FUNDAMENTALS OF CIRCUIT ANALYSIS

FUNDAMENTALS OF CIRCUIT ANALYSIS

with Applications to Electronics

James Fisk

Northern Essex Community College

Macmillan Publishing Company
New York

Collier Macmillan Publishers
London

To my wife, Ruth

Copyright © 1987, Macmillan Publishing Company, *a division* of Macmillan, Inc.

Printed in the United States of America

All rights reserved. No part of this book may be reproduced or transmitted in any form or by any means, electronic or mechanical, including photocopying, recording, or any information storage and retrieval system, without permission in writing from the publisher.

Macmillan Publishing Company
866 Third Avenue, New York, New York 10022

Collier Macmillan Canada, Inc.

Library of Congress Cataloging-in-Publication Data
Fisk, James,
   Fundamentals of circuit analysis.
   Includes index.
   1. Electric circuit analysis.   I. Title.
TK454.F54   1987   621.3815'3   86-8277
ISBN 0-02-337960-X

Printing: 1 2 3 4 5 6 7 8   Year: 7 8 9 0 1 2 3 4 5

ISBN 0-02-337960-X

PREFACE

Objectives and Features of This Text

Fundamentals of Circuit Analysis is aimed at students in Electronic Engineering Technology or Computer Technology programs in technical institutes and community colleges or those who need circuit analysis in nonelectrical engineering programs. It provides a thorough coverage of the concepts and analytical techniques essential to any well-written circuits text, but does so with some significant differences.

One of these differences is the book's emphasis on the understanding of circuits as well as the techniques of writing and solving equations. Training in the mathematics of circuit analysis is certainly fundamental to developing habits of logical, systematic thinking. But our graduates in the field seldom, if ever, write and solve systems of equations when troubleshooting or doing R&D on new systems. What they do need is the ability to make quick approximations and predictions of circuit response to changing conditions. This requires an understanding or "feel" for circuits that is equally as important as the ability to do detailed mathematical analyses.

While no book can teach a "feel" for circuits, a circuits text can at least introduce students to this kind of thinking that is so important in industry. *Fundamentals of Circuit Analysis* does this by showing readers how to estimate circuit performance, then stressing the use of such estimates to check the results of detailed analysis. It also emphasizes fundamentals and avoids memorized "format" techniques. The book includes extensive trouble-shooting sections as well as practical discussions of real-world components and how those components respond to low, medium, and high frequencies. Conventional current is used throughout. A prior or concurrent course in college algebra and trigonometry is assumed; no calculus is required.

Additional Features of The Text

The book also presents some of its topics in a different sequence than other circuits texts. For instance:

- The Thévenin source is a series circuit and is covered in the series This leads logically to discussions of maximum power transfer, loading, and voltage regulation. Likewise, the Norton circuit appears in the parallel chapter (Thévenin and Norton theorems and analysis are covered in an advanced chapter.)
- Superposition is found in the series-parallel chapter since a multi-source circuit is analyzed with series-parallel techniques once all sources but one have been turned off.

- Bridge circuits are discussed in the series-parallel chapter since balanced bridges can be analyzed with series-parallel techniques.
- In order to have students analyzing circuits early in the course, detailed discussions of resistance, temperature coefficients, etc., are saved for Chapter 8 (carbon composition resistors are discussed in Chapter 3 because they are used in lab). However, Chapter 8 may be presented after Chapter 2 if desired.

Fundamentals of Circuit Analysis emphasizes the terminology and procedures used in computers and electronics. For example, it introduces waveforms early and does analyses with them, describes transducers, digital meters, fanout, ceramic resonators, and the bargraph, speaks of the driving and driven circuit, and shows the various ways schematics are drawn in the field. The text also has many practical sections such as, "How to Find an Unknown Voltage in a Circuit" and "Measuring Phase and Frequency on the Oscilloscope." SI units are used throughout.

The book contains more than 1300 schematics, photos, and diagrams, 327 worked examples and over 1000 exercise problems ranging from the fundamental "plug and chug" to the more challenging Extra Credit level. The exercises are divided according to sections. Odd-numbered answers are located at the back of the book.

Instructor's Manual

The Instructor's Manual accompanying *Fundamentals of Circuit Analysis* is a bit unusual. Besides containing solutions to all exercises, it has a wealth of suggestions for presenting two or three semester courses in circuit analysis. Among these suggestions, gleaned over more than twenty years of teaching circuits, are demonstration ideas, areas to emphasize, key exercises, and lab assignments appropriate to the chapter material.

Laboratory Manual

The Laboratory Manual accompanying the text presents 36 student-tested experiments that teach the circuit construction, data taking, graphing, and percent error analysis skills fundamental to sound technical programs. Many of the experiments sharpen understanding by requiring students to predict or estimate circuit behavior first, then test their predictions with measurements.

Suggestions for Improvements

It will be sincerely appreciated if notification of errors and suggestions for improving the text and/or manuals be sent to the author at Northern Essex Community College, Haverhill, Mass. 01830.

Acknowledgements

The author wishes to thank all those who reviewed the text, especially Dr. Roger A. Mussell, William Rainey Harper College, Palatine, Illinois; Professor Joseph Farren, University of Dayton, Dayton, Ohio; and Professor Terry E. Schulz, Mt. Hood Community College, Gresham, Oregon. I would also like to thank my colleagues Sandra Meldrum and Professors Bill Arnold, Adrien Berthiaume, Charles Montgomery, and James Sullivan at Northern Essex Community College for their support and suggestions. Finally, my thanks and appreciation to all the people at Macmillan Publishing Company, especially David Johnstone and Edward Neve.

James H. Fisk W1HL

"Lo, these are part of His ways" Job XXVI 14

CONTENTS

1 Introduction 1

	Introduction	1
	Chapter Objectives	2
1.1	Mathematics: The Language of Electronics	2
1.2	Electronic Calculators	2
1.3	Scientific Notation	3
1.4	Systems of Units	6
1.5	Learning to Work with Units	6
1.6	Evaluating Math Expressions	9
	Exercises	10
	Chapter Highlights	12

2 Introduction to Electricity 13

	Introduction	13
	Chapter Objectives	14
2.1	The Atom	14
2.2	Electric Charge	14
2.3	Ions	15
2.4	Charge Transfer	15
2.5	The Coulomb	16
2.6	The Electric Field and Coulomb's Law	16
2.7	Energy	19
2.8	Electrical Energy	20
2.9	Electric Potential	21
2.10	The Ampere	22

2.11	Conductors and Insulators	23
2.12	Conduction by Electrons	23
2.13	Charge Movement by Positive Carriers	25
2.14	Conventional Current versus Electron Current	25
	Exercises	25
	Chapter Highlights	28

3

The Basic Circuit 29

	Introduction	29
	Chapter Objectives	30
3.1	The Basic Circuit	30
3.2	Energy Transfer in the Basic Circuit	31
3.3	Polarity	32
3.4	Resistance	32
3.5	Conductance	33
3.6	Carbon-Composition Resistors	33
3.7	Ohm's Law	36
3.8	Power	38
3.9	Efficiency of Energy Conversion	41
3.10	Short Circuits and Open Circuits	43
3.11	How Circuits Are Drawn in Electronics	44
3.12	Grounds	45
3.13	Waveforms	46
	Exercises	51
	Chapter Highlights	55

4

The Series Circuit 57

	Introduction	57
	Chapter Objectives	58
4.1	What Does "In Series" Mean?	58
4.2	Sources in Series	59
4.3	Kirchhoff's Voltage Law	60
4.4	Resistances in Series	62

Contents ix

4.5	Analyzing the Series Circuit	64
4.6	How to Find an Unknown Voltage in a Circuit	71
4.7	Controlling Current with Series Resistance	74
4.8	Voltage Dividers	75
4.9	The Thévenin Equivalent Source	77
4.10	Troubleshooting the Series Circuit	82
	Exercises	83
	Chapter Highlights	99

5

The Parallel Circuit 101

	Introduction	101
	Chapter Objectives	102
5.1	What Does "In Parallel" Mean?	102
5.2	Kirchoff's Current Law	103
5.3	Resistances in Parallel	105
5.4	Analyzing the Parallel Circuit	113
5.5	The Norton Equivalent Source	117
5.6	How to Convert from One Source Model to the Other	119
5.7	Current Sources in Series and Parallel: Millman's Theorem	121
5.8	Controlled Sources	123
5.9	Troubleshooting the Parallel Circuit	123
	Exercises	124
	Chapter Highlights	135

6

Series–Parallel Circuits 137

	Introduction	137
	Chapter Objectives	138
6.1	Resistances in Series–Parallel	138
6.2	Analyzing the Series–Parallel Circuit	140
6.3	The Loaded Voltage Divider	149
6.4	The Bridge Circuit	152
6.5	T–Π Transformations	156

6.6 Superposition	163
6.7 The Ladder Method	169
Exercises	**172**
Chapter Highlights	**181**

7

Advanced Techniques of Analysis   183

Introduction	**183**
Chapter Objectives	**184**
7.1 Review of Math	184
7.2 Loop Analysis	188
7.3 Nodal Analysis	197
7.4 Thévenin Analysis	206
7.5 Norton Analysis	214
Exercises	**220**
Chapter Highlights	**227**

8

Resistance   229

Introduction	**229**
Chapter Objectives	**230**
8.1 The Resistance of a Conductor	230
8.2 The Effects of Temperature on Resistance	232
8.3 Conductors	235
8.4 Printed-Circuit Boards	238
8.5 Resistors	240
8.6 VA Characteristics of Linear and Nonlinear Resistance	243
8.7 Resistive Transducers	246
8.8 Switches	246
Exercises	**248**
Chapter Highlights	**253**

9 Meters and Measuring 255

	Introduction	255
	Chapter Objectives	256
9.1	The Effects of Loading On Measuring and How to Minimize them	256
9.2	Analog Meters	258
9.3	The D'Arsonval Meter Movement	258
9.4	Sensitivity, Precision, and Accuracy of the D'Arsonval Movement	260
9.5	How to Make the D'Arsonval Movement Read Higher Currents	261
9.6	The D'Arsonval Movement as a Voltmeter	263
9.7	Voltmeter Input Resistance: The Ohms-per-Volt Rating	264
9.8	The D'Arsonval Movement as an Ohmmeter	265
9.9	How to Use an Ohmmeter	267
9.10	The VOM and FETVOM	267
9.11	Digital Meters	269
9.12	Comparing Analog and Digital Meters	271
9.13	Other Analog Meters	272
	Exercises	274
	Chapter Highlights	277

10 Capacitance 279

	Introduction	279
	Chapter Objectives	280
10.1	Capacitance	280
10.2	Capacitive Current/Voltage Relations	283
10.3	What Determines Capacitance?	286
10.4	The Role of the Dielectric	288
10.5	Stored Energy	291
10.6	Capacitors in Series and Parallel	292
10.7	Practical Capacitors	296
10.8	Stray Capacitance	299
10.9	Capacitive Transducers	300
10.10	Transient Analysis	300
10.11	Using Thévenin's Theorem in Transient Analysis	309

Exercises		**311**
Chapter Highlights		**316**

11

Magnetism 317

Introduction		**317**
Chapter Objectives		**318**
11.1	Magnetic Flux	318
11.2	Electromagnetism	321
11.3	The Magnetic Circuit	322
11.4	Magnetizing Force	324
11.5	The B–H Curve	324
11.6	Ampere's Circuital Law	327
11.7	Analyzing the Magnetic Circuit	328
11.8	Air Gaps	331
11.9	Electromagnetic Induction	333
11.10	Electromagnetic Devices	336
	Exercises	**338**
	Chapter Highlights	**342**

12

Inductance 343

Introduction		**343**
Chapter Objectives		**344**
12.1	Self-Inductance	344
12.2	Stored Energy	350
12.3	Some Comparisons Between R, L and C	350
12.4	Practical Inductors	350
12.5	Inductors in Series and Parallel	355
12.6	Components in the Real World	357
12.7	Inductive Transducers	359
12.8	Transient Analysis	359
12.9	Using Thévenin's Theorem in Transient Analysis	364
	Exercises	**366**
	Chapter Highlights	**370**

13
Introduction to Alternating Current 373

Introduction	373
Chapter Objectives	374
13.1 Vectors	374
13.2 Generating the Sine Wave	376
13.3 Measuring the Sine Wave	381
13.4 Sine Waves on DC Levels	388
13.5 The Average Value of a Sine Wave	389
13.6 The Effective Value of a Sine Wave	392
Exercises	397
Chapter Highlights	402

14
Phase and Phasors 403

Introduction	403
Chapter Objectives	404
14.1 Phase Relations	404
14.2 Sine-Wave Addition: When 2 + 2 Does Not Equal 4	412
14.3 Complex Numbers: The j Operator	413
14.4 Complex Numbers and Phasors	415
14.5 Math Operations with Complex Numbers	417
Exercises	423
Chapter Highlights	426

15
Reactance and Impedance 429

Introduction	429
Chapter Objectives	430
15.1 The Derivative of a Sine Wave	430
15.2 What Causes Phase Shift? Reactance	433

15.3 Impedance	443
15.4 Average Power in Resistance and Reactance	449
15.5 Reactive Power	452
15.6 Apparent Power	453
15.7 Power Factor	457
Exercises	**459**
Chapter Highlights	**463**

16

Analyzing Series and Parallel AC Circuits 465

Introduction	**465**
Chapter Objectives	**466**
16.1 Z_{total} of Series Impedances	466
16.2 Analyzing Series AC Circuits	467
16.3 The Voltage Division Formula	473
16.4 Response of *RC*, *RL* and *RLC* Series Circuits to Changing Frequency	474
16.5 *RL* and *RC* Filter Circuits	476
16.6 Z_{total} of Parallel Impedances: Susceptance and Admittance	480
16.7 Analyzing Parallel AC Circuits	486
16.8 Response of Parallel *RC*, *RL* and *RLC* Circuits to Changing Frequency	493
16.9 Series and Parallel Equivalent Circuits	495
Exercises	**502**
Chapter Highlights	**509**

17

Analyzing Series–Parallel AC Circuits 511

Introduction	**511**
Chapter Objectives	**512**
17.1 Analysis by Series–Parallel Reduction	512
17.2 AC Bridges	526
17.3 T-Π Transformations	531
17.4 Thévenin and Norton AC Sources	535

17.5	Superposition	542
17.6	The Ladder Method	545
	Exercises	547
	Chapter Highlights	554

18

Advanced Analysis of AC Circuits 555

	Introduction	555
	Chapter Objectives	556
18.1	Loop Analysis of AC Circuits	556
18.2	Nodal Analysis of AC Circuits	559
18.3	Thévenin Analysis of AC Circuits	564
18.4	Norton Analysis of AC Circuits	568
	Exercises	571

19

Resonant Circuits 579

	Introduction	579
	Chapter Objectives	581
19.1	The Series Resonant Circuit	581
19.2	The Quality Factor, Q	585
19.3	The Importance of Q	589
19.4	Bandwidth	591
19.5	The Parallel Resonant Circuit	597
19.6	Q of the Parallel Resonant Circuit	601
19.7	Bandwidth of the Parallel Resonant Circuit	605
19.8	The Tank Circuit as an AC Generator	608
19.9	Matching Impedance with Resonant Circuits: The L Network	609
19.10	Resonant Filter Circuits	614
19.11	Other Resonant Devices	620
	Exercises	622
	Chapter Highlights	626

20 Polyphase Circuits 629

	Introduction	629
	Chapter Objectives	630
20.1	Single and Two-Phase Circuits	630
20.2	Three-Phase Power	631
20.3	The Wye-Connected Generator	633
20.4	Loading the Wye-Connected 3-ϕ Generator: The Balanced Wye-Connected Load	636
20.5	The Unbalanced Wye-Connected Load	639
20.6	The Delta-Connected Load	642
	Exercises	645
	Chapter Highlights	648

21 Transformers 649

	Introduction	649
	Chapter Objectives	650
21.1	Iron-Core Transformer Theory	650
21.2	The Dot System	657
21.3	Analyzing Circuits Containing Ideal Transformers	659
21.4	Flux in the Nonideal Core	662
21.5	Mutual Inductance	665
21.6	Equivalent Circuit of the Iron-Core Transformer	670
21.7	Loosely Coupled Transformers	671
21.8	Other Types of Transformers	674
	Exercises	677
	Chapter Highlights	682

22 AC Measurements 683

	Introduction	683
	Chapter Objectives	684
22.1	Using the D'Arsonval Movement for AC Measurements	684

22.2 Other Types of AC Meters	687
22.3 AC Power Measurements	690
22.4 Basic AC Measurements on the Oscilloscope	691
22.5 Measuring Phase and Frequency on the Oscilloscope: The Lissajous Pattern	694
22.6 Frequency Counters	698
Exercises	**699**
Chapter Highlights	**701**

23

Periodic Non-Sinusoidal Waveforms: The Fourier Series 703

Introduction	**703**
Chapter Objectives	**704**
23.1 An Experiment	704
23.2 The Fourier Series	706
23.3 Symmetry	707
23.4 Fourier Series of Common Electronic Waveforms	709
23.5 RMS Values and Average Power	713
23.6 Using Fourier Series in Circuit Analysis	715
Exercises	**719**
Chapter Highlights	**721**
Appendix: Derivation of Π–to–T and T–to–Π Conversion Formulas of Sections 6.5 and 17.3	**723**
Answers to Odd-Numbered Exercises	**729**
Index	**739**

CHAPTER 1

Introduction

Any training in electronics must start with courses in circuit analysis because it is in circuits that transistors, integrated circuits, and other devices do the things that are so important to our lives.

What is a circuit? A circuit is a combination of electronic parts connected together in a way that makes it able to do a specific job. For example, a stereo amplifier is a circuit designed to amplify music from a record or tape and send that music to a loudspeaker. A computer is a circuit, too. However, it is not the same circuit as that of the stereo because it has a different job to do.

This book covers basic circuits, starting with the easiest and working up to the more complicated. An understanding of them is necessary in order to move to higher-level courses in solid-state devices, computer electronics, communications, and other areas.

Before we can begin studying circuits, however, there are two things to do. The first is to review some rules of mathematics and learn how to handle the units used in electronics. This is accomplished here in chapter one. The second is to get a solid understanding of the principles of electricity. This is necessary because a circuit is really nothing more than something we build to make electricity do what we want it to do. These principles are covered in Chapters 2 and 3.

> At the end of this chapter, you should be able to:
> - List the features students should look for when buying a calculator.
> - Use scientific notation so that large and small numbers can be handled easily.
> - Convert units more easily and with fewer errors.
> - Tell when variables are related by direct or inverse proportionality.
> - Explain why exponents have such enormous impact on the value of math expressions.

1.1 MATHEMATICS: THE LANGUAGE OF ELECTRONICS

Important concepts in electronics are usually described by mathematical equations. Mathematics is the language of electronics. So it is important that you develop the skills needed to handle math smoothly and accurately, to get a "feel" for it. This takes practice.

There are two ways to get that practice while using this book. First, understand each step in the text's explanation of a problem. Mark those steps you don't understand. Then ask your instructor about them at the next class meeting. Second, do the exercise problems at the end of each chapter.

Besides our work in algebra we will be doing ordinary calculations with addition, subtraction, and the like. Two important tools to help in these calculations, the electronic calculator and scientific notation, are discussed next.

1.2 ELECTRONIC CALCULATORS

A calculator is a necessary tool for anyone in electronics, not only while in school but throughout a career. Do not skimp when buying one. Look for the following minimum features:

- The four basic arithmetic operations: add, subtract, multiply, and divide.
- Full trigonometric capabilities: sine, cosine, tangent, arcsine, arccosine, and arctangent, in both degrees and radians.
- Specific functions: e^x; $\ln x$; $\log x$; 10^x; y^x; $1/x$; the square root of x; x^2; the constant, π.
- Rectangular-to-polar and polar-to-rectangular conversion capabilities. Indicated by R → P, P → R, on the keys.
- At least one memory location for storing intermediate solutions.

Watch for the words *scientific calculator* when shopping. They describe models that perform these functions.

Numbers are put into calculators in one of two ways. There are advantages and disadvantages to each. The *algebraic method,* used by Texas Instruments and others, is

1.3 SCIENTIFIC NOTATION

easier to learn and used on less expensive instruments. The *RPN* (reverse Polish notation) *method* requires a bit more learning time but, once learned, allows easier data entry. It is usually found on more expensive calculators, such as those made by Hewlett-Packard.

Do not put extra money into a programmable calculator unless there is strong interest in that capability.

No matter which make and model calculator you choose, it is important to study the instruction manual and learn to use the instrument. Go out of the way to find practice calculations such as determining team averages and gas mileages or adding long grocery bills. This develops confidence which pays off later when doing advanced calculations.

One other suggestion. Repeat the calculations in the text's example problems on your calculator. Make sure that your answer agrees with the one given. By doing so, you will get more practice and understand the thinking behind the analysis better.

1.3 SCIENTIFIC NOTATION

In electronics we work with extremely large and small numbers. Values in the billions and billionths are routine. *Scientific notation* is a method by which these numbers are changed to a form that is easier to work with. This form consists of a number multiplied by 10 raised to some power by an exponent.

Since exponents are so important to scientific notation, we need to review the laws of exponents.

$$X = X^1$$

Here the exponent 1 is understood

$$X^1 X^1 X^1 = X^3 \quad \text{or} \quad X^2 X^3 = X^5$$

When multiplying exponentialized numbers of the same base (in this case, X), add exponents.

$$(X^2)^3 = X^6$$

Pay particular attention to the difference between this example and the last.

$$\frac{X^5}{X^3} = \frac{X^2 X^3}{X^3} = X^2 = X^{5-3}$$

When dividing numbers of the same base, subtract the denominator exponent from the numerator exponent. (Remember: denominator down in a fraction)

$$1 = \frac{X}{X} = \frac{X^1}{X^1} = X^{1-1} = X^0$$

Any number to the zero power = 1.

$$\frac{1}{X^2} = \frac{X^0}{X^2} = X^{0-2} = X^{-2}$$

When a number is moved from the denominator of a fraction to the numerator, the sign of its exponent is changed.

Since X may be any number, let it be 10, the base of our numbering system. Then, using the laws of exponents on some powers of 10, we get

$$10 = 10^1 \quad \text{understood exponent}$$
$$10^2 = 10^1 \cdot 10^1 = 10 \cdot 10 = 100$$
$$10^3 = 10^1 \cdot 10^2 = 10 \times 100 = 1000$$
$$10^4 = 10{,}000 \quad 10^5 = 100{,}000 \quad 10^6 = 1{,}000{,}000 = 1 \text{ million}$$
$$10^9 = 1{,}000{,}000{,}000 = 1 \text{ billion}$$

It is much easier to write 1 billion as 10^9 than as a 1 followed by nine zeros. Better yet, trying writing a number as large as 10^{23}.

Going next to numbers less than zero:

$$\frac{1}{10} = 0.1 = \frac{10^0}{10^1} = 10^{0-1} = 10^{-1}$$

$$\frac{1}{100} = 0.01 = \frac{10^0}{10^2} = 10^{0-2} = 10^{-2}$$

$$\frac{1}{10^3} = 0.001 = 10^{-3}$$

$$\frac{1}{10^6} = 0.000001 = 10^{-6}$$

Note how numbers greater than 1 have positive exponents, whereas those less than 1 have negative exponents.

Example 1.1 shows how this system of writing powers of 10 may be used to write large and small numbers in a simpler form.

EXAMPLE 1.1

Write each number as another number between 1 and 10, multiplied by some power of 10.

a. 17,500.

The decimal point is to the right of the right-hand zero. Move it four places to the left to get the number between 1 and 10. This gives 1.75. This must be multiplied by 10 raised to an exponent equal to the number of places the decimal point was moved. Final answer: 1.75×10^4.

b. 873,000,000,000.

The decimal point is moved 11 places to the left to get 8.73, the number between 1 and 10. 8.73 must be multiplied by a huge number in order to retain the original value. Therefore, the final answer is: 8.73×10^{11}.

c. 0.037.

The number between 1 and 10 is 3.7. This must be divided by 100 to retain the original value. It could just as well be multiplied by 0.01 or by 10^{-2}. Final answer: 3.7×10^{-2}.

d. 0.00000000068.

Moving the decimal point 10 places to the right, we get 6.8. This must be multiplied by 10^{-10}, which effectively moves the point back to its original location to keep the original value. Final answer: 6.8×10^{-10}.

1.3 SCIENTIFIC NOTATION

So far the system has been used only to write large and small numbers in a more convenient form. It is also very useful when doing numerical calculations.

EXAMPLE 1.2

Do the following calculation, typical of that found in electronics.

$$\frac{8{,}370{,}000 \times 0.0024}{130}$$

Rewriting:

$$\frac{8.37 \times 10^6 \times 2.4 \times 10^{-3}}{1.3 \times 10^2} = \frac{8.37 \times 2.4 \times 10^3}{1.3 \times 10^2}$$

$$= \frac{20.1}{1.3} \times 10^1 = 15.452 \times 10^1 = 154.52$$

Rounding the answer to one place beyond the decimal point gives us 154.5. (We will usually round to one place unless doing so causes a large percentage of error.) ■

Many calculators will work in scientific notation. They have a key marked EE, EX, EXP, SCI, or something similar. The calculator's display shows a number such as 123,000,000 as 1.23 on the left and an 08 on the right, like this:

$$1.23 \qquad 08$$

The exponent is always for base 10. 0.0000000123 would be

$$1.23 \qquad -08$$

Students often have trouble determining the sign of the exponent when writing numbers in scientific notation. Consider our last example, 123,000,000, again. It was changed to 1.23, a number considerably less than 123 million. To keep the two forms of the number equal, 1.23 must be multiplied by a large number to increase it from something less than 10 to a value in the millions. 10^{-8} won't do the job because it is much less than 1 (0.00000001). 10^8, however, will do because it is a large number, 100 million.

Using the same thinking, we see that 1.23 must be multiplied by a small number, 10^{-8}, 0.00000001, to bring it back down to 0.0000000123.

EXAMPLE 1.3

Do the following calculation with scientific notation: $\dfrac{0.00000000045 \times 905}{0.077 \times 17{,}500}$

$$\frac{0.00000000045 \times 905}{0.077 \times 17{,}500} = \frac{4.5 \times 10^{-10} \times 9.05 \times 10^2}{7.7 \times 10^{-2} \times 1.75 \times 10^4}$$

$$= \frac{4.5 \times 9.05}{7.7 \times 1.75} \times \frac{10^{-8}}{10^2} = 302 \times 10^{-10} \quad ■$$

1.4 SYSTEMS OF UNITS

Besides numbers, we will be working with physical units such as seconds, watts, and volts. These units are called *parameters*.

There are several systems of units. The most familiar is the one we use in everyday life, the *English system*. In it, all parameters are defined in terms of the basic units, the foot, pound, and second.

The English system is not used in science and technology because it requires so many conversion factors, 36 inches to the yard, for example. Instead, we use the *Standard International* or *SI* system, in which the meter, kilogram, and second are the basic units. Conversions in this system are easily made by shifting the decimal point.

In certain areas of specialization, another system, called the CGS, or centimeter, gram, second, system is used.

In any system, standards must be set and carefully followed so that people refer to the same thing when they speak of a meter, a kilogram, or any other parameter. This is so important that an International General Conference on Weights and Measures with about 40 member countries meets regularly in France. Here, the closely guarded Standard International kilogram is kept. It is a rod of special alloy approximately 3.9 centimeters in height and diameter. Carefully checked copies of this and other standards are kept by other countries. Those of the United States are under the control of the National Bureau of Standards in Washington, D.C. The National Research Council of Canada in Ottawa, Ontario, maintains those of Canada.

The meter was originally defined by the French as one ten-millionth of the distance between the equator and either pole. This proved impractical and the meter was, until recently, based on the length of a rod of the same material as the weight standard. Now the SI unit of length is more precisely defined in terms of radiation from atoms.

The second was originally defined as part of the time it takes the earth to rotate. This proved impractical because of tiny changes in the globe's daily movements. In 1967, the standard was changed to an atomic clock whose accuracy is 1 second in about 3100 years, or 1 part in 10^{11}.

In summary, the SI system of units has the kilogram, meter, and second for the basic unit of mass, length, and time, respectively. Add the Standard International unit of electric current, the ampere, and we have the basis of every other unit used in electronics. The ampere will be defined later.

1.5 LEARNING TO WORK WITH UNITS

People in any field of technology must be able to convert from one unit to another. The following examples are designed to show an easy and dependable way to do this.

Let's look at a very familiar situation from a different angle, that is, from the standpoint of the units. The situation is a job paying $5 per hour. We wish to calculate pay received after 8 hours of work. Writing a math expression:

$$\text{pay} = 5 \frac{\text{dollars}}{\text{hour}} \times 8 \text{ hours} = 5 \times 8 \frac{\text{dollars}}{\cancel{\text{hour}}} \times \cancel{\text{hours}}$$

← cancel hours

$$= 40 \text{ dollars}$$

1.5 LEARNING TO WORK WITH UNITS

Note carefully how the hours cancel in the second step. This happens because hours appear both in the numerator (upper part of a fraction) and the denominator (lower part), and anything divided by itself equals 1. Be sure that this important point is clear.

The units of the answer are dollars. Are they correct? Yes, because we measure pay in dollars.

We'll do the calculation again, same rate per hour and number of hours, but this time the company pays a bonus of $12 for working evenings:

$$\text{pay} = 5 \frac{\text{dollars}}{\text{hour}} \times 8 \text{ hours} + 1 \text{ bonus} \times 12 \frac{\text{dollars}}{\text{bonus}}$$

hours cancel — bonus cancels

$$= 40 \text{ dollars} + 12 \text{ dollars} = 52 \text{ dollars}$$

Notice how units are added to the same units here.

Suppose that the bonus is changed to a steak dinner—what is the pay now? Forty dollars plus one steak dinner. Why? Because we cannot add dollars and steak dinners.

What would have to be done to get a final figure on pay? Study this carefully:

$$\text{pay} = 5 \frac{\text{dollars}}{\text{hour}} \times 8 \text{ hours} + 1 \text{ steak dinner} \times 12 \frac{\text{dollars}}{\text{steak dinner}}$$

steak dinner cancels

$$= 40 \text{ dollars} + 12 \text{ dollars} = 52 \text{ dollars}$$

For another example of handling units, we will figure pay after 6 months, assuming 40 hours of work per week and 4.33 weeks per month:

$$\text{pay} = 5 \frac{\text{dollars}}{\text{hour}} \times 40 \frac{\text{hour}}{\text{week}} \times 4.33 \frac{\text{weeks}}{\text{month}} \times 6 \text{ months}$$

$$= 5196 \text{ dollars}$$

Pay close attention to the placement of units in numerators and denominators so that they cancel as needed.

After all the canceling is done, what are the only units remaining on the right side? Dollars. It could not be hours because pay is not measured in hours.

This approach of placing units in both numerator and denominator so that they cancel is used when converting from one unit to another, for example, from centimeters to feet. To do this, however, certain conversion factors are needed.

EXAMPLE 1.4

Convert 17 centimeters to feet.

$$17 \text{ centimeters} \times \frac{1 \text{ inch}}{2.54 \text{ centimeters}} \times \frac{1 \text{ foot}}{12 \text{ inches}}$$

conversion factors

$$= \frac{17}{2.54 \times 12} \text{ feet} = 0.56 \text{ foot}$$

EXAMPLE 1.5

Convert 1 week into seconds.

$$1 \text{ week} \times \frac{7 \text{ days}}{\text{week}} \times \frac{24 \text{ hours}}{\text{day}} \times \frac{60 \text{ minutes}}{\text{hour}} \times \frac{60 \text{ seconds}}{\text{minute}}$$

$$= 6.05 \times 10^5 \text{ seconds}$$

EXAMPLE 1.6

Convert \$50/day to pennies/minute.

$$\frac{\$50}{\text{day}} \times \frac{100 \text{ cents}}{\text{dollar}} \times \frac{1 \text{ day}}{24 \text{ hours}} \times \frac{1 \text{ hour}}{60 \text{ minutes}}$$

$$= 3.47 \frac{\text{cents}}{\text{minute}}$$

Another system of units used in technology is one in which certain frequently used powers of 10 are given metric names. For example, radio announcers give a station's location on the dial in terms of kilohertz or megahertz. Kilo means thousands, and mega means millions.

Table 1.1 lists those terms regularly used in electronics.

TABLE 1.1 Powers of 10 and Their Metric Names

Name	Value	Letter abbreviation
giga	10^9	G
mega	10^6	M
kilo	10^3	k
milli	10^{-3}	m
micro	10^{-6}	μ (Greek letter, pronounced *mew*)
nano	10^{-9}	n
pico	10^{-12}	p
femto	10^{-15}	f

EXAMPLE 1.7

How many kilovolts are there in 34,000 volts?

$$34{,}000 \text{ volts} \times \frac{1 \text{ kilovolt}}{1000 \text{ volts}} = 34 \text{ kV}$$

EXAMPLE 1.8

Convert 2200 picofarads to microfarads.

$$2200 \text{ picofarads} \times \frac{1 \text{ microfarad}}{10^6 \text{ picofarads}}$$

$$= 2200 \times 10^{-6} \text{ microfarad} = 0.0022 \text{ microfarad}$$

1.6 EVALUATING MATH EXPRESSIONS

One of the best ways to gain understanding of a circuit is to write a math expression that describes how the circuit works. Once we have this, it is easy to determine the effect on circuit operation of a change in one of the parameters in that circuit. Simply put the change into the expression and note what the value of the expression does.

For example, the relation between distance traveled by a car, the car's speed, and time of travel is described by the formula

$$D = ST$$

where D = distance, in miles
S = speed, in miles per hour
T = time, in hours

S and T are called *independent variables*. They can have any value. D is also a variable, but its value depends on the values of S and T; therefore, D is called a *dependent variable*.

Application of the rules of algebra produces a different form of the equation

$$T = \frac{D}{S},$$

which tells us the time needed to travel a given distance at a certain speed.

Suppose that we want to go 500 miles (mi). At a speed of 50 miles per hour (mi/h), the trip takes 500/50 = 10 h. Double the distance to 1000 mi and time doubles to 1000/50 = 20 h (speed constant). In other words, T and D are tied together; a change in D causes exactly the same change in T. T is said to be *directly proportional* to D.

Let us examine next how T changes with changes in S.

At a 25-mi/h crawl our 500-mi trip takes 500/25 = 20 h. Double the speed to 2 × 25 or 50 mi/h and the time drops to 20/2 = 10 h. In a jet going 1000 mi/h trip time falls to just $\frac{1}{2}$ h.

Notice the connection between T and S. As speed, S, increases, time, T, decreases (with D held constant, of course). T is said to be *inversely proportional* to S.

To summarize these important points: T is directly proportional to D and D is in the numerator of the right side of the equation (directly proportional—numerator). T is inversely proportional to S and S is in the denominator of the right side (inversely proportional—denominator).

Another concept that helps evaluate math expressions is to understand the effect of exponents. Consider the expression

$$A = B$$

When the exponent of B is 1, as it is understood to be now, A increases directly with B (see the left-hand column of Table 1.2). If, however, the exponent is greater than 1, A increases at a much greater rate; the higher the exponent, the greater the rate. This important relationship is seen in the middle and right-hand columns of the table when the B exponent is 2 and 3, respectively. Study the three parts of the table. Note carefully how much more rapidly A rises when the exponent of B is increased by 1.

Exponents cause equally fast changes in the inverse direction. Consider

$$A = \frac{1}{B}$$

TABLE 1.2 Changes in a Variable with Increasing Exponents

Let $A = B$	Let $A = B^2$	Let $A = B^3$
When $B = 1$ $A = 1$	When $B = 1$ $A = 1$	When $B = 1$ $A = 1$
When $B = 2$ $A = 2$	When $B = 2$ $A = 4$	When $B = 2$ $A = 8$
When $B = 3$ $A = 3$	When $B = 3$ $A = 9$	When $B = 3$ $A = 27$
When $B = 5$ $A = 5$	When $B = 5$ $A = 25$	When $B = 5$ $A = 125$
When $B = 10$ $A = 10$	When $B = 10$ $A = 100$	When $B = 10$ $A = 1000$

and let $B = 5$. When the exponent of B is 1, $A = 1/5 = 0.2$. When the exponent is 2, A drops to $1/25 = 0.04$, and when the exponent is 3, A falls to $1/125 = 0.008$.

Learn to spot these all-important exponents in math expressions and be alert to their effect on results.

EXAMPLE 1.9

Given: the expression

$$M = \frac{KL^4}{B^2}$$

a. How does M vary with changes in the variables?

M varies directly with K, increases with the fourth power of L, and varies inversely with the square of B.

b. What happens to the value of M if K remains the same, L doubles, and B triples?

K causes no change. Doubling L causes M to increase 2^4 times $= 16$. Tripling B causes a reduction to $1/9$. Therefore, the final change is: M increases to $16/9 = 1.8$ times original value. ∎

EXERCISES

Section 1.3

1. Express each of the following numbers as a number between 1 and 10 times a power of 10.
   a. 57,000
   b. 9.23
   c. 0.0000037
   d. 1015
   e. 0.0685
   f. 186,000
   g. 0.00000000092
   h. 2,350,000
   i. 0.000077
   j. 1.56
   k. 755
   l. 0.0755
   m. 22
   n. 15,700
   o. 0.00053

2. Repeat Exercise 1, parts a through e, but express the number as a number between 10 and 100 times a power of 10.

3. Repeat Exercise 1, parts f through j, but express as a number between 0.0001 and 0.001 times a power of 10.

4. Repeat Exercise 1, parts k through o, but express as a number between 1000 and 10,000 times a power of 10.

5. Using a calculator, evaluate each of the following. (Express each answer as a number between 1 and 10 rounded to one place beyond the decimal point.)
   a. 53,000(415)
   b. 12.2(255)(942)
   c. 0.0078(0.013)(0.0000046)

EXERCISES

d. $\dfrac{186{,}000}{0.13}$ e. $\dfrac{10{,}700}{922{,}000}$ f. $\dfrac{100{,}092(158)}{17}$

g. $\dfrac{255(34)(0.0023)}{4800(0.066)}$ h. $155{,}000{,}000(0.0000044)$ i. $\dfrac{22}{0.0000000000000011}$

j. $\dfrac{505 \times 10^{-4}}{78 \times 10^{5}}$

6. Repeat Exercise 5 for the following, but round each answer to two places beyond the decimal point.

a. $0.0019(34)$ b. $17.2(1.8)(39)$

c. $0.092(47.4)(0.00071)$ d. $\dfrac{41{,}000}{0.0039}$

e. $\dfrac{1730(0.94)}{26}$ f. $\dfrac{47{,}500}{0.00073(6200)}$

g. $\dfrac{33(1.5)(0.064)}{19.8(0.1)}$ h. $\dfrac{782.7}{0.00000006}$

i. $\dfrac{23.3 \times 10^{7}}{9.1 \times 10^{-4}}$ j. $\dfrac{1.57 \times 10^{-23})(7.7 \times 10^{5})(83)}{14.6 \times 10^{2}(0.67)}$

Section 1.5

7. Convert each of the following as indicated: *Note:* Answers must include necessary units.
a. Millimeters to yards (2.54 centimeters = 1 inch).
b. Grams to tons (2.21 pounds = 1 kilogram).
c. $\dfrac{\$9000}{\text{year}}$ to $\dfrac{x \text{ cents}}{\text{minute}}$.
d. $\dfrac{15 \text{ gallons}}{\text{min}}$ to $\dfrac{x \text{ pints}}{\text{leap year}}$.
e. 15,000 microinches to inches.
f. 47,000 dollars to kilodollars.
g. 6800 milliliters to liters.
h. 14×10^{-8} bushels to picobushels.
i. 0.093 megaton to tons.
j. 107,000,000,000 dollars to gigadollars.

8. There are 43,560 square feet in an acre. How many acres are there in a square mile?

9. The "jiffy" has been defined by physicists as the time required for light to travel 1 centimeter. How long does this take? (Light travels 300,000,000 meters per second.) Express the answer in picoseconds.

10. Calculate the distance traveled by a car going 55 mi/h for 4.7 h. Show clearly how the dimensions are handled to get the correct units in the answer.

11. Assuming measurement in the metric system, prove that length, area, and volume have units of meters, meters squared, and meters cubed, respectively.

12. A friend says that he has 2 million microdollars in his pocket. How much is this?

13. How long would it take to spend a megadollar at the rate of a kilodollar per day? Express the answer in kilodays.

14. *Extra Credit.* Calculate the number of centimeters in a light-year. (A light-year is the distance traveled by light in 1 year. Light travels 3×10^{10} centimeters per second.)

Section 1.6

15. State what happens to the dependent variable on the left when the change indicated is made.
a. $P = I^2$ I increases to 5 times original value
b. $I = \dfrac{1}{R}$ R decreases to $\tfrac{1}{4}$ original value

c. $W = \dfrac{E^2}{2}$ E increases to 11 times original value

d. $J = \dfrac{1}{Q^3}$ Q increases to 4 times original value

e. $E = \sqrt{4KTBR}$ B decreases to $\tfrac{1}{9}$ original value. All other variables remain constant.

16. In your own words, define direct and indirect proportionality.
17. How are X and F related in the following two equations?   a. $X = 2\pi FL$; b. $X = 1/(2\pi FC)$.
18. In the equation $F = 1/2\pi\sqrt{LC}$, how does F vary with: a. L; b. C; c. π?
19. The volume of a globe or sphere is given by the formula $V = \tfrac{4}{3}\pi R^3$. By what percentage will the sphere's volume increase if the radius is increased by 10%?
20. The ability of a wire to carry electricity is directly proportional to the wire's diameter squared. If a given wire is replaced by one having $\tfrac{1}{4}$ the original diameter, by what factor has the carrying capacity changed? In which direction?
21. Above mach 1, the speed of sound, the drag or friction of air on a supersonic aircraft increases with the cube of the velocity. Assume that we wish to raise the speed of one of these planes 2.5 times. What increase in drag must the engines overcome?
22. We normally say we work for so many dollars per hour. Could we say the inverse, hours per dollar? Express $5 per hour as hours per dollar, days per dollar, and weeks per dollar.
23. One member of a crew lowering a piano from a fifth story lets go of the rope and the speed of descent doubles. What increase in kinetic energy must be controlled by the remaining members? The formula for kinetic energy is $KE = \tfrac{1}{2}MV^2$, where M = piano mass or weight and V = velocity.
24. Lay out two perpendicular axes on a sheet of paper or use graph paper. Make the vertical or Y axis 9 inches long and divide it into $\tfrac{1}{4}$-inch increments, each representing one unit. Make the X axis 6 inches long, each inch representing one unit. On this set of axes plot the following equations: $Y = X$, $Y = X^2$, and $Y = X^3$. Discuss the rate at which Y increases for each curve with increasing values of X.

CHAPTER HIGHLIGHTS

Time used to gain confidence with a calculator is time well spent.

    Skill in using scientific notation and the laws of exponents is necessary for handling large and small numbers with ease.

    Terms such as "kilo" and "pico" are as much a part of a number as the numerical value; they must be respected as such.

    Data usually come in two parts, the numerical value and the units (for example, 110 volts). Do not leave out the units.

    Units or dimensions must be handled according to the laws of mathematics and common sense. We cannot add gallons to feet or liters to meters.

    Strive to develop a "feel" for numbers and mathematical expressions. Pay special attention to exponents; they can have a powerful impact. Know the difference between direct and indirect proportionality.

CHAPTER 2

Introduction to Electricity

Anyone who has played with magnets knows that their poles push and pull one another. They do this under the familiar rule: Opposite poles attract, like poles repel.

Static electricity causes pushing and pulling just like magnets. On a dry winter's day, for example, we feel our hair rising to meet the comb.

Science explains the attracting and repelling forces of static electricity with a mathematical expression called Coulomb's law. This law explains many things in electricity and electronics. We will study it in this chapter.

After Coulomb's law, the chapter moves on to discussions of voltage and current, conductors and insulators.

At the end of this chapter, you should be able to:

- Explain the difference between static and dynamic electricity.
- Use Coulomb's law to calculate the forces of static electricity.
- Tell how energy can be put into and taken out of a system of charged bodies.
- Define the coulomb, volt, and ampere as well as voltage rise and drop.
- Explain how conductors differ from insulators.
- Describe how charge moves as a current.

2.1 THE ATOM

In earlier science and chemistry courses we learned that our world and everything in it is made of various combinations of *elements*. Among these elements are carbon, sodium, germanium, and so on, about 100 in all.

The smallest possible amount of a material such as sodium which has all the characteristics of that element is called an *atom*. Atoms are so small that they can be studied only by means of a model. The model we use was proposed by the Danish physicist Niels Bohr (1885–1962) in 1913. It is a microscopic copy of our solar system. At its center (see Figure 2.1) is the heaviest part of the atom, the *nucleus*. It corresponds to the sun. Spinning around the nucleus, much like our earth and other planets orbit the sun, are one or more *electrons*. Each element has a unique number of electrons in its atoms. Hydrogen, for example, has only 1, while sodium has 11 and uranium, 92. As tiny as a nucleus must be, it is itself made up of even smaller subatomic particles, such as *protons* and *neutrons*.

2.2 ELECTRIC CHARGE

Electrons and protons have a mysterious something called *electric charge*. To this day, charge is not fully understood, but it is known that electrons have as much of it as protons do. However, electron charge and proton charge act in opposite ways. Therefore, electrons are said to be negatively charged and protons, positively charged. Neutrons carry no charge, so they are neutral.

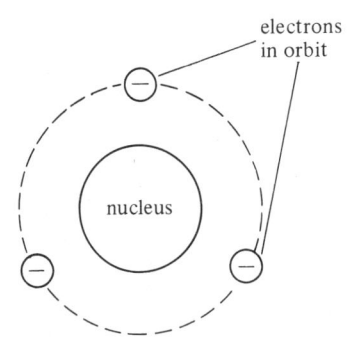

FIGURE 2.1 The Atom as Described by Niels Bohr.

2.3 IONS

A normal atom has as many electrons in orbit as it has protons in its nucleus. The negative charges thus cancel or balance out the positive charges, leaving the atom electrically neutral.

Suppose that an atom picked up an extra electron somehow. It is no longer neutral because it now has more negative charge than positive. As such, it acts like any other negatively charged body. On the other hand, let a normal atom lose an electron and it becomes positively charged because there is an excess proton in the nucleus.

An atom that has become a charged body because of lost or gained electrons is called an *ion*. It is a positive ion if electrons were lost, a negative ion if electrons were gained.

EXAMPLE 2.1

In its neutral state an atom has eight electrons. Answer the following questions.

a. How many protons are in the atom's nucleus?

| Since the atom is neutral, there must be eight protons in the nucleus.

b. The atom picks up two electrons. What kind of ion does it become?

| The atom is now a negative ion because there are not enough protons to neutralize the newcomers.

c. The ion loses five electrons. What is it now?

| It now has three too many protons and has become a positive ion. ■

2.4 CHARGE TRANSFER

Just as an atom becomes charged by gaining or losing electrons, so can other things. A pocket comb, for example, becomes negatively charged when rubbed with a piece of wool or fur because it acquires electrons. As such, it attracts bits of paper or makes our hair stand up. We say that this is caused by *static electricity*, that is, the effects of charge at rest (static). A more correct name is *electrostatic attraction*.

Charge also accumulates on our bodies as we cross a rug or slide over a car seat on a crisp winter day. The charge remains static until we touch something. Then it drains from our body to the uncharged object. As it does, the charge is no longer static electricity but *dynamic* electricity, that is, charge in motion. We feel our body's reaction to this moving charge as a shock.

It should be understood that charge is not created when a comb is rubbed with fur, just transferred from one body to the other. Only after rubbing, when the fur lost electrons to the comb, did it take on a positive charge and the comb a negative one.

Actually, rubbing is not necessary to cause charge transfer between bodies; they only need to touch. But rubbing increases the surface area contacted in a given time.

In the examples above, the same results would have been obtained if protons had been swapped between bodies instead of electrons. Nature has not seen fit to have this happen, however.

2.5 THE COULOMB

How do we measure charge? The charge of a single electron or proton could be used, but that amount is so small that is is impractical. Instead, we use the *coulomb*, named in honor of Charles Augustin de Coulomb of Paris (1736–1806), who made great contributions to the study of electricity.

1 coulomb (C) equals the charge of 6.24×10^{18} electrons.

Charge in coulombs is represented by the letter, Q.
We often deal in smaller units of charge such as the microcoulomb, 10^{-6} C.

EXAMPLE 2.2

Two coulombs of negative charge were transferred from body *A* to body *B*. Answer the following questions.

a. How many electrons changed bodies?

$$\text{number of electrons} = 2 \text{ C} \times \frac{6.24 \times 10^{18} \text{ electrons}}{\text{C}}$$

$$= 12.5 \times 10^{18} \text{ electrons}$$

b. What is the polarity of *A* with respect to *B* (assuming that both bodies had the same charge to begin with)?

Since *A* lost negative charge, it must be positively charged with respect to *B*. Conversely, *B* must be negatively charged with respect to *A*.

EXAMPLE 2.3

What charge in microcoulombs does a body acquire if it loses 26.2×10^{12} electrons?

Since the body lost electrons, it becomes positively charged.

$$Q = 26.2 \times 10^{12} \text{ electrons} \times \frac{1 \text{ C}}{6.24 \times 10^{18} \text{ electrons}} \times \frac{10^6 \text{ }\mu\text{C}}{\text{C}}$$

$$= 4.2 \text{ }\mu\text{C}$$

2.6 THE ELECTRIC FIELD AND COULOMB'S LAW

How is it possible for a charged comb to attract bits of paper? The answer to that question is not known. It is known, however, that charged bodies have a mysterious field all around them. This field is called an *electric field*.

By means of their electric fields, charged bodies push and pull each other over large distances. For example, a positively charge body attracts or pulls a negatively charged body toward it: *Unlike charges attract.* Two bodies charged the same, either both positive or negative, push away from each other: *Like charges repel.*

The pushing and pulling forces between charged bodies act a lot like those of gravity and magnetism. They are called *electrostatic forces*.

2.6 THE ELECTRIC FIELD AND COULOMB'S LAW

Coulomb studies these forces. He found they increased as the charge on the two bodies increased and decreased as the bodies got farther apart. The exact relation between force, charge, and distance is described by a mathematical statement known as *Coulomb's law*:

$$F = \frac{kQ_1Q_2}{d^2} \tag{2.1}$$

where F = force between bodies 1 and 2, in *newtons* [the SI unit of force; 1 newton (N) = 0.23 pounds (lb)]
k = constant needed to make the force come out in newtons
$= 9 \times 10^9 \dfrac{\text{newton-meters}^2}{\text{coulomb}^2}$
Q_1 and Q_2 = charge on body 1 and 2, respectively, in coulombs
d = distance between bodies, in meters

*** *A Word of Warning Before We Continue* ***

All calculations for Coulomb's law must be done using only the basic units in which the law is defined: namely, *newtons*, *coulombs*, and *meters*. Do not be fooled by problems that give these parameters in any other form (such as distance in feet or charge in microcoulombs). This warning holds for all definitions in later chapters.

The force, F, in Equation 2.1 has an algebraic sign. When that sign is negative, the force is one of attraction—the bodies try to move toward each other. A positive sign indicates repulsion—the bodies try to move away from each other.

EXAMPLE 2.4

Find the electrostatic force between body 1 charged to 6 and body 2 charged to −4 C. The distance between the two is 0.006 meter (m).

$$F = \frac{kQ_1Q_2}{d^2}$$
$$= \frac{9 \times 10^9 \times 6 \times (-4)}{(6 \times 10^{-3})^2}$$
$$= -6 \times 10^3 \text{ N}$$

Since the sign of the result is negative, the force is one of attraction. ■

EXAMPLE 2.5

Two charged bodies experience a repelling force between them of 3.2×10^2 N. One body has a charge of 3 μC. It is 15 millimeters (mm) from the second body. What is the magnitude and sign of the second charge?

Since we are looking for Q_2, the expression for Coulomb's law must be solved for Q_2 using algebra. (*Note:* It is usually easier to do as much work as possible in algebra first before putting numbers into an expression.)

$$F = \frac{kQ_1Q_2}{d^2}$$

Solving for Q_2, we have

$$Q_2 = \frac{Fd^2}{kQ_1} = \frac{3.2 \times 10^2 (15 \times 10^{-3})^2}{9 \times 10^9 \times 3 \times 10^{-6}}$$
$$= 2.7 \times 10^{-6} = 2.7 \ \mu C$$

The sign of Q_2 is positive since the force is repelling. ∎

The electrostatic forces described by Coulomb's law are found throughout electronics. In the CRT (*cathode ray tube*) found in oscilloscopes, for example, they are used to move, or *deflect*, a beam of electrons. Figure 2.2, part (a), shows a simplified diagram of a CRT. The electron gun shoots a beam of electrons at the front screen. The screen is coated with chemicals, called *phosphors*, that glow when hit by the electrons.

Note how the beam passes between two *vertical deflection plates*. If the upper plate is positively charged, it will attract the electrons in the beam. This causes the beam to bend, moving the light spot up to position *A* on the screen. Make the lower plate positive and the beam moves down to *B*. Vary the charge levels and the spot moves up and down. Vary the charge on the *horizontal deflection plates* [Figure 2.2), part (b)] and the beam moves horizontally. By changing the charge on both sets of plates, the spot of light can be moved anywhere on the screen.

Static electricity is important to other fields besides electronics. For instance, it helps in the fight against air pollution by removing dust particles from factory smoke. This is done in the *electrostatic precipitator* (also called the Cottrell precipitator after its inventor, Frederick Cottrell).

Smoke, heavy with solid particles, enters the precipitator, where it pases through a screen connected to a source of high voltage. The particles take on the same charge as the screen. They next pass over a collecting electrode which has a charge opposite to that of the screen. Because of the opposite charges the dust is attracted to the electrode, where it falls into a hopper for disposal.

Another use of static electricity is in the production of sandpaper. Here the bare paper, smeared with glue, is given one charge and the sand waiting for application the other. The attractive force between sand and paper causes the sand to spread evenly over the paper. The copy machine, which has become so important for photostating, also works on electrostatic principles.

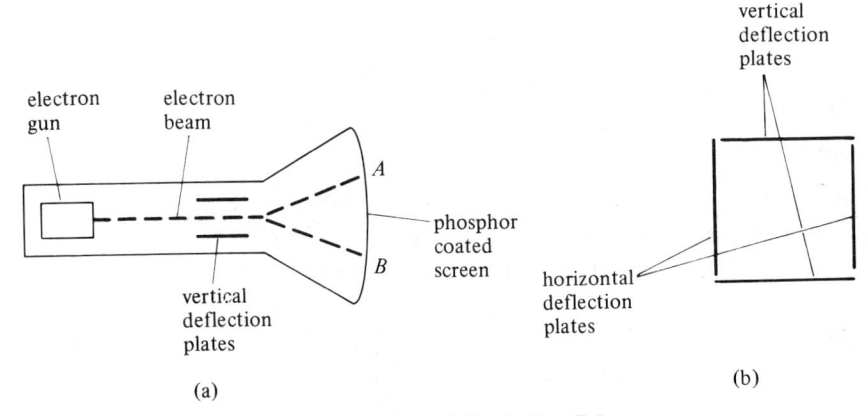

FIGURE 2.2 Coulomb's Law at Work in a Cathode Ray Tube.

2.7 ENERGY

Static electricity is not without problems, however. It destroys integrated-circuit "chips," attracts dust to high-voltage terminals such as TV picture tubes, and its sparks are a potential source of fire and explosion.

What is *energy*? Physics defines it as the ability to do *work*. Work is further defined as a force pushing a body over a distance. By this definition, a man can push a car with locked brakes all day and even though he is exhausted, he will not have done any work. But release the brakes, move the car, and the man has exerted energy; he has worked.

The formula for work or energy is

$$W = FD \qquad (2.2)$$

where W = work or energy expended in *joules* (pronounced *jewels*) in the SI system, foot-pounds in English
F = force, in newtons (SI) or pounds (English)
D = distance over which force was applied, in meters (SI) or feet (English)

The joule is named for an English physicist, James Joule (1818–1889). The foot-pound is the unit of energy in the English system. It equals the energy needed to raise a 1-pound object 1 foot. One foot-pound equals 1.36 joules (J).

EXAMPLE 2.6

A force of 17×10^3 N is required to push something 53 centimeters (cm). What energy was expended?

Here is that conversion trap again—centimeters must be converted to meters.

$$W = FD$$
$$= 17 \times 10^3 \text{ N} \times 53 \text{ cm} \times \frac{1 \text{ m}}{100 \text{ cm}}$$
$$= \frac{17 \times 10^3 \times 53}{100} = 17 \times 53 \times 10 \text{ J}$$
$$= 9 \times 10^3 \text{ J}$$

EXAMPLE 2.7

How many foot-pounds are expended by a 150-lb man climbing an 8-ft flight of stairs? Convert the answer to joules.

$$W = FD = 150 \text{ lb} \times 8 \text{ ft} = 1200 \text{ ft-lb}$$
$$= 1200 \text{ ft-lb} \times \frac{1.36 \text{ J}}{\text{ft-lb}}$$
$$= 1.6 \times 10^3 \text{ J}$$

Physics also teaches the principle of the *conservation of energy*:

| Energy cannot be made or destroyed, only changed from one form to another.

A car's engine, for example, converts chemical energy locked in gasoline to heat and mechanical energy.

2.8 ELECTRICAL ENERGY

Because of the pushing and pulling forces between charged bodies, energy can be carried in electrical form. To see how, suppose that we have two bodies each charged with +1 C. One of them is bolted down tightly; call it Q_f for fixed charge. The other is Q_m a movable charge. It will be placed miles from here, so far away that the force between the bodies is practically zero.

Now, suppose that you took Q_m and started bringing it closer to Q_f. Work must be done because you are bucking the natural repelling force of the electric field between bodies with the same charge. The distance decreases. You are struggling now. Coulomb's law says that the force quadruples each time the distance is cut in half. Imagine that you have wrestled Q_m right up next to Q_f. You are straining against the enormous force trying to push the bodies apart. Suddenly, you release Q_m. Imagine the result. Q_m leaps from your hands and disappears over the horizon from whence it came. As it does, the energy you added to the system in order to push the charges together is released.

Forcing like charges together is the electrical equivalent of compressing a spring (see Figure 2.3). It takes force to push the ends of the spring together. Free them and they return to their relaxed positions, doing work as they go, such as lifting a car or snapping a mousetrap.

Let's review the sequence of energy transfer in the system of charged bodies. When pushing Q_m toward Q_f, work was being done because a force, F, was needed to push Q_m a distance of D miles. It took force to overcome the natural repulsion between like charges. This is the same as pushing a car uphill against the force of gravity.

While pushing Q_m, energy was being stored in the system of charges. This energy is called *potential energy*. Since energy was being put in, we say that the system was going through a *potential rise*.

Then Q_m was released. During its return flight the stored potential energy was released in one form or another, perhaps as heat or light. This output of energy is called a *potential drop*.

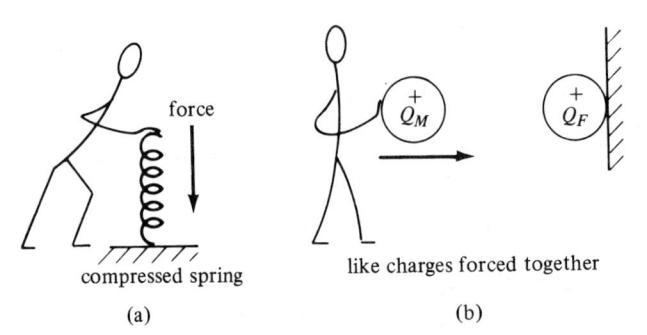

FIGURE 2.3 Energy Stored in (a) A Mechanical System and (b) An Electrical System.

2.9 ELECTRIC POTENTIAL   21

To summarize:

- A potential rise is putting energy into a system.
- A potential drop is taking energy out of a system.

Also, note carefully that energy was recovered only while Q_m was moving, not while it was static, standing still.

In electricity, charge in motion is called *current*:

To get energy out of an electrical system, there must be a current of moving charge. In other words, moving charge (current) has the capacity to transfer energy.

2.9 ELECTRIC POTENTIAL

In order to study electric potential, we return to our system of one coulomb charges, Q_f and Q_m (see Figure 2.4). As before, Q_f is fixed. Q_m lies at location A, which is so far from Q_f that the energy stored in the system is, for practical purposes, zero.

Again we force Q_m from point A toward Q_f, storing energy as we do; the system undergoes a potential rise.

Eventually, the one coulomb Q_m reaches a point (call it B) where its potential energy has risen 1 J above its value at point A. We say that the potential at point B is 1 joule per coulomb relative to point A:

A potential of 1 joule per coulomb is defined as 1 *volt* (V). The charge Q_m at point B therefore has a potential of 1 V relative to its potential at point A.

If we continue to push Q_m closer to Q_f, we arrive at point C, where 2 J is stored in the system. Q_m has now gone through a potential rise or *voltage rise* of 2 V with respect to A and 1 V with respect to B.

Release Q_m and it falls through a potential drop or *voltage drop* of 1 V in moving back to B. Let it continue through another drop of 1 V and it arrives in the zero energy position, A.

Another way to describe potential rises and drops is to call them *potential differences*. There is a potential difference of 1 V between points A and B and 2 V between A and C.

Thus we see that *voltage is the ratio of energy content to charge*. A 110-V home outlet supplies 110 J of energy for every coulomb that flows from it. A 12-V car battery, on the other hand, supplies only 12 J to the same coulomb.

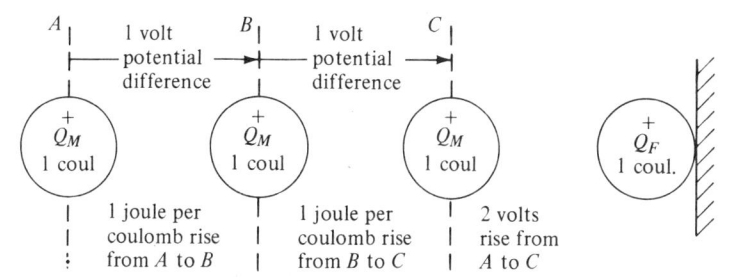

FIGURE 2.4 Voltage Rises.

Since voltage is the difference in energy levels between two points, we should be careful to define the points (see Exercise 23). Also, by its definition, voltage cannot move. People often say, "there's 12 volts coming out of that battery." This is incorrect. Charge is what flows from a battery. Voltage is the energy difference between two points that causes charge movement.

The formula for voltage is

$$V = \frac{W}{Q} \tag{2.3}$$

where V = volts
W = energy, in joules
Q = charge, in coulombs

The volt is named for an Italian scientist, Alessandro Volta (1745–1827).

Voltage has other names, such as potential, potential difference, EMF, (electromotive force), and high tension. By convention, the letter E represents a voltage rise, while V represents a voltage drop.

EXAMPLE 2.8

83.2 mC undergoes a voltage drop of 12 V. What energy loss took place?

Solving for W yields

$$W = VQ = 12 \times 83.2 \times 10^{-3}$$
$$= 998.4 \times 10^{-3} = 0.9984 \text{ J} \approx 1 \text{ J}$$

Note: The symbol $\approx$ means "is approximately equal to." ■

2.10 THE AMPERE

Current consists of charge moving under the pressure of a difference of electric potential, that is, a voltage. It is measured in units honoring the Frenchman André M. Ampère (1775–1836):

The SI unit, the *ampere* (A), is defined as the movement of 1 coulomb of charge past a point in 1 second.

The French speak of the *intensity* of charge flow, so the letter I represents current in amperes (or amps). The formula for current is

$$I = \frac{Q}{T} \tag{2.4}$$

where I = current, in amperes
Q = charge, in coulombs,
T = time, in seconds

Current is usually a flow of electrons, but sometimes it is moving positive carriers. It may consist of negative and positive carriers together.

EXAMPLE 2.9

If 3.4 C pass a point in 0.25 second (s), what current is flowing?

$$I = \frac{Q}{T} = \frac{3.4}{0.25} = 13.6 \text{ A}$$

EXAMPLE 2.10

How many charge carriers pass if 0.87 A flows for 215 milliseconds (ms)?

$$I = \frac{Q}{T}$$

Therefore,

$$Q = IT = 8.7 \times 10^{-1} \times 2.15 \times 10^{-1}$$
$$= 18.7 \times 10^{-2} \text{ C}$$
$$\text{carriers} = 18.7 \times 10^{-2} \text{ C} \times \frac{6.24 \times 10^{18} \text{ carriers}}{\text{C}}$$
$$= 11.7 \times 10^{17} \text{ carriers}$$

2.11 CONDUCTORS AND INSULATORS

We have seen that current consists of vast numbers of charged particles moved by voltage. These charges are usually *free electrons*, "free" meaning released, able to move.

Any material with lots of free electrons carries current easily and is called a *conductor*. Most metals, such as silver, aluminium, and copper, are excellent conductors.

On the other hand, glass, plastics, rubber, and many other materials have almost no free electrons and will not conduct electricity. They are called *insulators*.

Conductors, usually in the form of wires, carry current to where it is wanted. Insulators stop current from going where it is not wanted.

Solid-state electronics uses two *semiconductors*, germanium and silicon. These have fewer free electrons than conductors but more than insulators.

2.12 CONDUCTION BY ELECTRONS

What are free electrons? How does a material get them and why does one material have so many and another hardly any?

A free electron is one that has managed to break the electrostatic forces holding it to an atom. To do this, the electron must acquire energy from some source. That source is usually thermal energy (heat), which is everywhere, but light and other forms of radiation may be used.

How much energy an electron needs to break away is important because it determines whether a material is an insulator or conductor. For example, electrons in conductor atoms need almost no energy to break free. What little they need is easily obtained from heat; consequently, vast numbers are available to carry current. Insulator electrons, on the other hand, need so much energy to break loose that few manage to do so.

If extreme voltage is applied across an insulator, *breakdown* may occur. This is the voltage at which the forces of Coulomb's law become so great that electrons are ripped from their atoms. The material switches suddenly from being an insulator to a conductor, with results that can be disasterous.

The voltage required to cause breakdown varies between materials. It is one measure of insulator quality. A 1-milliinch layer of air, for example, can stand about 10 V before breakdown. The same thickness of ceramic will take about 100 V and some plastic films about 4 kilovolts (kV).

How do electrons travel in a conductor? Under the influence of heat and in the absence of an electric field, they move within a material in a completely haphazard manner [see Figure 2.5, part (a)]. As many go one way as the other. The opposite motions cancel one another, so net movement is zero; there is no transfer of charge.

Connect a voltage to the conductor and an electric field is set up within the conductor. This field causes electron motion to change in a very important way. While the aimless wanderings continue, there is now a net movement or *drift* toward the positive polarity of the voltage. This movement is a current.

The average velocity of an electron in a current has been calculated to be about 0.00045 mi/h. This surprises anyone who thought electricity moved at the speed of light. It does, almost. But individual electrons do not move that fast. What does, then?

Consider a line of billiard balls laid out as shown in Figure 2.6. A ball coming from the right hits the end ball and almost instantaneously another shoots out from the opposite end. It is not the same ball, but since they are all alike, the result is as though the first ball had traveled down the line in almost zero time.

Apply our billiard-ball analogy to a table lamp. An electron enters the conductor at the wall plug and repels other electrons already in the wire. They, in turn, repel their neighbors, and so on. This repulsion travels at nearly the speed of light. At the lamp, it forces an electron into the bulb, which repels another into the return line to the plug.

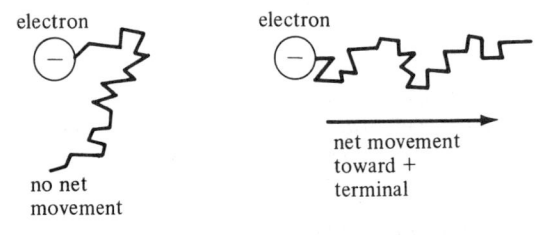

a. random motion

b. drift due to electric field

FIGURE 2.5 Random Motion and Drift of an Electron in a Conductor.

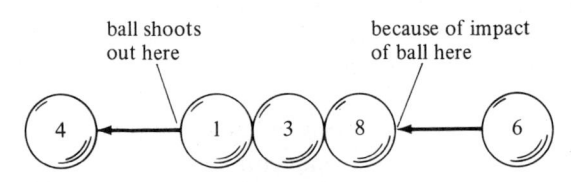

FIGURE 2.6 Billiard Ball Analogy.

2.14 CONVENTIONAL CURRENT VERSUS ELECTRON CURRENT

For the repelling effect to push charge, we see that a *continuous conductive path* from plug to bulb back to plug must exist. This is called a *closed path*. Break that path in any way, such as with a switch, and charge movement, current, comes to a halt. This is called an *open path*.

2.13 CHARGE MOVEMENT BY POSITIVE CARRIERS

Current can be the movement of positive carriers, too. One of these, the ion, has already been mentioned. It is the carrier when current flows through a liquid, or *electrolyte*, as in a lead-acid car battery.

Another positive carrier is the *hole*, so important to transistors and other semiconductor devices. Holes are studied in transistor courses.

2.14 CONVENTIONAL CURRENT VERSUS ELECTRON CURRENT

Since carriers of either or both polarity can be present at the same time, how do we define current, as the movement of positive or of negative carriers? Regardless of which choice is made, it will sometimes be wrong.

Current direction should have been defined in terms of the electron, for they are by far the more important carrier. Unfortunately, they had not been discovered when Benjamin Franklin (1706–1790) is said to have made the decision that current is the flow of positively charged bodies. Franklin's decision, although wrong, stuck.

> A movement of positively charged bodies is called a *conventional current* (it is sometimes called a *Ben Franklin current*). A movement of electrons is called an *electron current*. The two always travel in opposite directions.

Throughout this book, current will always be conventional current.

Actually, we seldom care about the charge nature of current. Our main concern is that we have the right current in the right place at the right time.

EXERCISES

Sections 2.1 through 2.5

1. Explain in your own words.
   a. Charge neutrality.
   b. Ionization.
   c. What determines whether a body is negatively or positively charged.
2. Do we create charge when rubbing a comb with fur? Explain.
3. Convert as indicated. (Refer to Table 1.1 for the values of kilo, pico, etc.)
   a. 13.2×10^3 C to electrons.
   b. 74.7×10^{-5} C to electrons.
   c. 50 μC to electrons.
   d. 15×10^{12} electrons to coulombs.
   e. 62 gigaelectrons to coulombs.
   f. 19.7×10^{11} megaelectrons to coulombs.
4. Repeat Exercise 3 for the following.
   a. 750×10^4 C to electrons.
   b. 12 mC to electrons.

c. 37.3 kC to electrons.
d. 17×10^{13} electrons to coulombs.
e. 4×10^{25} megaelectrons to coulombs.
f. 37.4 picoelectrons to femtocoulombs.

5. A glass rod, rubbed with a silk cloth, loses 50 C to the silk. What is the charge on the rod with respect to the silk, and vice versa?
6. Express the number of electrons in a coulomb in terms of megaelectrons, picoelectrons, kiloelectrons, microelectrons, and gigaelectrons.

Section 2.6

7. What is the magnitude and direction of the force between body A whose charge is 85 nC and body B charged to -65 nC? Distance between the bodies is 10^{-4} m.
8. Repeat Exercise 7 for $A = -16\ \mu$C and $B = -37\ \mu$C. The distance is 4 m.
9. Two bodies, separated by a distance of 3000 cm experience a force of 5 kN trying to bring them together. If one body is charged positively to 100 μC, find the charge on the other.
10. What is the new force between the two bodies of Exercise 9 if the distance between them is halved? Quartered?
11. State for each of the following whether the particles attract, repel, or do nothing.
   a. Two protons.
   b. Proton and electron.
   c. $(+)$ and $(-)$.
   d. Two electrons.
   e. Positive and negative.
   f. $(+)$ and $(+)$.
   g. Like charges.
   h. Positive and positive.
   i. Negative and negative.
   j. Unlike charges.
   k. Two neutrons.
12. How many inches separate two charged bodies, $A = 23$ mC and $B = 12$ mC, if a force of 0.005 N exists between them?
13. A car battery has $+$ and $-$ terminals. Toward which terminal will each of the following be attracted?
   a. Positive ion.
   b. Electron.
   c. Neutron.
   d. Negative ion.
   e. Proton.
14. *Extra Credit.* Explain how the charge level on the horizontal deflection plates of a CRT [see Figure 2.2, part (b)] should be varied to cause the electron beam to move from the left to the right side of the screen at a constant velocity.

Sections 2.7 and 2.8

15. We have investigated the storage of energy in a system of two like charges. Discuss the means whereby it could be done with two unlike charges.
16. Explain each term in your own words.
   a. Energy.
   b. Potential rise and drop.
   c. Stored energy.
17. Could energy be stored in a system of two magnets? Discuss.
18. What is the original source of all energy here on earth?
19. Three aircraft fly as follows: A is at 2000 f, B is at 2500 f, and C is at 3500 f. Tell whether a potential rise or drop is involved for each of the following.
   a. A moves to C.
   b. B moves to A.
   c. B moves to C.
   d. C moves to A.
20. What must charge be doing in order to recover its stored energy?
21. Given: a negatively charged body locked in place and several situations involving movable charges. What occurs in each case; is energy released or stored?
   a. Positive charge moved away.
   b. Negative charge moved away.
   c. Positive charge moved nearer.
   d. Negative charge moved nearer.
22. Repeat Exercise 21, stating whether the system underwent a potential rise or drop in each case.

Section 2.9

23. Which of the following statements is correct?
a. There is a signal of 17 mV at point M.

EXERCISES

b. There is a signal of 35 µV between the antenna terminals.
c. The newspaper reported that the victim was electrocuted when 10kV passed through him.

24. Explain in your own words.
a. The volt.
b. The terms "voltage rise" and "drop."
c. Why voltage should always be referred between two points.

25. Fill in the blanks.

	E(V)	W(J)	Q(C)
a.	4,110	0.037	
b.		105	3.9
c.	0.625		0.008
d.	535		0.63
e.	15,400	8,750	
f.		200,000	5×10^{-4}

26. 85×10^4 C releases 17×10^7 J while transferring between two terminals. What is the voltage?

27. Given: an EMF of 15 V between two terminals. How many joules are transferred by 1 C moving "downhill" from the high-energy terminal to the low-energy terminal? How many joules are transferred if the voltage is doubled? Discuss the effect on energy transfer of a voltage rise.

28. 37.4 nC undergoes a voltage drop of 59 mV. What energy was released?

29. What are some of the voltages you might encounter in a typical day?

30. Convert values as indicated.
a. 5 V to millivolts.
b. 1 V to microvolts.
c. 5×10^6 V to megavolts.
d. 3 MV to kilovolts.
e. 1000 mV to volts.

Section 2.10

31. Fill in the blanks (note the time units carefully).

	I(A)	Q(C)	T
a.		80×10^{-3}	2×10^{-3} s
b.	3.9		1897.4 s
c.	270×10^{-3}	85	s
d.		1,200	2 m
e.	18.5×10^{-6}		33.5×10^{-3} s
f.	10	72,000	h

32. A current of 4.73 A flows for 28.2 s. How much charge moves?

33. A 16-gage wire is rated to carry 15 A. Can it handle the movement of 9.25×10^4 C along it in 1.7×10^3 s?

34. How long does it take 55 µA to transport 37.6 pC?

35. What current flows as 21.4×10^{21} carriers move along 2.3 mi of wire in 1.79 min?

36. How long does it take 49×10^{27} carriers to move along 22.3 mi of wire, diameter 15.9 milliinches, at a current of 9.3 A? Express answer in days.

37. Convert each figure as indicated:
a. 0.006 A to milliamperes.
b. 5×10^7 µA to amperes.
c. 0.0075 mA to microamperes.

38. *Extra Credit.* Using the basic equations for voltage and current, do a dimensional analysis to prove that voltage × current × time = energy.

Sections 2.11 through 2.14

39. Explain each of the following in your own words:
a. Free charge carrier.
b. The characteristic of a material that determines whether the material will be a conductor, semiconductor, or insulator.
c. The billiard-ball analogy.
d. Why it is necessary to have one conductor from a wall socket to a lamp and another back to the socket.
e. A closed path and an open path.
40. Give one example each of a conductor, a semiconductor, and an insulator.
41. What is the function of a conductor? Of an insulator?
42. Name a device that depends on insulator breakdown to work properly. *Hint:* Think automobiles.
43. What is the difference between conventional and electron current?

CHAPTER HIGHLIGHTS

Charge has an electric field by which it can apply force over distance. The force attracts for opposite charges, repels for like charges

Charge is measured in coulombs. Charge movement, in coulombs per second, is measured in amperes.

Energy is the ability to do work. Work is force applied over a distance.

Coulomb's law describes the electrostatic forces between charged bodies. Because of these forces, energy can be transported by electricity.

An electric field from a battery or generator causes charge to move. Only moving charge can release energy.

Voltage causes charge movement and therefore, energy transfer.

Voltage tells the amount of energy carried by a given amount of charge. Charge moves through conductors, not voltage.

Energy is put onto charge in a voltage rise. Energy is taken off charge in a voltage drop.

The availability of free electrons determines whether a material will be a conductor or insulator.

The effects of electricity move at a speed close to that of light, but single charges move very slowly.

Formulas to Remember

1 C equals the charge on 6.24×10^{18} electrons.

$$F = \frac{kQ_1Q_2}{d^2} \quad \text{Coulomb's law}$$

$$W = FD \quad \text{work or energy}$$

$$V = \frac{W}{Q} \quad \text{voltage}$$

$$I = \frac{Q}{T} \quad \text{current}$$

CHAPTER 3

The Basic Circuit

In this chapter we use all the concepts of electricity from Chapter 2 to study the basic circuit.

At the end of this chapter, you should be able to:

- Name the parts of the basic circuit and explain their function.
- Discuss the energy changes taking place in a circuit.
- Explain what polarity means.
- Describe resistance and conductance.
- Read resistance values and tolerance by the EIA color code.
- Draw the symbols for fixed and variable resistances.
- Calculate I, E, and R by Ohm's law.
- Explain the difference between power and energy.
- Calculate power.
- List important resistor characteristics.
- Determine the efficiency of a system.
- Describe the ideal short and open circuit and tell why the short can be dangerous.
- Use the ground symbol in schematic diagrams.
- Identify some of the many waveforms used in electronics and do calculations with them.

3.1 THE BASIC CIRCUIT

There are four parts to the basic electric circuit:

- *Source:* changes another form of energy to electrical energy and puts it on moving charge. Has a continuous voltage rise between its terminals.
- *Load:* takes electrical energy from moving charge and converts it into a form we want, such as heat or sound. Has a voltage drop between its terminals.
- *Conductors:* carry current between source and load.

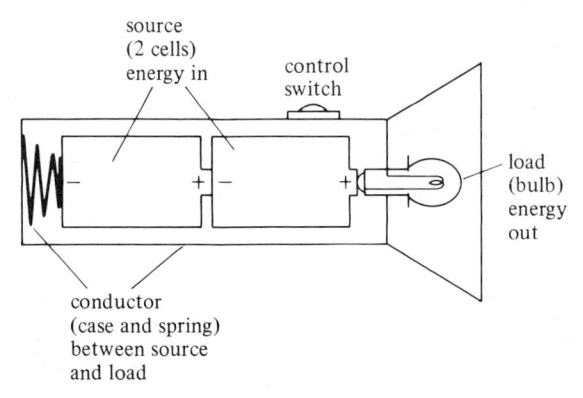

FIGURE 3.1 The Flashlight as a Basic Circuit.

3.2 ENERGY TRANSFER IN THE BASIC CIRCUIT

- *Control* and *safety devices:* control devices such as the switch provide a convenient way to open and close a circuit. Some provide a way to vary current. For example, a car's dashboard lights are dimmed by adjusting a *rheostat* (pronounced ree-o-stat). Safety devices such as the fuse and *circuit breaker* protect a circuit from damage due to excessive current.

A flashlight is a basic circuit (see Figure 3.1). Its source of one or more cells or batteries sends current through a control switch to a bulb, which is the load. Because one terminal of the bulb touches a battery terminal, only one conductor is needed between source and load. That is usually the flashlight case.

3.2 ENERGY TRANSFER IN THE BASIC CIRCUIT

Let's study the energy transfer taking place in the basic circuit.

Figure 3.2 shows a closed circuit. Note the closed switch and unbroken path between source and load. Charge moving along the bottom conductor has low energy. It enters the lower source terminal and takes on energy as it is "pushed" uphill in a voltage rise. Its energy replaced, the charge, moving in a current, travels along the upper conductor to the load. Here it "coasts downhill" through an energy or voltage drop, transferring all its energy to the load, which converts that energy into another form, such as light or heat. The charge then returns to the source via the lower conductor, where its energy is replaced.

All energy supplied by the source is completely converted by the load, that is, in any circuit, energy input = energy output. Say this another way, $E_{rise} = V_{drop}$, and we have an important law of electricity which will be studied in Chapter 4.

Actually, part of the source energy never arrives at the load. It is wasted in the conductors, which are not perfect current carriers. However, the amount is usually so small that we neglect it.

Two questions:

- To what energy level will the charge be raised by the source?

Depends on source voltage. Remember, voltage was defined as, $V = W/Q$. Therefore, $W = VQ$, that is, energy is proportional to voltage.

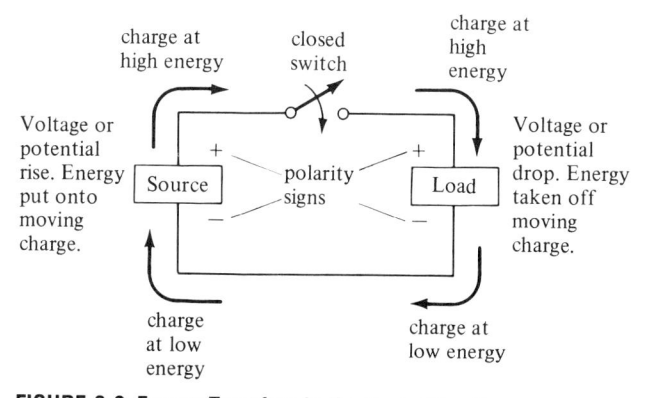

FIGURE 3.2 Energy Transfers in the Basic Circuit.

■ How long will the circuit operate?

As long as the source continues to have energy, it can convert to electrical energy. Flashlight batteries, for example, deliver energy for a time determined by the amount of chemicals they contain.

If the switch in Figure 3.2 is opened, current stops. There is still a voltage rise across the source, but no energy transfer because there is no charge moving. For the same reason, there is no voltage drop across the load.

3.3 POLARITY

The positive and negative signs shown with the voltages in Figure 3.2 are *polarity signs*. They indicate current direction and whether a component is a source or a load.

Remembering that we are using conventional current, here is how polarity is defined:

Terminals at which charge has *high energy* are marked *positive*. These are the source terminal from which current is leaving and the load terminal at which it is entering [see Figure 3.3, part (a)].
Terminals at which charge has *low energy* are marked *negative*. These are the source terminal at which current is entering and the load terminal from which current is leaving [see Figure 3.3, part (b)].

So, for a source, current leaves the positive terminal and enters the negative. For a load, current enters the positive terminal and leaves the negative.

The symbol for a battery shows polarity by using a longer line for the positive terminal [Figure 3.3, part (c)].

3.4 RESISTANCE

All materials oppose or resist the flow of current through them. This opposition is called *resistance*. It is represented by the letter R and measured in *ohms*. The Greek capital letter Ω (omega, pronounced o-may-guh) is the symbol for "ohms."

Insulating materials such as glass and plastics have *high resistance* and let very little current pass. On the other hand, *conductors*, such as most metals, have *low resistance* and let heavy currents pass.

(a) source — current enters negative terminal and leaves positive terminal — voltage rise

(b) load — current enters positive terminal and leaves negative terminal — voltage drop

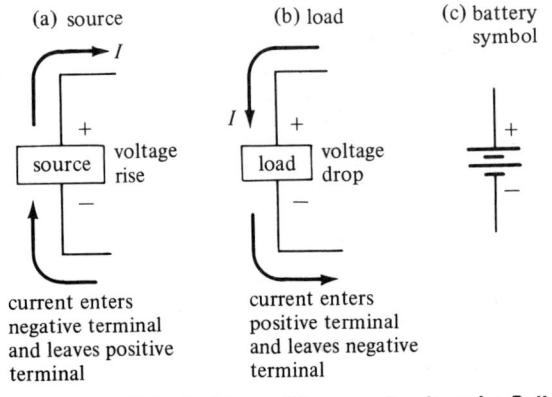

(c) battery symbol

FIGURE 3.3 Polarity Signs of Sources, Loads and a Battery.

3.6 CARBON-COMPOSITION RESISTORS

As friction causes heat in mechanical systems, resistance causes heat when current flows through a conductor. This heat is energy from the source that is lost before arriving at the load. Engineers call it *line loss*. To simplify analysis, we disregard line loss by considering conductors as being ideal; that is, they have zero resistance. Resistance is covered in greater detail in Chapter 8.

3.5 CONDUCTANCE

Conductance is the *inverse of resistance*. It is represented by the letter G and measured in SI units of *siemens* (pronounced *see-mens*), in honor of a German scientist, Werner von Siemens (1816–1892). Siemens (S) replaced the mho ("ohm" spelled backward), which is still found in many books.

$$G = \frac{1}{R} = R^{-1} \tag{3.1}$$

EXAMPLE 3.1

What is the conductance of a copper wire with a resistance of 3.9 Ω?

$$G = \frac{1}{R} = \frac{1}{3.9} = 0.26 \text{ S}$$

3.6 CARBON-COMPOSITION RESISTORS

Usually, we work hard to get rid of resistance in a circuit. Other times it is deliberately added to get certain results. This is done by wiring in a convenient bundle of resistance called a *resistor*.

Resistors are probably the most common component in electronic equipment. Their resistance values range from less than an ohm to millions of ohms, and they are sold in a wide variety of styles. The most popular of these is the *carbon-composition resistor*, shown in Figure 3.4. It is discussed here; others are described in Chapter 8.

The resistance of a carbon-composition resistor is obtained by squeezing a mixture of carbon dust, clay, and glue into a rod. The rod is covered with a protective, insulating layer and provided with connecting wires called *leads*, or *pigtails*, with which the unit is soldered into a circuit. Various resistance values are obtained by changing the ratio of carbon to clay.

Resistors are sold individually or by the hundreds mounted on long strips of paper for computer-controlled parts placement on printed-circuit boards.

Two of the many resistor characteristics we need to know are:

- *Resistance:* Given in ohms, kiloohms, and megohms (ohms is usually dropped from the last two and they are called simply, k and megs). Composition resistors are sold in values ranging from 2.7 ohms to 22 megohms.
- *Tolerance:* Resistors cannot be made with exact resistance values. Instead, a manufacturer guarantees the measured resistance of its products to fall within a certain percentage of the specified value. For example, a 10-k 10% unit will have a resistance between 11 and 9 k (10% above and below). This percentage is called the *tolerance*. Carbon-composition resistors are sold in 5, 10, and 20% tolerances; the higher the tolerance value, the lower the cost.

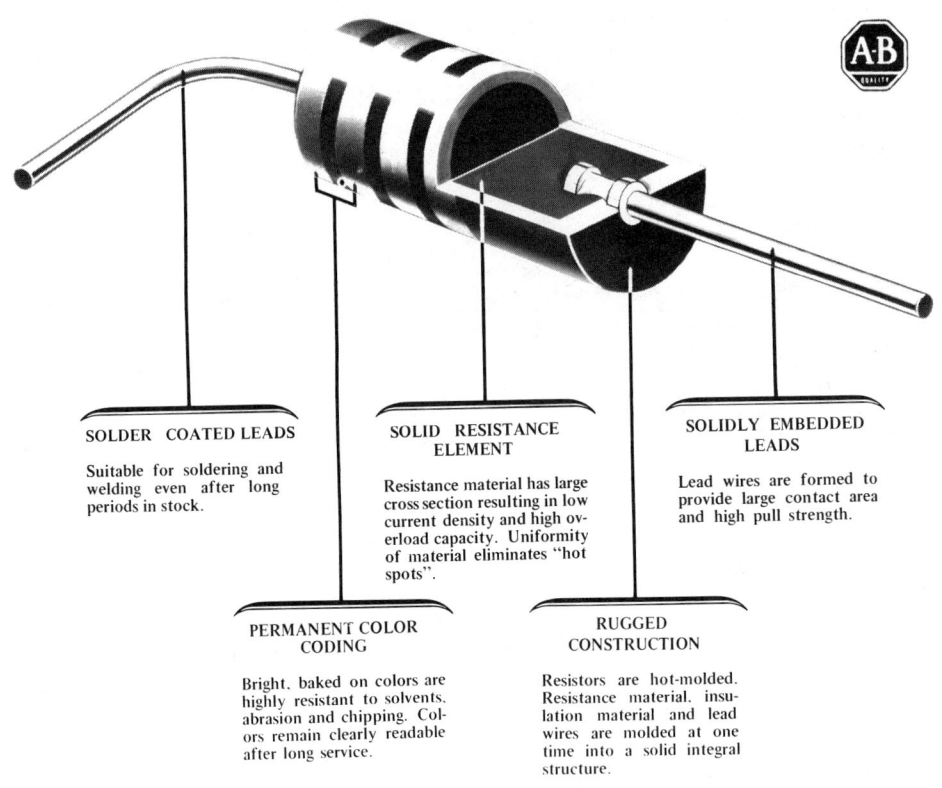

SOLDER COATED LEADS
Suitable for soldering and welding even after long periods in stock.

SOLID RESISTANCE ELEMENT
Resistance material has large cross section resulting in low current density and high overload capacity. Uniformity of material eliminates "hot spots".

SOLIDLY EMBEDDED LEADS
Lead wires are formed to provide large contact area and high pull strength.

PERMANENT COLOR CODING
Bright, baked on colors are highly resistant to solvents, abrasion and chipping. Colors remain clearly readable after long service.

RUGGED CONSTRUCTION
Resistors are hot-molded. Resistance material, insulation material and lead wires are molded at one time into a solid integral structure.

FIGURE 3.4 Carbon-Composition Resistor Construction and Symbol. (Courtesy Allen-Bradley, Inc.)

EXAMPLE 3.2

A 47 k, 20% tolerance resistor is measured on an ohmmeter and reads 37,450 Ω. Is this within tolerance?

$$20\% \text{ of } 47,000 \text{ is } 9400 \text{ }\Omega$$

The lowest value this reistance can have and still be within tolerance is

$$47,000 - 9400 = 37,600 \text{ }\Omega$$

Since the measured value is less than this, the unit is out of tolerance. ∎

Resistance and tolerance values are indicated by a group of three to five colored bands painted around the body of a composition resistor (see Figure 3.5). Notice that the bands are nearer one end. Placing that nearer end to our left, read the bands from left to right.

Bands 1 and 2 tell the significant value of the resistance. Band 3 is a multiplier. It indicates how many times the significant value is multiplied by 10. Band 4 (if there is one) states the tolerance. A fifth band indicates reliability, that is, the expected percentage of failures per 1000 hours of operation. It is found only on military-specification (called mil-spec) resistors.

3.6 CARBON-COMPOSITION RESISTORS

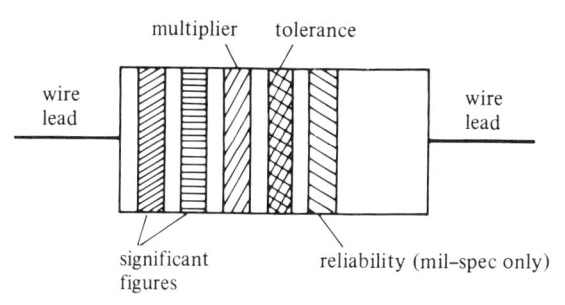

FIGURE 3.5 EIA Color Code on Carbon-Composition Resistor.

The numerical rating of each band is given by its color according to a code adopted by the EIA (Electronics Industry Association). The code is given in Table 3.1.

TABLE 3.1 EIA Color Code

Color	Band 1 First digit	Band 2 Second digit	Band 3 Color	Band 3 Multiplier
Black	0	0	Black	0
Brown	1	1	Brown	10
Red	2	2	Red	100
Orange	3	3	Orange	1,000
Yellow	4	4	Yellow	10,000
Green	5	5	Green	100,000
Blue	6	6	Blue	1,000,000
Violet	7	7	Silver	0.01
Gray	8	8	Gold	0.1
White	9	9		

Color	Band 4 Tolerance (%)	Color	Band 5 Reliability (%) (military only)
Gold	5	Brown	1
Silver	10	Red	0.1
No band	20	Orange	0.01
		Yellow	0.001

EXAMPLE 3.3

Given: resistors with color bands as indicated. Find their resistance and tolerance range.

a. Yellow, violet, orange, silver.

The first two bands tell us that the significant values are 4 and 7. The orange band says that the 47 is multiplied by 10 three times, that is, by 1000. Therefore, resistance is 47,000 Ω or 47 k. Band 4 indicates a tolerance of 10%, which means that the manufacturer guarantees the actual resistance will be 47,000 plus or minus 4700 Ω, or between 42.3 and 51.7 k.

b. Brown, black, black.

Bands 1 and 2 show a significant value of 10. The third band is black. This indicates that the significant value is multiplied by 10 zero times. Therefore, the resistance is 10 Ω. Tolerance is 20% since there is no fourth band. Final resistance value falls between 8 and 12 Ω.

c. Orange, blue, gold, gold, orange.

Bands 1 and 2 indicate 36. The third is a multiplier of 0.1; therefore, the resistance is 3.6 Ω. Tolerance, 5%. Final range, 3.42 to 3.78 Ω. The fifth band indicates a reliability rating of 0.01%, which says that the military expects 0.01% of these resistors to fail during the first 1000 hours of operation. ∎

3.7 OHM'S LAW

A German physicist, Georg Simon Ohm (1787–1854), studied the relation between voltage, current, and resistance. He found that current through a test sample was directly proportional to the voltage across it. This meant that the ratio of voltage to the current caused by that voltage was constant for a particular piece of material.

Ohm based his *law of constant proportionality* on that fact. Written as a formula, it is

$$\frac{E}{I} = k$$

where k = a constant for a particular piece of material.

Ohm noticed that other materials had the same constant voltage versus current relation but that the value of k would be different. He found that for the same voltage, current was low through materials of high k values and high through those with low k. In other words, whatever it was in the material that determined the k value acted to oppose or resist the movement of charge.

The k characteristic of a material was named *resistance*, which, as we know, is measured in ohms.

A resistance of 1 ohm has a voltage drop of 1 volt across it when a current of 1 ampere passes through it.

We state, then, *Ohm's law*, a fundamental relation between voltage, current, and resistance in a circuit that must be second nature to anyone in electronics:

$$\frac{E}{I} = R \tag{3.2}$$

where E = voltage, in volts
I = current, in amperes
R = resistance, in ohms

Using algebra, the law can be written in two other forms:

$$E = IR$$

and

$$R = \frac{E}{I}$$

3.7 OHM'S LAW

Let's examine each form to see what it is telling us about circuits.

- $E/I = R$: If a voltage of E volts is applied to a circuit and we want I amps to flow, R ohms must be in the circuit.

EXAMPLE 3.4

How much resistance is needed to limit current to 37 mA in a circuit with an applied EMF of 15 V?

$$R = \frac{E}{I} = \frac{15}{37 \times 10^{-3}} = 406 \ \Omega$$

- $E = IR$: This equation says two things: (1) to get I amperes to flow through R ohms, apply E volts; and (2) the voltage drop across R ohms when I amps flows is E volts.

EXAMPLE 3.5

What voltage is needed to push 120 μA through 47 k?

$$E = IR = 120 \times 10^{-6} \times 47 \times 10^3 = 5640 \times 10^{-3} = 5.6 \ V$$

EXAMPLE 3.6

How many joules per coulomb are lost as heat when 2.38 A flows in 46.22 Ω?

$$E = IR = 2.38 \times 46.22 = 110 \ J/C \ (\text{or volts})$$

- $I = E/R$: When R ohms is connected across E volts, I amperes will flow. Note carefully that I is proportional to E and inversely proportional to R.

EXAMPLE 3.7

Perform the following steps.

a. A toaster of 10.7 Ω is plugged into a 120-V outlet. What current will be drawn?

$$I = \frac{E}{R} = \frac{120}{10.7} = 11.2 \ A$$

b. What I flows when the toaster is connected to a 240-V outlet?

Since I varies directly with E and E doubled, I must double to 22.4 A.

c. The toaster is reconnected to the 120-V outlet, but its resistance has dropped 50% to 5.35 Ω. What I flows now?

Since R was halved, I doubles to 22.4 A, twice its value in part a. Note carefully that current can be increased by either an increase in E or a decrease in R.

If G is substituted for R in Ohm's law, we get the three equations in conductance form:

$$G = \frac{I}{E} \qquad E = \frac{I}{G} \qquad I = EG$$

To summarize the important points about Ohm's law:

- High resistance (low conductance) means low current.
- Low resistance (high conductance) means high current.
- Current can be increased by raising voltage or dropping resistance (raising conductance).

3.8 POWER

Although some people think *power* and *energy* are the same, they are not. A lawn-mower engine installed in a race car and given enough time will get the car around the track, but it is not likely to win any races because it trickles energy out. What is needed is a huge engine that pours out energy in a tremendous, winning blast. The big engine is more powerful than the lawn-mower engine, not because it delivers more energy (it doesn't), but because it delivers the same energy in a fraction of the time.

The definition of power:

> Power is the rate at which energy is being transformed from one form to another. It is represented by the letter P and measured in *watts*. One watt means that 1 joule of energy is being converted in 1 second. In math form,

$$P = \frac{W}{T} \tag{3.3}$$

where P = power, in watts
W = energy, in joules
T = time, in seconds

One horsepower (hp) = 746 watts (W).

EXAMPLE 3.8

300 megajoules are supplied by a machine in 1.8 min. What power is developed?

$$P = \frac{W}{T}$$

$$= \frac{3 \times 10^8 \text{ J}}{1.8 \text{ min} \times 60 \text{ s/min}}$$

$$= \frac{3 \times 10^8}{1.8 \times 60}$$

$$= 2.8 \times 10^6 \text{ W}$$

$$= 2.8 \text{ MW}$$

∎

EXAMPLE 3.9

a. How long does it take a 60 W light bulb to convert 3 kilojoules of electric energy to light energy (assuming 100% of the energy goes into light and not heat)?

$$P = \frac{W}{T}$$

Therefore,

$$T = \frac{W}{P} = \frac{3 \times 10^3 \text{ J}}{60 \text{ J/s}} = 50 \text{ s}$$

b. What is the horsepower of the bulb?

$$60 \text{ W} \times \frac{1 \text{ hp}}{746 \text{ W}} = 0.08 \text{ hp}$$ ∎

Power in electrical terms is

$$P = EI \tag{3.4}$$

where P = power, in watts
E = voltage, in volts
I = current, in amperes

Using Ohm's law, we can express the power formula in two other ways:

1. Substituting $I = E/R$ into Equation 3.4:

$$P = EI = E \times \frac{E}{R} = \frac{E^2}{R} \tag{3.5}$$

2. Substituting $E = IR$ into Equation 3.4:

$$P = EI = IR \times I = I^2 R \tag{3.6}$$

To summarize the three power equations:

$$P = EI \qquad P = \frac{E^2}{R} \qquad P = I^2 R$$

EXAMPLE 3.10

How many watts are being dissipated by the resistor in this circuit? (*Remember:* Conductors between source and load are considered ideal; that is, their resistance is zero.)

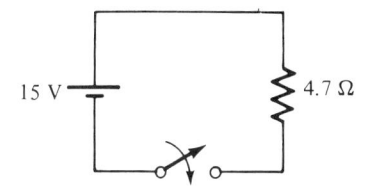

$$P = \frac{V^2}{R} = \frac{15^2}{4.7} = \frac{225}{4.7} = 47.9 \text{ W}$$ ∎

EXAMPLE 3.11

What current flows in the circuit of Example 3.10?

There are two ways to do this problem:

1. By Ohm's law:

$$I = \frac{E}{R} = \frac{15}{4.7} = 3.2 \text{ A}$$

2. Solving the power equation for I, we have

$$P = I^2 R$$

therefore,

$$I^2 = \frac{P}{R} \quad \text{and} \quad I = \sqrt{\frac{P}{R}}$$

Substituting values yields

$$I = \sqrt{\frac{47.9}{4.7}} = \sqrt{10.19} = 3.2 \text{ A} \quad \blacksquare$$

The heat generated when current flows through resistance is measured in watts. Carbon composition resistors get rid of their heat by providing a surface area that transfers it to the surrounding air. They come in five power ratings: $\frac{1}{8}, \frac{1}{4}, \frac{1}{2}$, 1, and 2 W. Figure 3.6 gives a comparison of sizes.

Resistors are often *derated*; that is, they are operated below their maximum power ratings in order to increase reliability.

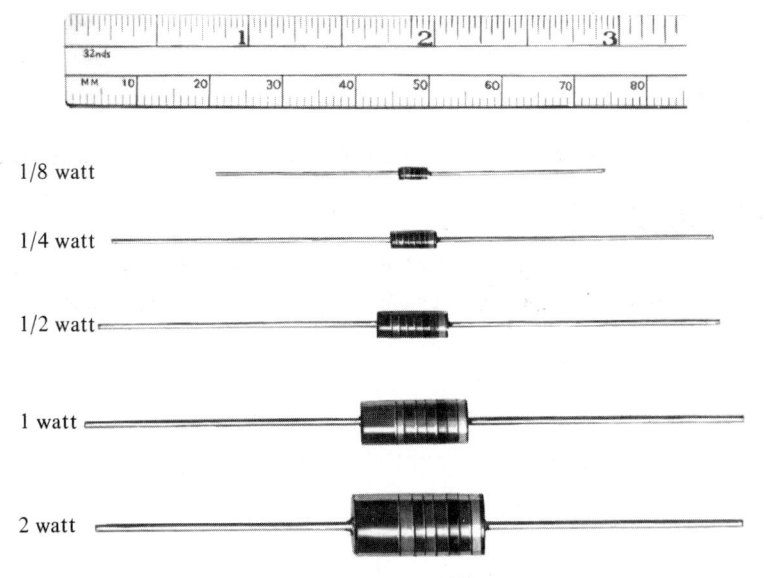

FIGURE 3.6 Power Rating and Sizes of Carbon Resistors.

3.9 EFFICIENCY OF ENERGY CONVERSION

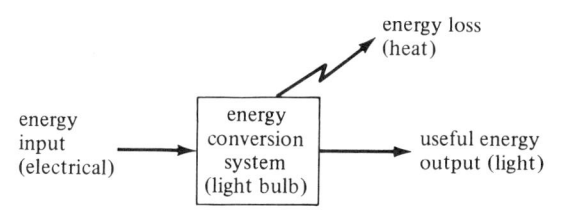

```
                    energy loss
                    (heat)
                        ↗
energy              ┌──────────┐
input               │  energy  │
(electrical)   ──→  │conversion│  ──→  useful energy
                    │  system  │       output (light)
                    │(light bulb)│
                    └──────────┘
```

FIGURE 3.7 Losses in Energy-Conversion System.

3.9 EFFICIENCY OF ENERGY CONVERSION

When we supply energy to a conversion system such as a light bulb, we would like to get all that energy out as useful work, in this case, as light. Unfortunately, much of the input energy is lost as heat (see Figure 3.7).

The useful energy out of a piece of equipment compared to the amount put in is called the *efficiency* of the device. Represented by the Greek lowercase letter, η, (eta; pronounced *ay-da*), percent efficiency is defined as

$$\%\eta = \frac{W_{out}}{W_{win}} \times 100 \tag{3.7}$$

Note: Multiplication by 100 converts to percentage form.

Equation 3.7 can be put in more practical form by rewriting it in terms of power. This is easily done using the fact that $W = PT$ from Equation 3.3. Substituting this into Equation 3.7 gives us

$$\%\eta = \frac{P_{out}\cancel{T}}{P_{in}\cancel{T}} \times 100$$

$$= \frac{P_{out}}{P_{in}} \times 100 \tag{3.8}$$

Because power output from a device equals power input minus power loss, the efficiency equation may be rewritten

$$\%\eta = \frac{P_{in} - P_{loss}}{P_{in}} \times 100 \tag{3.9}$$

Note that efficiency has no units.

EXAMPLE 3.12

A radio transmitter has a power input of 15 kW and a power output of 8.2 kW.

a. What is its percent efficiency?

$$\%\eta = \frac{P_{out}}{P_{in}} \times 100$$

$$= \frac{8.2 \times 10^3}{15 \times 10^3} \times 100$$

$$= 54.7\%$$

b. What power is lost as heat?

$$\begin{aligned}\text{Power loss} &= P_{in} - P_{out} \\ &= (15 - 8.2) \times 10^3 \\ &= 6.8 \text{ kW}\end{aligned}$$

Another way the losses could be determined is by solving Equation 3.9 for P_{loss}:

$$\eta = \frac{P_{in} - P_{loss}}{P_{in}}$$

$$\eta P_{in} = P_{in} - P_{loss}$$

Therefore,

$$\begin{aligned}P_{loss} &= P_{in} - \eta P_{in} \\ &= P_{in}(1 - \eta) \\ &= 15(1 - 0.547) \\ &= 15(0.453) \\ &= 6.8 \text{ kW}\end{aligned}$$ ■

Most energy-conversion systems are made up of parts called *subsystems*. A car, for example, consists of the engine, transmission, axle, and so on. Each subsystem has its own efficiency (see Figure 3.8).

How is a system's total efficiency determined when the efficiency of each subsystem is known? An equation will be derived. Notice the following in Figure 3.8:

$$P_{in_{total}} = P_{in_1}$$
$$P_{out_1} = P_{in_2}$$

and

$$P_{out_2} = P_{out_{total}}$$

Now

$$P_{out_1} = \eta_1 P_{in_1} = \eta_1 P_{in_{total}} = P_{in_2}$$

and

$$P_{out_2} = \eta_2 P_{in_2} = \eta_2 \eta_1 P_{in_1}$$

Substituting $P_{in_{total}}$ for P_{in_1} and $P_{out_{total}}$ for P_{out_2} yields

$$P_{out_{total}} = \eta_2 \eta_1 P_{in_{total}}$$

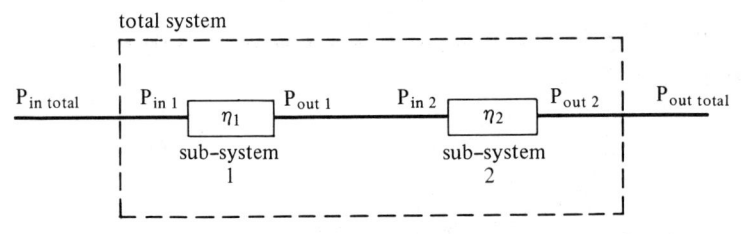

FIGURE 3.8 Finding Total Efficiency of a System Made-up of Sub-Systems.

3.10 SHORT CIRCUITS AND OPEN CIRCUITS

Divide both sides by $P_{in_{total}}$:

$$\frac{P_{out_{total}}}{P_{in_{total}}} = \eta_1 \eta_2$$

But

$$\eta_{total} = \frac{P_{out_{total}}}{P_{in_{total}}}$$

Therefore,

$$\eta_{total} = \eta_1 \eta_2$$

If the formula works for two subsystems, it will work for n subsystems. Therefore,

$$\eta_{total} = \eta_1 \eta_2 \eta_3 \cdots \eta_n \tag{3.10}$$

where η_n is the efficiency of the last subsystem.

EXAMPLE 3.13

The radio transmitter of Example 3.12 is made up of four subsystems:

P_{in} — [sub-system 1: $\eta_1 = 0.92$] — [sub-system 2: $\eta_2 = 0.88$] — [sub-system 3: $\eta_3 = 0.73$] — [sub-system 4: $\eta_4 = 0.926$] — P_{out}

What is the total system efficiency?

$$\eta_{total} = \eta_1 \eta_2 \eta_3 \eta_4 = 0.92 \times 0.88 \times 0.73 \times 0.926$$
$$= 0.547 \quad \text{or} \quad 54.7\%$$

This is the same answer as that found in Example 3.12.

People dream of building a perpetual-motion machine with an efficiency greater than 100%, one that puts out more energy than was put in. Such machines violate the laws of physics and cannot be built.

3.10 SHORT CIRCUITS AND OPEN CIRCUITS

To short-circuit (or short) two points, connect them with a conductor having zero ohms resistance [see Figure 3.9, part (a)]. With no resistance there can be current but no voltage between the points.

To open-circuit (or open) two points, leave them unconnected with infinite ohms between them [Figure 3.9. part (b)]. With infinite resistance there can be voltage but no current between the points.

These are the *ideal short* and *open*. Like all things ideal, they do not exist. Any conductor has some resistance and it is impossible to get infinite ohms. But the concepts are useful in circuit analysis, as will be seen in later chapters.

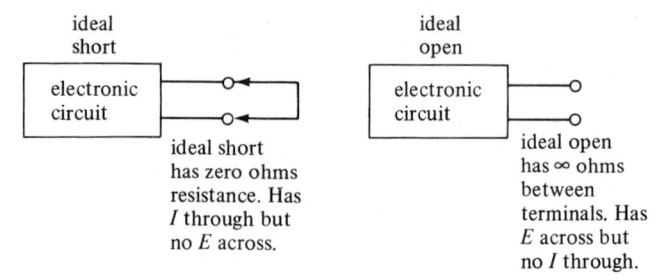

FIGURE 3.9 Ideal Short Circuit and Open Circuit.

ideal short has zero ohms resistance. Has I through but no E across.

ideal open has ∞ ohms between terminals. Has E across but no I through.

In real life, the term *short circuit* refers to a breakdown of insulation that bypasses the load in a circuit. This drops circuit resistance, causing current to rise sharply. Heat generation by the conductors between source and load rises even faster because it goes up with the square of the current ($P = I^2R$). The normally cool wires can get hot enough to cause a fire.

Fuses provide protection against the high currents of short circuits. They are connected so that source current passes through them. If that current rises above the fuse's protection value, a special conducting material melts and opens the circuit. Fuses are sold with capacities ranging from a fraction of an ampere to hundreds of amperes.

3.11 HOW CIRCUITS ARE DRAWN IN ELECTRONICS

Even though the diagrams of Figure 3.10 differ, they are electrically equal. Each shows a different way to draw the same schematic diagram.

Style (a) needs little comment; it has been used many times already. Style (b) leaves out the source and shows the 15 V applied between the terminals. Style (c) introduces the *ground symbol*. It is seen at the bottom of both source and load. The thing to remember about grounds is:

| All points in a circuit that have ground symbols are connected together.

In other words, the ground connections in circuit (c) do the same job as the lower conductor in circuit (a), connecting the bottom ends of the battery and resistor: *The two circuits are electrically identical.*

Styles (d) and (e) show how circuits are drawn most often in electronics. Their only difference is in the choice of ground symbols (the various symbols are explained in Section 3.12). Circuit (f) shows another symbol used less frequently.

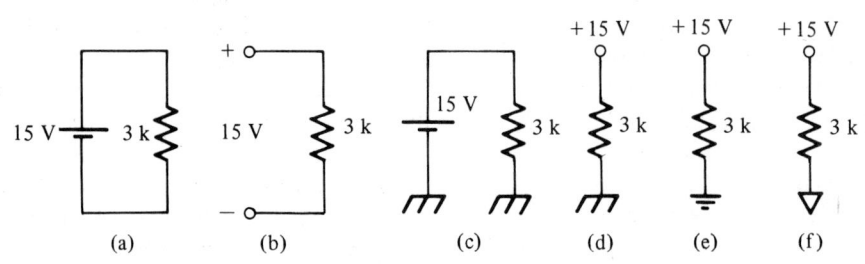

FIGURE 3.10 Six Ways to Draw the Same Circuit.

3.12 GROUND

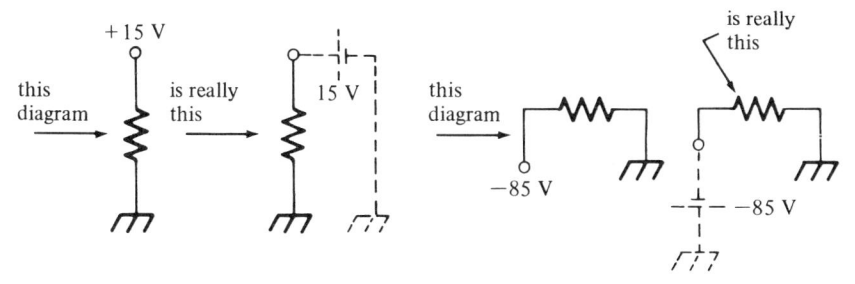

FIGURE 3.11 Understood Source Between a Terminal and Ground.

Note the terminal marked +15 V in diagrams (d), (e), and (f). Where is the source supplying this potential, and where is its negative terminal?

A terminal marked with only a voltage and a polarity is understood to have a source of that voltage connected to it. The source terminal of opposite polarity (unseen) is grounded.

Figure 3.11 shows two examples of this type of schematic. Note carefully the polarity of the grounded source terminal in each.

3.12 GROUNDS

Proper grounding techniques are extremely important to the proper operation of electronic circuits and in the reduction of electrical noise. There are two types: earth grounds and chassis grounds.

An *earth ground* is literally a connection to the earth. It is obtained by connecting to a system of buried wires or to one or more metal rods driven into the dirt. Used mostly by the power and radio communications industries, these grounds are important for safety reasons and for providing reference points of zero voltage and resistance. The earth-ground symbol is shown in Figure 3.10, part (e).

A *chassis ground* is a connection to the chassis (usually aluminum) on which a circuit's components are mounted. Besides acting as a return conductor such as that of Figure 3.10, part (c), chassis grounds also provide zero-voltage reference points. Their symbol is that of Figure 3.10, parts (c) and (d); the symbol of part (f) is seen on occasion. Many books use the earth-ground symbol for chassis ground.

To ground a point in electronics means to connect that point to the chassis. To measure the voltage across a component with one side grounded, place one voltmeter terminal on the "hot" ungrounded point and the other on ground, the chassis. The procedure is illustrated in Figure 3.12.

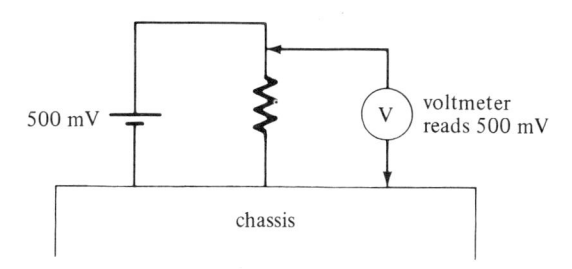

FIGURE 3.12 Reading Voltage Across a Component with One Side Grounded.

3.13 WAVEFORMS

So far, we have talked about the basic circuit in which a switch is closed, current flows, and a voltage drop is developed across a load.

Consider next a car's turn signal, in which current is switched on and off, over and over. Figure 3.13 shows the circuit. If the timer closes the switch for 1 s, opens it for the next, and so on, current flows in 1-s bursts—on one second, off the next. Graph this on–off relation between current and time and the result is a *current waveform*.

Figure 3.14 shows such a waveform. Note the following carefully:

- Time is plotted on the *X* or *horizontal* axis. The parameter plotted versus time, current in this case, is on the *Y* or *vertical axis*.
- Each axis has an arrow at the far end, indicating the direction of increasing value.
- The parameter being plotted and its units are shown clearly along each axis. For example, the *Y* axis indicates current, in amps.
- Each axis is divided into equal *increments*. The value of each increment is indicated clearly.

Next, study the plot or waveform itself, noting that:

- The switch is open, current equals zero, and the bulb is off whenever the plot lies on the *X* axis, for instance, between 0 s and the first second.
- When time equals 1 s, the switch closes and the waveform jumps to an amplitude of 6 A. The signal is on.

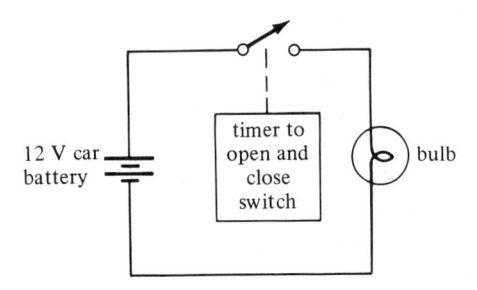

FIGURE 3.13 Circuit for Car's Turn Signal.

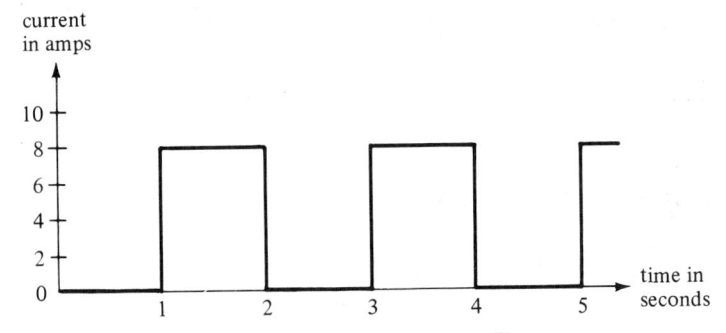

FIGURE 3.14 Plot of Turn-Signal Current Versus Time.

3.13 WAVEFORMS

- When time equals 2 s, the switch opens and current drops to zero. The signal is off.
- Current comes in 1-s bursts called *pulses*.
- The off-time (1 s) plus the on-time (1 s) is called the *period* of the waveform. In other words, the period is the *time needed to complete one entire off–on operation*. For this waveform the period is 2 s.
- All the pulses have the same *amplitude* and period. Therefore, the wave is said to be *periodic*. A periodic waveform is one that repeats over and over, always the same.
- The full name of the turn-signal current waveform is a *periodic, pulsed waveform*.

Plot the current pulses to an automatic door opener at a supermarket and they would not be evenly spaced, like those in Figure 3.14. In this circuit current flows only when a customer enters, and customers enter at random. As a result, the pulses would not have a fixed period, meaning that the waveform would not be periodic. It would be an *aperiodic* (pronounced *ay-periodic*) waveform.

Waveforms are drawn for either current, voltage, or power. There are many important ones in electronics. The simplest is *dc* or *direct current* from a battery. Dc is only used to power circuits. It has a dull waveform that rises to some fixed value and sits there (see Figure 3.15).

The most important wave is the *sine wave*, which will be studied in Chapter 13. There are two reasons for its importance. First, the electricity that runs the world travels in sine-wave form as *ac* or *alternating current*. Second, all periodic waveforms can be built out of some combination of sine waves. We have more to say on these points in later chapters. The sine wave is shown in Figure 3.15.

Other periodic waveforms used in electronics are shown in Figure 3.16. Note that time is always plotted on the X axis and either voltage, current, or power on the Y axis. Students of electronics study the circuits that generate these waveforms.

Aperiodic waveforms are important, too. They are made in a number of ways. For example, a microphone generates a weak voltage, or *signal*, that changes according to the sound entering it. A TV camera generates a signal that varies with the darks and lights of the scene being televised.

But regardless of the type of waveform, there is one point that must be remembered:

No matter how a voltage, current, or power may change, the rules of electricity hold at all times.

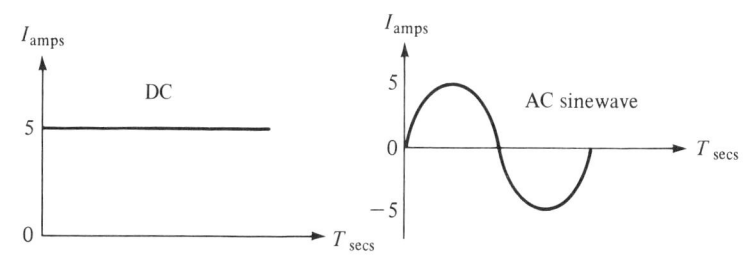

FIGURE 3.15 The DC Waveform and the Sine Wave.

I ▲ sawtooth → *T*

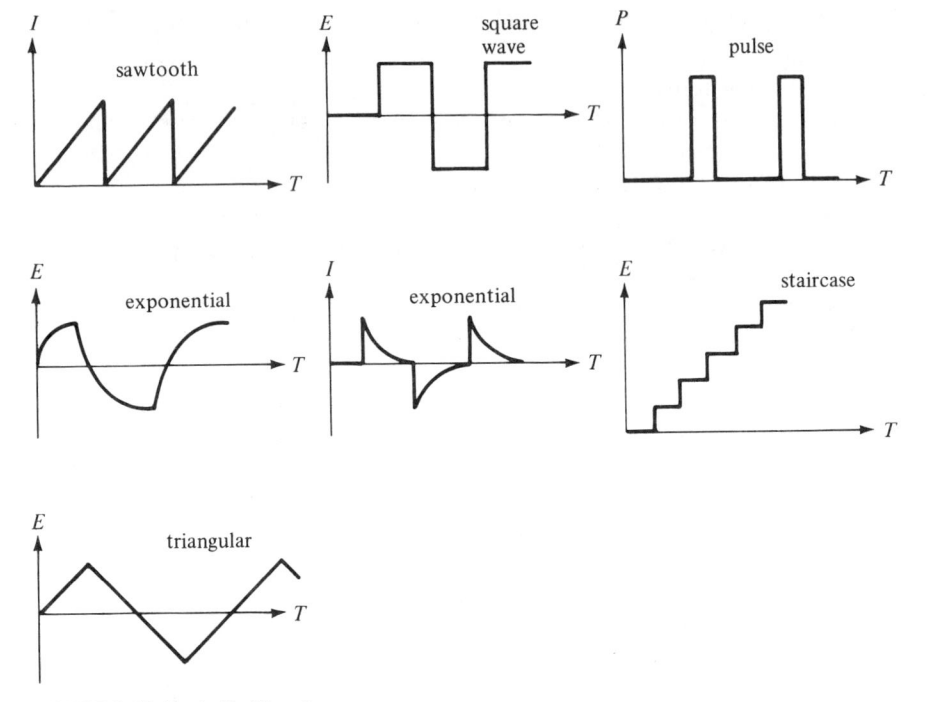

E ▲ square wave → *T*

P ▲ pulse → *T*

E ▲ exponential → *T*

I ▲ exponential → *T*

E ▲ staircase → *T*

E ▲ triangular → *T*

FIGURE 3.16 Periodic Waveforms.

Some examples illustrating this important point follow.

EXAMPLE 3.14

The switch in each circuit spends 5 s in position *A* and 3 s in position *B*, then return to *A* to repeat the cycle. Plot the load current waveform for each case.

a.

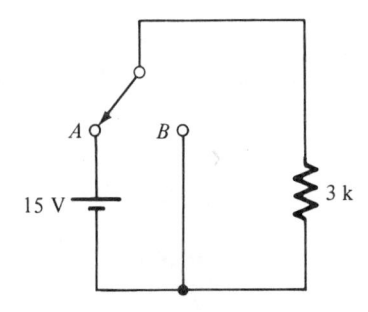

While the switch is in position *A*, the current is

$$I = \frac{E}{R} = \frac{15}{3000} = 5 \text{ mA}$$

3.13 WAVEFORMS

The current equals zero while the switch is in position B. The load current waveform is

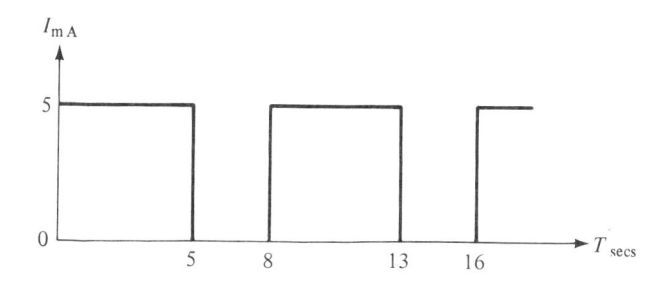

b. Current will flow in both directions in this circuit because the batteries have opposite polarities. If we define the clockwise direction as positive, current from the 15-V source will be negative. Study the plot carefully as to this point.

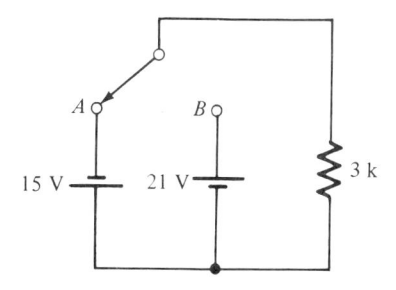

$$I \text{ from battery } A = \frac{-15}{3 \text{ k}} = -5 \text{ mA}$$

$$I \text{ from battery } B = \frac{21}{3 \text{ k}} = 7 \text{ mA}$$

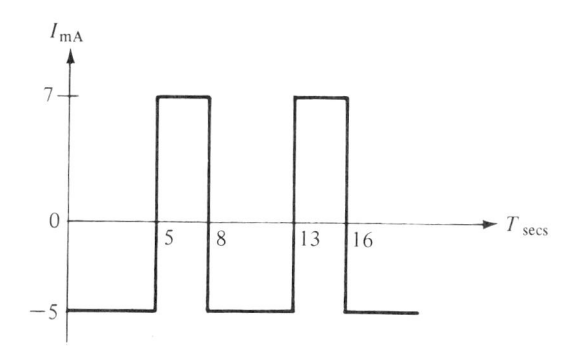

The load current waveform is ■

Here is another example, one using a sawtooth voltage waveform. Note how lowercase letters, *e* and *i*, are used to represent changing values. Capital (uppercase) letters indicate fixed values.

EXAMPLE 3.15

The sawtooth of voltage shown here is applied to the 2-Ω resistor. Draw the current waveform.

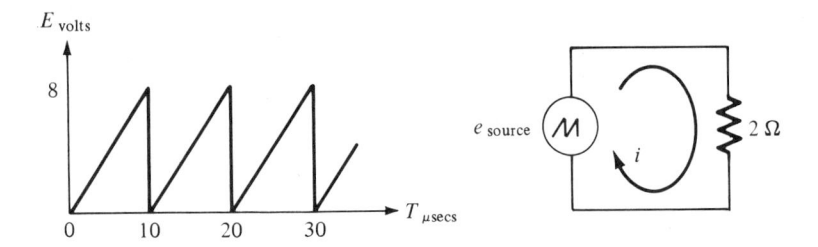

Note: There are other sources besides batteries. Their symbol is a circle; sometimes the waveform generated is drawn inside.

Four things are needed to draw the current waveform: its minimum and maximum values, how long it takes to go between them, and its shape.

$$i_{minimum} = \frac{e_{minimum}}{R}$$

$e_{minimum}$ from the plot = 0 V

Therefore,

$$i_{minimum} = \frac{0}{2} = 0 \text{ A}$$

$$i_{maximum} = \frac{e_{maximum}}{R}$$

$e_{maximum}$ from the plot = 8 V

Therefore,

$$i_{maximum} = \frac{8}{2} = 4 \text{ A}$$

The plot shows the voltage rising from 0 to 8 V in 10 μs. Therefore, current must rise from 0 to 4 A in the same time. And it will rise in sawtooth form as the voltage did. The current waveform is

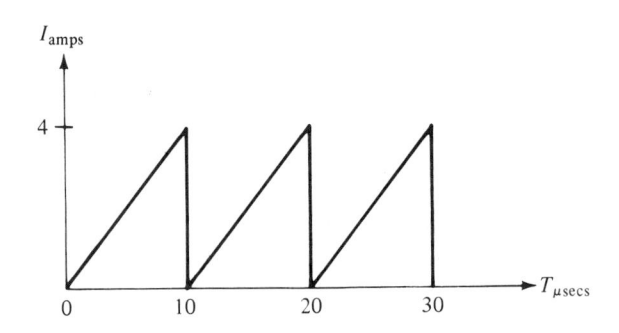

EXERCISES

Sections 3.1 through 3.3

1. Name the four components of the basic circuit and describe the function of each.
2. Name sources other than the battery. What type of energy is being changed to electrical energy in each?
3. When is a car battery a source, and when is it a load? Discuss.
4. Explain the following in your own words:
a. The difference between a source and a load.
b. Closed circuits and open circuits.
c. Polarity.
5. Which of these is a source? Which is a load?

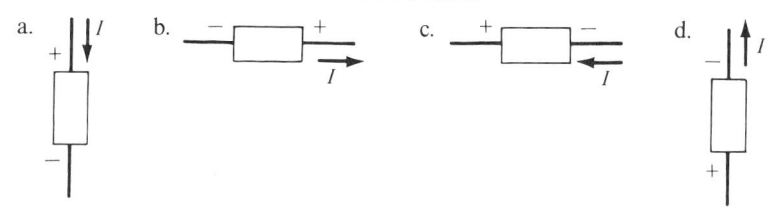

6. What does the law that voltage rise equals voltage drop mean in terms of energy? Discuss.
7. Our definition of polarity is based on a body having energy at one location and lacking it at another. With this in mind, how should each of the following be polarized?
a. The input and output of a water pump.
b. An elevator at the ground floor and the fifteenth floor.
c. A bullet at the muzzle of a rifle and at the target.
d. A space vehicle on the ground and in orbit.
8. Can a source have voltage between its terminals if there is no current flowing through it? Discuss.

Sections 3.4 and 3.5

9. State whether each of the following represents a conductor or an insulator:
a. High resistance.
b. Poor ability to conduct moving charge.
c. Glass.
d. Good conductivity.
e. Aluminum.
f. Low resistance.
g. Rubber.
h. Electrons easily separated from parent atom.
10. Think of some factors that might cause the resistance of a resistor to change.
11. Change each of these by replacing k or M with zeros: a. 10 k; b. 47 k; c. 330 k; d. 560 k; e. 1.2 M; f. 4.7 M; g. 2.2 M.
12. Convert as indicated:
a. 330,000 Ω to kilohms.
b. 1,500,000 Ω to megohms.
c. 22,000 Ω to kilohms.
d. 1500 Ω to kilohms.
13. Find the conductance of each resistance in Exercise 12.
14. Which resistor has the most resistance?

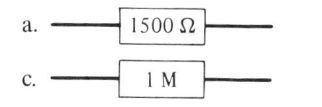

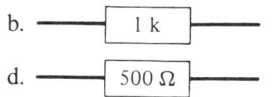

a. 1500 Ω
b. 1 k
c. 1 M
d. 500 Ω

Section 3.6

15. Which of these resistors has a resistance within tolerance?
a. A 2.7-k, 10%, which reads 2480 Ω.
b. An 82-Ω, 20%, which reads 99 Ω.
c. A 150,000-Ω, 5%, which reads 155 k.

16. Explain tolerance in your own words.
17. State the resistance of each of the following. Then calculate maximum and minimum values based on given tolerance.
   a. Yellow, orange, brown, gold.
   b. Green, blue, red, silver.
   c. Red, red, orange.
   d. Green, brown, gold, silver, brown.
   e. Orange, white, green, gold.
   f. Grey, red, black, silver, orange.
18. State the color code to be found on each of the following resistors. a. 7500 Ω, 5%; b. 10 Ω, 20%; c. 33 k, 10%; d. 910 k, 5%; e. 560 Ω, 10%; f. 470 k, 20%; g. 6.8 M, 10%.

Section 3.7

19. In order to change current in a circuit, we can (true or false):
   a. Increase EMF.
   b. Hold E and R constant.
   c. Decrease R.
   d. Increase E and reduce R.
   e. Increase R while reducing E.
20. The amount of current flowing in a circuit depends on (true or false):
   a. The device used to measure I.
   b. Where in the circuit I is measured.
   c. The voltage applied to the circuit.
   d. The resistance in the circuit.
21. When the voltage applied to a circuit is increased (true or false):
   a. I will increase if R is held constant.
   b. I will decrease if R is held constant.
   c. I will increase if R is decreased.
22. Given: a circuit in which the following changes are made. State what I will do in each case.
   a. R held constant, E doubled.
   b. R doubled, E halved.
   c. R halved, E doubled.
   d. R doubled, E doubled.
   e. R doubled, E held constant.
   f. R halved, E halved.
23. Fill in the blanks.

	E(V)	I(A)	R(Ω)
a.	15 V	23 mA	
b.		8.1 A	0.03 Ω
c.	282 mV		83 k
d.	1.5 V	65 μA	
e.	37 V		4.7 k
f.		500 mA	15 Ω

24. Which of the following show Ohm's law correctly?
   a. R held constant, E increased, I decreased.
   b. R increased, E held constant, I increased.
   c. E held constant, R decreased, I increased.
   d. R held constant, E increased, I increased.
   e. E held constant, R decreased, I decreased.
   f. R held constant, E decreased, I increased.
   g. R held constant, E decreased, I decreased.
   h. E held constant, R increased, I decreased.
25. Find the missing quantity in each of the following.
   a. +10 V, 330 k, +0.3 V, $I = ?$
   b. +15 V, 270 k, $I = 50\ \mu A$, $E = ?$
   c. −20 V, 120 k, −33 V, $I = ?$
   d. −12 V, $R = ?$, +8 V, $I = 10\ \mu A$
26. Ohm's law teaches that I goes down as R goes up. Based on this, explain why there is no current in an open circuit. *Hint:* What is the resistance of an open switch?

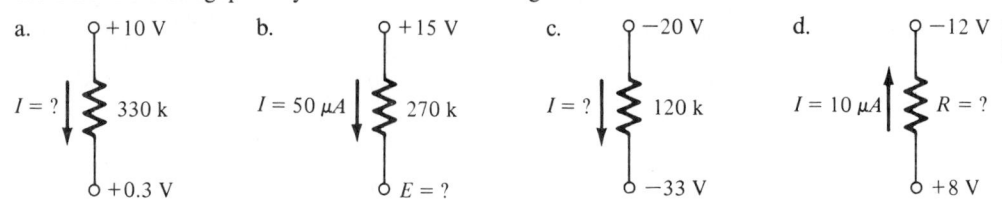

EXERCISES

27. Fill in the blanks.

	E(V)	I(A)	G(S)
a.	3.5 V	21 μA	
b.		4.16 A	0.028 S
c.	40 mV		10^{-7} S

28. Which of the following show Ohm's law correctly?
a. G increases, E held constant, I decreases.
b. G held constant, E increases, I decreases.
c. G increases, E held constant, I increases.
d. G decreases, E held constant, I decreases.
e. G increases, E held constant, I held constant.
f. G held constant, E increases, I increases.

Section 3.8

29. Find the current drawn by a 60-, a 75-, a 100-, and a 200-W light bulb when connected to 110 V. How does the current vary with power rating? Does this agree with the equation $P = EI$? Repeat the problem finding the resistance of each bulb. Compare resistance values. Do the variations in R agree with the equation $P = E^2/R$?

30. Multiply the basic units of voltage and current to prove that the units of their product are joules per second (or watts).

31. Fill in the blanks.

	P(W)	E(V)	I(A)	R(Ω)
a.	1600 W	40 V		
b.	81 mW		27 mA	
c.		15 μV		300 Ω
d.	1 kW			72 Ω

32. In what form does wasted electrical energy usually show up?

33. The most common conductor used to wire homes is 14-gage copper wire. It has a resistance of 2.6 Ω per 1000 ft. How many watts will be lost by, and what is the voltage drop across, 35.7 ft of this wire carrying 12.8 A?

34. Derive the three power equations in terms of G rather than R.

35. What are the units of each expression?   a. $\left(\dfrac{E}{R}\right)^2 R$;   b. $\dfrac{(IR)^2}{R}$;   c. $\left(\dfrac{P}{R}\right)^{1/2}$.

36. A 5-hp motor requires how many watts?

Section 3.9

37. A jet engine burns fuel to generate 55×10^{18} J. Of these, 33.55×10^{18} J is actually available to push an aircraft. What is the engine's efficiency?

38. What are the losses in a motor that takes in 1.7×10^4 J and has an efficiency of 79.4%?

39. A transistorized stereo amplifier has 67 W input power and an efficiency of 48%.
a. What is its output power?
b. What power is wasted by the amplifier?
c. How does this wasted power show up?

40. A power supply must lose no more than 13 W; otherwise, its temperature climbs and reliability falls. The supply's output power is 92 W.
   a. What is the maximum input power?
   b. What is the percent efficiency?

41. What is the missing efficiency in each case?
   a. $\eta_1 = 0.75, \eta_2 = 0.63, \eta_3 = 0.91, \eta_{total} = ?$
   b. $\eta_1 = \eta_2 = \eta_3 = 0.82, \eta_{total} = ?$
   c. $\eta_1 = 0.65, \eta_2 = 0.43, \eta_{total} = 0.25, \eta_3 = ?$

42. *Extra Credit.* An electrical device operates with an energy input of 83 W and an efficiency of 74.2%. The operating characteristics are improved, so that for the same input, the output increases 17.3%. What is the new efficiency?

Section 3.10

43. In your own words, describe ideal shorts and opens.

44. Explain how a short can occur in a lamp cord.

45. Why is the exponent of I in the power equation $P = I^2R$, important to the short of Exercise 44?

46. Explain how a fuse works.

Sections 3.10 and 3.11

47. How are grounded points in a circuit related?

48. List three jobs performed by an electronic chassis.

49. Find the current through each resistance and its direction.

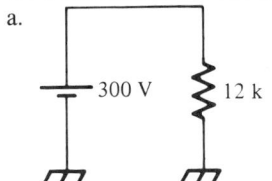

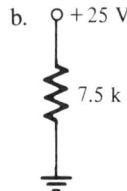

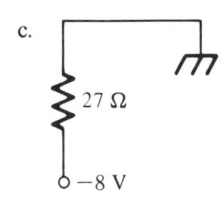

   a. [circuit: 300 V, 12 k]
   b. [circuit: +25 V, 7.5 k]
   c. [circuit: 27 Ω, −8 V]

50. The +25-V terminal of Exercise 49, circuit b, is said to be, "25 V above ground," in the language of electronics. Remembering that ground is 0 V, how is the −8-V terminal in circuit c referred to?

Section 3.13

51. In your own words, explain the difference between periodic and aperiodic waveforms.

52. Besides turn signals on a car, there are many other periodic events in daily life. List some. Repeat for aperiodic events.

The remaining exercises ask that waveforms be sketched. "Sketch" means a freehand plot without using a straight edge or careful measuring. Just draw the waveforms neatly, get the right shape, and show important values on both axes; keep it simple.

53. Sketch the load current waveform for each circuit.
a. The switch spends equal times in A and B.

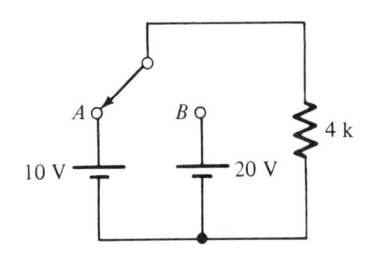

[circuit: switch positions A and B, 10 V, 20 V, 4 k]

EXERCISES

b. The switch spends 10 s in A, 1 s in B, then repeats.

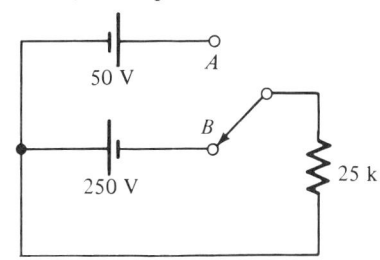

c. The switch is in A for 100 μs, then in B for 500 μs, then repeats.

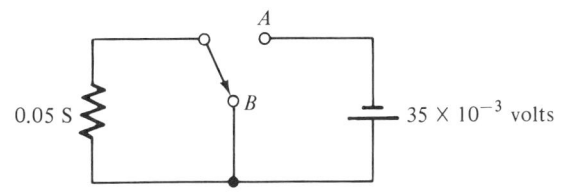

54. Repeat Exercise 53, drawing the load power dissipation for each circuit.

55. This switch spends 1 s in each position, back and forth. Sketch source current waveform. *I* shows direction of positive current.

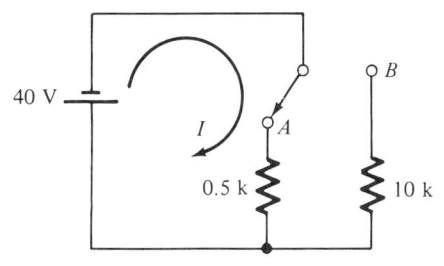

56. Sketch the load current waveform.

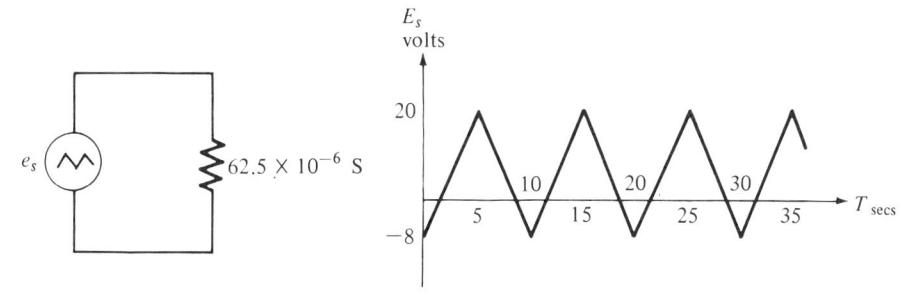

CHAPTER HIGHLIGHTS

Electric circuits consist of:

- *Source:* converts some form of energy to electrical energy.
- *Load:* converts electrical energy to another form.
- *Control devices:* provide control over circuit parameters.
- *Conductors:* carry current from source to load.

Polarity tells where charge has energy. At the positive side of a source or load, energy is high. At the negative side, energy is low.

To tell a source from a load, compare current direction with polarity.

Since energy in must equal energy out, voltage rise must equal voltage drop.

Resistance is a material's opposition to the movement of charge through it.

Conductance is the reciprocal of resistance. It indicates how well a material will carry current.

Three most important things to know about a resistor are its resistance, tolerance, and wattage rating.

The EIA color code indicates resistor data with a group of colored bands.

Ohm's law is the most frequently used tool in circuit analysis. It must be known in all three forms.

Power is the rate at which energy is being transformed. Efficiency tells how much energy put into a system comes out in a useful form.

Short circuits or shorts occur when circuit resistance is bypassed, causing current to jump to high levels. Fuses protect against shorts by opening a circuit.

Grounded points are connected together.

Waveforms are graphs showing how voltage, current, and power change with time.

The laws of electricity hold at all times.

The result of circuit analysis is an understanding of how a circuit works. This provides the all-important ability to predict what a circuit will do under different conditions.

Formulas to Remember

$$G = \frac{1}{R} \qquad \text{conductance}$$

$$E = IR \quad I = \frac{E}{R} \quad R = \frac{E}{I} \qquad \text{Ohm's law in resistance form}$$

$$E = \frac{I}{G} \quad I = EG \quad G = \frac{I}{E} \qquad \text{Ohm's law in conductance form}$$

$$P = \frac{W}{T} \qquad \text{basic formula for power}$$

$$P = EI = I^2 R = \frac{E^2}{R} \qquad \text{three formulas for power}$$

$$\eta = \left[\frac{P_{\text{out}}}{P_{\text{in}}}\right] \times 100 \text{ percent efficiency}$$

CHAPTER 4

The Series Circuit

Now that the basics of electricity and circuits have been covered, we are ready to move on to the first of the more complex circuits, the series circuit.

At the end of this chapter, you should be able to:

- Tell when components are connected in series.
- Calculate the total voltage from sources in series.
- Use Kirchhoff's voltage law to understand and analyze series circuits.
- Calculate the total resistance of a group of resistances in series.
- Do a rough analysis of a circuit when estimates are sufficient.
- Find an unknown voltage between two points in a circuit.
- Use the voltage-divider formula.
- Determine the series resistance needed to keep current to some level.
- Calculate loading, percent voltage regulation, and maximum power transfer for a Thévenin source.
- Troubleshoot series circuits.

4.1 WHAT DOES "IN SERIES" MEAN?

Note that all the source current in Figure 4.1 must flow through R_1 and R_2; there is no other path. R_1, R_2, and the source are said to be connected in *series*.

| Components are connected in series when they have the same current through them.

For contrast, the circuits of Figure 4.2 are not series circuits. In circuit (a), the source current splits at the top of R_1 into two parts, I_{R_1} and I_{R_2}. Although these may have the same value, they cannot possibly be the same current. Therefore, R_1 and R_2 are not in series. They are in *parallel*, a connection studied in Chapter 5.

Figure 4.2, circuit (b) shows both series-connected and parallel-connected components combined in a *series–parallel* circuit (described in Chapter 6). R_2 and R_3 are in parallel. R_1, R_4, R_5, and the source are in series since current in one must flow through the others.

The fact that R_1 is in series with R_5 in circuit (b) illustrates an important point:

| Components do not have to be next to one another to be in series.

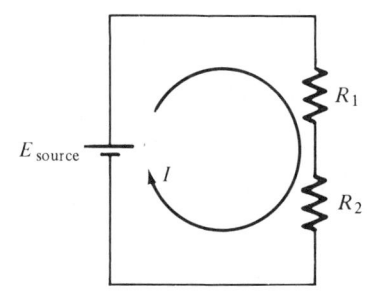

FIGURE 4.1 Series Circuit.

4.2 SOURCES IN SERIES

(a) parallel circuit

(b) series–parallel circuit

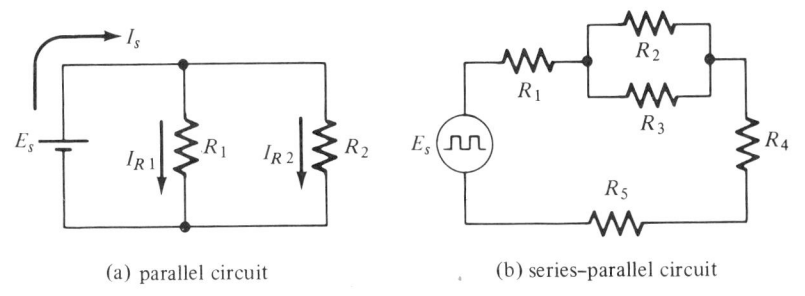

FIGURE 4.2 Nonseries Circuits.

Students must develop the ability to tell how components are connected because series circuits are analyzed with math techniques different from those used on parallel circuits.

4.2 SOURCES IN SERIES

When sources are connected in series, their voltages aid or oppose one another, depending on polarity. Figure 4.3 is an example of *series aiding*. Current along the lower conductor enters the 4-V battery at the negative terminal and leaves at the positive. This is a 4-V rise in potential. It does the same thing for a 7-V rise in the next source. By the time the current leaves the positive terminal of the 7-V source, it has gone through a total rise of 11 V. In other words, the total rise equals the sum of the parts. Sources are aiding when their terminals are connected in a minus-to-plus, minus-to-plus sequence.

Figure 4.4 shows the opposite effect, that of sources in *series opposition*. The 4-V source forces current clockwise, while the 7-V source does just the opposite. The 7-V

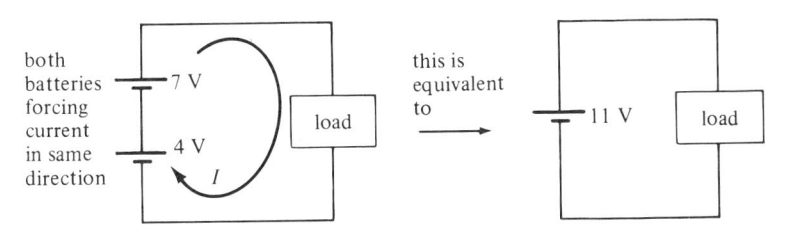

FIGURE 4.3 Sources in Series Aiding.

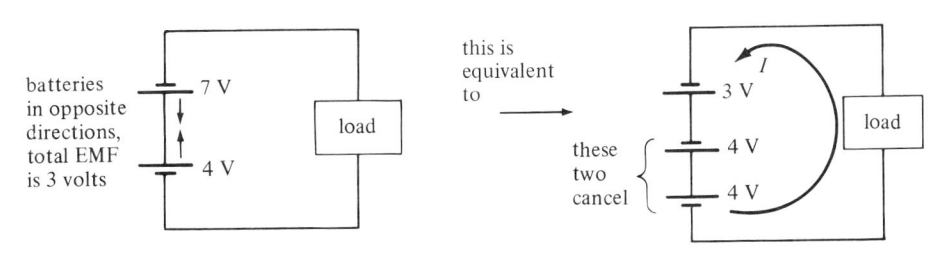

FIGURE 4.4 Sources in Series Opposition.

potential cancels the 4-V rise, leaving a final 3 V delivering current in the counterclockwise direction.

If the 4-V source was increased to, say, 15 V, the result would be an 8-V battery forcing current clockwise.

| When two sources are connected with opposing polarities, the one having the greater voltage determines final current direction.

EXAMPLE 4.1

What current flows in this circuit?

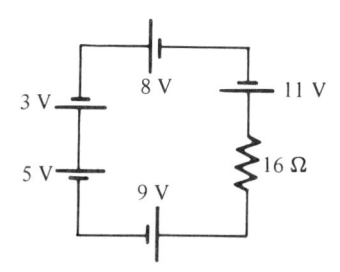

Total voltage forcing current clockwise = 11 + 5 = 16 V.
Total in the counterclockwise direction = 8 + 3 + 9 = 20 V.
Final voltage = 20 − 16 = 4 V.

$$I = \frac{E}{R} = \frac{4}{16} = 0.25 \text{ A counterclockwise}$$

Final direction determined by the 20-V combined source.

4.3 KIRCHHOFF'S VOLTAGE LAW

Figure 4.5 shows a 12-V car battery consisting of six 2-V cells in series aiding. It is connected to a string of three resistors, each with a voltage drop. Starting at point A, we will follow around the entire closed loop, keeping track of its energy rises and drops as we go.

Heading upward from point A, the charge undergoes a potential (voltage) rise of 2 J/C (volts) in the first cell. It picks up an additional 10 J/C in the other five cells and leaves the battery with a total of 12 J/C (how many coulombs flow depends on source voltage and total resistance). The charge then travels the upper conductor to R_1 and loses 3 J/C in a 3-V drop. Of the 9 J/C remaining, 7 J/C are lost in R_2 and 2 J/C in R_3. Charge arrives at the bottom conductor with no energy left and returns to point A.

The point of this example is very important. It is:

| The sum of the voltage rises around a closed loop must equal the sum of the voltage drops.

4.3 KIRCHHOFF'S VOLTAGE LAW

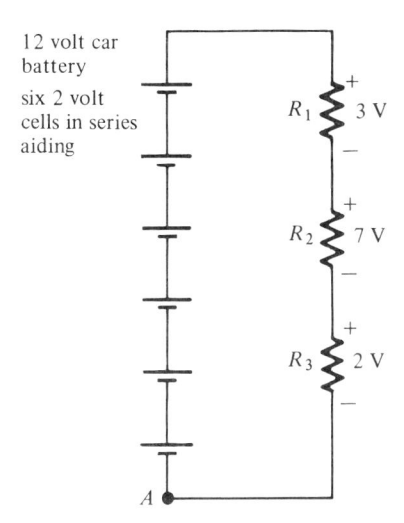

FIGURE 4.5 Series Circuit Demonstrating Kirchhoff's Voltage Law.

A German physicist, Gustov Kirchhoff (1824–1887), studied this relation between voltage rises and drops. It is called *Kirchhoff's voltage law* in his honor and abbreviated KVL.

In the shorthand of mathematics, KVL is written

$$\sum E_{\text{rises}\circlearrowleft} = \sum V_{\text{drops}\circlearrowleft} \qquad (4.1)$$

where $\sum$ = the Greek capital letter sigma, which stands for "the sum of"
$\circlearrowleft$ = around a closed loop

EXAMPLE 4.2

What is the source voltage in this circuit?

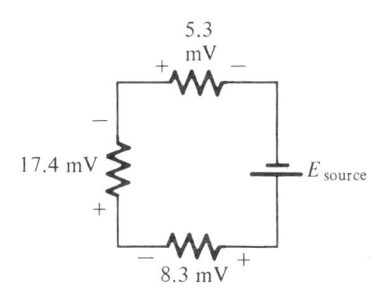

I By KVL,

$$E_{\text{source}} = (5.3 + 17.4 + 8.3) \text{ mV}$$
$$= 31 \text{ mV}$$

EXAMPLE 4.3

Find the unknown voltage across R_1.

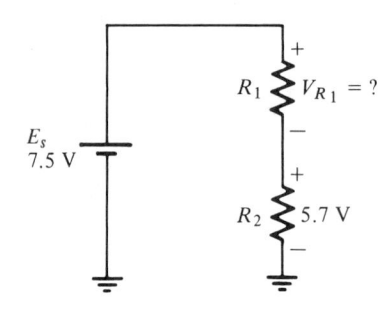

By KVL, the sum of the drops must equal the rise, or

$$7.5 = 5.7 + V_{R_1}$$

Solving for V_{R_1}

$$V_{R_1} = 7.5 - 5.7 = 1.8 \text{ V}$$

4.4 RESISTANCES IN SERIES

Suppose that we connect several resistances in series. What is the total resistance of the combination? Kirchhoff's voltage law will help us find the answer to that question.

In the circuit of Figure 4.6, we know that

$$E_{\text{source}} = V_{R_1} + V_{R_2} + V_{R_3}$$

by KVL. We also know that

$$E_{\text{source}} = IR_{\text{total}}$$

and

$$V_{R_1} = IR_1, \, V_{R_2} = IR_2 \quad \text{and} \quad V_{R_3} = IR_3$$

by Ohm's law. Substituting for each voltage in the KVL expression:

$$IR_{\text{total}} = IR_1 + IR_2 + IR_3$$

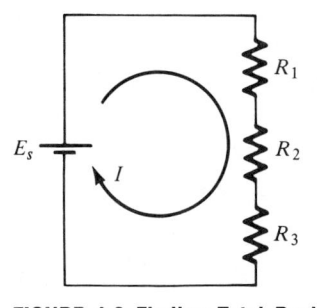

FIGURE 4.6 Finding Total Resistance of Series-Connected Resistances.

4.4 RESISTANCES IN SERIES

Factoring I from the right side:

$$IR_{total} = I(R_1 + R_2 + R_3)$$

Dividing both sides by I, we have

$$R_{total} = R_1 + R_2 + R_3$$

which says:

| The total resistance of resistances in series equals the sum of the individual resistances.

The math statement can be put in a more general form that works for any number of series resistances:

$$R_{total} = R_1 + R_2 + \cdots + R_n$$
$$= \sum R \qquad (4.2)$$

where R_n is the nth resistance.

NOTE: The group of math steps just finished is called a *derivation*. Using the laws of electricity and algebra, we derived a simple, yet powerful math statement that covers an infinite number of resistance combinations. You are strongly urged to study this derivation and others that follow carefully, step by step. This is an excellent way to learn how mathematics is used to analyze circuits.

EXAMPLE 4.4

What is the total resistance of a 10-Ω, a 50-Ω, and a 90-Ω series combination?

$$R_{total} = \sum R = 10 + 50 + 90 = 150 \; \Omega \qquad \blacksquare$$

NOTE: Example 4.4 tells something very important about series-connected resistances:

Adding resistance in series always increases total resistance.

EXAMPLE 4.5

What is the value of the missing resistance?

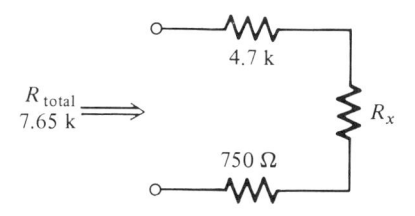

$$R_{total} = 7.65 \, k = 4.7 \, k + R_x + 0.75 \, k$$

Solving for R_x

$$R_x = 7.65 \, k - (4.7 \, k + 0.75 \, k) = 7.65 \, k - 5.45 \, k = 2.2 \, k \qquad \blacksquare$$

To find R_{total} for a series string of resistances all of the same value, multiply the value of one by the number in the string. In equation form this relation is

$$R_{total} = NR \tag{4.3}$$

where N = number of series-connected resistances of same value
R = value of one resistance

The next example is worded in the language of electronics, the way someone in the field would say it.

EXAMPLE 4.6

What is the resistance seen looking into these terminals?

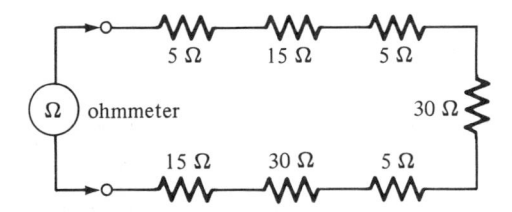

[Circuit diagram: an ohmmeter connected to a series loop containing resistors of 5 Ω, 15 Ω, 5 Ω, 30 Ω, 5 Ω, 30 Ω, 15 Ω]

The words *see* and *look* mean: What resistance would be measured by an ohmmeter connected to the terminals?

$$R_{total} = (3 \times 5) + (2 \times 15) + (2 \times 30)$$
$$= 15 + 30 + 60 = 105 \; \Omega \quad \blacksquare$$

4.5 ANALYZING THE SERIES CIRCUIT

Since series circuits can have any number of components, an analysis often involves many voltages, currents, and powers. As a result, the chance for errors grows. One way to reduce these errors is to spend a moment checking the answers of a problem to see if they make sense in light of the circuit from which they came. The next example shows how to do this.

EXAMPLE 4.7

Find the voltage drop across and power dissipated by each resistor in this circuit.

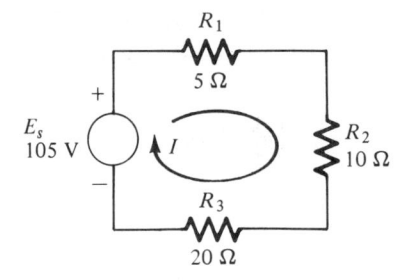

[Circuit diagram: Source $E_s = 105$ V with current I, $R_1 = 5\;\Omega$, $R_2 = 10\;\Omega$, $R_3 = 20\;\Omega$ in series]

4.5 ANALYZING THE SERIES CIRCUIT

The first thing to do is find I:

$$I = \frac{E_s}{R_{\text{total}}}$$
$$= \frac{105}{(5 + 10 + 20)}$$
$$= \frac{105}{35}$$
$$= 3 \text{ A}$$

This is the current through each resistance. It will be used frequently, so save time by storing it in calculator memory. Using Ohm's law to find the drops:

$$V_{R_1} = IR_1 = 3(5) = 15 \text{ V}$$
$$V_{R_2} = IR_2 = 3(10) = 30 \text{ V}$$

By KVL,

$$V_{R_3} = E_S - (V_{R_1} + V_{R_2})$$
$$= 105 - (15 + 30)$$
$$= 105 - 45 = 60 \text{ V}$$

For review, all three power equations are used:

$$P_{R_1} = V_{R_1} I_{R_1}$$
$$= 15(3) = 45 \text{ W}$$

Note the voltage used. V_{R_1} is the only voltage that can determine P_{R_1}.

$$P_{R_2} = \frac{V_{R_2}^2}{R_2}$$
$$= \frac{30^2}{10} = 90 \text{ W}$$

$$P_{R_3} = I^2 R_3$$
$$= 3^2(20) = 180 \text{ W}$$

Here are some "checking questions" to use when checking results:

- Does the sum of the three drops equal the source voltage?
- Since R_2 is twice R_1, V_{R_2} should be twice V_{R_1}. Is it?
- Compare R_2 and R_3, then V_{R_2} and V_{R_3}. Do the voltages make sense?
- What about V_{R_3} and V_{R_1}?
- Compare the powers. Do they make sense?
- Does the sum of the three output powers equal that supplied by the source? ■

A useful tool in the analysis of series circuits is the *voltage-divider formula*. It tells how source voltage is divided among series resistances. The formula is

$$V_{R_n} = \frac{E_{\text{source}} R_n}{R_{\text{total}}} \tag{4.4}$$

where V_{R_n} = voltage across resistance N
R_n = resistance n

The formula does not say anything new. We know that E_{source}/R_{total} equals circuit current, I; and I times R_n equals V_{R_n}. This is how the last example was done. However, many books refer to the formula, so you should be familiar with it.

EXAMPLE 4.8

Find the drop across the 13 K using the voltage-division formula.

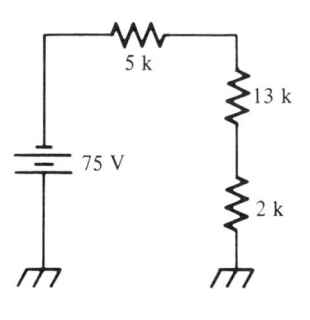

$$V_{13 k\Omega} = \frac{E_s 13\ k}{R_{total}}$$

$$= \frac{75(13\ k)}{5\ k + 13\ k + 2\ k}$$

$$= \frac{75(13)}{20}$$

$$= 48.8\ V$$

The next two circuits are drawn as they might appear in electronic schematic diagrams.

EXAMPLE 4.9

Find the voltage between terminals A and B for each circuit.

a.

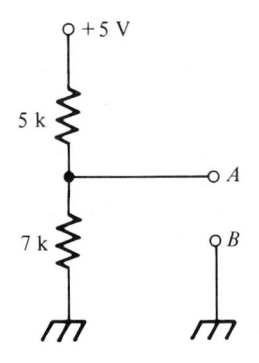

4.5 ANALYZING THE SERIES CIRCUIT

Remember: The negative source terminal is grounded. The circuit is actually this:

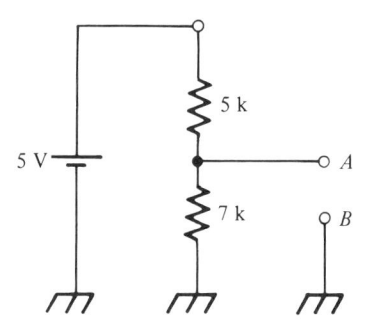

The terminals are across the 7 k therefore, by voltage division:

$$V_{7k} = \frac{5(7\ k)}{5\ k + 7\ k}$$

$$= \frac{35\ k}{12\ k}$$

$$= 2.9\ V$$

b.

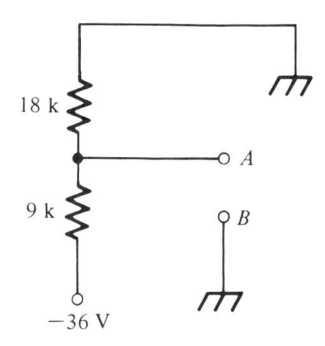

At first glance it appears we are to find the drop across the 9 k, but look carefully at the terminals. They are really across the 18 k. Therefore,

$$V_{18k} = -\frac{36(18\ k)}{18\ k + 9\ k}$$

$$= -\frac{36(18\ k)}{27\ k}$$

$$= -36\left(\frac{2}{3}\right)$$

$$= -24\ V$$

Point *A* is 24 V below ground.

Checking question:
    18 k is two-thirds of R_{total}. Therefore, it should have two-thirds of E_{source} across it. Does it? ■

EXAMPLE 4.10

Determine the power waveform for the 33-Ω resistance.

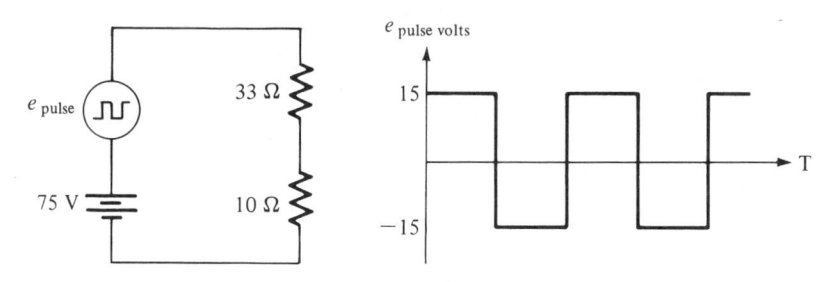

Equivalent diagrams show the effects of the changing polarity of the pulse generator. When $e_{\text{Pulse}} = 15$ V:

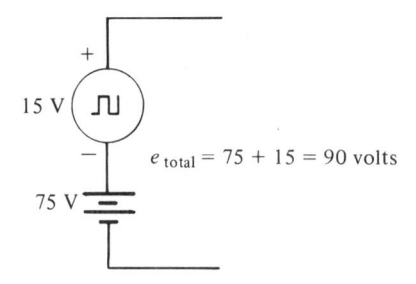

$$e_{\text{total}} = 75 + 15 = 90 \text{ volts}$$

When $e_{\text{pulse}} = -15$ V:

$$e_{\text{total}} = 75 - 15 = 60 \text{ volts}$$

Using the voltage-division formula to find the drop across the 33 Ω for both conditions:

$$e \text{ at } +15 \text{ V} = \frac{90(33)}{33 + 10} = 69.1 \text{ V}$$

$$e \text{ at } -15 \text{ V} = \frac{60(33)}{33 + 10} = 46.1 \text{ V}$$

$$p \text{ at } +15 \text{ V} = \frac{e^2}{R} = \frac{69.1^2}{33} = 144.6 \text{ W}$$

$$p \text{ at } -15 \text{ V} = \frac{e^2}{R} = \frac{46.1^2}{33} = 64.3 \text{ W}$$

4.5 ANALYZING THE SERIES CIRCUIT

Plot the power waveform:

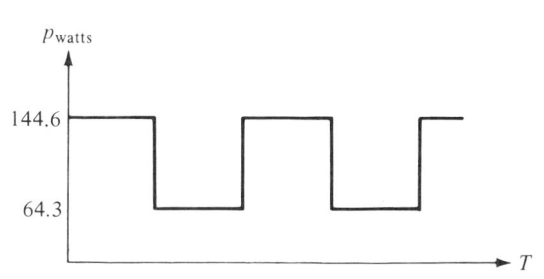

[Power waveform plot showing p_{watts} vs T, with values 144.6 and 64.3]

Data needed to solve a problem are not always given but must be calculated using the information supplied. This often takes some hard thinking. Because each circuit is different, no one can tell you how to do this, but a few general guidelines might help:

- Study the problem carefully to see what is given and what is needed. Plan the steps you will take to get needed data from given data.
- If you cannot think of a plan for a solution, do what can be done with available data. Calculate currents, powers, whatever. While doing this, a plan may show itself.
- The resistance with two of its four parameters (E, I, R, P) known is often the key to the solution of series circuits; look for it.
- Find errors by asking checking questions.

EXAMPLE 4.11

Find the resistance of R_1.

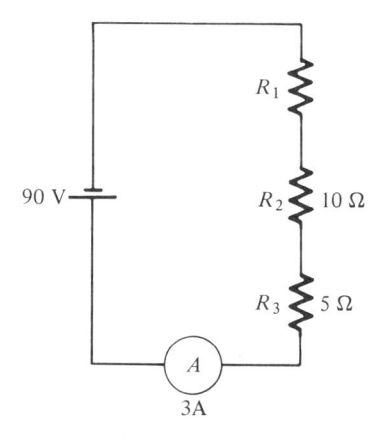

[Circuit diagram: 90 V source in series with R_1, $R_2 = 10\,\Omega$, $R_3 = 5\,\Omega$, and ammeter A reading 3A]

By Ohm's law:

$$R_1 = \frac{V_{R_1}}{I} = \frac{V_{R_1}}{3}$$

By KVL:

$$V_{R_1} = E_s - (V_{R_2} + V_{R_3})$$

By Ohm's law:
$$V_{R_2} + V_{R_3} = I(R_2 + R_3) = 3(10 + 5)$$
$$= 3(15) = 45 \text{ V}$$

Finally, by Ohm's law:
$$R_1 = \frac{45}{3} = 15 \text{ }\Omega$$ ∎

Sometimes "ballpark" estimates of voltage or current are sufficient for the job at hand. This is especially true in troubleshooting.

EXAMPLE 4.12

Approximate the drop across the resistance in each of the following circuits.

a.

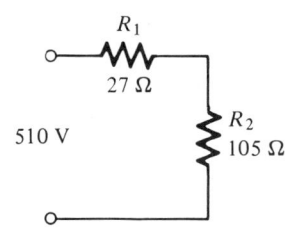

In round numbers, $R_1 = 25 \text{ }\Omega$, $R_2 = 100 \text{ }\Omega$, and $E_{\text{source}} = 500$ V. If there are x volts across R_1, there must be $4x$ volts across R_2.

$$4x + x = 5x = 500 \text{ V}$$
$$x = 100 \text{ V}$$

Thus, there are about 100 V across R_1 and 400 V across R_2. (The answers by analysis are 104.3 and 405.7 V, respectively. Both errors are less than 5%.)

b.

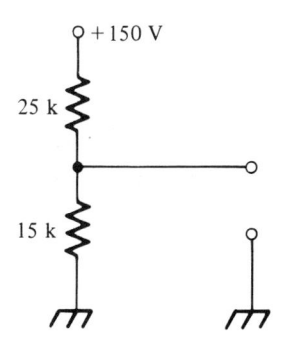

Note that each resistance is divisable by 5 k; three times 5 k on the bottom, five times 5 k on top. Therefore, for every $3x$ volts across the 15 k there must be $5x$ volts across the 25 k.

$$3x + 5x = 8x = 150 \text{ V}$$

4.6 HOW TO FIND AN UNKNOWN VOLTAGE IN A CIRCUIT

Change the 150 to 160 for easy division by 8, and get

$$x = \frac{160}{8} = 20\ V$$

Then

$$V \text{ across the } 15\ k = 3(20) \approx 60\ V$$

and

$$V \text{ across the } 15\ k = 5(20) \approx 100\ V$$

The answers by analysis are 56.3 and 93.8 V. Our approximations are wrong but they are good ballpark figures; both are less than 10% in error.

A troubleshooting reading of 57.9 V between the terminals is close enough to the estimated 60 V to assume that the circuit is working properly. On the other hand, a reading around 30 V indicates the need for closer checking. ∎

Another good estimating tool is the *10-to-1 rule*. It says:

| If R_1 and R_2 are series connected and $R_1 \geq 10R_2$, $R_{total} \approx R_1$ and $V_{R_1} \approx E_{source}$.

Remember this rule; we will see it again.

EXAMPLE 4.13

Suppose that R_2 in Example 4.12, part a, is changed to 405 Ω.

a. Approximate the new values of R_{total} and V_{R_2}:

| Since $405 > 10(27)$, $R_{total} \approx 405\ \Omega$ and $V_{R_2} \approx 510\ V$.

b. Calculate the new values of R_{total} and V_{R_2}:

$$R_{total} = 405 + 27 = 432\ \Omega$$
$$V_{R_2} = \frac{405 \times 510}{432} = 478.1\ V$$

| Both estimates are about 6.7% in error. ∎

4.6 HOW TO FIND AN UNKNOWN VOLTAGE IN A CIRCUIT

Kirchhoff's voltage law (Equation 4.1) says that the sum of the voltage rises must equal the sum of the voltage drops around any closed loop. If we look on a drop as being a negative rise, KVL can be written in an easier form:

| The sum of the voltages around a closed loop equals zero.

In math shorthand this is

$$\sum E_\circlearrowright = 0 \qquad (4.5)$$

where $\circlearrowright$ means "around a closed loop."

The math statement adding all the voltages around a loop is called a *loop equation*. Loop equations are an important tool in circuit analysis. The next example shows how to write them.

EXAMPLE 4.14

Write a loop equation for this circuit.

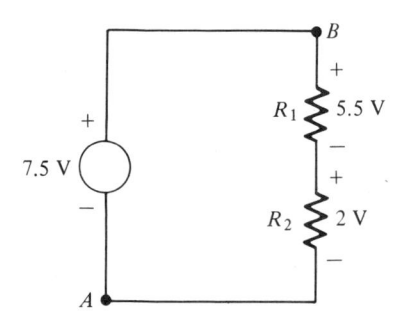

To write a loop equation:

- Start anywhere in the loop and go around it in either direction.
- When a voltage is met at its negative sign, enter it in the equation as a negative quantity.
- When a voltage is met at its positive sign, enter it as a positive quantity.
- Go around the entire loop entering each and every voltage until you return to the starting point.
- Finish making an equation by setting the statement equal to zero.

Follow these steps as a loop equation is written for this example.
Start at point A and go CW (clockwise). The first sign and voltage is -7.5 V, then $+5.5$ V, and finally $+2$ V. Then we are back at A and done. Putting this in an equation:

$$-7.5 + 5.5 + 2 = 0$$

which is true; our loop equation is correct.

To prove that starting point and direction make no difference, start at point B and go CCW (counterclockwise). The same equation results. ∎

Loop equations help us determine unknown voltages. Follow these steps:

- Chose a polarity for the unknown potential (do not worry about your choice; if it is wrong, the answer is simply negative).
- Write a loop equation that includes the unknown voltage.
- Solve for the unknown voltage.

EXAMPLE 4.15

What is in the box, a source or a load?

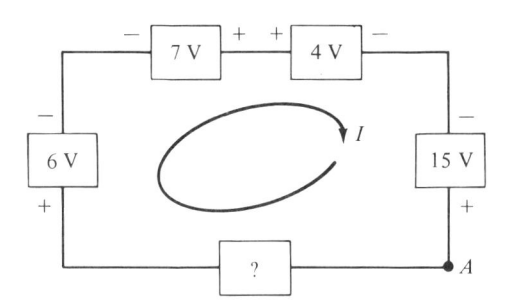

To determine the contents of the box, we must first calculate the magnitude and polarity of the voltage across the box. Assume that the unknown polarity is + on the left. Writing a loop equation in the CCW direction starting from point A:

$$+15 - 4 + 7 - 6 + E_x = 0$$

Solving for E_x:

$$22 - 10 + E_x = 0$$
$$12 + E_x = 0$$
$$E_x = -12 \text{ V}$$

The answer is negative; our choice of E_x polarity was wrong. The unknown voltage is 12 V, positive on the right. (As an exercise, assume opposite polarity, start the equation at another point, and go CW; the same results must be obtained.) Since current enters the positive terminal, we know that the box is a load with 12 V drop across it. ■

EXAMPLE 4.16

Find the magnitude and polarity of the voltage between points *A* and *B*.

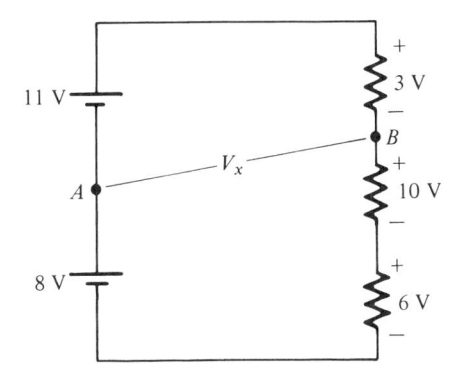

There are two loops containing V_x. Either may be used. The loops are:

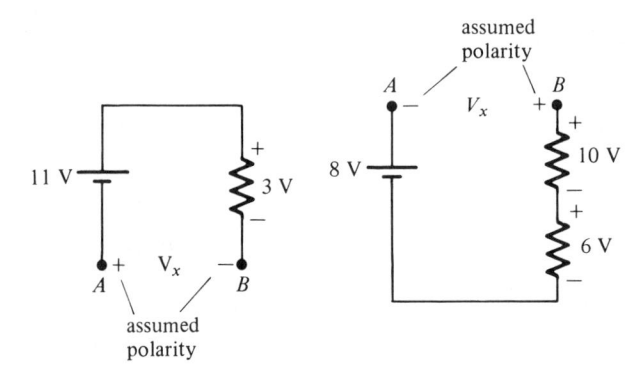

After assuming a polarity for V_x, the equation is written going CCW from B on the right-hand circuit:

$$V_x + 8 - 6 - 10 = 0$$
$$V_x - 8 = 0$$
$$V_x = 8 \text{ V}$$

The unknown is 8 V, + on the B side. ∎

4.7 CONTROLLING CURRENT WITH SERIES RESISTANCE

If a source voltage is too high, series resistance must be added between source and load; otherwise, excessive current might damage a device. The amount of resistance needed is determined by analysis.

EXAMPLE 4.17

How much resistance must be added in series with a 9-V transistor radio that draws 30 mA so that it can be run off a 12-V car battery? What power rating must the resistance have?

The resistance will have $12 - 9 = 3$ V across it and 30 mA through it:

$$R = \frac{V}{I} = \frac{3}{0.03} = 100 \text{ }\Omega$$

$$P = VI$$
$$= 3(0.03)$$
$$= 0.09 \text{ W or } 90 \text{ mW}$$ ∎

Rheostats are variable resistances that allow us to adjust current to vary the speed of a model train or dim dashboard lights. One end of the rheostat's shaft has a knob. The other end has a *wiper* which rubs across, or, *wipes*, a band of resistive material (see Figure 4.7). As the wiper rotates clockwise, more of this material lies between the two connections and the resistance increases. The line with the arrowhead in the symbol represents the wiper. Arrowheads in electronic symbols indicate variability.

4.8 VOLTAGE DIVIDERS

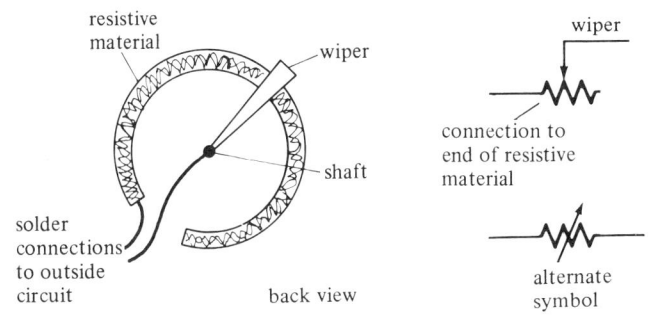

FIGURE 4.7 Rheostat and its Symbol.

Series-connected resistances divide an applied voltage into any number of lower values. By grounding one terminal of such a *voltage divider*, potentials above and below ground are obtained.

EXAMPLE 4.18

Find the voltage with respect to ground for each terminal of this voltage divider.

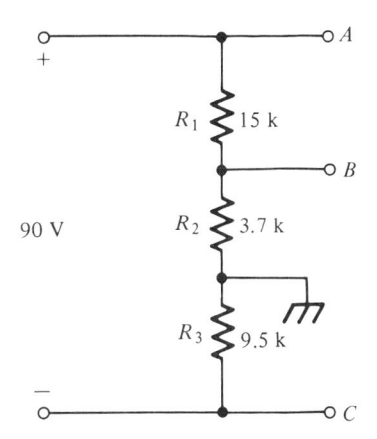

The drop across each resistance is found with the voltage-division formula:

$$V_{R_1} = \frac{90(15\ \text{k})}{15\ \text{k} + 3.7\ \text{k} + 9.5\ \text{k}} = \frac{1350\ \text{k}}{28.2\ \text{k}} = 47.9\ \text{V}$$

$$V_{R_2} = \frac{90(3.7\ \text{k})}{28.2\ \text{k}} = 11.8\ \text{V}$$

$$V_{R_3} = \frac{90(9.5\ \text{k})}{28.2\ \text{k}} = 30.3\ \text{V}$$

Point A is above ground by the combined drops across R_1 and R_2. Therefore,

$$V_a = 47.9 + 11.8 = 59.7\ \text{V above ground}$$

Point *B* is above ground by the drop across R_2:

$$V_b = 11.8 \text{ V above ground}$$

The top of R_3 is grounded and therefore at zero potential. The bottom is 30.3 V negative with respect to the top. Therefore, point *C* is 30.3 V below ground, or negative. ■

A word of caution: Output voltages from a divider such as that in Example 4.18 are obtained only under *no-load* or *open-circuit conditions*; that is, no current is being drawn from any terminal.

Special resistors, both fixed and variable, are made for voltage division. The fixed type has *taps* positioned to provide the desired potentials. The variable unit is the *potentiometer* (pronounced *puh-tench-ee-ahm-uter*), or *pot*. Pots are very common. They are the *volume* and *tone* controls in radios and stereo sets as well as *constrast* and *brightness* controls on TV sets.

Figure 4.8 shows that pots work much the same as rheostats. However, they have three terminals because both ends of the resistive material are brought out to connections. The center terminal is always the wiper. The outer terminals connect the ends of the resistive material so that maximum resistance may be measured between them. One of them is usually grounded.

Figure 4.9, part (a), shows how the wiper divides the pot's resistive material into a two-resistor voltage divider. Each resistor lies between one end of the resistive material

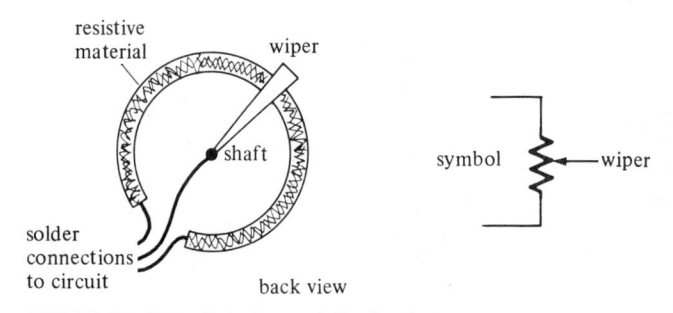

FIGURE 4.8 Potentiometer and its Symbol.

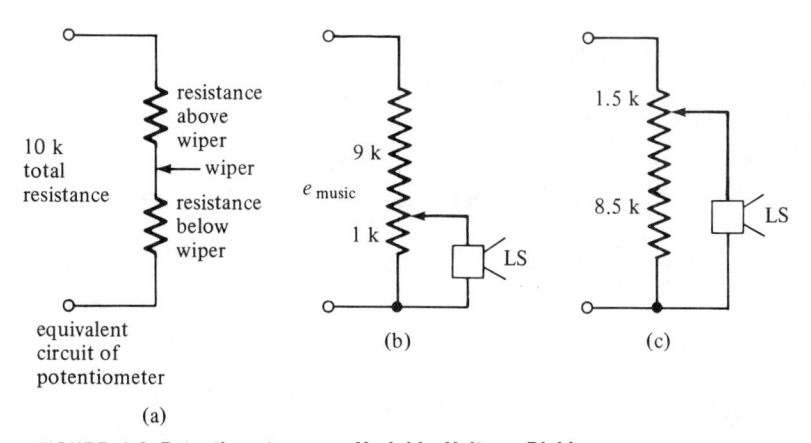

FIGURE 4.9 Potentiometer as a Variable Voltage Divider.

and the wiper. In Figure 4.9 (b), a voltage, changing with music, has been applied between the pot's outer terminals and a loudspeaker connected between wiper and ground.

When the wiper is in the 1 k–9 k position [part (b)], the speaker voltage is one-tenth of the music voltage because the speaker only sees the drop across the 1 k between wiper and ground. In part (c), however, the speaker gets the higher voltage developed across 8.5 k and the music is louder. The pot is thus a volume control.

4.9 THE THÉVENIN EQUIVALENT SOURCE

Every source has internal resistance because of the way it is made. The copper wire in a car's alternator, for instance, has resistance. So do the chemicals within a flashlight battery. Because it is scattered through the source, internal resistance is difficult to measure and often changes value under different operating conditions.

For these reasons, real sources are difficult to analyze. However, a good understanding can be obtained by studying an approximation or model called the *Thévenin equivalent source*, named after a French engineer, M. L. Thévenin (1857–1926).

Thévenin's equivalent source has two components, representing parts inside a real source. They are: an *ideal voltage source* in series with an *internal resistance* (see Figure 4.10). The subscript oc on the source voltage stands for "open circuit." It will be explained.

An ideal voltage source supplies the same voltage no matter what current is drawn from it, that is, its internal resistance is zero. Of course, such sources are only an idea; they do not exist. Nevertheless, that idea helps us understand and analyze real devices and circuits.

Writing a loop equation for the circuit of Figure 4.10, we find

$$E_{oc} = IR_{int} + IR_{load}$$

But

$$IR_{load} = V_o, \text{ the output voltage}$$

Substituting:

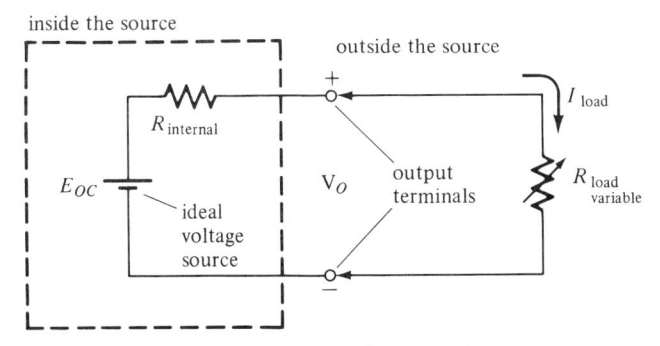

$$E_{oc} = IR_{int} + V_o$$

FIGURE 4.10 Thévenin Equivalent Source with Load.

Solving for V_o:

$$V_o = E_{oc} - IR_{int} \qquad (4.6)$$

Solving the first equation for I, we get:

$$I = \frac{E_{oc}}{R_{int} + R_{load}} \qquad (4.7)$$

Let us vary R_{load} from a minimum of zero ohms (output shorted) to infinite ohms (output open). As we do, the changes in V_o and I will be noted:

- For $R_{load} = 0\ \Omega$ (short-circuit conditions):

$$I = \frac{E_{oc}}{R_{int} + 0}$$
$$= \frac{E_{oc}}{R_{int}} \text{ amperes}$$

This is the most current the circuit will deliver under any conditions. It is called I_{sc} for *short-circuit current*.

$V_o = 0$ V, the output is shorted. All of E_{oc} is being dropped across R_{int}.

Note that maximum output current occurs with minimum output voltage.

- For $R_{load} = \infty\ \Omega$ (open-circuit conditions):

$I = 0$ A; the output is open.

$V_o = E_{oc} - IR_{int} = E_{oc} - 0 = E_{oc}$ volts

E_{oc}, the *open-circuit voltage*, is the most voltage the source can deliver under any conditions. It is the value of the Thévenin ideal source.

Note that minimum output current occurs with maximum output voltage.

- For $0 < R_{load} < \infty$:

As R_{load} increases from $0\ \Omega$ (short) to $\infty\ \Omega$ (open), V_o increases from zero to E_{oc} and I decreases from I_{sc} to zero. Find V_o and I by regular series circuit analysis.

Figure 4.11 summarizes the output voltage/load current variations for the Thévenin source model.

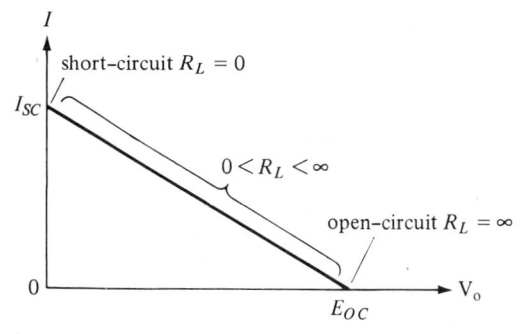

FIGURE 4.11 Output Characteristics of the Thévenin Equivalent Source.

4.9 THE THÉVENIN EQUIVALENT SOURCE

EXAMPLE 4.19

Calculate V_o and I of this Thévenin source for each value of R_{load}.

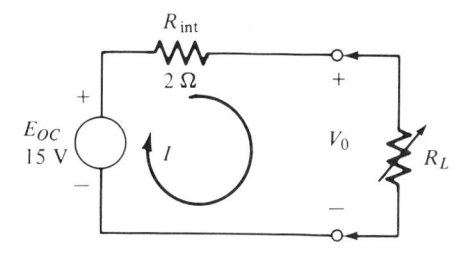

a. R_{load} = infinite ohms.

Open-circuit conditions.

$$V_o = E_{oc} = 15 \text{ V}$$
$$I = 0 \text{ A (no drop across } R_{int})$$

b. $R_{load} = 4 \, \Omega$.

$$V_o = \frac{E_{oc} R_{load}}{R_{load} + R_{int}}$$
$$= \frac{15(4)}{4 + 2}$$
$$= \frac{60}{6} = 10 \text{ V}$$

$$I = \frac{E_{oc}}{R_{load} + R_{int}}$$
$$= \frac{15}{4 + 2} = \frac{15}{6} = 2.5 \text{ A}$$

c. $R_{load} = 0 \, \Omega$.

Short-circuit conditions:

$$V_o = 0 \text{ V (all of } E_{oc} \text{ is dropped across } R_{int})$$
$$I = \frac{E_{oc}}{R_{int}}$$
$$= \frac{15}{2} = 7.5 \text{ A} = I_{sc}$$

■

Note carefully how the load and internal resistance of the Thévenin circuit create a voltage divider. As in any divider, when R_{load} decreases, its share of E_{oc} decreases. In other words, *the output voltage decreases as the load resistance is reduced*. In Example 4.19, for instance, V_o fell from 15 V to 10 V when R_{load} was reduced from an open to 4 Ω, a drop in output of 33%.

The output voltage of a real source drops when a load is connected just as it does for the Thévenin model. When it drops, we say that R_{load} is *loading*, or *loading down*, the source. Since loading is caused by voltage division between R_{load} and source internal resistance:

| Loading can be minimized by making $R_{load} \gg R_{internal}$.

The symbol $\gg$ means "very much greater than." How much is "very much?" In electronics, the 10-to-1 rule is used, which says that 10 times is very much greater.

The fact that loading causes output voltage decrease is a constant problem in electronics. To prevent these drops requires a "stiff" source, that is, a source whose terminal voltage remains constant as output current increases. Such a source is said to have good *voltage regulation*.

Percent voltage regulation is defined:

$$\% \text{ voltage regulation} = \frac{E_{\text{no-load}} - E_{\text{full-load}}}{E_{\text{full-load}}} \times 100 \qquad (4.8)$$

$E_{\text{no-load}}$ is the source output voltage when R_{load} is infinite—no current is being drawn. $E_{\text{full-load}}$ is the output voltage when the source is operating at its rated current; R_{load} is at its operating value.

Voltage regulation provides a "figure of merit" by which sources can be compared. Needless to say, those with the lowest, and therefore best, regulation, cost the most.

Finally, let us consider the power delivered to a load by a source. It will be zero when R_{load} is zero, rise to a maximum for some value of load, and fall back to zero when R_{load} is infinite.

What value of R_{load} draws that maximum power from the source? That question is easily answered with higher math. For now, accept the following *maximum power transfer theorem*:

| A source delivers maximum power to a load when the load has the same resistance as the internal resistance of the source.

Some of the most common situations in electronics involve steps taken to ensure maximum power transfer between a source and a load.

EXAMPLE 4.20

Answer the questions for this Thévenin source.

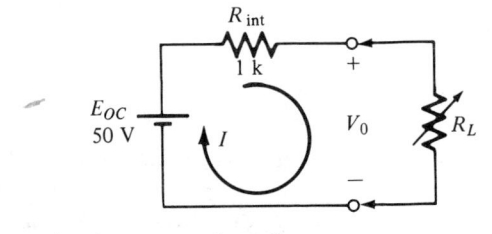

a. What is the no-load output voltage?

$$R_{load} = \text{infinity}$$

Therefore,

$$V_o = 50 \text{ V}$$

4.9 THE THÉVENIN EQUIVALENT SOURCE

b. How badly does a resistance of 5 k load the source?

$$V_o = \frac{E_{oc} R_{load}}{R_{load} + R_{int}} = \frac{50(5\ k)}{5\ k + 1\ k} = 41.7\ V$$

Loading caused the terminal voltage to drop 8.3 V.

c. What is the value of R_{load} for minimum loading?

There is no set figure; the higher, the better. The 10-to-1 rule says anything greater than 10(1 k), or 10 k.

d. What is the percent voltage regulation when $R_{load} = 5\ k$?

$$\%\ \text{voltage regulation} = \frac{E_{no\text{-}load} - E_{full\text{-}load}}{E_{full\ load}} \times 100$$

$$= \frac{50 - 41.7}{41.7} \times 100 = 20\%$$

e. What value of R_{load} will draw maximum power?

For maximum power transfer, R_{load} must equal $R_{internal}$, or 1 k. ∎

Two or more Thévenin circuits may be combined in series. When doing so, pay close attention to source polarities.

EXAMPLE 4.21

Reduce this circuit to a single Thévenin circuit.

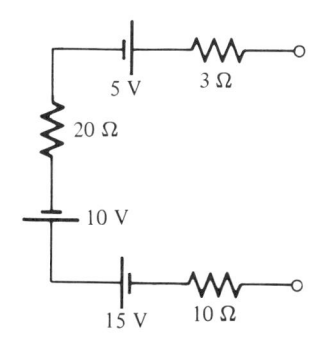

The three resistances add:

$$3 + 10 + 20 = 33\ \Omega = \text{final } R_{internal}$$
$$E_{oc}\ \text{final} = (5 + 15) - 10 = 10\ V$$

Draw the final circuit:

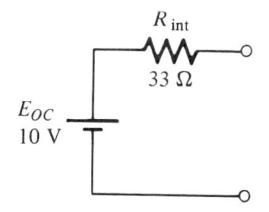

∎

4.10 TROUBLESHOOTING THE SERIES CIRCUIT

Troubleshooting electronic equipment requires logical thinking that starts with the questions: What are the indications that something is wrong? What could happen to cause these indications? The answers to these questions are clues that tell the repairperson where to start looking. Measurements are taken and tests performed to find the problem. Then repairs are made.

In order to know what tests and measurements to take and to be able to interpret their results, the repairperson must know:

- The theory behind the circuit being repaired. The greater the knowledge, the quicker the results.
- How to use test equipment.
- The limitations of each piece of test equipment, such as its accuracy.
- How the test equipment might affect the circuit being tested.

Most series-circuit troubleshooting can be done with a voltmeter and/or an ohmmeter (meters are discussed in Chapter 9). Currents are seldom read because the circuit must be cut in order to connect the ammeter in series. The same information can usually be obtained with less work by using voltage readings and Ohm's law.

Before starting troubleshooting, make sure that the equipment is on. Valuable time is lost searching for voltages that are not there because the equipment is not plugged in or there's a blown fuse. Also, be sure that the test equipment is on and working. Check it against other equipment known to be functioning.

The easiest problems to find are shorts and opens.

The Short Circuit

When a short occurs, total resistance goes down, current goes up and voltages are not normal. Voltage readings taken across components located between source and short will be higher than usual because current is high. Readings taken beyond the short will be low or zero, depending on the severity of the short. The correct voltages are often printed on schematics supplied by the equipment manufacturer.

The short itself is found at the point where there is either no voltage or a voltage below normal. Look for breaks in insulation, defective components, solder, or other conductive material in the wrong places: anything that could cause unusually low resistance.

EXAMPLE 4.22

Troubleshoot this circuit:

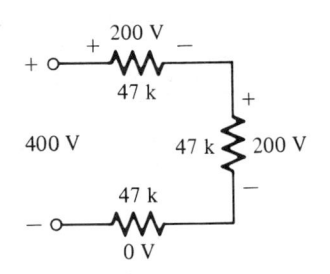

```
        + 200 V −
  + o───WWW───┐
         47 k │ +
              │
  400 V   47 k ≷ 200 V
              │
         47 k │ −
  − o───WWW───┘
          0 V
```

4.10 TROUBLESHOOTING THE SERIES CIRCUIT

With three equal resistances, we would expect the 400 V to divide evenly at 133 V across each. Instead, we see 200 V across two of them. Therefore, there is current in the circuit; the problem is not an open. The 0V across one resistance signals a short there. Check for a defective resistor, frayed wires, or shorted connections. ■

The Open Circuit

An open circuit means infinite or unusually high resistance and zero or very low current. Consequently, the voltage across all components through which that current normally flows will be zero or very low. With one exception: The actual break, which is the open, looks like a very high resistance. It will therefore have most, if not all, the source voltage across it. Look for broken connections, a switch not closing properly, a blown fuse or tripped circuit breaker, dirty or corroded connections, anything that interrupts current.

Do not be afraid to poke, wiggle, and prod components, cables, and connections while troubleshooting. Opens and shorts often announce themselves by sudden changes in meter readings when this is done.

EXAMPLE 4.23

Troubleshoot this circuit:

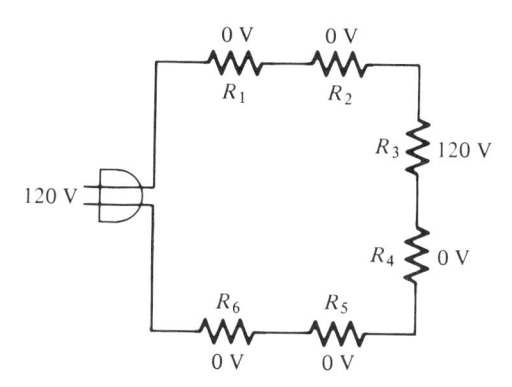

Since there is no drop across all but one of the components, there must be no current through them. The presence of the source voltage across R_3 indicates that resistance is open. Check for a defective component or broken connections. ■

The ohmmeter is especially useful as a *continuity checker* in searching for opens. By "continuity" is meant a continuous path of low resistance between two points. For example, if ohmmeter leads are connected to each end of a wire, the resistance should read very low. If it reads high or infinite, there is no continuity; there is a break in the line.

EXERCISES

Section 4.1

1. Each of three resistors has 5 mA through it. Are the three in series? Under what conditions would they be in series? Draw a circuit for this. Under what conditions would they not be in series? Draw a circuit for this.

84 THE SERIES CIRCUIT

2. What is the current at each of points A, B, and C? Discuss.

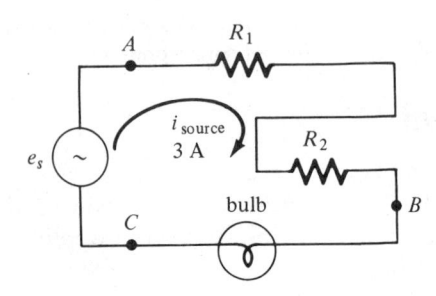

3. Which of these characteristics does each of the following circuits have?

   a. More than one path for current flow.
   b. Series circuit.
   c. Only one path for current flow.

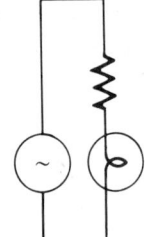

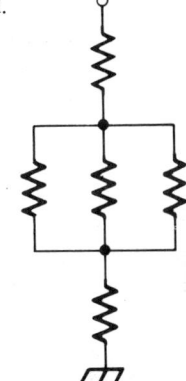

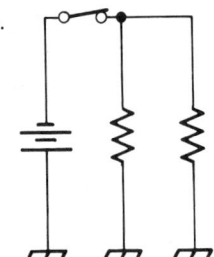

4. Fuses are connected in series with the circuit being protected. This forces all current to pass through them. At what point, A, B, or C, could a fuse be connected in the circuit of Exercise 2? Discuss.

5. What components in each circuit are in series? Do not overlook the source.

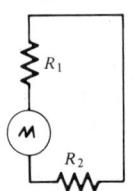

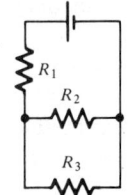

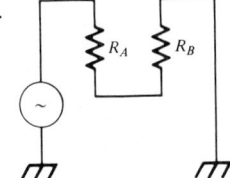

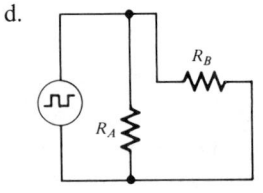

EXERCISES

85

e.

f.

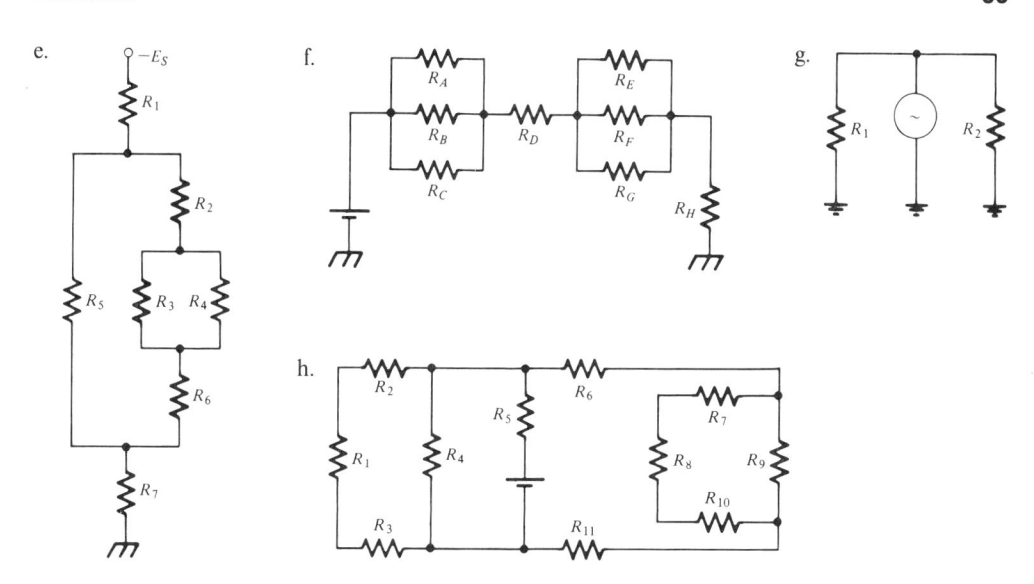

g.

h.

6. Current is measured by ammeters which are connected in series. The terminals of the meter are marked with + and − signs. Current entering the positive terminal causes a positive reading. Current entering the negative terminal causes a negative reading. Which of these ammeters reads positively?

a.

b.

7. Circuit B from Exercise 5 is redrawn with ammeters included. Tell which meter(s) will read the current requested.

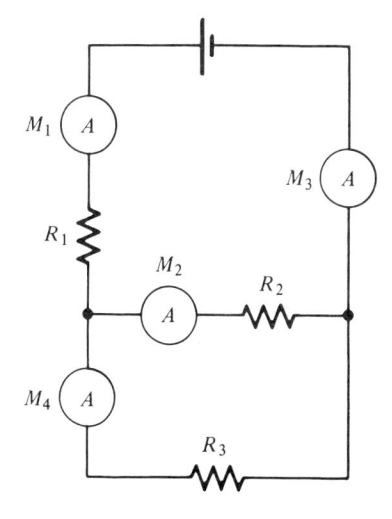

a. Source current.   b. Current through R_2.   c. Current through R_3.

8. Repeat Exercise 7 for circuit E for the following currents.

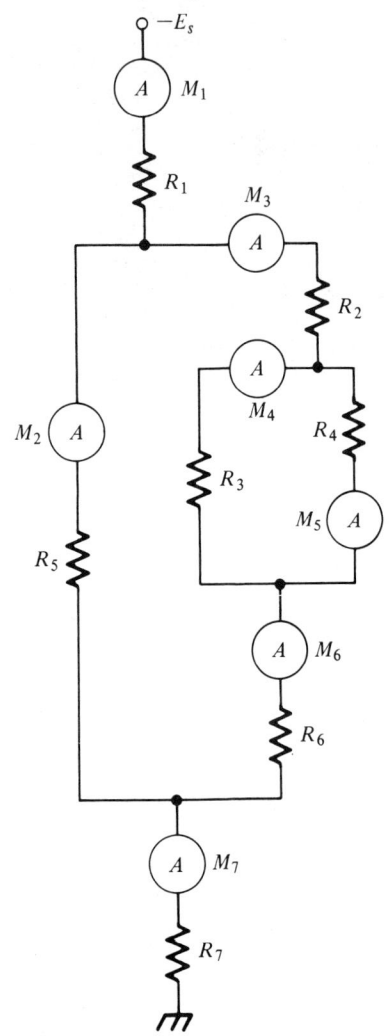

a. Source current.
b. Current through R_2.
c. Current through R_4.
d. Current through R_5.

Section 4.2

9. Given: 5-V, 6-V, 8-V and 12-V sources, one of each. Connect all of them in series so that the total provides each of the following. a. 3 V; b. 9 V; c. 7 V.

10. How should sources be connected for series aiding? Explain why this is so from an energy standpoint. Repeat for series opposing.

11. What is the terminal voltage of each circuit? State which terminal is +.

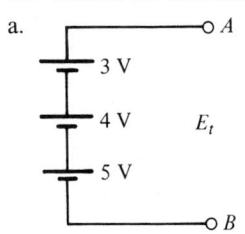

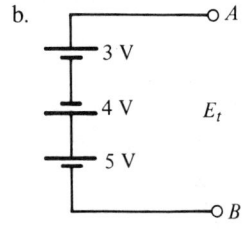

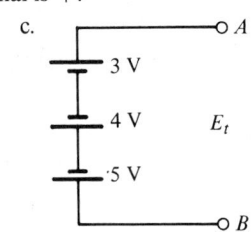

EXERCISES

d.

[Circuit diagram: 10 V, 25 V, 5 V cells in top row; 12 V cell on left, 2 V cell on right; terminals A and B at bottom, E_t labeled]

e.

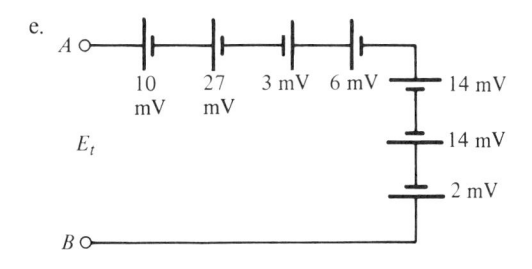

[Circuit diagram: A at top-left; cells 10 mV, 27 mV, 3 mV, 6 mV, 14 mV, 14 mV, 2 mV; E_t labeled; B at bottom-left]

12. a. Why must flashlight cells (batteries) be inserted with the center terminal of one against the bottom terminal of the next?
b. What would happen if the center terminals of a two-cell flashlight were touching?

13. Find the following voltages for this circuit. a. V_{ab}; b. V_{bc}; c. V_{cd}; d. V_{da}; e. V_{ac}; f. V_{bd}. Indicate the positive terminal in each case.

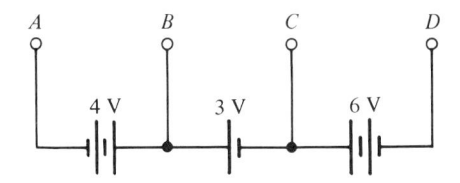

[Circuit diagram showing terminals A, B, C, D with 4 V, 3 V, 6 V cells between them]

14. Show how seven 2-V cells can be connected in series to supply 10 V.

Section 4.3

15. Calculate the source voltage in each of the following circuits.

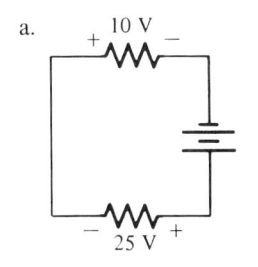

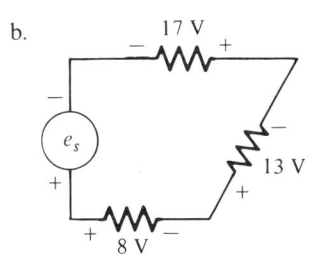

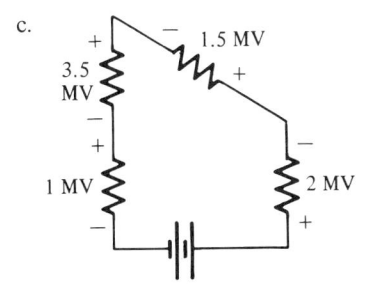

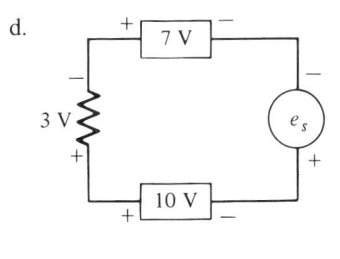

a. [Circuit with 10 V resistor (top), battery on right, 25 V resistor (bottom)]

b. [Circuit with e_s source, 17 V, 13 V, 8 V resistors]

c. [Circuit with 3.5 MV, 1.5 MV, 1 MV, 2 MV resistors and battery source]

d. [Circuit with 7 V, e_s source, 3 V, 10 V resistors]

16. Explain Kirchhoff's voltage law in your own words, first from an energy standpoint, then from a voltage standpoint.

17. Calculate the unknown voltage in each of the following circuits.

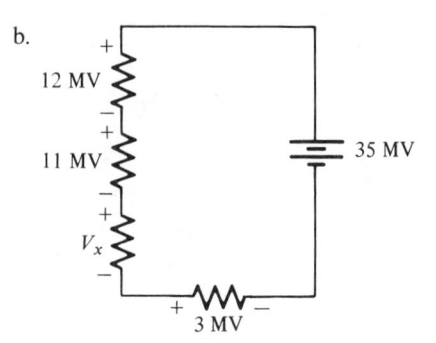

a. [Circuit with $e_s = 110$ V, $+75$ V drop, v_x, and -8 V$+$]

b. [Circuit with $+12$ MV, $+11$ MV, V_x, $+3$ MV, and 35 MV source]

18. Which statements about this circuit are true?

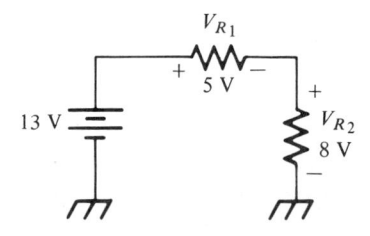

[Circuit with 13 V source, $V_{R_1} = 5$ V, $V_{R_2} = 8$ V]

a. The polarity of V_{R_1} is correct.
b. The $E_{\text{rise}} = $ the sum of the drops.
c. Total potential drop is 26 V.
d. Total potential rise is 13 V.
e. The drop across R_2 is greater than the drop across R_1.

Section 4.4

19. Find the total resistance for each circuit.

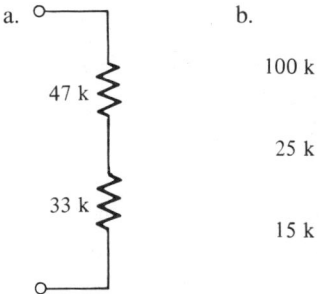

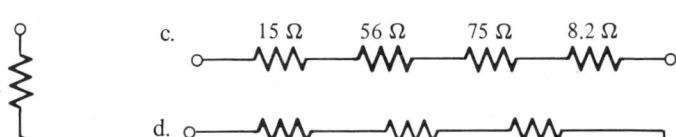

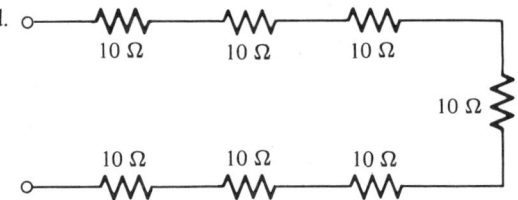

a. [47 k, 33 k in series]

b. [100 k, 25 k, 15 k in series]

c. [15 Ω, 56 Ω, 75 Ω, 8.2 Ω in series]

d. [10 Ω, 10 Ω, 10 Ω, 10 Ω, 10 Ω, 10 Ω, 10 Ω in series loop]

20. Find the unknown resistance in each circuit.

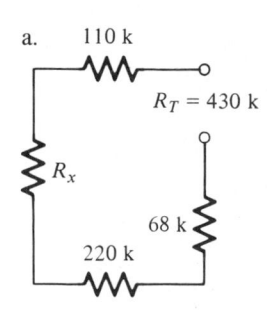

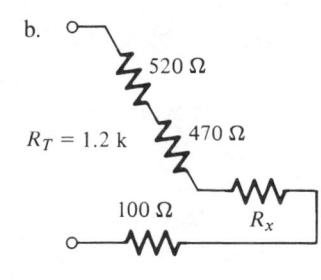

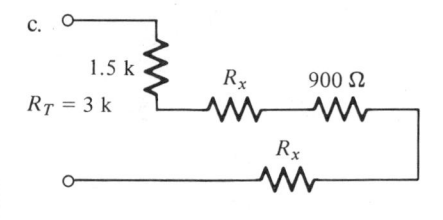

a. [110 k, R_x, 220 k, 68 k; $R_T = 430$ k]

b. [520 Ω, 470 Ω, 100 Ω, R_x; $R_T = 1.2$ k]

c. [1.5 k, R_x, 900 Ω, R_x; $R_T = 3$ k]

EXERCISES

21. Answer the questions for this circuit.

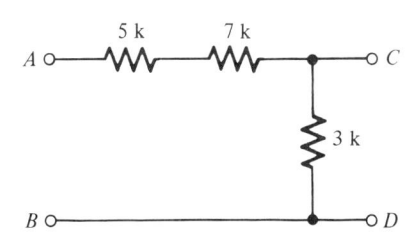

a. What is the resistance looking into terminals A and B?
b. What is the resistance looking into terminals C and D?

22. Repeat Exercise 21 for this circuit.

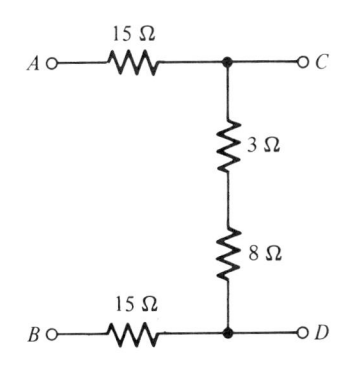

23. Find the current in each circuit.

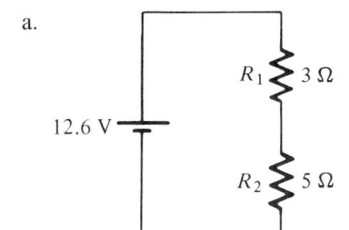

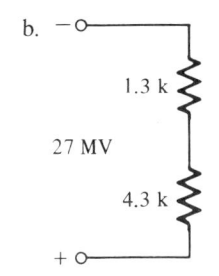

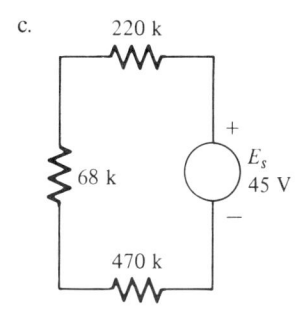

24. What voltage must be applied to each circuit to force the indicated current?

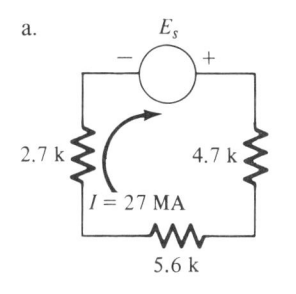

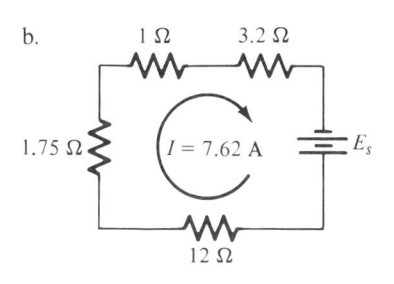

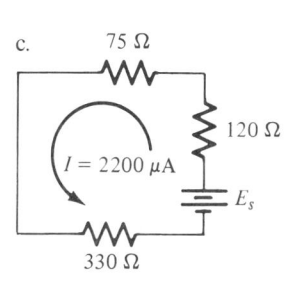

Section 4.5

25. Answer the questions for this circuit:

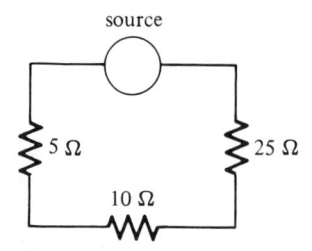

a. Which resistance will have the greatest voltage drop no matter what the source voltage?
b. Which will have the least drop?
c. How will the drop across the 25 Ω compare to that across the 5 Ω?
d. How will the drop across the 10 Ω compare to that across the 5 Ω?
e. How will the drop across the 25 Ω compare to that across the 10 Ω?
f. Will the relations of parts a through e hold regardless of source voltage?

26. Answer the questions for this circuit:

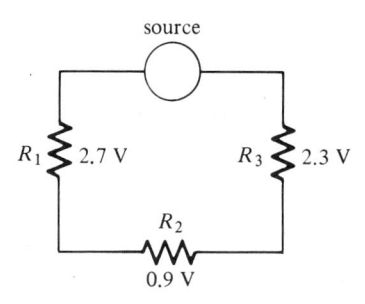

a. Which resistor has the highest resistance?
b. Which has the lowest?
c. How does R_2 compare to R_1?

27. Determine the drop across each resistance in both circuits, then answer the questions.

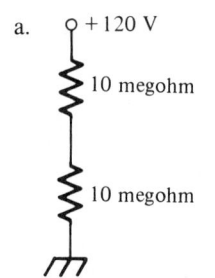

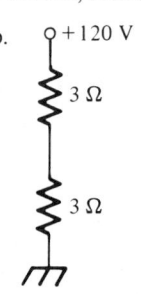

a. How do the drops across the 10-M resistances compare? Why is this so?
b. How do the 3-Ω drops compare?
c. What conclusion can be made about equal-valued resistances in series?
d. How would the voltages compare in either circuit if the source voltage was doubled?
e. What does affect how the drops compare?

28. a. Two resistances of the same value are connected in series and placed across a 50 V source. What is the drop across each?
b. Repeat for 40 equal-valued resistances across 2 kV.
c. Repeat for n equal-valued resistances across E volts.

EXERCISES

29. What must the resistance of R_x be in order that V_{R_x} be each of the following?

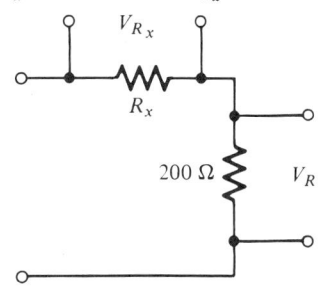

a. $V_{R_x} = 4V_R$ b. $V_{R_x} = 1.5V_R$ c. $V_{R_x} = V_R$
d. $V_{R_x} = 0.5V_R$ e. $V_{R_x} = 0.15V_R$

30. Using the circuit of Exercise 29, what must the resistance of R_x be so that V_{R_x} is: a. 50% of E_{source};   b. 67% of E_{source};   c. 33% of E_{source};   d. $\approx E_{source}$.

31. Analyze each circuit. Use calculator memory where useful. Remember to ask checking questions.

a.

b.

c.

d.

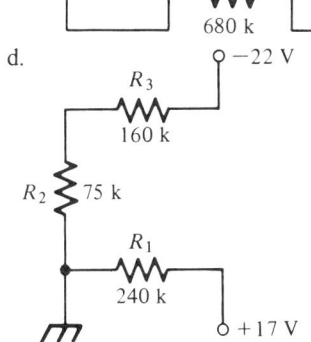

32. What does each math statement say about this circuit?

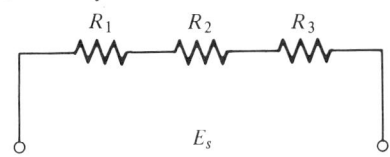

a. $\dfrac{E_{source}}{R_1 + R_2 + R_3}$

b. $\dfrac{E_{source} R_2}{R_1 + R_2 + R_3}$

c. $\left(\dfrac{E_s}{R_1 + R_2 + R_3}\right)^2 R_1$

d. $\left(\dfrac{E_s R_3}{R_1 + R_2 + R_3}\right)^2 \dfrac{1}{R_3}$

33. Calculate the value of R_x in each circuit.

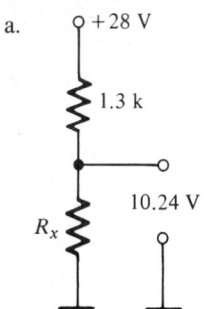

a. [Circuit: +28 V source, 1.3 k resistor with 10.24 V tap, R_x to ground]

b. [Circuit: E_s 90 MV, R_x with 24 MV, 120 k resistor]

34. Find the items indicated for each circuit.

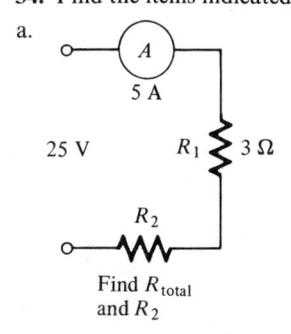

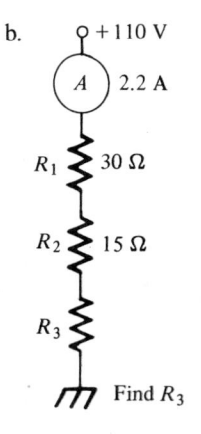

a. [Circuit: 25 V, ammeter A reading 5 A, $R_1 = 3\ \Omega$, R_2]
Find R_{total} and R_2

b. [Circuit: +110 V, ammeter A reading 2.2 A, $R_1 = 30\ \Omega$, $R_2 = 15\ \Omega$, R_3]
Find R_3

35. Find the items indicated for each circuit.

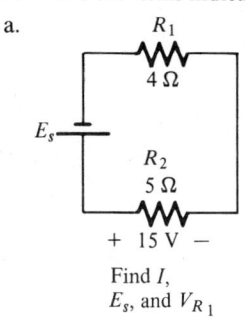

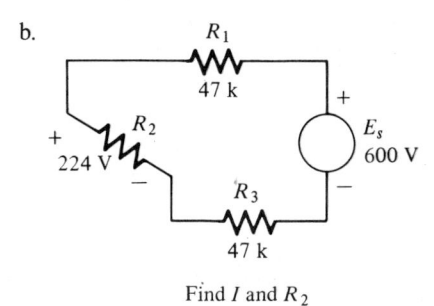

a. [Circuit: E_s, $R_1 = 4\ \Omega$, $R_2 = 5\ \Omega$, +15 V−]
Find I, E_s, and V_{R_1}

b. [Circuit: $R_1 = 47\ k$, $E_s = 600$ V, $R_3 = 47\ k$, R_2, 224 V]
Find I and R_2

36. The ideal ammeter has 0 Ω resistance. Real meters, however, have resistance that increases total circuit resistance. Perform the indicated operations for this circuit.

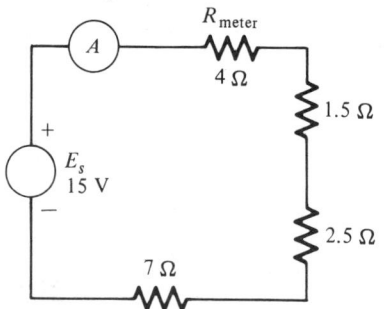

[Circuit: ammeter A, $R_{meter} = 4\ \Omega$, 1.5 Ω, 2.5 Ω, 7 Ω, $E_s = 15$ V]

a. Assume that the ammeter has 0 Ω of resistance. Calculate I.
b. Include the meter's 4 Ω and recalculate I.
c. Find the error caused by the meter.

EXERCISES

37. Draw the waveforms of the current and both drops in this circuit:

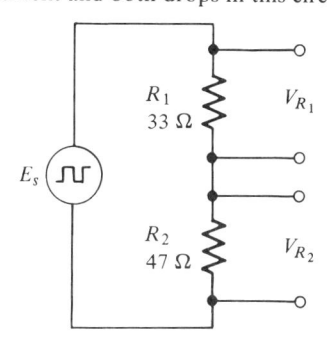

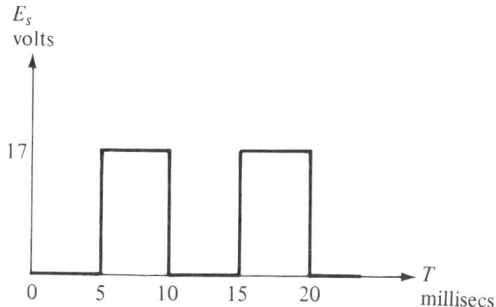

38. Repeat Exercise 37 with this source-voltage waveform:

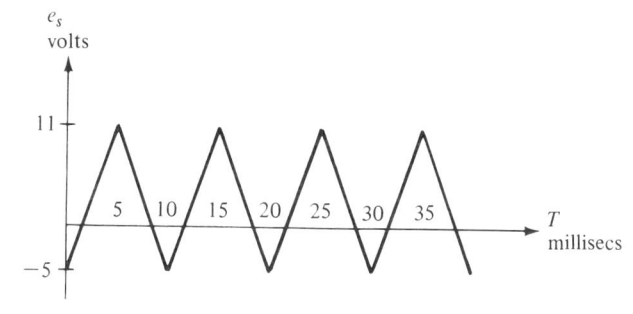

39. Two resistances, a 9.1 k and an 11 k are connected in series. The source voltage is the following sawtooth waveform. Sketch the waveform of the voltage drop across the 9.1 k.

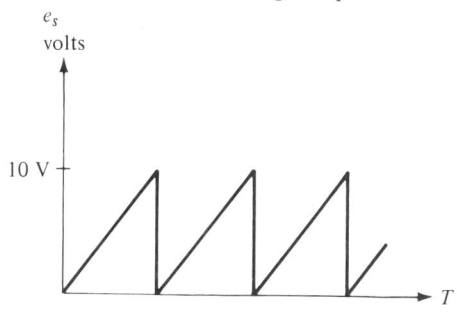

40. Two 20% tolerance resistances, a 5.6 k and a 4.7 k, are connected in series across an 80-V source. Perform the following steps.
a. Estimate each voltage drop.
b. Calculate exact drops. Compare results with those of part a.
c. Let the 5.6 k be at its maximum tolerance value and the 4.7 k at its minimum. Recalculate exact voltage drops.

THE SERIES CIRCUIT

41. Estimate the current in each circuit without using a calculator. Then do an exact analysis and compare results.

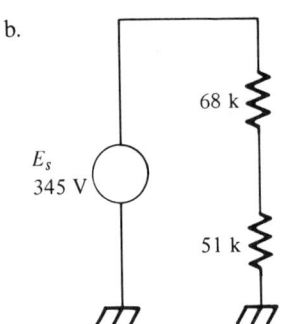

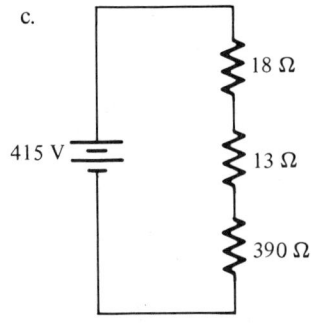

a. 19.5 V, 4 Ω, 5 Ω

b. E_s = 345 V, 68 kΩ, 51 kΩ

c. 415 V, 18 Ω, 13 Ω, 390 Ω

42. Repeat Exercise 41 for the voltage drop across each resistance in these circuits.

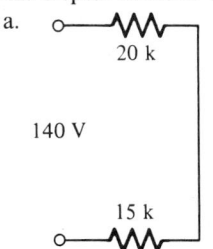

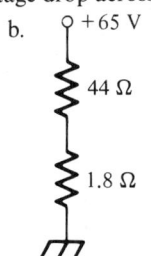

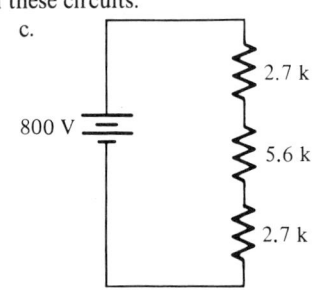

a. 140 V, 20 k, 15 k

b. +65 V, 44 Ω, 1.8 Ω

c. 800 V, 2.7 k, 5.6 k, 2.7 k

Section 4.6

43. Find the unknown voltage, E_x, for each circuit and its polarity.

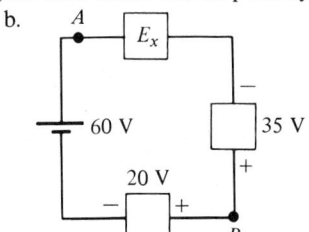

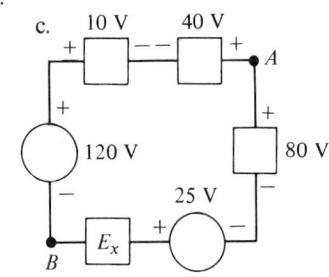

a. 5 V, 15 V, E_x (points A, B)

b. E_x, 60 V, 35 V, 20 V (points A, B)

c. 10 V, 40 V, 120 V, 80 V, 25 V, E_x (points A, B)

44. Redo Exercise 43 by starting the loop equation at a different location and going in the opposite direction. Compare equations.

45. Find the voltage between points A and B of each circuit in Exercise 43. State which terminal is positive.

46. What is inside the X box in each circuit, a resistance, a source, or whatever?

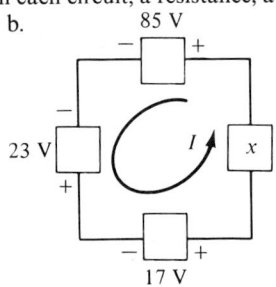

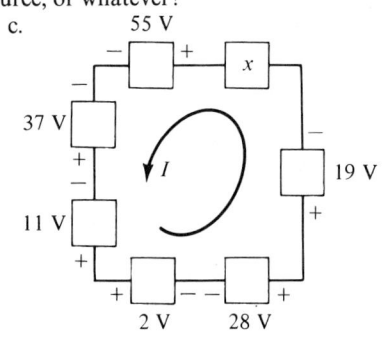

a. 15 V, x, 9 V, 3 V

b. 85 V, 23 V, x, 17 V

c. 55 V, x, 37 V, 11 V, 2 V, 28 V, 19 V

EXERCISES

47. Find the items indicated for each circuit.

a. V_{R_2}, E_{source}, R_{total}, and V_{ab}. The ammeter resistance is zero.

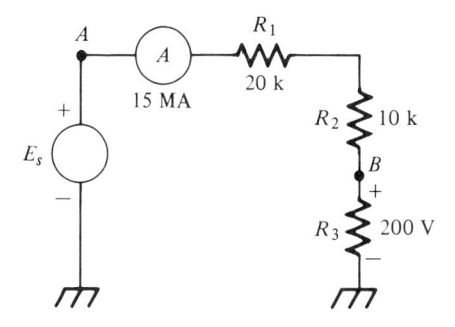

b. E_{source}.

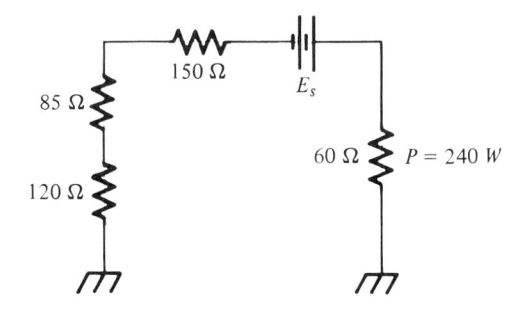

48. Find the items indicated for each circuit.

a. If $R_{total} = 200\ k\Omega$, find R_2, I, V_{R_1}, V_{R_2}, E_{source}, and V_{mp}.

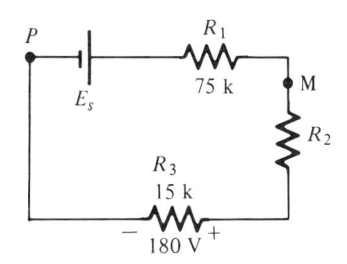

b. R_1, R_2, and I.

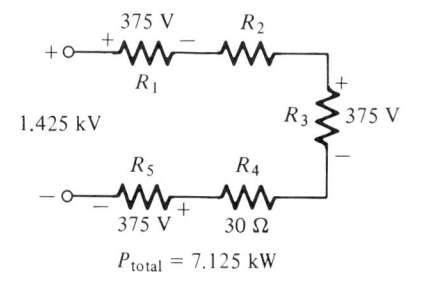

Section 4.7

49. What resistance must be added in series with a 4.5-V bulb that draws 75 mA so that it can be powered by a 6-V source? Could a $\frac{1}{4}$-W resistor do the job?

50. A model race car motor draws 1.65 A at maximum speed on the straightaways. This must drop to 0.92 A in the curves or the car leaves the track. If the applied voltage is 14 V and no resistance is used on the straightaways, how much resistance must be added to maintain control in the curves?

51. How much resistance must be placed in series with a 60-W, 120-V bulb so that it will give the same light when connected to a 220-V line? What is the power rating of this resistor?

52. Series resistances can divide a voltage down. Can they ever multiply one up? Discuss.

Section 4.8

53. For no-load conditions, find the voltage at each point with respect to ground for each circuit.

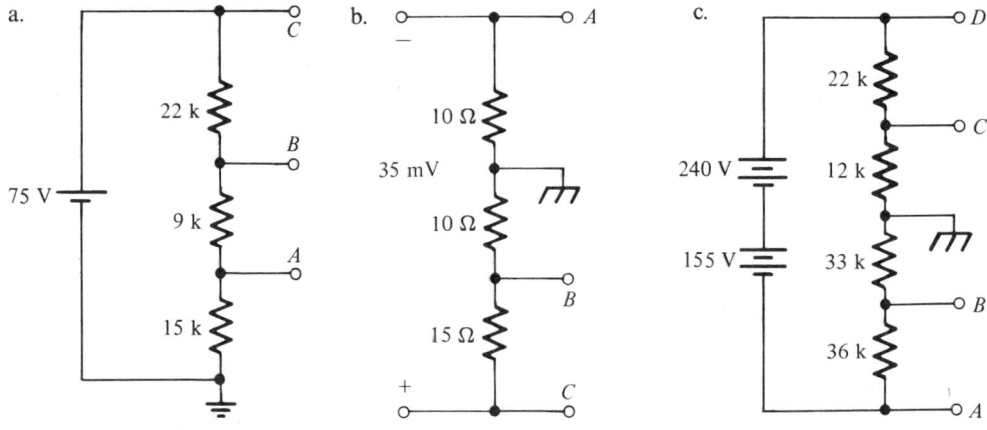

54. Repeat Exercise 53 for these circuits.

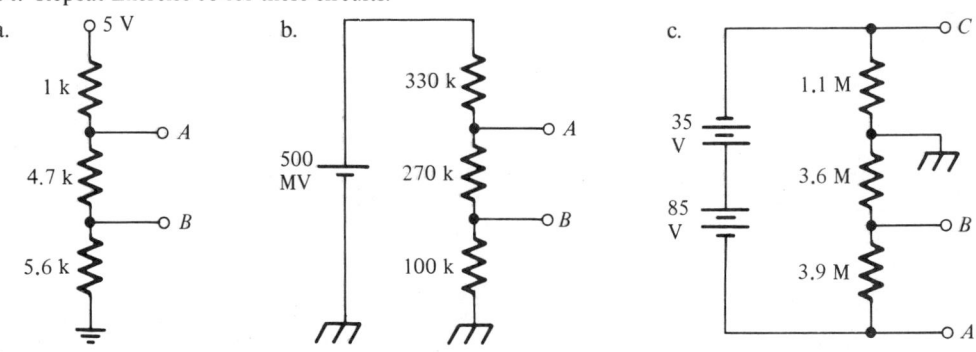

55. The wiper on this pot is one-fourth of the way from one end as shown.
a. Between what two terminals will each of these resistances be measured: 5 k, 15 k, and 20 k?
b. Where would the wiper have to be in order to read 10 k between W and A?

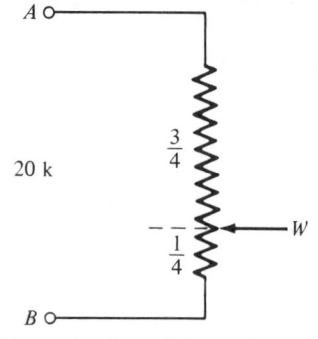

56. What is meant by open-circuit or no-load conditions when referring to voltage dividers?

EXERCISES

Section 4.9

57. What is meant by open-circuit and short-circuit conditions when referring to a source? Should a car battery be operated under open-circuit conditions? Under short-circuit conditions? Explain.

58. Explain why the open-circuit voltage is the most voltage that can possibly be obtained from a source.

59. Find V_o and I for each circuit under open- and short-circuit conditions.

a. R_{int} 0.5 Ω, E_{OC} 200 V, V_0

b. R_{int} 410 Ω, E_{OC} 18.3 mV, V_0

60. Let R_{load} of circuit a in Exercise 59 be 1.5 Ω. Solve for V_o, I, P_{load}, and % voltage regulation.

61. A source with an E_{oc} of 215 V delivers 10.8 A to a 14-Ω load. What is the internal resistance?

62. The Thévenin source model with a load is a two-resistor voltage divider. On that basis, show why loading occurs.

63. A Thévenin circuit has an $E_{oc} = 100$ V and an $R_{int} = 0.5$ Ω. Calculate V_o and I for each value of R_{load}: a. 0.1 Ω; b. 0.25 Ω; c. 0.5 Ω; d. 2 Ω; e. 10 Ω. f. Comment on the changes in V_o as load resistance increases.

64. Given: a source with an $R_{internal}$ of 500 Ω. Without doing any calculations, tell what values of R_{load} would cause each of the following (use the 10-to-1 rule).
a. Negligible loading.   b. Moderate loading.   c. Severe loading.

65. A standard D-size flashlight cell (battery) has an E_{oc} of 1.5 V and an $R_{internal}$ of approximately 0.25 Ω.
a. What is the voltage across a 6-Ω bulb connected to the cell?
b. The cell ages and $R_{internal}$ climbs to 3 Ω. What is the new E_{oc} for the battery? What is the new voltage across the bulb?
c. Based on these results, why is it a waste of time to test a cell under open-circuit conditions?

66. Find the % voltage regulation when a Thévenin source of $E_{oc} = 18.3$ mV and an internal resistance of 410 Ω is loaded by 350 Ω.

67. Calculate power delivered to the load of Exercise 66. Then increase R_{load} from 350 to 450 Ω in 10-Ω increments, calculating load power for each. What value of R_{load} drew maximum power? How does this value compare to $R_{internal}$? Calculate P_{load} for $R_{load} = 409$ and 411 Ω. Were both of these values less than P_{load} for 410 Ω? What value of R_{load} draws maximum power from the source of Exercise 63?

68. *Extra Credit.* What value of $R_{internal}$ should a source have in order to transfer maximum power to a load? Be careful with this one.

69. Reduce to a single Thévenin source.

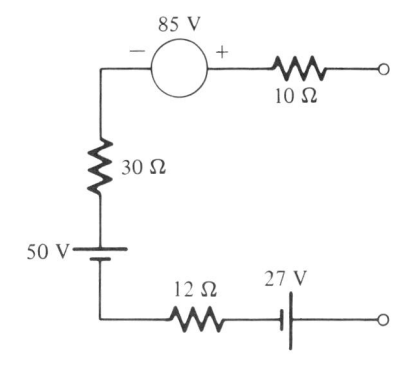

85 V, 10 Ω, 30 Ω, 50 V, 12 Ω, 27 V

70. Reduce to a single Thévenin source.

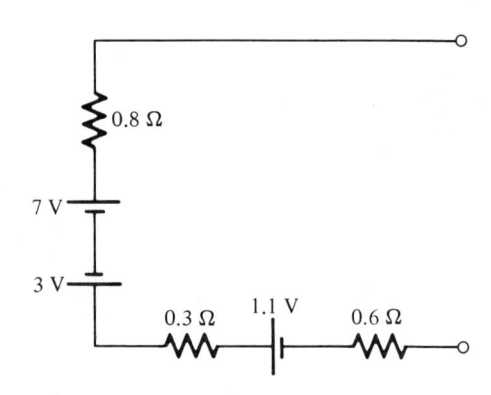

Section 4.10

71. Why is current seldom measured when troubleshooting?

72. What is meant by continuity in a circuit?

73. A normal switch is either all on or all off. The resistance between the contacts of a particular switch under test reads a different value each time the switch is closed. What does this indicate?

74. Troubleshoot each circuit.

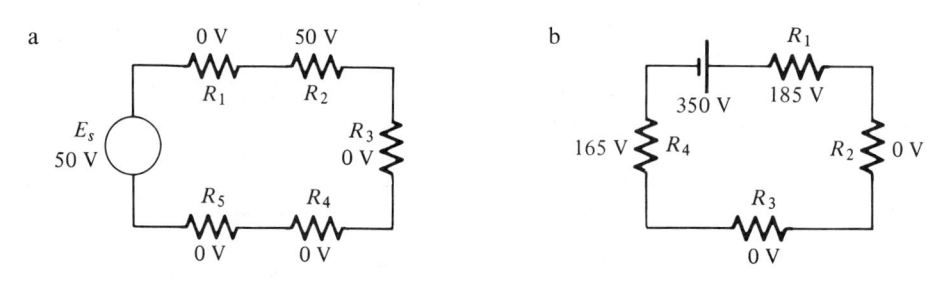

75. What will the voltage be between the output terminals of this circuit for each of the following conditions?

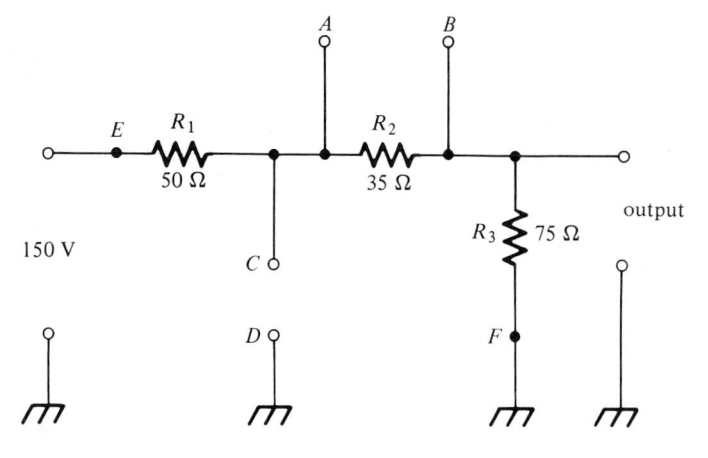

a. A short between points A and B.
c. A break at point E.
b. A short between points C and D.
d. A break at point F.

EXERCISES

76. Four bulbs are wired as shown. Indicate what will happen to each of the following if bulb 3 opens. Use an I for increase, D for decrease, and NC for no change. Repeat for bulb 3 shorting.

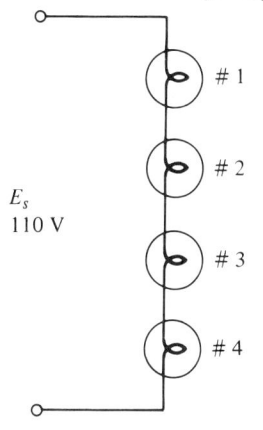

E_s 110 V

\#1
\#2
\#3
\#4

a. Current.
c. E_{source}.

b. Brightness of bulb 4.
d. R_{total}.

CHAPTER HIGHLIGHTS

Elements in series have the same current through them.
    Check polarities to see if series-connected sources are adding or subtracting their potentials.
    Adding series resistance always increases total resistance.
    Examine the results of analysis to see if they are reasonable.
    To calculate power dissipation, voltage drop, current, or resistance for a component, use only the parameters associated with that specific component.
    Loop equations sum all voltages around a closed loop. They can be used to find an unknown voltage between two points in a circuit.
    The Thévenin source model explains much about source operation, including loading, voltage regulation, and maximum power transfer.
    Troubleshooting requires logical thinking, knowledge of circuit being repaired, and knowledge of test equipment.

Formulas to Remember

loop equations

$\sum E \text{ rises} = \sum V \text{ drops}$ Kirchhoff's voltage law

$R_{total} = \sum R$ total resistance of series-connected resistances

$E \text{ across } R_n = \dfrac{E_{source} R_n}{R_{total}}$ voltage-divider formula

$\sum E_\circlearrowleft = 0$ loop equation

$V_o = E_{oc} - I(R_{internal})$ output voltage of Thévenin source

$\dfrac{E_{no\text{-}load} - E_{full\text{-}load}}{E_{full\text{-}load}} \times 100$ % voltage regulation

$R_{load} = R_{internal}$ maximum power transfer

CHAPTER 5

The Parallel Circuit

Having completed our study of series circuits, we move to the parallel circuit.

At the end of this chapter, you should be able to:

- **Tell when components are connected in parallel.**
- **Use Kirchhoff's current law to analyze parallel circuits.**
- **Calculate total resistance of a group of resistances in parallel.**
- **Analyze circuits in terms of conductance as well as resistance.**
- **Convert between Norton and Thévenin source models.**
- **Use Millman's theorem to combine sources in parallel.**
- **Troubleshoot parallel circuits.**

5.1 WHAT DOES "IN PARALLEL" MEAN?

A *parallel* circuit is shown in Figure 5.1. It has an upper and a lower *node*, each signified by a black ball indicating a connection between conductors. A node is defined as a point where three or more conductors are connected.

Between the nodes are three *branches*. A branch is defined as a part of a circuit between two nodes that contains one or more components. The right-hand branch, for example, contains R_2. Current flowing in a branch is called a *branch current*.

Note how the total current leaves the source in Figure 5.1, travels to the upper node and splits, part heading down through R_1, the rest going over to R_2. The two branch currents meet at the lower node, recombine to form I_{total}, and head back to the source. The effect is the same as river water splitting to pass an island on both sides, then rejoining at the downstream end. Because of the split at the upper node, I_{R_1} cannot be the same current as I_{R_2}. Therefore, R_1 and R_2 are not in series.

Remembering that the conductors are considered ideal (zero resistance), R_1 and the source must have the same voltage across them because there is nothing between them to cause different voltages. R_1 is said to be in parallel or in *shunt* with the source. For the same reason, R_2 is also in parallel with R_1 and the source; all three components are in shunt with each other.

| Components are said to be in parallel when they have the same voltage across them.

Be sure that the difference between series and parallel is clear:

- Elements in series have the same current through them.
- Elements in parallel have the same voltage across them.

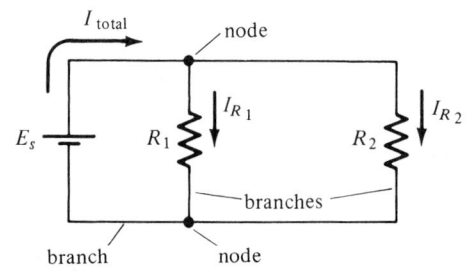

FIGURE 5.1 Parallel Circuit.

5.2 KIRCHHOFF'S CURRENT LAW

(a) series circuit

(b) series–parallel circuit

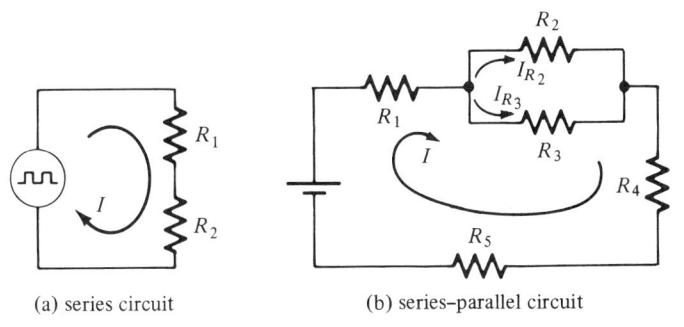

FIGURE 5.2 Nonparallel Circuits.

For contrast, Figure 5.2 shows two nonparallel circuits. Circuit (a) is the familiar series circuit from Chapter 4. Circuit (b), a series–parallel circuit (also seen in Chapter 4), shows the same current flowing through the source, R_1, R_4, and R_5. They are, therefore, in series. But R_2 and R_3 cannot have the same current. They do, however, have the same voltage across them and are in parallel.

The parallel connection is used far more often than the series connection. The outlets in our homes, for example, are all in parallel, as are all the houses in an area. In fact, a high-tension line feeding several towns sees each of them as a parallel load.

5.2 KIRCHHOFF'S CURRENT LAW

Note how the 6-A I_{total} going toward the upper node in Figure 5.3 equals the sum of the 4-A and 2-A currents going away from the node. This demonstrates the second of Kirchhoff's laws that are so important to electronics, *Kirchhoff's current law*, abbreviated *KCL*. It says:

| The sum of the currents entering a node equals the sum of the currents leaving the node.

In the shorthand of mathematics, KCL is

$$\sum I_{\text{entering a node}} = \sum I_{\text{leaving the node}} \qquad (5.1)$$

If a current leaving is considered a negative entering current, KCL may be simplified to

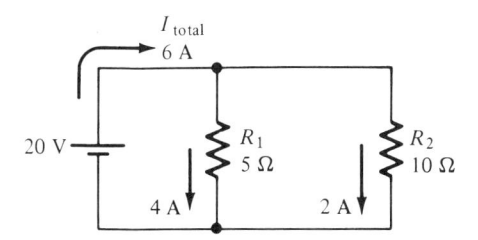

$$\sum I = 0 \qquad (5.2)$$

FIGURE 5.3 Parallel Circuit to show Kirchhoff's Current Law.

Any equation based on Kirchhoff's current law that sums all the currents entering and leaving a node is called a *nodal equation*.

To contrast KVL and KCL:

- In a loop equation, KVL says that the sum of the voltages around a closed loop equals zero.
- In a nodal equation, KCL says that the sum of the currents entering and leaving a node equals zero.

Loop and nodal equations are the basis for the two most powerful techniques of circuit analysis. They are discussed in Chapter 7.

EXAMPLE 5.1

Use a nodal equation to find the unknown current.

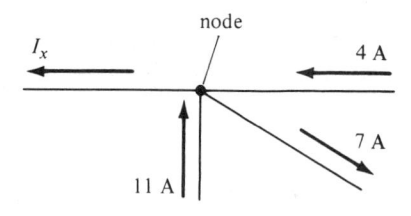

Use KCL and write a nodal equation:

$$\sum I_{\text{entering}} = \sum I_{\text{leaving}}$$
$$4 + 11 = 7 + I_x$$
$$I_x = 15 - 7 = 8 \text{ A}$$

EXAMPLE 5.2

Find the magnitude and direction of each unknown current.

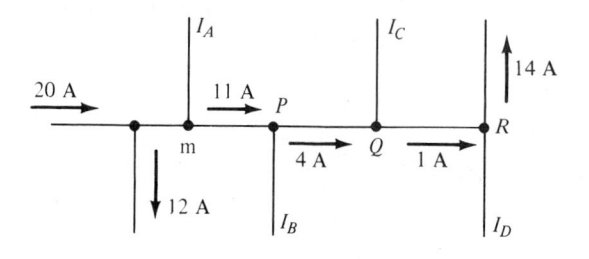

Nodal equation at node M:
I_A is assumed to be entering the node. If this assumption is wrong, the sign will be negative.
I_C

$$I_A + (20 - 12) = 11$$
$$I_A = 11 - 8 = 3 \text{ A entering } M$$

At node P:
$$11 = 4 + I_B$$
$$I_B = 11 - 4 = 7 \text{ A leaving } P$$

At node Q:
$$4 = I_C + 1$$
$$I_C = 4 - 1 = 3 \text{ A leaving } Q$$

At node R:
$$14 = I_D + 1$$
$$I_D = 14 - 1 = 13 \text{ A entering } R \qquad \blacksquare$$

5.3 RESISTANCES IN PARALLEL

Because of the open switch in the circuit of Figure 5.4, part (a), no current passes through R_2 and I_{total} is the 4 A through R_1. R_{total} seen by the source is 20 V/4 A = 5 Ω, the resistance of R_1.

In Figure 5.4, circuit (b), the switch is closed. R_2 draws 2 A, causing I_{total} to increase to 6 A. R_{total} is now 20/6 = 3.3 Ω.

Note carefully that adding R_2 caused R_{total} to decrease:

| Total resistance decreases whenever parallel resistance is added.

A moment's thought shows the logic in this statement. Adding a parallel resistance, no matter what its value, provides another current path, which causes I_{total} to increase. Therefore, R_{total} must have decreased. This is just the opposite of series circuits, in which adding resistance causes R_{total} to increase.

Using KCL and Ohm's law, we will derive an equation that tells the total resistance of a group of parallel-connected resistances. Consider two resistances in shunt with a source as we saw in Figure 5.1. KCL tells us that

$$I_{\text{total}} = I_{R_1} + I_{R_2}$$

But Ohm's law says that

$$I_{\text{total}} = \frac{E_{\text{source}}}{R_{\text{total}}}$$

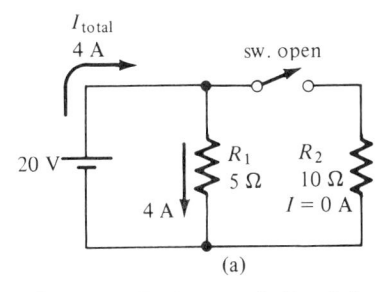

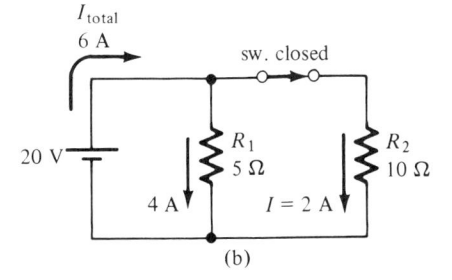

FIGURE 5.4 Resistances in Parallel.

It also says that

$$I_{R_1} = \frac{E_{source}}{R_1}$$

and

$$I_{R_2} = \frac{E_{source}}{R_2}$$

Substitution into our original KCL equation yields

$$\frac{E_{source}}{R_{total}} = \frac{E_{source}}{R_1} + \frac{E_{source}}{R_2}$$

Dividing both sides of the equation by E_{source}, we have

$$\frac{1}{R_{total}} = \frac{1}{R_1} + \frac{1}{R_2}$$

Reciprocating both sides, we get

$$R_{total} = \frac{1}{1/R_1 + 1/R_2}$$

For n resistances the equation becomes

$$R_{total} = \frac{1}{1/R_1 + 1/R_2 + \cdots + 1/R_n} \tag{5.3}$$

where R_n is the Nth resistance.

EXAMPLE 5.3

Find R_{total} looking into the terminals of this circuit.

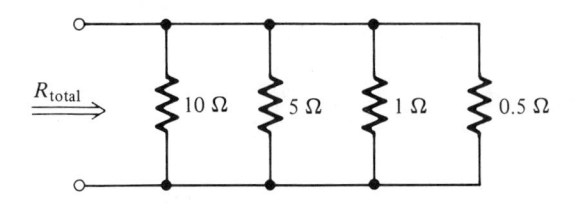

$$R_{total} = \frac{1}{1/10 + 1/5 + 1/1 + 1/0.5}$$
$$= \frac{1}{0.1 + 0.2 + 1 + 2}$$
$$= \frac{1}{3.3}$$
$$= 0.3 \ \Omega$$

∎

5.3 RESISTANCES IN PARALLEL

Here is a quick test to use when calculating R_{total} for resistances in parallel:

The total resistance of a group of shunt resistances will be less than the lowest resistance in the group.

In Example 5.3, R_{total} equals 0.3 Ω, which is less than 0.5 Ω, the lowest resistance in the circuit. Although this test will not tell if an answer is right, it will signal an error if the total is not less than the lowest.

Note that each reciprocal of resistance in the denominator of Equation 5.3 is a conductance. The equation could therefore be rewritten

$$R_{total} = \frac{1}{1/R_1 + 1/R_2 + \cdots + 1/R_n}$$
$$= \frac{1}{G_1 + G_2 + \cdots + G_n}$$
$$= \frac{1}{G_{total}}$$

There is an important point being made here. It is:

$$G_{total} = \sum G \qquad (5.4)$$

Add parallel-connected conductances to get total conductance.

EXAMPLE 5.4

Calculate R_{total} for this circuit:

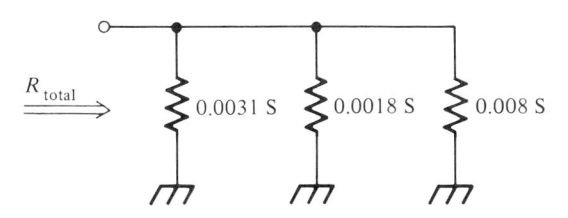

$R_{total} \rightarrow$ | 0.0031 S | 0.0018 S | 0.008 S

The conductances are added directly to get G_{total}:

$$G_{total} = 0.0031 + 0.0018 + 0.008$$
$$= 0.0129 \text{ S}$$
$$R_{total} = \frac{1}{G_{total}} = \frac{1}{0.0129} = 77.5 \text{ Ω}$$

The lowest resistance in this problem is the one with the highest conductance. Therefore, reciprocating 0.008, we get

$$\frac{1}{0.008} = 125 \text{ Ω}$$

which is greater than R_{total}.

EXAMPLE 5.5

Calculate R_{total} for this circuit:

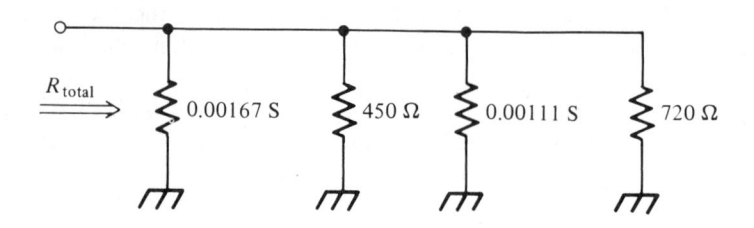

[Circuit diagram: four parallel branches with values 0.00167 S, 450 Ω, 0.00111 S, and 720 Ω, with R_{total} indicated at the input.]

The resistances are converted to conductances because all four must be in the same units for addition:

$$R_{total} = \frac{1}{\underbrace{1/450}_{\substack{\text{convert} \\ R \text{ to } G}} + \underbrace{0.00167 + 0.00111}_{\substack{\text{these two} \\ \text{already in} \\ G \text{ form}}} + \underbrace{1/720}_{\substack{\text{convert} \\ R \text{ to } G}}}$$

$$= \frac{1}{0.0022 + 0.00167 + 0.00111 + 0.00139}$$

$$= \frac{1}{0.00637}$$

$$= 157 \; \Omega$$

There are several shortcuts that help find R_{total}.

Shortcut A: Finding R_{total} on a Calculator

Follow these steps with calculators using *algebraic notation*:

Step number		
1	Enter R_1	} converts R_1 to G_1
2	$\frac{1}{x}$	
3	+	
4	Enter R_2	} converts R_2 to G_2
5	$\frac{1}{x}$	
6	=	gives G_{total}
7	$\frac{1}{x}$	reciprocates G_{total} to get R_{total}

Steps 3, 4, and 5 may be repeated for any number of resistances.

5.3 RESISTANCES IN PARALLEL

Follow these steps with calculators using *RPN notation*:

Step number		
1	Enter R_1	
2	$\dfrac{1}{x}$	converts R_1 to G_1
3	Enter R_2	
4	$\dfrac{1}{x}$	converts R_2 to G_2
5	$+$	add G_1 to G_2 to get G_{total}
6	$\dfrac{1}{x}$	reciprocates G_{total} to get R_{total}

Steps 3, 4, and 5 may be repeated for any number of resistances.

Shortcut B: Finding R_{total} for Several Resistances of the Same Value

Suppose that three resistances of the same value are put into Equation 5.3:

$$R_{total} = \frac{1}{1/R + 1/R + 1/R}$$
$$= \frac{1}{3/R} = \frac{R}{3}$$

When N resistances of the same value are connected in parallel, R_{total} equals the resistance of one of them divided by N.

EXAMPLE 5.6

Find R_{total} for this circuit:

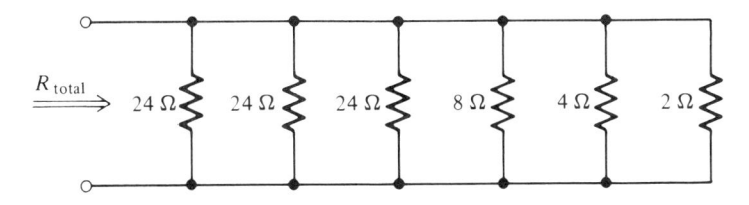

$R_{total} \rightarrow$ 24 Ω ∥ 24 Ω ∥ 24 Ω ∥ 8 Ω ∥ 4 Ω ∥ 2 Ω

The three 24 Ω reduce to a single 8 Ω. The circuit is now

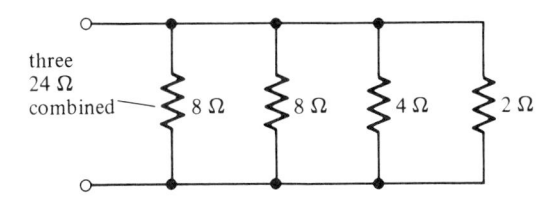

three 24 Ω combined → 8 Ω ∥ 8 Ω ∥ 4 Ω ∥ 2 Ω

The two 8 Ω reduce to 4 Ω:

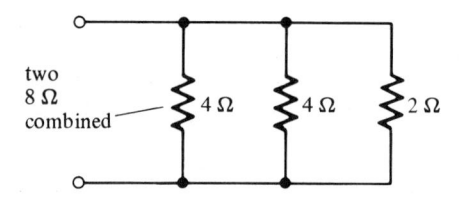

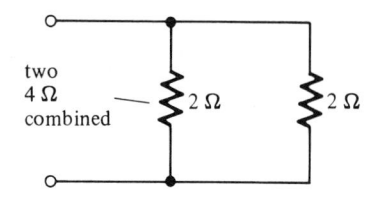

The two 4 Ω reduce to 2 Ω:

which reduces to a single 1-Ω resistance.

EXAMPLE 5.7

Calculate R_{total} looking into terminals F and H.

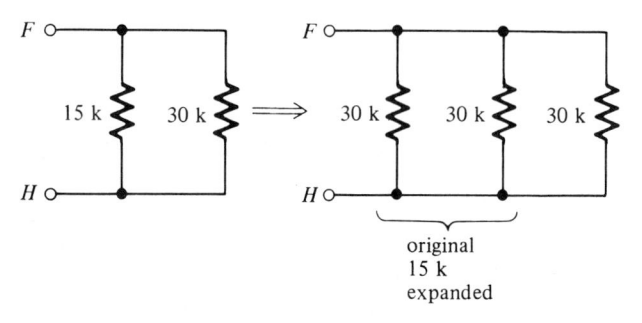

The 15 k can be considered two 30 k in shunt. Do this and the circuit becomes three 30 k in parallel. R_{total} is then found easily:

$$R_{total} = \frac{30 \text{ k}}{3} = 10 \text{ k}$$

This approach can be used with many combinations, for example:

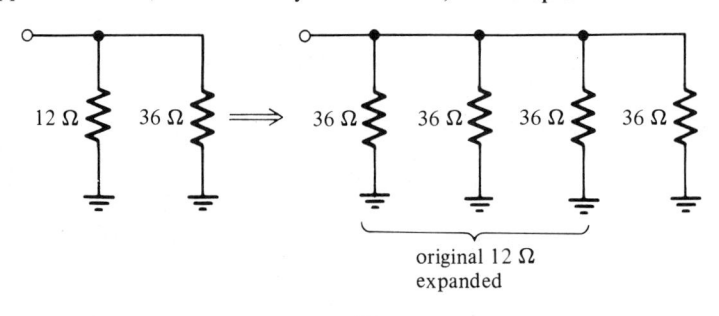

$$R_{total} = \frac{36}{4} = 9 \text{ Ω}$$

5.3 RESISTANCES IN PARALLEL

Shortcut C: Two resistances of Any Value in Parallel

Use Equation 5.3:

$$R_{total} = \frac{1}{\dfrac{1}{R_1} + \dfrac{1}{R_2}}$$

Substitute the lowest common denominator:

$$R_{total} = \frac{1}{\dfrac{R_2}{R_1 R_2} + \dfrac{R_1}{R_1 R_2}} = \frac{1}{\dfrac{R_2 + R_1}{R_1 R_2}}$$

$$R_{total} = \frac{R_1 R_2}{R_1 + R_2} \tag{5.5}$$

This is called the *product-over-sum* (*POS*) rule.

EXAMPLE 5.8

Find R_{total} for this configuration:

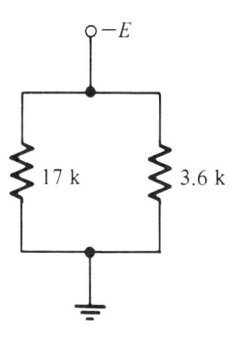

$$R_{total} = \frac{R_1 R_2}{R_1 + R_2}$$
$$= \frac{17\,k(3.6\,k)}{17\,k + 3.6\,k}$$
$$= 3\,k \qquad ■$$

· The POS rule can be used on circuits of more than two resistances by following these steps:

- Combine any two resistances by POS or shortcut B. This reduces the circuit by one resistance.
- Use the R_{total} from the first step together with any other resistance to find a new R_{total} by POS or shortcut B. The circuit is reduced again.
- Continue these steps until finished.
- Before starting, check the circuit to see if any combinations can be easily reduced in your head.

EXAMPLE 5.9

Find R_{total} looking into terminals *M* and *P*.

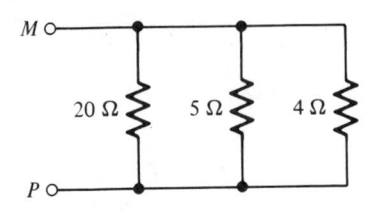

Combine the 20- and 5-Ω units by POS:

$$R_{total} = \frac{20(5)}{20 + 5}$$

$$= \frac{100}{25} = 4 \, \Omega$$

Combine the two 4 Ω by shortcut B:

$$R_{total} = \frac{4}{2} = 2 \, \Omega$$

Check: 2 Ω is less than 4 Ω.

Shortcut D: Using Approximations and Ballpark Figures

Disregard high-valued resistances when they are in parallel with much lower values. By the 10-to-1 rule:

If R_1 and R_2 are parallel-connected and $R_1 \geq 10R_2$, $R_{total} \approx R_2$ and $I_{R_2} \approx I_{total}$.

EXAMPLE 5.10

What is the approximate R_{total} for this combination?

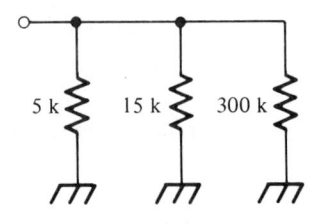

Disregard the 300 k

$$R_{total} \approx \frac{15 \, k(5 \, k)}{15 \, k + 5 \, k}$$

$$= 3.8 \, k$$

The value by analysis is 3.7 k. The error caused by the approximation is 2.7%.

EXAMPLE 5.11

Find the approximate R_{total} for this circuit:

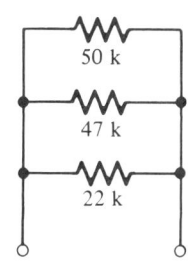

[Circuit diagram: three resistors 50 k, 47 k, 22 k in parallel]

The 50 k and 47 k are about the same, so their total is about 25 k. This is in shunt with another approximate 25 k; therefore R_{total} is about 25 k/2 or 12.5 k. The answer by analysis is 11.5 k. Our approximation is off by 8.7%. ∎

5.4 ANALYZING THE PARALLEL CIRCUIT

The pointers for analyzing series circuits in Chapter 4 apply to parallel circuits, too:

- Think of a plan to use known data to find unknowns.
- Take time to consider whether results make sense.
- Make ballpark estimates and compare to calculated figures.
- Ask checking questions.

EXAMPLE 5.12

Analyze this circuit:

[Circuit diagram: E_s = 120 V source in parallel with R_1 = 10 Ω, R_2 = 30 Ω, R_3 = 60 Ω; branch currents I_{R_1}, I_{R_2}, I_{R_3}; total current I_{total}]

Find the branch currents:

$$I_{R_1} = \frac{E_{source}}{R_1} = \frac{120}{10} = 12 \text{ A}$$

$$I_{R_2} = \frac{E_{source}}{R_2} = \frac{120}{30} = 4 \text{ A}$$

$$I_{R_3} = \frac{E_{source}}{R_3} = \frac{120}{60} = 2 \text{ A}$$

$$I_{total} = \sum I = 12 + 4 + 2 = 18 \text{ A}$$

THE PARALLEL CIRCUIT

Checking questions: Does the lowest-valued resistance have the highest current? Since R_2 is three times R_1, I_{R_2} should be one-third I_{R_1}. Is it? What about I_{R_1} and I_{R_3}? I_{R_2} and I_{R_3}?

Next, R_{total} will be calculated two ways. The results must be the same.

Method 1:

$$R_{\text{total}} = \frac{E_{\text{source}}}{I_{\text{total}}} = \frac{120}{18} = 6.7\ \Omega$$

Method 2:

$$R_{\text{total}} = \frac{1}{1/10 + 1/30 + 1/60}$$
$$= \frac{1}{0.1 + 0.033 + 0.0167}$$
$$= \frac{1}{0.15}$$
$$= 6.7\ \Omega$$

Calculate the power dissipation of each resistance:

$$P_{R_1} = \frac{E_{\text{source}}^2}{R_1} = \frac{120^2}{10} = 1440\ \text{W}$$
$$P_{R_2} = (I_{R_2})^2 R_2 = 4^2(30) = 16(30) = 480\ \text{W}$$
$$P_{R_3} = E_{\text{source}} I_{R_3} = 120(2) = 240\ \text{W}$$

The sum of the three dissipations is

$$1440 + 480 + 240 = 2160\ \text{W}$$

Power delivered by source is

$$E_{\text{source}} I_{\text{total}} = 120(18) = 2160\ \text{W}$$

Power supplied = power dissipated, as it must. ∎

Note carefully that I_{R_2} was used to calculate P_{R_2} in Example 5.12 and I_{R_3} to calculate P_{R_3}. Although this point has been made before, it is worth repeating:

> To calculate any parameter for a particular component, use only the current, voltage, resistance, and power associated with that specific component.

The analysis of parallel circuits is simplified by the *current-division formula*. It tells how total current entering a group of parallel-connected resistances divides among the individual resistances. The formula is

$$I_n = \frac{I_{\text{total}} R_{\text{total}}}{R_n} \tag{5.6}$$

where I_n = current through branch n
R_n = resistance of branch n
R_{total} = total shunt resistance

5.4 ANALYZING THE PARALLEL CIRCUIT

The formula does not say anything new. A look at the math shows that the numerator equals the voltage across the parallel combination. Division of that voltage by R_n yields the current through R_n. However, many books refer to it, so you should be familiar with it.

For two parallel resistances R_1 and R_2, the current division formula reduces to

$$I_{R_1} = \frac{I_{\text{total}} R_2}{R_1 + R_2} \tag{5.7}$$

and

$$I_{R_2} = \frac{I_{\text{total}} R_1}{R_1 + R_2} \tag{5.8}$$

Note carefully how R_2 is in the numerator for I_{R_1}, and vice versa.

EXAMPLE 5.13

Calculate the current in each resistance.

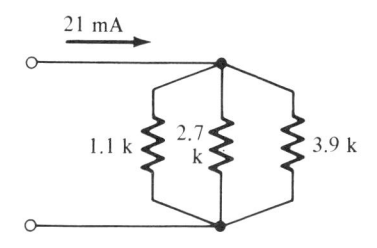

Calculate R_{total}:

$$R_{\text{total}} = \frac{1}{1/1.1\text{ k} + 1/2.7\text{ k} + 1/3.9\text{ k}}$$

$$= \frac{1}{909 \times 10^{-6} + 370 \times 10^{-6} + 256 \times 10^{-6}}$$

$$= \frac{1}{1.54 \times 10^{-3}}$$

$$= 651 \; \Omega$$

By current division, current through the 1.1 k is

$$I_{1.1\text{k}} = \frac{21\text{ mA}(651)}{1.1\text{ k}}$$

$$= 12.4\text{ mA}$$

Subtracting $I_{1.1\text{k}}$ from I_{total} gives the combined current through the 2.7 k and 3.9 k:

$$= 21\text{ mA} - 12.4\text{ mA}$$

$$= 8.6\text{ mA}$$

Calculating current through the 2.7 k using Equation 5.7:

$$I_{2.7k} = \frac{8.6 \text{ mA}(3.9 \text{ k})}{2.7 \text{ k} + 3.9 \text{ k}}$$
$$= 5.1 \text{ mA}$$

Calculating current through the 3.9 k by KCL:

$$I_{3.9k} = 8.6 \text{ mA} - 5.1 \text{ mA}$$
$$= 3.5 \text{ mA}$$

Checking question: Does the sum of the three branch currents equal I_{total}? ∎

As in series circuits, source voltage in parallel circuits can change according to various waveforms.

EXAMPLE 5.14

Determine the waveform of i_{R_2}.

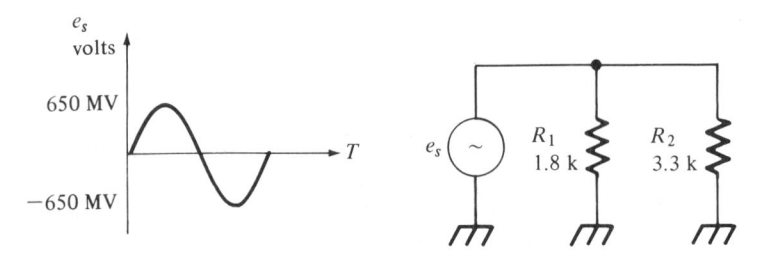

Since the positive maximum has the same absolute value as the negative minimum, i_{R_2} need only be calculated once:

$$i_{R_2} = \frac{e_{\text{maximum}}}{R_2} = \frac{650 \times 10^{-3}}{3.3 \times 10^3}$$
$$= 197 \text{ }\mu\text{A}$$

There is no need to calculate i for the three points when e_{source} is zero (when it crosses the X axis). Draw the waveform:

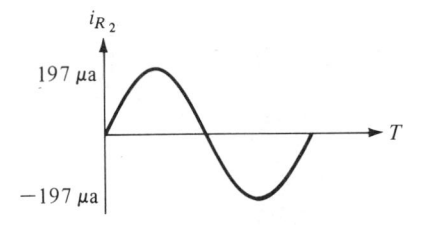

∎

Skill at estimating parameters in parallel circuits is just as important as it is in series circuits.

EXAMPLE 5.15

While testing this circuit, 10.2 mA is read on an ammeter in series with the 5.6 k. Without doing exact calculations, tell whether this is a reasonable figure; I_{total} is known to be correct.

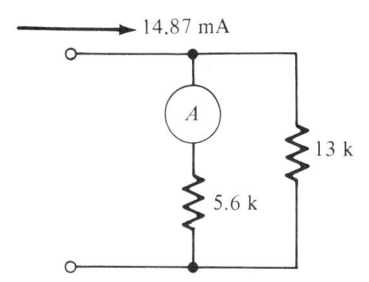

13 k is about two times 5.6 k. Therefore, for every unit of current through the 13 k there should be about 2 units through the 5.6 k. One unit plus 2 units equals 3 units. Consider I_{total} as 15 mA. Dividing it into 3 units makes 5 mA per unit. $I_{through\ the\ 5.6k} = 2$ units or 10 mA. Since the measured value is close to this approximation, the meter reading is reasonable. The answer by analysis is 10.4 mA.

5.5 THE NORTON EQUIVALENT SOURCE

The parallel equivalent of the Thévenin source model is the *Norton equivalent source*, named for Edward Norton (b. 1898) of Bell Laboratories.

Like the Thévenin model, there are two parts to the Norton circuit. These are an *ideal constant-current source* in shunt with an *internal resistance*. The model is shown with an external load resistance in Figure 5.5. The letters sc near the current source symbol stand for short circuit. This will be explained. The arrow in the symbol shows current direction.

Like its ideal voltage-source counterpart, an ideal current source is only a mathematical idea that helps us analyze electronic circuits and devices. It supplies the same current no matter what load resistance is connected. In other words, it has infinite internal resistance.

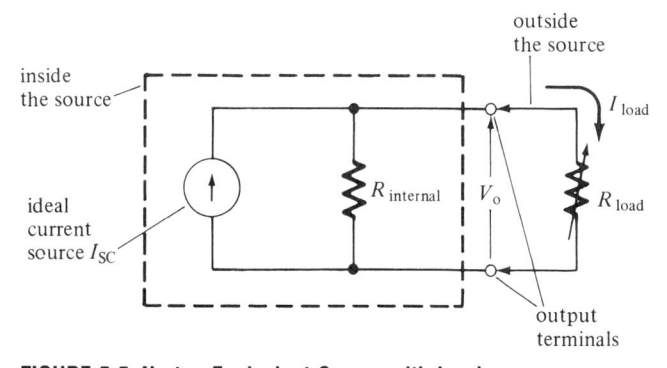

FIGURE 5.5 Norton Equivalent Source with Load.

By current division:

$$I_{\text{load}} = \frac{I_{\text{sc}} R_{\text{int}}}{R_{\text{int}} + R_{\text{load}}} \qquad (5.9)$$

The output voltage, V_o, equals the source current times R_{total}, the total resistance seen by the source:

$$V_o = I_{\text{sc}} R_{\text{total}}$$
$$= \frac{I_{\text{sc}} R_{\text{int}} R_{\text{load}}}{R_{\text{int}} + R_{\text{load}}} \qquad (5.10)$$

As we did for the Thévenin source, the values of V_o and I_{load} will be calculated as R_{load} is varied from zero ohms (output shorted) to infinite ohms (output open):

- For $R_{\text{load}} = 0\,\Omega$ (output shorted):

    All of I_{sc} is through the short, none flows through R_{int}. This is the most current that can be drawn from the circuit. That is why the source is called, I_{sc}.
    $V_o = 0$ V
    Note that maximum output current occurs with minimum output voltage.

- For $R_{\text{load}} = \infty\,\Omega$ (open circuit conditions):

    $I_{\text{load}} = 0$ A. All of I_{sc} passes through R_{internal}. R_{total} is at its highest value; therefore, V_o is highest.
    $V_o = I_{\text{sc}}(R_{\text{int}})$
    Note that minimum output current occurs with maximum output voltage.

- For $0 < R_{\text{load}} < \infty$:

    Find I_{load} and V_o by regular parallel circuit analysis.

The discussions on loading, percent voltage regulation, and maximum power transfer made in Chapter 4 for the Thévenin model are true for the Norton model because the two are equivalent. Similarly, the Norton V_{out} versus I_{load} characteristics are the same as those of the Thévenin equivalent (see Figure 4.11).

EXAMPLE 5.16

Answer the questions for this circuit:

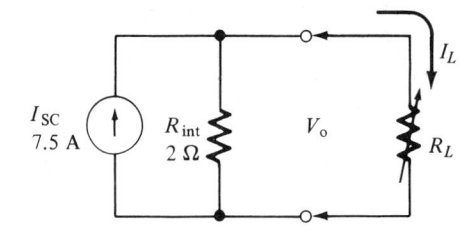

I_{SC} 7.5 A, R_{int} 2 Ω, V_o, R_L, I_L

a. What are I_{load} and V_o for $R_{\text{load}} = 0\,\Omega$?

Short-circuit conditions.

$$I_{\text{load}} = I_{\text{sc}} = 7.5 \text{ A, its highest value.}$$
$$V_o = E_{\text{short-circuit}} = 0 \text{ V, its lowest value.}$$

b. What are I_{load} and V_0 for $R_{\text{load}} = $ infinite ohms?

5.6 HOW TO CONVERT FROM ONE SOURCE MODEL TO THE OTHER

Open-circuit conditions.

$$I_{load} = 0 \text{ A, its lowest value.}$$
$$V_o = E_{open\text{-}circuit} = I_{sc}R_{int}$$
$$= 7.5(2) = 15 \text{ V, its highest value}$$

c. What are I_{load} and V_o for $R_{load} = 4 \, \Omega$?

$$I_{load} = \frac{I_{sc}R_{int}}{R_{int} + R_{load}} = \frac{7.5(2)}{2+4} = 2.5 \text{ A}$$
$$V_o = I_{load}R_{load} = 2.5(4) = 10 \text{ V}$$

d. What is the loading and percent voltage regulation for $R_{load} = 4 \, \Omega$?

V_o dropped from 15 V (R_{load} open) to 10 V ($R_{load} = 4 \, \Omega$); therefore, the loading was 5 V.

$$\% E_{regulation} = \frac{E_{no\text{-}load} - E_{full\text{-}load}}{E_{full\text{-}load}} \times 100$$
$$= \frac{15 - 10}{10} \times 100 = 50\%$$

e. What value of R_{load} will draw maximum power from this Norton model?

$$R_{load} = R_{internal} = 2 \, \Omega \text{ for maximum power transfer.}$$

5.6 HOW TO CONVERT FROM ONE SOURCE MODEL TO THE OTHER

Because the Norton and Thévenin models are equivalent, one can be converted to the other. Here is how:

- When going from Norton to Thévenin, the value of E_{oc} is needed. Get it by open-circuiting the Norton circuit (R_{load} = infinite ohms). $E_{oc} = I_{sc}R_{int}$.
- When going from Thévenin to Norton, the value of I_{sc} is needed. Get it by short-circuiting the Thévenin circuit ($R_{load} = 0 \, \Omega$). $I_{sc} = E_{oc}/R_{int}$.
- $R_{internal}$ is the same in either case. Put it in series with the ideal source, E_{oc}, in the Thévenin model, in shunt with the ideal source, I_{sc}, in the Norton model.
- Draw the new circuit.

EXAMPLE 5.17

Convert as indicated.

a. Convert this Thévenin source to its Norton equivalent.

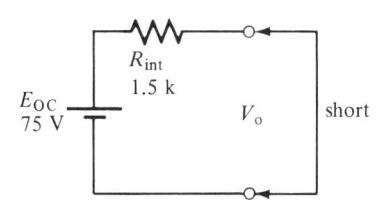

Short the terminals and calculate I_{sc}:

$$I_{sc} = \frac{E_{oc}}{R_{int}} = \frac{75}{1.5 \text{ K}} = 50 \text{ mA}$$

Draw the Norton equivalent:

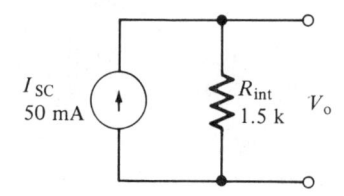

I_{SC} 50 mA, R_{int} 1.5 k, V_o

b. Convert this Norton source to its Thévenin equivalent:

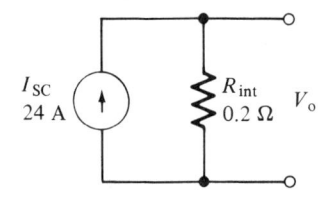

I_{SC} 24 A, R_{int} 0.2 Ω, V_o

Leave the terminals open and calculate E_{oc}:

$$E_{oc} = I_{sc} R_{int} = 24(0.2) = 4.8 \text{ V}$$

Draw the Thévenin equivalent:

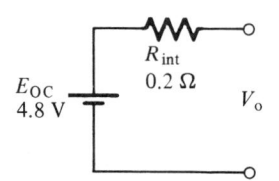

E_{OC} 4.8 V, R_{int} 0.2 Ω, V_o

EXAMPLE 5.18

Show that the two circuits of Example 5.17, part b, are equivalent.

If the circuits are equivalent, they must have the same V_o and I_{load} for a given R_{load}. Let $R_{load} = 1 \text{ Ω}$.
For the Norton circuit:

$$I_{load} = \frac{I_{sc} R_{int}}{R_{int} + R_{load}} = \frac{24(0.2)}{0.2 + 1} = \frac{4.8}{1.2} = 4 \text{ A}$$

$$V_o = I_{load} R_{load} = 4(1) = 4 \text{ V}$$

For the Thévenin circuit:

$$I_{load} = \frac{E_{oc}}{R_{load} + R_{int}} = \frac{4.8}{1 + 0.2} = \frac{4.8}{1.2} = 4 \text{ A}$$

$$V_o = \frac{E_{oc} R_{load}}{R_{load} + R_{int}} = \frac{48(1)}{1 + 0.2} = \frac{48}{1.2} = 4 \text{ V}$$

The circuits are equivalent.

5.7 CURRENT SOURCES IN SERIES AND PARALLEL: MILLMAN'S THEOREM

Millman's theorem lets us combine parallel-connected Norton sources.

EXAMPLE 5.19

Combine these Norton sources into a single source.

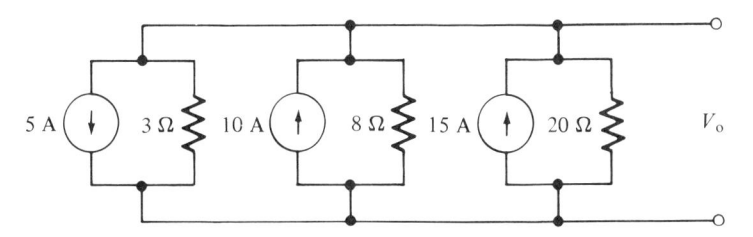

The circuit can be redrawn without changing any electrical characteristics:

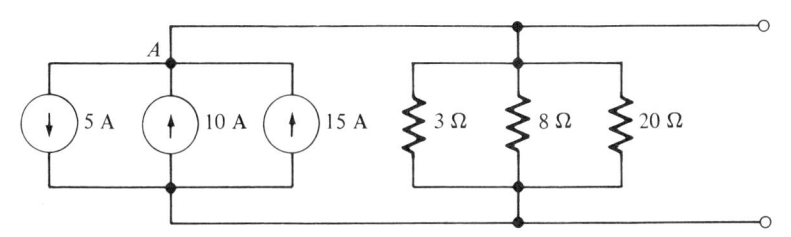

A nodal equation at point A indicates there is a 20-A current leaving the node. That is, the three sources combine into a single 20-A source. Similarly, the resistances combine to form a single $R_{internal}$:

$$R_{internal} = \frac{1}{\frac{1}{3} + \frac{1}{8} + \frac{1}{20}}$$

$$= \frac{1}{0.51} = 2\,\Omega$$

Draw the final reduced circuit:

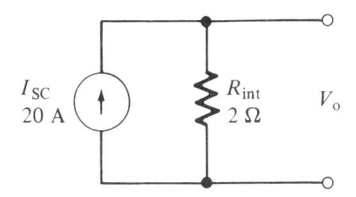

Of course, if the theorem works on the three generators of this example, it works on any number.

Thévenin circuits can be combined in parallel by first converting each to its Norton equivalent, then following the procedure just shown. A word of caution, however. Do not connect real sources in parallel unless they have the same voltage and internal resistance. If they do not, one source may be absorbing energy rather than delivering it to the load.

EXAMPLE 5.20

Large banks of batteries are used by the phone company as backups in the event of a power outage. By connecting these batteries in parallel, it is possible to get more current through a load.

a. Suppose that each battery has an open-circuit voltage of 12 V and an internal resistance of 0.03 Ω. Calculate the current through a 0.1-Ω load connected to one of them.

$$I_{load} = \frac{E_{oc}}{R_{int} + R_{load}} = \frac{12}{0.1 + 0.03}$$
$$= 92.3 \text{ A}$$

b. Shunt two of the batteries and calculate the new I_{load}.

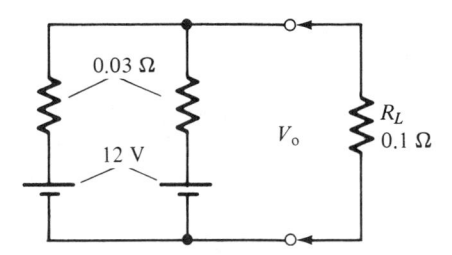

Internal resistances are not in parallel and cannot be combined. Nor can the generators. But convert each Théveinin source to its Norton equivalent of a 400-A source ($E_{oc}/R_{int} = 12/0.03 = 400$ A) in shunt with 0.03 Ω, and we have:

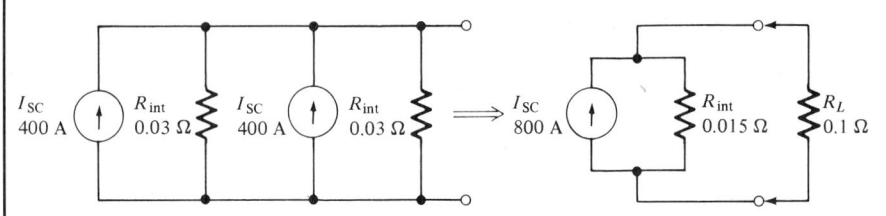

This combines into the circuit on the right. I_{load} now equals

$$\frac{I_{sc}R_{int}}{R_{int} + R_{load}} = \frac{800(0.015)}{0.015 + 0.1}$$
$$= 104.3 \text{ A}$$

c. Calculate the percent increase in I_{load} from paralleling two sources.

$$\% \text{ increase} = \frac{\text{amount of increase}}{\text{starting value}} \times 100$$
$$= \frac{104.3 - 92.3}{92.3} \times 100$$
$$= 13.1\%$$

To combine series-connected Norton sources, change each to its Thévenin equivalent first. Then follow the procedure for Thévenin generators.

5.8 CONTROLLED SOURCES

A *controlled* source supplies a voltage or current which is a multiple of another voltage or current elsewhere in a circuit. Such sources are also called *dependent* sources since their output depends on another current or voltage.

EXAMPLE 5.21

What is the current supplied by the current source for each value of E_{in}?

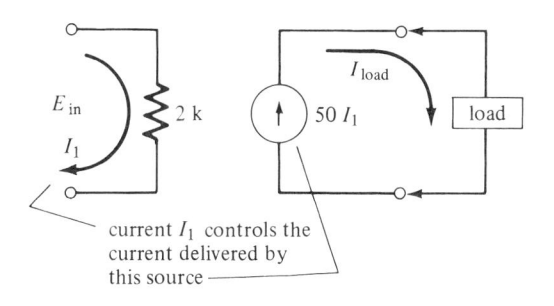

current I_1 controls the current delivered by this source

a. $E_{in} = 8$ V:

$$I_1 = \frac{E_{in}}{2\text{ k}} = \frac{8}{2\text{ k}} = 4 \text{ mA}$$

The controlled source supplies a current which is 50 times I_1. Therefore,

$$I_{load} = 50I_1 = 50(4 \text{ mA}) = 200 \text{ mA}$$

b. E_{in} increases to 20 V

$$I_1 = \frac{20}{2\text{ k}} = 10 \text{ mA}$$

$$I_{load} = 50I_1 = 50(10 \text{ mA}) = 500 \text{ mA}$$ ∎

Many electronic devices act as though they have controlled or dependent sources inside them. A small input current to a transistor, for example, seems to control an internal controlled source which multiplies the input to provide a larger output current. Multiplying a small input signal to get a larger output is called *amplifying* in electronics.

There are four kinds of controlled sources. The one in Example 5.21 was a current-controlled current source. There are also voltage-controlled current sources. The other two are current and voltage-controlled voltage sources.

5.9 TROUBLESHOOTING THE PARALLEL CIRCUIT

Just as they are for series circuits, shorts and opens are the main reasons parallel circuits stop working. But they act in different ways. For example, an open in a series circuit kills (turns off) the entire circuit; everything goes dead. This is not necessarily so in the parallel case. An open could kill the whole circuit or just a part, depending on its location. A short across a series resistance takes out only that resistance and raises the

EXAMPLE 5.22

Show how a short or open at each of the indicated points will affect circuit operation.

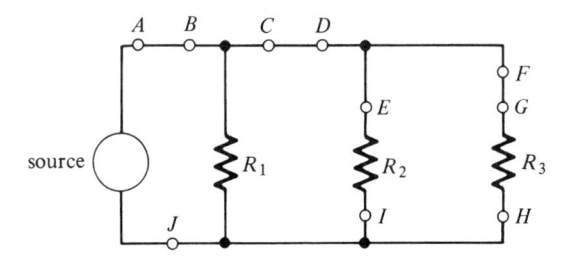

a. A short between *A* and *J*.

| A short here kills the entire circuit. Same as in series case.

b. A short between *E* and *I* or between *G* and *H*.

Although each is a short across a single resistance, the result is the same as a short across the source; therefore, the entire circuit goes dead. This is a major difference between series and parallel hookups.

c. An open between *A* and *B*.

| As in the series circuits, this kills the entire circuit because all current flows from *A* to *B*.

d. An open between *C* and *D*.

Current continues to flow in R_1 but not in R_2 or R_3, but an open at *E* stops current to R_2 only. In contrast, open a series circuit anywhere and it turns off the entire hookup. ∎

EXERCISES

Section 5.1

1. Two resistances are in parallel.
a. Can they have the same current through them?
b. Can they have the same value of current through them?
2. Answer the questions for this circuit:

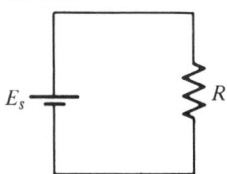

a. Are the two components in series? Why?   b. Are they parallel? Why?

EXERCISES

3. Consider these two circuits:

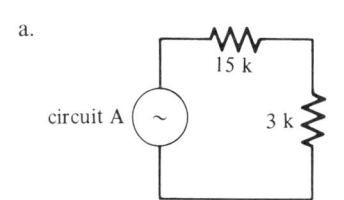

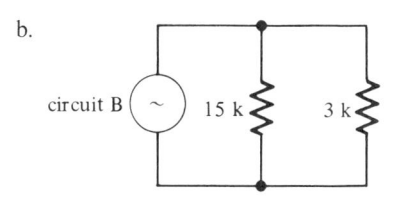

a.

circuit A — 15 k, 3 k

b.

circuit B — 15 k ∥ 3 k

a. Which resistance has the larger voltage drop in circuit 1?
b. Which resistance has the larger voltage drop in circuit 2?
c. Which resistance has the larger current in circuit 1?
d. Which resistance has the larger current in circuit 2?
e. Discuss the reason behind the answers in parts a through d.

4. What components in each of the following circuits are in parallel? Do not overlook the source.

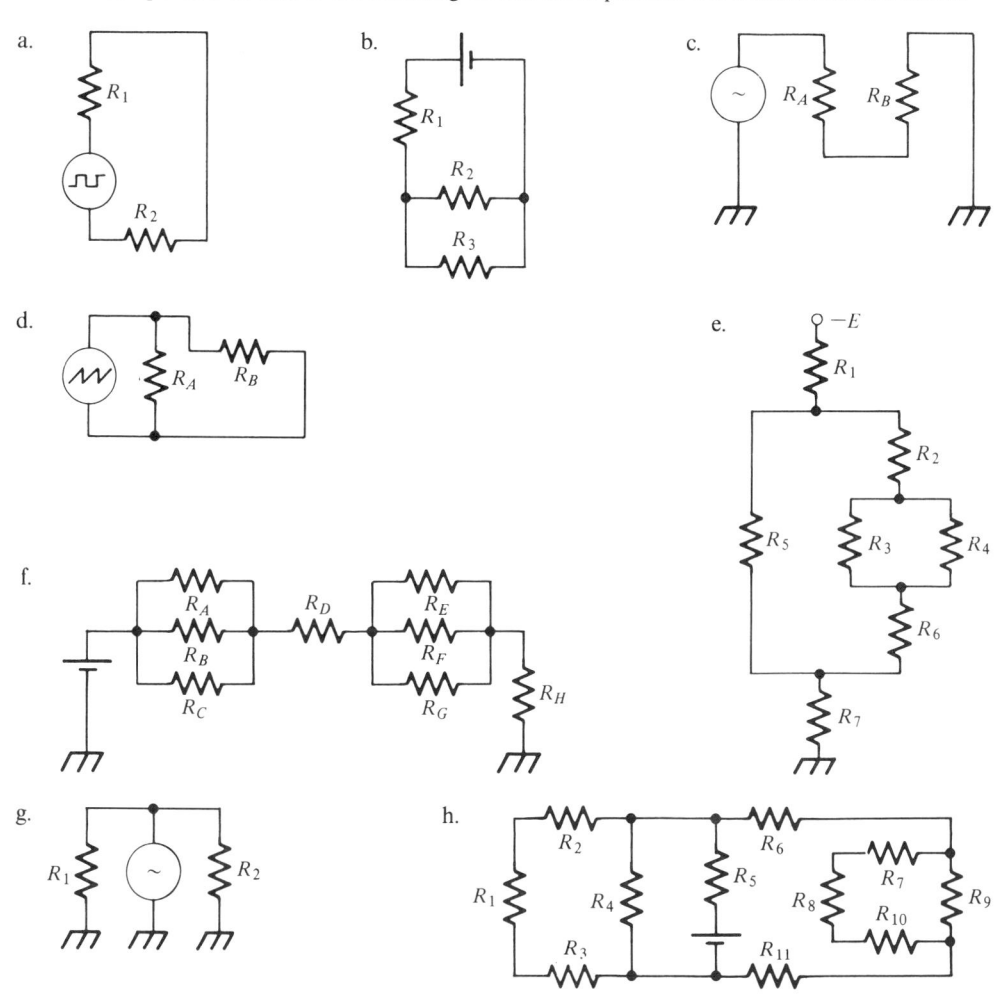

a. R_1, R_2 with source

b. R_1, R_2, R_3 with battery

c. R_A, R_B with source

d. R_A, R_B with source

e. $-E$, R_1, R_2, R_3, R_4, R_5, R_6, R_7

f. R_A, R_B, R_C, R_D, R_E, R_F, R_G, R_H

g. R_1, source, R_2

h. R_1, R_2, R_3, R_4, R_5, R_6, R_7, R_8, R_9, R_{10}, R_{11}

5. How many nodes are there in each of the circuits in Exercise 4? (*Remember:* All grounded points in any circuit are connected together.) How many branches are there in each circuit?

6. Voltmeters measure voltage. They are connected in shunt with the component whose voltage is being measured. The meter terminals are marked with polarity signs so that the + terminal may be

connected to the + end of a voltage and the − terminal to the − end. Connected this way, the meter gives a positive reading. Reverse the connections and the reading will be negative. What is the reading on each of these meters?

a.

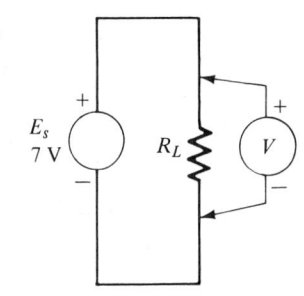

E_s 7 V, R_L, V

b.

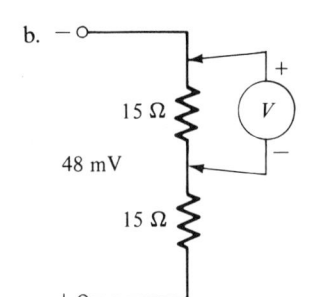

15 Ω, 48 mV, 15 Ω, V

7. Circuits b and e from Exercise 4 are redrawn here with various voltmeters included. Answer the questions for each.

b.

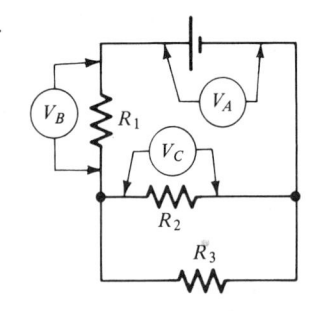

e.

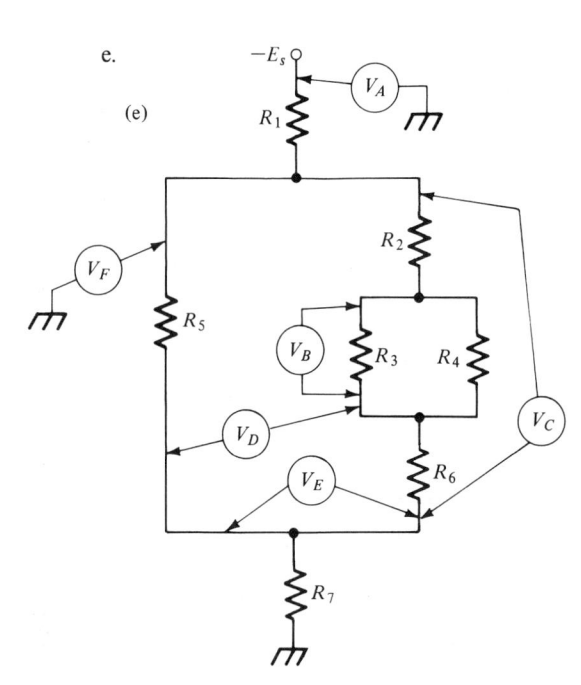

Circuit b:
a. What meter reads E_{source}?
b. What meter reads V_{R_1}?
c. What meter reads V_{R_3}?
d. What does the sum of V_c and V_b equal?

Circuit e:
a. What meter reads E_{source}?
b. What meter reads V_{R_4}?
c. What meter reads V_{R_5}?
d. What meter reads V_{R_6}?
e. What is the reading on V_e?
f. What does $V_c - (V_b + V_d)$ equal?
g. What does $V_a - V_f$ equal?
h. What readings would give V_{R_7}?

8. Keeping in mind the basic definitions of series and parallel connections, tell how these resistances are connected in circuits of Exercise 4.
a. In circuit b, R_1 and R_2.
b. In circuit e, R_1 and R_2. Then R_5 and R_6.

EXERCISES

Section 5.2

9. Use nodal equations to solve for both the magnitude and direction of each unknown current.

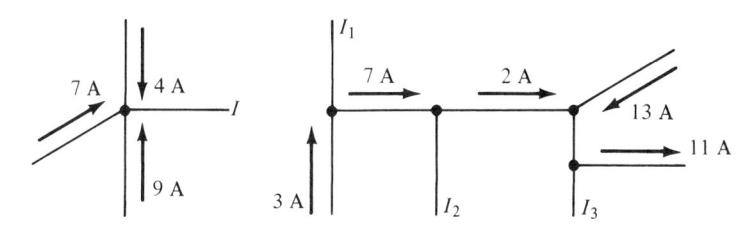

10. What is the drop across the 7 k in each circuit?

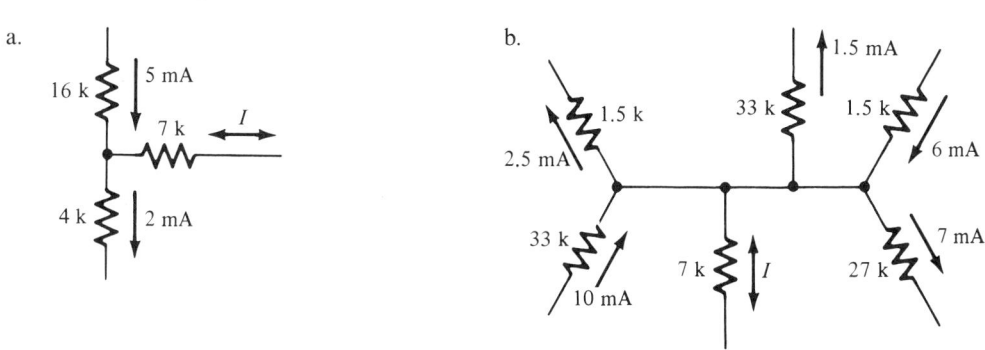

a.

b.

Section 5.3

11. Fill in the blanks with I (increase) or D (decrease):
a. When resistance is added in series total resistance _____ and source current _____ .
b. When resistance is added in parallel total resistance _____ and source current _____ .

12. a. When can resistances be added directly?
b. When can conductances be added directly?

13. Calculate R_{total} for each circuit. Do not use approximations. Be sure to test answers.

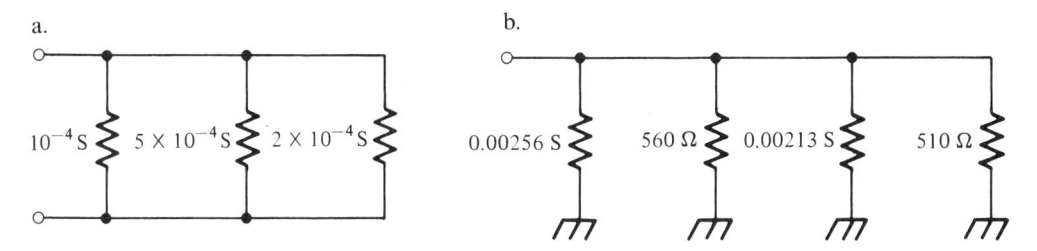

a.

b.

14. Find R_{total} for each circuit. Watch for shortcuts. Test answers.

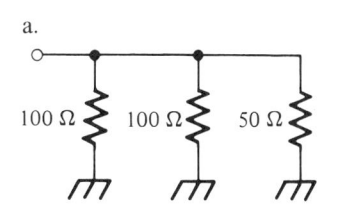

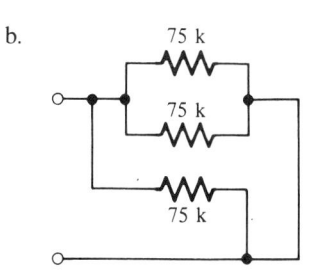

a.

b.

128 THE PARALLEL CIRCUIT

c.

d.

e.

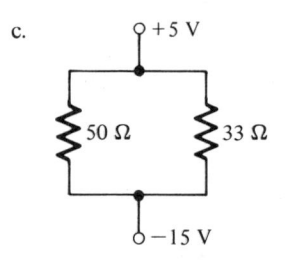

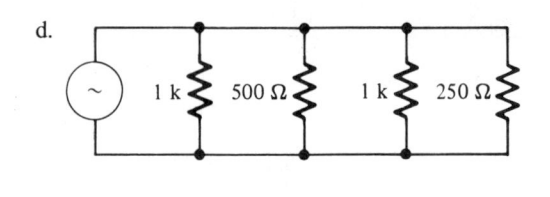

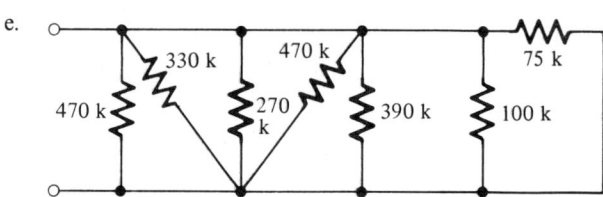

15. Answer the questions for this circuit:

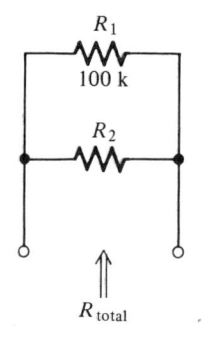

a. What value of R_2 could be put in to make $R_{total} \approx R_2$?
b. What value of R_2 could be put in to cause $R_{total} \approx 100$ k?
c. What value of R_2 could be put in to cause $R_{total} = 50$ k?
d. What value of R_2 would cause $R_{total} > 100$ k?

16. Approximate R_{total} for each circuit. Then find R_{total} by analysis. Compare results.

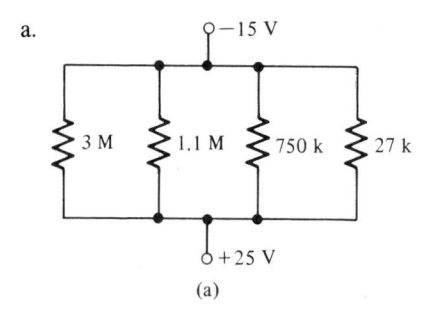

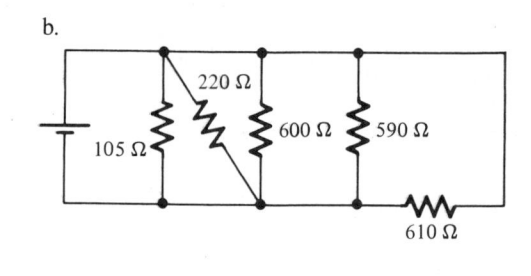

a.

b.

(a)

Advanced Problems

17. Two resistances in parallel have an $R_{total} = 148.5$ k. Find their values if the resistance of $R_1 = 82\%$ of R_2.

18. Two parallel resistances have an R_{total} of 400 Ω. The same two, connected in series, have an R_{total} of 1.8 k. What are the values of the two resistances?

EXERCISES

Section 5.4

19. The ammeters in this circuit are ideal. What will each one read?

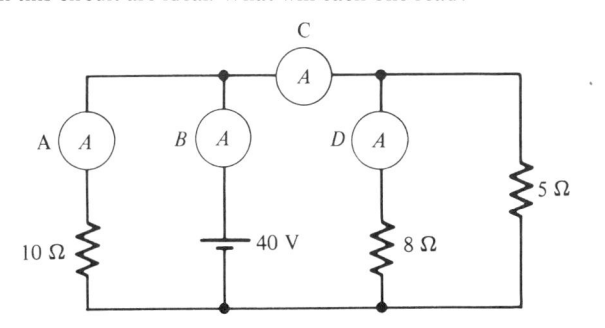

20. Find the current through and power dissipated by each resistance. Use calculator memory where possible. Do not forget to check results.

a.

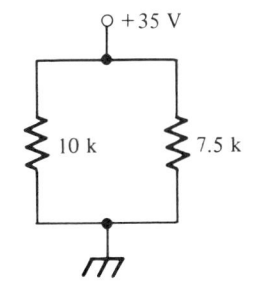

b.

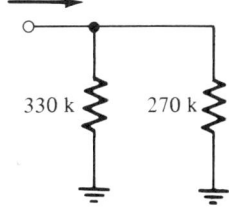

21. Repeat Exercise 20 for these circuits.

a.

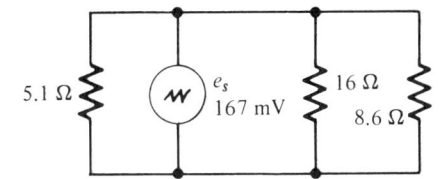

b.

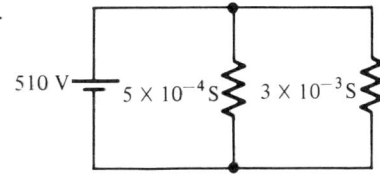

22. The resistances in Exercise 20, circuit a, are 20% tolerance units. Calculate the minimum and maximum current in each branch.

23. Which resistance is most likely to fail in this circuit?

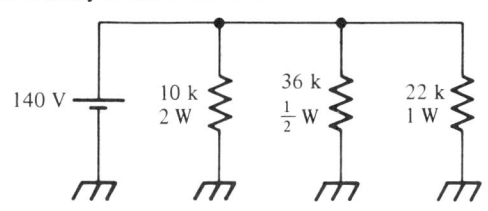

24. What does each expression tell about this circuit?

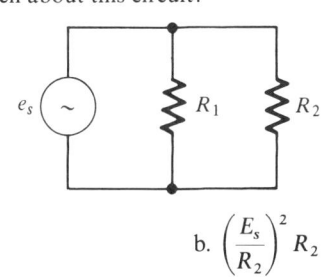

a. $\dfrac{E_s(R_1 + R_2)}{R_1 R_2}$

b. $\left(\dfrac{E_s}{R_2}\right)^2 R_2$

c. $\left[\dfrac{E_s R_2}{\dfrac{R_1 R_2}{R_1 + R_2}(R_1 + R_2)} \right]^2 R_1$

d. $\dfrac{E_s^2 (R_1 + R_2)}{R_1 R_2}$

25. Draw the current waveform for each resistance.

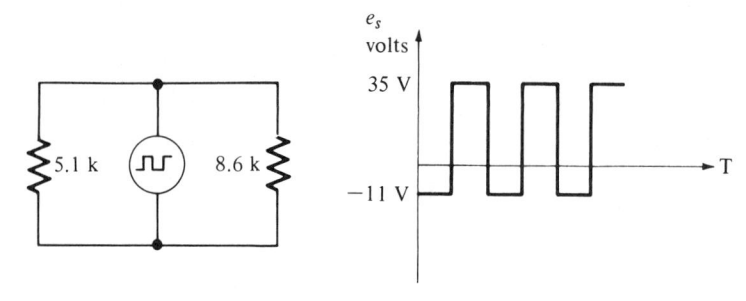

Advanced Problems

26. Show how to connect a group of 5-Ω 1-W resistors to make a 25-Ω 5-W resistance.

27. What is the resistance of R_1 and R_2 in this circuit?

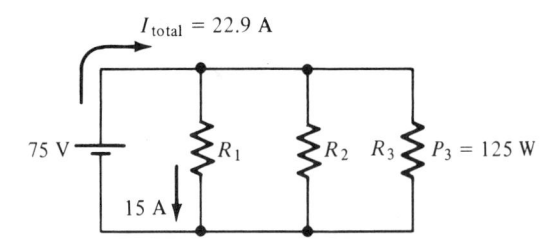

28. Find each of the four unknown currents in this circuit:

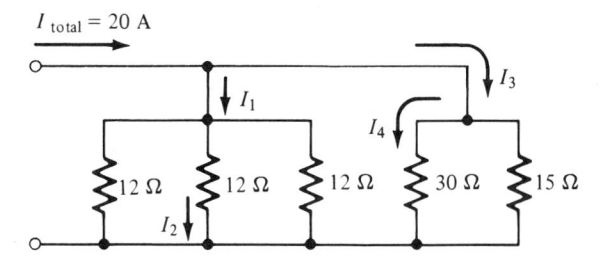

29. Find I_{total} for this circuit:

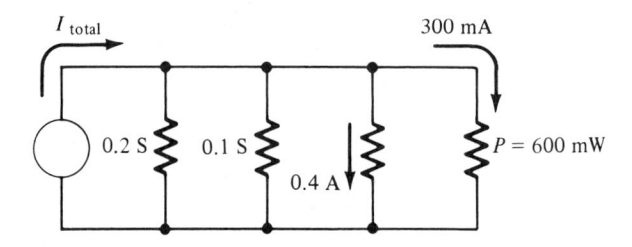

30. Imagine a 1 megohm resistance in parallel with a 1-Ω resistance. People say "current takes the path of least resistance." Does that mean that all the current flows through the 1 Ω? Will there be any current through the million-ohm resistance? Discuss.

EXERCISES

Section 5.5

31. a. Can the ideal E_{source} be operated under short-circuit conditions and the ideal I_{source} under open-circuit conditions? Comment.
b. Can the Thévenin model be operated under short-circuit conditions and the Norton model under open-circuit conditions? Comment.
c. Why the difference between parts a and b?

32. Why can't the Thévenin equivalent source be built like this:

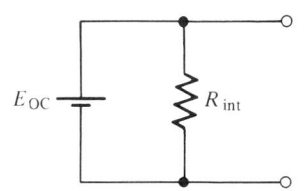

and the Norton like this?

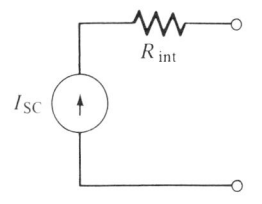

Hint: Consider Exercise 31, part b.

33. A Norton equivalent circuit has an ideal current source, I_{sc}, of 5 A in shunt with an internal resistance of 3300 Ω. Answer the following questions.
a. What are I_{load}, V_o, and P_{load} for $R_{load} = $ infinity?
b. Repeat part a for $R_{load} = 0$ Ω.
c. Repeat part a for $R_{load} = 5.6$ k.
d. For what value of R_{load} will maximum power be transferred from source to load?
e. For what range of R_{load} values will I_{load} be approximately equal to I_{sc}?
f. For what range of R_{load} values will V_o be approximately equal to $I_{sc}R_{int}$?

34. Calculate the loading and percent voltage regulation for the circuit of Exercise 33 when R_{load} goes from an open to 5.6 k.

Section 5.6

35. Convert each model to its opposite form.

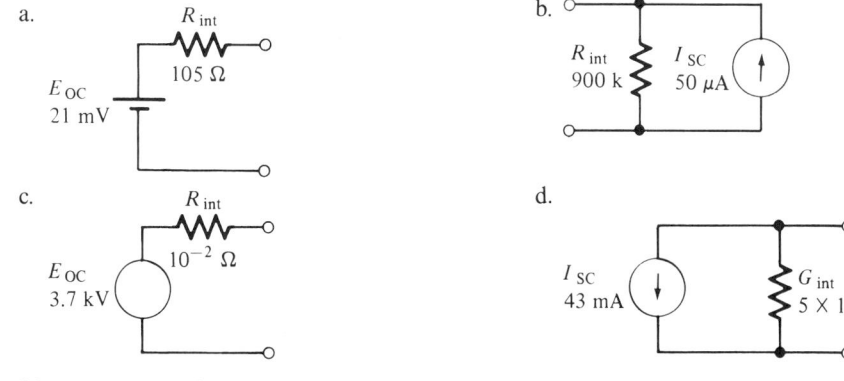

a. R_{int} 105 Ω, E_{OC} 21 mV

b. R_{int} 900 k, I_{SC} 50 μA

c. R_{int} 10^{-2} Ω, E_{OC} 3.7 kV

d. I_{SC} 43 mA, G_{int} 5×10^{-4} S

36. For an R_{load} of 70 Ω, prove that the Thévenin circuit of Exercise 35, part a, is equal to its Norton equivalent.

THE PARALLEL CIRCUIT

Extra Credit Assignment

37. a. Write the equation for I_{load} for a Thévenin circuit.
b. Let $R_{internal} \gg R_{load}$ and show that the equation changes to the approximation $I_{load} \approx E_{oc}/R_{int}$.
c. Holding $R_{internal} \gg R_{load}$ and keeping in mind the approximation of part b, what happens to I_{load} as R_{load} is varied?
d. Fill in the blanks: I_{load} is approximately constant for changing R_{load} when _____ is very much greater than _____.
e. Repeat parts a and b for the Norton circuit, showing that $I_{load} \approx I_{sc}$ under the same conditions.
f. Based on your answers in part d, what should the internal resistance of a practical source be in order that the source approximate a constant-current generator?

38. a. Write the equation for V_0 of a Thévenin circuit as a voltage divider between R_{load} and R_{int}.
b. Let $R_{internal} \ll R_{load}$ and show that the equation changes to the approximation $V_0 \approx E_{oc}$.
c. Holding $R_{internal} \ll R_{load}$ and keeping in mind the approximation of part b, what happens to V_0 as R_{load} is varied?
d. Fill in the blanks: V_0 is approximately constant for changing R_{load} when _____ is very much less than _____.
e. Repeat parts a and b for the Norton circuit, showing that $V_0 \approx I_{sc} R_{int}$ under the same conditions.
f. Based on your answers to part d, what should the internal resistance of a practical source be in order that the source approximate a constant-voltage generator?

Section 5.7

39. Combine the generators in each circuit into an equivalent source of the type indicated. Pay close attention to polarities.

a. Into a Thévenin equivalent.

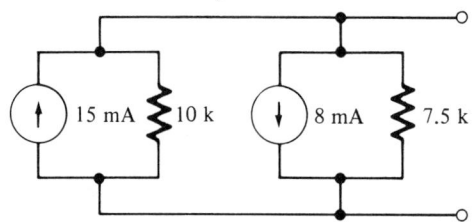

b. Into a Thévenin equivalent.

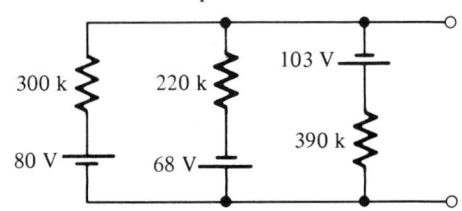

c. Into a Norton equivalent.

d. Into a Thévenin equivalent.

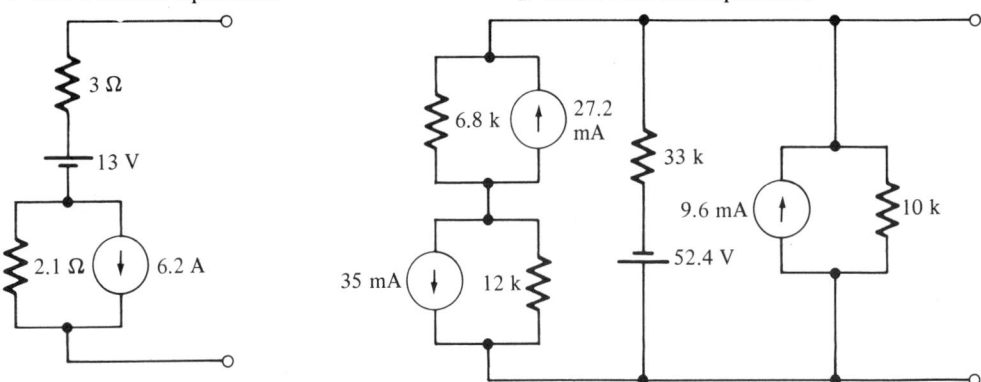

40. A battery having an open-circuit terminal voltage of 250 V and internal resistance of 1.5 k, is connected to a 5.1 k load.
a. Find I_{load}.
b. Find I_{load} when three of these batteries are paralled and connected to the same load.
c. What is the percent increase in I_{load} gained by paralleling the three?

EXERCISES

Section 5.8

41. Calculate E_{out} for each circuit as indicated.

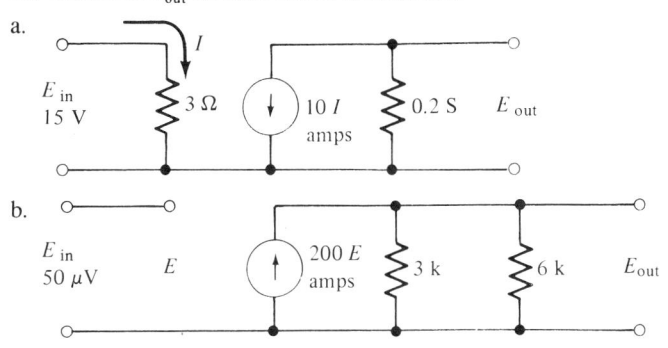

a. $E_{in} = 15$ V, $3\ \Omega$, $10I$ amps, 0.2 S, E_{out}

b. $E_{in} = 50\ \mu\text{V}$, E, $200E$ amps, 3 k, 6 k, E_{out}

42. Calculate I_{out} or E_{out} for each circuit as indicated.

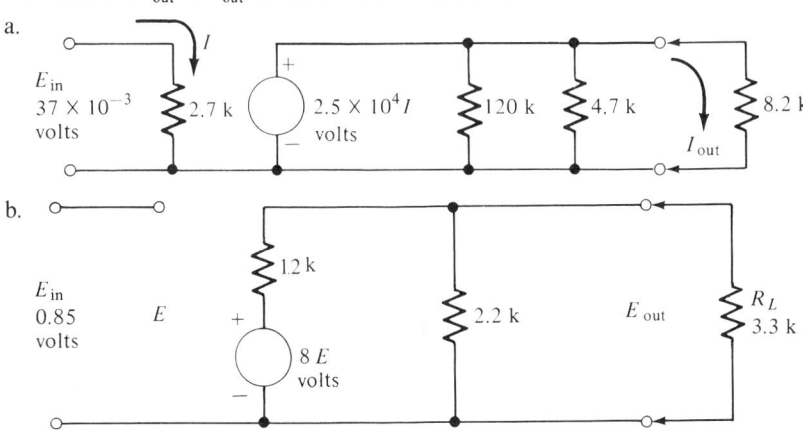

a. $E_{in} = 37 \times 10^{-3}$ volts, 2.7 k, $2.5 \times 10^4 I$ volts, 120 k, 4.7 k, I_{out}, 8.2 k

b. $E_{in} = 0.85$ volts, E, 1.2 k, $8E$ volts, 2.2 k, E_{out}, $R_L = 3.3\text{ k}$

Section 5.9

43. Using NC for no change, I for increase, and D for decrease, indicate what will happen to the reading on each ammeter if an open occurs at point X.

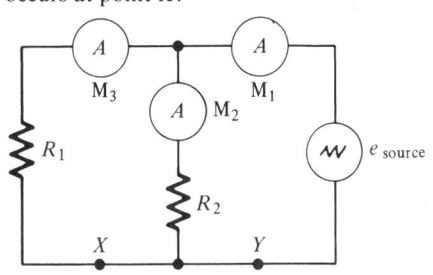

44. Close the open at X in the circuit of Exercise 43 and repeat the exercise for an open at Y.

45. Use NC, I, or D to indicate what will happen to each of the parameters when an open occurs at point X. a. e_{source}; b. i_{total}; c. i_{R_1}; d. v_{R_2}; e. i_{R_3}; f. v_{R_4}.

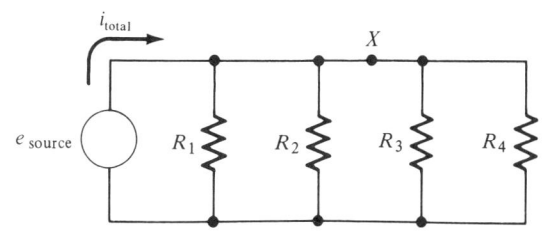

134   THE PARALLEL CIRCUIT

46. Use NC, I, or D to indicate what will happen to the reading on each ameter if a short occurs between points A and B.

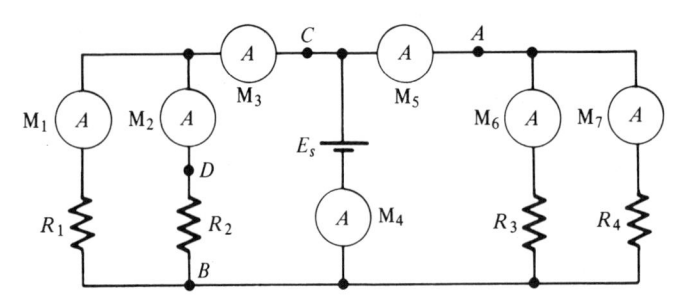

47. Repeat Exercise 46 for a short between C and D (the short between A and B having been removed).

48. Find the problem in this circuit:

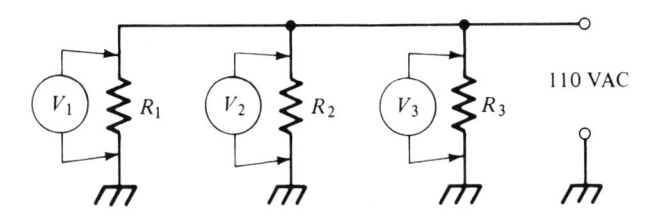

	Correct reading should be (V):	Trouble-shooting reading is (V):
V_1	110	110
V_2	110	0
V_3	110	110
V_4	110	110

49. Find the problem in this circuit:

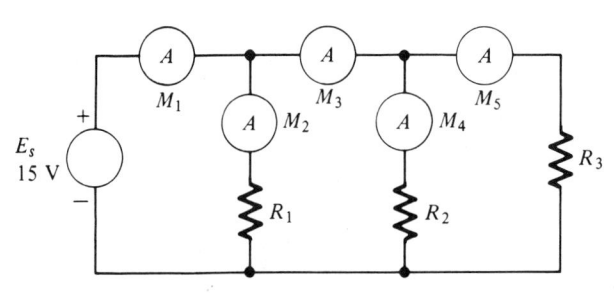

	Correct reading should be (A):	Troubleshooting reading is (A):
M_1	12	11
M_2	4	4
M_3	8	7
M_4	1	0
M_5	7	7

CHAPTER HIGHLIGHTS

Components in series have the same current through them.
Components in parallel have the same voltage across them.
Node: point at which three or more conductors are connected.
Branch: portion of a circuit between two nodes that contains one or more circuit components.
Adding resistance in parallel always causes a reduction in total resistance. This reduction causes total current to increase.
Conductances in parallel may be added directly, like resistances in series.
Total resistance of a group of shunt-connected resistances must be smaller than the lowest value.
There are many ways to determine total resistance of a group of parallel-connected resistances.
Asking "checking questions" is an excellent method for finding errors when doing problems.
To calculate E, I, R, or P for any component, use only the E, I, R, or P associated with that specific component.
The Norton source is to parallel circuits what the Thévenin is to series circuits. They are equal and each is easily changed to the other.
The voltage or current supplied by a dependent source is controlled by a voltage or current from some other place in a circuit.
Shorts and opens often act much different in series circuits than they do in parallel circuits. Find them quickly by combining knowledge of circuit operation with logical thinking.

Formulas to Remember

nodal equations

$$\sum I \text{ entering a node} = \sum I \text{ leaving the node} \quad \text{KCL}$$
$$\sum I = 0 \quad \text{KCL}$$

resistances in parallel:

$$R_{\text{total}} = \frac{1}{1/R_1 + 1/R_2 + \cdots + 1/R_n}$$

$$G_{\text{total}} = G_1 + G_2 + \cdots + G_n$$

$$R_{\text{total}} = \frac{1}{G_{\text{total}}}$$

$$R_{\text{total}} = \frac{R}{N} \qquad \text{for } N \text{ resistances all the same value}$$

$$R_{\text{total}} = \frac{R_1 R_2}{R_1 + R_2} \qquad \text{product-over-sum rule for two resistances in parallel}$$

$$R_{\text{total}} \approx R_1 \qquad \text{if } R_1 \text{ is in parallel with } R_2 \text{ and } R_2 \geq 10\, R_1$$

Quick check for all methods: R_{total} must be less than the lowest value of resistance in the parallel circuit.

$$I_{R_n} = \frac{I_{\text{total}} R_{\text{total}}}{R_n} \qquad \text{current division}$$

CHAPTER 6

Series–Parallel Circuits

In this chapter series and parallel connections of resistances are combined to make circuits that are more complicated than any studied so far. To analyze these *series-parallel* circuits, we need all the techniques covered earlier plus new ones to be studied next.

At the end of this chapter, you should be able to:

- Calculate the total resistance of a group of resistances connected in series–parallel.
- Analyze series–parallel circuits.
- Calculate the output voltage of a loaded voltage divider.
- Identify bridge circuits and tell when they are balanced.
- Indentify Π and T networks and convert from one to the other.
- Use the superposition technique to analyze circuits with more than one source.
- Analyze circuits by the ladder method.

6.1 RESISTANCES IN SERIES–PARALLEL

Perhaps the first step in a series–parallel analysis is to find the circuit's total resistance. Follow this procedure:

- Combine resistances that are obviously in series.
- Combine resistances that are obviously in parallel.
- Repeat these steps until all resistances have been reduced to one R_{total}.

EXAMPLE 6.1

Reduce each circuit to a single resistance.

a.

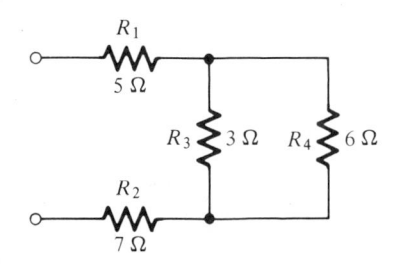

R_3 and R_4 are in parallel. Their combined value is

$$\frac{R_3 R_4}{R_3 + R_4} = \frac{3 \times 6}{3 + 6} = \frac{18}{9} = 2 \, \Omega$$

R_1 and R_2 are in series. Their combined value is

$$R_1 + R_2 = 5 + 7 = 12 \, \Omega$$

138

6.1 RESISTANCES IN SERIES–PARALLEL

Redraw the circuit:

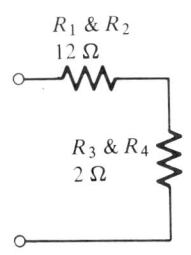

R_1 & R_2
12 Ω

R_3 & R_4
2 Ω

The 2-Ω equivalent of R_3 and R_4 is in series with the 12 Ω from R_1 and R_2. Combining them yields

$$R_{\text{total}} = 2 + 12 = 14 \text{ Ω}$$

b.

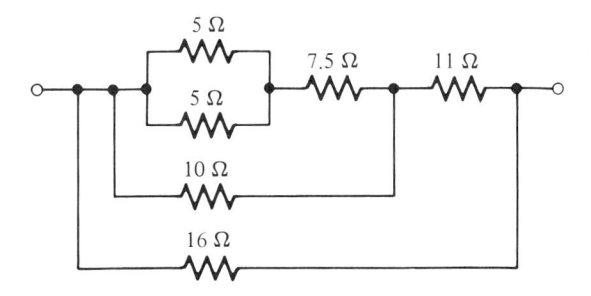

5 Ω
5 Ω
7.5 Ω
11 Ω
10 Ω
16 Ω

The two 5 Ω in shunt combine for 2.5 Ω. This is in series with the 7.5 Ω, which adds to 10 Ω. The circuit has been reduced to

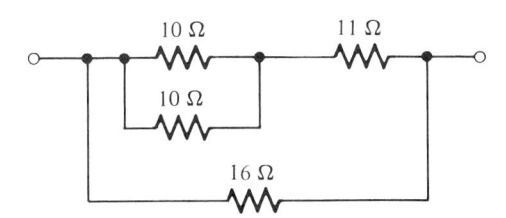

10 Ω
10 Ω
11 Ω
16 Ω

The two 10 Ω in shunt combine for 5 Ω. This is in series with the 11 Ω, which adds to 16 Ω. The circuit is now

16 Ω
16 Ω

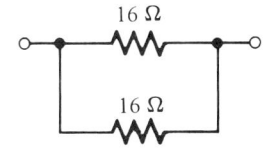

Finally, $R_{\text{total}} = 16/2 = 8 \text{ Ω}$.

Sometimes the connection between components changes when we look at the circuit from a different direction.

Example 6.2

This circuit is called a *two-port* or *four-terminal network*. Find its input and output resistances, R_{in} and R_{out}.

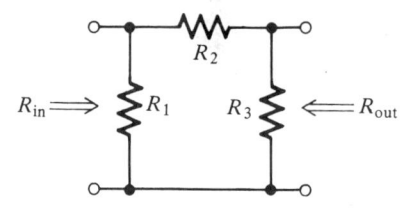

R_{in} sees R_2 and R_3 in series. Their combination shunts R_1. Therefore,

$$R_{in} = \frac{R_1(R_2 + R_3)}{R_1 + R_2 + R_3}$$

R_{out} sees R_1 and R_2 in series and their combination in shunt with R_3. This gives us

$$R_{out} = \frac{R_3(R_1 + R_2)}{R_1 + R_2 + R_3}$$ ∎

Note carefully how R_1 of Example 6.2 was in series with R_2 for R_{out} and in parallel with the series combination of R_2 and R_3 for R_{in}. The important point here is:

A component's connection in a circuit may change according to the way we analyze that circuit.

6.2 ANALYZING THE SERIES–PARALLEL CIRCUIT

The method of series–parallel analysis we will use most often requires that we first reduce the circuit to a single R_{total} and a single source. We will call this method, *series–parallel reduction* and abbreviate it *SPR*.

A circuit may be analyzed by SPR techniques if it meets two requirements:

- It has only one source, *or*, a number of sources that can be combined into one (for example, two batteries in series), *and*
- All resistances are in series or parallel combinations that can be reduced to a single R_{total}.

Follow these steps when doing an SPR analysis:

- Combine resistances into R_{total}.
- Combine sources into a single source.
- Find I_{total} by Ohm's law.
- Work away from the source finding voltage drops by Ohm's law and KVL. Find currents by Ohm's law and KCL.
- Calculate powers.

6.2 ANALYZING THE SERIES–PARALLEL CIRCUIT

EXAMPLE 6.3

Analyze this circuit:

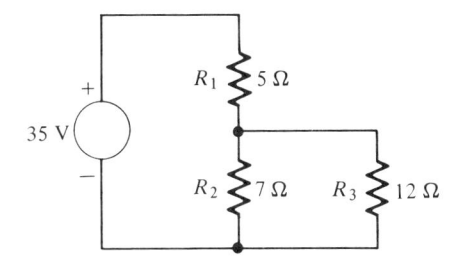

This circuit has only one source and its resistances can be combined; it can be analyzed by SPR.

Finding R_{total}:

$$R_{total} = R_1 + \frac{R_2 R_3}{R_2 + R_3}$$

$$= 5 + \frac{7(12)}{7 + 12}$$

$$= 5 + \frac{84}{19}$$

$$= 9.4 \; \Omega$$

Find I_{total}:

$$I_{total} = \frac{E_{total}}{R_{total}}$$

$$= \frac{35}{9.4}$$

$$= 3.7 \; A$$

Finding the voltage drop and current for individual resistances:

I_{total} flows through R_1; therefore,

$$V_{R_1} = I_{total} R_1 = 3.7(5) = 18.5 \; V$$

$$V_{R_2} \text{ by KVL} = E_{source} - V_{R_1} = 35 - 18.5 = 16.5 \; V$$

$$I_{R_2} = \frac{V_{R_2}}{R_2} = \frac{16.5}{7} = 2.4 \; A$$

$$I_{R_3} \text{ by KCL} = I_{total} - I_{R_2} = 3.7 - 2.4 = 1.3 \; A$$

Finding powers:

$$P_{R_1} = V_{R_1} I_{total} = 18.5(3.7) = 68.5 \; W$$

$$P_{R_2} = (I_{R_2})^2 R_2 = (2.4)^2 (7) = 40.3 \; W$$

$$P_{R_3} = \frac{(V_{R_3})^2}{R_3} = \frac{(16.5)^2}{12} = 22.7 \; W$$

Sum output powers:
$$P_{total} = 68.5 + 40.3 + 22.7$$
$$= 131.5 \text{ W}$$

Power delivered by source:
$$E_{source}I_{total} = 35(3.7) = 129.5 \text{ W}$$

Error between two power calculations is less than 2%. ∎

Watch for the circuit of Example 6.3 in following chapters. We will be analyzing it again by other techniques to illustrate the differences between techniques and show that there is no one "right way" to analyze a particular circuit.

EXAMPLE 6.4

Find V_{R_1}, I_{R_4}, P_{R_6} for this circuit:

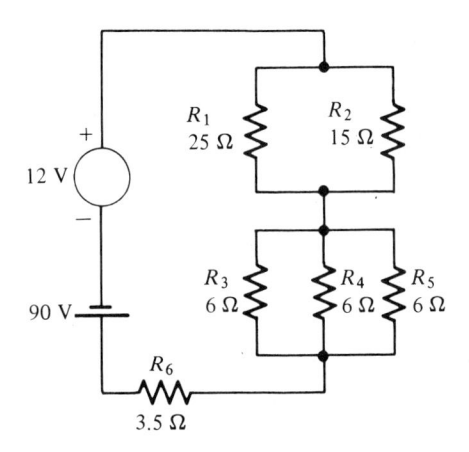

The sources can be combined into a single 78-V source forcing current CCW. The resistances can be combined into R_{total}. Therefore, SPR will work.

$$R_{total} = \frac{25 \times 15}{25 + 15} + \frac{6}{3} + 3.5$$
$$= 9.4 + 2 + 3.5 = 14.9 \text{ }\Omega$$

$$I_{total} = \frac{E_{source}}{R_{total}} = \frac{78}{14.9} = 5.2 \text{ A}$$

$$V_{R_1} = I_{total}\frac{R_1 R_2}{R_1 + R_2}$$
$$= 5.2\left[\frac{25(15)}{25 + 15}\right]$$
$$= 5.2(9.4)$$
$$= 48.9 \text{ V}$$

6.2 ANALYZING THE SERIES–PARALLEL CIRCUIT

$$I_{R_4} = \frac{I_{\text{total}}}{3}$$
$$= \frac{5.2}{3}$$
$$= 1.73 \text{ A}$$
$$P_{R_6} = \frac{(V_{R_6})^2}{R_6}$$

By KVL we know that $V_{R_6} = E_{\text{source}} - (V_{R_1} + V_{R_4})$.

$$V_{R_4} = I_{R_4} R_4 = 1.73(6)$$
$$= 10.4 \text{ V}$$

Substituting for V_{R_1} and V_{R_4}:

$$V_{R_6} = 78 - (48.9 + 10.4) = 78 - 59.3$$
$$= 18.7 \text{ V}$$

Therefore,

$$P_{R_6} = \frac{18.7^2}{3.5} = 99.9 \text{ W}$$

EXAMPLE 6.5

Answer the questions for this circuit:

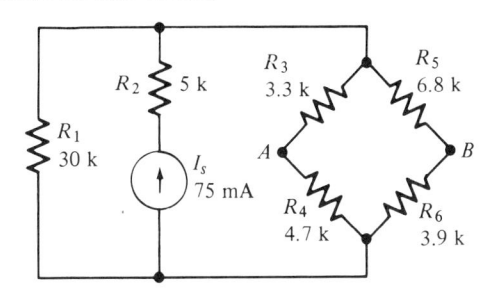

The circuit can be analyzed using SPR techniques. Since I_{total} is given, we skip the first two steps of the technique.

a. Find the voltage across each resistance.

$$V_{R_2} = I_{\text{total}} R_2 = 75 \text{ mA}(5 \text{ k})$$
$$= 375 \text{ V}$$

We need I_{R_1} to find V_{R_1}. To get it, current division will be used on I_{total} to see how I_{total} splits between R_1 and the total resistance of the diamond-shaped group on the right.

Finding R_{total} of the diamond-shaped group (call it $R_{3,4,5,6}$):

$$R_{3,4,5,6} = (R_3 + R_4) \text{ in parallel with } (R_5 + R_6)$$
$$R_3 + R_4 = 3.3 \text{ k} + 4.7 \text{ k} = 8 \text{ k}$$
$$R_5 + R_6 = 6.8 \text{ k} + 3.9 \text{ k} = 10.7 \text{ k}$$
$$R_{3,4,5,6} = 8 \text{ k in parallel with } 10.7 \text{ k}$$
$$= \frac{8 \text{ k}(10.7 \text{ k})}{8 \text{ k} + 10.7 \text{ k}} = 4.6 \text{ k}.$$

The circuit is now:

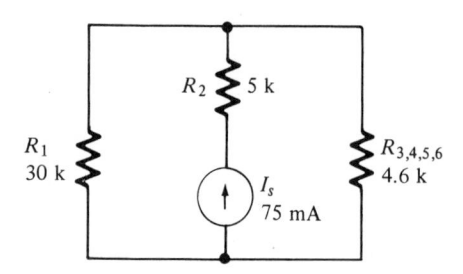

By current division,

$$I_{R_1} = \frac{I_{\text{total}} R_{3,4,5,6}}{R_{3,4,5,6} + R_1}$$

$$= \frac{75 \text{ mA}(4.6 \text{ k})}{4.6 \text{ k} + 30 \text{ k}}$$

$$= \frac{345}{34.6 \text{ k}} = 10 \text{ mA}$$

$$V_{R_1} = I_{R_1} R_1 = 10 \text{ mA}(30 \text{ k})$$
$$= 300 \text{ V}$$

Looking at the original circuit we see that R_1 is in parallel with $(R_3 + R_4)$. Therefore, knowing V_{R_1} means that we also know the voltage across $(R_3 + R_4)$. Voltage division tells how much of that voltage is dropped across R_3:

$$V_{R_3} = \frac{300 R_3}{R_3 + R_4}$$

$$= \frac{300(3.3 \text{ k})}{3.3 \text{ k} + 4.7 \text{ k}} = 123.8 \text{ V}$$

By KVL,

$$V_{R_4} = V_{R_{3,4}} - V_{R_3}$$
$$= 300 - 123.8$$
$$= 176.2 \text{ V}$$

The voltage across $(R_5 + R_6)$ is known because this combination is also in parallel with R_1. Using voltage division again we find that V_{R_5} and V_{R_6} equal 190.7 V and 109.3 V respectively.

b. What is the source voltage?

To find E_s write an equation around any loop that includes the source. There are three. We will choose the one around the source, R_2 and R_1.

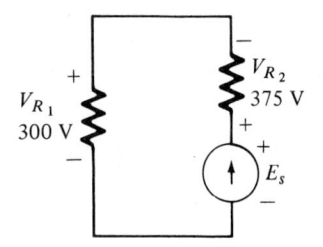

6.2 ANALYZING THE SERIES-PARALLEL CIRCUIT

Start at the lower end of the source and go CCW:

$$E_s + 375 + 300 = 0$$
$$E_s = -375 - 300$$
$$E_s = 675 \text{ V}$$

c. Prove KVL by writing a loop equation around R_1, R_5, and R_6.

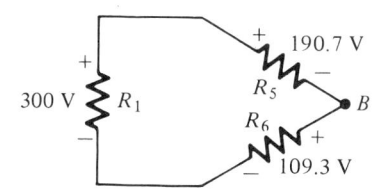

Start at B and go CW:

$$109.3 - 300 + 190.7 \stackrel{?}{=} 0$$
$$-300 + 300 = 0$$

Practice this important point by testing other loops.

d. Find V_{BA}.

Pick a polarity for V_{BA}. Then write an equation around any loop containing V_{BA}, for example, the loop V_{BA}, R_3, and R_5.

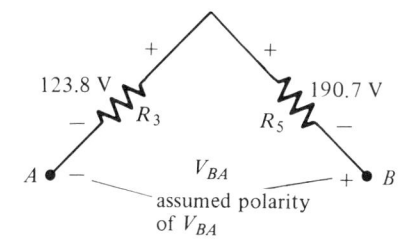

Start at point A and go CW:

$$-123.8 + 190.7 + V_{BA} = 0$$
$$66.9 + V_{BA} = 0$$
$$V_{BA} = -66.9 \text{ V}$$

The negative sign says that the assumed polarity is backward.

e. Find P_{R_3}.

The power equation, $P = E \times I$, will be used.
Question: What E and what I must be used?
Answer: Only the voltage across R_3 times the current through R_3.

$$V_{R_3} = 123.8 \text{ V (from work done before)}$$

I_{R_3} is part of $I_{R_{3,4,5,6}}$, the total current through the diamond-shaped group.

$$I_{R_{3,4,5,6}} = I_{\text{total}} - I_{R_1}$$
$$= 75 \text{ mA} - 10 \text{ mA}$$
$$= 65 \text{ mA}$$

This current splits. We want the part through R_3.

By current division:

$$I_{R_{3,4}} = \frac{I_{R_{3,4,5,6}} R_{5,6}}{R_{3,4} + R_{5,6}}$$

$$= \frac{65 \text{ mA}(10.7 \text{ k})}{10.7 \text{ k} + 8 \text{ k}}$$

$$= \frac{695.5}{18.7 \text{ k}}$$

$$= 37.2 \text{ mA}$$

Finally,

$$P_{R_3} = V_{R_3} I_{R_3}$$
$$= 123.8(37.2 \text{ mA})$$
$$= 4.6 \text{ W}$$ ∎

EXAMPLE 6.6

This problem will not be done using reduction techniques. It is different in that some information is given. Good thinking, Ohm's law, KCL, and KVL will find the rest. Ammeters have 0 Ω resistance.

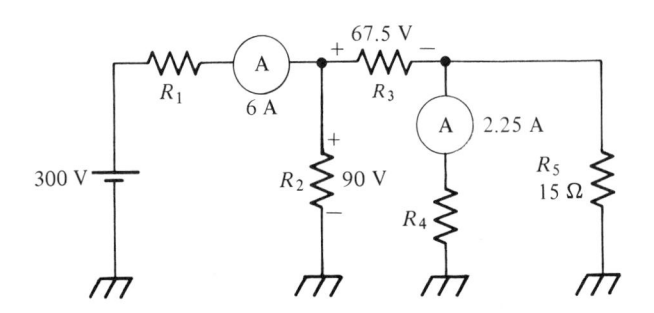

a. Find V_{R_1}.

Go around the loop R_1, R_2, and source:

$$\sum E = 0$$
$$-300 + V_{R_1} + 90 = 0$$
$$V_{R_1} = 300 - 90$$
$$= 210 \text{ V}$$

b. Find R_1.

By Ohm's law,

$$R_1 = \frac{V_{R_1}}{I_{R_1}}$$
$$= \frac{210}{6}$$
$$= 35 \text{ Ω}$$

6.2 ANALYZING THE SERIES–PARALLEL CIRCUIT 147

c. Find I_{R_2}.

By KCL, $I_{R_2} = I_{R_1} - I_{R_3}$. But I_{R_3} is unknown at this point. Therefore, leave I_{R_2} for now.

d. Find R_3.

By Ohm's law, $R_3 = V_{R_3}/I_{R_3}$. But I_{R_3} is still unknown. So leave R_3 for now.

e. Find V_{R_4}.

By KVL around the loop R_2, R_3, and R_4:

$$-90 + 67.5 + V_{R_4} = 0$$
$$V_{R_4} = 90 - 67.5$$
$$= 22.5 \text{ V}$$

f. Find I_{R_5}.

$V_{R_5} = 22.5$ V because R_5 is in parallel with R_4. Therefore,

$$I_{R_5} = \frac{V_{R_5}}{R_5} = \frac{22.5}{15}$$
$$= 1.5 \text{ A}$$

g. Find R_4.

$$R_4 = \frac{V_{R_4}}{I_{R_4}}$$

I_{R_4} given at 2.25 A, Therefore,

$$R_4 = \frac{22.5}{2.25}$$
$$= 10 \text{ }\Omega$$

Now we have the data needed to do parts c and d.

d. R_3.

By KCL,

$$I_{R_3} = I_{R_4} + I_{R_5}$$
$$= 2.25 + 1.5 = 3.75 \text{ A}$$

Therefore,

$$R_3 = \frac{V_{R_3}}{I_{R_3}}$$
$$= \frac{67.5}{3.75}$$
$$= 18 \text{ }\Omega$$

c. I_{R_2}.

By KCL,

$$I_{R_2} = I_{R_1} - I_{R_3}$$
$$= 6 - 3.75$$
$$= 2.25 \text{ A}$$

■

Example 6.7

Write an equation for the voltage, V_{R_d}, in this circuit:

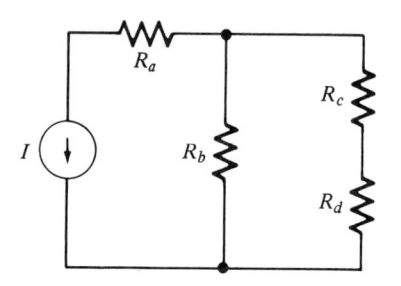

$V_{R_d} = I_{R_d} R_d$, but I_{R_d}, by current division, is

$$\frac{IR_b}{R_b + (R_c + R_d)}$$

Therefore,

$$V_{R_d} = \frac{IR_b}{R_b + (R_c + R_d)} R_d$$

$$= \frac{IR_b R_d}{R_b + R_c + R_d}$$

Doing a units check on this last expression:

$$\text{volts} = \frac{\text{amps} \times \text{ohms} \times \cancel{\text{ohms}}}{\cancel{\text{ohms}}}$$

$$= \text{volts} \qquad \blacksquare$$

Example 6.8

What does the following statement tell us about the circuit in Example 6.7?

$$\frac{\left[\dfrac{IR_b(R_c + R_d)}{R_b + (R_c + R_d)}\right]^2}{R_b}$$

Inside the numerator brackets.

$$\frac{R_b(R_c + R_d)}{R_b + R_c + R_d}$$

is the product-over-sum calculation of R_{total} for the two parallel branches on the right. Note that R_{total} is multiplied by I. This gives us the voltage across the branches. One of these branches is R_b. Therefore, the entire part inside the brackets is the voltage across R_b.

6.3 THE LOADED VOLTAGE DIVIDER

Squaring the voltage across R_b and dividing by R_b, we see that the statement tells the power dissipated by R_b. Checking by units analysis to be sure that the statement really does result in watts:

$$\left[\frac{\text{amps} \times \text{ohms} \times \cancel{\text{ohms}}}{\cancel{\text{ohms}}}\right]^2 \bigg/ \text{ohms} = \frac{\text{volts}^2}{\text{ohms}} = \text{watts} \qquad \blacksquare$$

6.3 THE LOADED VOLTAGE DIVIDER

Electronic circuits do not work alone. They are connected to form circuits, which are often called *systems*.

Figure 6.1 shows three circuits wired together. Note how the output of one connects to the input of the next. This is called the *cascade* connection. We say that circuit 1 *drives* circuit 2; circuit 1 is the *driver*, circuit 2 is *driven*. In a similar manner, circuit 3 is driven by circuit 2.

An important point to remember when circuits are cascaded is:

The input resistance of a driven circuit loads the output of the driver.

In Figure 6.1, for example, the input resistance of circuit 2 loads the output of circuit 1. Similarly, circuit 3 loads circuit 2. We know from earlier work that *loading causes the output voltage of any circuit to fall*. This is a constant problem in electronics, one of which the circuit designer is always aware.

Figure 6.2 shows a voltage divider in both unloaded and loaded states. The output of the unloaded divider is easily calculated with the voltage-division formula. But this will not work for the loaded circuit. What was a series circuit is now series–parallel and SPR techniques are needed to find V_{out}.

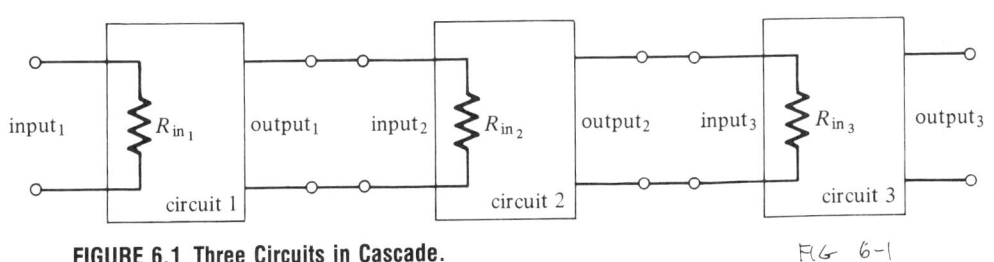

FIGURE 6.1 Three Circuits in Cascade.

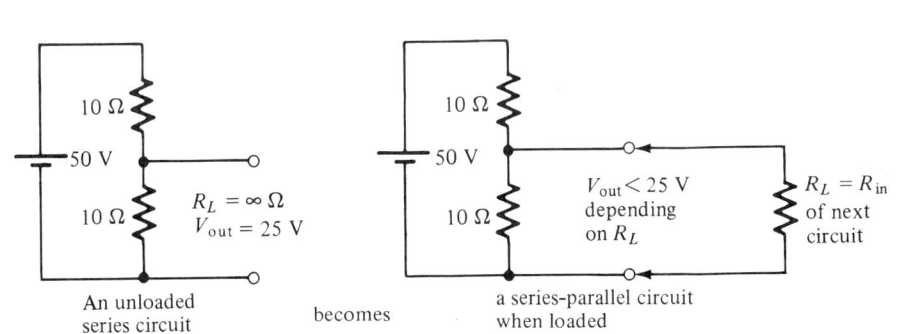

FIGURE 6.2 Voltage Dividers: Unloaded and Loaded.

EXAMPLE 6.9

Answer the questions for this voltage divider:

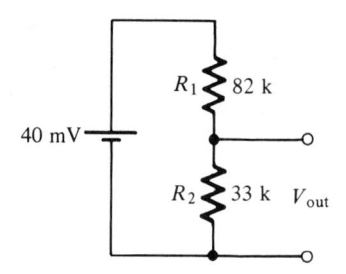

a. What is the output voltage of this divider?

$$V_{out} \text{ by voltage division} = \frac{E_s R_2}{R_1 + R_2}$$

$$= \frac{40 \text{ mV}(33 \text{ k})}{82 \text{ k} + 33 \text{ k}}$$

$$= 11.5 \text{ mV}$$

b. Another circuit with an input resistance of 150 k is connected to the terminals of this divider. What is the new V_{out}? By what percentage has V_{out} dropped?

As far as the source is concerned, the circuit now looks like this:

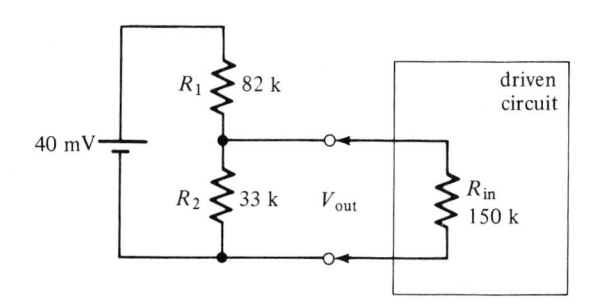

Note how the resistance that develops V_{out} has dropped from 33 k to 27 k (33 k in shunt with 150 k). It follows that the output must also fall.

$$V_{out} = \frac{40 \text{ mV}(27 \text{ k})}{82 \text{ k} + 27 \text{ k}}$$

$$= 9.9 \text{ mV}$$

$$\% \text{ loading} = \frac{\text{Change}}{\text{original value}} \times 100$$

$$= \frac{11.5 - 9.9}{11.5} \times 100$$

$$= 13.9\%$$

6.3 THE LOADED VOLTAGE DIVIDER

c. What value of R_{in} would cause even worse loading?

| Any value < 150 k

d. How could loading be reduced?

| By making R_{in} of the driven circuit $\gg 150$ k.

EXAMPLE 6.10

Find the output voltages of this loaded dual divider:

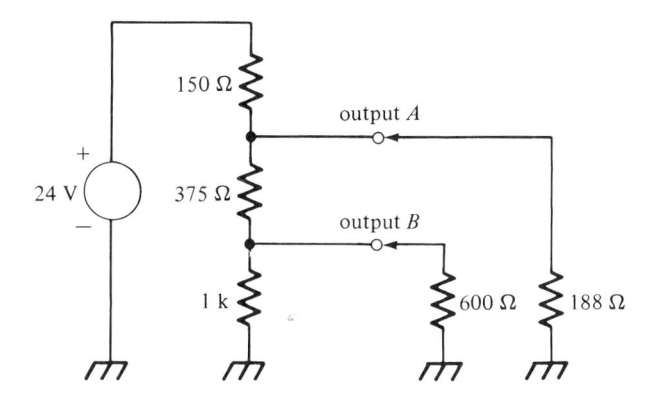

Combine resistances:
The 1 K and 600 Ω in parallel reduce to

$$\frac{1\,K(600)}{1\,K + 600} = 375\,\Omega$$

This is in series with the 375 Ω:

$$375 + 375 = 750\,\Omega$$

This is in parallel with the 188 Ω. The combination reduces to

$$\frac{750(188)}{750 + 188} = 150\,\Omega$$

The circuit now looks like this:

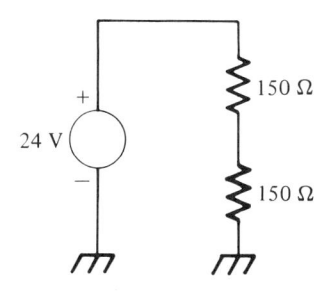

Because they are equal, each resistance has $24/2 = 12$ V across it. Output A is taken at the junction of these two, therefore:

$$V_a = 12 \text{ V}$$

The other 12 V is dropped across this part of the circuit:

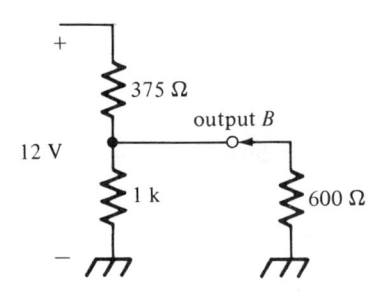

[Circuit diagram: 12 V source with 375 Ω resistor on top, 1 k resistor on bottom left, 600 Ω resistor on bottom right, output B taken between them]

We know that the 1 k and 600 Ω reduce to 375 Ω from previous calculations. Therefore, the 12 V is being divided evenly by two 375-Ω resistances. Output B is taken at the junction of these two; therefore,

$$V_b = 6 \text{ V}$$

6.4 THE BRIDGE CIRCUIT

One of the most useful circuits in electronics is the *bridge*. They are used regularly in many applications, especially in test equipment, where they are the basic circuit in meters that measure many different parameters.

Figure 6.3 shows the basic bridge in three different conditions. In all of them, point A is fixed at 25 V above ground by the drop across R_2. Point B, on the other hand, has different voltages because R_4 is variable.

In Figure 6.3, part (a), point B is 20 V above ground. This puts it 5 V below point A. If there was a connection between A and B, current would flow "downhill" from A, the point of higher potential, to B.

In part (b), R_4 has been increased. Its voltage drop is now 30 V. This puts B 5 V above A. Current would now flow from B to A if there was a conductor between the points.

In part (c), R_4 is set so that its drop is 25 V. This means that:

Point B is at the same potential as A. Since there is no difference in potential between A and B, no current would flow between them if they were connected. When this happens, the bridge is said to be *balanced*.

The ability to be balanced is what makes the bridge so useful. So that it will be easy to see in complicated diagrams, its four legs are drawn in the shape of a diamond.

The bridge is not a series–parallel circuit. Therefore, we cannot analyze it in its unbalanced condition using the series–parallel reduction techniques of this chapter. However, we are only interested in the balanced bridge. This is a series–parallel circuit because its center leg is effectively gone; SPR can be used.

Using the fact that V_{R_4} must equal V_{R_2} for balance, a balance equation can be derived:

$$V_{R_2} = V_{R_4}$$

6.4 THE BRIDGE CIRCUIT

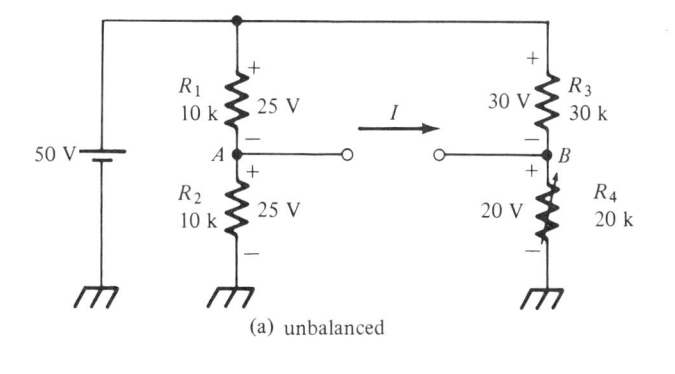

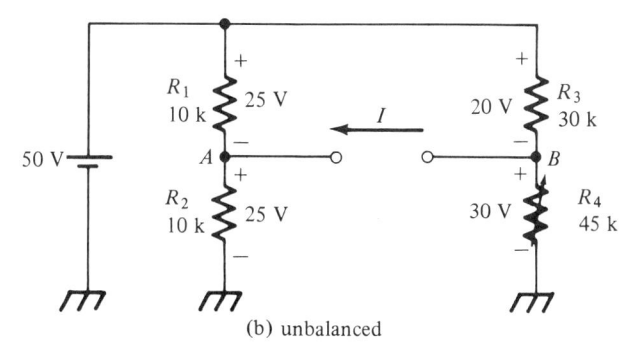

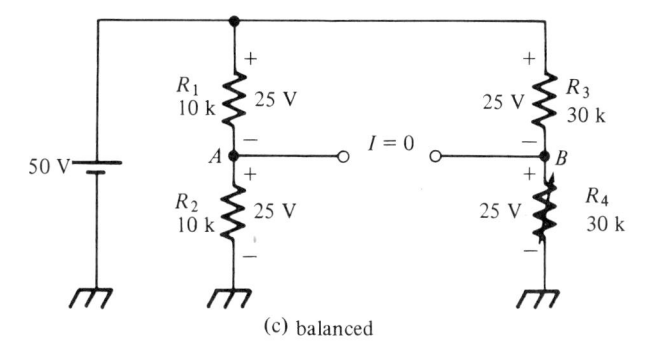

FIGURE 6.3 Basic Bridge; three conditions.

But, by voltage division,

$$V_{R_2} = \frac{E_s R_2}{R_1 + R_2}$$

and

$$V_{R_4} = \frac{E_s R_4}{(R_3 + R_4)}$$

Setting the two expressions equal, we have

$$\frac{E_s R_2}{R_1 + R_2} = \frac{E_s R_4}{R_3 + R_4}$$

Dividing both sides by E_s gives us

$$\frac{R_2}{R_1 + R_2} = \frac{R_4}{R_3 + R_4}$$

Cross-multiplying:

$$R_2(R_3 + R_4) = R_4(R_1 + R_2)$$
$$R_2 R_3 + R_2 R_4 = R_1 R_4 + R_2 R_4$$

Subtracting the common term, $R_2 R_4$, from both sides:

$$R_2 R_3 = R_1 R_4$$

Divide both sides by $R_2 R_4$:

$$\frac{R_2 R_3}{R_2 R_4} = \frac{R_1 R_4}{R_2 R_4}$$

and we get

$$\frac{R_3}{R_4} = \frac{R_1}{R_2}$$

or

$$\frac{R_1}{R_2} = \frac{R_3}{R_4} \qquad \text{the balance equation} \qquad (6.1)$$

EXAMPLE 6.11

Are these bridges balanced?

a.

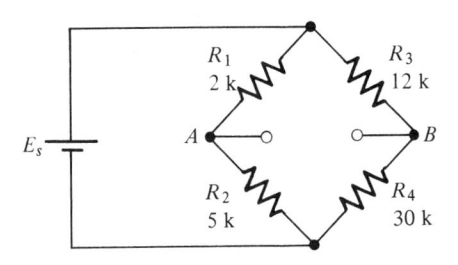

For balance, the ratio R_1/R_2 must equal the ratio R_3/R_4. In this case

$$\frac{R_1}{R_2} = \frac{2\text{ k}}{5\text{ k}} = 0.4$$

$$\frac{R_3}{R_4} = \frac{12\text{ k}}{30\text{ k}} = 0.4$$

$$0.4 = 0.4$$

The bridge is balanced. There is no potential difference between points A and B.

6.4 THE BRIDGE CIRCUIT

b.

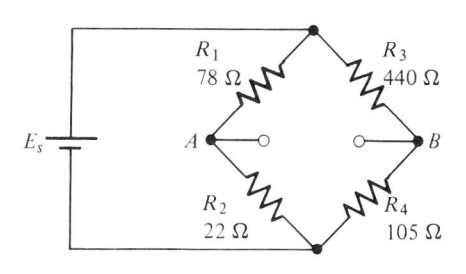

$$\frac{R_1}{R_2} = \frac{78}{22} = 3.55$$

$$\frac{R_3}{R_4} = \frac{440}{105} = 4.19$$

Since the two ratios are not equal, the bridge is not balanced. Which way will current flow if a conductor is placed between points A and B? Point A is at

$$V_a = \frac{E_s(22)}{78 + 22} = 0.22E_s$$

Point B is at

$$V_b = \frac{E_s(105)}{440 + 105} = 0.19E_s$$

Point A is at the higher potential. Therefore, current will flow from A to B. ■

Notice E_s was canceled early in the derivation, so it does not appear in the balance equation. This is an extremely important point. It is telling us that *bridge balance is not affected by changes in E_s*. If balance was affected by E_s variations, the bridge circuit would be worthless since it is not possible to hold any source voltage perfectly constant. Changing source voltage is one of the big headaches in electronics.

The equations does say that four resistances determine bridge balance. If they are of high quality so that their values are stable through temperature, humidity, and time changes, any bridge made of them will be just as stable.

Let's consider two examples that show how useful the bridge is: If a freezer breaks down and nobody notices, the contents spoil. To prevent this, a heat-sensitive resistor called a *thermistor* is connected as one of the legs of a bridge and placed in the freezer. The bridge is balanced. As long as the temperature is correct, the bridge remains balanced; there is no current through the center branch. But if the temperature rises, the thermistor changes resistance, upsetting the bridge. Current flows in the center branch and sounds an alarm. Since there is no reaction to changes in source voltage, the alarm will not go off every time there is a dip on the 110-V line.

Many people got their names in the history book of electricity by inventing a new bridge that does a particular job well. Probably the best known of these is Sir Charles Wheatstone, who, in 1843, designed a very accurate and precise ohmmeter that came to be known as the *Wheatstone bridge* (see Figure 6.4). The Wheatstone bridge circuit has a precision variable resistor as R_2 and an extra-sensitive ammeter called a *galvanometer* in the center branch. R_4 is removed. In its place is a set of terminals to which the unknown R_x is connected. By making R_1 and R_3 equal, the balance equation

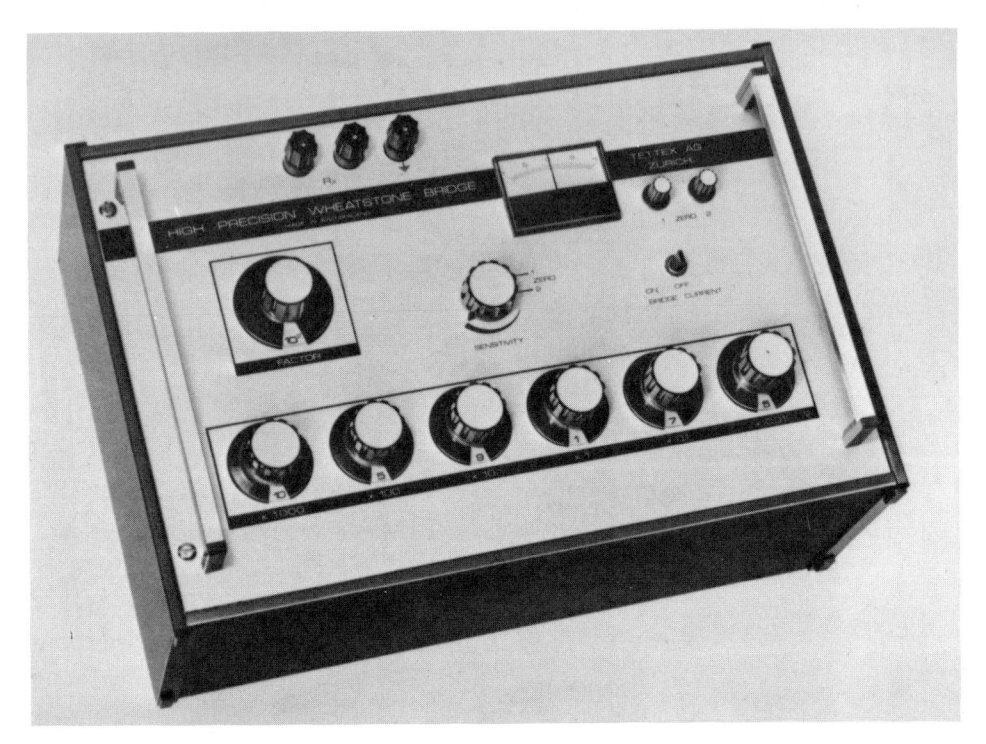

FIGURE 6.4 Wheatstone Bridge. (Courtesy of TETTEX Instruments, Inc.)

becomes
$$\frac{R_1}{R_2} = \frac{R_1}{R_x}$$

Dividing both sides of the equation by R_1, we get
$$\frac{1}{R_2} = \frac{1}{R_x}$$
or
$$R_x = R_2$$

To use the bridge, the unknown is connected and R_2 adjusted until the galvanometer reads zero. When it does, the bridge is balanced and the value of R_x can be read from the dial of R_2.

6.5 T–Π TRANSFORMATIONS

T and Π *networks* are groups of components connected to form a circuit having the shape of the letter (see Figure 6.5). They are found frequently in electronics and in the electric power industry, where the T is called the *Y* (or *wye*) or *star* connection and the Π is called the *delta* connection because of its similarity to the shape of the Greek capital letter delta.

Each network can be converted or *transformed* into an equivalent network of the opposite type. For example, transform a T network into its Π equivalent and the circuit outside the network cannot tell the difference. This often allows us to change a circuit requiring high-level analysis into a simpler form that yields to SPR analysis. The next example shows how this can happen.

6.5 T–Π TRANSFORMATIONS

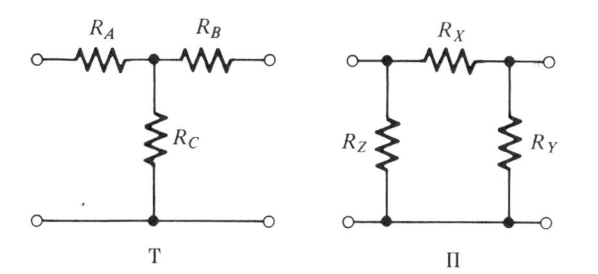

FIGURE 6.5 T and Π Networks.

EXAMPLE 6.12

The resistances in this circuit cannot be reduced to a single R_{total} because none of them are in series or in parallel.

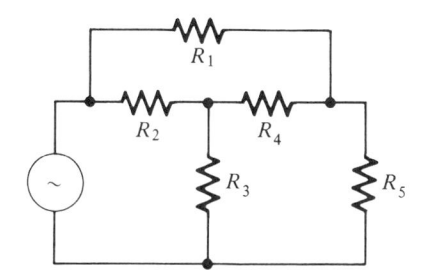

a. Find the T and Π networks hiding in this circuit.

T circuits

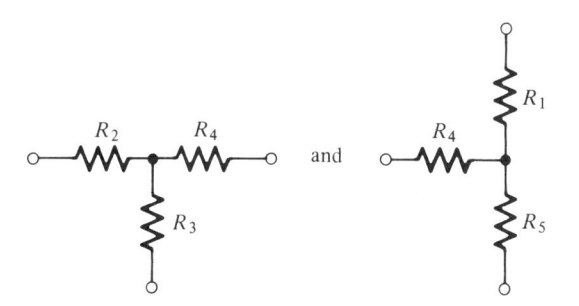

R_2 — R_4 with R_3 and R_4 with R_1 and R_5

Π circuits

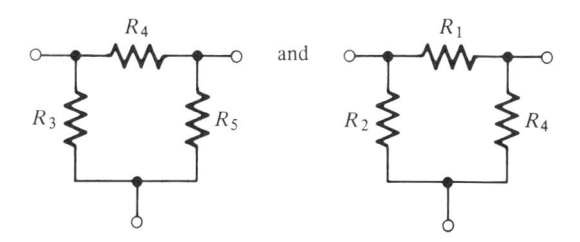

R_4 with R_3 and R_5 and R_1 with R_2 and R_4

Note the small circles indicating terminals or connections. It is extremely important that the equivalent circuit be placed so that its three connections are wired to these same points. Be very careful on this.

b. Replace the T consisting of R_2, R_3, and R_4 with its Π equivalent.

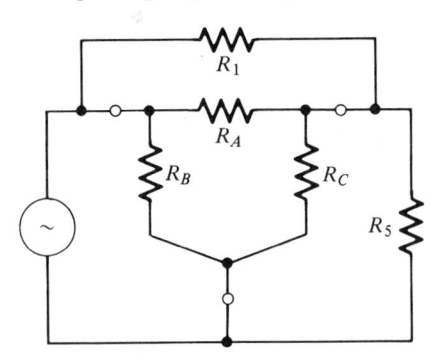

The Π of R_a, R_b, and R_c replaces the T of R_2, R_3, and R_4. Note that the circuit's resistances are in series and parallel. They can now be reduced to a single R_{total}.
The T of R_1, R_4, and R_5 could also have been replaced by its Π equivalent.

c. Replace the Π consisting of R_3, R_4, and R_5 with its T equivalent.

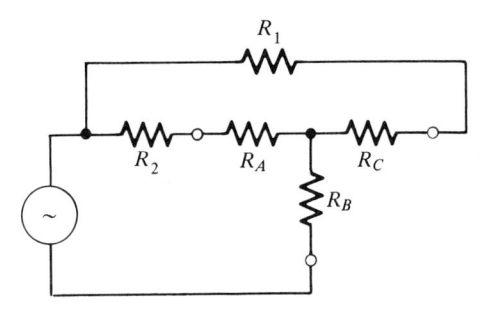

The T of R_a, R_b, and R_c replaces the Π. The circuit can be redrawn with a single R_{total} now.
The Π of R_1, R_2, and R_4 could also have been replaced by its T equivalent.

The formulas for converting a Π to its equivalent T are listed in Figure 6.6 (the formulas are derived in Appendix A). Note their characteristics carefully:

$$R_A = \frac{R_x R_z}{R_x + R_y + R_z} \quad (6.2)$$

$$R_B = \frac{R_x R_y}{R_x + R_y + R_z} \quad (6.3)$$

$$R_C = \frac{R_y R_z}{R_x + R_y + R_z} \quad (6.4)$$

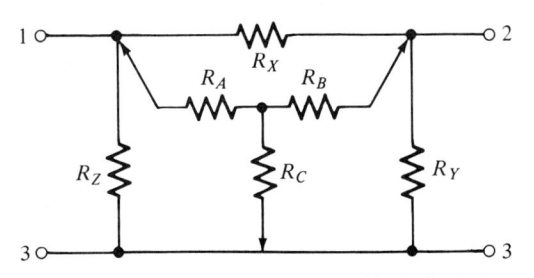

FIGURE 6.6 Converting a Π to its Equivalent T.

6.5 T-Π TRANSFORMATIONS

- The denominators are the same, namely, the summation of all three Π components. Calculate it once and store in memory.
- One end of each T resistance and two Π resistances meet at the same terminal. The product of the two Π resistances is the numerator in the formula for that T resistance. For example, Figure 6.6 shows T resistance, R_a, and Π resistances, R_x and R_z, meeting at terminal 1. $R_x R_z$ is the numerator in the formula for R_a.

EXAMPLE 6.13

R_{total} of this circuit cannot be found because elements are not in series or parallel. Convert the 500-Ω, 200-Ω, and 300-Ω Π within the terminals to its equivalent T and solve for R_{total}.

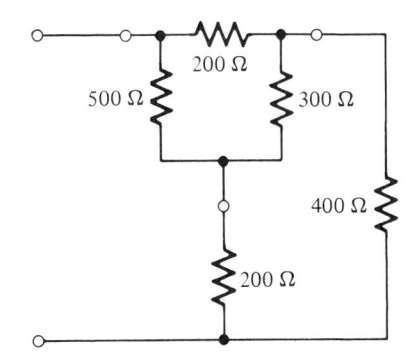

The Π network is removed from the circuit. Its T equivalent is connected to the same terminals:

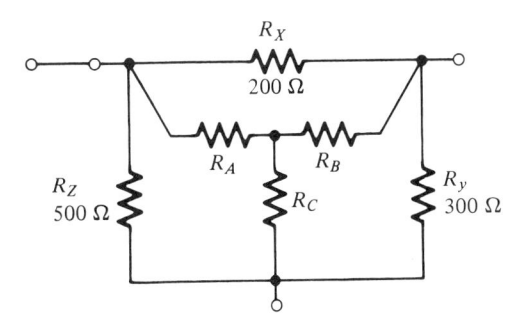

The denominator, D, in all cases is the sum of the three Π elements:

$$D = 500 + 200 + 300 = 1000$$

$$R_a = \frac{R_x R_z}{D} = \frac{200(500)}{1000} = 100 \text{ Ω}$$

$$R_b = \frac{R_x R_y}{D} = \frac{200(300)}{1000} = 60 \text{ Ω}$$

$$R_c = \frac{R_y R_z}{D} = \frac{300(500)}{1000} = 150 \text{ Ω}$$

The new circuit with the T in place is:

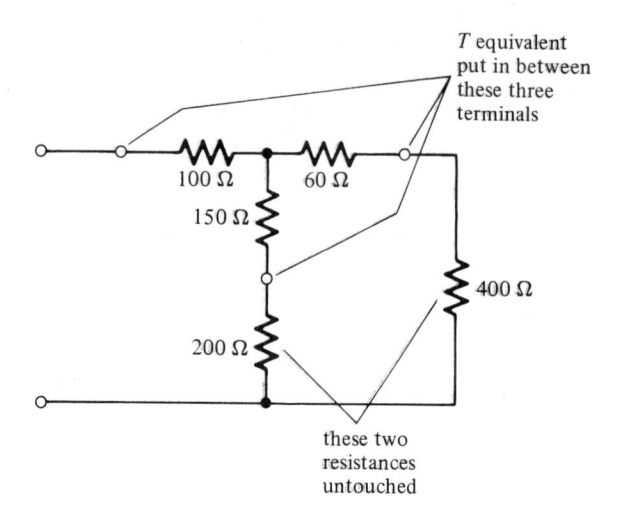

T equivalent put in between these three terminals

100 Ω, 60 Ω, 150 Ω, 200 Ω, 400 Ω

these two resistances untouched

Calculating R_{total}:
Add series resistances:

$$60 + 400 = 460 \text{ Ω}$$
$$150 + 200 = 350 \text{ Ω}$$

These two are in parallel:

$$\frac{460(350)}{460 + 350} = 199 \text{ Ω}$$

This is in series with the 100 Ω.

Final $R_{total} = 199 + 100 = 299$ Ω. ∎

* * * A Word of Warning * * *

No information can be obtained about the original components involved in a transformation because they have vanished. Therefore:

> If information is needed about a component, do not bury that component in a network transformation.

If this means that transform techniques cannot be used, the analysis must be done with methods covered in Chapter 7.

Figure 6.7 lists the formulas (also derived in Appendix A) for converting from T to Π. Note their characteristics carefully:

$$R_x = \frac{R_A R_B + R_B R_C + R_C R_A}{R_C} \tag{6.5}$$

$$R_y = \frac{R_A R_B + R_B R_C + R_C R_A}{R_A} \tag{6.6}$$

$$R_z = \frac{R_A R_B + R_B R_C + R_C R_A}{R_B} \tag{6.7}$$

6.5 T–Π TRANSFORMATIONS

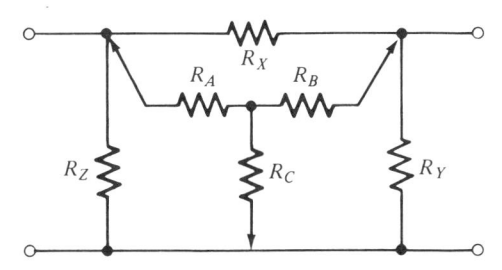

FIGURE 6.7 Converting a T to its Equivalent Π.

- The numerators are the same, namely, the sum of three products. Calculate it once. Then store it.
- The denominators are always a single resistance. Here is how to tell which one: Notice the node at which all three T elements meet in Figure 6.7. An imaginary line drawn along one of the T resistances and through the node hits a Π element. That T element is the denominator. For example, a line along R_a, through the node, hits R_y. R_a is the denominator in the formula for R_y.

EXAMPLE 6.14

The circuit of Example 6.13 is redrawn to point out the 500-Ω, 200-Ω, and 300-Ω T in it. Convert this to its equivalent Π and calculate R_{total}.

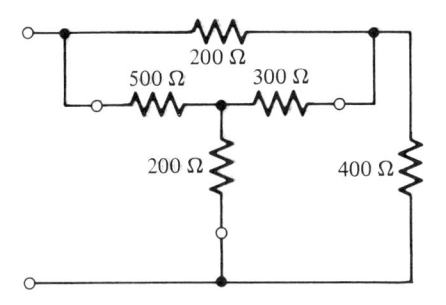

The numerator

$$N = 500(200) + 200(300) + 300(500)$$
$$= (10 \times 10^4) + (6 \times 10^4) + (15 \times 10^4)$$
$$= 31 \times 10^4$$

$$R_x = \frac{N}{R_c} = \frac{31 \times 10^4}{200}$$
$$= 1550 \text{ Ω}$$

$$R_y = \frac{N}{R_a} = \frac{31 \times 10^4}{500}$$
$$= 620 \text{ Ω}$$

$$R_z = \frac{N}{R_b} = \frac{31 \times 10^4}{300}$$
$$= 1030 \text{ Ω}$$

Pull out the T and put in its equivalent Π. The new circuit is

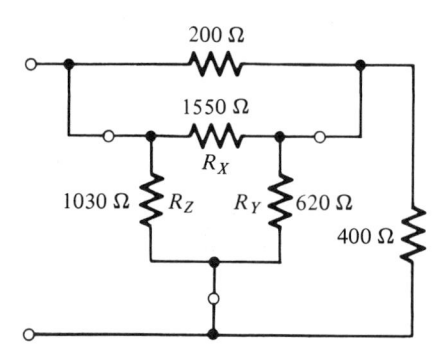

Calculating R_{total}:
Calculate parallel values:

$$\frac{620(400)}{620 + 400} = 243 \; \Omega$$

$$\frac{1550(200)}{1550 + 200} = 177 \; \Omega$$

The circuit is now

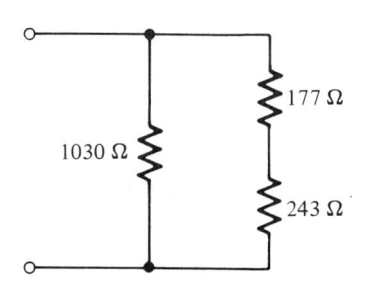

Add series resistances:

$$177 + 243 = 420 \; \Omega$$

Finally,

$$R_{\text{total}} = \frac{1030(420)}{1030 + 420}$$
$$= 298.5 \; \Omega$$

The answer by Π–T transformation in Example 6.13 was 299 Ω. ∎

When the three resistances in a network have the same value, the transform equations simplify to

$$R_T = \frac{R_\Pi}{3} \tag{6.8}$$

and

$$R_\Pi = 3R_T \tag{6.9}$$

6.6 SUPERPOSITION

If the sources in a circuit cannot be combined into one, series–parallel reduction techniques will still provide a solution if a technique called *superposition* is used.

Superposition is a simple concept; we see examples of it every day. For instance, a diving board droops 2 in. for one person and 1.5 in. for another. Superposition tells us that it will deflect 3.5 in. if both are on together.

The diving board illustrates a *linear cause-and-effect relation*. Weight, the *cause*, produced a proportional *effect*, board droop. Such a linear relation between cause and effect is necessary for superposition to work.

The cause in electricity is usually voltage. Its effect is current. Any component that has a linear voltage–current relation can have its total current determined by superposition, that is, by adding the individual currents. The component must also be *bilateral*; that is, it must have the same E versus I relation for either current direction.

Resistance is both linear and bilateral (as are all the components studied in this book except one). Voltage across it causes the same current regardless of polarity and we know from Ohm's law that

$$I = \frac{1}{R}E$$

a linear equation. Therefore, any circuit consisting of sources and resistances can be analyzed by superposition.

Here is how to use superposition:

1. Turn off all the sources except one. This changes the circuit to an ordinary series–parallel case.
2. Using SPR techniques, calculate the current caused by the "on" source through the desired component. Label results clearly, being sure to pay close attention to current direction.
3. Turn source 1 off and source 2 on. Repeat step 2.
4. Continue until the contribution of every source has been found.
5. Add the currents in one direction. Add those in the other direction. Subtract the smaller from the larger. The result is I_{total}.

Step 1 mentioned turning sources off. How is this done?

In earlier chapters we studied Thévenin and Norton circuits. Each has an ideal source in series or shunt with an internal resistance.

> Sources must be shut off in such a way that their internal resistances remain in the circuit. To do this requires that:
> The voltage source in the Thévenin circuit be short-circuited.
> The current source in the Norton circuit be open-circuited.

Figure 6.8 illustrates this important point.

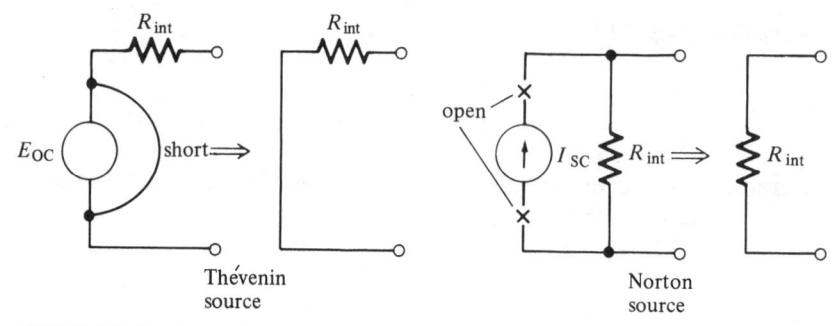

FIGURE 6.8 Turning Off Thévenin and Norton Sources.

Note carefully: Only the ideal source is affected—$R_{internal}$ remains in the circuit.

EXAMPLE 6.15

Find the current through R_2 by superposition.

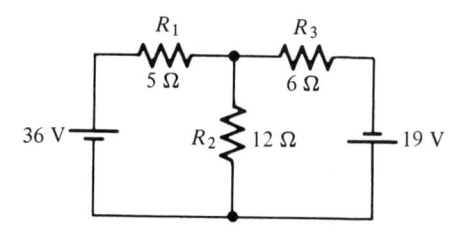

Turn off the 19-V source by shorting it. The circuit becomes

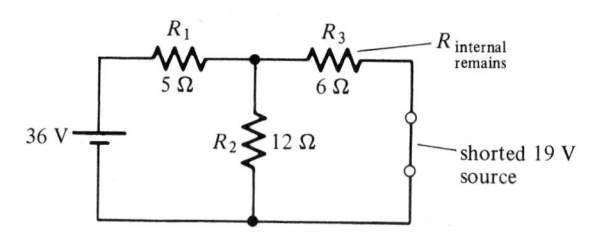

Using SPR, find $I(36\ V)$, the current supplied by the 36-V source:

$$R_{total} = 5 + \frac{12(6)}{12 + 6}$$

$$= 5 + \frac{72}{18}$$

$$= 5 + 4$$

$$= 9\ \Omega$$

$$I_{total} = \frac{36}{9}$$

$$= 4\ A$$

6.6 SUPERPOSITION

This is the total current from the 36-V source. It splits—part goes through R_2, the rest through R_3. We want the part through R_2. By current division:

$$I_{R_2} = \frac{4R_3}{R_2 + R_3}$$

$$= \frac{4(6)}{12 + 6}$$

$$= \frac{24}{18} = 1.3 \text{ A}$$

This is the current magnitude. We need its direction as well.

Current from the 36-V source goes CW; therefore, it travels downward through R_2. We indicate the current as

$$I_{36\,V} = 1.3 \text{ A down}$$

Next, turn off the 36-V source, turn on the 19-V source and do another analysis. The circuit becomes

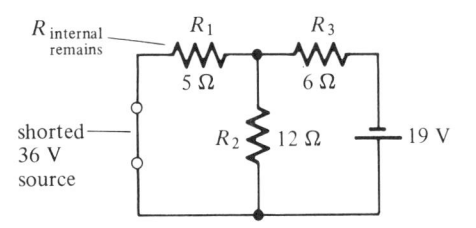

R_{internal} remains — R_1 $5\,\Omega$ — R_3 $6\,\Omega$

shorted 36 V source

R_2 $12\,\Omega$ — 19 V

$$R_{\text{total}} = 6 + \frac{12(5)}{12 + 5} = 6 + \frac{60}{17}$$

$$= 6 + 3.5 = 9.5\,\Omega$$

$$I_{\text{total}} = \frac{19}{9.5} = 2 \text{ A}$$

How much of this passes through R_2? By current division,

$$I_{R_2} = \frac{2R_1}{R_1 + R_2}$$

$$= \frac{2(5)}{5 + 12}$$

$$= \frac{10}{17}$$

$$= 0.6 \text{ A. The direction is up.}$$

Finally,

$$I_{R_2} = I_{36\,V} + I_{19\,V}$$
$$= 1.3 \text{ A down} + 0.6 \text{ A up}$$
$$= 0.7 \text{ A down}$$

The accuracy of calculations can be checked by using Ohm's law, KVL, and KCL to find every voltage drop and current in the circuit. Then sum the voltages around two or more loops. If the sums are close to zero, the results are probably correct. Do not expect perfection, though, because there has been some rounding off. Experience will help tell when a sum is too far from zero.

The procedure is illustrated in the next example.

EXAMPLE 6.16

Check the results of Example 6.15.

Redraw the circuit with I_{R_2} indicated:

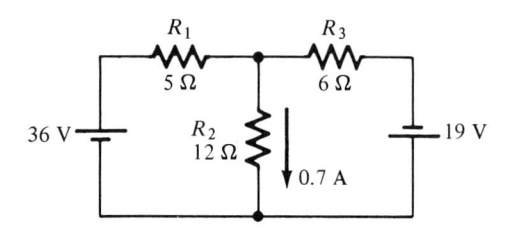

$$V_{R_2} = 0.7(12) = 8.4 \text{ V}$$

Draw the loop containing R_2, R_3, and the 19-V source and assume a polarity for V_{R_3}:

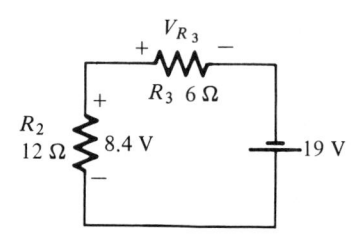

By KVL,

$$V_{R_2} + 19 - V_{R_3} = 0$$
$$8.4 + 19 - V_{R_3} = 0$$
$$V_{R_3} = 27.4 \text{ V}$$
$$I_{R_3} = \frac{V_{R_3}}{R_3} = \frac{27.4}{6}$$
$$= 4.6 \text{ A from left to right}$$

Consider the node at the top of R_2:

6.6 SUPERPOSITION

By KCL,

$$I_{R_1} = I_{R_2} + I_{R_3}$$
$$= 0.7 + 4.6 = 5.3 \text{ A}$$

Then

$$V_{R_1} = 5.3(5) = 26.5 \text{ V}$$

The circuit with all voltages and their polarities is

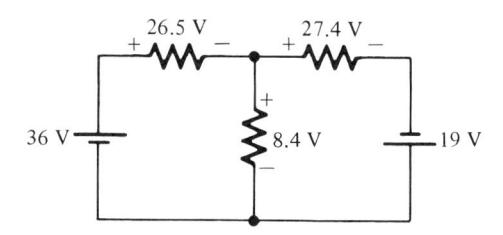

Be sure to understand how the polarity of each voltage was determined.
Check by KVL around all three loops:
The left loop: $-36 + 26.5 + 8.4 = -36 + 34.9$. 36 is not equal to 34.9 but the error is only about 3%.
The right loop: $-8.4 + 27.4 - 19 = -27.4 + 27.4 = 0$.
Around the outside: $-36 + 26.5 + 27.4 - 19 = -55 + 53.9$. The difference is only about 2%. ∎

When doing superposition, do not assume that all the current from a source passes through a resistance. Check and be sure.

EXAMPLE 6.17

Find the current through R_3 by superposition.

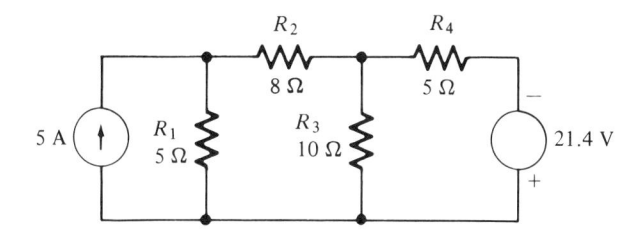

Turn off the 21.4-V source by shorting it. The circuit is now:

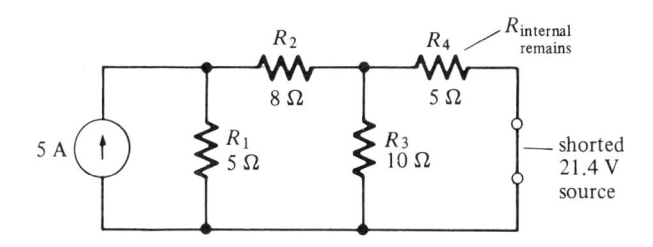

To determine how much of the 5 A passes through R_3, we must first find how much passes through R_2. To do this, we need R_{total} of $R_2 + R_3 \| R_4$ ($\|$ means "in shunt with"):

$$R_3 \| R_4 = \frac{10(5)}{10+5} = \frac{50}{15} = 3.3 \, \Omega$$

$$R_3 \| R_4 + R_2 = 3.3 + 8 = 11.3 \, \Omega$$

Use current division to determine I_{R_2}:

$$I_{R_2} = \frac{I_{total} R_1}{R_1 + R_2 + R_3 \| R_4}$$

$$= \frac{5(5)}{5 + 11.3} = \frac{25}{16.3}$$

$$= 1.5 \, A$$

Use current division again to find out how much of I_{R_2} passes through R_3:

$$I_{R_3} = \frac{I_{R_2} R_4}{R_3 + R_4}$$

$$= \frac{1.5(5)}{10 + 5}$$

$$= \frac{7.5}{15} = 0.5 \, A$$

I from the 5-A source $= 0.5$ A; the direction is down.

Next, turn off the current source, turn on the 21.4-V source, and do another analysis. The circuit is now:

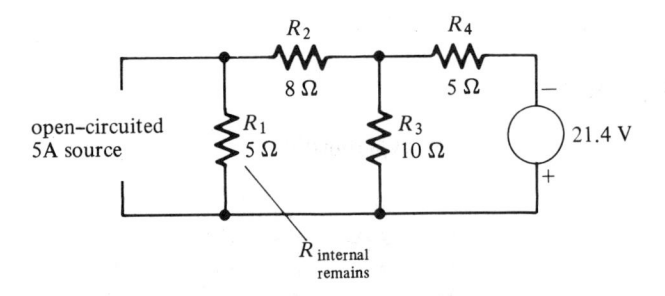

Doing an SPR analysis to find I_{R_3}:
Find R_{total}:

$$R_1 + R_2 = 5 + 8 = 13 \, \Omega = R_{1,2}$$

$$R_{1,2} \| R_3 = \frac{13(10)}{13 + 10}$$

$$= \frac{130}{23} = 5.7 \, \Omega = R_{1,2,3}$$

$$R_{1,2,3} + R_4 = 5.7 + 5 = 10.7 \, \Omega$$

$$I_{total} = \frac{E_s}{R_{total}} = \frac{21.4}{10.7}$$

$$= 2 \, A$$

This current splits at the node at the bottom of R_3:

$$I_{R_3} = \frac{I_{total}R_{1,2}}{R_{1,2} + R_3}$$

$$= \frac{2(13)}{13 + 10}$$

$$= \frac{26}{23} = 1.1 \text{ A}$$

I from the 21.4-V source = 1.1 A; the direction is up.

$$I_{total} = \sum I$$
$$= I \text{ from 5 A} + I \text{ from 21.4 V}$$
$$= 0.5 \text{ A down} + 1.1 \text{ A up}$$
$$= 0.6 \text{ A up}$$

Checking is left to the reader. ∎

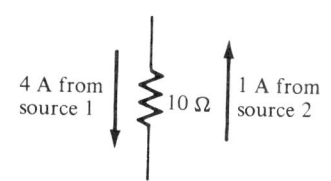

FIGURE 6.9 Using Superposition Improperly.

Here is the wrong way to use superposition. Suppose that a resistance had two currents calculated as shown in Figure 6.9. Find power dissipation by superposition:

$$P_{down} = I^2 R = 4^2(10) = 16(10) = 160 \text{ W}$$
$$P_{up} = I^2 R = 1^2(10) = 1(10) = 10 \text{ W}$$
$$P_{total} = 160 - 10 = 150 \text{ W}$$

This approach may seem logical but it is totally wrong because superposition holds only for linear relationships and

$$P = I^2 R$$

with the exponent, 2, is not linear. The calculation must be done this way:

$$I_{total} = I_{down} - I_{up} = 4 - 1 = 3 \text{ A}$$
$$P = I^2 R = 3^2(10) = 9(10) = 90 \text{ W}$$

6.7 THE LADDER METHOD

In the *ladder method* (so named because it works well on circuits shaped like a ladder), a current, *I*, is assumed to flow in the components farthest from the source. *The equations for all voltages and currents throughout the circuit are written in terms of this I.* Watch carefully how this is done in the next example.

EXAMPLE 6.18

Perform the following steps for this circuit:

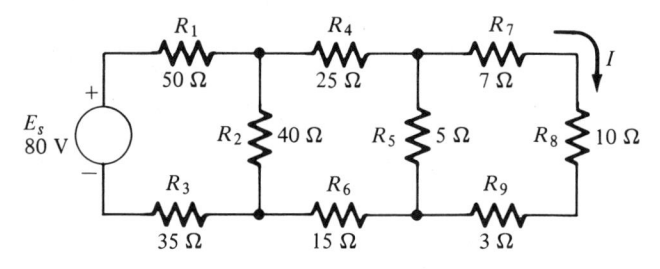

a. Write the ladder method equations and solve for I.

Note the current, I, flowing in R_7, R_8, and R_9. By KVL,

$$V_{R_5} = V_{R_7} + V_{R_8} + V_{R_9}$$
$$= IR_7 + IR_8 + IR_9$$
$$= 7I + 10I + 3I$$
$$= 20I$$

(This voltage, like all others, has been written in terms of I.) By Ohm's law,

$$I_{R_5} = \frac{V_{R_5}}{R_5} = \frac{20I}{5}$$
$$= 4I$$

(This current, like all others, has been written in terms of I.) By KCL,

$$I_{R_4} = I_{R_6} = I_{R_5} + I$$
$$= 4I + I$$
$$= 5I$$

By KVL,

$$V_{R_2} = V_{R_4} + V_{R_5} + V_{R_6}$$
$$= I_{R_4}R_4 + V_{R_5} + I_{R_6}R_6$$
$$= 5IR_4 + V_{R_5} + 5IR_6$$
$$= 5I(25) + 20I + 5I(15)$$
$$= 125I + 20I + 75I$$
$$= 220I$$

By Ohm's law,

$$I_{R_2} = \frac{V_{R_2}}{R_2} = \frac{220I}{40}$$
$$= 5.5I$$

By KCL,

$$I_{R_1} = I_{R_3} = I_{R_2} + I_{R_4}$$
$$= 5.5I + 5I$$
$$= 10.5I$$

6.7 THE LADDER METHOD

By KVL,

$$\begin{aligned}E_s &= V_{R_1} + V_{R_2} + V_{R_3}\\ &= I_{R_1}R_1 + V_{R_2} + I_{R_3}R_3\\ &= 10.5IR_1 + V_{R_2} + 10.5IR_3\\ &= 10.5I(50) + 220I + 10.5I(35)\\ &= 525I + 220I + 367.5I\\ &= 1112.5I\end{aligned}$$

But E_s is given as 80 V; therefore,

$$I = \frac{E_s}{1112.5} = \frac{80}{1112.5} = 72 \text{ mA}$$

Now that all voltages and currents have equations written in terms of I, it is easy to find the value of any of them.

b. What is the voltage across R_2 and the current through R_6?

From the ladder equations,

$$V_{R_2} = 220I = 220(72 \text{ mA}) = 15.8 \text{ V}$$
$$I_{R_2} = 5I = 5(72 \text{ mA}) = 360 \text{ mA}$$

Using SPR techniques, a change in source voltage means that an entire analysis must be repeated to find new currents. Not so using the ladder method.

c. Suppose that the source voltage was increased to 127.6 V. What is the new value of I_{R_1}?

From the equations,

$$E_s = 1112.5I$$

Therefore,

$$I = \frac{E_s}{1112.5} = \frac{127.6}{1112.5} = 115 \text{ mA}$$

From the equations,

$$I_{R_1} = 10.5I = 10.5(115 \text{ mA}) = 1.2 \text{ A}$$

Here is another example of the flexibility of the ladder method.

d. Suppose that a current of 57 mA was needed through R_5. What source voltage would be required?

The ladder equation for I_{R_5} is $I_{R_5} = 4I$, but I_{R_5} must be 57 mA. Therefore,

$$57 \text{ mA} = 4I \quad \text{and} \quad I = \frac{57 \text{ mA}}{4} = 14.3 \text{ mA}$$

The source-voltage equations is

$$E_s = 1112.5I$$

Substitute the new value of I and solve for E_s:

$$E_s = 1112.5(14.3 \text{ mA}) = 15.9 \text{ V}$$

EXERCISES

Section 6.1

1. Find R_{total} for each circuit.

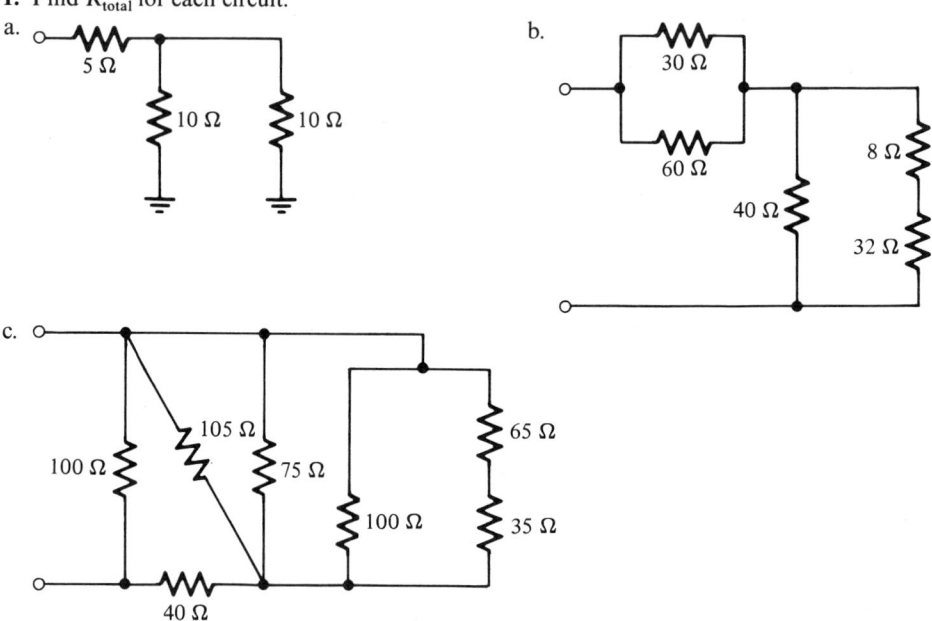

a.

b.

c.

2. What R_{load} would draw maximum power from the terminals of this circuit? *Hint:* Reduce to a Thévenin equivalent source.

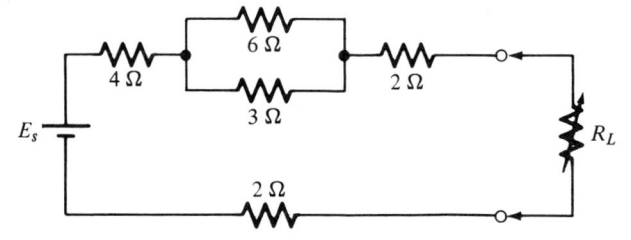

3. Two circuits, called two-port or four-terminal networks, appear quite often in electronics, the Π and T, so named because of their shapes. The two left-hand terminals of these networks are called the input terminals, or the input. An input signal (a voltage or a current) is applied to the input, changed in some way by the network, and appears between the output terminals on the right. For this T network, calculate the following.

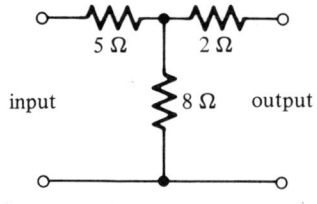

a. R_{in} (resistance looking into the input terminals) with the output open.
b. R_{in} with output shorted.
c. R_{in} with output loaded by 6 Ω.
d. R_{in} with output loaded by 100 k.

EXERCISES

e. R_{out} (resistance looking back into output terminals) with input open.
f. R_{out} with input shorted.

4. Repeat Exercise 3 for this Π network:

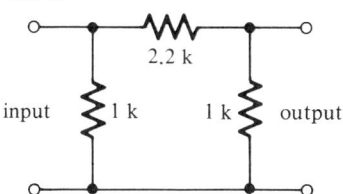

Section 6.2

5. a. Does R_1 have any affect on the voltage across or current through R_3 in this circuit? Discuss.

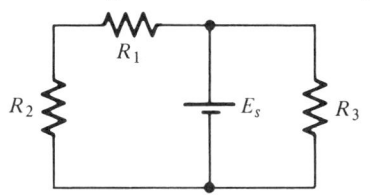

b. Redo part a when R_1 is moved as shown.

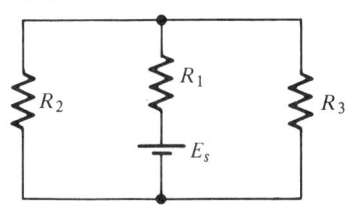

6. Analyze each circuit. Remember to ask checking questions.
   a.
   b. $I = 25$ A

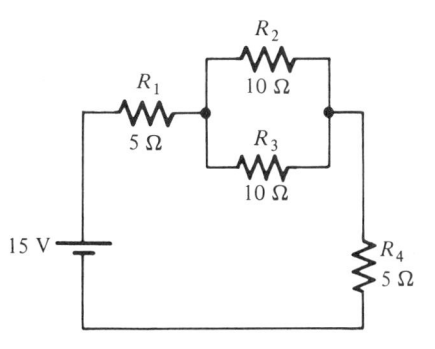

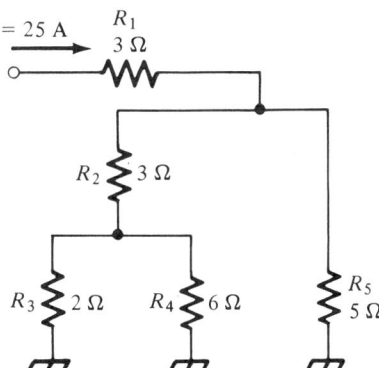

7. Analyze each circuit.
   a.

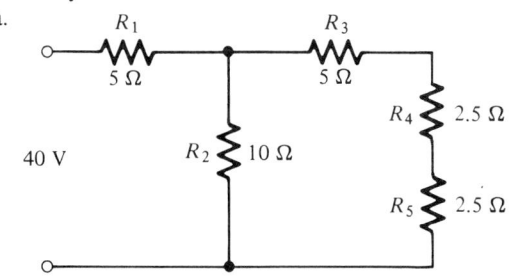

b.

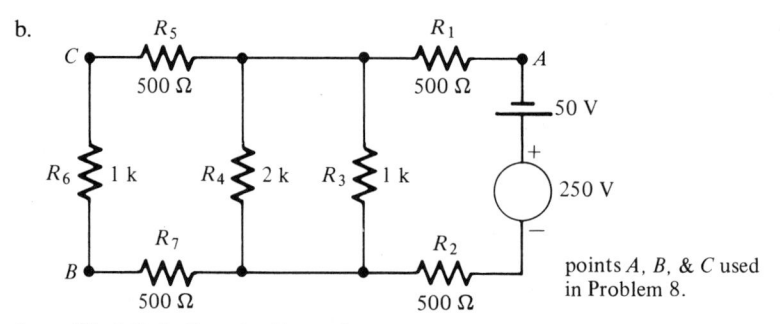

[Circuit diagram with: $R_5 = 500\,\Omega$, $R_1 = 500\,\Omega$ along top between C and A; 50 V and 250 V sources on right; $R_6 = 1\,k$, $R_4 = 2\,k$, $R_3 = 1\,k$ vertical resistors; $R_7 = 500\,\Omega$, $R_2 = 500\,\Omega$ along bottom to B; points A, B, & C used in Problem 8.]

8. a. Find V_{ab} in Exercise 7, part b.

b. Ground point C in Exercise 7, part b. What is the potential at point A with respect to ground? Of point B?

9. The voltmeter across the source reads how many volts?

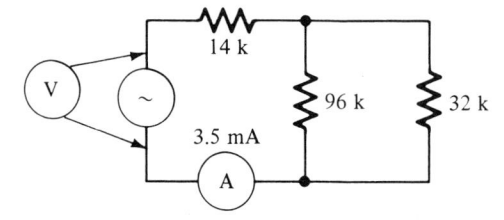

[Circuit with voltmeter V, AC source, ammeter reading 3.5 mA, 14 k resistor in series, 96 k and 32 k resistors in parallel.]

10. Find I_{R_7}. *Hint:* Use Millman's theorem.

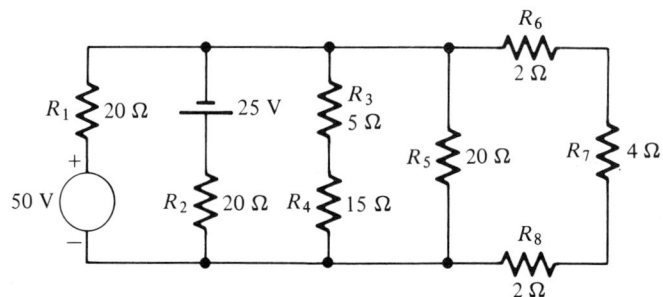

[Circuit with 50 V source, $R_1 = 20\,\Omega$, 25 V source, $R_2 = 20\,\Omega$, $R_3 = 5\,\Omega$, $R_4 = 15\,\Omega$, $R_5 = 20\,\Omega$, $R_6 = 2\,\Omega$, $R_7 = 4\,\Omega$, $R_8 = 2\,\Omega$.]

11. Find V, I, and R for each resistance in this circuit. To make sure that nothing is missed, set up a table with four columns, like this:

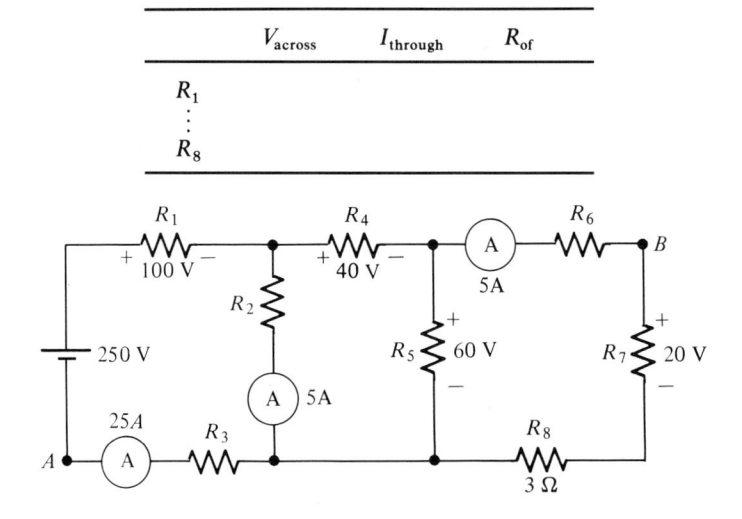

	V_{across}	$I_{through}$	R_{of}
R_1			
$\vdots$			
R_8			

[Circuit diagram: 250 V source, R_1 with 100 V across, R_4 with 40 V across, ammeter reading 5A, R_6 to point B, R_2, R_5 with 60 V, R_7 with 20 V, $R_8 = 3\,\Omega$, ammeter showing 25A, R_3, ammeter showing 5A from point A.]

EXERCISES

12. Perform the indicated operations for the circuit of Exercise 11.
a. Prove KVL around the loop R_2, R_4, R_6, R_7, and R_8.
b. Find V_{ab}. Check results by rewriting the loop equation around two other loops.

13. Perform the indicated operations for this circuit:

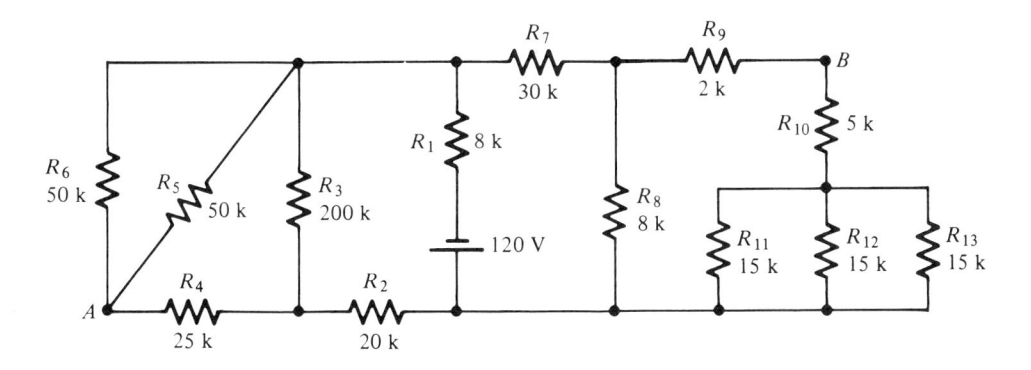

a. Find P_{R_8}.
b. Find I_{R_3}.
c. Find V_{ba}.
d. Prove KVL around R_5, R_4, R_2, R_{11}, R_{10}, R_9, and R_7.

14. What does each of the following tell about this circuit? Check answers by units analysis.

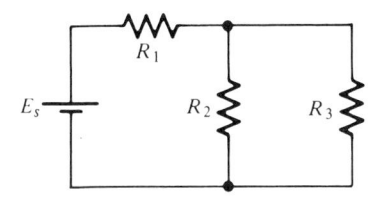

a. $\dfrac{E_s}{R_1 + \dfrac{R_2 R_3}{R_2 + R_3}}$

b. $\dfrac{E_s R_1}{R_1 + \dfrac{R_2 R_3}{R_2 + R_3}}$

c. $\dfrac{E_s R_2}{\left(R_1 + \dfrac{R_2 R_3}{R_2 + R_3}\right)(R_2 + R_3)}$

d. $\left[\dfrac{E_s R_3}{\left(R_1 + \dfrac{R_2 R_3}{R_2 + R_3}\right)(R_2 + R_3)}\right]^2 R_2$

Section 6.3

15. Using I, D, and NC (increase, decrease, and no change, respectively), indicate what happens to each of the following when the switch on this circuit is closed. a. I_{total}; b. V_{R_1}; c. V_{R_2}; d. $V_{R_{input}}$.

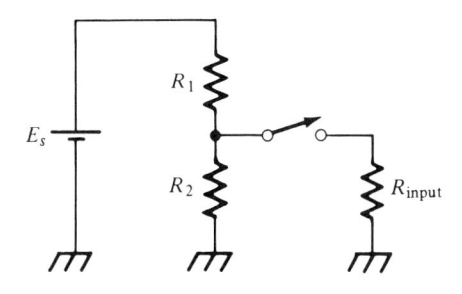

176   SERIES–PARALLEL CIRCUITS

16. A stereo system has three main parts: turntable, amplifier, and loudspeaker.
    a. Which part drive which?
    b. Which part loads which?
17. Answer the questions for this circuit:

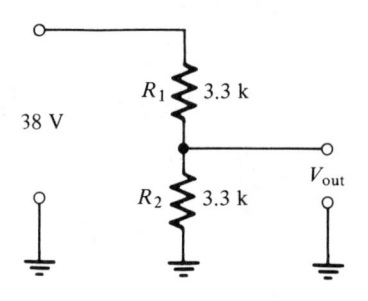

   a. What is V_{out} for open-circuit conditions?
   b. What is V_{out} when the divider drives a circuit with an input resistance of 4.7 k?
   c. Without doing any calculations, tell what V_{out} in part b would do if the input resistance was increased to 47 k.
18. Find V_1 and V_2 for each of the following conditions.

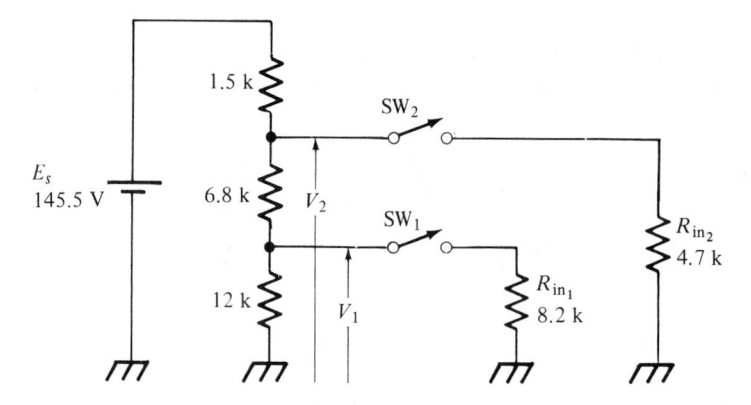

   a. Under no-load conditions.
   b. Only SW_1 closed.
   c. Only SW_2 closed.
   d. Both switches closed.
19. What must R_1 be to hold V_1 to 180 V?

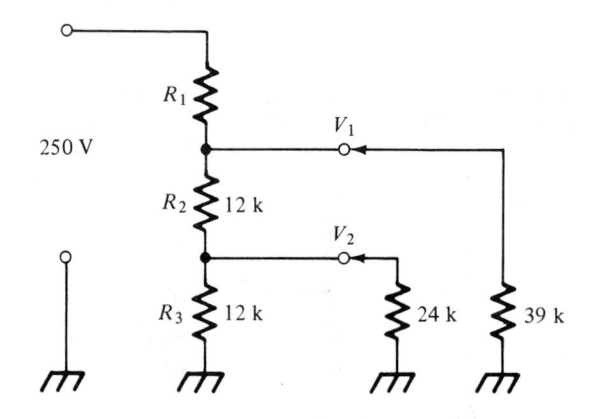

EXERCISES

20. *Extra Credit.* Each additional driven input places another 180-Ω resistance across the driver output in this circuit. This increases output loading, causing E_{out} to decrease. The number of driven circuits that may be connected without E_{out} falling below some critical level is called the driver's *fanout*.

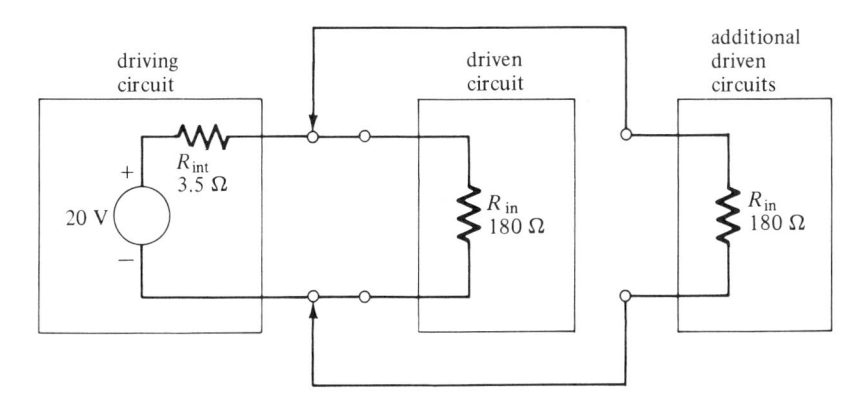

a. What is the fanout of this driver if the driven inputs require an E_{in} of 16.3 V?
b. What can be done to R_{int} to increase fanout?
c. What can be done to driven circuit input resistance to increase fanout?

Section 6.4

21. An engineer is designing a battery-operated weather monitor to be parachuted to a remote mountain peak.
a. What characteristic of bridge circuits is important in this application?
b. Why should the design call for high-valued resistances?
c. Describe how a bridge could be made to detect changes in barometric pressure.

22. Explain what is meant by the term "balance" when referring to a bridge. How is it achieved?

23. Are these bridges balanced?

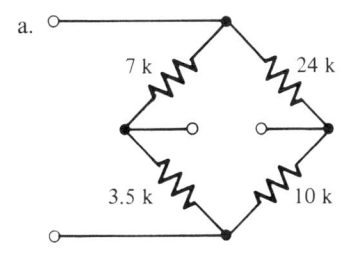

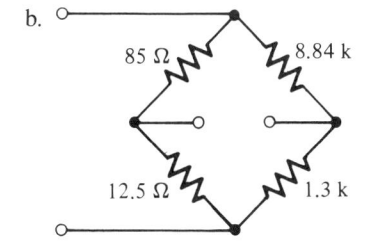

24. In which direction will current flow through the meter in each Wheatstone bridge?

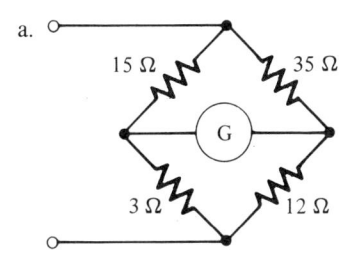

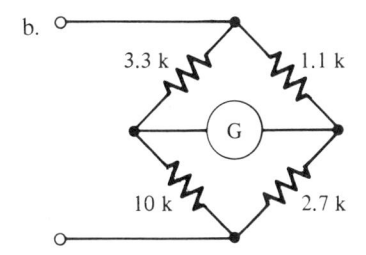

Section 6.5

25. Find R_{total} for this circuit by converting a T to a Π. Check the answer by converting a Π to a T. Be sure to circle connections so that the equivalents can be connected properly.

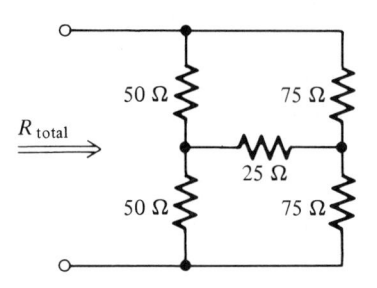

26. Find I_{total} for this circuit:

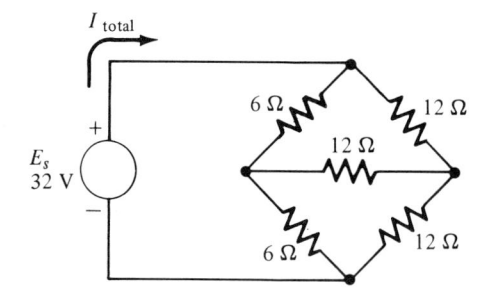

27. Find the power dissipated by the 20 k. *Remember:* Do not bury the component for which data are wanted.

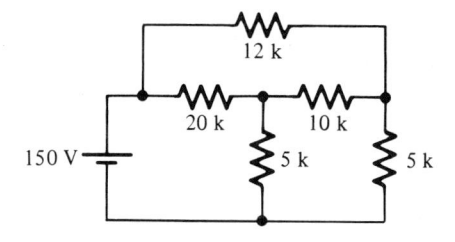

28. Find R_{total} of this circuit:

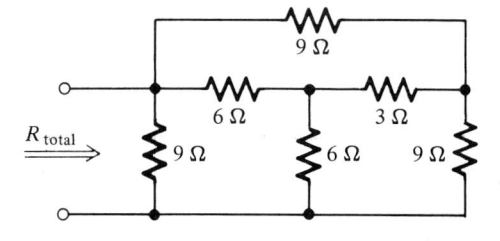

Section 6.6

29. When would the cause-and-effect relation between weight and droop for the diving board become nonlinear?

EXERCISES

30. Find the magnitude and direction of the current through the center branch by superposition.

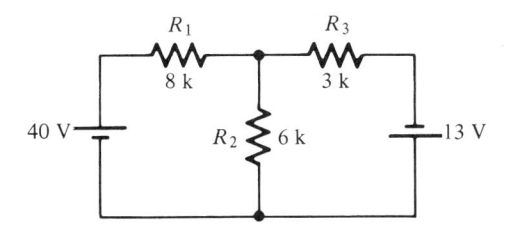

31. Find the current I through R_4 and its direction by superposition.

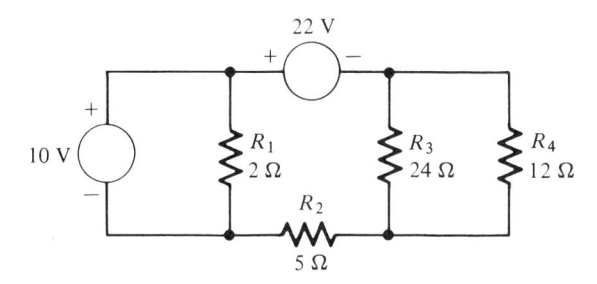

32. By superposition, find I through the 60 k and its direction.

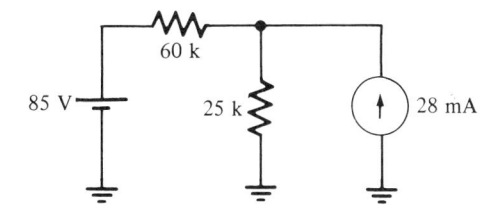

33. Using superposition, find the power dissipated by the 9-Ω resistance.

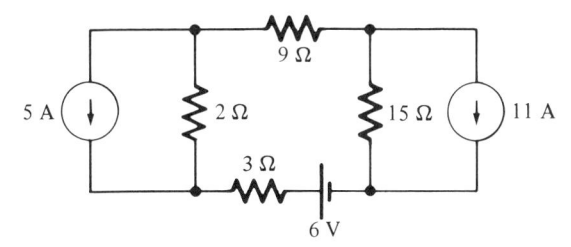

34. Using superposition, find the power dissipated by R_2.

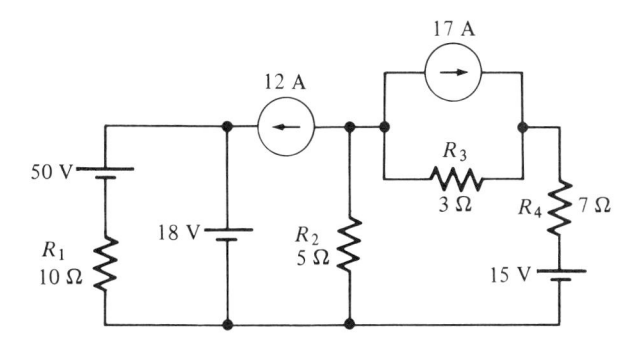

Section 6.7

35. Write the ladder equations for this circuit. Then find I.

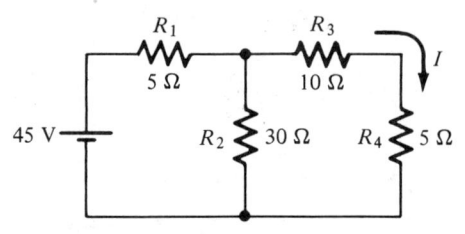

36. a. Use the ladder method to find the voltage across R_4.
b. If E_s is increased to 62 V, what is the new value of V_{R_4}?

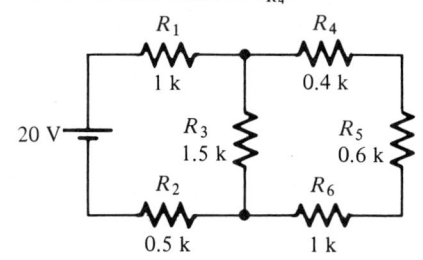

37. Answer the questions for this circuit:

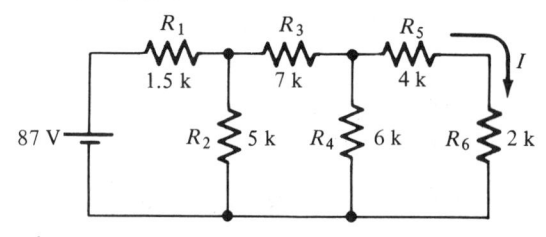

a. Write the ladder equations.
b. Find I.
c. What are V_{R_2} and I_{R_3}?
d. What source voltage would be needed to push 7.5 mA through R_2?
e. Prove KVL around the loop R_2, R_3, R_5, and R_6 using the original value of I.

38. Use the ladder method to tell what I_{total} equals when I is 2 A.

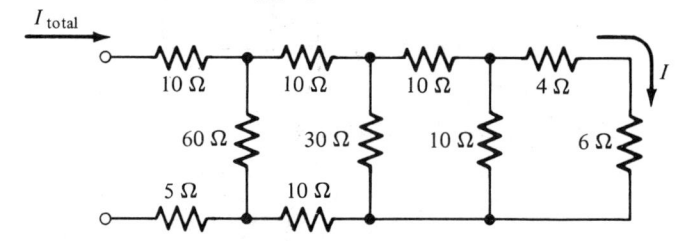

39. Answer the questions for this circuit:

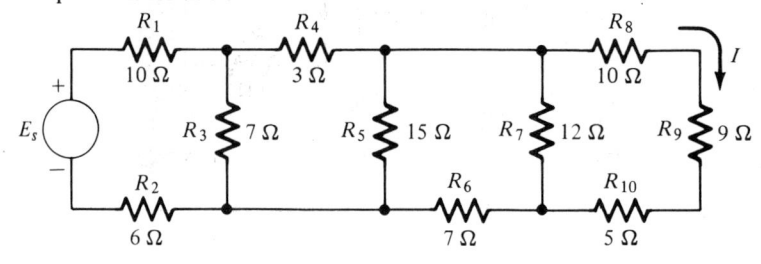

EXERCISES

a. If $I = 4$ A, what is E_s?
b. If $E_s = 285$ V, what is I?
c. If $V_{R_5} = 115$ V, what is E_s?
d. If $I_{R_3} = 65$ A, what is I?
e. What E_s is needed to force 45 A through R_6?

Troubleshooting

40. Answer the questions for this circuit:

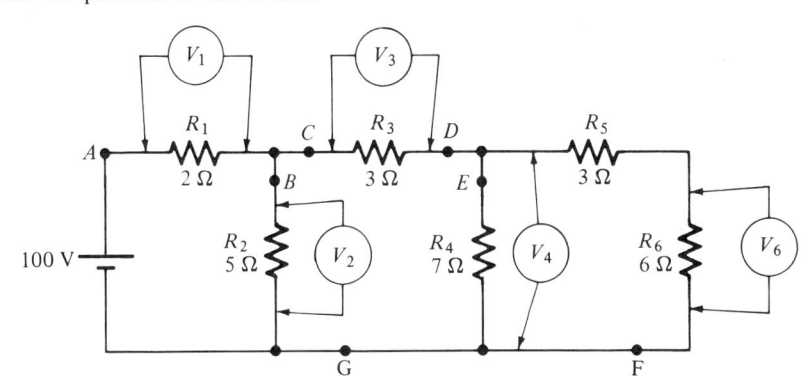

a. What will V_4 read if R_6 opens?
b. What will V_6 read if R_4 opens?
c. What will V_3 read if R_5 is shorted?
d. What will V_3 read if R_4 is shorted?
e. What will V_3 read if R_2 is shorted?
f. What will V_2 read if R_3 opens?
g. At what point should an ammeter be placed to read the total current through R_3, R_4, R_5, and R_6?
h. At what point should a fuse be placed to protect R_4?

41. Indicate by a letter what happens to each lamp under the stated condition. Let D represent gets dimmer, B for brightens, G for goes out, and U for unchanged. All bulbs are the same wattage.

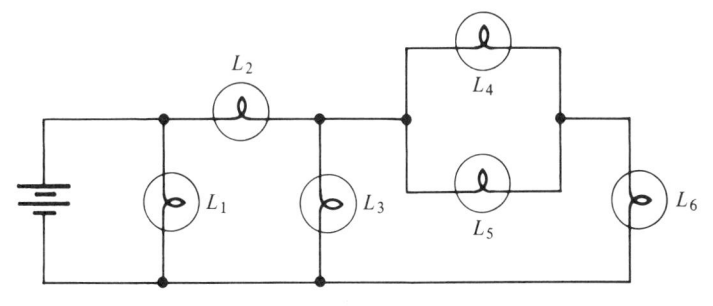

a. L_4 if L_3 opens.
b. L_6 if L_3 shorts.
c. L_5 if L_2 shorts.
d. L_6 if L_2 shorts.
e. L_6 if L_2 opens.
f. L_4 if L_1 shorts.
g. L_6 if L_1 opens.
h. L_5 if L_4 shorts.
i. L_2 if L_5 shorts.
j. L_2 if L_3 opens.

CHAPTER HIGHLIGHTS

Resistances must be in series or in parallel to reduce them into a single R_{total}. Both connections are found in series–parallel circuits.

Whether components are in series or parallel can change depending on how a circuit is viewed.

Series–parallel reduction (SPR) is the most common way to analyze series–parallel circuits. It can only be used when a circuit can be reduced to a single source and load.

SPR steps: Find R_{total}, solve for I_{total}, use I_{total} with KVL, KCL, and Ohm's law to complete analysis.

In advanced electronics, the form of an equation is often more important than its solution.

Circuits do not work alone. They are connected to, or drive, other circuits and are, in turn, driven. Any driven circuit loads its driver; that is, driver output voltage decreases.

Besides their many applications in electronics, bridges react to physical parameters by choice of resistive element in one leg. Bridge balance condition not affected by changes in source voltage.

Sometimes network transformation allows a difficult connection of resistances to be combined for R_{total}.

Do not bury the resistance being analyzed by a network transformation.

Superposition says: If cause-and-effect relations are linear, total effect equals the sum of individual effects.

Superposition: Find the contribution of each source with others turned off. Add parts to get the total. Sources turned off with internal resistance remaining—Thévenin: short E_{oc}, Norton: open I_{sc}.

Ladder equations written in terms of one current. With them, a circuit's voltages and currents are easily determined for changing conditions.

Formulas to Remember

$$\frac{R_1}{R_2} = \frac{R_3}{R_4} \quad \text{bridge balance}$$

$$R_a = \frac{R_x R_z}{D} \quad \text{Π-to-T transformations}$$

$$R_b = \frac{R_x R_y}{D}$$

$$R_c = \frac{R_y R_z}{D} \quad \text{where } D = R_x + R_y + R_z$$

$$R_x = \frac{N}{R_c} \quad \text{T-to-Π transformations}$$

$$R_y = \frac{N}{R_a}$$

$$R_z = \frac{N}{R_b} \quad \text{where } N = R_a R_b + R_b R_c + R_c R_a$$

CHAPTER 7

Advanced Techniques of Analysis

In this chapter we call on our "first-string players," the methods of analysis that work on the most complicated circuits. The first two are *loop analysis*, based on KVL, and *nodal analysis*, based on KCL. Because these techniques require higher mathematics than used so far, the chapter begins with a discussion of the math tools we will be using.

After loop and nodal analysis, we study powerful techniques based on Thévenin's and Norton's theorems. These reduce complicated circuits to simple Thévenin or Norton sources.

At the end of this chapter you should be able to:

- Use Cramer's rule to solve systems of equations by determinants.
- Analyze circuits by means of loop and nodal analysis.
- Find the Thévenin or Norton equivalent of the circuit driving any component.

7.1 REVIEW OF MATH

In both loop and nodal analysis we write a group, or *system*, of linear equations, one equation for each unknown current and voltage. The system is then solved for the unknowns and the analysis completed.

There are three math techniques for solving systems of linear equations. They are *substitution, addition/subtraction,* and *determinants.* The first two are covered in high school algebra and will only be reviewed here, substitution in Example 7.6 and addition/subtraction in Example 7.7. On the other hand, some students will not have seen determinant solutions, so they are discussed.

Readers needing reviews of the substitution and addition/subtraction methods are urged to turn to Examples 7.6 and 7.7 now and study the appropriately marked portions (no knowledge of the electrical theory behind the equations is necessary for these reviews). In-depth coverage of the subject may be found in any college algebra text in the section on solution of simultaneous linear equations.

Solution by Determinants

Substitution and addition/subtraction work fine for two equations in two unknowns. They get very complicated for larger systems, however, and determinants should be used. A *determinant* is a special arrangement of numbers. When processed according to certain rules, a single number results. It is the *value* of the determinant.

Figure 7.1 shows a determinant. It is called a *second-order* determinant because it has two horizontal rows and two vertical columns. Lines on the sides outline the determinant; they do not mean absolute value.

Figure 7.1 also shows the method by which the value of the determinant is calculated.

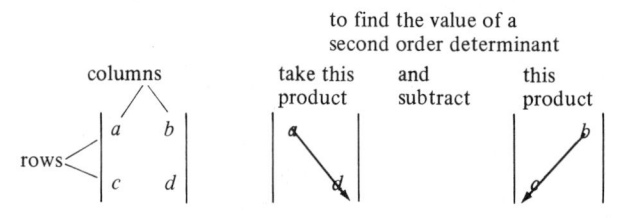

FIGURE 7.1 Second-Order Determinant and its Value.

EXAMPLE 7.1

Evaluate this second-order determinant:

$$\begin{vmatrix} 3 & -5 \\ 2 & 7 \end{vmatrix} = \begin{vmatrix} 3 & \searrow \\ & 7 \end{vmatrix} - \begin{vmatrix} & -5 \\ 2 & \end{vmatrix}$$

$$= (3 \times 7) - (2 \times -5)$$
$$= 21 - (-10)$$
$$= 31$$

Note how each arrow passes through two numbers in a second-order determinant. It must be clearly understood that the negative sign between multiplications is required; it has nothing to do with the signs of the numbers. Do not forget to put it in. ■

A system of equations is solved with determinants by following *Cramer's rule*. The technique is best described by an example.

EXAMPLE 7.2

Solve the following set of equations using determinants and Cramer's rule.

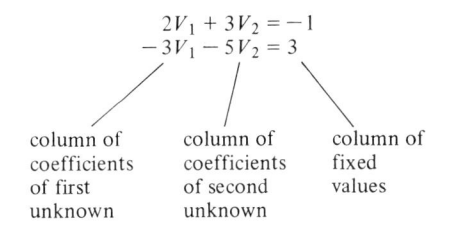

$$2V_1 + 3V_2 = -1$$
$$-3V_1 - 5V_2 = 3$$

column of coefficients of first unknown

column of coefficients of second unknown

column of fixed values

Cramer's rule says that the value of V_1 is found by dividing two determinants in the following manner:

because we're solving for unknown 1, column 1 is the column of fixed values

column 2 is the same in both numerator and denominator

$$V_1 = \frac{\begin{vmatrix} -1 & 3 \\ 3 & -5 \end{vmatrix}}{\begin{vmatrix} 2 & 3 \\ -3 & -5 \end{vmatrix}} = \frac{(-1 \times -5) - (3 \times 3)}{(2 \times -5) - (3 \times -3)} = \frac{5 - 9}{-10 - (-9)} = \frac{-4}{-1} = 4$$

denominator determinant is the four coefficients of V_1 and V_2 copied from the equations

The only change needed to solve for V_2 is to place the column of fixed values in the numerator's second column:

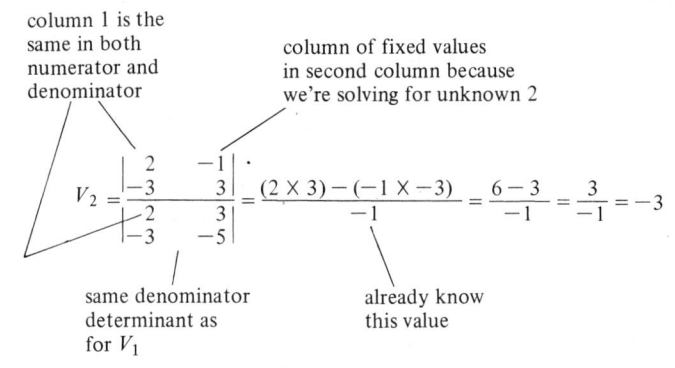

column 1 is the same in both numerator and denominator

column of fixed values in second column because we're solving for unknown 2

$$V_2 = \frac{\begin{vmatrix} 2 & -1 \\ -3 & 3 \end{vmatrix}}{\begin{vmatrix} 2 & 3 \\ -3 & -5 \end{vmatrix}} = \frac{(2 \times 3) - (-1 \times -3)}{-1} = \frac{6-3}{-1} = \frac{3}{-1} = -3$$

same denominator determinant as for V_1

already know this value

As with any system of equations, solutions are checked by substituting them into the original equations:

$$2(4) + 3(-3) = -1$$
$$-3(4) - 5(-3) = 3$$

Solutions are correct. ∎

Third-order determinants have three rows and three columns. Their value is calculated in much the same way as for second-order determinants. But now each arrow passes through three numbers instead of two. To ensure that the arrow goes through the correct three, the determinant's first two columns are repeated to the right.

EXAMPLE 7.3

Evaluate the third-order determinant:

$$\begin{vmatrix} 3 & 3 & -1 \\ 2 & 1 & -1 \\ 1 & 1 & -4 \end{vmatrix}$$

Redraw the determinant with the first two columns repeated and diagonals drawn:

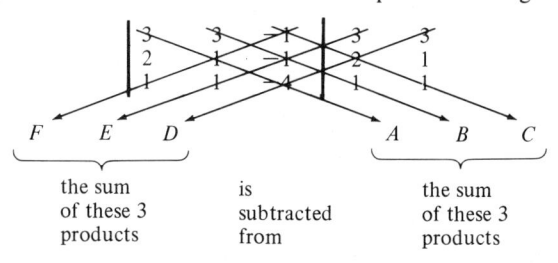

$$\begin{array}{cccccc} 3 & 3 & -1 & 3 & 3 \\ 2 & 1 & -1 & 2 & 1 \\ 1 & 1 & -4 & 1 & 1 \end{array}$$

F   E   D   A   B   C

the sum of these 3 products

is subtracted from

the sum of these 3 products

The value of the determinant is determined by

$$(A + B + C) - (D + E + F)$$

where each letter signifies the product of the three numbers on that diagonal:

$$[(3)(1)(-4) + (3)(-1)(1) + (-1)(2)(1)]$$
$$- [(3)(2)(-4) + (3)(-1)(1) + (-1)(1)(1)]$$
$$= (-12 - 3 - 2) - (-24 - 3 - 1)$$
$$= (-17) - (-28) = 11$$

∎

7.1 REVIEW OF MATH

Although determinants work for any number of equations, the procedure becomes very lengthy for perhaps five or more unknowns and a computer should be used. Our work will be limited to systems of three equations in three unknowns. They are solved with determinants and Cramer's rule in the same way that second-order systems were.

EXAMPLE 7.4

Solve this set of equations using determinants:

$$3I_1 + 3I_2 - I_3 = 4$$
$$2I_1 + I_2 - I_3 = 7$$
$$I_1 + I_2 - 4I_3 = -6$$

The determinant denominator was calculated in Example 7.3. Its value is 11. In the numerator determinant for I_1, column 1 is replaced by the column of fixed values:

$$I_1 = \frac{\begin{vmatrix} 4 & 3 & -1 \\ 7 & 1 & -1 \\ -6 & 1 & -4 \end{vmatrix}}{\begin{vmatrix} 3 & 3 & -1 \\ 2 & 1 & -1 \\ 1 & 1 & -4 \end{vmatrix}} = \frac{\begin{vmatrix} 4 & 3 & -1 \\ 7 & 1 & -1 \\ -6 & 1 & -4 \end{vmatrix} \begin{matrix} 4 & 3 \\ 7 & 1 \\ -6 & 1 \end{matrix}}{11}$$

repeat columns 1 and 2 ↗ ↘

11 ← already known

$$= \frac{[(4)(1)(-4) + (3)(-1)(-6) + (-1)(7)(1)] - [(3)(7)(-4) + (4)(-1)(1) + (-1)(1)(-6)]}{11}$$

$$= \frac{(-16 + 18 - 7) - (-84 - 4 + 6)}{11} = \frac{(-5) - (-82)}{11} = 7$$

For I_2, the second numerator column is replaced by the fixed values:

$$I_2 = \frac{\begin{vmatrix} 3 & 4 & -1 \\ 2 & 7 & -1 \\ 1 & -6 & -4 \end{vmatrix}}{11} = \frac{\begin{vmatrix} 3 & 4 & -1 \\ 2 & 7 & -1 \\ 1 & -6 & -4 \end{vmatrix} \begin{matrix} 3 & 4 \\ 2 & 7 \\ 1 & -6 \end{matrix}}{11}$$

repeat columns 1 and 2

$$= \frac{[(3)(7)(-4) + (4)(-1)(1) + (-1)(2)(-6)] - [(4)(2)(-4) + (3)(-1)(-6) + (-1)(7)(1)]}{11}$$

$$= \frac{(-84 - 4 + 12) - (-32 + 18 - 7)}{11} = \frac{(-76) - (-21)}{11} = -5$$

Since I_1 and I_2 are known, I_3 can be found by substitution:

$$I_1 + I_2 - 4I_3 = -6$$
$$7 - 5 - 4I_3 = -6$$
$$2 - 4I_3 = -6$$
$$-4I_3 = -6 - 2$$
$$I_3 = \frac{-8}{-4} = 2$$

Solutions are $(7, -5, 2)$. Checking is left to the reader. ∎

7.2 LOOP ANALYSIS

Loop analysis is a powerful method for analyzing circuits that gets results when other techniques cannot. Its name comes from its use of loop equations summing voltages around closed loops.

The loop was defined in Chapter 4 as any closed electrical path. That definition was good enough then, but it needs to be tighter for the work we are now starting:

A loop is a closed path in a circuit which passes through no branch or node more than once.

A branch of mathematics called *topology* says that the number of equations needed to analyze a circuit by loop analysis is found by the formula

$$\text{Number of loop equations} = B - N + 1 \tag{7.1}$$

where B = number of circuit branches
N = number of circuit nodes

EXAMPLE 7.5

Answer the questions about this circuit:

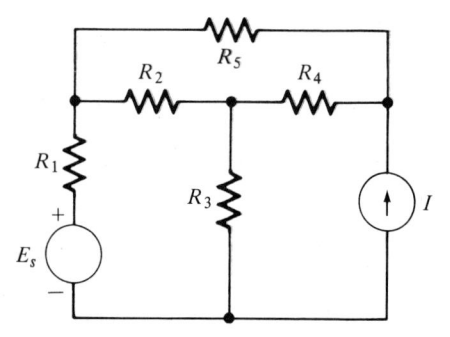

a. Show all possible loops. This is done using a simplified form of the circuit:

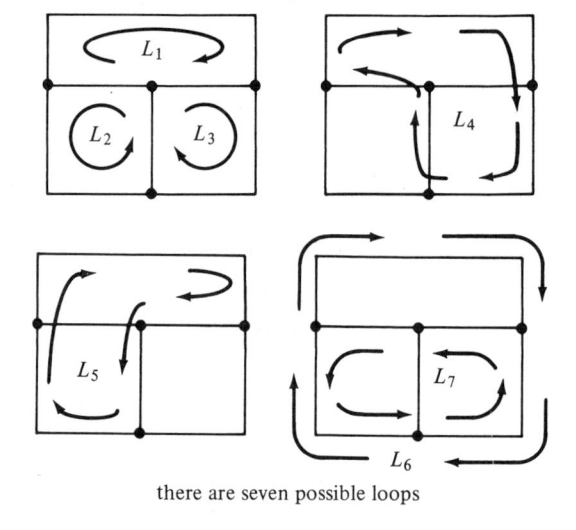

there are seven possible loops

7.2 LOOP ANALYSIS

b. What is wrong with these loops?

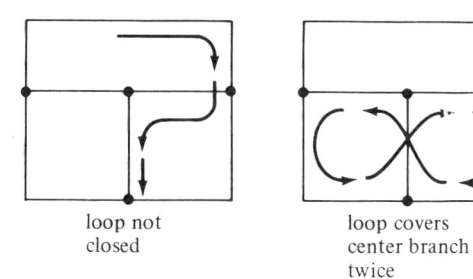

loop not closed

loop covers center branch twice

c. Indicate the nodes and count them.

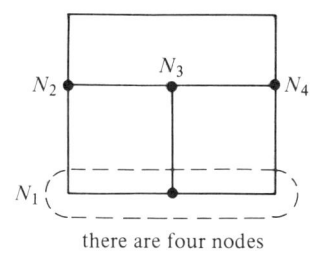

there are four nodes

Note how the entire bottom line is a single node. This point is clearly seen if the circuit is redrawn as follows:

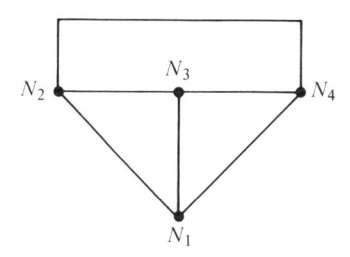

d. Indicate branches and count them.

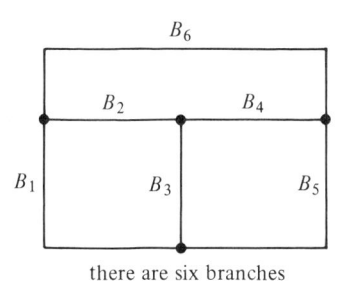

there are six branches

e. How many loop equations are needed to analyze the circuit?

$$\text{number of loop equations} = B - N + 1$$
$$= 6 - 4 + 1 = 3$$

Three loop equations are needed.

Here is how to write the equations and do loop analysis:

- Choose the $(B - N + 1)$ loops around which the equations will be written. Make sure that every branch is included in at least one loop.
- Draw a *loop current* flowing completely around each loop. Indicate the direction of flow with an arrowhead. These currents tell two things: first, they show voltages included in each loop equation, second, their direction indicates the voltage-drop polarity. Do not worry about choice of direction; if it is wrong, the answer is negative. Identify each current with a subscript such as I_3 or I_b.
- At first it might be helpful to write in voltage-drop polarity. Remember that current enters the positive end of a drop and leaves the negative.
- Go all the way around a loop summing all voltages in a loop equation. Travel in either direction, with the arrow or against it. Repeat for remaining loops.
- The result of these steps is a system of equations. Solve the system for the unknown currents.
- Check answers using known currents to calculate all voltage drops. Apply KVL to any loop. Repeat on other loops until satisfied with the results. Check currents at nodes with KCL.

EXAMPLE 7.6

Find the voltage drops in this circuit by loop analysis. (This problem was analyzed by SPR in Example 6.3.)

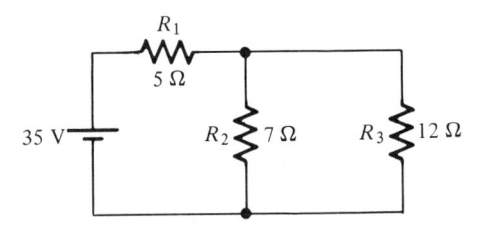

R_1 5 Ω, 35 V, R_2 7 Ω, R_3 12 Ω

There are three branches and two nodes. Therefore, two loop equations are needed $(B - N + 1 = 3 - 2 + 1 = 2)$. How many possible loops are there?

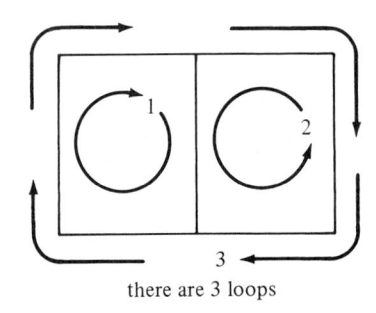

there are 3 loops

The two inner loops (1 and 2) include every branch and will be used for our analysis.

7.2 LOOP ANALYSIS

Redraw the circuit with loop currents added and labeled:

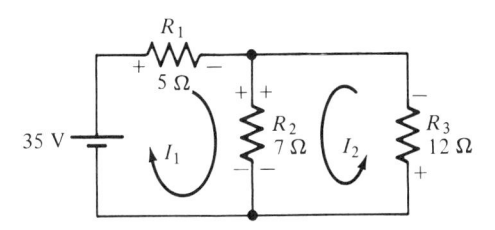

Voltage-drop polarities are shown. Note carefully that R_2 has two drops across it because both I_1 and I_2 pass through it.

To write the first loop equation, we start at the negative terminal of the 35-V battery and follow the arrow:

$$\underset{\uparrow}{\text{battery}}$$
$$-35 + \underset{\underset{R_1}{\text{across}}}{\underset{\text{drop}}{\uparrow}} 5I_1 + \underbrace{7I_1 + 7I_2}_{\substack{\text{two drops}\\ \text{across } R_2}} = 0$$

Combine terms:

$$12I_1 + 7I_2 = 35$$

The second equation starts at the bottom of R_3 and goes with the arrow:

$$12I_2 + \underbrace{7I_2 + 7I_1}_{\substack{\text{two drops}\\ \text{across } R_2}} = 0$$
$$\underset{\underset{R_3}{\text{across}}}{\underset{\text{drop}}{\uparrow}}$$

Combine terms and rewrite:

$$7I_1 + 19I_2 = 0$$

We now have a system of two equations in two unknowns:

$$\text{(a) } 12I_1 + 7I_2 = 35$$
$$\text{(b) } 7I_1 + 19I_2 = 0$$

Here is a quick test: If the equations are written correctly, the drop from I_2 in Equation (a) must be the same as the drop from I_1 in Equation (b). This always happens when analyzing circuits with linear, bilateral components. This system passes the test.

(Begin Substitution Technique Review)
Solving for I_1 by substitution: Solve Equation (b) for I_2:

$$19I_2 = -7I_1$$
$$\text{(c) } I_2 = \frac{-7}{19} I_1$$

Substitute Equation (c) in equation (a):

$$12I_1 + 7\left(\frac{-7I_1}{19}\right) = 35$$

$$12I_1 - \frac{49I_1}{19} = 35$$

$$I_1(12 - 2.6) = 35$$

$$I_1 = \frac{35}{9.4} = 3.7 \text{ A}$$

Substituting I_1 into Equation (b) and solving for I_2:

$$7(3.7) + 19I_2 = 0$$

$$I_2 = \frac{-25.9}{19} = -1.4 \text{ A}$$

(End Review)

The negative sign says that the choice of current direction in loop 2 was backwards. There are two ways to find the total current through R_2:

- Stay with the original choice of direction for I_2. Both I_1 and I_2 flow down through R_2. Therefore,

$$I_{R_2} = I_1 + I_2$$
$$= 3.7 + (-1.4)$$
$$= 2.3 \text{ A}$$

- Reverse the direction of I_2. It will then be $+1.4$ A. Now I_1 and I_2 flow in opposite directions through R_2; see the diagram:

$$I_{R_2} = I_1 - I_2$$
$$= 3.7 - (1.4)$$
$$= 2.34$$

It is important to understand these two approaches.
Calculating voltage drops:

$$V_{R_1} = I_1 R_1 = 3.7(5) = 18.5 \text{ V}$$
$$V_{R_2} = I_{R_2} R_2 = 2.3(7) = 16.1 \text{ V}$$
$$V_{R_3} = I_2 R_3 = 1.4(12) = 16.8 \text{ V}$$

With I_2 reversed from its original direction.
The SPR results for this problem in Chapter 6 were

$$V_{R_1} = 18.5 \text{ V} \qquad V_{R_2} = V_{R_3} = 16.5 \text{ V}$$

Differences are due to rounding. ■

7.2 LOOP ANALYSIS

Because of the choice of loops in Example 7.6, it was necessary to calculate both I_1 and I_2 in order to determine I_{R_2}. Here is an easier way. Choose either inner loop plus the outer loop. Now there is only one current through R_2 to calculate. In other words, a wise choice of loops reduces the work.

EXAMPLE 7.7

Find I_{R_2} in the circuit of Example 7.6.

Redrawing the circuit with different loops:

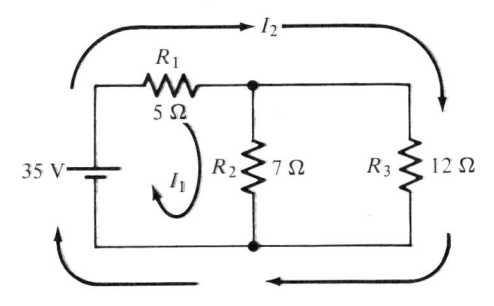

Note how R_2 has only I_1 through it now.
For loop 1:

$$-35 + 5I_1 + 5I_2 + 7I_1 = 0$$

or

$$12I_1 + 5I_2 = 35$$

For loop 2:

$$-35 + 5I_1 + 5I_2 + 12I_2 = 0$$

or

$$5I_1 + 17I_2 = 0$$

Our system of equations is

(a) $12I_1 + 5I_2 = 35$
(b) $5I_1 + 17I_2 = 0$

As a check, note the $5I_2$ in loop 1 and the $5I_1$ in loop 2.
(Begin Addition/Subtraction Technique Review)
Solving the system of equations for I_1:
Multiply Equation (a) by 17:

(c) $204I_1 + 85I_2 = 595$

Multiply Equation (b) by 5:

(d) $25I_1 + 85I_2 = 175$

The I_2 coefficients are now the same.
Subtract Equation (d) from (c):

$$179I_1 + 0I_2 = 420$$

Therefore,

$$I_1 = \frac{420}{179} = 2.4 \ A$$

(End Review)
Example 7.6 has two currents through R_2, for a total of 2.3 A. ∎

EXAMPLE 7.8

Determine the voltage across each resistance in this circuit:

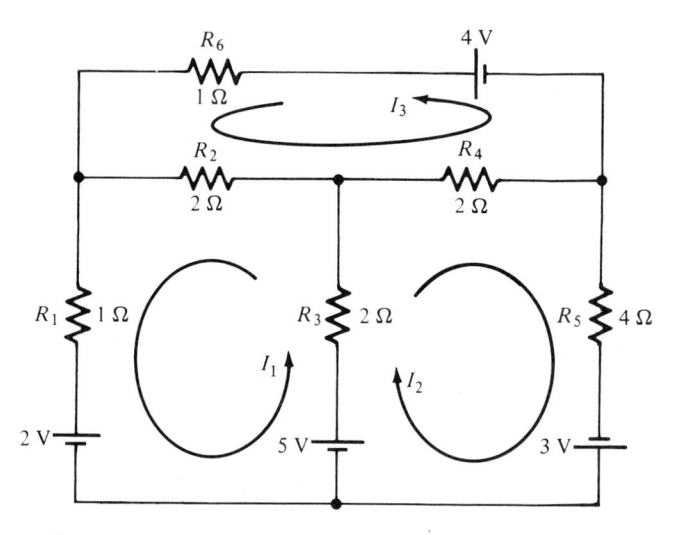

To calculate voltages, we need currents. Loop analysis will be used to find them.
There are six branches in the circuit and four nodes ($B - N + 1 = 6 - 4 + 1 = 3$); three equations are needed. The loops for these equations are seen in the schematic. Loop equation 1 starts at the bottom of the 5-V source and goes with the arrow:

$$-5 + 2I_1 + 2I_2 + 2I_1 - 2I_3 + I_1 + 2 = 0$$

positive because I_2 in same direction as I_1

negative because I_3 flows against I_1

Combine terms:

$$5I_1 + 2I_2 - 2I_3 = 3$$

Loop 2 starts at the same point as loop 1 and also follows its arrow:

$$-5 + 2I_1 + 2I_2 + 2I_2 + 2I_3 + 4I_2 - 3 = 0$$

positive because same direction as I_2

Combine terms:

$$2I_1 + 8I_2 + 2I_3 = 8$$

7.2 LOOP ANALYSIS

It is left as an exercise for the reader to determine where loop 3 begins and its direction:

$$-4 + I_3 + 2I_3 - 2I_1 + 2I_2 + 2I_3 = 0$$

Combine terms:

$$-2I_1 + 2I_2 + 5I_3 = 4$$

The three equations are

$$5I_1 + 2I_2 - 2I_3 = 3$$
$$2I_1 + 8I_2 + 2I_3 = 8$$
$$-2I_1 + 2I_2 + 5I_3 = 4$$

Test questions:
Is the drop from I_3 in loop 1 the same as the drop from I_1 in loop 3?
Is the drop of I_2 in loop 1 the same as that of I_1 in loop 2? And three in loop 2 versus two in loop 3?
Using determinants to solve for the unknowns:

$$I_1 = \frac{\begin{vmatrix} 3 & 2 & -2 \\ 8 & 8 & 2 \\ 4 & 2 & 5 \end{vmatrix}}{\begin{vmatrix} 5 & 2 & -2 \\ 2 & 8 & 2 \\ -2 & 2 & 5 \end{vmatrix}} = \frac{76}{112} = 0.7 \text{ A} \qquad I_2 = \frac{\begin{vmatrix} 5 & 3 & -2 \\ 2 & 8 & 2 \\ -2 & 4 & 5 \end{vmatrix}}{112} = \frac{70}{112} = 0.6 \text{ A}$$

$$I_3 = \frac{\begin{vmatrix} 5 & 2 & 3 \\ 2 & 8 & 8 \\ -2 & 2 & 4 \end{vmatrix}}{112} = \frac{92}{112} = 0.8 \text{ A}$$

From the schematic:

$$I_{R_1} = I_1 = 0.7 \text{ A}$$
$$V_{R_1} = I_1 R_1 = 0.7(1) = 0.7 \text{ V}$$
$$I_{R_2} = I_3 - I_1 \quad \text{since } I_3 > I_1$$
$$= 0.8 - 0.7$$
$$= 0.1 \text{ A}$$
$$V_{R_2} = I_{R_2} R_2 = 0.1(2) = 0.2 \text{ V}$$
$$I_{R_3} = I_1 + I_2 \quad \text{since both arrows in the same direction}$$
$$= 0.7 + 0.6 = 1.3 \text{ A}$$
$$V_{R_3} = I_{R_3} R_3 = 1.3(2) = 2.6 \text{ V}$$
$$I_{R_4} = I_2 + I_3 = 0.6 + 0.8 = 1.4 \text{ A}$$
$$V_{R_4} = I_{R_4} R_4 = 1.4(2) = 2.8 \text{ V}$$
$$I_{R_5} = I_2 = 0.6 \text{ A}$$
$$V_{R_5} = I_{R_5} R_5 = 0.6(4) = 2.5 \text{ V}$$
$$I_{R_6} = I_3 = 0.8 \text{ A}$$
$$V_{R6} = I_{R_6} R_6 = 0.8(1) = 0.8 \text{ V}$$

Redraw the circuit with voltages indicated:

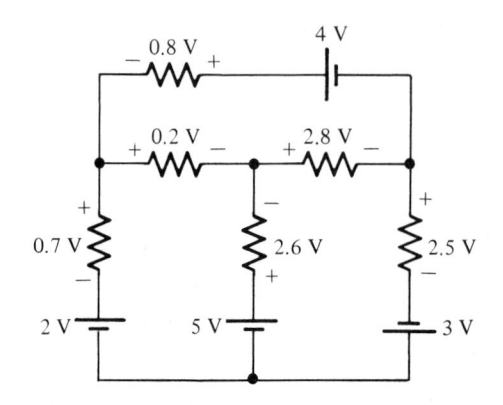

[Circuit diagram showing resistors with voltage drops: 0.8 V, 0.2 V, 2.8 V, 4 V, 0.7 V, 2.6 V, 2.5 V, and sources 2 V, 5 V, 3 V]

Check results by KVL around any loop. For example, start at the negative terminal of the 4-V source and go CCW around the outside:

$$-4 + 0.8 + 0.7 + 2 + 3 - 2.5 = 0 \text{ V} \qquad \blacksquare$$

Norton sources may be changed to their Thévenin equivalents to eliminate unnecessary loops. A current source with no parallel resistance cannot be converted. However, this often simplifies an analysis because the source current is known. These points are illustrated in the next example.

EXAMPLE 7.9

Find the current in R_3 by loop analysis.

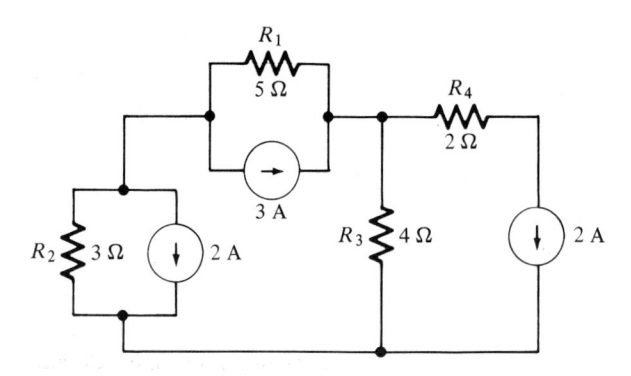

[Circuit diagram with $R_1 = 5\,\Omega$, $R_4 = 2\,\Omega$, 3 A source, 2 A source, $R_2 = 3\,\Omega$, $R_3 = 4\,\Omega$, 2 A source]

Convert Norton sources to their Thévenin equivalents:

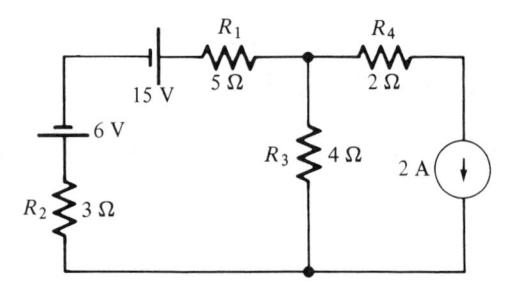

[Circuit diagram with 15 V source, $R_1 = 5\,\Omega$, $R_4 = 2\,\Omega$, 6 V source, $R_3 = 4\,\Omega$, 2 A source, $R_2 = 3\,\Omega$]

Combine Thévenin sources and draw loop currents:

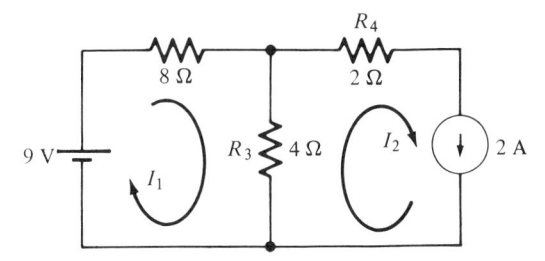

Write an equation around loop 1:
$$-9 + 8I_1 + 4I_1 - 4I_2 = 0$$
Combine terms:
$$12I_1 - 4I_2 = 9$$
But I_2 is known to equal 2 A; therefore,
$$12I_1 - 4(2) = 9$$
$$12I_1 = 9 + 8 = 17$$
Therefore,
$$I_1 = \frac{17}{12} = 1.4 \text{ A}$$

7.3 NODAL ANALYSIS

In nodal analysis, another powerful way of analyzing circuits, an unknown voltage is assumed to exist at each node in a circuit. These voltages cause currents to flow. The currents are added together in nodal equations and the equations solved for the unknown voltages.

How many nodal equations are needed to analyze a circuit? Topology says that $N - 1$ equations are required (where N = the number of nodes in the circuit).

The next example shows how node currents can be described in terms of node voltages.

EXAMPLE 7.10

What is the current leaving point A in each of the following?

In each case, the current leaving point A passes through a resistance. To describe that current with an Ohm's law statement, find the voltage across the resistance in terms of the voltage at either end, then divide this voltage by the resistance.

a.

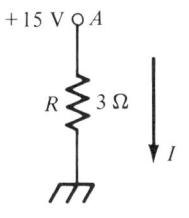

One end of the resistance, point A, is 15 V above some reference point. That point is ground. The other end of the resistance is at reference ground, or 0 V. The drop across the resistance is therefore 15 − 0 = 15 V. Current flows down from *A* to ground. Its value is 15/3 = 5 A.

b.

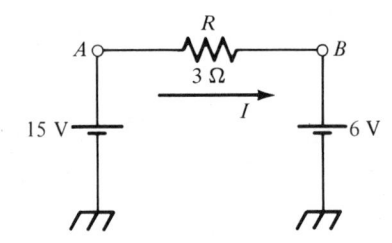

Figure: Circuit with 15 V source on left, connected through resistor R = 3 Ω between nodes A and B, with 6 V source on right. Current I flows from A to B.

Point *A* is 15 V above reference ground. Point *B* is only 6 V above ground. It follows that *A* must be 15 − 6 = 9 V above *B*. It also follows that this 9 V is the drop across *R*. Since *A* is at the higher potential, current flows from *A* to *B*. Its value is

$$I = \frac{V_a - V_b}{R} = \frac{15 - 6}{3} = \frac{9}{3} = 3 \text{ A}$$

Be sure to understand how the drop across *R* and the current through it were obtained. Let's review the steps:

- Find the voltage at each end of *R* with respect to the reference point. This will be one of the nodes in the circuit.
- Calculate the drop across *R* by subtracting the smaller "end" voltage from the larger.
- Divide the drop by *R* to get the current, which is what is needed for the nodal equation.

Watch for these steps.

c.

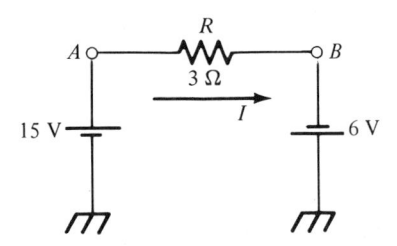

Figure: Circuit with 15 V source on left, resistor R = 3 Ω between nodes A and B, and 6 V source on right with reversed polarity.

Point *A* is still 15 V above ground. But point *B* is now 6 V below ground. Therefore, *A* is 15 − (−6) = 21 V above *B*.
The current from *A* to *B* is

$$I = \frac{V_a - V_b}{R} = \frac{15 - (-6)}{3}$$

$$= \frac{21}{3} = 7 \text{ A}$$

Next, we let V_a be an unknown, the way it will be in the nodal equations. All unknown voltages are assumed positive with respect to the reference point. If that assumption is wrong, the node voltage comes out negative.

7.3 NODAL ANALYSIS

d.

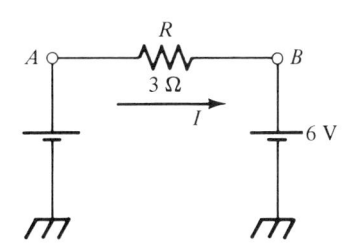

If we assume current from point A to point B, V_a must be more positive than V_b. The difference between them is the drop across R. Therefore,

$$I = \frac{V_a - V_b}{R} = \frac{V_a - 6}{3} \quad \text{amperes}$$

e.

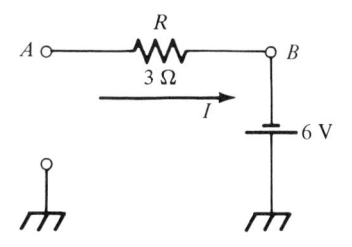

Point B is now 6 V below ground. Point A is V_a volts above ground. The current is

$$I = \frac{V_a - V_b}{R}$$
$$= \frac{V_a - (-6)}{3}$$
$$= \frac{V_a + 6}{3} \quad \text{amperes}$$

f. Suppose that the potential at both ends of the resistance is unknown.

If we assume current flowing from point A to B, V_a must be greater than V_b. The difference between them is the voltage drop across R:

$$I = \frac{V_a - V_b}{R} = \frac{V_a - V_b}{3} \quad \text{amperes}$$

This is how currents are described in terms of unknown voltages in nodal equations.

On the other hand, if $V_b > V_a$, current flows from B to A and

$$I = \frac{V_b - V_a}{R} = \frac{V_b - V_a}{3} \quad \text{amperes}$$

g. Consider this last example.

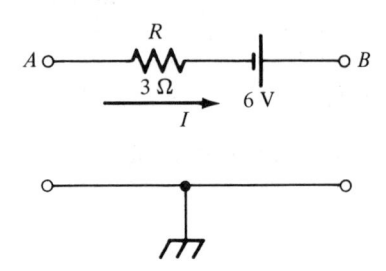

Remember: V_a and V_b are assumed positive with respect to the reference point, ground. Here are the voltages in the circuit if we say that current is from A to B:

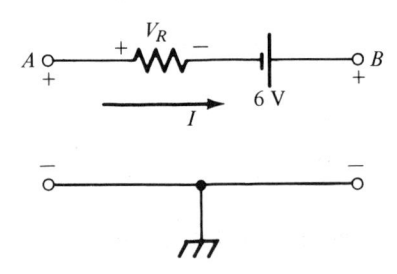

We need V_R in order to divide it by R to get I. How do we find an unknown voltage? As always, write a loop equation:

$$-V_a + V_R - 6 + V_b = 0$$

or

$$V_R = V_a + 6 - V_b$$

Therefore,

$$I = \frac{V_R}{R} = \frac{V_a + 6 - V_b}{3} \quad \text{amperes} \qquad \blacksquare$$

Here are the steps for nodal analysis:

- Choose one node in the circuit to be the zero-potential or reference node. This can be any node but is usually the bottom one. Identify the node in some way, with a "O" perhaps or a ground symbol. All unknown node voltages are treated as being positive with respect to this reference.
- Label the unknown voltage at each of the remaining nodes: for example, V_a or V_4.
- Write a nodal equation summing currents at each of the $N - 1$ remaining nodes. Do not worry about the direction of unknown currents. Let them all leave a node. Those actually going the opposite direction will have negative answers.

7.3 NODAL ANALYSIS

- Solve the system of equations for unknown node voltages.
- Use node voltages to find the drop across each resistance.
- Check your work by applying KVL around loops. Calculate currents and check KCL at nodes.

The first problem done by nodal analysis is the same circuit analyzed by loop analysis in Example 7.6 and by SPR in Example 6.3.

EXAMPLE 7.11

Find the current in each resistance by nodal analysis.

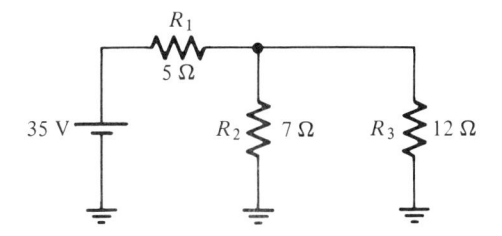

Of the two nodes, the bottom one, ground, is chosen as the reference node. A single nodal equation will be written for the node at which R_1, R_2, and R_3 meet. The unknown in the equation is the voltage, V, of the node with respect to reference.

Since there are three branches meeting at the upper node, there are three currents in the equation. Treat all three as leaving. Negative signs indicate currents going the opposite way. The three currents are:

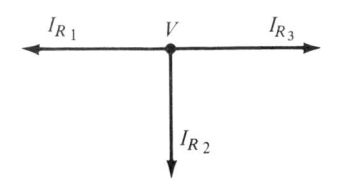

$$I_{R_1} = \frac{V - 35}{R_1}$$

$$I_{R_2} = \frac{V}{R_2}$$

and

$$I_{R_3} = \frac{V}{R_3}$$

Put these into a nodal equation:

$$\frac{V - 35}{5} + \frac{V}{7} + \frac{V}{12} = 0$$

Multiply both sides of the equation by the lowest common denominator (5 × 7 × 12), to eliminate denominators:

$$(5 \times 7 \times 12)\left(\frac{V-35}{5} + \frac{V}{7} + \frac{V}{12}\right) = (5 \times 7 \times 12) \times 0$$
$$(7 \times 12)(V-35) + (5 \times 12)V + (5 \times 7)V = 0$$
$$84(V-35) + 60V + 35V = 0$$
$$84V - 2940 + 60V + 35V = 0$$

Combine and rewrite:

$$179V = 2940$$
$$V = \frac{2940}{179}$$
$$= 16.4 \text{ V}$$

Calculate the currents:

$$I_{R_1} = \frac{V-35}{R_1} = \frac{16.4-35}{5} = \frac{-18.6}{5}$$
$$= -3.7 \text{ A}$$

The negative sign (and common sense) says that the current is entering the node, not leaving it.

$$I_{R_2} = \frac{V}{R_2} = \frac{16.4}{7}$$
$$= 2.3 \text{ A}$$
$$I_{R_3} = \frac{V}{R_3} = \frac{16.4}{12}$$
$$= 1.4 \text{ A}$$

These are the same results that we got using SPR in Example 6.3 and loop analysis in Example 7.6. ∎

Notice that it took one nodal equation to analyze Example 7.11; loop analysis needed two. Nodal analysis often gets results with less math than loop analysis.

EXAMPLE 7.12

Find I_{R_3} by nodal analysis.

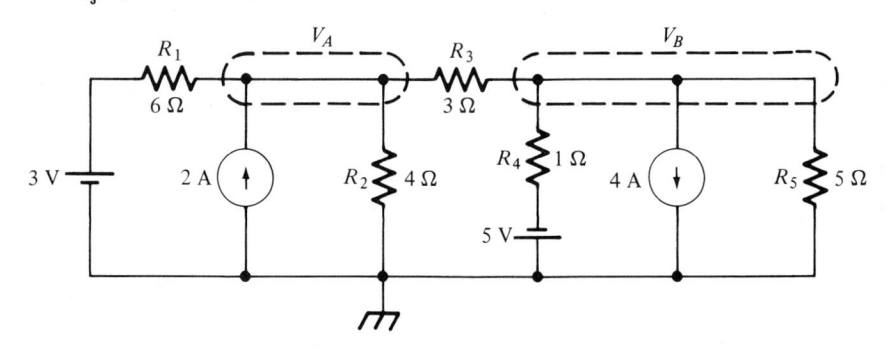

The dashed line at V_a shows that this entire area is node A; similarly for node B.

7.3 NODAL ANALYSIS

Write the equation at A:

$$\text{(a)} \quad \frac{V_A - 3}{6} - 2 + \frac{V_A}{4} + \frac{V_A - V_B}{3} = 0$$

The 2-A current is negative because it enters the node.
The equation at B is

$$\text{(b)} \quad \frac{V_B - V_A}{3} + \frac{V_B + 5}{1} + 4 + \frac{V_B}{5} = 0$$

The 4-A current is positive because it is leaving the node. Multiply equation (a) by 12:

$$2(V_A - 3) - 24 + 3V_A + 4(V_A - V_B) = 0$$

Combine:

$$9V_A - 4V_B = 30$$

Multiply equation (b) by 15:

$$5(V_B - V_A) + 15(V_B + 5) + 60 + 3V_B = 0$$

Combine:

$$-5V_A + 23V_B = -135$$

Finally, the two equations are

$$\text{(c)} \quad 9V_A - 4V_B = 30$$
$$\text{(d)} \quad -5V_A + 23V_B = -135$$

Solve for V_a by Cramer's rule:

$$V_A = \frac{\begin{vmatrix} 30 & -4 \\ -135 & 23 \end{vmatrix}}{\begin{vmatrix} 9 & -4 \\ -5 & 23 \end{vmatrix}} = \frac{690 - 540}{207 - 20} = \frac{150}{187} = 0.8 \text{ V}$$

Substitute V_a in equation (c) to get V_b:

$$9(0.8) - 4V_B = 30$$
$$-4V_B = 30 - 7.2 = 22.8$$
$$V_B = \frac{22.8}{-4} = -5.7 \text{ V}$$

Find I_{R_3}:

$$\underset{0.8 \text{ V}}{\overset{V_A}{\circ}} \underset{3\,\Omega}{\overset{R_3}{-\!\!\!\text{W}\!\!\!-}} \underset{-5.7 \text{ V}}{\overset{V_B}{\circ}}$$

$$I_{R_3} = \frac{V_{R_3}}{R_3} = \frac{0.8 - (-5.7)}{3}$$

$$= \frac{6.5}{3} = 2.2 \text{ A, left to right}$$

EXAMPLE 7.13

Use nodal analysis to determine the voltage across R_3.

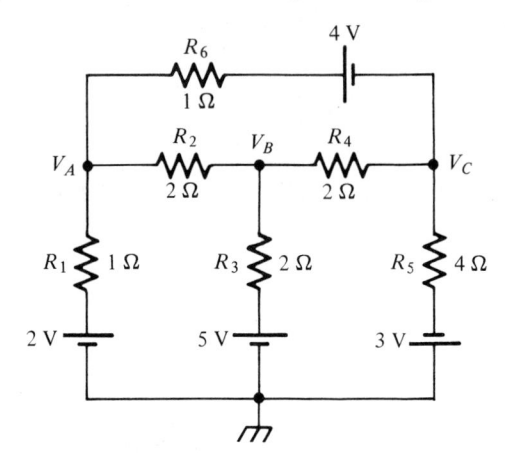

At node A:

$$\text{(a)} \quad \frac{V_A - 2}{1} + \frac{V_A - V_B}{2} + \frac{V_A - 4 - V_C}{1} = 0$$

At node B:

$$\text{(b)} \quad \frac{V_B - V_A}{2} + \frac{V_B - 5}{2} + \frac{V_B - V_C}{2} = 0$$

At node C:

$$\text{(c)} \quad \frac{V_C + 4 - V_A}{1} + \frac{V_C - V_B}{2} + \frac{V_C + 3}{4} = 0$$

The voltage was found to be 2.6 V by loop analysis in Example 7.8. Writing the equations:

Multiply equation (a) by 2:

$$2(V_A - 2) + (V_A - V_B) + 2(V_A - 4 - V_C) = 0$$

Combine:

$$5V_A - V_B - 2V_C = 12$$

Multiply equation (b) by 2:

$$(V_B - V_A) + (V_B - 5) + (V_B - V_C) = 0$$

Combine:

$$-V_A + 3V_B - V_C = 5$$

Multiply equation (c) by 4:

$$4(V_C + 4 - V_A) + 2(V_C - V_B) + (V_C + 3) = 0$$

Combine:

$$-4V_A - 2V_B + 7V_C = -19$$

7.3 NODAL ANALYSIS

The three equations, ready for solution, are

$$\text{(d)} \quad 5V_a - V_b - 2V_c = 12$$
$$\text{(e)} \quad -V_a + 3V_b - V_c = 5$$
$$\text{(f)} \quad -4V_a - 2V_b + 7V_c = -19$$

Solve for V_b by determinants:

$$V_b = \frac{\begin{vmatrix} 5 & 12 & -2 \\ -1 & 5 & -1 \\ -4 & -19 & 7 \end{vmatrix}}{\begin{vmatrix} 5 & -1 & -2 \\ -1 & 3 & -1 \\ -4 & -2 & 7 \end{vmatrix}} = \frac{134}{56} = 2.4 \text{ V}$$

Draw the circuit for R_3 and assume polarity for V_{R_3}:

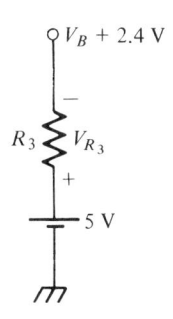

Write a loop equation:

$$-5 + V_{R_3} + 2.4 = 0$$

Therefore,

$$V_{R_3} = 2.6 \text{ V}$$

the same results as those obtained in Example 7.8.

EXAMPLE 7.14

Find the value of R_1 needed to hold the current into the transistor, I_{in}, to 1 mA. The voltage across the transistor input terminals is 0.7 V.

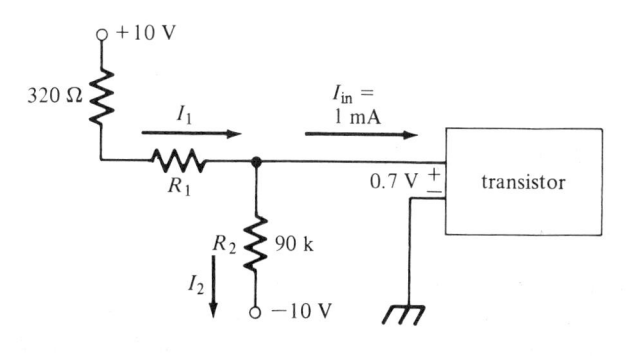

One equation will be written for the node at the top of R_2. Note that the node voltage is known in this case. It is the 0.7 V at the transistor input. The node equation is

$$\text{(a)} \quad -I_1 + I_2 + I_{in} = 0$$

Therefore,

$$\text{(b)} \quad I_1 = I_2 + I_{in}$$

I_1 is the current through R_1 and the 320-Ω resistor. To find it, we need the voltage across the series combination of R_1 and the 320 Ω. Writing a loop equation:

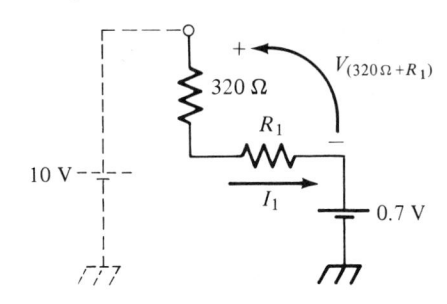

$$-10 + V_{(320\Omega + R_1)} + 0.7 = 0$$
$$V_{(320\Omega + R_1)} = 10 - 0.7 = 9.3 \text{ V} = I_1(320 \text{ Ω} + R_1)$$

Therefore,

$$I_1 = \frac{9.3}{320 + R_1}$$

For I_2, the voltage across the 90 k is needed. The loop equation is

$$-0.7 + I_2 R_2 - 10 = 0$$
$$I_2 R_2 = 10.7$$
$$I_2 = \frac{10.7}{R_2} = \frac{10.7}{90 \text{ k}} = 0.12 \text{ mA}$$

Substitute for $I_1, I_2,$ and I_{in} in equation (b):

$$\frac{9.3}{320 + R_1} = 0.12 \text{ mA} + 1 \text{ mA} = 1.12 \text{ mA}$$

Solve for R_1:

$$\frac{9.3}{1.12 \text{ mA}} = 320 + R_1$$
$$8.3 \text{ k} = 320 + R_1$$
$$R_1 = 8300 - 320 \approx 8 \text{ k}$$

∎

7.4 THÉVENIN ANALYSIS

Suppose that we did the following things to the circuit in Figure 7.2, part (a).

1. Measure the open-circuit voltage, V_{oc}, between terminals A and B.
2. Turn off all sources in the circuit just as we did in superposition. That is, short-circuit voltage sources, open-circuit current sources [see Figure 7.2, part (b)].
3. With all sources off, determine the equivalent resistance seen looking into terminals A and B.

7.4 THÉVENIN ANALYSIS

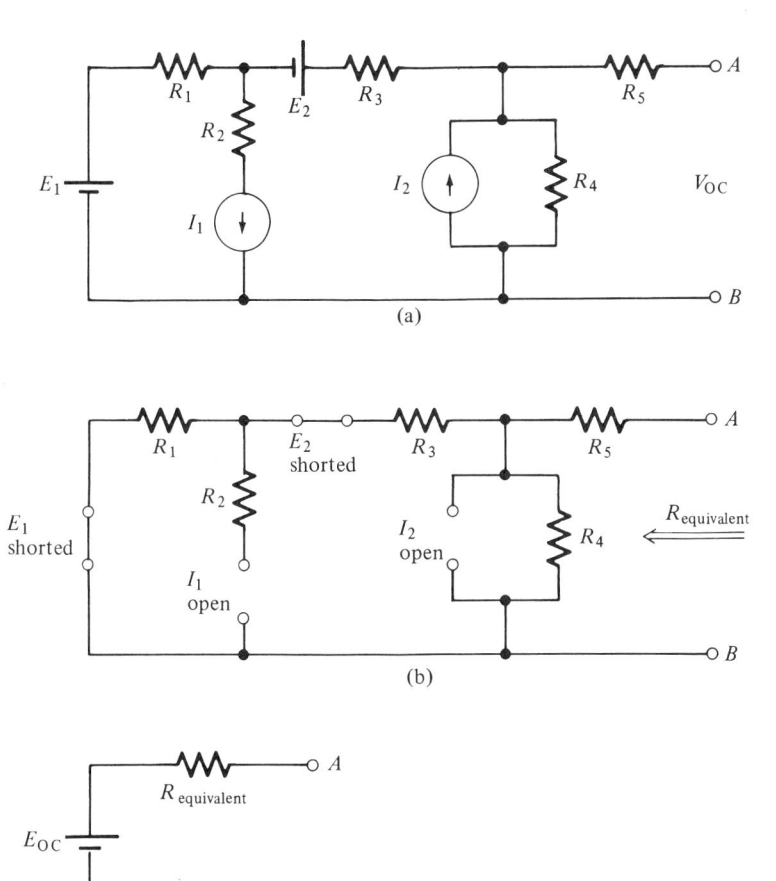

FIGURE 7.2 Finding the Thévenin Equivalent of a Circuit.

Next, we build a new circuit, one having just two parts [see Figure 7.2, part (c)]. Its source voltage we set equal to the V_{oc} measured in step 1. Its resistance is set to the $R_{equivalent}$ of step 3.

Now, here is the interesting part. Connect any component to terminals A and B of circuit (c), and it will have the same voltage, current, and power as it does when connected to the original circuit, (a). In other words, *the component cannot tell whether it is wired to circuit (a) or (c); the two circuits are equivalent.*

Notice that the circuit of Figure 7.2, part (c), is the familiar Thévenin source, E_{oc}, in series with $R_{internal}$, studied in Chapter 4. Circuit (c) is said to be the *Thévenin equivalent* of circuit (a).

The fact that a circuit with two parts can be equivalent to a much more complicated circuit is described in one of the most important ideas in electricity, *Thévenin's theorem*. It says:

> When looking into a set of terminals at a circuit made up of sources and linear, bilateral components, the entire circuit can be replaced by a source in series with a resistance if:
>
> The source voltage equals the open-circuit voltage across the terminals, and
>
> The resistance equals the resistance seen looking into the terminals with all sources turned off.

Every student of electronics must be able to look into a pair of terminals and determine the two parts of the Thévenin equivalent circuit seen there. Learning to do this is one of the most important skills acquired in circuit analysis.

Let's review the steps:

- To find the Thévenin source voltage, E_{oc}: Calculate the open-circuit voltage between terminals using any of the methods of analysis studied so far.
- To find the Thévenin internal resistance, R_{int}: Turn off all sources just as we did in superposition. Look into the terminals and calculate the equivalent resistance.
- Draw the Thévenin equivalent circuit, E_{oc}, in series with R_{int}.

EXAMPLE 7.15

Théveninize this circuit, then connect the load and determine I_{load}.

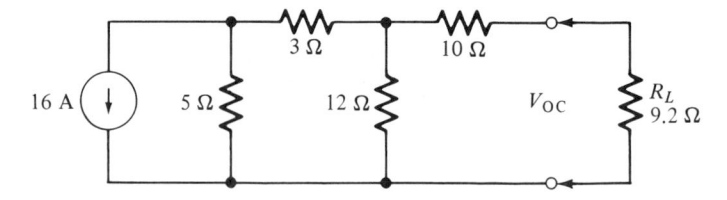

a. Finding V_{oc}.

The open-circuit terminal voltage is the drop across the 12-Ω resistance (there is no current through the 10 Ω). Use current division to determine how much of I_{total} passes through the 12 Ω:

$$I \text{ through } 12\,\Omega = \frac{16(5)}{5 + (12 + 3)}$$

$$= \frac{80}{20} = 4 \text{ A}$$

$$V_{oc} = 4(12) = 48 \text{ V}$$

b. Find the Thévenin resistance, R_{int}.

With the source turned off (opened), the circuit is

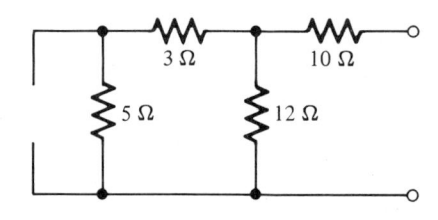

$$R_{int} = 10 + [12 \text{ in parallel with } (3 + 5)]$$

$$= 10 + \frac{12(8)}{12 + 8} = 10 + 4.8$$

$$= 14.8\,\Omega$$

7.4 THÉVENIN ANALYSIS

c. Draw the circuit:

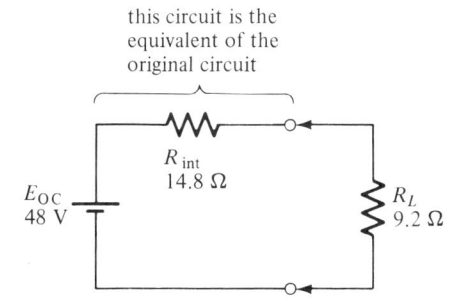

this circuit is the equivalent of the original circuit

E_{OC} 48 V, R_{int} 14.8 Ω, R_L 9.2 Ω

d. Find the current through an R_{load} of 9.2 Ω.

$$I_{load} = \frac{E_{oc}}{R_{int} + R_{load}}$$
$$= \frac{48}{14.8 + 9.2}$$
$$= 2 \text{ A}$$

The Thévenin equivalent circuit concept is a powerful tool for analyzing complicated circuits. Consider what happens if we were to cut a component out of a circuit. A gap or hole is left behind. The ends of the wires on either side of the gap can be considered terminals. It could be said that the component was originally connected to these terminals.

What do we see looking into these terminals? An open-circuit voltage and an equivalent resistance. In other words, behind the two terminals is the Thévenin equivalent of the circuit that drove the component we removed. Find it and the component current, voltage and power are easily determined by analyzing a simple series circuit.

EXAMPLE 7.16

Find the current through R_3 by Théveninizing the circuit driving R_3.

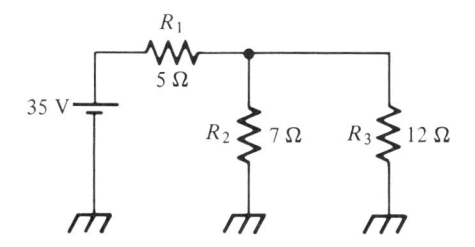

35 V, R_1 5 Ω, R_2 7 Ω, R_3 12 Ω

Note: This circuit was analyzed by SPR in Example 6.3, by loop analysis in Example 7.6, and by nodal analysis in Example 7.11. In all cases, the answer has been approximately 1.4 A.

Step 1: Divorce R_3 from the circuit:

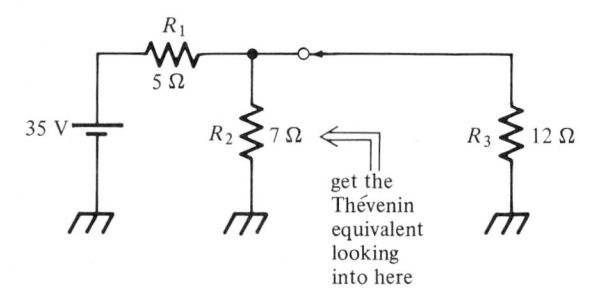

Step 2: Calculate V_{oc} between the terminals created when R_3 was removed:

$$V_{oc} = V_{R_2} = \frac{35(7)}{5+7} = 20.4 \text{ V}$$

Step 3: Find R_{int}: With the source turned off (shorted),

$$R_{int} = \frac{5(7)}{5+7} = \frac{35}{12} = 2.9 \text{ }\Omega$$

Step 4: Draw the Thévenin circuit seen by R_3. Connect R_3 to the circuit and calculate I_{R_3}:

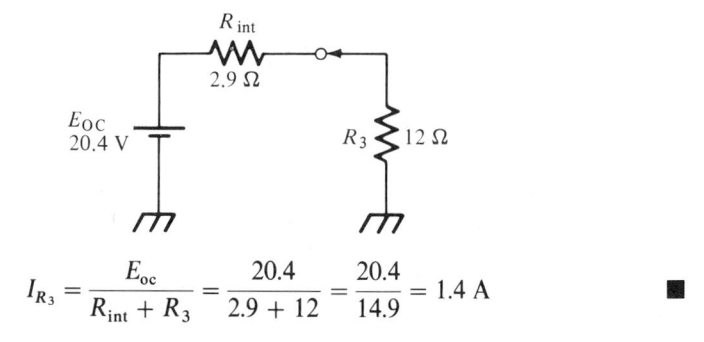

$$I_{R_3} = \frac{E_{oc}}{R_{int} + R_3} = \frac{20.4}{2.9 + 12} = \frac{20.4}{14.9} = 1.4 \text{ A}$$ ∎

EXAMPLE 7.17

Find the Thévenin equivalent of the circuit seen by R_4, then calculate I_{R_4}. What value of R_4 draws maximum power from the original circuit?

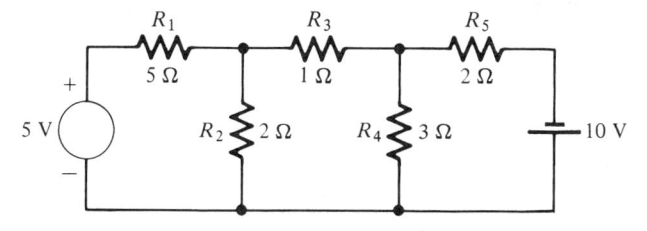

Step 1: Divorce R_4 from this circuit. This leaves:

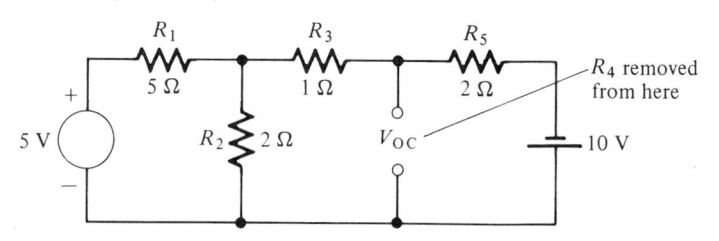

7.4 THÉVENIN ANALYSIS

Step 2: Find the open-circuit voltage, V_{oc}. Finding V_{oc} is the point that troubles students the most in Thévenin analysis. Chapter 4 said: To find an unknown voltage, write a loop equation containing that voltage, then solve for the unknown. There are several loops containing V_{oc}, for example, the loop around R_5 and the 10-V source, or the one around R_3, R_1, and the 5-V source. Since there is no one "right" loop, we choose the one around V_{oc}, R_5, and the 10-V source. But to use this loop, the drop across R_5 must be known. How are we to find V_{R_5}?

Forget for the moment that we are in the middle of a Thévenin analysis. Ask yourself: "How can I find V_{R_5}?" Study these possibilities (remember that R_4 is gone):

- Two loop equations—solve for I_{R_5}.
- Superposition to get I_{R_5}.
- A single nodal equation would give the voltage at the junction of R_1, R_2, and R_3. Subtract the 10 V from this voltage to get the drop across R_3 and R_5. Use voltage division to find V_{R_5}.
- Théveninize the 5 V, R_1, and R_2 at the left. Calculate I_{R_5}.

For no special reason, superposition will be used.
Find I from the 5-V source (10-V source shorted):

$$R_{\text{total}} = 5 + 2 \| (1 + 2) = 5 + \frac{2(3)}{2 + 3}$$

$$= 5 + \frac{6}{5}$$

$$= 6.2 \, \Omega$$

$$I_{\text{total}} = \frac{E}{R_{\text{total}}}$$

$$= \frac{5}{6.2}$$

$$= 0.81 \, \text{A}$$

Finding I_{R_5} by current division:

$$I_{R_5} = \frac{0.81(2)}{2 + 3}$$

$$= \frac{1.6}{5}$$

$$= 0.32 \, \text{A to the right}$$

Finding I from the 10-V source (5-V source shorted):

$$R_{\text{total}} = 2 + 1 + 2 \| 5 = 2 + 1 + \frac{2(5)}{2 + 5}$$

$$= 3 + \frac{10}{7}$$

$$= 4.4 \, \Omega$$

$$I_{\text{total}} = \frac{E}{R_{\text{total}}} = I_{R_5} \text{ for 10-V source}$$

$$= \frac{10}{4.4}$$

$$= 2.3 \, \text{A to the right}$$

$$I_{R_5} = \sum I = 0.32 + 2.3$$
$$= 2.6 \text{ A, left to right}$$
$$V_{R_5} = I_{R_5} R_5 = 2.6(2) = 5.2 \text{ V}$$

Now that V_{R_5} is known, it is placed in the V_{oc}, V_{R_5}, 10-V source loop. The loop is

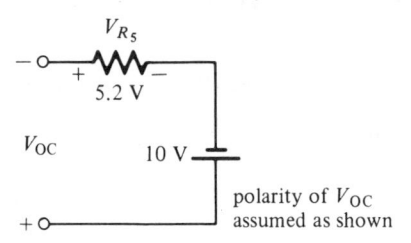

[circuit diagram showing $V_{R_5} = 5.2$ V, 10 V source, V_{OC} terminals; polarity of V_{OC} assumed as shown]

Write the loop equation to find V_{oc}:

$$V_{oc} + 5.2 - 10 = 0$$
$$V_{oc} = 10 - 5.2 = 4.8 \text{ V}$$

Step 3: Next, turn off all sources. Look back into the terminals and calculate the Thévenin resistance:

$$R_{int} = 2 \| (1 + 5 \| 2)$$

resistance to right of R_4 terminals ↗
resistance to the left of R_4 terminals

$$= 2 \| \left(1 + \frac{5(2)}{5+2}\right) = 2 \| \left(1 + \frac{10}{7}\right)$$
$$= 2 \| 2.4 = \frac{2(2.4)}{2 + 2.4}$$
$$= \frac{4.8}{4.4}$$
$$= 1.1 \, \Omega$$

Step 4: Draw the Thévenin equivalent of the circuit seen by R_4:

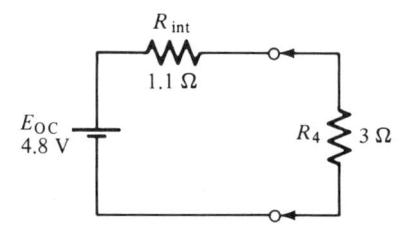

[circuit: $R_{int} = 1.1\,\Omega$, $E_{OC} = 4.8$ V, $R_4 = 3\,\Omega$]

Step 5: Calculate I_{R_4} and determine R_4 for maximum power transfer:

$$I_{R_4} = \frac{E_{oc}}{R_{int} + R_4} = \frac{4.8}{1.1 + 3} = 1.2 \text{ A}$$

Suppose that R_4 changed values. Using loop or nodal techniques, each I_{R_4} requires a separate analysis. But once the Thévenin equivalent is calculated, the currents are determined with simple series circuit/Ohm's law calculations.

For maximum power transfer, R_4 must have the same resistance as R_{int} or, $1.1\,\Omega$. ∎

7.4 THÉVENIN ANALYSIS

EXAMPLE 7.18

Find the current through R_3 by Théveninizing the circuit driving R_3.

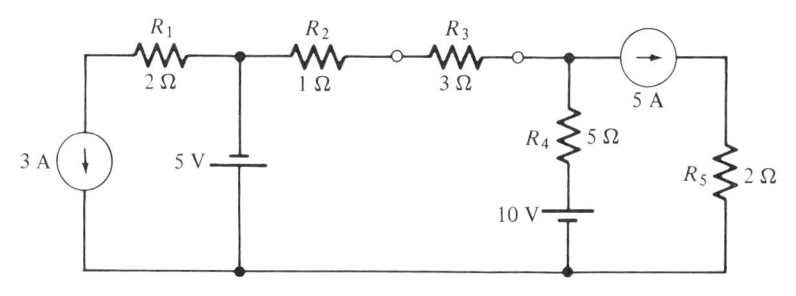

Step 1: Remove R_3. The circuit to be Théveninized is

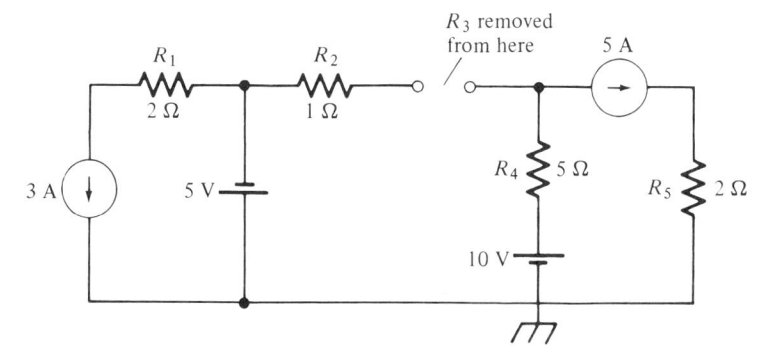

Step 2: Find the Thévenin equivalent voltage E_{oc}:
Ground the bottom line in the circuit to show that it is being used as a 0-V reference. The left terminal of V_{oc} is 5 V below ground because of the battery. Note that R_2 has no affect on V_{oc} since there is no current through it. Finding the voltage on the right terminal of V_{oc} is a bit harder. Drawing the circuit to the right of the V_{oc} terminals:

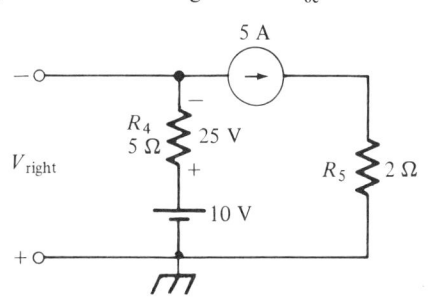

R_4 has the 5 A of the current source through it.
Therefore,

$$V_{R_4} = 5(5) = 25 \text{ V}$$

Write a loop equation to find the voltage, V_{right} (assuming the upper terminal is negative with respect to ground):

$$V_{\text{right}} - 25 + 10 = 0$$
$$V_{\text{right}} = 15 \text{ V}$$

The right-hand terminal of V_{oc} is thus 15 V below ground. Therefore, it is 10 V below the

left terminal, and
$$E_{oc} = 10 \text{ V}$$

Step 3: Find the Thévenin resistance. With all sources off, we have:

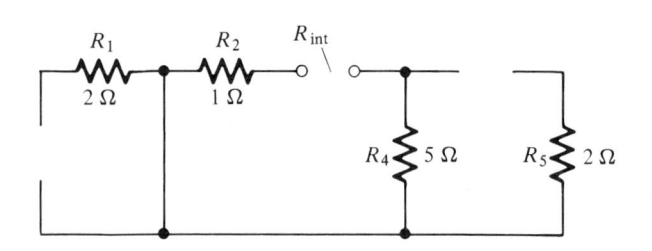

$$R_{int} = R_2 \text{ in series with } R_4$$
$$= 1 + 5$$
$$= 6 \, \Omega$$

Step 4: Draw the equivalent circuit:

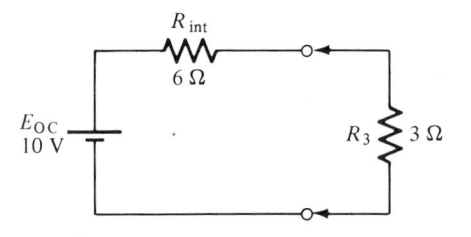

Finding I_{R_3}:

$$I_{R_3} = \frac{E_{oc}}{R_{int} + R_3}$$
$$= \frac{10}{6 + 3} = \frac{10}{9} = 1.1 \text{ A}$$ ∎

7.5 NORTON ANALYSIS

Norton's theorem is worded about the same as Thévenin's:

> When looking into a pair of terminals at a circuit made up of sources and linear, bilateral components, the entire circuit can be replaced by an ideal current source in shunt with a resistance if:
> The source current equals the current through a short between the terminals, and
> The resistance equals the resistance seen looking into the terminals with all sources turned off.

Circuits are analyzed by the Norton method just as they were by the Thévenin except that $I_{\text{short circuit}}$ is found rather than $V_{\text{open circuit}}$.

Here is Example 7.16 again, done by Norton analysis. Study carefully the similarities and differences between the two techniques.

7.5 NORTON ANALYSIS

EXAMPLE 7.19

Find the current through R_3 by getting the Norton equivalent of the circuit driving R_3.

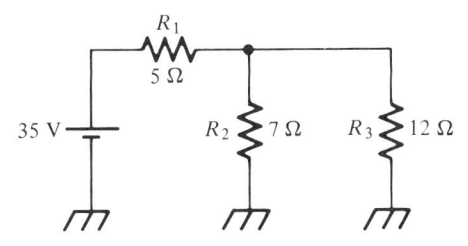

This circuit was done by SPR in Example 6.3, loop analysis in Example 7.6, nodal analysis in Example 7.11, and by Thévenin in Example 7.16.

Step 1: Divorce R_3 from the circuit.

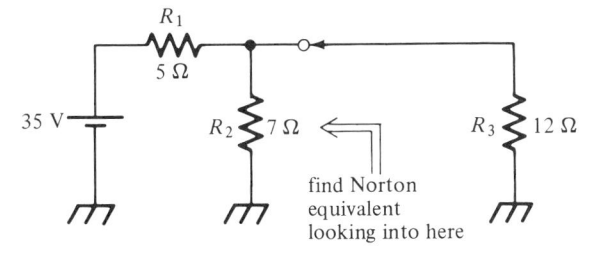

Step 2: Short the terminals and solve for I_{sc}.

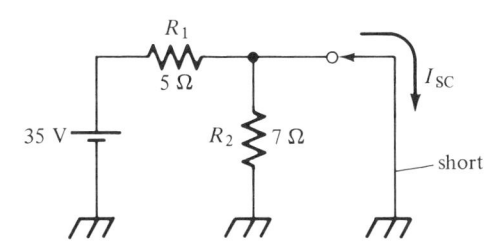

R_2 is shorted out.

$$I_{sc} = \frac{E_s}{R_1}$$
$$= \frac{35}{5}$$
$$= 7 \text{ A}$$

Step 3: Remove the short. Determine $R_{internal}$.

$$R_{int} = 7\,\Omega \| 5\,\Omega$$
$$= \frac{7(5)}{7+5}$$
$$= \frac{35}{12}$$
$$= 2.9\,\Omega$$

Step 4: Draw the Norton equivalent circuit seen by R_3. Connect R_3 to it. Calculate I_{R_3}.

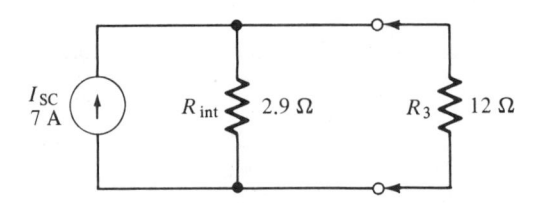

By current division,

$$I_{R_3} = \frac{I_{sc} R_{int}}{R_{int} + R_3}$$

$$= \frac{7(2.9)}{2.9 + 12}$$

$$= \frac{20.3}{14.9}$$

$$= 1.4 \text{ A}$$ ∎

EXAMPLE 7.20

Find the current through the 3-Ω resistance by Norton analysis.

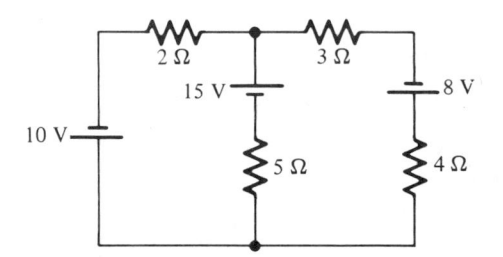

The 3 Ω is removed, a pair of terminals formed, and a short placed between them. We must solve for the current through this short. How? For no special reason, loop analysis will be used.

Draw the circuit with the resistance removed, a short placed between the terminals, and loop currents indicated:

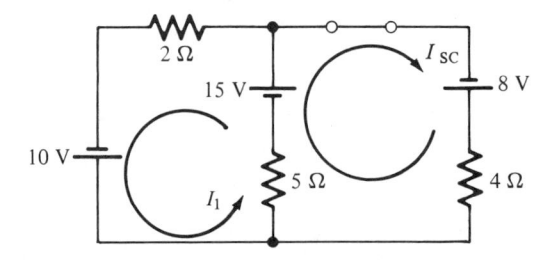

Write loop equations:

$$-10 + 5I_1 + 5I_{sc} - 15 + 2I_1 = 0$$
$$5I_1 + 5I_{sc} - 15 - 8 + 4I_{sc} = 0$$

7.5 NORTON ANALYSIS

Combine and rewrite:

$$7I_1 + 5I_{sc} = 25$$
$$5I_1 + 9I_{sc} = 23$$

Solve for I_{sc}:

$$I_{sc} = \frac{\begin{vmatrix} 7 & 25 \\ 5 & 23 \end{vmatrix}}{\begin{vmatrix} 7 & 5 \\ 5 & 9 \end{vmatrix}} = \frac{36}{38} = 0.95 \approx 1 \text{ A}$$

Remove the short (do not forget to do this!) and calculate $R_{internal}$:

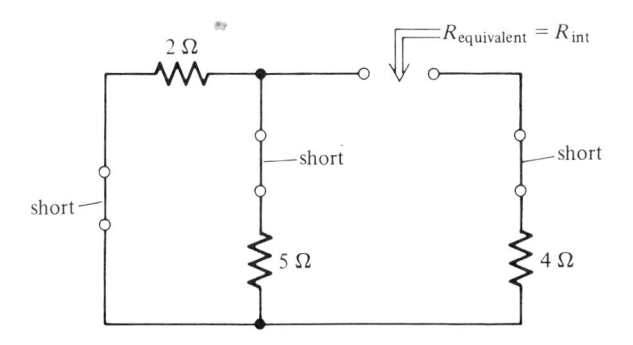

$$R_{int} = 4 + 2 \| 5$$
$$= 4 + \frac{2(5)}{2 + 5}$$
$$= 5.4 \; \Omega$$

Draw the Norton equivalent, connect the 3-Ω resistance, and calculate the current through it.

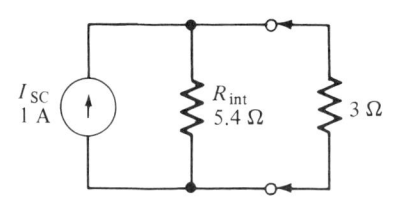

By current division

$$I_{3\Omega} = \frac{I_{sc} R_{int}}{R_{int} + 3}$$
$$= \frac{1(5.4)}{5.4 + 3}$$
$$= \frac{5.4}{8.4}$$
$$= 0.64 \text{ A}$$

EXAMPLE 7.21

Find the Norton equivalent of the circuit driving R_3. Calculate I_{R_3}.

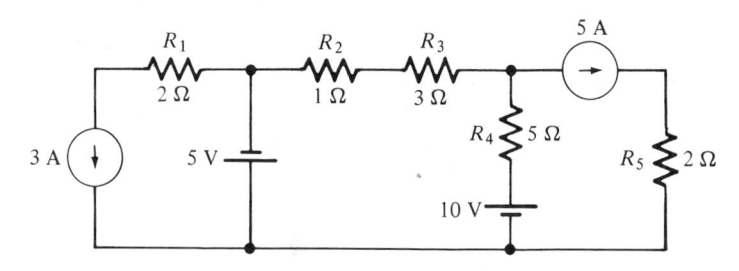

This problem was done by Thévenin analysis in Example 7.18. Removing R_3 and replacing it with a short:

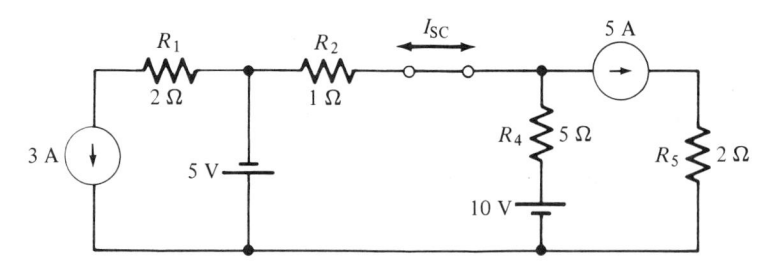

Find I_{sc}. How?

Suppose that the circuit to the right of the short—R_4, R_5, and the 5-A and 10-V sources—was Théveninized. Then the circuit to the left of the short—R_2, R_1, and the 3-A and 5-V sources—was also Théveninized. The result would be this:

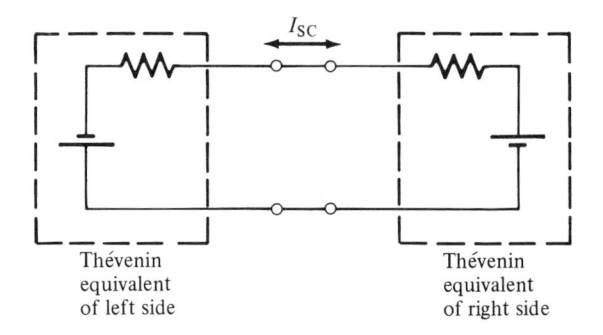

Thévenin equivalent of left side

Thévenin equivalent of right side

The problem becomes a simple series circuit analysis.
Reducing the right side:

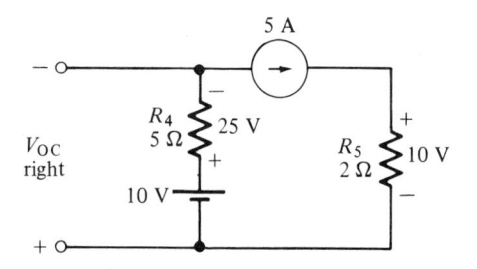

7.5 NORTON ANALYSIS

The 5 A flows through R_4, developing a 25-V drop. Assume polarity of V_{oc} and write a loop equation:

$$V_{oc} - V_{R_4} + 10 = 0$$
$$V_{oc} - 25 + 10 = 0$$
$$V_{oc} = 15 \text{ V}$$

With the 5-A source opened and the 10-V source shorted, $R_{internal}$ becomes simply R_4, or 5 Ω. The right Thévenin circuit is

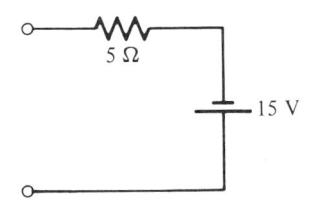

The circuit is to the left of the terminals is

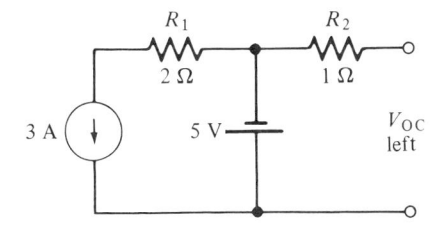

Getting Thévenin equivalent: $V_{oc} = 5$ V (there is no drop across R_2). Find R_{int}:

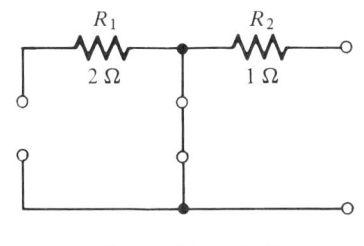

$$R_{int} = R_2 = 1 \text{ Ω}$$

The left-hand Thévenin circuit is

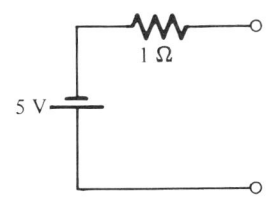

Combining the two Thévenin circuits into the final equivalent circuit:

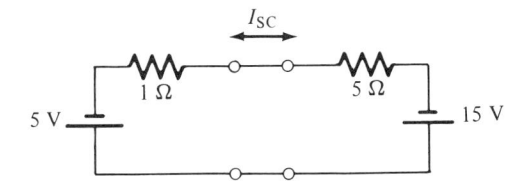

Solve for the Norton current, I_{sc}:

$$I_{sc} = \frac{15-5}{5+1} = \frac{10}{6}$$

$$= 1.7 \text{ A}$$

We have determined the value of I_{sc}. Half the job is done. Next, go back to the original circuit, remove the short, turn off all sources, and solve for the Norton equivalent resistance. This is done the same way that it was for the Thévenin circuit; therefore, the work will not be repeated. From Example 7.18, $R_{int} = 6\,\Omega$. Draw the final Norton circuit:

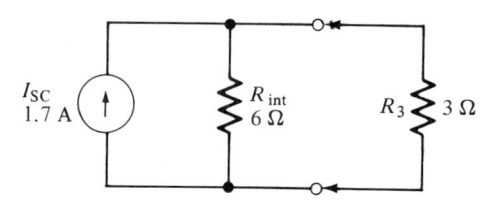

Calculate I_{R_3} by current division:

$$I_{R_3} = \frac{I_{sc} R_{int}}{R_{int} + R_3}$$

$$= \frac{1.67(6)}{6+3}$$

$$= \frac{10}{9}$$

$$= 1.1 \text{ A}$$

This equals the value obtained from Thévenin analysis. ∎

EXERCISES

Section 7.1

1. Calculate the value of each second-order determinant.
   a. $\begin{vmatrix} 1 & 2 \\ 3 & 4 \end{vmatrix}$ b. $\begin{vmatrix} 1 & 2 \\ -2 & 5 \end{vmatrix}$ c. $\begin{vmatrix} 3 & -1 \\ -2 & 4 \end{vmatrix}$ d. $\begin{vmatrix} 3 & 4 \\ -2 & 5 \end{vmatrix}$

2. Solve the following systems of equations by determinants and Cramer's rule.
   a. $V_1 + V_2 = 3$
      $3V_1 - 2V_2 = 14$
   b. $2I_1 + I_2 = 1$
      $-I_1 + I_2 = 4$
   c. $3V_1 - 2V_2 = 19$
      $V_1 + V_2 = 23$

3. Solve the following systems of equations by determinants and Cramer's rule.
   a. $4I_1 - 6I_2 = 2$
      $30I_1 - 42I_2 = 17$
   b. $3V_1 + 5V_2 = 9$
      $\dfrac{-7V_1}{2} + 5V_2 = -4$
   c. $\dfrac{I_1}{3} - \dfrac{I_2}{4} = 2$
      $\dfrac{I_1}{4} - \dfrac{I_2}{2} = 7$

4. Check solutions to Exercise 3 by redoing each system by substitution and then by addition/subtraction.

EXERCISES

5. Evaluate each third-order determinant.

a. $\begin{vmatrix} -1 & 0 & 5 \\ 7 & 3 & 2 \\ 0 & 5 & 9 \end{vmatrix}$
b. $\begin{vmatrix} 3 & 2 & 4 \\ -2 & 0 & 1 \\ 3 & 0 & 5 \end{vmatrix}$
c. $\begin{vmatrix} 8 & 2 & -3 \\ 1 & -1 & 4 \\ 2 & -2 & 1 \end{vmatrix}$

6. Repeat Exercise 5 for these determinants.

a. $\begin{vmatrix} 0 & 0 & 2 \\ 0 & 2 & 5 \\ 1 & 2 & 2 \end{vmatrix}$
b. $\begin{vmatrix} 3 & 5 & 1 \\ 1 & -4 & -2 \\ 7 & -5 & -4 \end{vmatrix}$
c. $\begin{vmatrix} 1 & 1 & 1 \\ 0 & 0 & 1 \\ 3 & 4 & -3 \end{vmatrix}$

7. Solve these systems of three equations by determinants. *Note:* The denominator determinant for each part was evaluated in Exercise 5, parts b and c.

a. $3I_1 + 2I_2 + 4I_3 = -5$
   $-2I_1 + 0I_2 + I_3 = 0$
   $3I_1 + 0I_2 + 5I_3 = -13$

b. $8V_1 + 2V_2 - 3V_3 = 7$
   $V_1 - V_2 + 4V_3 = 9$
   $2V_1 - 2V_2 + V_3 = 11$

8. Repeat Exercise 7 for these systems of equations. *Note:* The denominator determinant for each part was evaluated in Exercises 6, parts b and c.

a. $3I_1 + 5I_2 + I_3 = 16$
   $I_1 - 4I_2 - 2I_3 = -8$
   $7I_1 - 5I_2 - 4I_3 = -4$

b. $V_1 + V_2 + V_3 = -3$
   $0V_1 + 0V_2 + V_3 = 2$
   $3V_1 + 4V_2 - 3V_3 = -21$

Section 7.2

9. Answer the following questions for this circuit:

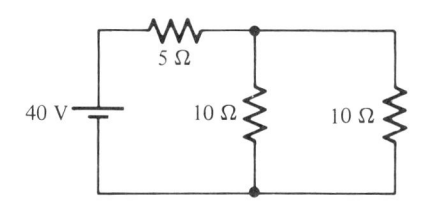

a. How many branches are there?
b. How many nodes are there?
c. How many loop equations are needed to analyze the circuit?
d. Write the equations. Then solve them to find the current through each resistance.
e. Redo the exercise by SPR. Compare results with those of part d.

10. Find the current through each resistance. Write loop equations around E_{s_1}, R_1, and R_2, and E_{s_2}, R_2, and R_3. Calculate the voltage across each resistance. Do a KVL summation around all three loops to check results.

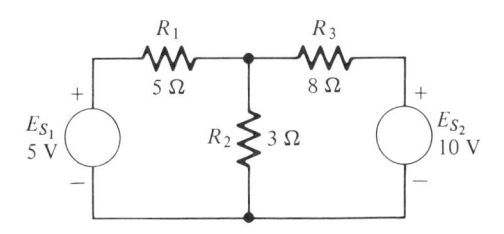

11. Repeat Exercise 10, but write loop 1 around E_{s_1}, R_1, R_3, and E_{s_2}. Write loop 2 around E_{s_1}, R_1, and R_2. Compare the results to those of Exercise 10.

12. Repeat Exercise 10 with loop 1 around E_{s_1}, R_1, R_3, and E_{s_2}. Write loop 2 around E_{s_2}, R_2, and R_3. Compare the results to those of Exercises 10 and 11.

222   ADVANCED TECHNIQUES OF ANALYSIS

13. Using loop analysis, calculate the input current to this T network and the current through the 3-Ω element. Redo the exercise by superposition and compare the results.

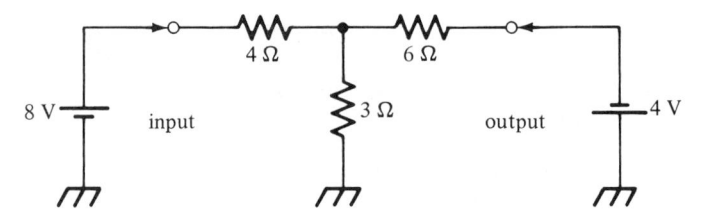

14. Using loop analysis, find the input and output currents for this Π network:

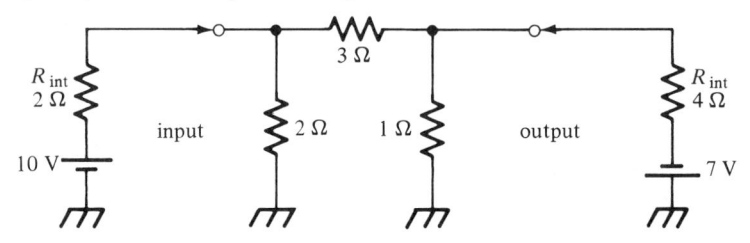

15. Find the current through the 35-Ω resistance. Do this by writing loop equations so that just one current passes through the 35-Ω resistance.

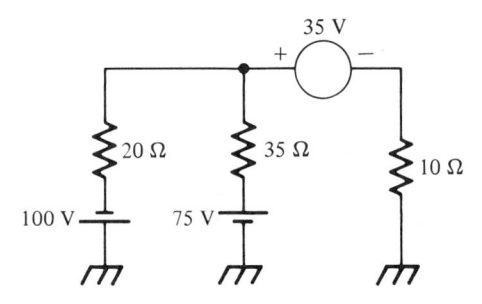

16. Perform the indicated operations for this circuit:

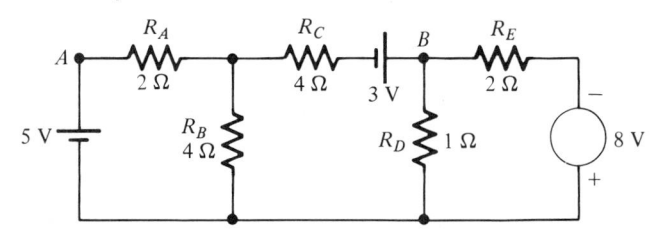

   a. Write loop equations as follows: around the 5-V source, R_a, and R_b; around R_b, R_c, 3-V source, and R_d; and around R_d, R_e, and the 8-V source.
   b. Solve for the current through each resistance.
   c. Calculate the voltage across each resistance.
   d. Check results by KVL.
   e. Calculate the power dissipated by each resistance. Add them to get the total power output.
   f. Calculate the power supplied by each source. Add them to get the total input power.
   g. Compare input power to output power.
   h. Find V_{ab}.

17. Do Exercise 16, parts a and b again, but write the loop equations this way: around the 5-V source, R_a, and R_b; around the 5-V source, R_a, R_c, the 3-V source, and R_d; and around the 5-V source, R_a, R_c, 3-V source, R_e, and the 8-V source.
Compare results.

EXERCISES

18. Using the loops indicated, solve for the current in each resistance.

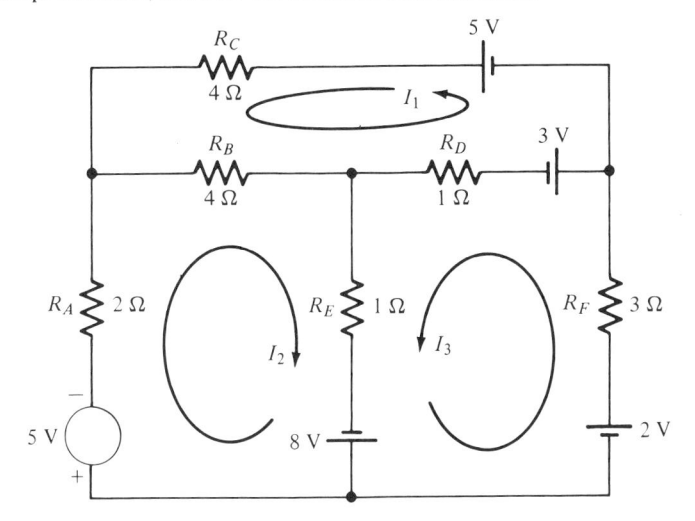

19. Solve for the current through R_2. Check the results by redoing the exercise by superposition.

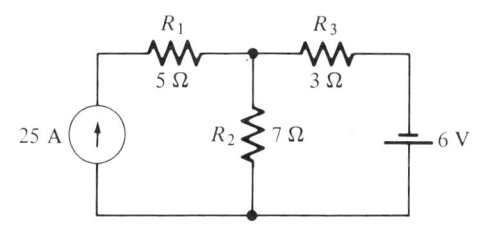

20. Reduce this circuit to one that can be solved by two loop equations. Then find the power delivered by the 7-V source using loop analysis.

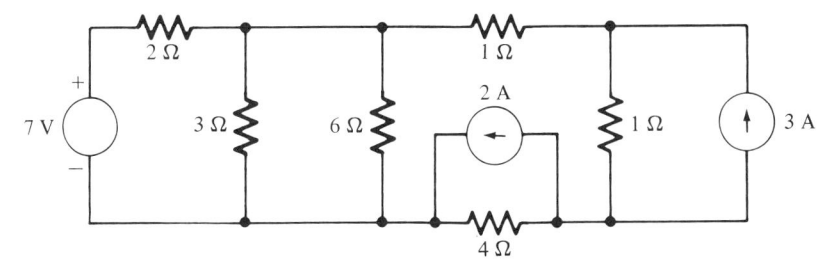

Section 7.3

21. What is the current from point A to point B in each circuit? Define a positive current as one from A to B.

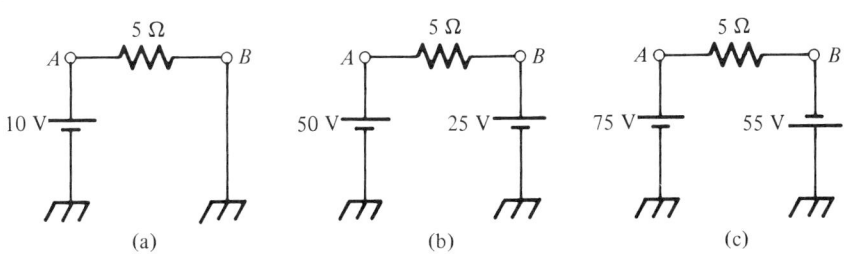

(a)   (b)   (c)

22. Repeat Exercise 21 for the current from B to A. Define positive currents as from B to A.

23. Write a math statement that describes the current from point A to point B in each circuit. This direction is positive. Assume that unknown voltages are positive with respect to ground.

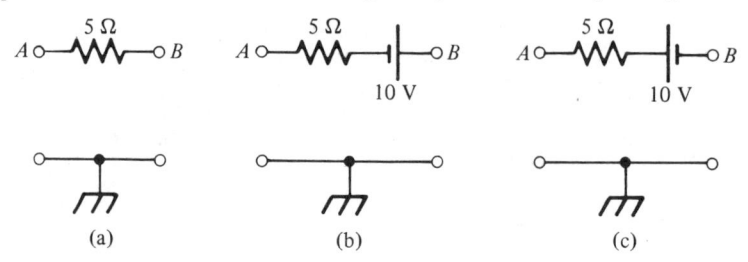

(a) (b) (c)

24. Repeat Exercise 23 for currents from B to A. This direction is positive.

25. Answer the questions for this circuit:

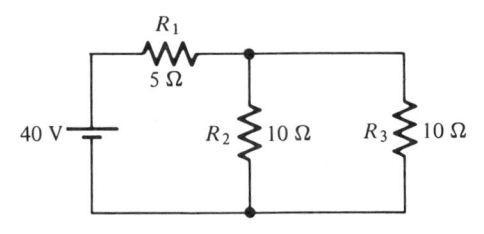

a. How many nodes are there?
b. How many nodal equations are necessary for a solution by nodal analysis?
c. Ground the bottom node. Write a nodal equation at the top node.
d. Solve for the current through each resistance.
e. Redo the exercise by SPR. Compare the results.

26. Convert the 40-V 5-Ω source in Exercise 25 to its Norton equivalent. Redo Exercise 25, parts c and d. Compare the results.

27. Find the current through R_2 by nodal analysis. Redo the exercise by superposition. Compare the results.

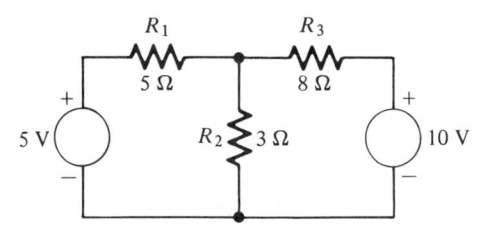

28. Perform the indicated operations for this circuit:

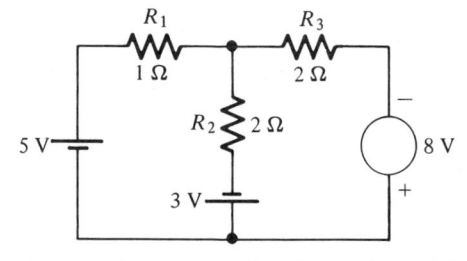

a. Find the voltage drop and its polarity across each resistance by nodal analysis.
b. Check the results by KVL around the loops.
c. Calculate total P_{out}.
d. Calculate total P_{in}.
e. Compare P_{out} to P_{in}.

EXERCISES

29. Find I_{R_3} by nodal analysis.

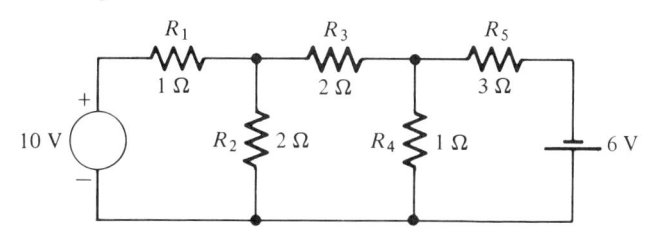

30. Convert the 10-V 1-Ω and 6-V 3-Ω sources in Exercise 29 to their Norton equivalents. Write nodal equations for this new circuit. Compare these equations and those of Exercise 29 closely. Are they the same?

31. Find: a. the current through R_6; b. the voltage across R_1. Use nodal analysis.

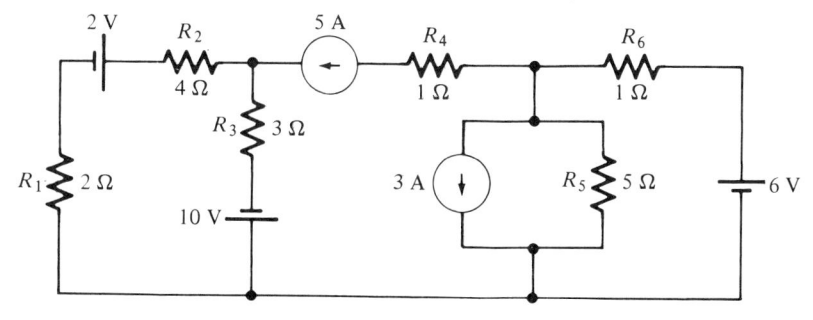

32. Using nodal analysis, find: a. the voltage across source 3; b. I_{R_2}; c. V_{R_4}.

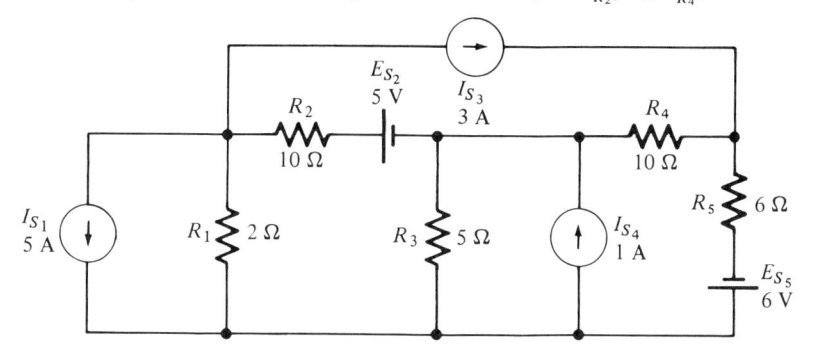

33. a. What must R_1 be to hold I_{in} to 2 mA?
b. What is the drop across the 400-Ω resistance?
c. Will a $\frac{1}{4}$-W dissipation resistor do for R_1?

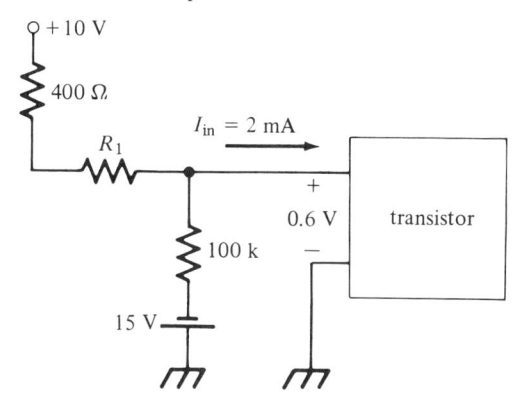

34. What must R_1 be to hold I_{in} to 1.8 mA?

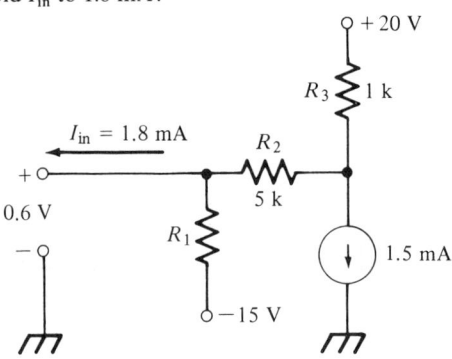

Section 7.4

35. For each of the circuits below:

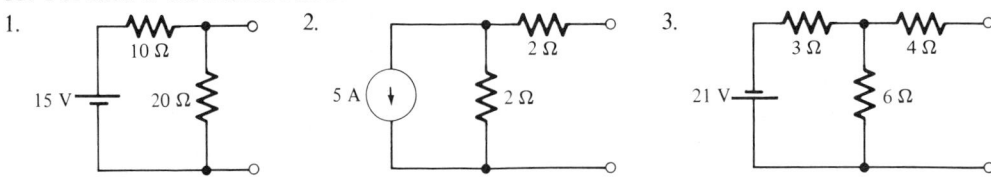

1. [circuit: 15 V source, 10 Ω series, 20 Ω shunt]
2. [circuit: 5 A source, 2 Ω series, 2 Ω shunt]
3. [circuit: 21 V source, 3 Ω series, 4 Ω series, 6 Ω shunt]

a. Determine the Thévenin equivalent circuit.
b. Load each Thévenin circuit with a 10-Ω resistance and calculate V_o and I_{load}.

36. Repeat Exercise 35 for the following circuits, but change the load to 10 k.

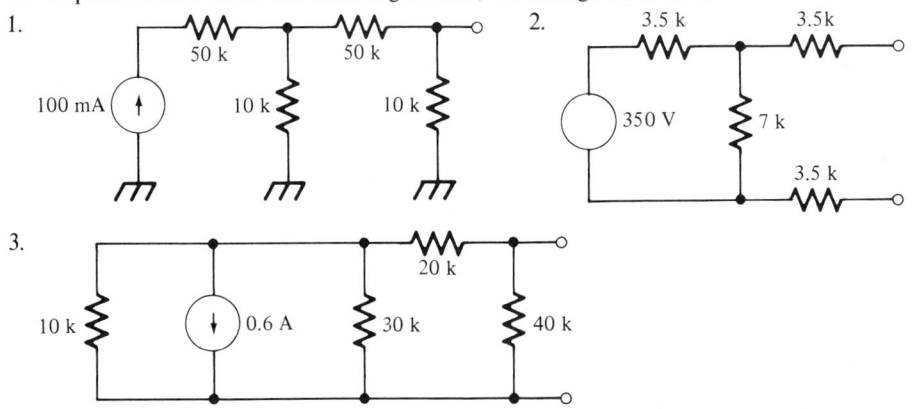

1. [circuit: 100 mA source, 50 k, 50 k series, 10 k, 10 k shunts]
2. [circuit: 350 V source, 3.5 k, 3.5 k, 7 k, 3.5 k]
3. [circuit: 10 k, 0.6 A source, 30 k, 20 k, 40 k]

37. What resistance draws maximum power from each circuit in Exercise 35?

38. Follow the steps for this circuit:

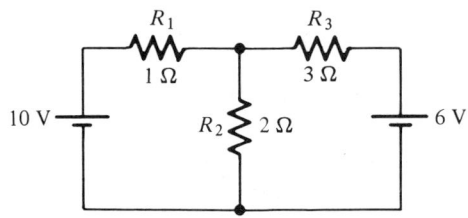

[circuit: 10 V source, $R_1 = 1\,\Omega$, $R_2 = 2\,\Omega$, $R_3 = 3\,\Omega$, 6 V source]

a. Find the Thévenin equivalent of the circuit driving R_1. Put R_1 across the Thévenin output. Calculate I_{R_1}.
b. Repeat part a for R_2.
c. Repeat part a for R_3.

EXERCISES

d. Repeat part a for the 6-V source.
e. Using loop equations, solve for the current through R_1, R_2, R_3, and the 6-V source. Compare these currents to those obtained from the Thévenin circuits.
f. Convert the 10 V/R_1 and 6 V/R_3 connections to their Norton equivalents. Use Millman's theorem to combine the two Nortons into one. Convert this into its Thévenin equivalent. Place R_2 across the Thévenin terminals. Calculate I_{R_2}. Compare the results with those of parts b and e.

Section 7.5

39. Determine the Norton equivalent of each circuit in Exercise 35. Do not convert the Thévenin circuits determined in Exercise 35 to do this.

40. Determine the Norton equivalent of each circuit in Exercise 36. Do not convert the Thévenin circuits determined in Exercise 36 to do this.

41. Follow the steps for this circuit:

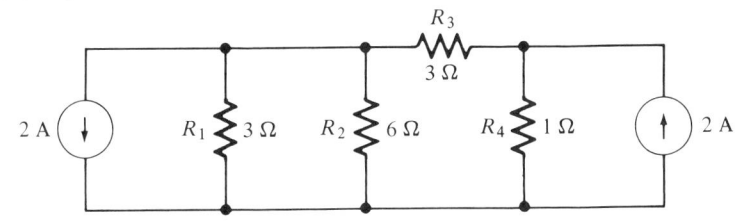

a. Find the Thévenin equivalent of the circuit driving R_3. Put R_3 across the Thévenin output. Calculate I_{R_3}.
b. Repeat part a for R_2.
c. Repeat part a for R_4.
d. Write nodal equations for the nodes at each end of R_3. Solve for the current through R_2, R_3, and R_4. Compare the results with those of parts a, b, and c.

42. Follow the steps for this circuit:

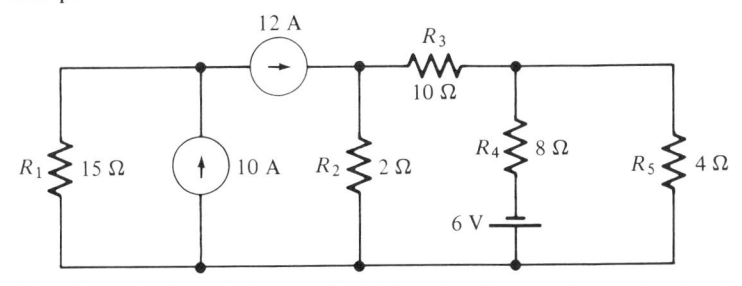

a. Find the Thévenin equivalent of the circuit driving R_3. Connect R_3 to the circuit. Calculate its current.
b. Find the Norton equivalent of the circuit driving R_3. Do not do this by converting the Thévenin circuit of part a. Calculate the current through R_3. Compare to the current of part a.
c. Repeat parts a and b for R_5.

CHAPTER HIGHLIGHTS

Save work in loop analysis by choosing loops with care.
    Loop equations sum voltages around a loop per Kirchhoff's voltage law. The unknowns in these equations are currents.
    Nodal equations sum currents at a node per Kirchhoff's current law. The unknowns are voltages.
    Find the open-circuit output voltage and the equivalent circuit resistance looking into a pair of terminals with all sources turned off. Put these together in a Thévenin equivalent circuit.

Find the current through a short between two terminals. Then find the equivalent resistance looking into the same terminals with the short removed and all sources off. Put these together in a Norton equivalent circuit.

Formulas to Remember

Value of a determinant:

value of a determinant

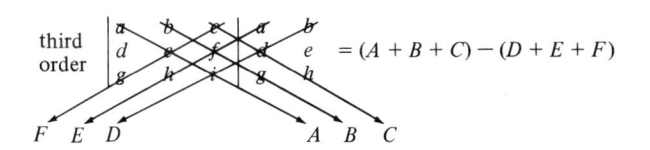

$$\text{second order} \quad \begin{vmatrix} a & b \\ c & d \end{vmatrix} = ad - bc$$

$$\text{third order} \quad \begin{vmatrix} a & b & c & a & b \\ d & e & f & d & e \\ g & h & i & g & h \end{vmatrix} = (A + B + C) - (D + E + F)$$

$$ F \quad E \quad D A \quad B \quad C$$

Cramer's rule:

$B - N + 1$ Number of loop equations needed for loop analysis solution
$N - 1$ Number of nodal equations needed for nodal analysis solution

$R_{load} = R_{internal}$ of Thévenin or Norton circuit for maximum power transfer to a load from an equivalent circuit.

CHAPTER 8

Resistance

In Chapter 3 we learned that resistance the opposition to current flow, that materials with high resistance allow little current to pass and are called insulators, and that materials with low resistance allow lots of current to pass and are called conductors. There is much more to learn about resistance.

At the completion of this chapter, you should be able to:

- Tell what determines the resistance of a piece of material.
- Explain what a temperature coefficient is and use it to find resistance at different temperatures.
- Calculate the cross-sectional area of a conductor in circular mills.
- List conductor characteristics important in electronics.
- Use the military code for determining resistance.
- Describe how printed-circuit boards are made and list their advantages and disadvantages.
- Explain how resistance values are determined.
- State the difference between precision and accuracy.
- Define a linear relationship.
- Read and interpret VA characteristics.
- Explain dynamic resistance and how to tell if a component has it.
- Tell what a transducer does and give examples of them.
- List important characteristics of resistors and switches.

8.1 THE RESISTANCE OF A CONDUCTOR

Four factors determine the resistance of a conductor. One of these, temperature, is discussed in Section 8.2. Two of the remaining three depend on the size of the conductor. They are:

- *Length:* Resistance is directly proportional to conductor length. Think of a wire as many short pieces hooked together, similar to links of a chain (see Figure 8.1). Each piece has resistance. These resistances are in series. The more pieces, the more resistors in series, and the higher the total resistance.
- *Area:* Resistance is inversely proportional to cross-sectional area. Think of the wire as being sliced into many strips (see Figure 8.1). Each strip has resistance. These resistances are in parallel. The thicker the wire, the greater the number of parallel-connected strip resistances, and the lower the total resistance.

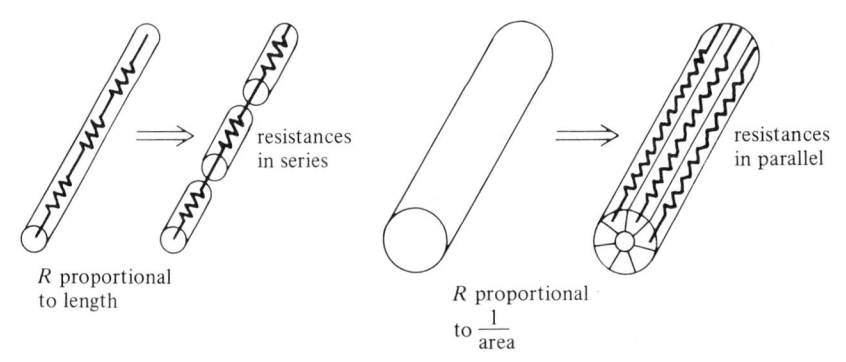

FIGURE 8.1 Effects of Length and Cross-Sectional Area on the Resistance of a Piece of Material.

8.1 THE RESISTANCE OF A CONDUCTOR

The third factor determining conductor resistance is the ease with which the conductor's atoms release electrons to become free charge carriers. In Chapter 2, we learned that atoms of low-resistance materials give up carriers easily, whereas those of high-resistance materials do so with difficulty. This atomic characteristic is so important that it is given a name, *resistivity*. Resistivity is represented by the Greek lowercase letter ρ (rho, pronounced *row*).

The resistivities of some common materials are listed in Table 8.1.

TABLE 8.1 Resistivity of Common Materials

Materials	Resistivity (Ω-meters) at 20°C
Silver	1.65×10^{-8}
Copper	1.7×10^{-8}
Gold	2.4×10^{-8}
Aluminum	2.8×10^{-8}
Tungsten	5.5×10^{-8}
Iron	9.8×10^{-8}
Nichrome	100×10^{-8}
Carbon	3500×10^{-8}
Glass	10^{10} to 10^{14}
Mica	10^{11} to 10^{15}

Note carefully how the resistivities of the conductors, copper and aluminum, are much less than 1, while those of the insulators, such as glass, are in the billions.

These three factors—resistivity, length, and cross-sectional area—determine the resistance of a piece of material according to the equation:

$$R = \frac{\rho L}{A} \tag{8.1}$$

where R = resistance, in ohms
ρ = resistivity, in ohm-meters
L = length, in meters
A = cross-sectional area, in square meters

EXAMPLE 8.1

What is the resistance of this iron bar for current flowing in the direction of the arrow?

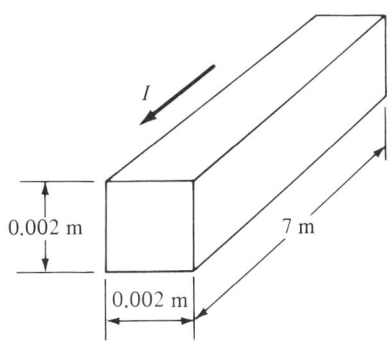

$$R = \frac{\rho L}{A} \qquad \rho = 9.8 \times 10^{-8}\,\Omega\text{-}m \text{ from Table 8.1}$$

$$= \frac{9.8 \times 10^{-8} \times 7}{(0.002)^2} = 17.2 \times 10^{-2}\,\Omega$$

EXAMPLE 8.2

What is the resistance of 14 kilometers of copper wire 2 cm in diameter?

Find the cross-sectional area of the wire:

$$\text{area of a circle} = \pi R^2 \text{ or } \frac{\pi D^2}{4}$$

The diameter in centimeters must be converted to meters. Therefore,

$$\text{area} = \frac{\pi D^2}{4} = \frac{\pi (2 \times 10^{-2})^2}{4}$$

$$= \pi(10^{-4})$$

$$= 3.14(10^{-4})$$

$$R = \frac{\rho L}{A}$$

$$= \frac{(1.7 \times 10^{-8})(14 \times 10^3)}{3.14 \times 10^{-4}}$$

$$= 0.76\,\Omega$$

EXAMPLE 8.3

What is the resistance of the conductor of Example 8.2 if the material is changed to silver?

The hard way to do the problem is to repeat all the calculations. Here is the easy way: The only thing that changed was the value of ρ. Therefore, the new resistance must be

$$R_{new} = \frac{\rho_{silver}}{\rho_{copper}}(R_{copper})$$

$$= \frac{1.65 \times 10^{-8}}{1.7 \times 10^{-8}}(0.76) = 0.97(0.76) = 0.74\,\Omega$$

8.2 THE EFFECTS OF TEMPERATURE ON RESISTANCE

Good circuit design pays close attention to the effects of temperature because electrical characteristics change as materials get hot or cold. The resistivity of most materials, for example, increases as the temperature goes up. Thus temperature is the fourth factor, besides length, area, and resistivity, that determines the resistance of a piece of material.

How much the resistivity of a material changes for a given temperature shift is called the *resistivity temperature coefficient*. The coefficient is represented by the Greek lowercase letter α (alpha, pronounced *al-fuh*). It is positive if resistivity goes up as the temperature goes up. It is negative if resistivity goes down as the temperature goes up.

8.2 THE EFFECTS OF TEMPERATURE ON RESISTANCE

The tungsten filament in an incandescent light bulb, for example, has a positive temperature coefficient. Its resistance undergoes about a 10-to-1 increase when switching from cold (bulb off) to hot (bulb on). On the other hand, the thermistor, which we met in bridge circuits in Chapter 6, usually has a negative coefficient. It is used to cancel the positive coefficient of other devices, such as transistors. This reduces the changes in a circuit's operation due to temperature increases.

Table 8.2 lists the resistivity temperature coefficient of common materials. Used in Equations 8.2, these values allow us to calculate the resistance of a material at any temperature.

TABLE 8.2 Resistivity Temperature Coefficient of Common Materials

Material	Resistivity temperature coefficient, α, per °C at 20°C
Silver	+0.0038
Copper	+0.00393
Gold	+0.0034
Aluminum	+0.00391
Tungsten	+0.0045
Iron	+0.006
Nichrome	+0.0004
Carbon	−0.0005

Note how Equation 8.2 starts with resistance at 20°C (68°F), and adds or subtracts an amount determined by the temperature coefficient of Table 8.2 multiplied by the temperature change:

$$R_{new} = R_{20°}(1 + \alpha \Delta T) \tag{8.2}$$

where R_{new} = resistance at the new temperature
$R_{20°}$ = resistance at 20°C
α = Temperature coefficient from Table 8.2
ΔT = change in temperature from 20°C; the Greek letter Δ means "change in"

EXAMPLE 8.4

A relay coil wound of copper wire has a resistance 3.7 Ω at 20°C. Find its resistance at 140°C.

First, find ΔT:

$$\Delta T = T_{new} - 20°C = 140° - 20° = 120°$$
$$R \text{ at } 140°C = R \text{ at } 20°C(1 + \alpha \Delta T)$$
$$= 3.7(1 + [0.00393(120)])$$
$$= 3.7(1 + 0.47) = 3.7(1.47)$$
$$= 5.44 \text{ Ω}$$

Note how the positive temperature coefficient of resistivity caused the coil's resistance to increase as the temperature rose.

EXAMPLE 8.5

A carbon rod has a resistance of 37 × 10³Ω at 20°C. It is heated to 350°C. Find its new resistance.

Find ΔT:

$$\Delta T = 350° - 20° = 330°$$

α for carbon from Table 8.2 $= -0.0005$.

$$\begin{aligned}
R \text{ at } 350°C &= R \text{ at } 20°C(1 + \alpha \Delta T) \\
&= (37 \times 10^3)[1 + (-0.0005)(330)] \\
&= (37 \times 10^3)(1 - 0.165) \\
&= (37 \times 10^3)(8.35 \times 10^{-1}) \\
&= 30.9 \times 10^3 \, \Omega
\end{aligned}$$

Carbon's negative temperature coefficient caused rod resistance to decrease with a temperature increase. ∎

Another technique for determining temperature effects works directly on a unit's resistance rather than on the resistivity of the material from which the unit is made. It requires a *temperature coefficient of resistance* (often called TCR, or temco, on data sheets). This coefficient is given as a percent change in resistance per degree Celsius temperature change.

EXAMPLE 8.6

A resistor has a resistance of 150 k at 20°C and a temperature coefficient of resistance of 0.03%/°C. What is its resistance at $T = 165°C$?

$$\begin{aligned}
R \text{ at } 165°C &= R \text{ at } 20°C[1 + \text{TCR}(\Delta T)] \\
&= 150 \text{ k}[1 + 0.0003(165 - 20)] \\
&= 150 \text{ k}[1 + 0.0003(145)] \\
&= 150 \text{ k}[1 + 0.0435] \\
&= 156.5 \text{ k} \quad\quad\quad\quad\quad\quad ∎
\end{aligned}$$

Industry often expresses the % change/°C coefficient in another form, "parts per million change per degree C temperature change" (ppm/°C). Let us see how. (*Note:* The word "parts" in ppm/°C is used because the technique has other applications in electronics besides resistance. Read "parts" as "ohms" here.)

EXAMPLE 8.7

Redo Example 8.6 but convert the TCR into ppm/°C form first.

Multiplying the TCR percentage times 1 million converts it into ppm/°C form:

$$0.03\%(10^6) = 0.0003(10^6) = 300 \text{ parts per million (ppm)}$$

The resistance of Example 8.6 changes 300 ppm/°C, that is, it changes 300 Ω for every 1 million ohms it has for a 1°C temperature change. How many million ohms does the 150-k resistance have?

$$150 \times 10^3 \text{ parts (ohms)} \times \frac{1 \text{ million}}{10^6} = 0.15 \text{ million ohms}$$

The resistance change is then

$$(0.15 \text{ million ohms}) \times \frac{300 \text{ }\Omega/\text{million ohms}}{°C} (145°C) = 6.5 \times 10^3 \text{ }\Omega$$

The final resistance is

$$R_{\text{original}} + R_{\text{change}} = 150 \text{ k} + 6.5 \text{ k}$$
$$= 156.5 \text{ k} \quad \text{the same answer as before}$$

EXAMPLE 8.8

A 3.3 megohm resistor has a temperature coefficient of 125 ppm/°C. How much does it resistance increase when its temperature rises 25°C?

The positive temperature coefficient says: For every million ohms the resistor has, its resistance increases 125 Ω for a 1°C rise. The resistor has

$$3.3 \times 10^6 \text{ }\Omega \times \frac{1 \text{ million}}{10^6} = 3.3 \text{ million ohms}$$

Resistance therefore changes:

$$3.3 \text{ million ohms} \left(\frac{125 \text{ }\Omega}{\text{million ohms}} \right) \text{per °C} = 412.5 \text{ }\Omega \text{ per °C}$$

The temperature rises 25°C. Therefore, the resistance rise is

$$\frac{412.5 \text{ }\Omega}{°C} (25°C) = 10{,}313 \text{ }\Omega$$

Pay close attention to the way units cancel when the problem is written as a math expression.

8.3 CONDUCTORS

Most of the conductors used in electronics are insulated copper wires. Here are some of the things that electronics personnel need to know about wires:

- *Insulation:* Usually plastic, but silk, cotton, and varnish are also used. Insulation is colored so that conductors can be traced through the maze of wires found in equipment.
- *Stranded and solid:* Stranded wires are made up of many thin wires for flexibility. Solid conductors are used where little or no movement is expected.
- *Tinning:* A shiny coating of tin/lead alloy put on wire to make soldering easier.
- *Shielding:* Wires running through equipment act as antennas that pick up other signals or electrical noise which can mask legitimate signals. To prevent this,

the wire is wrapped with a layer of conducting "cloth" woven of many fine wires. Called *shielding*, this layer prevents unwanted signals from getting to the wire inside; it is normally connected to chassis ground. Cables connecting microphones and phonograph pickups to amplifiers, for example, are always shielded to prevent hum.

- *Single and multiconductor:* Besides the usual single-wire lines, or *leads*, as they are called, there are cables made of many wires and available in a wide variety of styles.

Wire diameters are measured in AWG (American Wire Gage) numbers; the smaller the *gage number*, the thicker the wire. Table 8.3 lists common sizes and some of their characteristics. Houses are wired with 12- and 14-gage conductors, while the sizes from approximately 18 to 32 gage are used for electronic "hookup" wire.

TABLE 8.3 Copper Wire Table

Wire size	Diameter in (mils)	Cross-sectional area in circular mils (CM)	Ohms per 1000 ft at 25°C	Ampacity (continuous duty current capacity), open air in amps
10	102	10,380	1.018	55
12	80.8	6,529	1.619	41
14	64.1	4,109	2.575	32
16	50.8	2,580	4.094	22
18	40.3	1,624	6.510	16
20	32	1,024	10.35	11
22	25.3	640	16.46	—
24	20.1	404	26.17	—
26	15.9	254	41.6	—
30	10.0	101	105.2	—
36	5.0	25	423	—

A few comments about Table 8.3:

- Column 2 lists wire diameters in mils. These are milliinches, 10^{-3} in.
- Note carefully how diameter, and therefore cross-sectional area, decrease as gage size increases.
- Column 4 shows the effects of this drop in area; as area decreases, resistance increases, as Equation 8.1 says it must. The results of increasing resistance are demonstrated in column 5 by a drop in current capacity.
- Square any gage diameter in column 2. Note that the result is found in the next column under cross-sectional area and that the units of this area are *circular mils*.
- This tells us:

A wire's cross-sectional area in circular mils is equal to its diameter in mils (milliinches) squared. In math form, this is

$$A, \text{ area in circular mils} = (D, \text{ diameter in mils})^2 \qquad (8.3)$$

8.3 CONDUCTORS

EXAMPLE 8.9

What is the cross-sectional area in circular mils of a wire 8 mils in diameter?

$$\begin{aligned} \text{Area, in CM} &= (\text{diameter, in mils})^2 \\ &= (8)^2 \\ &= 64 \text{ CM} \end{aligned}$$

It should be clearly understood that this area is not the wire's cross-sectional area in square mils. Using circular-mil area is a shortcut that removes the necessity for multiplying diameter squared by $\pi/4$ to get area. ∎

EXAMPLE 8.10

A wire has a cross-sectional area of 2580 CM. What is its diameter in mils and in inches? What gage wire is this?

Since

$$\text{area, in CM} = (\text{diameter, in mils})^2$$

it follows that

$$\sqrt{\text{area, in CM}} = \text{diameter, in mils}$$

Therefore,

$$\begin{aligned} \text{diameter} &= \sqrt{\text{area}} = \sqrt{2580} \\ &= 50.8 \text{ mil} \end{aligned}$$

The diameter in inches is

$$50.8 \text{ mil} \left(\frac{1 \text{ inch}}{1000 \text{ mils}} \right) = 0.0508 \text{ in.}$$

What gage wire is this?
A check of Table 8.3 shows this to be 16-gage wire. ∎

TABLE 8.4 Resistivity of Common Materials in English Units

Material	Resistivity (Ω-CM/ft) at 20°C
Silver	9.8
Copper	10.37
Gold	14.7
Aluminum	17
Tungsten	26.3
Iron	59
Nichrome	602
Carbon	21,070
Glass	10^{18}–10^{22}
Mica	10^{19}–10^{23}

The resistance formula (Equation 8.1) is rewritten in more familiar English units:

$$R = \frac{\rho L}{A} \qquad (8.4)$$

where ρ = resistivity, Ω-CM/ft
L = length, in feet
A = cross-sectional area, in circular mils

The ρ values of Table 8.1 are converted from metric to English units for use in Equation 8.4. They are listed in Table 8.4.

EXAMPLE 8.11

What is the resistance of 14 × 10³ ft of 4 gage aluminum wire 0.2043 in. in diameter?

The diameter in mills is

$$0.2043 \text{ in.} \frac{1000 \text{ mil}}{\text{in.}} = 0.2043(1000) = 204.3 \text{ mils}$$

From Table 8.4, ρ for aluminum = 17 Ω-CM/ft. Substitute into equation 8.4:

$$R = \frac{\rho L}{A} = \frac{17(14 \times 10^3)}{(204.3)^2}$$
$$= 5.7 \text{ }\Omega$$

8.4 PRINTED-CIRCUIT BOARDS

After wires, the conductors used most often in electronics are the copper strips found on *printed-circuit boards* (PCBs). Figure 8.2 illustrates an example of this style of wiring.

High-quality board material (FR-4) is made from as many as eight layers of fiberglass glued into a sheet $\frac{1}{16}$ in. thick with flame-retardant epoxy. It is used in military electronics, computers, communications equipment, industrial controls, and test equipment. At about half the cost are the paper-based boards (such as FR-2 and XXXP) used in automotive electronics, TV games, calculators, and other consumer goods.

Printed circuit boards are coated with a layer of copper about 3 mil thick. It is from this layer that the conductors are formed. Here is one way it is done:

The desired circuit is drawn on the copper with a special ink that resists acid. The board is then placed in an acid bath. This eats away all copper not protected by the ink. What is left after the bath is a pattern of copper strips matching the inked circuit. These strips interconnect the resistors, transistors, integrated circuits, and other circuit components mounted on the board. This entire process can be automated for mass production.

Many PC boards are double-sided; they have copper on both sides. This allows more electronics to be packed on the board. There are also flexible PC boards as well as multilayer boards having 20 or more conductor layers for interconnecting complex circuits.

8.4 PRINTED-CIRCUIT BOARDS

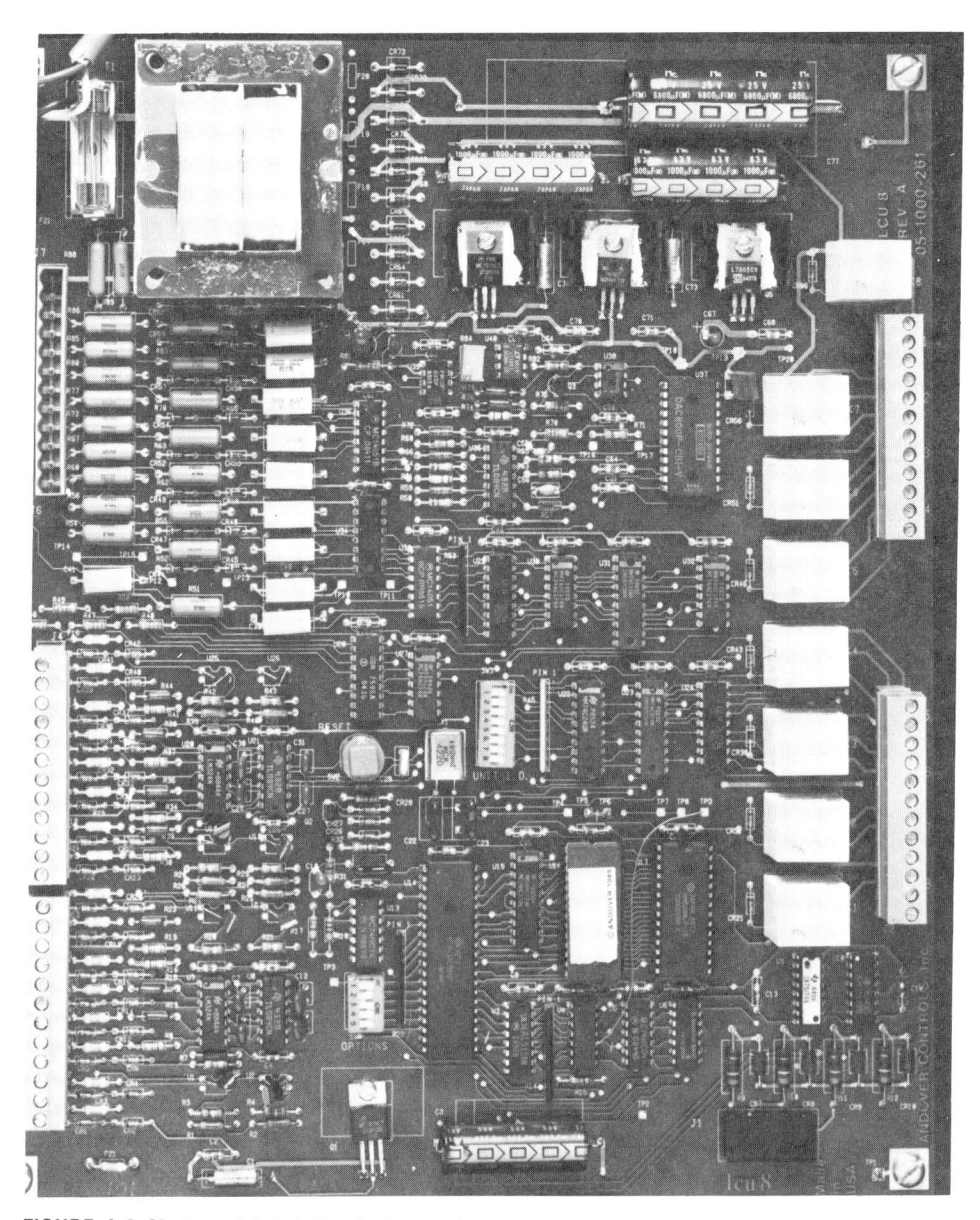

FIGURE 8.2 Modern Printed-Circuit Board. (Courtesy of Andover Controls.)

The advantages of PC boards over regular wiring are:

- Greater reliability, neater, more compact.
- Cheaper when made in quantities.
- Can be mass produced.
- Work very well with robots and other machines for automatic placement of components.

The disadvantages are:

- High layout and design costs for small quantities.

- Difficult to change finished circuit.
- Electrical characteristics of board material may be different for high-frequency signals. This can make it difficult to predict how a circuit built on the board will work.

8.5 RESISTORS

The carbon resistor and color code were introduced in Chapter 3 so that they could be used in the lab. There is much more to know about resistors.

General Characteristics:

The following characteristics apply to all types of resistors:

- *Resistance, power rating,* and *temperature coefficient.*
- *Noise generation:* Switch a TV to a blank channel, turn up the volume, and a loud, rushing sound is heard. In electronics, that sound is called *noise*. All resistances generate tiny noise voltages. These are no problem in a stereo system but could drown out weak signals from space. Circuit designers working with such signals must know the noise characteristics of each type of resistor.
- *Cost:* Often the most important parameter.
- *Fixed* versus *variable.*
- *How* is the resistor *connected* into the circuit? Does it have leads (pigtails)? If so, are they intended for automatic parts placement on a PC board or to be wrapped around a terminal and soldered? Or is the resistor designed to be plugged into a *SIP* (single-in-line package) or *DIP* (dual-in-line package) socket? Figure 8.3 shows each of these popular types of packaging.
- *Precision* and *accuracy:* People often confuse these two. *Precision*, in engineering, refers to the size of the smallest measuring unit relative to the size of the parameter being measured. For example, a ruler marked in feet measures the

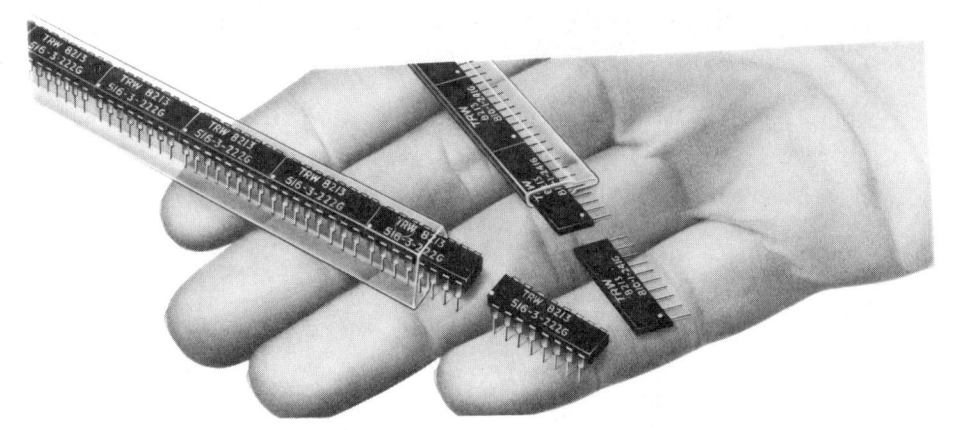

FIGURE 8.3 SIP and DIP Resistor Networks. (Courtesy of Resistive Products Division, TRW Electronic Components Group.)

distance to the moon with very high precision and the thickness of a hair with very low precision. *Accuracy*, on the other hand, tells how close the feet on the ruler are to being exactly 1 foot long. (In electronics, we use the word *tolerance* when speaking of accuracy.) Precision resistors have precise values of resistance, such as 49.9 Ω. Their tolerances may be as low as 0.01%. Needless to say, high precision and tight tolerances are expensive.

- *Reliability:* A figure based on tests done by the manufacturer.
- *Stability:* Stable resistors keep the same resistance as they age, get wet, or are heated during soldering or by heavy current.

Types of Resistors

Carbon Composition

An old favorite but losing ground to film types. Available with values from 2.7 Ω to 22 M power ratings of $\frac{1}{8}, \frac{1}{4}, \frac{1}{2}$, 1, and 2 W, and tolerances of 5, 10, and 20%. Higher price than film resistors. Poor stability. Noisy. Poor temperature coefficient.

Carbon Film

Made by depositing film of fine carbon particles on rod of insulating material, then covering with an insulating, protective jacket. Resistance value determined by length and width of film. Similar to carbon composition resistors in size, noise, and temperature coefficient, but more stable and reliable and cheaper. Available from 1 Ω to 22 M in 2 and 5% tolerances, $\frac{1}{8}$ to 2 W. Uses the same color code as composition resistors.

Metal Film

Low-power, high-reliability, and high-precision resistors. Made by depositing film of metal particles on insulating rod, then encasing in protective cover. Final resistance value set by trimming film with diamond-edged cutter or laser. Available in standard sizes from less than an ohm to several gigaohms. Excellent accuracy (as low as 0.005%) and stability. Lower noise and temperature coefficient than carbon types but higher cost. Popular for new designs. Used heavily in military equipment. Military coding system discussed below.

Wirewounds

Made by winding special resistance wire around an insulating spool. For reasons explained in a later chapter, wirewounds must not be used with high-frequency signals. Two main types:

1. *High precision.* As wire is wound on a spool, its resistance is checked until the desired value is reached. Wirewounds therefore have high precision and accuracy (to 0.01%), and if wire having a temperature coefficient near zero is used, their temperature stability may be as good as 0.001%. Expensive. Maximum value approximately 1 k.
2. *High power dissipation.* High-resistivity wire is wound around a ceramic core able to stand high temperatures. Core often hollow to provide "chimney effect" for better cooling. Resistance values range from a few ohms to about 100 k. Wattage ratings from about 1 to 250 W.

Thin Film

Microscopic resistors built into integrated circuits. Made of metal film less than one-millionth of a meter thick. Range from 20 Ω to about 50 k. Final value set by laser trimming or acid process similar to that used in making PC boards. Resistance depends on length and width of conducting path. In some cases, ρ can also be adjusted.

Thick Film

Resistance material greater than 1 millionth of a meter thick is mounted on insulating slab. Resistance set by path size and material resistivity. Well suited for forming several resistors on the same slab to make resistor networks (see Figure 8.3 and Exercises 38 and 39). Sold in SIP and DIP packages. Values range from about 10 Ω to 1 M, depending on the company. A popular type is CERMET (CERamic-METal), in which the resistivity is controlled by mixing insulating ceramic and conducting metal particles. Power dissipation capability only a few tenths of a watt. Becoming increasingly popular as SMD (surface-mounted device) "chip" resistors (see Figure 8.4).

How Resistance Values Are Determined

You may have noticed that resistor values do not change in a nice 10, 20, 30, 40, ... pattern. Instead, the significant numbers follow a special sequence. For 20% tolerance units, that sequence is: 10, 15, 22, 33, 47, 68, and 100. Here is how it works: A 10-Ω, 20% tolerance resistor may have a maximum resistance of 1.2(10) or 12 Ω. A 15-Ω resistor (the next number in the sequence) may have a minimum value of 0.8(15) or 12 Ω. Note how the maximum resistance of one value meets the minimum resistance of the next-higher value.

The thinking behind the choice of values is this: It is too costly to make resistors of every resistance. Instead, certain specific values are produced and the tolerance is expected to fill the gaps between them.

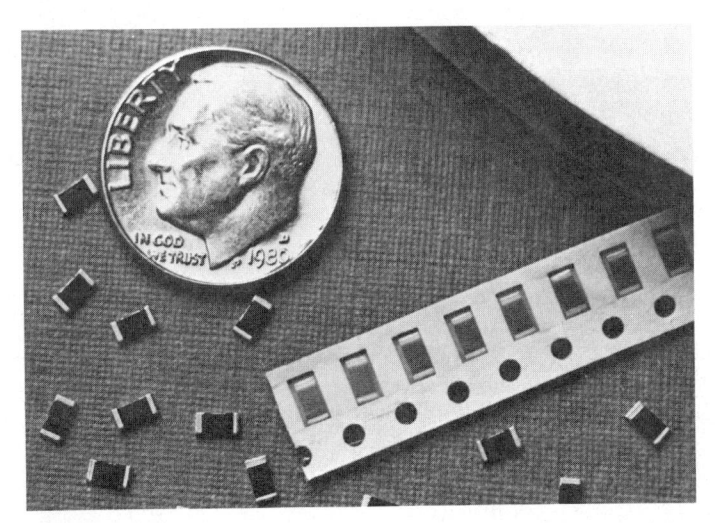

FIGURE 8.4 Chip Resistors for Surface Mounting. (Courtesy of PACCOM Electronics.)

Should a resistance be needed that is not one of the specific values, do this: Find the value whose tolerance covers the resistance wanted. Measure a number of these on an ohmmeter until one is found that is close enough to the desired value. For example, a 42-k resistor is needed. 33-k units will not cover this because their maximum value is 33 k(1.2) = 39.6 k. But 47 k will, because its minimum is 0.8(47 k) = 37.6 k. Therefore, measure 47 k's until a 42 k is found.

The values in the 10% sequence are: 10, 12, 15, 18, 22, 27, 33, 39, 47, 56, 68, 82, and 100. In the 5% sequence they are: 10, 11, 12, 13, 15, 16, 18, 20, 22, 24, 27, 30, 33, 36, 39, 43, 47, 51, 56, 62, 68, 75, 82, 91, and 100. The student is strongly urged to calculate several maximum and minimum tolerance values for adjacent numbers in each sequence to show how well the gaps are covered.

The Military Coding Method

For 2% and 5% tolerance units, the military code uses a three-digit number followed by a letter. The first two digits are the significant numbers in the resistance. The third is a multiplier. It tells the number of zeros that follow the significant figures. The letter tells the tolerance; K indicates a 10% accuracy, J shows 5%, G is for 2%, and F, 1%. For example, a unit marked 154G has a resistance of 15 followed by four zeros, or 150,000 Ω, or 150 k. The tolerance is 2%. If the resistance is less than 10 Ω, there is no multiplier. Instead, the letter R is used to indicate a decimal point. For example, 5R0 means 5.0 Ω, and, 3R6 means 3.6 Ω. A four-digit number is used by the military system for 1% tolerance units. In this case, 3402 F means 340 followed by two zeros; the final value is 34,000 Ω or 34 k.

8.6 VA CHARACTERISTICS OF LINEAR AND NON-LINEAR RESISTANCE

The resistance we have worked with so far has always had a *linear relation* between the variables, V across it and I through it. This means that if the voltage applied is doubled, the current through doubles, or halve the voltage and the current drops 50%, and so on. In other words:

> Two variables have a linear relation when a change in one causes the same amount of change in the other.

The word "linear" also means that a graph of one variable versus the other will be a straight line; "linear" means "line." This is seen on the left side of Figure 8.5, in which current versus voltage data for a linear component, a resistance, have been plotted. To the right is the plot of data from a nonlinear device, the transistor. This graph is not a straight line.

These plots are called *volt-ampere characteristics*, or simply *VA characteristics*.

- The VA characteristic of a linear device is a straight line.
- The VA characteristic of a nonlinear device is not a straight line.

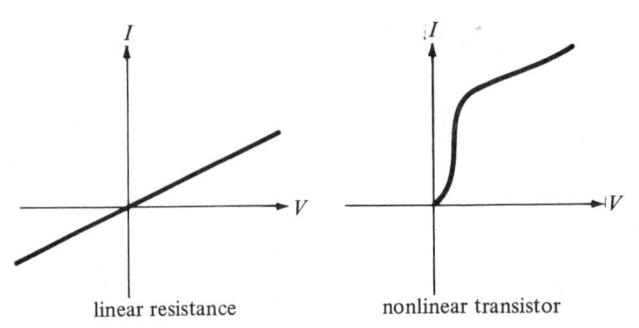

FIGURE 8.5 VA Characteristics of Linear and Non-linear Devices.

Note how current is plotted on the Y axis and voltage on the X axis.

Circuit designers are very interested in the VA characteristics of a component because these graphs tell how that component is going to act in a circuit. The plots do this in two important ways:

1. They provide current and voltage data. That is, the graph tells what current flows through the device when a particular voltage is applied.

    We do not need a graph to get this information for a linear component such as a resistance. There is a much more accurate tool for the purpose, Ohm's law. But most electronic components have *nonlinear* volt-ampere relations that are not described by mathematical equations. For example, there is no equation that tells the current through a transistor for a given voltage. For these nonlinear devices we are forced to use VA characteristics to find the data. That is why these plots are so important in solid-state work. In fact, they are used so much that special oscilloscopes, called *curve tracers*, are made that display VA characteristics on the screen of a cathode ray tube.
2. The plots tell the resistance of the device. This is no help in the case of resistors. They have a fixed value of resistance which is easily determined by color code or ohmmeter. But transistors and other solid-state devices have *dynamic resistance*, that is, resistance that changes depending on applied voltage. Dynamic resistance cannot be measured on an ohmmeter. It is, however, easily obtained from the *slope* of a VA characteristic.

We learn from mathematics that slope tells how steep a line is. It is defined as

$$M = \frac{\Delta Y}{\Delta X} \qquad (8.5)$$

where M = slope
ΔY = change in the vertical or Y direction
ΔX = change in the horizontal or X direction

Because current is plotted on the Y axis and voltage on the X axis, the slope of a VA characteristic is in units of amperes/volts. The inverse slope, therefore, will be in volts/amperes, which is, of course, ohms. In other words, and here is the important point:

> The inverse slope ($\Delta X / \Delta Y$) of a component's VA characteristic tells the resistance of the component for the range over which ΔX and ΔY were measured.

8.6 VA CHARACTERISTICS OF LINEAR AND NON-LINEAR RESISTANCE

EXAMPLE 8.12

Given: the VA characteristics of two components.

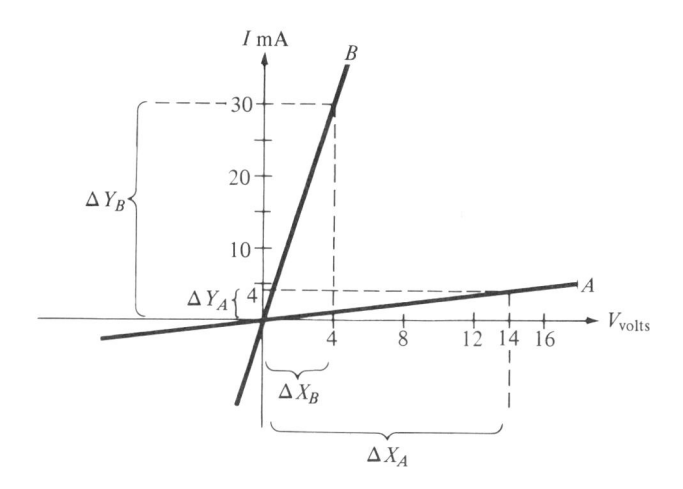

a. What are the two components?

| Since both have linear VA characteristics, they must be resistances.

b. What is the resistance of each?

Calculate the inverse slope for each:

$$R_A = \frac{\Delta V}{\Delta I} = \frac{14 - 0}{(4 - 0) \times 10^{-3}} = \frac{14}{4 \times 10^{-3}} = 3.5 \text{ k}$$

$$R_B = \frac{\Delta V}{\Delta I} = \frac{4 - 0}{(30 - 0) \times 10^{-3}} = \frac{4}{30 \times 10^{-3}} = 133.3 \text{ }\Omega$$ ∎

By measuring the changing slope of a component's VA characteristics, we can determine whether that component acts as a high or low resistance in a circuit. For example, study the transistor VA characteristics in Figure 8.6. When the voltage across the transistor is low, the transistor operates in region A. Here we see a large ΔI for a small ΔV; the inverse slope is low. The transistor is therefore acting as a low resistance.

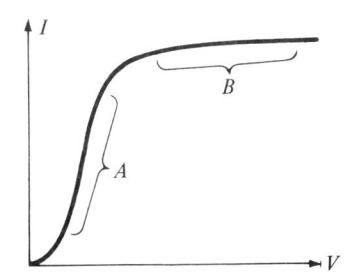

FIGURE 8.6 Transitor VA Characteristics Showing Regions of Low and High Resistance.

Let V increase and operation moves along the curve to region B. Now there is large ΔV for little ΔI; the inverse slope is high. The transistor now acts as a high resistance. To review:

> A device acts as a high resistance when the inverse slope of its VA characteristic is high. It acts as a low resistance when the inverse slope is low. Resistance that changes depending on how a component is operated is called dynamic resistance.

8.7 RESISTIVE TRANSDUCERS

There are many physical parameters, such as temperature, that are regularly measured by electronic equipment. But electronic equipment only responds to electrical signals. Therefore, there must be a device at the input to the electronics which supplys a voltage or current that changes in step with the changes in the parameter being answered. Such a device is called a *transducer*.

Resistive transducers take advantage of the ability of certain physical parameters to change one of the three factors (ρ, L, and A) that determine resistance. Here are some examples:

Resistive Transducers That Respond to Changes in Temperature

- *Thermistors:* THERMal-resISTOR made of semiconductor material whose resistivity is inversely proportional to temperature.
- *Positive-temperature-coefficient* (PTC) *resistors:* made from semiconductor material treated to cause positive temperature coefficient.
- *Platinum resistance thermometers:* wirewound resistance made of platinum. Very popular due to stability and accuracy.

Resistive Transducers That Respond to Changes in Light Level:

Photoconductive cells. Certain semiconductor materials have reduced resistivity in strong light.

Resistive Transducers That Respond to Changes in Strain:

The *strain gage*. Resistor made of very thin wire wound in special way. When subjected to strain, that is, stretched, length increases and cross-sectional area decreases, causing resistance to increase. About the size of a postage stamp, they are glued to bridge girders to measure bending under load or to airplane wing for ground testing under forces equaling those expected in flight.

8.8 SWITCHES

Switches make, or close, a circuit by bringing two contacts together and break, or open, the circuit by pulling those contacts apart. Since there is a great deal that can be said about these components, our remarks will be limited to some of the characteristics that apply to all types:

Voltage Rating

An open switch has the source voltage across it. The maximum voltage rating tells how high this voltage may be without an arc jumping between contacts and ruining the switch.

8.8 SWITCHES

Current Rating

Maximum current that may flow through switch resistance.

Poles and Throws

A pole is the movable part of a switch which is thrown against fixed contacts to complete circuits. The most common switch, the single-pole, single-throw (SPST), has a single pole that can be thrown to just one location. The single-pole, double-throw, on the other hand, has a single pole that can be thrown to two locations. The symbols of these and other common switch types are shown in Figure 8.7.

Contact Material

Contacts of high-quality, high-reliability switches, often gold or silver plated to prevent corrosion. Mercury switches in thermostats operate when tilted.

Mechanical Force to Operate

The "push" that operates a switch comes from various sources: *Manual:* Manually operated switches include toggle switches, rotary switches changed by turning a shaft, pushbuttons, slide switches often found on stereo sets, and rocker switches. *Mechanical:* Switches activated by levers, push-rods, cams, and plungers, to name a few possibilities. *Magnetic:* Usually, electromagnetism developed by relay coil, which forces movable arm to close switch contacts.

Figure 8.8 shows a variety of switches, including DIP (bottom), pushbutton (lower left), keyboards, rotary (left center), and solid state (upper right).

Normally Open and Normally Closed Contacts.

When a relay is relaxed, that is, there is no current through its coil, its switch contacts are said to be in their *normal* position. If this is an open position, the contacts are called *normally open*. It follows that they close when the relay is activated. The opposite of normally open is *normally closed*. The two are abbreviated NO and NC, respectively.

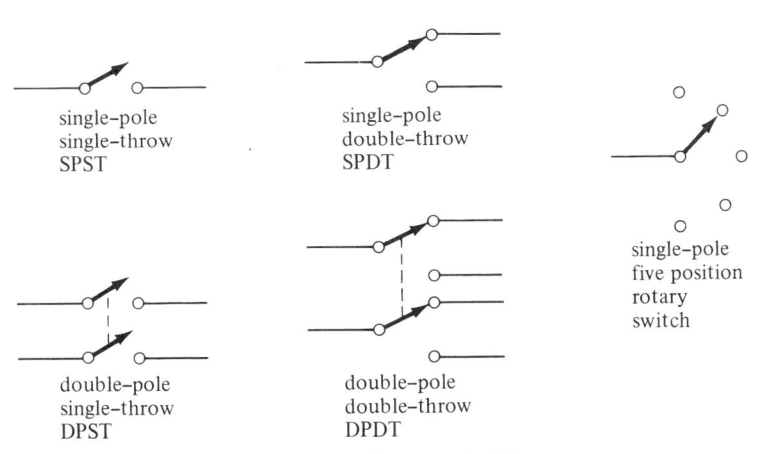

FIGURE 8.7 Schematic Symbols for Common Switches.

FIGURE 8.8 Types of Switches. (Courtesy of Grayhill, Inc.)

EXERCISES

Section 8.1

1. List the four factors that influence the resistance of a piece of material.

2. For each of the following, state whether the resistance of a conductor is directly proportional, indirectly proportional, or not proportional.
 a. To cross-sectional area. b. To resistivity of material. c. To circuit current.
 d. To conductor length. e. To applied EMF.

3. What happens to a conductor's resistance under each of the following conditions?
 a. Double the length.
 b. Halve the cross-sectional area.
 c. Halve the length and halve the cross-sectional area.
 d. Double the length and halve the cross-sectional area.
 e. Triple the diameter.
 f. Halve the diameter and quadruple the length.

4. Fill in the blanks.

	R (Ω)	ρ ($\Omega - m$)	Length (m)	Area (m^2)
a.		1.65×10^{-8}	3.5	3.5×10^{-3}
b.	39.2		4.8×10^3	1.2×10^{-5}
c.	0.35	3.5×10^{-5}		10^{-6}
d.	21.3×10^{-3}	1.7×10^{-8}	275	
e.		2.8×10^{-8}	3.5×10^5	78.5×10^{-6}

EXERCISES

5. An underground telephone cable is protected by an aluminum sheath whose dimensions are as shown. What is the resistance of 11.5 kilometers of this covering?

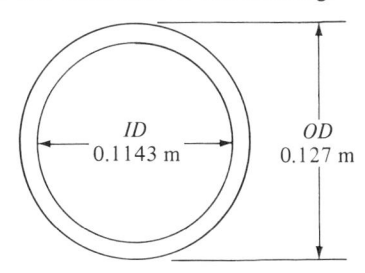

ID 0.1143 m
OD 0.127 m

6. A printed-circuit board (PCB) is covered with the copper layer shown below. Calculate the resistance of this layer in the direction of arrow *A*. Repeat for arrow *B*.

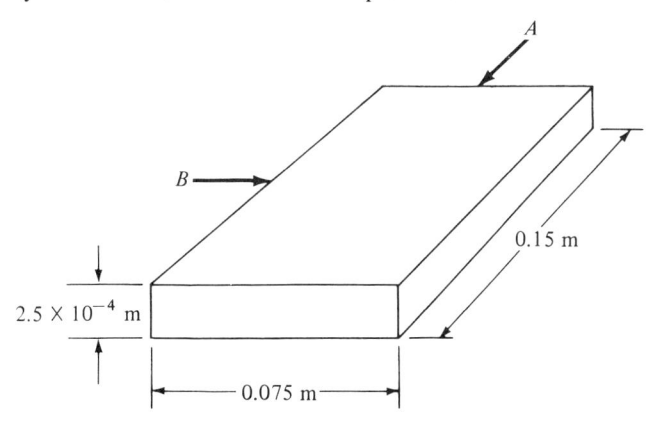

0.15 m
2.5×10^{-4} m
0.075 m

7. An aluminum wire $\frac{1}{4}$ in. in diameter is 300 ft long. Finds its resistance.

8. Which of the variables of Equation 8.1 is being changed when the resistance of a rheostat or potentiometer is adjusted? See Figures 4.7 and 4.8, respectively.

Section 8.2

9. Explain the term "temperature coefficient" in your own words. Then explain the difference between positive and negative temperature coefficients.

10. If a parameter such as resistance is to be stable under changing temperature conditions, should it have a high- or a low-temperature coefficient? Discuss.

11. A copper rod is 10 m long and 1 cm in diameter.
a. What is its resistance at 20°C? b. What is its resistance at 600°C?

12. Nichrome wire is used in the heating elements of toasters. If an element has a resistance of 12.1 Ω at 20°C, what is its resistance at 400°C?

13. The tungsten filament of a 100-W 110-V incandescent bulb has 12 Ω resistance when the bulb is off (filament at 20°C.)
a. The bulb is turned on. What is the instantaneous power drawn before the filament heats up?
b. Filament resistance increases with temperature so that the bulb draws 100 W when hot. What is the resistance now?
c. At what temperature is the filament operating now?

14. a. What is the ampacity (continuous-duty current capacity) of 14-gage wire in open air? (Find the value in Table 8.3.)
b. The value is listed for continuous duty. Could the wire handle greater current if the current was pulsed, on and off? Discuss.
c. The value is given for open air. What do you suppose happens to maximum current capacity when the wire is enclosed in an insulated wall? Discuss.

RESISTANCE

15. Find the new resistance in each case.

	$R_{original}$	TC (ppm/°C)	ΔC	R_{new}
a.	1.2 M	100	50	
b.	150 k	50	110	
c.	47 k	−220	67.7	
d.	0.75 k	1250	45	

16. Calculate the temperature coefficient of resistance as a percent change per degree Celsius for each part of Exercise 15.

Section 8.3

17. Copper wire resistance per 1000 ft doubles (approximately) as the gage increases three sizes. The resistance of 10 gage is 1.018 Ω.
a. Calculate the approximate resistance of 16 gage per 1000 ft. Compare to the value in Table 8.3.
b. If 18 gage has 6.51 Ω per 1000 ft, what is the approximate value for 30 gage?
c. Compare your results to those listed in Table 8.3.

18. Fill in the blanks.

	Diameter (mil)	Diameter (in.)	Area (CM)
a.		0.185	
b.			1,050,625
c.	7.2		

19. Fill in the blanks.

	Resistance (Ω)	ρ (Ω–CM/ft)	Length (ft)	Diameter (mil)
a.		10.37	39600	400
b.	0.03		0.4	15
c.	81.5	17		600
d.	0.01	9.8	0.003	

20. Electrical energy is shipped 72 mi. from Niagara Falls, Ontario, to Toronto, over an aluminum wire $\frac{1}{2}$ in. in diameter. What is the round-trip wire resistance?

21. A car's taillight uses 18-gage copper wire for one path and the steel chassis for the other. Assume that both paths are 15 ft long and that the ρ for steel is the same as that for iron. Also, assume that the effective area of the chassis equals 2.25×10^6 cm. Compare the resistances of the two paths. Which path has the smaller resistance, and why?

22. *Extra Credit.* Use metric-to-English conversions to show that each ρ value in Table 8.1 is multiplied by 6×10^8 to get the corresponding value in Table 8.4. *Hints:* 2.54 cm = 1 in. $\frac{\pi}{4}$ included in the metric case.

Section 8.4

23. Explain how the leads are formed in the copper layer of a PC board.
24. List the advantages and disadvantages of PC boards over regular wiring.

EXERCISES

Section 8.5

25. What considerations should be given to the choice of resistors when designing a radio receiver to process signals from a space vehicle photographing Mars? What types of resistors should be used? What types avoided? Why?

26. Repeat Exercise 25 for the design of an electronic toy.

27. a. What effect does humidity have on the resistance of a 10-M resistor? *Hint:* Think of resistors in parallel.
b. Can you see a practical limit to the resistance of a resistor? *Hint:* Think of the outer insulating layer.

28. Why would it not be practical to make resistors of very low resistance, say, $0.005 \, \Omega$?

29. Does each marked value in this group have accuracy and/or precision?

	Marked value (Ω)	True value (Ω)
a.	45,357.672 Ω	45,354.591 Ω
b.	45 k	39 k
c.	45 k	45.013 k
d.	45,357.672 Ω	39,200 Ω

30. Discuss the difference between precision and accuracy.

31. What happens to the cost of a resistor when its reliability, precision, and/or accuracy go up? Why?

32. In what applications is it worth spending the extra money to get high reliability?

33. Why are there more resistance values in the 5% sequence than in the 10%?

34. Describe how a wirewound resistor is made. How is its resistance determined?

35. Describe how the resistance of a film resistor can be adjusted by trimming.

36. What is the difference between thick- and thin-film resistance? Where is each type used?

37. Resistors are to be made with 15% tolerance. The lowest value is $10 \, \Omega$. What are the next four whole values so that there are no gaps?

38. Shown here is a resistive network of the type made in thick film and sold in SIP form. Suppose that E volts were applied across the input terminals. Calculate the voltage between each output terminal and ground. Make up a name that describes the function of the network.

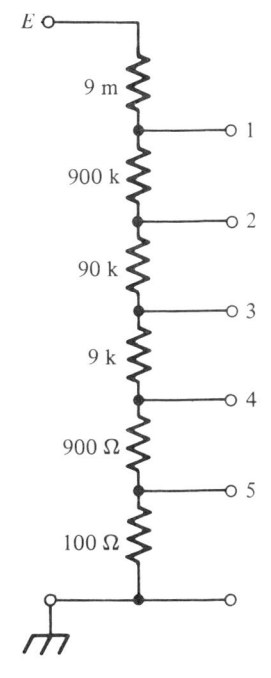

39. *Extra Credit.* The following resistor network, called an R/2R ladder, is part of a D/A (digital-to-analog) converter. Perform the following steps.

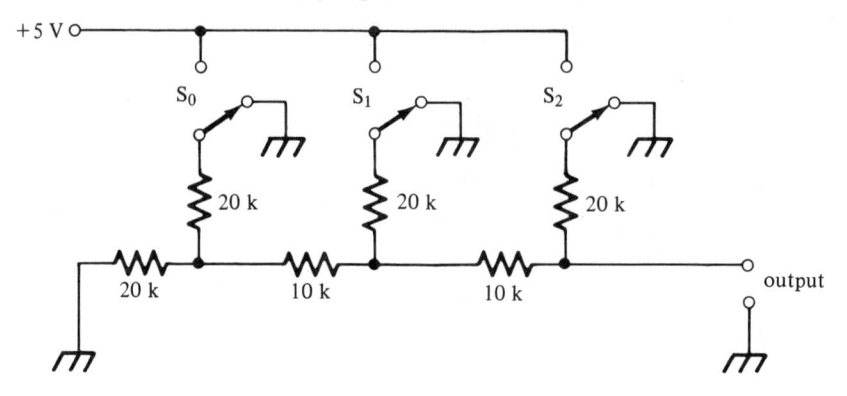

a. With S_2 in the 5-V position and S_0 and S_1 in the grounded position, determine the Thévenin equivalent circuit looking back into the output.
b. Repeat part a with S_1 in the 5-V position and S_2 and S_0 in the ground position.
c. Repeat part a with S_0 in the 5-V position and S_1 and S_2 in the ground position.
d. Compare the Thévenin circuits of parts a, b, and c. What is the internal resistance in each case? What fraction of 5 V is the open-circuit output voltage in parts a, b, and c? Compare the three Thévenin circuits and discuss what the circuit is doing.

40. Every value of 10% tolerance EIA-coded carbon-composition resistors is available in its own drawer. In which drawer should we look for each of the following? a. 13.9 k; b. 53 Ω; c. 1.63 M; d. 349.5 k; e. 9.4 k.

41. The following resistors are marked with military code. What is the resistance and tolerance of each? a. 103K; b. 915G; c. 5R7F; d. 762J.

42. What is the maximum resistance that each of the resistors of Exercise 41 could have and still be within tolerance?

Section 8.6

43. What does it mean when two variables are said to have a linear relationship?

44. Given: three lines—one horizontal, one vertical, and one slanting. Which line has: a. both ΔX and ΔY; b. ΔX but no ΔY; c. ΔY but no ΔX?

45. Shown here are the VA characteristics of a solid-state device called a *diode*. In what region does the diode: a. have a large ΔI for a given ΔV; b. have a small ΔI for a given ΔV; c. exhibit high resistance; d. exhibit low resistance?

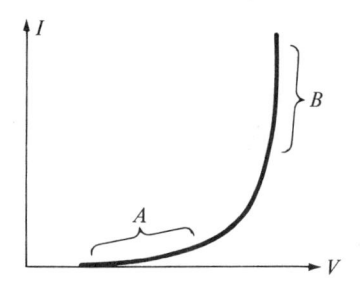

46. Define slope mathematically. What is the slope of a horizontal line? Of a vertical line? Of a slanting line?

47. Lightning strikes, causing a microsecond kilovolt burst on a 110 V power line. *Transients*, as these bursts are called, destroy solid-state components. For protection, another solid-state device, the *varistor* (from VARiable-resISTOR), is connected across the line. At 110 V, varistor resistance is

EXERCISES

megohms, but when a transient comes, resistance drops to a few ohms in nanoseconds, absorbing the transient. Draw the VA characteristics of this device. Indicate areas of low and high resistance.

48. What is dynamic resistance? Does a resistor have it? What components do have it?

Section 8.7

49. What is the function of a transducer? What do we call the transducer that does each of the following?
   a. Converts electricity to light
   b. Converts sound to electricity
   c. Converts electricity to sound
   d. Converts electricity to heat

50. Describe how a potentiometer could be used as a transducer to indicate the direction pointed to by a rotatable TV antenna.

51. Describe how a resistance could be used as a transducer to indicate fluid height in a tank.

52. Describe how a thermistor could be used as a transducer.

Section 8.8

53. What is meant by the terms "pole" and "throw" when referring to switches. What do the letters SPST mean? Draw the schematic symbol of this component. Repeat for SPDT, DPST, and DPDT.

54. Switch contacts are often plated with gold or silver to prevent corrosion. Why? What is wrong with corrosion on switch contacts?

55. Analyze this circuit and tell where it might be found in everyday life. Remember that there are only two positions for either switch—they do not stop in the middle.

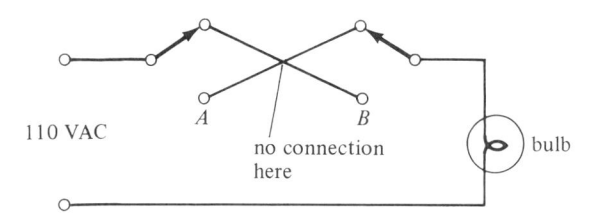

CHAPTER HIGHLIGHTS

Four factors affecting resistance: length, cross-sectional area, resistivity, and temperature.

Temperature coefficient of resistance tells how a component's resistance changes with temperature. Positive TC resistance increases as temperature increases. Negative TC resistance decreases as temperature increases.

Wire sizes are given in AWG (American Wire Gage) numbers. The smaller the number, the larger the wire.

Printed circuit boards provide a reliable, compact, neat, and low-cost (in quantities) method of wiring circuits.

Reliability is especially important in military electronics and other areas involving life and/or high cost.

Precision is how close we can make a measurement. Accuracy is how close that measurement is to the truth.

Two variables have a linear relationship when a change in one causes equal change in the other.

Logic behind resistance values— high reading of one value is close to or overlaps low reading of the next.

By reading the inverse slope of VA characteristics, circuit designers determine component resistance in a specific voltage range.

Static resistance: same resistance for all voltages.
Dynamic resistance: varying resistance for different voltages.
Transducers provide voltage or current that changes as some physical parameter changes.

Formulas to Remember

$$R = \frac{\rho L}{A} \qquad \text{resistance of a piece of material}$$

Two ways to determine the effects of temperature change:

R at some new temperature $= R$ at $20°C(1 + \alpha \Delta T)$
ppm/°C

Wire's cross-sectional area in circular mils:

A, in CM $= (D$, diameter in mils$)^2$

The inverse slope, $\Delta X/\Delta Y$, of VA characteristics is $\Delta V/\Delta I$, which provides resistance information.

CHAPTER 9

Meters and Measuring

A large portion of your career in electronics will be spent working with meters of one form or another. They will be used to calibrate, adjust, and troubleshoot circuits and systems, monitor equipment operation, and measure the performance of new designs. To get the best results from these meters, you must know their strengths and weaknesses and how to use them properly.

This chapter covers those points. It begins by describing the effects of loading on measuring, continues with a discussion of the two major types of meters, digital and analog, and ends with descriptions of other dc meters. Ac meters are described in Chapter 22.

At the completion of this chapter, you should be able to:

- State the question that must be asked whenever test equipment is connected to a circuit.
- Choose the meter that minimizes loading errors.
- Describe differences between analog and digital meters.
- Explain how the D'Arsonval meter movement works.
- Convert the basic ammeter to read voltage, resistance, and higher current.
- Use a voltmeter's ohms-per-volt rating to calculate meter loading.
- State the difference between a VOM and a FETVOM.
- Explain the basics of digital meters and state what a "half-digit" is.
- Discuss the strengths and weaknesses of digital and analog meters.

9.1 THE EFFECTS OF LOADING ON MEASURING AND HOW TO MINIMIZE THEM

A jogger friend asks you to design an electronics package to measure her performance. You go to work and when finished, proceed to strap 100 lb of batteries and equipment on her back. What happens? The weight of the electronics slows the runner so badly that any measurements are worthless. In other words, *the measuring equipment changed that which was being measured.*

Remember this rule:

Whenever you connect a piece of test equipment to a circuit, ask yourself: Will this equipment affect the data I want to measure?

A great deal of time and embarrassment can be saved by doing so.

Just as the jogger's speed drops when she is loaded down, so will output voltage when a circuit is loaded. We studied this in Chapter 6. You will recall the statement that driver circuit output falls when loaded by driven circuit input resistance.

The same thing happens whenever a meter is connected to a circuit to read a voltage (which is the most frequent reading we make). The circuit becomes the driver, the meter becomes the driven, and meter input resistance loads driver output. If loading is severe, the meter readings will not mean what you think they mean; the measuring equipment has changed that which was being measured.

Chapter 6 also showed how to reduce loading to acceptable levels by making driven circuit input resistance as high as possible compared to driver output resistance. This leads to an important point, one that electronic personnel must keep in mind when taking voltage readings:

To get the least loading and the best accuracy when measuring voltage, always use the meter with the highest input resistance.

EXAMPLE 9.1

Given: the following voltage divider:

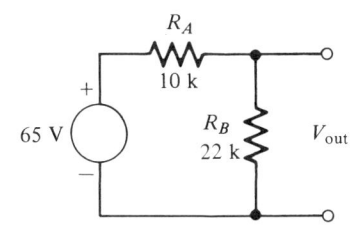

a. Calculate output voltage under no-load conditions—or, look at it another way—measure V_{out} with a meter having infinite internal resistance.

Using the voltage-divider formula gives us

$$V_{out} = \frac{ER_B}{R_A + R_B}$$

$$= \frac{65(22\ k)}{10\ k + 22\ k}$$

$$= 44.7\ V$$

b. What would the voltage be if read by a meter having an internal resistance of 50 k?

The circuit becomes

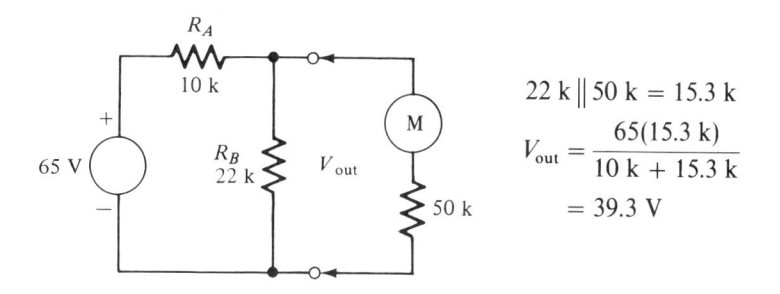

$22\ k \parallel 50\ k = 15.3\ k$

$$V_{out} = \frac{65(15.3\ k)}{10\ k + 15.3\ k}$$

$$= 39.3\ V$$

V_{out} has dropped from an expected 44.7 V to 39.3 V because of loading.

c. Repeat part b for a meter having an internal resistance of 3 megohms:

$$3\ \text{megohms} \parallel 22\ k \approx 22\ k$$

$$V_{out} \approx \frac{65(22\ k)}{10\ k + 22\ k} \approx 44.7\ V$$

Measured voltage is now essentially the same as the open-circuit value. Pay particular attention to the way that measured voltage comes closer to the true value as meter resistance becomes very much greater than circuit resistance.

d. Change the 22 k to 2.2 M and the 10 k to 1 M. Remeasure V_{out} with the meter of part c.

With this change, meter resistance is no longer very much greater than circuit resistance; therefore, loading should be severe again:
The circuit is now

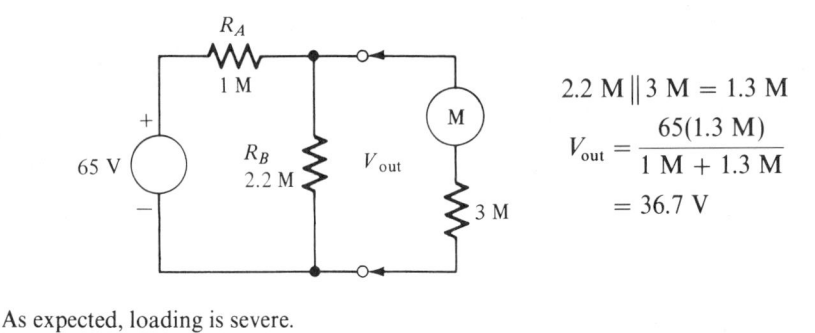

$2.2 \text{ M} \| 3 \text{ M} = 1.3 \text{ M}$

$V_{out} = \dfrac{65(1.3 \text{ M})}{1 \text{ M} + 1.3 \text{ M}}$

$= 36.7 \text{ V}$

As expected, loading is severe. ∎

9.2 ANALOG METERS

A group of analog meters is shown in Figure 9.1. They are called *panel meters* because they are built to be mounted on the panel of a piece of equipment (most meters in electronics are panel type). Note the parameter being measured as well as the variety of scales.

Analog meters have a *needle* which moves in front of a scale of numbers. The operator reads data from the meter by noting needle position relative to the scale. Thus needle position is ANALOGous to the value of the parameter being measured.

9.3 THE D'ARSONVAL METER MOVEMENT

The part of any analog meter that makes the pointer move is called the *meter movement*. Although there are several types of movements, by far the most common is the *D'Arsonval*, named after its inventor. To explain how it works, we must discuss magnetism first.

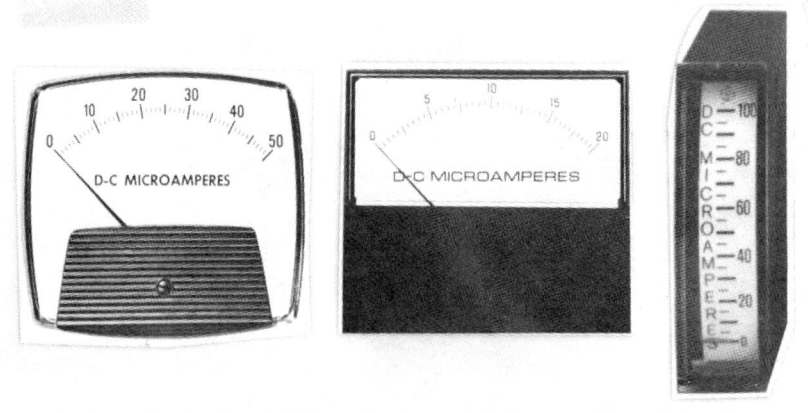

FIGURE 9.1 Analog Panel Meters. Courtesy of Yokogawa Corp. of America

9.3 THE D'ARSONVAL METER MOVEMENT

There is something special about the region between a magnet's north and south poles called a *magnetic field*. Any iron object in this field is pushed and pulled by invisible forces of *magnetism*. These forces act on other magnets, too. For example, the north poles of two magnets repel one another, as do the two south poles. However, the north pole of one magnet attracts the south pole of another. (Here again is the familiar rule: Opposites attract, likes repel.)

A current-carrying conductor has a magnetic field around it. Therefore, it will be pushed and pulled by magnetic forces if placed in a magnetic field. This is the principle behind the electric motor and the D'Arsonval movement.

A horseshoe-shaped permanent magnet provides a fixed magnetic field for the D'Arsonval movement (see Figure 9.2). Suspended inside the field is a coil of fine wires mounted on a shaft so that it can rotate. At the ends of the shaft are springs that limit shaft rotation and return the shaft to its resting position.

The current measured by the meter passes through the coil and develops another magnetic field. Interaction between the two fields creates a force that rotates the coil about the shaft; the stronger the current, the greater the rotation. Since the pointer is also mounted on the shaft, it rotates too, coming to rest at a spot determined by coil current and the counterforce of the springs.

The direction of the force turning the coil depends on the direction of the current. Therefore, polarity must be observed; otherwise, the needle travels down scale rather than up. To show polarity, positive and negative signs are located near a meter's terminals.

There are two ways to suspend the coil in the magnetic field:

The Pivot-and-Jewel (P & J) Suspension

In P & J suspension, the shaft on which the coil is mounted is made of hardened steel and called the *pivot*. In high-quality meters pivot ends ride in low-friction jewels as in a fine watch. Mounted at the ends of the pivot are the springs mentioned above. Besides counteracting the magnetic forces, they act as conductors to carry current to the coil. P & J instruments have good accuracy and repeatability (always give the same reading for the same current). They are very popular and stand up well to vibration but are somewhat more expensive and their jewels can break from the shock of a fall.

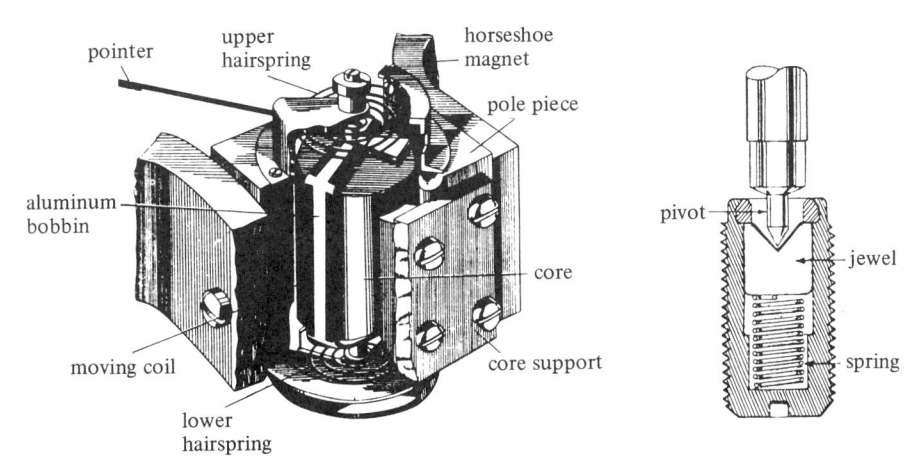

FIGURE 9.2 Inside View of the D'Arsonval Meter Movement with Pivot and Jewel Suspension.

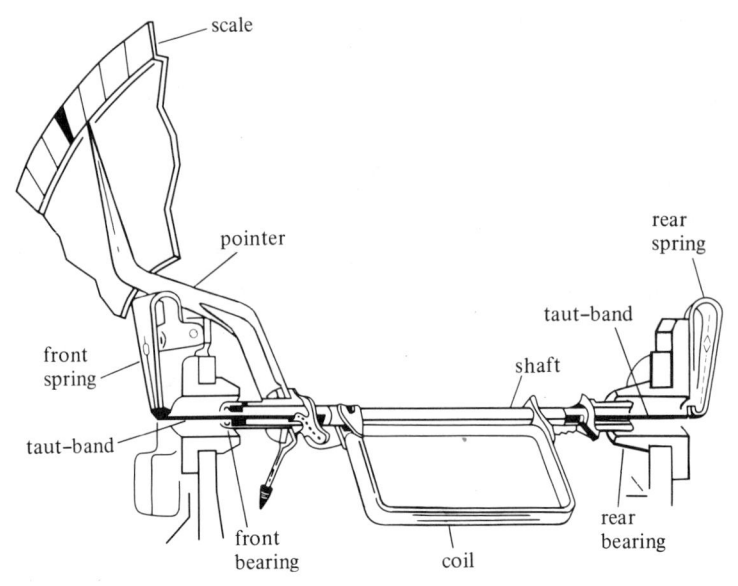

FIGURE 9.3 Taut-Band Suspension System. Courtesy of Yokogawa Corp. of America.

The taut-band Suspension

Each end of the meter's shaft has a fine steel ribbon or band fixed to it (see Figure 9.3). The bands are pulled taut and their other ends clamped so that the bands twist when coil and shaft rotate. This twisting acts as a spring (much as a torsion bar in a car) providing the counterforce to the magnetic force. The technique eliminates the friction of the pivot-and-jewel system. As a result, taut-band meters have greater sensitivity and are more rugged than P & J units but do not stand vibration well.

9.4 SENSITIVITY, PRECISION, AND ACCURACY OF THE D'ARSONVAL MOVEMENT

The current necessary to cause pointer swing to the far end of a D'Arsonval scale is called the *full-scale deflection current*. It is abbreviated I_{fsd}. A movement with an I_{fsd} of 1 mA is often referred to as a 1-mil movement.

The lower the value of I_{fsd}, the greater the sensitivity of the movement. Inexpensive meters, for example, take 1 mA to swing the pointer full scale, while expensive units require only 20 μA. The high-quality meter is thus 50 times more sensitive.

The precision of an analog meter is determined by the number of places that can be read from the scale. It is limited by two factors. The first of these is human—the care and skill that goes into a measurement. The second factor depends on scale size. The larger the scale, the greater the number of dividing lines drawn on it, therefore, the smaller the unit of measure. Typically, an analog panel meter can be read to two places, with the operator able to estimate a third.

The usual accuracy of analog panel meters is 2%. But watch it; there's a trap. It is natural to think the number means that any reading is accurate to within plus or minus 2%. It doesn't. The number really means that any reading is accurate to within 2% of full-scale value. So a reading anywhere on the scale of a 100-V, 2% meter may be off as much as 2 V. This is less of a problem at the high end of the scale. A reading of 95 V, for example, could read anywhere from 93 to 97 V, an error of about 2.2%. At 15 V, however, at the low end of the scale, the reading falls anywhere between 13 and 17 V, a maximum error of 13.3%. At 5 V, possible error has climbed to 40%. The point is: Any error becomes a higher percentage of the actual reading as that reading moves down scale. Therefore, do not read 5 V on a 100-V meter. Use the meter whose full-scale value is closest to the value being read. Remember:

| Choose analog meters so that readings fall in the upper third of the scale.

There is another factor that affects analog accuracy. The meter's pointer is not actually on, but a bit in front of, the scale. Therefore, reading the meter from the side causes an error. For example, if your eye is to the right of the pointer, the pointer appears to lie over a lower scale value. When looking from the left, the value appears higher. The difference between the true value under the pointer and the value that appears to be under the pointer is called a *parallax error*.

To prevent parallax errors, an antiparallax mirror is built into higher-quality meters. To use it, the operator moves his or her eye until the pointer covers its own mirror image. This places the eye directly above the needle, thereby preventing parallax error.

9.5 HOW TO MAKE A D'ARSONVAL MOVEMENT READ HIGHER CURRENTS

If a current greater than I_{fsd} is sent through a meter movement, the pointer swings off scale to the far right, stopping when it hits the inside of the case. When this happens, we say that the meter is *pinned*. In severe cases, pinning is so violent that the needle slams against the case and is bent or broken.

A D'Arsonval movement can be modified to handle currents greater than I_{fsd}. This is done by connecting a resistance in parallel with the movement. Called a *shunt*, this resistance carries the excess current greater than I_{fsd}. Let us derive a formula for finding R_{shunt}.

Figure 9.4 shows a movement in parallel with a shunt resistance. Current through the shunt is

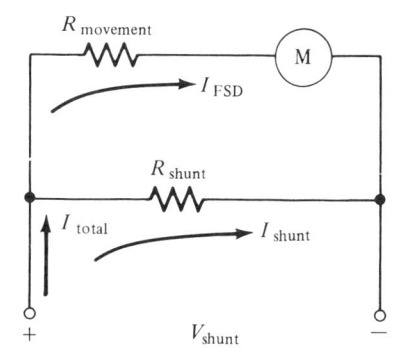

$$I_{shunt} = I_{total} - I_{fsd}$$

FIGURE 9.4 Shunt Used to Raise Current Capacity of D'Arsonval Movement.

The voltage across the shunt equals that across the movement. Therefore,

$$V_{\text{shunt}} = I_{\text{fsd}} R_m$$

where R_m = movement resistance.

By Ohm's law,

$$R_{\text{shunt}} = \frac{V_{\text{shunt}}}{I_{\text{shunt}}}$$

$$= \frac{I_{\text{fsd}} R_m}{I_{\text{total}} - I_{\text{fsd}}} \qquad (9.1)$$

If $I_{\text{total}} \gg I_{\text{fsd}}$, the equation reduces to an approximation:

$$R_{\text{shunt}} \approx \frac{I_{\text{fsd}} R_m}{I_{\text{total}}}$$

Ammeters often come with several shunt resistances, which may be switched to provide a variety of current ranges.

EXAMPLE 9.2

Modify a 10-mA, 1.8-Ω D'Arsonval movement to measure 35 mA full scale. Calculate the total meter resistance.

$$R_{\text{shunt}} = \frac{I_{\text{fsd}} R_m}{I_{\text{total}} - I_{\text{fsd}}}$$

$$= \frac{10 \text{ mA}(1.8)}{35 \text{ mA} - 10 \text{ mA}}$$

$$= \frac{1.8 \times 10^{-2}}{25 \times 10^{-3}}$$

$$= 0.72 \text{ Ω}$$

$$R_{\text{total}} = R_{\text{shunt}} \text{ in parallel with } R_m$$

$$= \frac{0.72(1.8)}{0.72 + 1.8}$$

$$= 0.51 \text{ Ω}$$

The circuit diagram is

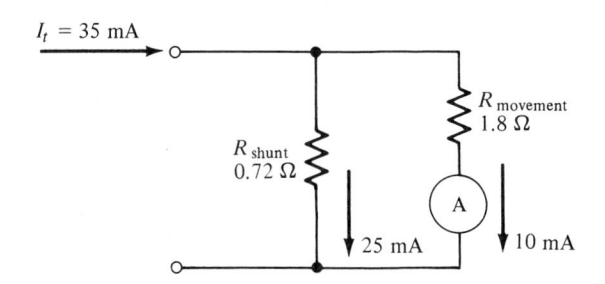

9.6 THE D'ARSONVAL MOVEMENT AS A VOLTMETER

What voltage must be applied between the terminals of a D'Arsonval movement to force an I_{fsd} of 50 mA through an R_m of 0.5 Ω? By Ohm's law, $V = IR =$ 50 mA(0.5 Ω) = 25 mV. So whenever there is 25 mV between terminals, the needle reads full scale. Reduce V to 12.5 mV, the current drops to 25 mA and the needle falls to half scale, and so on. The point is: The needle responds to voltage changes. Therefore, remove the milliamp scale, put in a millivolt scale, and what was a 50 milliammeter is now a 25 millivoltmeter.

Twenty-five millivolts will never win prizes as the most useful voltage. One hundred volts is a lot more practical. But we can't put that potential across a 50-mil movement with 0.5 Ω resistance; 200 A will flow (but not for long). Something must be done to limit current to I_{fsd}.

What do we always do to limit current when voltage can't be reduced? We add resistance. R_m must be increased by adding series resistance. How much? Let us derive an equation for determining this series resistance.

Figure 9.5 shows a meter movement in series with a resistance that we will call $R_{series\,add}$. We will also call total meter resistance R_{input}.

$$R_{input} = R_{movement} + R_{series\,add}$$

Therefore,

$$R_{series\,add} = R_{input} - R_{movement}$$

But

$$R_{input} = \frac{\text{range voltage}}{I_{fsd}}$$

Therefore,

$$R_{series\,add} = \frac{V_{range}}{I_{fsd}} - R_{movement} \tag{9.2}$$

If $R_{input} \gg R_{movement}$,

$$R_{series\,add} \approx \frac{V_{range}}{I_{fsd}}$$

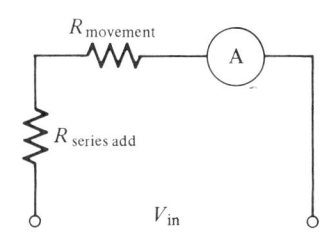

FIGURE 9.5 Adding Series Resistance to Ammeter to Make a High-Range Voltmeter.

EXAMPLE 9.3

In Example 9.2, a 10 mA, 1.8-Ω D'Arsonval movement was converted to a higher-range ammeter by adding shunt resistance.

a. With the shunt resistance removed, what voltage causes I_{fsd} to flow?

$$V_m = I_{fsd} R_m$$
$$= (10 \times 10^{-3})(1.8)$$
$$= 18 \text{ mV}$$

b. Add series resistance to the same movement to make a 100-mV voltmeter.

$$R_{series\,add} = \frac{V_{range}}{I_{fsd}} - R_{movement} = \frac{100 \text{ V}}{10 \text{ mA}} - 1.8$$
$$= 10 - 1.8$$
$$= 8.2 \text{ Ω}$$

EXAMPLE 9.4

Modify the movement of Example 9.3 to make a 10-V voltmeter. Draw the circuit of a three-range voltmeter for 18 mV, 100 mV, and 10 V.

$$R_{series\,add} = \frac{V_{range}}{I_{fsd}} - R_{movement} = \frac{10 \text{ V}}{10 \text{ mA}} - 1.8$$
$$= 1000 - 1.8$$
$$\approx 1000 \text{ Ω}$$

The circuit diagram is

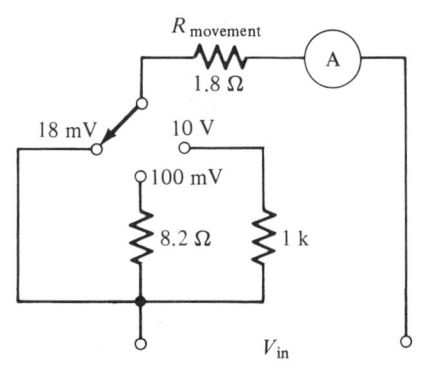

9.7 VOLTMETER INPUT RESISTANCE: THE OHMS-PER-VOLT RATING

It was pointed out earlier that loading is minimized by choosing the voltmeter with highest input resistance. How is this value determined?

We have seen that the formula for input resistance of a voltmeter is

$$R_{input} = \text{range voltage} \frac{1}{I_{fsd}}$$

Note the factor $1/I_{fsd}$ in the equation. Its units must be ohms per volt because by Ohm's law, we know that the units of I_{fsd} are volts per ohm.

The reciprocal of a meter's full-scale deflection current is $1/I_{fsd}$. This is called the meter's *ohms-per-volt rating*.

Remove I_{fsd} from the equation for voltmeter input resistance and replace it with a meter's ohms-per-volt rating. The result is the industry formula for finding voltmeter input resistance:

$$R_{input} = \text{range voltage} \times \text{ohms-per-volt rating} \qquad (9.3)$$

EXAMPLE 9.5

What is the ohms-per-volt rating of a 50-μA D'Arsonval movement?

$$\begin{align}\text{ohms-per-volt rating} &= (I_{fsd})^{-1}\\ &= (50 \times 10^{-6})^{-1}\\ &= 20{,}000 \; \Omega/V\end{align}$$

∎

EXAMPLE 9.6

If the movement of Example 9.5 was made into a 100-V voltmeter, what would its input resistance be? Repeat for a 500-V scale.

$$\begin{align}R_{input} &\text{ for 100-V scale}\\ &= \text{ohms-per-volt rating (range voltage)}\\ &= \frac{20 \text{ k ohms}}{\text{volt}}(100 \text{ V})\\ &= 20 \times 10^5 = 2 \text{ M}\end{align}$$

$$\begin{align}R_{input} &\text{ for 500-V scale}\\ &= \frac{20 \text{ k ohms}}{\text{volt}}(500 \text{ V})\\ &= 10^7 = 10 \text{ M}\end{align}$$

∎

9.8 THE D'ARSONVAL MOVEMENT AS AN OHMMETER

Figure 9.6 shows the diagram of a basic ohmmeter built around a D'Arsonval movement. The scale of the movement is marked in ohms and the resistance being measured, R_x, is connected to the terminals.

When the unknown is 0 Ω (terminals shorted), the movement reads:

$$\begin{align}\frac{E}{R_{series} + R_{movement}} &= \frac{10 \text{ V}}{9990 + 100}\\ &= \frac{10 \text{ V}}{10 \text{ k}}\\ &= 1 \text{ mA} \quad \text{or} \quad I_{fsd}\end{align}$$

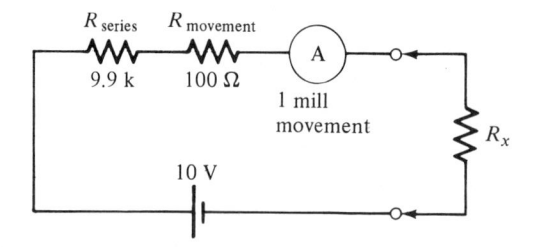

FIGURE 9.6 Simple Ohmmeter Using a D'Arsonval Movement.

Therefore, "zero ohms" is printed at the right end of the meter's scale. For $R_x = 10$ k, series-circuit analysis tells us that

$$I = \frac{E}{R_x + R_{series} + R_{movement}}$$

$$= \frac{10}{10 \text{ k} + 9.9 \text{ k} + 100 \text{ }\Omega}$$

$$= \frac{10}{20 \text{ k}}$$

$$= 0.5 \text{ mA}$$

The needle stops at half scale because circuit resistance has doubled. A 10 representing 10 k is printed here (the k for kilo is omitted for reasons given below.)

When the resistance is infinite (terminals open) the needle does not move and the ∞ sign is printed at the far left end.

Using the same technique, other values of R_x are printed. When all points have been plotted, a scale such as that in Figure 9.7 results (currents are included to help you understand how the scale is drawn; they do not appear on actual ohmmeters).

Notice how the scale is non-linear. It also reads backwards. This happens because the meter only "understands" current and zero current flows when resistance is maximum.

Practical ohmmeters are equipped with several switch-selectable ranges to cover a wider span of resistances. These ranges are all read on one scale. To make this possible, that scale is printed with numbers only; the k for kilo is omitted. The range switch indicates a multiplier for each setting. The needle reading is multiplied by that multiplier to obtain the final resistance. For example, a reading of 5 with the switch in the $R \times 100$ setting stands for 500 Ω. The same 5 in the $R \times 100$ k setting represents a 500 k resistance, and so forth.

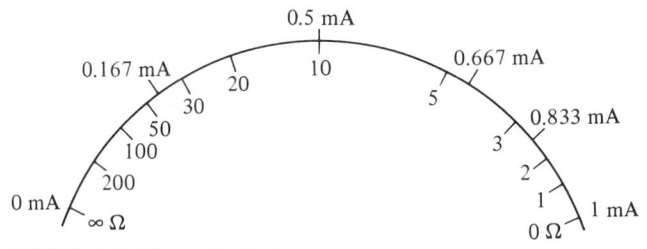

FIGURE. 9.7 Ohmmeter Scale.

Our ohmmeter has one remaining problem. As the battery ages, its voltage drops causing the meter to read below full scale when R_x is zero ohms. To prevent this the battery voltage is made a bit too high. Then a potentiometer called the *zero adjust* or *ohms adjust* pot is connected in series with the battery. Before measuring a resistance short the ohmmeter terminals and adjust the pot so the needle sits on zero ohms. This is called *zeroing the meter* (when the meter can no longer be zeroed, change the battery). This technique also cancels any resistance of the leads connecting the ohmmeter to the component under test.

9.9 HOW TO USE AN OHMMETER

There are four rules to remember when using an ohmmeter:

- Zero the meter whenever a range change is made.
- Make sure that the circuit being tested is off.

An ohmmeter is really an ammeter fooled into thinking that it is measuring resistance. Any outside voltage from a "hot" circuit causes current which pins the needle.

- Measure only one resistance at a time.

More often than not, a resistor is shunted by other circuit components. These can affect readings. Check the schematic. If in doubt, lift (unsolder) one end of the resistor before taking a reading.

- Know that some components can be ruined by testing with an ohmmeter.

An ohmmeter measures current that its internal battery forces through a component. Some components can be damaged by this current or voltage. Be extra careful with solid-state devices, especially integrated circuits.

9.10 THE VOM AND FETVOM

Combine a D'Arsonval movement with the resistances and batteries needed to make it read volts, ohms, and milliamperes and a switch to make all the right connections, and we get one of the most useful tools in electronics, the VOM, or as it is also called, a multimeter.

A typical VOM (see Figure 9.8) has the following ranges: six for dc voltage, six for ac voltage, five for dc current, and four ohmmeter ranges. It has a 20,000-Ω/V rating on dc and 5000 Ω/V on ac.

While the model described above has six dc voltage ranges, there are only three voltage scales on the face of the movement. These are 0 to 10, 0 to 50, and 0 to 250. The 0-to-10 scale serves for both the 10- and 1000-V ranges, the 0-to-50 scale for the 50- and 5000-V ranges, and the 0-to-250 scale for the 2.5- and 250-V ranges. The operator supplys the decimal point. Remember: When using a VOM, the setting of the range switch is as important as the reading from the scale.

FIGURE 9.8 Typical VOM. (Courtesy of Triplett Corp.)

A few other pointers about using a VOM:

- Avoid pinning the needle when reading an unknown voltage or current by starting on the highest range and working down.
- If the meter is connected backward, the needle pins in the negative direction. To correct this, either reverse the leads or change the *polarity-reversal switch* included on most VOMs.
- Remember the rules for using the ohmmeter.
- Ac scales look the same as dc scales. Make sure to read the correct scale.
- Be extra careful when using high-voltage ranges (usually 1 kV and 5 kV). Make all connections to the circuit while it is off and don't touch it when it is on.

Low-cost VOMs have an ohms-per-volt rating of 1000 on dc. Quality units have ratings of 20,000 or 50,000 Ω/V. Even these might cause too much loading in some cases. When this happens, use a FETVOM.

The voltage or current being measured by a FETVOM drives a solid-state device called a *field-effect transistor*, or *FET*. The FET input is nearly an open circuit; hence the FETVOM has a very high input resistance, reaching 10^{14} Ω in some models. And the resistance does not vary with range changes as it does for a regular VOM.

The FET's output, which is proportional to the input, drives the meter circuit. Therefore, the meter loads the FET output but not the circuit under test. In other words, the FET isolates the circuit from the meter.

A typical FETVOM is shown in Figure 9.9.

FIGURE 9.9 FETVOM. (Courtesy of Triplett Corp.)

9.11 DIGITAL METERS

Digital meters supply their data in actual numbers. There is no need to read needle position over a scale. And being electronic, they have some interesting abilities not found in their analog counterparts.

There are six parts to a basic digital meter:

- *Clock generator*: circuit in any digital device whose function is to generate many voltage pulses per second. The pulse-per-second rate is called the *clock frequency*.
- *Gate*: lets clock pulses pass when open and stops them when closed.
- *Counter*: counts clock pulses passed by the gate.
- *Display*: shows the number of pulses counted by the counter. Usually in *seven-segment* form, either *LED* (light-emitting diodes) or *LCD* (liquid-crystal display).
- *Ramp generator*: generates a ramp of voltage which rises X volts per second.
- *Comparator*: compares ramp voltage to the unknown voltage, E_x, being measured. When the ramp is less than E_x, the comparator opens the gate. When the ramp equals E_x, the comparator shuts the gate.

A block diagram showing these parts connected as a basic digital voltmeter is given in Figure 9.10.

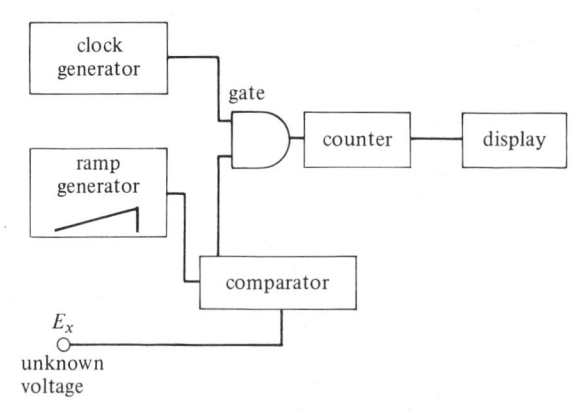

FIGURE 9.10 Block Diagram of Basic Digital Voltmeter.

To make our explanation of circuit operation easier, the clock frequency is set to 1000 pulses per second, the ramp climbs at 1 volt/per second and the unknown voltage is 3.5 V.

Both the unknown and ramp signals drive the comparator. The ramp starts climbing from zero. As long as it is less than E_x, the comparator leaves the gate open and clock pulses are counted by the counter. But as soon as the ramp reaches E_x, the comparator shuts the gate. The count freezes. At what value? Let's calculate it: The ramp climbs 1 V/s. The unknown, E_x, is 3.5 V; therefore, it must takes the ramp 3.5 s to reach E_x. During those 3.5 s, the open gate lets 1000 pulses per second pass. How many pulses get to the counter to be counted?

$$3.5 \text{ s} \times \frac{1000 \text{ pulses}}{\text{second}} = 3500 \text{ pulses}$$

Put a decimal point between the 3 and the 5, print volts on the front panel, and we have a digital voltmeter.

Digital meters are described by the number of 9s they can display. For example, a unit that can show 999 is called a three-digit meter. By using a range switch, these 9s can mean maximum readings of 999, 99.9, 9.99, or 0.999.

For the 999 range, the least significant digit (LSD) changes in steps of 1; that is, we could read 998 or 999 but not 998.5. We say that the range has a *resolution* of 1. Similarly, the resolution on the 99.9 range is 0.1, and so on. If better resolution is needed a four- or five-digit meter would be necessary.

EXAMPLE 9.7

Can a voltage of 72.5 V be read to within 5mV with a four-digit DPM (digital panel meter)?

The DPM would be set on the 99.99-V range in order to read 72.5 V. The LSD on this range is 0.01 V (two places to the right of the decimal point), which is 10 mV. Therefore, this meter will not read to the desired resolution. A five-digit DPM must be used. ∎

For the small cost of some extra electronics, the range of a digital meter can be doubled. This means that the maximum reading of a three-digit unit goes from 999 to

**FIGURE 9.11 Digital Multimeter.
(Courtesy of Beckman Industrial Corp.)**

1999, or that of a four-digit from 9999 to 19,999. Since the most significant digit (MSD) can only be a 1 or blank, it is called a *half-digit*.

EXAMPLE 9.8

What is the maximum reading of a $5\frac{1}{2}$-digit DPM?

The maximum reading is a 1 followed by five 9s, or 199999. ∎

The digital equivalent of the VOM is the DMM (digital multimeter). A typical example is shown in Figure 9.11. Besides the usual abilities of a VOM to read current, voltage, and resistance, many DMM models have automatic ranging and polarity features.

9.12 COMPARING ANALOG AND DIGITAL METERS

Digital meters, being electronic, have very high input resistance which is constant with range changes. They work well with computers and have excellent accuracy, much better than that of analog devices. Typical values of 0.1%. That figure means percentage of actual reading, not of full-scale value.

Other advantages of digital meters are: short response time (no needle to move), easy to read, no parallax error, not affected by magnetic fields, and can stand more physical abuse. Finally, digital meters have excellent *set-point control*. This means that they can be set to trigger an alarm should the reading go above or below some value.

Analog meters have strong advantages, too. Besides being inexpensive and popular, they have excellent immunity from electrical noise (a frequent problem with digital devices) and require no power source.

There are also important human advantages to analog devices. Data are not only obtained from a meter's needle position. Sometimes the speed and direction of a change in that position supplys important "trend" information. This is difficult to do with digital displays; they show change in a blur of numbers.

Another human factor is that often a glance at a needle's position is all that is needed from an analog meter. The same glance at a digital readout tells nothing until the operator compares the reading against a known value.

9.13 OTHER ANALOG METERS

There are two other analog meters to be discussed, the *electrodynamometer* and the *bargraph*.

The D'Arsonval movement, you will recall, has a coil of wire mounted between permanent-magnet poles. When current flows through the coil, torque, proportional to current, turns the coil.

The electrodynamometer works much the same way except that an electromagnet supplys the magnetic field rather than a permanent magnet. Figure 9.12 shows two coils forming the electromagnet. They are locked in place. Inside is a movable coil which has the meter's pointer mounted on it.

The electrodynamometer may be used as a wattmeter because of its two-coil system. One coil has circuit current through it. The other is connected across the source voltage. The result is a force proportional to power.

This meter is being phased out rapidly in favor of less expensive ways to measure power with electronics. It is mentioned because many of them are still in use.

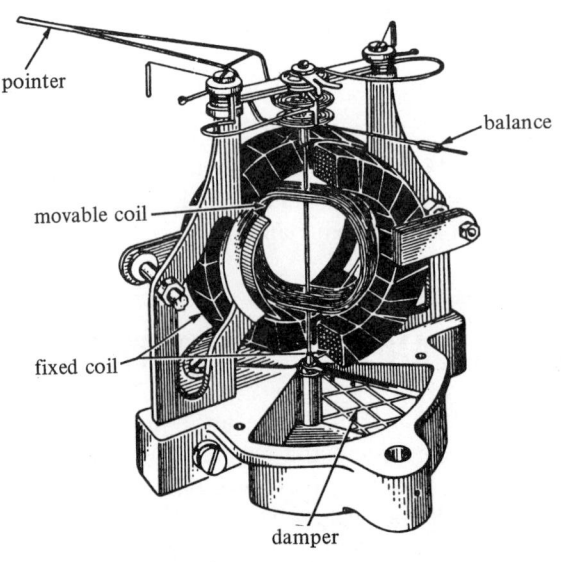

FIGURE 9.12 Electrodynamometer.

9.13 OTHER ANALOG METERS

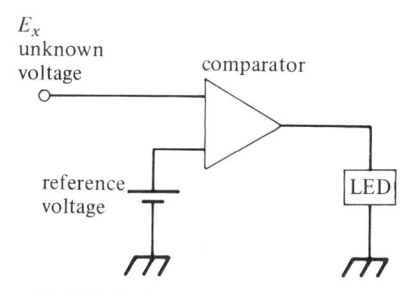

FIGURE 9.13 Comparator Circuit: The Heart of the Bargraph.

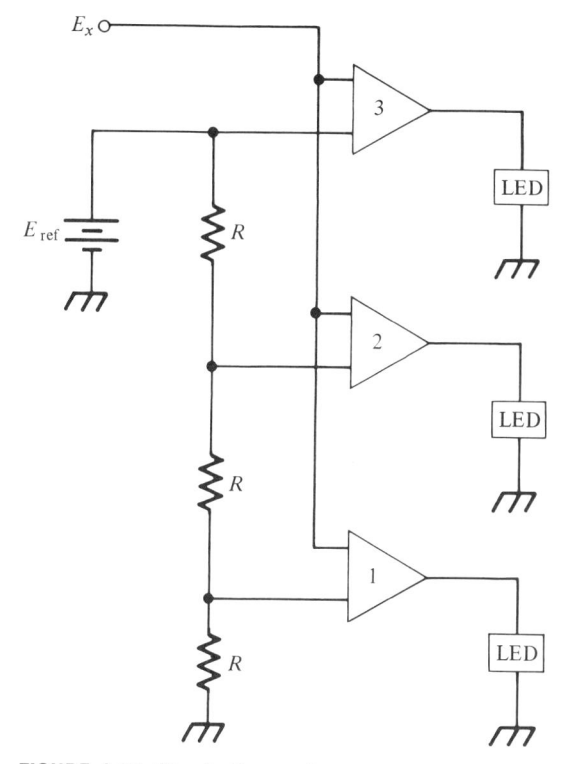

FIGURE 9.14 Simple Bargraph.

The bargraph is a solid-state analog meter. To understand how it works, we first review the comparator circuit seen earlier. Figure 9.13 shows a comparator connected to a LED. Notice the two inputs to the comparator. $E_{\text{reference}}$ is fixed at the battery voltage. E_x varies. When E_x is greater than E_{ref}, the LED is lit. When E_x is less than E_{ref}, the LED is out.

Next, connect three comparators as shown in Figure 9.14. Pay special attention to the voltage divider in the figure. It divides $E_{\text{reference}}$ into thirds. One-third goes to comparator 1, two-thirds to 2, and three-thirds to 3. Note also how all E_x inputs are tied together.

If E_x is 0 V, none of the LEDs is lit because reference input to every comparator is greater than E_x input. But as E_x increases, it eventually exceeds one-third of E_{ref}. When

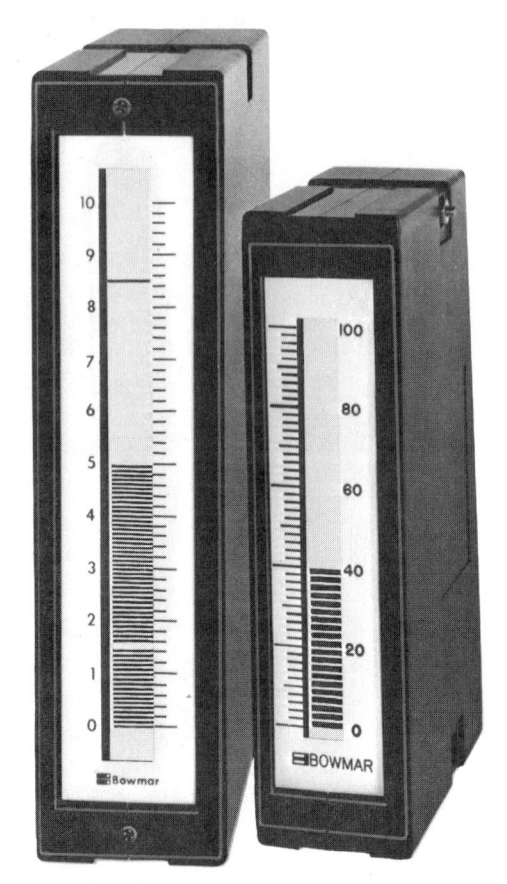

FIGURE 9.15 Modern Bargraph. (Courtesy of Bowmar/Ali Inc.)

it does, LED 1 turns on. LED 2 comes on when E_x is greater than two-thirds E_{ref}, and 3 lights up when E_x exceeds E_{ref}. In other words,

| The number of LEDs lit is an analog of the voltage being measured.

Put 50 or 100 comparators and LEDs in the circuit for good resolution and the result is a rugged, reliable instrument not affected by vibration, shock, or magnetic fields. Bargraphs have excellent computer compatibility and set-point control and make fine trend indicators. In addition, they cannot be pinned and can be seen in the dark. However, they are a bit more expensive than other types of meters and suffer from a rather low input resistance of about 100 k. A modern bargraph is shown in Figure 9.15.

EXERCISES

Section 9.1

1. One hundred pounds of test equipment upset the jogger's performance. Would the same weight influence the speed tests of an ocean liner? Write a rule that tells when test-equipment weight affects data and when it does not.

EXERCISES

2. Rewrite your rule of Exercise 1 in terms of voltmeter resistance compared to circuit resistance.

3. Measure the output of this circuit with: a. a meter with 100-k input resistance; b. a meter with 10-M input resistance. Comment on differences.

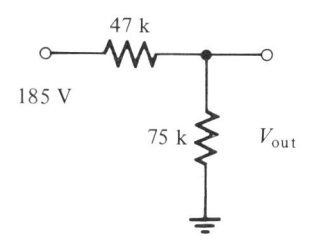

4. State the input-resistance rule for choosing voltmeters.

5. We have talked about the loading effects of voltmeters, but ammeters can also upset circuits. Explain how their resistance does this.

6. a. Calculate the true current flowing in this circuit.
b. Calculate the circuit current measured on an ammeter with 0.5-Ω resistance.
c. Repeat part b for a 10-Ω meter.

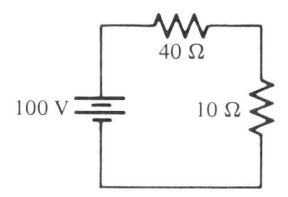

7. a. Using the 10-to-1 rule, what is R_{total} of a circuit whose readings are seriously affected by a 75-Ω ammeter?
b. Repeat part a for a circuit not seriously affected by the meter.

8. Rewrite your rule of Exercise 1 for ammeters.

Section 9.3

9. Explain how the D'Arsonval movement works.

10. Compare the pivot-and-jewel and taut-band suspensions.

Section 9.4

11. A 50-V meter has an accuracy of 5% of full-scale value. The needle rests over 42 V.
a. What is the maximum percent error that this reading could be off?
b. What is the minimum percent error?
c. Repeat part a when the needle lies over 12 V.

12. How is the sensitivity of a D'Arsonval movement described?

13. State the rule for choice of meter scale compared to the value of the parameter being measured.

14. Explain how parallax errors occur. How can they be prevented?

Section 9.5

15. Given: a 50-mil movement with $R_{meter} = 0.5\ \Omega$. Calculate the shunt resistance needed to allow the movement to read each of the following: a. 175 mA; b. 450 mA; c. 875 mA.

16. Given: a 2.5 mil movement with $R_{meter} = 2.6\ \Omega$. Calculate the shunt resistance needed to allow the movement to read each of the following: a. 10 mA; b. 140 mA.

17. Study the results of Exercise 15 and/or 16 and describe what happens to the value of R_{shunt} as I_{total} goes up. Do you see a practical limit to how high I_{total} can go? Discuss.

18. A 300-mA D'Arsonval movement has a resistance of 0.167 Ω. What value of shunt should be used to create a meter capable of reading 10 A? What must be done to the scale of the 300-mA movement after this change is made?

Section 9.6 and 9.7

19. Given: a 20-μA movement with an R_{meter} of 3.4 k.
   a. What is its ohms-per-volt rating?
   b. What is the voltage needed to cause I_{fsd} to flow?
   c. What is the series resistance needed to make a 0.5-V voltmeter?
   d. What is the series resistance needed to make a 50-V voltmeter?
   e. Draw the diagram of the movement connected as a voltmeter which can switch to these three ranges.
20. A meter movement has a 5000-Ω/V rating. What is its full-scale deflection current, I_{fsd}?
21. A meter movement has a 2000-Ω/V rating. What is its input resistance when made into each of these voltmeters?   a. 1 V;   b. 25 V;   c. 100 V;   d. 500 V.
22. Two D'Arsonval movements, one with an I_{fsd} of 50 μA, the other of 20 μA, are each converted to a 500-V voltmeter. Which meter loads the least? Why?

Section 9.8

23. Explain why the ohmmeter scale reads backwards compared to other VOM scales.
24. What are the steps to take when "zeroing" an ohmmeter?
25. What resistance should a switch have when checked with an ohmmeter? What readings on the meter would indicate that the switch is defective?
26. Why is it necessary to connect a pot in series with the battery in a VOM?
27. What two pieces of information are needed to get a resistance reading from a VOM?
28. What four precautions must be taken when using an ohmmeter in a circuit?

Section 9.10

29. What is the main advantage of a FETVOM over a regular VOM?
30. Discuss the similarities and differences between VOMs and FETVOMs. What do the acronyms stand for?

Section 9.11

31. Describe the role of each of the six parts of the basic digital meter. Then describe how the meter works.
32. What is a half-digit when referring to a digital meter? Where are half-digits always found on a digital display? What is the effect of a half-digit?
33. What are the possible maximum voltages on each range of: a. a five-digit DVM (digital volt meter);   b. a $5\frac{1}{2}$-digit DVM?
34. How close can a voltage of 0.1445 be read on: a. a four-digit DVM; b. a $4\frac{1}{2}$-digit DVM?

Section 9.12

35. What is meant by: a. set-point control;   b. computer compatibility?
36. Why are digital meters not as good as other meter types for trend information?

Section 9.13

37. How is the dynamometer able to read watts? How does this meter differ from the D'Arsonval movement?
38. Explain how a bargraph works.

CHAPTER HIGHLIGHTS

Be alert to possible effects of test equipment on the circuit being tested.

If loading is severe, meter readings will not mean what they say.

For least loading, choose the voltmeter with the highest input resistance (highest ohms-per-volt rating).

Readings are taken from analog meters by noting the needle position relative to a scale. The needle position is analogous to the value of parameter being measured.

Polarity must be observed with analog meters.

Use the analog scale that puts a reading closest to the top.

Avoid parallax errors.

Shunt resistances enable meters to handle greater current, while series resistors allow greater voltage.

FETVOMs are not necessarily more accurate than VOMs but have higher input resistance that is constant on all scales.

Digital meters work by counting voltage pulses.

No single meter is best for every job. Each type has good and bad points.

Formulas to Remember

$$\frac{1}{I_{fsd}} \quad \text{ohms-per-volt rating}$$

$R_{input} = (\text{range voltage})(\text{ohms per volt rating of meter}) \quad \text{voltmeter input resistance}$

CHAPTER 10

Capacitance

All electronic components have three characteristics that limit current in a circuit. So far, we have only talked about one of them, resistance. In this chapter we discuss the second, *capacitance*.

Capacitance is much like resistance in that it is always present in circuits. Like resistance, we add it in one circuit, then work hard to eliminate it in the next.

But resistance and capacitance are different in some important and interesting ways. For instance, capacitance causes voltage and current to change over a period of time. These changes are called *transients*. We study transient analysis later in the chapter.

At the end of this chapter, you should be able to:

- Explain what capacitance is.
- Describe the factors that determine capacitance.
- State the equation for capacitance.
- Discuss the voltage/current characteristics of capacitance.
- Calculate total capacitance of capacitors in series and parallel.
- Describe the major types of capacitors.
- Analyze transient circuits.

10.1 CAPACITANCE

Figure 10.1, part (a), shows two parallel metal sheets separated by a layer of insulating material called the *dielectric*. Note that each sheet is connected to a battery terminal.

Will there be any current when the switch is closed?

Our training so far tells us there will be no current because of the insulating dielectric. Due to something called *capacitance*, however, there will be a current. Let's see how this happens.

Before the switch is closed, the plates are electrically neutral [Figure 10.1, part (a)], that is, both have the same charge.

After switch closure, the positive battery terminal attracts electrons from the upper plate. Their departure along the upper conductor causes the sheet to become positively charged. At the same time, the negative battery terminal repels electrons along the bottom conductor. They crowd together on the bottom plate, causing it to become negatively charged.

By our definition of Chapter 2, this electron movement from upper to lower plate means that current flowed in the opposite direction, from the lower to the upper plate [see Figure 10.1, part (b)].

The surge of electrons between plates lasts only an instant. Here's why: Every electron leaving the upper plate causes that plate to become a bit more positively charged. As long as the battery's positive terminal has greater attraction than the plate's charge, electrons continue to leave. Similarly, every electron arriving on the bottom plate causes that plate to become more negatively charged. As long as the

(a) switch open
capacitor
uncharged

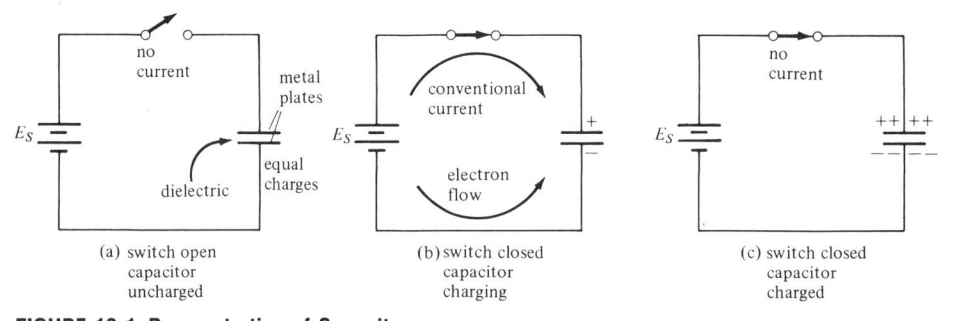

(b) switch closed
capacitor
charging

(c) switch closed
capacitor
charged

FIGURE 10.1 Demonstration of Capacitance.

10.1 CAPACITANCE

battery's negative terminal has greater repulsion than the plate's charge, electrons continue to arrive.

Eventually, the Coulomb's law forces of the charged plates equal those of the battery terminals. When this happens, electron movement (current) stops [Figure 10.1, part (c)].

It should be noted that charge movement (and therefore current) is heaviest the first instant after switch closure, when plate charges are lowest. It decreases as those charges grow.

Battery energy was needed to move carriers against the plate forces. That energy is now stored in a stretched electrical "spring," namely, the charge difference between plates. If the battery is disconnected, the charge difference remains and so does the energy. In other words, our device of metal sheets separated by a dielectric has the ability or capacity to store energy. That ability is called *capacitance*, and the device is called a *capacitor*.

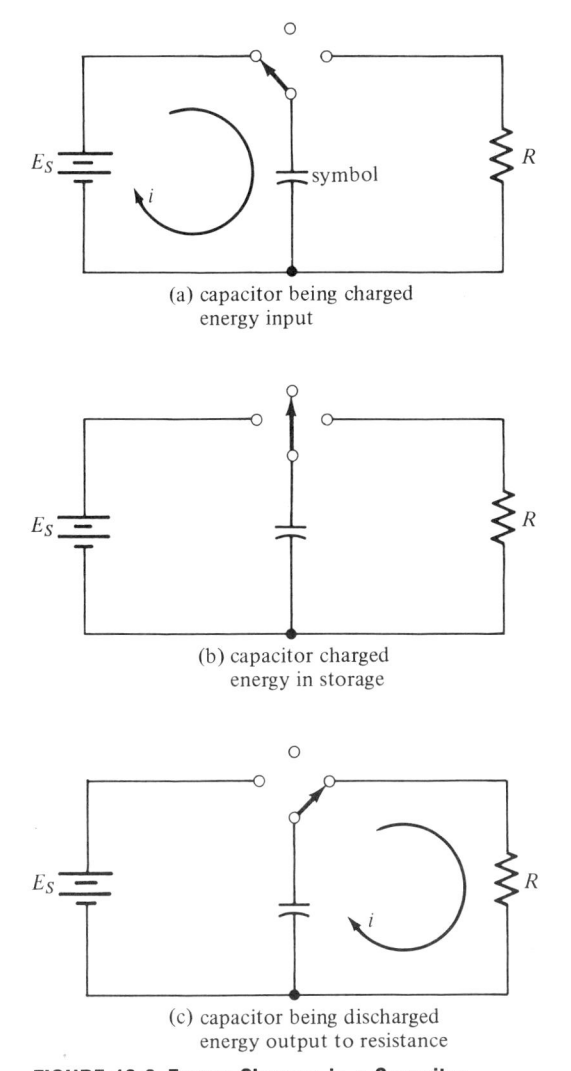

FIGURE 10.2 Energy Changes in a Capacitor.

A capacitor containing stored energy is said to be *charged*. Because of the charge difference between plates, a charged capacitor has a potential difference between its terminals (plates) just as a battery does. Connect the terminals with a wire and that voltage causes electron movement from the lower to the upper plate, which equalizes the charges. The stored energy is released in the process and the voltage drops to zero. The capacitor is *discharged*.

Figure 10.2 shows a capacitor passing through the energy changes occurring during charge and discharge. Note the capacitor symbol.

How much charge can be moved between the plates of a given capacitor? That depends on the voltage between the plates and the capacitance of the capacitor. The relation between the three is described by the equation:

$$Q = CV \qquad (10.1)$$

where Q = charge moved, in coulombs
C = capacitance, in farads
V = voltage, in volts

Capacitance is measured in *farads*, in honor of an English physicist, Michael Faraday (1791–1867):

> One farad (F) is defined as that capacitance for which a 1-volt potential between plates causes a charge transfer of 1 coulomb:
>
> $$1 \text{ farad} = \frac{1 \text{ coulomb}}{1 \text{ volt}}$$

EXAMPLE 10.1

A 30 microfarad capacitor has 450 V across it. What charge has been moved between plates?

$$\begin{aligned} Q = CV &= 30 \times 10^{-6}\,(450) \\ &= 135 \times 10^{-4} \\ &= 13.5 \text{ milliCoulombs} \end{aligned}$$

■

The farad is a huge amount of capacitance. We usually work in microfarads (10^{-6} F), abbreviated μF, and picofarads (10^{-12} F), abbreviated pF.

EXAMPLE 10.2

Convert 22000 pF to μF.

$$\begin{aligned} 22{,}000 \text{ pF} &\times \frac{1\ \mu\text{F}}{10^6\ \text{pF}} \\ &= 22{,}000 \times 10^{-6}\ \mu\text{F} \\ &= 0.022\ \mu\text{F} \end{aligned}$$

■

10.2 CAPACITIVE CURRENT/VOLTAGE RELATIONS

Let us examine capacitor current and voltage more closely. If Equation 10.1 is solved for V we find that:

$$V = \frac{Q}{C} \tag{10.2}$$

which tells us that capacitor voltage is zero until charge has been transferred between plates. Since charge can not be moved in zero time:

| Voltage across capacitance cannot change instantaneously.

Current and voltage waveforms for a resistor/capacitor circuit driven by a battery are shown in Figure 10.3. Time, t, is zero at the instant of switch closure and measured from then.

Notice how the current jumps immediately to its maximum value at $t = 0$. As we saw earlier, this is the time of greatest charge movement. However, very little charge has yet moved so the voltage is approximately zero. Later, when the capacitor is fully charged, current is zero and voltage maximum. Note carefully that *capacitive current is not proportional to capacitor voltage*. In other words, Ohm's law does not apply to capacitance as it does for resistance. Then how is capacitor current affected by applied voltage?

Note when i_c is maximum in Figure 10.3. Not when v_c is maximum as might be expected but at the instant when v_c has just begun rising above 0 volts. This is the point where v_c is changing the fastest, that is, when the slope of v_c is highest. As v_c increases, its waveform slope decreases; the curve is not rising as steeply as before. At the same time, current decreases.

Eventually the waveform levels off, voltage stops changing, and slope drops to zero. This point must be understood; *voltage did not drop to zero; it is at maximum value. But its slope, its rate of change with time, did.* Note that current is now zero. From this we conclude:

| Capacitor current is proportional to how fast capacitor voltage is changing.

This point can be shown mathematically. If V in Equation 10.1 is allowed to change, Q must change also (C is fixed). That is,

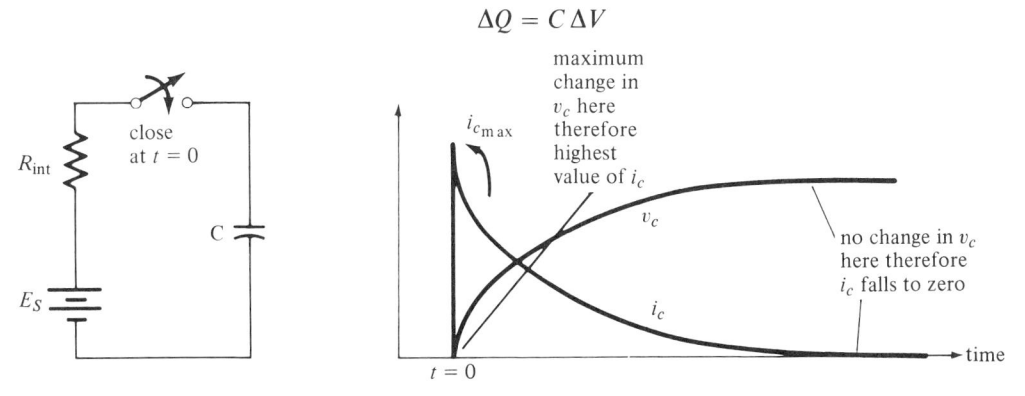

$$\Delta Q = C \Delta V$$

FIGURE 10.3 Capacitor Current is Proportional to the Rate of Change of Capacitor Voltage.

Divide ΔQ and ΔV by Δt to get change per unit time:

$$\frac{\Delta q}{\Delta t} = \frac{C \Delta v}{\Delta t}$$

(switch to lower case letters because parameters now changing with time.)

But $\dfrac{\Delta q}{\Delta t}$ is current.

Substituting and rewriting:

$$i = C \frac{\Delta v}{\Delta t} \tag{10.3}$$

where i = capacitor current, in amperes
$\quad\quad C$ = capacitance, in farads
$\Delta v/\Delta t$ = rate of change of voltage with time

Note carefully how this equation and the waveforms in Figure 10.3 say the same thing.

It should be clearly understood that the current waveform of Figure 10.3 is caused by the changing voltage in that figure. Apply a voltage that varies some other way and we get a different current waveform whose values are calculated with Equation 10.3.

EXAMPLE 10.3

The voltage shown in this waveform is applied to a 4500 μF capacitor. Draw the current waveform.

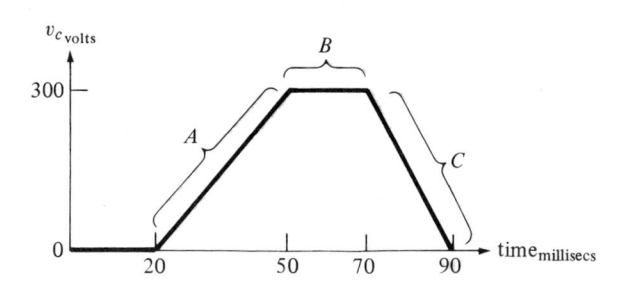

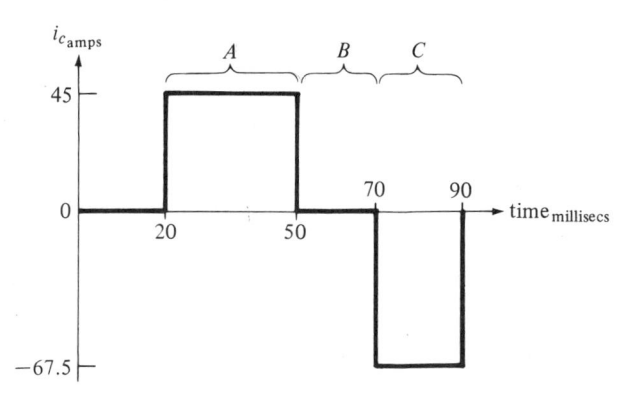

10.2 CAPACITIVE CURRENT/VOLTAGE RELATIONS

For part A of the waveform:

$$i_c = \frac{C \Delta v}{\Delta t}$$

$$= \frac{C(\text{final } v - \text{starting } v)}{\text{final time} - \text{starting time}}$$

$$= \frac{(4500 \times 10^{-6})(300 - 0)}{50 \times 10^{-3} - 20 \times 10^{-3}}$$

$$= \frac{(4.5 \times 10^{-3})(3 \times 10^2)}{30 \times 10^{-3}}$$

$$= 45 \text{ A}$$

Note how current jumps to 45 A showing that while capacitor voltage cannot change instantaneously, current can. Current is constant during time A because the voltage rate of change is constant; the voltage rises in a straight line.

For the flat part of the voltage waveform, part B, $i_c = 0$ A because there is no $\Delta v / \Delta t$.

For part C:

$$i_c = C \Delta v / \Delta t$$

$$= \frac{(4500 \times 10^{-6})(0 - 300)}{90 \times 10^{-3} - 70 \times 10^{-3}}$$

$$= \frac{(4.5 \times 10^{-3})(-300)}{20 \times 10^{-3}}$$

$$= -67.5 \text{ A}$$

Notice two things about the current during part C:

- It is negative because the slope of the voltage waveform is negative (contrast to part A).
- Its absolute value is greater than during time A because equal charge must be moved in shorter time. ■

When the switch in the circuit of Figure 10.3 is closed, current jumps to maximum, then decreases, or *decays*, to zero. During this time, voltage does the opposite. It starts at zero and climbs to its maximum, the source potential.

Two observations of these changes:

- Because capacitor current reaches its maximum first, we say:

| Current through a capacitor *leads* voltage across.

- A resistor has constant resistance but a capacitor is a short (current but no voltage) at the instant of switch closure, and an open (voltage but no current) when charged.

EXAMPLE 10.4

What is the current through each resistance at the indicated time?

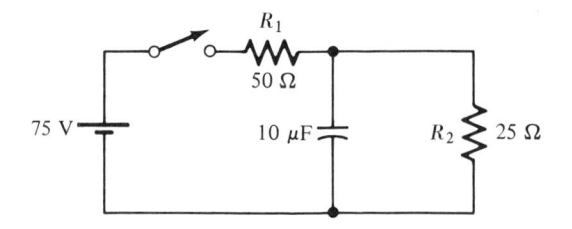

(Circuit: 75 V source, switch, $R_1 = 50\,\Omega$ in series; then parallel branch of $10\,\mu F$ capacitor and $R_2 = 25\,\Omega$.)

a. The first instant after switch closure?

At the instant of switch closure the capacitor is a short and the circuit acts like this:

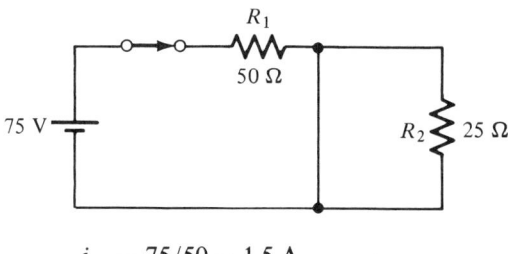

(Circuit: 75 V, switch, $R_1 = 50\,\Omega$, with $R_2 = 25\,\Omega$ shorted.)

$$i_{R_1} = 75/50 = 1.5 \text{ A}$$
$$i_{R_2} = 0 \text{ A (shorted by capacitor)}$$

b. A long time after switch closure?

After the capacitor is charged, it acts as an open and the circuit looks like this:

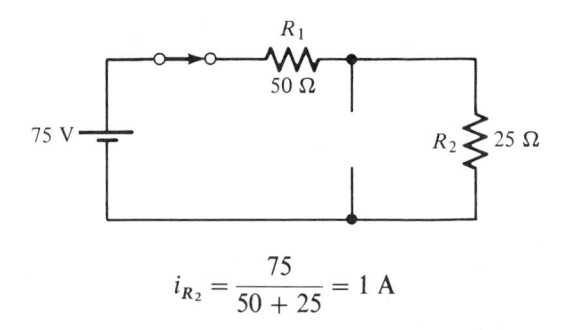

(Circuit: 75 V, switch, $R_1 = 50\,\Omega$ in series with $R_2 = 25\,\Omega$.)

$$i_{R_2} = \frac{75}{50 + 25} = 1 \text{ A}$$

10.3 WHAT DETERMINES CAPACITANCE?

When we studied Coulomb's law in Chapter 2 we learned that charged bodies are able to push and pull on each other by means of an electric field between them. Faraday described these fields as an area in which lines of force existed between charges. The higher the number of lines through a given area, the greater the *field intensity*.

These lines are shown in Figure 10.4. Their direction is always given from the positive charge to the negative.

A charged capacitor has an electric field between its plates because of the charges on them. The intensity of this field can be shown to be proportional to the

10.3 WHAT DETERMINES CAPACITANCE?

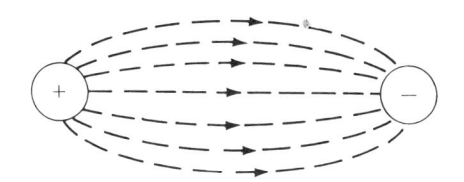

FIGURE 10.4 Lines of Force of an Electric Field.

charge moved between plates and inversely proportional to plate area. The exact relationship is

$$\mathscr{E} = \frac{1}{\varepsilon_0} \frac{Q}{A} \tag{10.4}$$

where $\mathscr{E}$ = electric field intensity, in volts per meter
ε_0 = dielectric permittivity of vacuum or dry air = 8.9×10^{-12} C^2/N-m
Q = charge moved between plates, in coulombs
A = area of either plate, in meters^2

Field intensity can also be calculated by the equation

$$\mathscr{E} = \frac{V}{d} \tag{10.5}$$

where V = voltage between plates, in volts
d = distance between plates, in meters

Substituting Equation 10.2 for V in Equation 10.5:

$$\mathscr{E} = \frac{Q}{Cd}$$

Set this equal to Equation 10.4:

$$\frac{Q}{Cd} = \frac{1}{\varepsilon_0} \frac{Q}{A}$$

Solve for C:

$$C = \frac{\varepsilon_0 A}{d} \tag{10.6}$$

where C = capacitance, in farads
ε_0 = dielectric permittivity of vacuum
A = area of either plate, in meters^2
d = distance between plates, in meters

EXAMPLE 10.5

What is the capacitance of two aluminum plates, each with an area of 0.03 m^2 separated by a layer of air 0.001 m thick?

$$C = \frac{\varepsilon_0 A}{d} = \frac{(8.9 \times 10^{-12})(3 \times 10^{-2})}{10^{-3}}$$

$$= 26.7 \times 10^{-11} \text{ F} = 267 \text{ pF}$$

10.4 THE ROLE OF THE DIELECTRIC

So far we have considered only dielectrics of air or vacuum. What happens if another material is used? To answer that question, we first examine the effects of an electric field on dielectric molecules.

The protons and electrons of dielectric molecules in an uncharged capacitor lie in a jumbled mess; there is no field to force them to do otherwise. The effect is shown in Figure 10.5, part (a) (for simplicity, each molecule is shown with just two positive and two negative charges).

Create a field by charging the capacitor and the particles behave much differently. If the dielectric was made of conducting material, there would be free electrons. These would drift toward the positively charged plate. But this is a dielectric, an insulator. It has no free electrons. Nevertheless, the force of the field still pushes and pulls charges within the molecule and they move what little they can. The result is a concentration of positive charges on one side of the molecule, negative on the other [see Figure 10.5, part (b)]. The molecule is now said to be *polarized*. Being polarized, it is called a *dipole*.

Of course, there are billions of dipoles in the dielectric, but let's consider just one layer of them (see Figure 10.6). Note how positive charge on the right side of one dipole neutralizes negative charge on the left side of its neighbor (each pair has a dashed circle around it). Being neutralized, these charges no longer have any effect. But the charges on the far ends, those closest to the plates, have no opposite dielectric charge nearby. They are not neutralized.

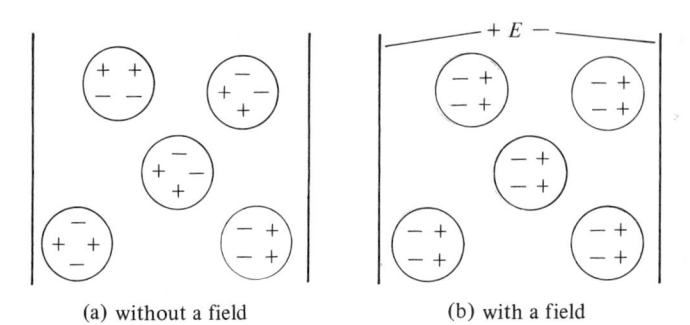

(a) without a field   (b) with a field

FIGURE 10.5 Action of Dielectric Molecules With and Without an Electric Field.

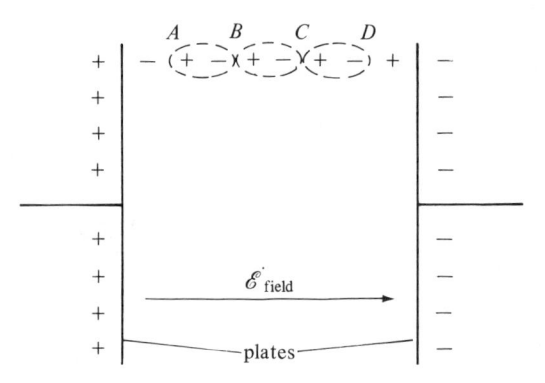

FIGURE 10.6 Dipole Layer showing Neutralization.

10.4 THE ROLE OF THE DIELECTRIC

If we expand our thinking from one layer of dipoles to the entire dielectric, we see a strip of unneutralized positive charges on the right edge and a negative one on the left (Figure 10.7). These are called *induced charges*. The strips of induced charges have their own electric field. Seen in Figure 10.7, its direction is opposite to the direction of the field caused by the applied voltage. The induced charge field therefore cancels part of the applied field. As a result, *the total field between plates is reduced*. Say this in equation form:

$$\mathscr{E}_{\text{total}} = \mathscr{E}_{\text{applied}} - \mathscr{E}_{\text{induced}} \tag{10.7}$$

By Equation 10.5,

$$\mathscr{E} = \frac{V}{d}$$

With $\mathscr{E}$ having dropped, one of the right-hand variables must change to maintain equality. Since d, the distance between plates, is fixed, V must decrease (the capacitor was disconnected from the source once it was charged). Substitute this drop into Equation 10.1 and a very important point is seen.

Equation 10.1 tells us that $Q = CV$. The charge was put into the capacitor during charging; it cannot change. Therefore, a drop in voltage must go hand in hand with a rise in capacitance. In other words:

> Use any material for a dielectric other than air or vacuum and the capacitance of a capacitor is increased.

Divide this increased capacitance with a material dielectric by the capacitance of the same set of plates but with air dielectric, and a number greater than 1 results. This number is the *dielectric constant* of the material (also called the *dielectric coefficient*). It tells how much better the material is as a dielectric than air and is represented by the letter K. Equation 10.8 defines K:

$$K \text{ of a material} = \frac{C \text{ with material dielectric}}{C \text{ with air dielectric}} \tag{10.8}$$

Note that K has no units.

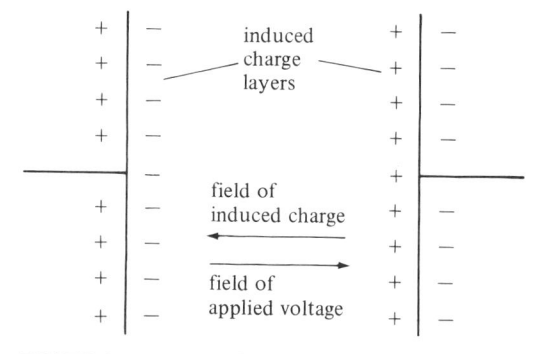

FIGURE 10.7 Induced Charges in a Dielectric.

Table 10.1 lists the dielectric constant for several popular materials.

TABLE 10.1 Dielectric constant and breakdown voltage for common dielectrics

Dielectric	K	Breakdown voltage (V/mil)
Air	1	10
Plastic films	2.3–3.1	4000–7000
Oil	4	400
Mica	5.5	5000
Ceramics	5000–10,000	75–100
Special ceramics	18,000	100

The parameter ε_0, dielectric permittivity, was included in Equation 10.6. For a vacuum its value is 8.9×10^{-12}. The dielectric permittivity for a given material is obtained by multiplying ε_0 by the K value for that material. When this is done, Equation 10.6 becomes

$$C = \frac{8.9 \times 10^{-12} \, KA}{d} \tag{10.9}$$

where C = capacitance, in farads
K = dielectric constant for a given material
A = plate area, in meters^2
d = distance between plates, in meters

EXAMPLE 10.6

750 V is applied to a capacitor whose plates are 0.01 by 0.015 m. The mica dielectric is 0.001 m thick.

a. Find the capacitance.

$$C = \frac{(8.9 \times 10^{-12})KA}{d}$$

$$= \frac{(8.9 \times 10^{-12})(5.5)(0.01 \times 0.015)}{0.001}$$

$$= 7.3 \text{ pF}$$

b. Find the electric field intensity in the dielectric.

$$\mathscr{E} = \frac{V}{d}$$

$$= \frac{750}{0.001}$$

$$= 7.5 \times 10^5 \text{ V}/m$$

10.5 STORED ENERGY

c. Find the charge moved between plates.

$$Q = CV$$
$$= (7.3 \times 10^{-12})750$$
$$= 5.5 \times 10^{-9} = 5.5 \text{ nC} \blacksquare$$

EXAMPLE 10.7

The mica dielectric in Example 10.6 is replaced with a ceramic material having a dielectric constant of 7300. What is the new capacitance?

Do not redo the problem. Just multiply the original answer by $K_{ceramic}/K_{mica}$:

$$C_{new} = C_{original} \frac{7300}{5.5} = 7.3 \text{ pF}(1330)$$
$$= 9690 \text{ pF} = 0.00969 \text{ }\mu\text{F} \blacksquare$$

There is a limit to the voltage that can be applied across a dielectric. Exceed that *breakdown* potential and field intensity becomes so great that electrons are torn from dielectric atoms. An arc jumps between plates. For most dielectric materials, that arc welds the plates together, shorting the capacitor, which must be replaced. Such a material is called *nonhealing*. Other materials, such as air and oil, are *self-healing*. They repair themselves after the excess voltage is turned off. Dielectric breakdown voltages are listed in Table 10.1.

Current through a broken-down dielectric creates I^2R heat that generates gases. These build up pressure because capacitors are usually sealed to keep out dirt and humidity. When pressure gets too high, the case may burst with a terrifying explosion.

EXAMPLE 10.8

A mica dielectric is 8 mils thick. What is its breakdown voltage?

From Table 10.1, breakdown voltage for mica is 5000 V/mil. Therefore,

$$E_{breakdown} = [5000 \text{ V/mil}]8 \text{ mils}$$
$$= 40,000 \text{ V} = 40 \text{ kV} \blacksquare$$

If a charged capacitor is discharged, the potential between the plates drops to zero, as expected. Measure the voltage a while later, however, and it may read as much as 10% of its earlier charged value. This mysterious characteristic is called *dielectric absorbtion* or *soaking* (as though the dielectric material has soaked up charge). Although studied by Faraday and Franklin, soaking is still a puzzle today. It can be a real problem in certain electronic circuits.

10.5 STORED ENERGY

Equation 10.1, $Q = CV$, tells us that charge moved between capacitor plates is proportional to voltage. If that voltage increases in ramp form, charge must do likewise (see Figure 10.8).

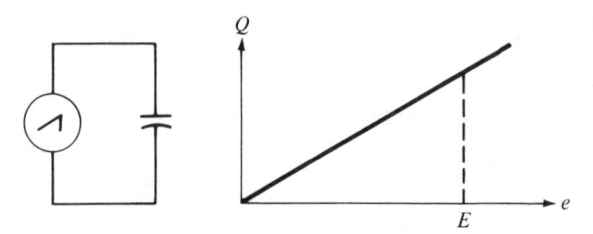

FIGURE 10.8 Variation of Charge with Changing Voltage.

Let the voltage increase to E, then calculate the area of the triangle under the plot. The base is E units wide, the height is Q. Area of a triangle = $\frac{1}{2}$ base × height = $\frac{1}{2}EQ$.

The units of this area are

$$\frac{1}{2}EQ = \frac{\text{joules}}{\text{coulombs}} \times \text{coulombs} = \text{joules}$$

Since joules are the units of energy, *the area under our QV waveform tells the energy stored in the capacitor*;

$$W = \frac{1}{2}QV \quad \text{joules} \tag{10.10}$$

Using Equation 10.1, the equation can be rewritten in terms of C and V:

$$W = \frac{1}{2}CV^2 \quad \text{joules} \tag{10.11}$$

or, in terms of Q and C:

$$W = \frac{1}{2}\frac{Q^2}{C} \quad \text{joules} \tag{10.12}$$

A capacitor's stored energy is useful in many applications. For example, it is discharged in thousandths of a second in a photographer's electronic flash, and farad-sized capacitors supply stored energy to prevent data loss from computer memory during short-term power outages. On the other hand, high-capacitance, high-voltage capacitors store lethal energy. Experienced electronics personnel short such capacitors before starting repairs.

10.6 CAPACITORS IN SERIES AND PARALLEL

Like resistors, capacitors in series and parallel can be combined by following certain rules.

The parallel capacitors in Figure 10.9 have the same voltage across them, and by KCL, total charge equals the sum of the individual charges, or

$$Q_{\text{total}} = Q_1 + Q_2$$

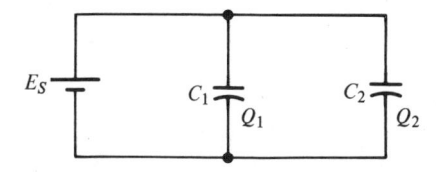

FIGURE 10.9 Capacitors in Parallel.

10.6 CAPACITORS IN SERIES AND PARALLEL

But
$$Q = CE_s$$

Therefore,
$$C_{total}E_s = C_1 E_s + C_2 E_s$$
$$= E_s(C_1 + C_2)$$

Cancelling E_s, we see that:
$$C_{total} = C_1 + C_2$$

For *n* capacitors in parallel:
$$C_{total} = C_1 + C_2 + \cdots + C_n \tag{10.13}$$

where *n* is any whole number. (Note the similarity with the calculation for R_{total} of series resistances.)

EXAMPLE 10.9

For this circuit:

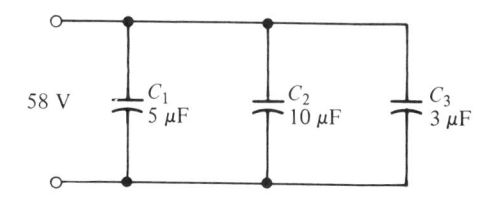

58 V | C_1 5 μF | C_2 10 μF | C_3 3 μF

a. Find C_{total}.
$$C_{total} = C_1 + C_2 + C_3 = (5 + 10 + 3)\,\mu F$$
$$= 18\,\mu F$$

b. Find the voltage across C_1.
$$V_{C_1} = E_{source} = 58\,V$$

c. Find the charge in C_2.
$$Q_{C_2} = C_2 E_{source}$$
$$= 10\,\mu F(58\,V)$$
$$= 580\,\mu C$$

d. Find the energy stored in C_3.
$$W_{C_3} = \tfrac{1}{2} C_3 E_s^2$$
$$= \tfrac{1}{2}(3 \times 10^{-6})58^2$$
$$= 5\,mJ \quad \blacksquare$$

For the series circuit of Figure 10.10:
$$E_s = V_{C_1} + V_{C_2}$$

by KVL.

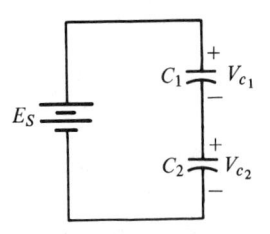

FIGURE 10.10 Capacitors in Series.

From Equation 10.2 we know that:

$$E_s = \frac{Q_{total}}{C_{total}}$$

and

$$V_{c_1} = \frac{Q_{c_1}}{C_1}$$

and

$$V_{c_2} = \frac{Q_{c_2}}{C_2}$$

Substitute into the loop equation:

$$\frac{Q_{total}}{C_{total}} = \frac{Q_{c_1}}{C_1} + \frac{Q_{c_2}}{C_2}$$

Since this is a series circuit,

$$Q_{total} = Q_{c_1} = Q_{c_2}$$

Therefore,

$$\frac{Q_{total}}{C_{total}} = \frac{Q_{total}}{C_1} + \frac{Q_{total}}{C_2}$$

Divide both sides by Q_{total}:

$$\frac{1}{C_{total}} = \frac{1}{C_1} + \frac{1}{C_2}$$

Solve for C_{total}:

$$C_{total} = \frac{1}{1/C_1 + 1/C_2}$$

For *n* capacitors in series:

$$C_{total} = \frac{1}{1/C_1 + 1/C_2 + \cdots + 1/C_n} \quad (10.14)$$

C_{total} of series capacitors is calculated the same way as R_{total} of parallel resistances.

For two series-connected capacitors, Equation 10.14 can be rewritten in the familiar product-over-sum form:

$$C_{total} = \frac{C_1 C_2}{C_1 + C_2} \quad (10.15)$$

10.6 CAPACITORS IN SERIES AND PARALLEL

EXAMPLE 10.10

For the following circuit, determine the indicated values.

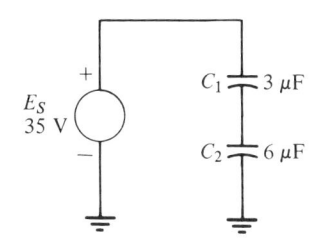

a. C_{total}.

$$C_{total} = \frac{C_1 C_2}{C_1 + C_2} = \frac{(3\,\mu F)(6\,\mu F)}{3\,\mu F + 6\,\mu F} = 2\,\mu F$$

b. Q_{total}.

$$Q_{total} = E_s C_{total} = 35(2 \times 10^{-6}) = 70\,\mu C$$

c. Q_{c_1} and Q_{c_2}.

$$Q_{c_1} = Q_{c_2} = Q_{total} = 70\,\mu C$$

d. V_{c_1}.

$$V_{c_1} = \frac{Q_{c_1}}{C_1} = \frac{70\,\mu C}{3\,\mu F} = 23.3\,V$$

e. V_{c_2}.

$$V_{c_2} = \frac{Q_{c_2}}{C_2} = \frac{70\,\mu C}{6\,\mu F} = 11.7\,V$$

Check by KVL: $23.3 + 11.7 = 35\,V$ ∎

Note carefully how the smaller capacitance in Example 10.10 had the greater voltage across it. This is the opposite of series-connected resistances.

Series–parallel capacitor circuits are analyzed much the same way as are their resistor counterparts.

EXAMPLE 10.11

For the following circuit:

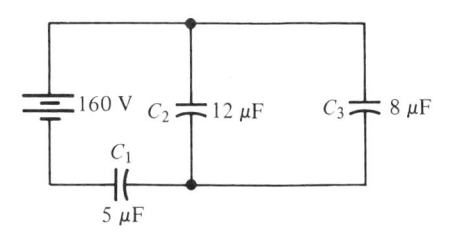

a. Find C_{total}.

$$C_2 + C_3 = 12\,\mu F + 8\,\mu F$$
$$= 20\,\mu F$$
$$C_{total} = \frac{C_1(C_2 + C_3)}{C_1 + (C_2 + C_3)}$$
$$= \frac{5\,\mu F(20\,\mu F)}{5\,\mu F + 20\,\mu F}$$
$$= 4\,\mu F$$

b. Find Q_{total}.

$$Q_{total} = C_{total}E_s = 4\,\mu F(160\,V) = 6.4 \times 10^{-4}\,C$$

c. Find Q_{c_1}.

$$Q_{c_1} = Q_{total} = 6.4 \times 10^{-4}\,C$$

d. Find Q_{c_3}.

$$Q_{c_2} = 1.5\,Q_{c_3}$$
$$Q_{c_2} + Q_{c_3} = Q_{total}$$
$$1.5\,Q_{c_3} + Q_{c_3} = 2.5\,Q_{c_3} = Q_{total}$$
$$Q_{c_3} = \frac{Q_{total}}{2.5} = \frac{6.4 \times 10^{-4}}{2.5} = 2.6 \times 10^{-4}\,C$$

e. Find Q_{c_2}.

$$Q_{c_2} = Q_{total} - Q_{c_3} = 6.4 \times 10^{-4} - 2.6 \times 10^{-4} = 3.8 \times 10^{-4}\,C$$

f. Find V_{c_1}.

$$V_{c_1} = \frac{Q_{c_1}}{C_1} = \frac{6.4 \times 10^{-4}}{5 \times 10^{-6}} = 128\,V$$

g. Find V_{c_2}.

$$V_{c_2} = E_s - V_{c_1} = 160 - 128 = 32\,V$$

10.7 PRACTICAL CAPACITORS

Some of the general characteristics that apply to all types of capacitors are:

- *Cost, reliability,* and *breakdown voltage.*
- *Temperature coefficient:* in parts per million per degree Celsius.
- *Tolerance:* typical values 0.1% to 20%. Often different in positive and negative directions, for example, +20%, −56%.
- *Stability:* ability to hold constant capacitance with age, applied voltage, and temperature. Usually determined by dielectric material.
- *Polarized* and *nonpolarized:* because of their dielectric, polarized capacitors must have voltage applied according to + and − signs printed on them. Nonpolarized capacitors may be wired either way.

10.7 PRACTICAL CAPACITORS

- *Leakage:* measure of how well capacitor holds charge. With an ideal dielectric (one having infinite resistance) charge held indefinitely. Practical dielectrics act as high resistance between plates through which charge "leaks", gradually discharging capacitor.

Most capacitors types are named after their dielectric material. Here is a description of major types together with their strengths and weaknesses:

Mica Capacitors

Mica capacitors are made of alternate layers of metal foil and mica. Capacitance range, approximately 1 pF to 0.01 μF. Stability, reliability, and accuracy good. Low leakage. Excellent for high frequency applications.

Ceramic Capacitors

Ceramic capacitors are the most popular type of capacitor. Ceramic is a claylike mixture of barium and other materials. Manufacturers vary mixtures to obtain dielectrics with desired characteristics, such as temperature coefficient or stability. K values from about 150 to 10,000, with special formulas achieving 18,000. High capacitance for size, inexpensive, low leakage, good accuracy and stability. (see Figure 10.11). Three basic types:

1. *Disc ceramic.* Ceramic disc coated with metal film on both sides and covered with plastic or epoxy shell. Capacity range about 10 pF to 1 μF.
2. *Tubular ceramic.* Ceramic material mixed for specific temperature characteristics. As temperature compensators, they cancel temperature changes of other components. Typical temperature coefficients: N500 (negative 500 ppm), P100 (positive 100 ppm), or NP0 (zero change with temperature); see Exercise 29. Capacity range about 0.5 to 680 pF.
3. *Monolithic ceramic.* Layers of metalized ceramic stacked to increase capacitance (often referred to as MLC, multilayer ceramic, capacitors). Stack

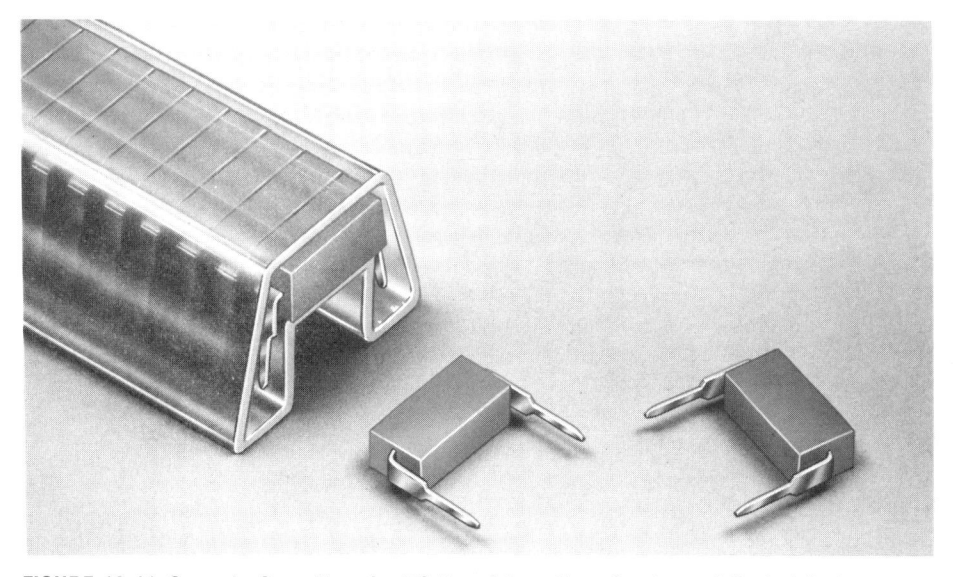

FIGURE 10.11 Ceramic Capacitors for PC Board Insertion. Courtesy of Centralab, Inc.

baked at high temperature into *monolithic* (Latin: *mono*, one; *lithim*, stone) piece and coated with plastic, glass, or epoxy shell. Leads may be left off to make "chip" capacitors. Well suited for automated parts placement. Capacity range about 1 pF to 20 μF.

Plastic Film Capacitors

Plastic films coated with metal film. Many layers stacked to form MLF (multilayer film) capacitor. Low cost, popular, but film's low K value makes them large for their capacitance. Wide temperature range. Low soakage. Able to stand high voltage for their size. Capacity range about 10 pF to 10 μF.

Electrolytic Capacitors

An *electrolyte* is a fluid that carries current by ion flow. An *electrolytic capacitor* uses special paper soaked with electrolyte for one plate. The other is aluminium foil with thousands of laser-etched grooves in its surface. These increase capacitance by increasing surface area. Foil and soaked paper rolled into tube and packaged with leads or terminals (see Figure 10.12). Voltage applied between plates forms aluminium oxide layer on foil surface. This oxide is the dielectric. It is very thin, so plates are close, which increases capacitance. Electrolytics provide highest capacitance for size of any type but have terrible leakage, short life, and are not stable; only used in certain applications. Polarized. Operation with reversed polarity destroys dielectric and capacitor. Capacity range about 0.1 μF to 0.5 F.

Tantalitic Capacitors

Tantalitic capacitors are also electrolytic. Plates made of *tantalum* metal. Dielectric tantalum pentoxide. Expensive, but provide high capacitance for size, with very low leakage and extremely long life. Polarized. Capacity range about 0.1 μF to 22,000 μF.

FIGURE 10.12 Electrolytic Capacitors. Courtesy of Mallory, Inc.

Oil-Filled Capacitors

Oil-filled capacitors are layers of aluminum foil rolled together with dielectric of plastic film or treated paper and inserted in can filled with special oil. Highest voltage rating. Very long life. Capacity range about 0.1 to 20 μF.

Air Dielectric Capacitors

Air dielectric capacitors are used in variable capacitors (see Figure 10.13). By turning shaft with a screwdriver, area between plates changes. Self-healing. Voltage rating set by distance between plates. Typical values 10 to 500 pF.

10.8 STRAY CAPACITANCE

Like resistance, capacitance is everywhere in circuits. Anytime that two conductors are separated by an insulator, unwanted or "stray" capacitance exists between them. A resistor's leads, for instance, act as capacitor plates (Figure 10.14); the resistive core is the dielectric. The figure shows other examples. Stray capacitance causes serious problems, particularly in high-frequency circuits.

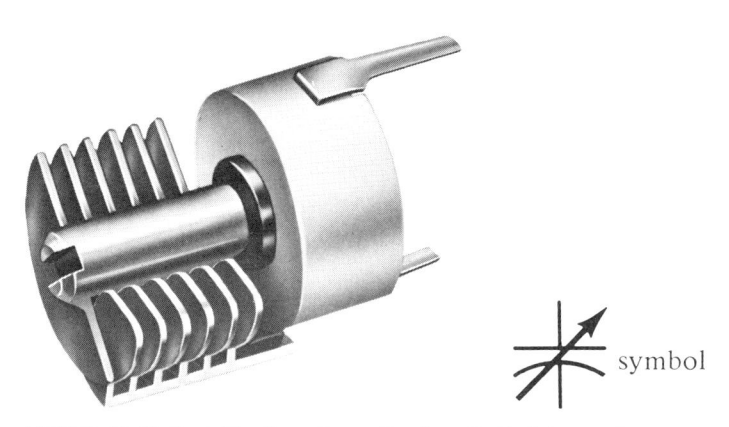

FIGURE 10.13 Variable Capacitors. Courtesy E. F. Johnson Co.

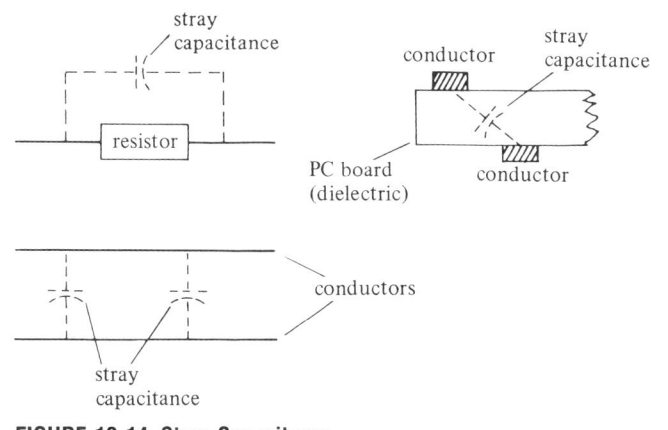

FIGURE 10.14 Stray Capacitance.

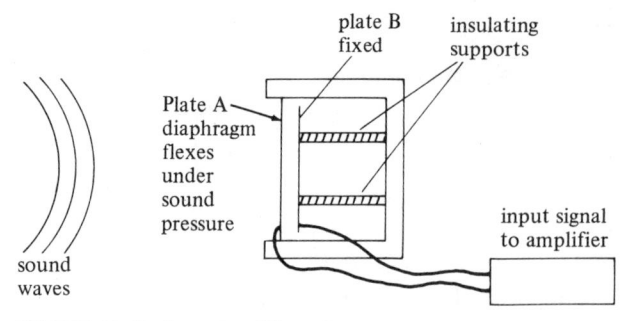

FIGURE 10.15 Capacitor Microphone.

10.9 CAPACITIVE TRANSDUCERS

Capacitive transducers vary plate area or separation according to some physical parameter. For example, the capacitor microphone is a capacitor with one plate fixed and the other movable. Sound pressure waves push and pull the movable plate, thereby changing plate separation (Figure 10.15). This varies capacitance, which develops a voltage changing in step with the sound.

10.10 TRANSIENT ANALYSIS

When switches are thrown in circuits containing resistance and capacitance, voltages and currents change with time. These changes are called *transients*, a word describing something that lasts a short time. (*Note:* Because they change, transient voltages and currents are indicated by lowercase letters.)

To help us analyze transients, we name the last instant of time before a switch is thrown, $t-$. Similarly, $t+$ is the first instant after a switch is thrown (the word "instant" means the shortest possible length of time). The instant the switch is thrown is t_0.

At t_B^-, the switch in Figure 10.16 part (a) has not yet left position A, the circuit is open, and there is no current or voltage across R or C.

At t_B^+, the switch has just arrived in position B. The state of v_R, v_C, and i at this instant is shown in the waveforms of Figure 10.16, part (b). v_C is still zero because no charge has moved yet. Since it is zero, the entire source voltage is dropped across the resistance; v_R is at its highest value, E_s. This means that i_R must also be maximum. Its value is E_s/R.

As time passes, v_C rises and i falls. With current dropping v_R does likewise. These waveforms are shown in Figure 10.16, part (c).

After enough time, v_C has risen to equal E_s and i and v_R have decayed to zero; the transients are over. At any instant,

$$v_C + v_R = E_s \tag{10.16}$$

by KVL. This equation will be solved for v_C to give us an equation by which capacitor voltage can be determined at any instant.

By Ohm's law,

$$v_R = iR \tag{10.17}$$

10.10 TRANSIENT ANALYSIS

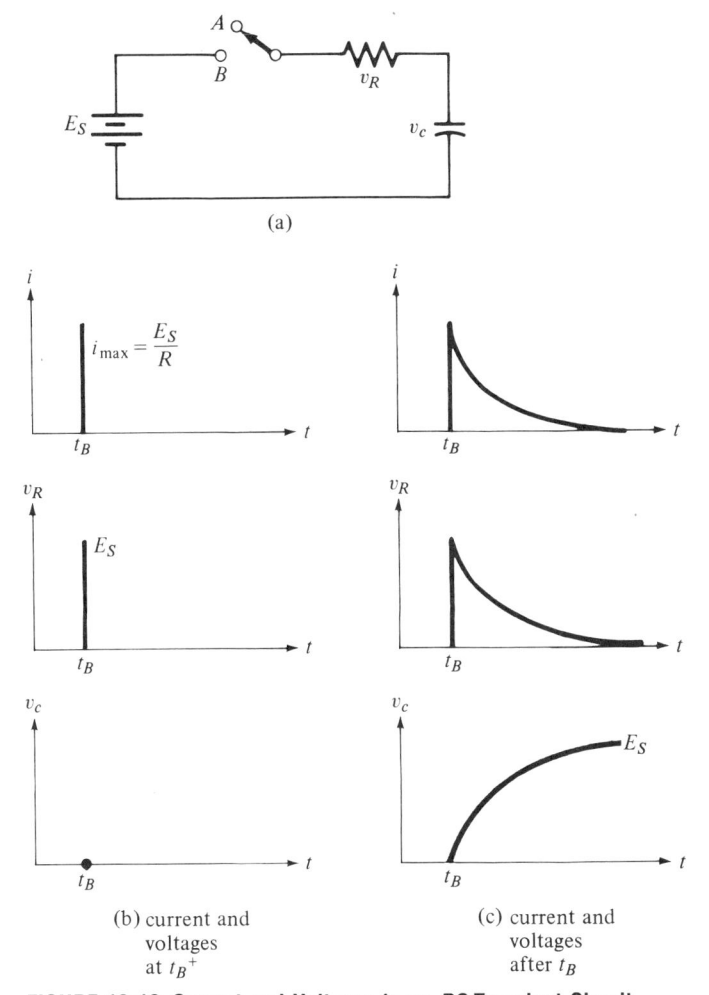

FIGURE 10.16 Current and Voltages in an *RC* Transient Circuit.

(a)

(b) current and voltages at $t_B{}^+$

(c) current and voltages after t_B

and

$$i = i_C$$
$$= C\frac{\Delta v}{\Delta t}$$

from Equation 10.3.

However, the delta sign applies to straight-line changes, and this current waveform is curved. So we use the symbol *d* from calculus, in which a curved line is considered to consist of an infinite number of infinitesimal straight lines. Thus we have

$$i_C = C\frac{dv}{dt} \tag{10.18}$$

Substituting for *i* in Equation 10.17 gives us:

$$v_(\, v_R = RC\frac{dv}{dt}$$

Substituting for v_R in Equation 10.16:

$$v_C + RC\frac{dv}{dt} = E_s$$

We cannot solve this equation for v_C with algebra because it contains the time-varying term $RC(dv/dt)$. Calculus must be used. When it is, the following equation results:

$$v_C = E_s(1 - e^{-t/\tau}) \text{ volts} \tag{10.19}$$

where v_C = instantaneous capacitor voltage at time t
  E_s = source voltage
  e = base of the natural logs, 2.7182818...,
  t = time, in seconds
  τ = circuit time constant, in seconds [τ is the Greek lowercase letter, tau (pronounced *tau*, rhymes with *cow*)]

The factor $(1 - e^{-t/\tau})$ contains several new items. Let us examine it in detail.

Note the t in the exponent of e. It makes $e^{-t/\tau}$ change with time, which, in turn, causes v_C to vary with time. By inserting an instantaneous t into the equation, we determine v_C at that instant.

A calculator purchased as suggested in Chapter 1 has a key marked ln and/or e^x. It does calculations with a special number called e, which equals 2.7182818.... Study your calculator manual until the use of this key is learned thoroughly.

The exponent of e, $-t/\tau$, will always be negative in our work.

τ is the circuit time constant. It is defined

$$\tau = RC \text{ seconds} \tag{10.20}$$

The units of τ must be seconds in order that the exponent of e be dimensionless. That they are seconds is easily shown:

$$RC = \frac{\text{volts}}{\text{ampere}} \cdot \frac{\text{coulombs}}{\text{volt}} = \frac{\text{coulombs}}{\text{ampere}}$$

$$= \frac{\text{coulombs}}{\frac{\text{coulombs}}{\text{second}}} = \text{seconds}$$

EXAMPLE 10.12

For this circuit:

83 V, 10 k, 2200 µF

10.10 TRANSIENT ANALYSIS

a. Write the equation for v_C:

$$\tau = RC = 10 \text{ k} (2200 \times 10^{-6}) = 22 \text{ s}$$

Substituting τ and the value of E_s into the general equation 10.19, we get a specific equation for this circuit:

$$v_C = 83(1 - e^{-t/22}) \quad \text{volts}$$

Study this equation carefully. With it, v_C can be determined at any instant.

b. What is v_C 5.3 seconds after switch closure?

Note how this specific time goes into the numerator of the exponent of e:

$$\begin{aligned} v_C &= 83(1 - e^{-5.3/22}) \\ &= 83(1 - e^{-0.24}) \\ &= 83(1 - 0.79) \\ &= 83(0.21) \\ &= 17.8 \text{ V} \end{aligned}$$

■

Using calculus again, it can be shown that transient current in RC circuits is

$$i = I_{\max}(e^{-t/\tau}) \quad \text{amperes} \tag{10.21}$$

Since $I_{\max} = \dfrac{E_s}{R}$, the equation can also be written:

$$i = \frac{E_s}{R} e^{-t/\tau} \quad \text{amperes} \tag{10.22}$$

where i = instantaneous current at time t.

EXAMPLE 10.13

Calculate the instantaneous current in the circuit of Example 10.12 at $t = 17.4$ seconds.

We first write the equation describing current in this circuit:

$$\begin{aligned} i &= \frac{E_s}{R} e^{-t/\tau} \quad \text{amperes} \\ &= \frac{83}{10 \text{ k}} e^{-t/22} \\ &= (83 \times 10^{-4}) e^{-t/22} \quad \text{A} \end{aligned}$$

Solve for the instantaneous current at $t = 17.4$ s by substituting that time in the equation:

$$\begin{aligned} i &= (83 \times 10^{-4}) e^{-17.4/22} \\ &= (83 \times 10^{-4}) 0.45 \\ &= 37.4 \times 10^{-4} \\ &= 3.74 \text{ mA} \end{aligned}$$

Current was a maximum 8.3 mA at $t = 0$ (instant of switch closure). In 17.4 s it has dropped to 3.74 mA.

■

To understand transients we must study how the factors $e^{-t/\tau}$ and $1 - e^{-t/\tau}$ change as time increases. We do this by giving values to t that are exact multiples of τ, that is, $t = 1\tau, 2\tau, 3\tau$, and so on. In this way the exponents become whole numbers. For example, let $\tau = 3t$. $e^{-t/\tau} = e^{-3\tau/\tau} = e^{-3}$.

Table 10.2 shows the values of the two factors as t varies from 0τ to 6τ. Be sure to understand the relation between these values and the waveform under the column.

TABLE 10.2 Values of $e^{-t/\tau}$ and $1 - e^{-t/\tau}$ for Various Values of t

t	$e^{-t/\tau}$	$1 - e^{-t/\tau}$
0τ	1.000	0.000
1τ	0.368	0.632
2τ	0.135	0.865
3τ	0.050	0.950
4τ	0.018	0.982
5τ	0.007	0.993
6τ	0.002	0.998
	Waveshape for these numbers	Waveshape for these numbers

Note carefully how the $e^{-t/\tau}$ column starts at 1 when $t = 0\tau$. Then, during the time from 0τ to 1τ, the value drops sharply to 0.368. In other words, the wave drops from 100% of maximum value to 36.8% during the time of one time constant.

During the next time constant the value drops at a slower rate, from 36.8% to 13.5%. Note how the wave is not dropping as steeply during this period.

The values continue to fall during succeeding time constants, but the amount of fall is smaller for each one. The waveform shows this by leveling out.

The value of $1 - e^{-t/\tau}$ jumps from 0 to 63.2% of maximum during the first time constant. The waveform reflects this by climbing very steeply during this period (note how maximum current occurs here, where there is maximum dv/dt). It climbs at a slower rate during succeeding time constants, rising to 86.5% at the end of the third, and 95% after the fourth, until it is almost level in the fifth and sixth. Since both waveforms change according to the value of the exponent of e, they are called *exponential waveforms*.

10.10 TRANSIENT ANALYSIS

The $e^{-t/\tau}$ wave never will fall to zero, nor will the $1 - e^{-t/\tau}$ ever reach 1. However, both are within 1% of those goals at the end of five time constants. For these reasons:

I Transients are considered finished after five time constants.

EXAMPLE 10.14

The switch is thrown from A to B at time $t_B = 0$. Answer the following questions.

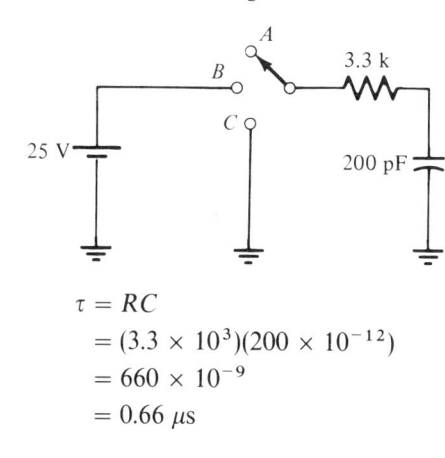

$$\tau = RC$$
$$= (3.3 \times 10^3)(200 \times 10^{-12})$$
$$= 660 \times 10^{-9}$$
$$= 0.66 \ \mu s$$

a. What is the instantaneous capacitor voltage at $t = 3\tau$?

From Table 10.2, when $t = 3\tau$,

$$v_C = 0.95 E_s$$
$$= 0.95(25)$$
$$= 23.8 \ V$$

This could also have been done by substituting $t = 3\tau = 3(0.66 \times 10^{-6}) = 1.98 \ \mu s$ into Equation 10.19.

b. How long will it take the capacitor to charge to 25 V?

$E_s = 25$ V. Therefore, when $v_C = 25$ V, the capacitor is fully charged. This requires five time constants.

$$5\tau = 5(0.66 \ \mu s)$$
$$= 3.3 \ \mu s$$

c. What is the equation for circuit current?

$$i = \frac{E_s}{R} e^{-t/\tau}$$
$$= \frac{25}{3.3 \ k} e^{-t/0.66 \mu s}$$
$$= 7.6 e^{-t/0.66 \mu s} \quad mA$$

d. With the switch held in position B for 3.3 μs (5τ), draw the waveforms of v_C, v_R, and i.

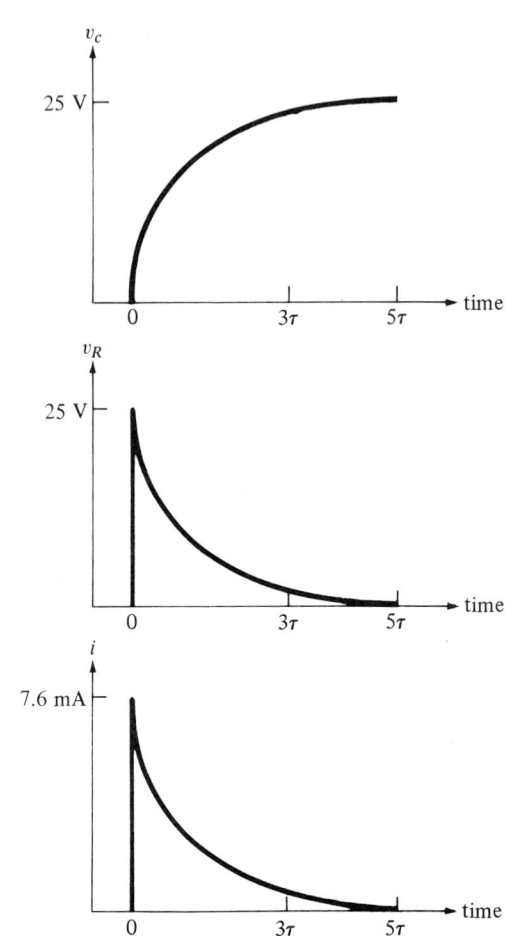

EXAMPLE 10.15

The switch in Example 10.14 has been held in position B long enough to complete charging. It is then thrown to position C to discharge the capacitor. Answer the following questions.

a. What is v_R at t_C^+?

Resistor current is left to right during charging, and right to left during discharging. Therefore, the polarity is reversed.

$$v_R \text{ at } t_C^+ = -25 \text{ V}$$

b. What is the current at t_C^+?

$$i = \frac{v_C}{R} = \frac{-25}{3300} = -7.6 \text{ mA}$$

Current is also negative because of opposite direction.

10.10 TRANSIENT ANALYSIS

c. Draw the waveforms of v_C, v_R, and i during discharge.

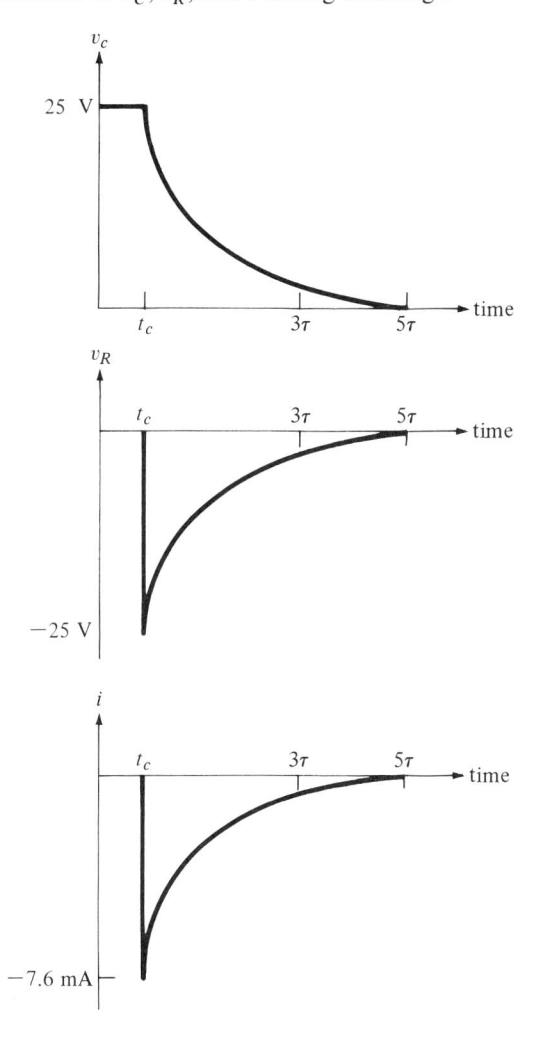

d. Write the equations describing v_C, v_R, and i during discharge.

It must be clearly understood that a brand new transient starts at time, $t_C = 0$, when the switch lands in position C. In other words, the time in these three equations is not the same time as in the equations describing circuit action during transient B (Example 10.14, part c).

At $t_C = 0$, v_C, v_R, and i are all at maximum value. They decay to zero as time increases. Therefore, each equation must include $e^{-t/\tau}$ because its value starts at 1 (when $t = 0$) and decreases to zero with time.

$$v_C = 25e^{-t/0.66\,\mu s} \quad \text{volts}$$
$$v_R = -25e^{-t/0.66\,\mu s} \quad \text{volts}$$
$$i = -7.6e^{-t/0.66\,\mu s} \quad \text{mA}$$

Electronic circuits often change time constants when a switch (transistor) opens or closes. It is quite common, for example, to see a capacitor charge quickly through a low resistance, then be switched to discharge slowly through a high resistance.

EXAMPLE 10.16

Answer the questions for this circuit.

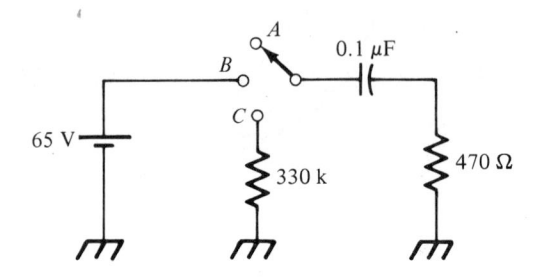

a. The switch is thrown from position A to position B at $t = 0$. What are i, v_C, and $v_{470\Omega}$ 283 μs later?

$$\tau = RC = 470(0.1 \times 10^{-6}) = 47 \; \mu s$$
$$5\tau = 5(47 \; \mu s) = 235 \; \mu s$$

283 μs is greater than 235 μs. Therefore, the transient is dead.

$$v_C = 65 \text{ V} \qquad i = 0 \text{ mA} \qquad v_R = 0 \text{ V}$$

b. The switch is thrown to position C. What is the voltage across the 330 k at t_C^+?

Since i is a maximum and 330 k $\gg$ 470 Ω,

$$v_{330k} \approx 65 \text{ V}$$

c. What is i at t_C^+?

$$i \approx \frac{v_C}{330 \text{ k}}$$
$$= \frac{65}{330 \text{ k}}$$
$$= 197 \; \mu A$$

But the direction is opposite to that when the capacitor was charging; therefore, this is $-197 \; \mu A$.

d. How long does it take v_{330k} to decay to zero?

$$\tau_{discharge} = R_{total}C$$
$$= (330 \text{ k} + 470 \; \Omega)0.1 \; \mu F$$
$$\approx 330 \text{ k}(0.1 \times 10^{-6})$$
$$= 330 \times 10^{-4}$$
$$= 33 \text{ ms}$$

Five time constants are needed for v_{330k} to decay to 0. Therefore,

$$5\tau = 5(33 \text{ ms}) = 165 \text{ ms}$$

10.11 USING THÉVENIN'S THEOREM IN TRANSIENT ANALYSIS

e. Draw the waveforms of v_C and i for the entire period that the switch was in position B then C:

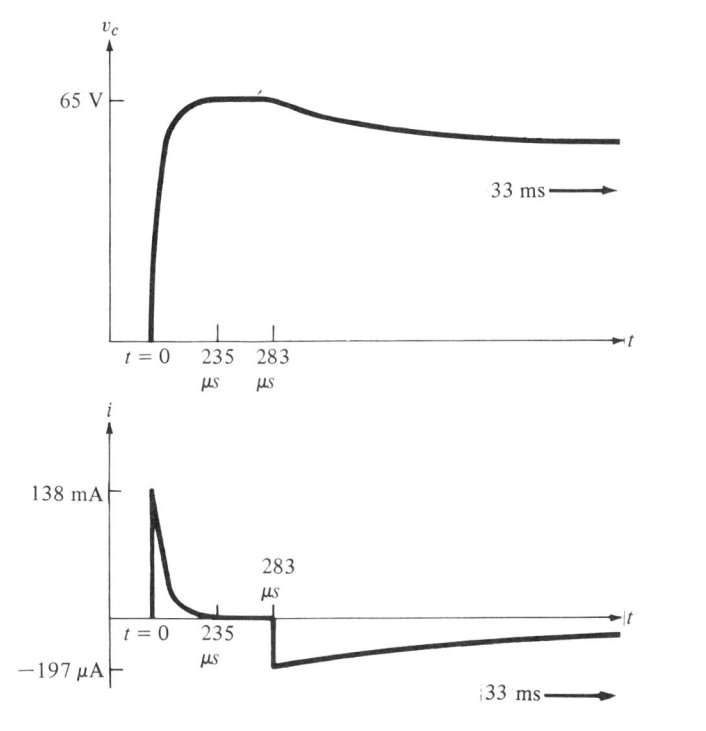

[Graph of v_C vs t: rises to 65 V, with markings at $t = 0$, 235 μs, 283 μs, and 33 ms]

[Graph of i vs t: peaks at 138 mA, drops to -197 μA, with markings at $t = 0$, 235 μs, 283 μs, and 33 ms]

■

Time constants give us control of time in electronics. For example, the motor in a delay-type windshield wiper is turned on by pulses from a speed-control circuit. Reducing a variable resistance reduces the time constant in that circuit. This allows a capacitor to charge to the switching voltage faster, which results in a shorter time between wipes.

10.11 USING THÉVENIN'S THEOREM IN TRANSIENT ANALYSIS

RC transient analysis offers an excellent chance to show the power of Thévenin's equivalent circuit.

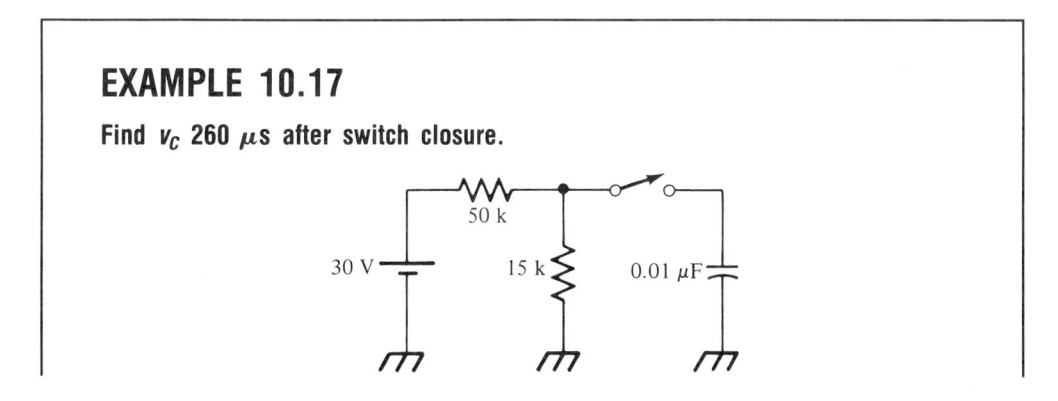

EXAMPLE 10.17

Find v_C 260 μs after switch closure.

[Circuit diagram: 30 V source, 50 kΩ resistor, 15 kΩ resistor, switch, and 0.01 μF capacitor]

How can the time constant be calculated when there is no single resistance through which the capacitor charges? To what voltage will the capacitor charge? Both of these questions are answered by Théveninizing the circuit driving the capacitor:

Finding R_{int}: Short out the source. This leaves the two resistances in parallel.

$$R_{int} = \frac{50\text{ k}(15\text{ k})}{50\text{ k} + 15\text{ k}}$$
$$= 11.5\text{ k}$$

Finding E_{oc}: E_{oc} equals the drop across the 15 k:

$$E_{oc} = \frac{30(15\text{ k})}{50\text{ k} + 15\text{ k}}$$
$$= 6.9\text{ V}$$

Redraw the circuit with the Thévenin equivalent driving the capacitor:

$$\tau = RC$$
$$= 11.5\text{ k}(0.01 \times 10^{-6})$$
$$= 11.5 \times 10^{-5}$$
$$= 115\text{ }\mu\text{s}$$

Write the equation for v_C:

$$v_C = E_s(1 - e^{-t/\tau})$$
$$= 6.9(1 - e^{-t/115\,\mu s})\quad\text{V}$$

Solve for v_C at $t = 260$ μs:

$$v_C = 6.9(1 - e^{-260\,\mu s/115\,\mu s})$$
$$= 6.9(1 - 0.1)$$
$$= 6.9(0.9)$$
$$= 6.2\text{ V}$$ ∎

The conclusion to be drawn from Example 10.17:

Thévenize the circuit driving the capacitor in a transient analysis if that circuit does not consist of a source and series resistance or cannot be reduced to that form.

EXAMPLE 10.18

What is i_C 10 seconds after switch closure?

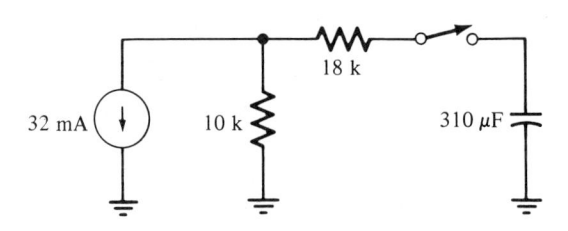

EXERCISES

Looking back from the capacitor, we do not see a source in series with a resistance; therefore, the circuit driving the capacitor must be Théveninized.

$$E_{oc} = (32 \times 10^{-3})10^4$$
$$= 320 \text{ V}$$
$$R_{int} \text{ (with current source opened)} = 10 \text{ k} + 18 \text{ k} = 28 \text{ k}$$

Draw the equivalent circuit:

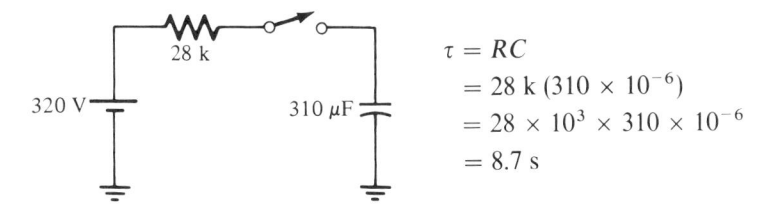

$$\tau = RC$$
$$= 28 \text{ k} (310 \times 10^{-6})$$
$$= 28 \times 10^3 \times 310 \times 10^{-6}$$
$$= 8.7 \text{ s}$$

The equation for the capacitor current is

$$i_C = I_{max} e^{-t/\tau}$$
$$= \frac{320}{28 \text{ k}} e^{-t/8.7}$$
$$= 11.4 e^{-t/8.7} \text{ mA}$$

Solve for instantaneous current at $t = 10$ s:

$$i_C = 11.4 e^{-10/8.7}$$
$$= 11.4 e^{-1.15}$$
$$= 11.4(0.32)$$
$$= 3.6 \text{ mA}$$

EXERCISES

Section 10.1

1. To the outside circuit, current appears to flow in one end of a capacitor and out the other, as it does with a resistance. Is this true? Explain.
2. What is a dielectric? Where is it found in a capacitor?
3. Explain how energy is stored in a capacitance. What must be done to release it?
4. Why is it important that a dielectric have a very high resistance?
5. Fill in the blanks.

	Voltage (V)	Q (C)	C (F)
a.		10^{-7}	150 pF
b.	46		65,000 pF
c.	185	3700 μC	

6. Convert each value as indicated.
   a. 30 F to picofarads.　　　b. 10^7 pF to microfarads.　　　c. 0.01 μF to pF.

Section 10.2

7. Find the current through and voltage across R_1 and R_2 in each circuit.
  a. At the first instant of switch closure (capacitor uncharged).
  b. Much later (capacitor charged).

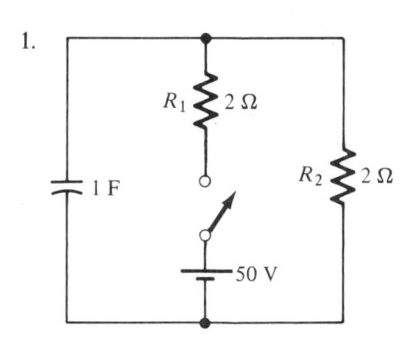

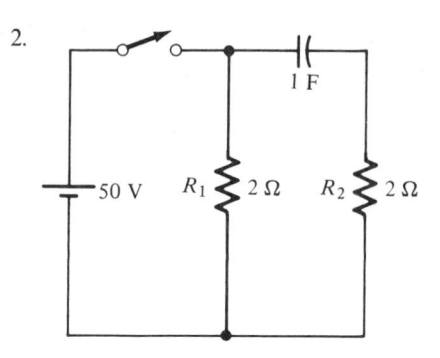

8. a. Can resistance have voltage across it with no current through it? Repeat for capacitance.
  b. Can capacitance have current through it with no voltage across it? Repeat for resistance.

9. Which voltage change causes the greater current through a capacitor, and why?
  a. Changes from 5 V to 55 V in 10 μs.   b. Changes from 47,000 V to 47,050 V in 10 μs.

10. The voltage waveform below is applied to a 10 μF capacitor. Plot the current waveform.

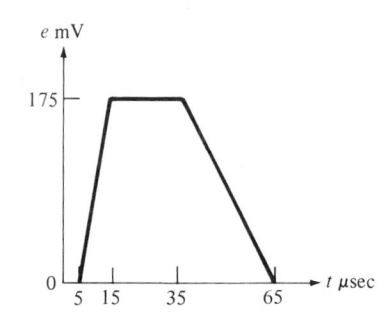

11. Repeat Exercise 10 for the voltage waveform:

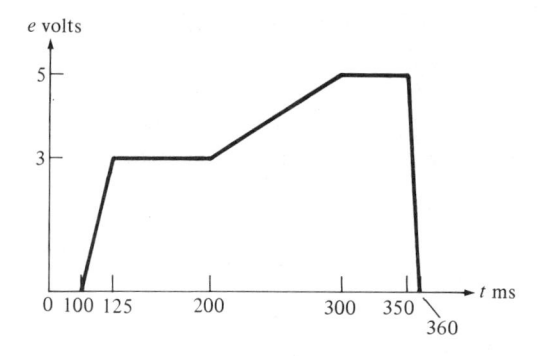

12. A current of 500 μA charges a capacitor to 300 mV in 10 s. What is the value of the capacitance.

Section 10.3

13. What is the capacitance of two silver plates each 0.2 by 0.05 m, separated by an air gap 10^{-4} m wide?

14. Repeat Exercise 13 for two aluminium plates 5 in. by 0.6 in., separated by an air gap $\frac{1}{16}$ in. thick.

EXERCISES

Section 10.4

15. In your own words, explain each of the following.
a. The development of induced charge in a dielectric.
b. How the field within a dielectric is reduced because of induced charge.
c. Why capacitance increases when a dielectric other than air is inserted between capacitance plates.
16. What K value is needed to raise the 140 pF of an air dielectric capacitor to 0.0033 μF?
17. Three hundred volts is applied across a capacitor having plates 4 cm by 60 cm separated by a ceramic dielectric 2×10^{-3} m thick with a K of 7000. Find: a. the capacitance; b. field intensity between plates; c. stored charge.
18. Each of two metal plates has an area of 3×10^{-2} m^2. When 1200 V is applied between them, the field intensity is 4×10^6 V/m. If $K = 6.8$, find the capacitance.
19. *Engineering Trade-off.* Increasing dielectric thickness raises breakdown voltage, but what does this do to capacitance?
20. What voltage punctures a dielectric $\frac{1}{8}$ in. thick with a breakdown of 960 V/mil?

Section 10.5

21. Fill in the blanks.

	W (J)	C (F)	Q (C)	V (V)
a.		2200 μF		35
b.			13×10^{-5}	3.4 kV
c.	78	0.15		
d.	108×10^{-15}		4.3×10^{-12}	

22. What energy is stored in a 350-μF capacitor with 900 V across it?

Section 10.6

23. Find C_{total} for each circuit.

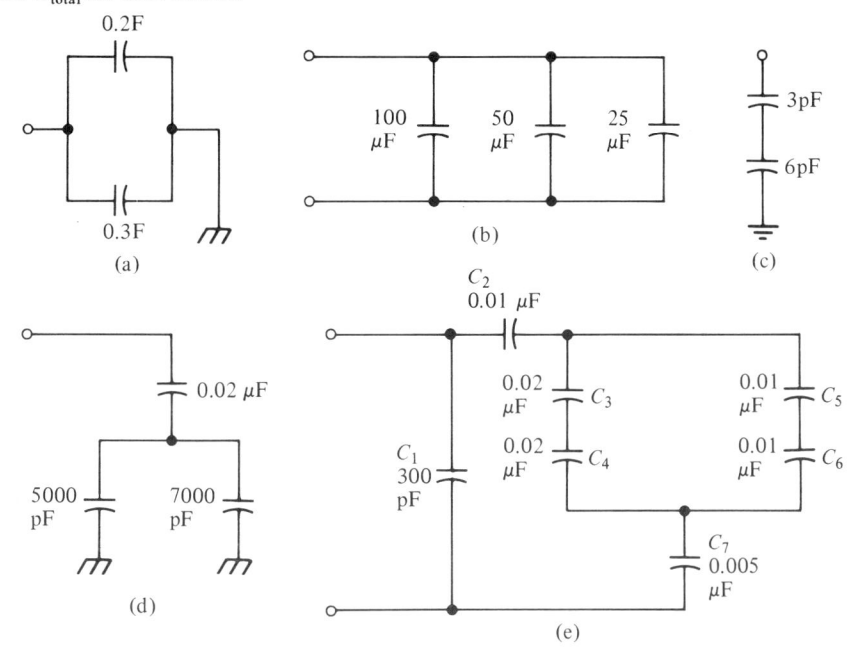

24. Approximate R_{total} for circuit a and C_{total} for circuit b. Discuss differences.

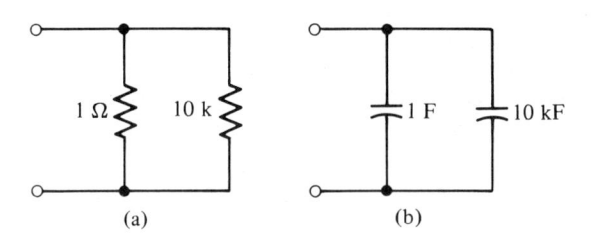

25. Repeat Exercise 24 for these circuits:

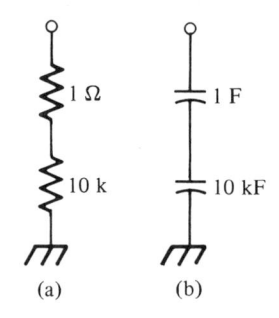

26. For this circuit:

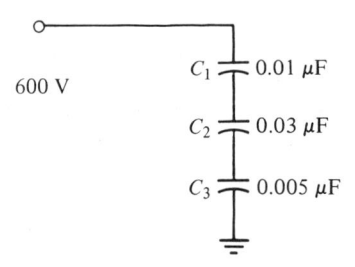

Find: a. Q_{total}; b. V_{c_1}; c. W_{c_2}; d. Q_{c_3}.

27. Circuit e of Exercise 23 has 75 V applied between terminals. What is the voltage across each capacitor?

28. Two capacitors, a 10 pF and a 75 pF, are series-connected across 1.2 kV. What dielectric breakdown rating must the 10 pF have?

Section 10.7

29. A monolithic ceramic capacitor has a capacitance of 200 pF at 20°C and a +TC of 100 ppm/°C.
a. What is its capacitance at 155°C?
b. The capacitor is replaced by a 175-pF capacitor having the same TC. In parallel with it is a temperature-compensating ceramic tubular capacitor, $C = 25$ pF at 20°C, TC = N750 (negative 750 ppm/°C). What is the capacitance of this combination at 20°C?
c. What is the capacitance of the combination at 155°C? Discuss the effect of the N750 unit.

30. A 5-μF tantalitic capacitor has a tolerance of +15%, −37%. What are the limits of its capacitance?

31. What precaution must be taken when connecting a polarized capacitor?

32. Explain how electrolytic capacitors are constructed to get high capacitance in a given volume.

EXERCISES

Section 10.10

33. Answer the following questions for this circuit:

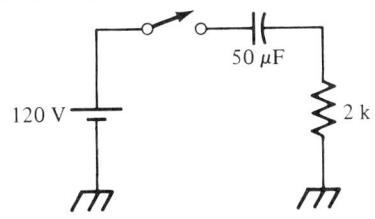

a. What is τ?
b. Write the equation for v_C, v_R, and i.
c. Calculate the current at $t = 55$ ms after switch closure.
d. Without doing any calculations, estimate what v_C should be at $t = 382$ ms. Then calculate the value. Compare the two.
e. What is v_R 600 ms after $t = 0$?
f. Sketch the waveforms of v_C, v_R, and i from $t = 0\tau$ to $t = 5\tau$.

34. Repeat Exercise 33 with the capacitor increased to 600 μF.

35. The switch is thrown to position B at $t = 0$.

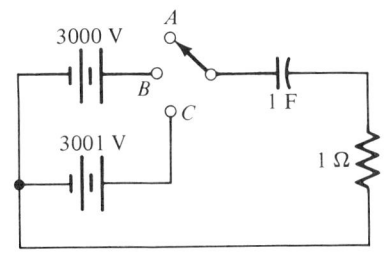

a. How long does it take the capacitor to charge to the source voltage?
b. The switch stays in B for the time determined in part a and is then thrown to position C. What will the capacitor do?
c. How long will it take to do so?
d. Having spent 5 s in position C, the switch is returned to position B. What will the capacitor do, and how long will it take to do it?
e. To what conclusion do you arrive about the charge time of a capacitor?

36. Answer the questions for this circuit:

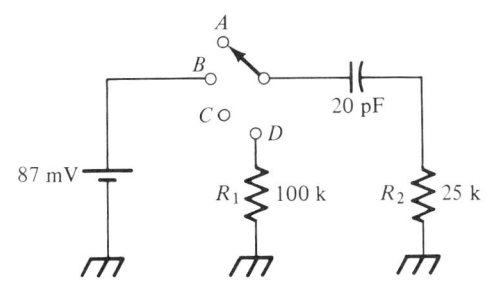

a. The switch is in position A. At $t = 0$ it is thrown to B. What is v_C 0.83 μs later?
b. What is the current 0.15 μs after $t = 0$?
c. The switch stays in position B for 4 μs. It is then switched to position C, where it remains for 5 μs, then switched to D. If the capacitor has no leakage, what is capacitor current at t_D^+?
d. What is v_{R_2} at t_D^+?
e. What is v_C 7.1μs after t_D?
f. Sketch the waveforms of v_C, i, and v_{R_2} for the period covering switch positions **B** and C and 12.5 μs in D.

Section 10.11

37. Answer the following questions for this circuit:

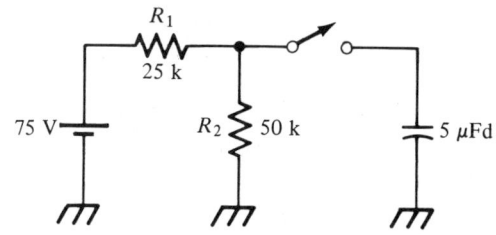

[Circuit diagram: 75 V source, $R_1 = 25\ k$ in series with switch, $R_2 = 50\ k$, and $5\ \mu Fd$ capacitor]

a. What is v_C 100 ms after $t = 0$?
b. What is i_C 223 ms after $t = 0$?

38. Answer these questions about the circuit of Exercise 37.

a. What is v_{R_1} at $t-$?
b. What is v_{R_2} at $t-$?
c. What is v_{R_2} at $t+$?
d. What is v_{R_1} at $t+$?
e. What is i_C at $t+$?
f. What is i_{R_2} at $t+$?
g. Sketch the waveforms of v_{R_1}, v_{R_2}, and i_{R_2} for time before and after $t = 0$.

CHAPTER HIGHLIGHTS

Capacitance, like resistance, is everywhere. Any two conductors separated by an insulator have capacitance between them.

A charged capacitor has stored energy and a potential between its plates.

Voltage across a capacitor cannot change instantaneously.

Resistor current is proportional to voltage. Capacitor current is proportional to the rate of change of voltage.

Do not exceed dielectric breakdown voltage.

No type of capacitor is right for every application.

Transients are voltages and currents that change with time. They are considered over after five time constants.

Know which transient equation describes which exponential waveform.

Thévenin's equivalent circuit must be used when we cannot see a single source and resistance in series with a capacitor.

Formulas to Remember

$$Q = CV \quad \text{charge moved between plates of a capacitor}$$

$$i = C\frac{dv}{dt} \quad \text{current through a capacitance}$$

$$C = \frac{8.9 \times 10^{-12}\ kA}{d} \quad \text{capacitance of a capacitor}$$

$$W = \frac{1}{2}CV^2 = \frac{1}{2}QV = \frac{1}{2}\frac{Q^2}{C} \quad \text{energy stored in a capacitor}$$

$$C_{total} = C_1 + C_2 + \cdots + C_n \quad \text{total capacitance of } n \text{ capacitors in parallel}$$

$$C_{total} = \frac{1}{1/C_1 + 1/C_2 + \cdots + 1/C_n} \quad \text{total capacitance of } n \text{ capacitors in series}$$

$$v_C = E_s(1 - e^{-t/\tau}) \quad \text{transient voltage across a capacitor}$$

$$\tau = RC \quad \text{time constant}$$

$$i_C = I_{max}e^{-t/\tau} = \frac{E_s}{R}e^{-t/\tau} \quad \text{transient current through a capacitor}$$

CHAPTER 11

Magnetism

In Chapter 2, we studied the electrostatic forces of charge at rest. In this chapter we study the electromagnetic forces of charges in motion.

Magnetism plays an important part in electronics. It is the basis of components such as meters, relays, and transformers, as well as tape recording of audio and video (TV) signals and magnetic data storage for computers. It is also fundamental to radio communications and the generation of electric power.

At the end of this chapter, you should be able to:

- Describe the magnetic field, its flux lines, and poles.
- List the parameters used in the study of magnetism.
- State the difference between magnetic and nonmagnetic materials.
- Discuss factors determining strength and direction of an electromagnetic field.
- Use the right-hand rule to determine field direction.
- Compare electric circuits and magnetic circuits.
- Use *B–H* curves and Ampère's circuital law to analyze magnetic circuits.
- State what air gaps are for and analyze magnetic circuits containing them.
- Explain the effects of Lenz's law on electrical generation.
- Describe the operation of several magnetic devices important to electronics.

11.1 MAGNETIC FLUX

Magnetism is still not completely understood. We do know that a magnet exerts force on distant objects by means of a *magnetic field* surrounding it. Scientists believe this field comes from the superposition of billions of tiny fields of individual electrons spinning in the magnet's atoms. Materials in which this happens are called *ferromagnetic materials*. They are used to make permanent magnets such as that in the D'Arsonval meter movement. Other materials, called *diamagnetic* and *paramagnetic*, are, for all practical purposes, nonmagnetic. They will not combine electron fields and cannot be used in magnetic devices.

A magnetic field acts as though it consists of many lines, called *flux lines*. Figure 11.1 shows the flux lines forming the fields of a bar magnet and a horseshoe magnet. Note the following:

- Flux lines have direction. A compass needle orients itself in this direction [Figure 11.1, part (a)].

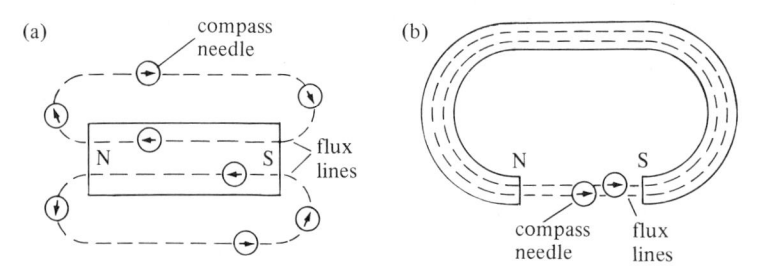

FIGURE 11.1 Magnetic Fields of Bar and Horseshoe Magnets.

11.1 MAGNETIC FLUX

- Flux lines are continuous loops. They do not begin and end as electrostatic field lines do. Instead, they leave a magnet at a location called the *north pole*, reenter at the *south pole*, and continue through the entire length of the magnet.
- Flux lines fill the volume between poles.
- Flux lines repel one another and do not cross.
- Flux lines arrange themselves to be as short as possible.

Flux is represented by the Greek lowercase letter ϕ (phi, rhymes with *pie*). It is measured in webers (Wb), an SI unit named in honor of German physicist Wilhelm E. Weber (1804–1891).

Flux density is the number of flux lines passing through a plane perpendicular to a magnetic field. It is represented by the letter B and defined:

$$B = \frac{\phi}{A} \quad \text{Wb/m}^2 \tag{11.1}$$

where ϕ = flux, in webers
A = area, in square meters

Besides the familiar SI units of meters, kilograms, and seconds, magnetic designers work in the *CGS* (*centimeter-gram-second*) system. In it, flux is measured in *lines* or *maxwells*. One weber equals 10^8 flux lines or maxwells. Flux density is measured in gauss; 1 gauss (G) equals 1 line or maxwell/centimeter^2. 1 Wb/m^2 = 10^4 G.

EXAMPLE 11.1

2.6×10^8 lines of flux pass through the face of this bar magnet.

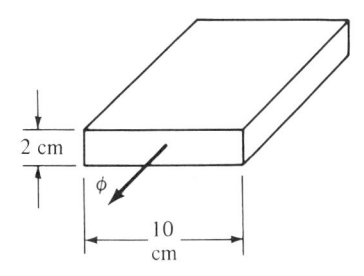

a. What is the flux density in gauss?

$$B = \text{lines/cm}^2 = \frac{2.6 \times 10^8 \text{ lines}}{10 \text{ cm} \times 2 \text{ cm}}$$

$$= \frac{2.6 \times 10^8 \text{ lines}}{20 \text{ cm}^2} = 13 \times 10^6 \text{ G}$$

b. Find the flux density in Wb/m^2.

$$(2.6 \times 10^8 \text{ lines}) \frac{1 \text{ Wb}}{10^8 \text{ lines}} = 2.6 \text{ Wb}$$

$$B = \frac{\phi}{A}$$
$$= \frac{2.6}{(10 \times 10^{-2})(2 \times 10^{-2})}$$
$$= \frac{2.6}{20 \times 10^{-4}}$$
$$= 1300 \text{ Wb/m}^2$$

This answer could have been obtained by converting the gauss of part a to Wb/m²:

$$(13 \times 10^6 \text{ G})\left(\frac{10^{-4} \text{ Wb/m}^2}{\text{gauss}}\right) = 1300 \text{ Wb/m}^2 \qquad \blacksquare$$

The ease with which flux lines may be established within a material is called the material's *permeability*. It is represented by the Greek lowercase letter μ. (Note: μ does not mean "micro" in this application.) The units of permeability are Webers/Amp-meter.

The permeability of a vacuum, represented by μ_0, is $4\pi \times 10^{-7}$ Wb/A-m. For other materials, permeability is given as a multiple of μ_0, called *relative permeability*, μ_r. The material's actual permeability is obtained by

$$\mu = \mu_r \mu_0 \tag{11.2}$$

EXAMPLE 11.2

What is the actual permeability of a material having a relative permeability of 40,000?

$$\mu = \mu_r \mu_0 = 40{,}000(4\pi \times 10^{-7})$$
$$= 50.3 \times 10^{-3} \text{ Wb/A-m} \qquad \blacksquare$$

Diamagnetic and paramagnetic materials such as aluminum, glass, copper, and other nonmagnetic substances have constant relative permeabilities of about 1. The ferromagnetics, iron, nickel, cobalt, and their alloys, on the other hand, have very high μ_r values. Iron, for example, ranges from 6000 to 8000, while permalloy, an alloy of iron and nickel, can go as high as 80,000. As we shall see, these values vary widely.

If a diamagnetic or paramagnetic substance is placed in a magnetic field, little happens [Figure 11.2, part (a)]. But a ferromagnetic material causes distortion of the field because flux lines bend to pass through the path of higher permeability [Figure 11.2, part (b)].

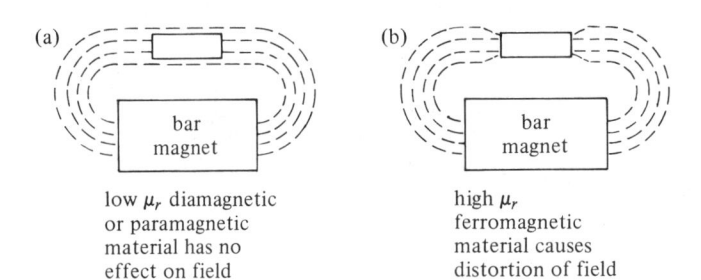

FIGURE 11.2 Effects of Permeability.

11.2 ELECTROMAGNETISM

A Danish physicist, Johann C. Oersted (1777–1851), was the first to report what scientists had long suspected, namely, a relation between electricity and magnetism. He noticed a compass needle position itself at right angles to a current-carrying conductor. André Ampère, whom we met in Chapter 2, and others, studied this effect and found that:

> Any current-carrying conductor is encircled by closed magnetic flux lines. The number of lines is proportional to current magnitude; their direction is set by current direction.

The direction of these flux lines can be determined with the help of the *right-hand rule*:

> The fingers of the right hand will point in the direction of magnetic flux lines if they are wrapped around a current-carrying conductor so that the thumb points in the direction of current (see Figure 11.3).

If the conductor is bent into a circle, the field inside is concentrated because all flux lines are in one direction, those outside are in the opposite direction (see Figure 11.4). Wind the conductor into a coil of many circles, or *turns*, and an enormous concentration of flux creates an electromagnet called a *solenoid*. Solenoids have all the

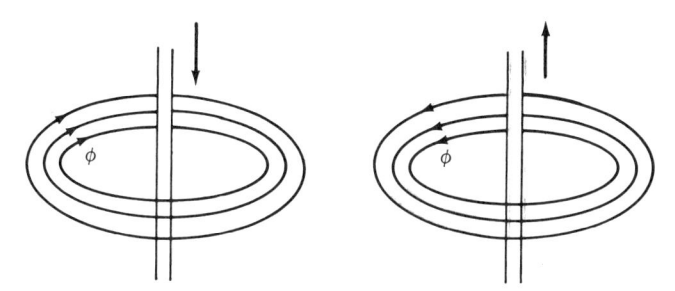

FIGURE 11.3 Using Right-Hand Rule to Determine Flux Direction Around Current-Carrying Conductor.

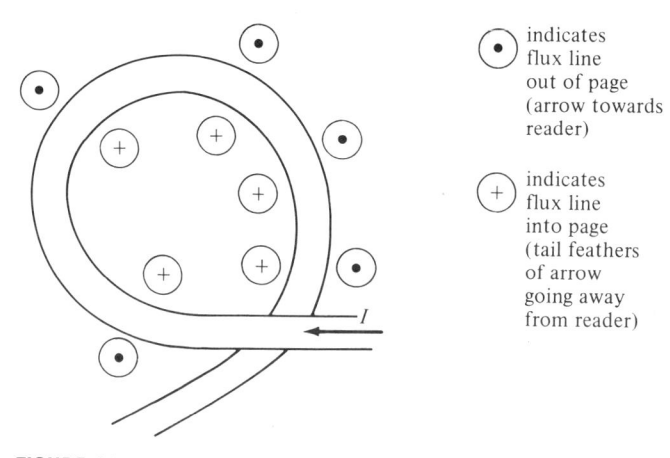

⊙ indicates flux line out of page (arrow towards reader)

⊕ indicates flux line into page (tail feathers of arrow going away from reader)

FIGURE 11.4 Flux Line Concentration Around Coiled Conductor.

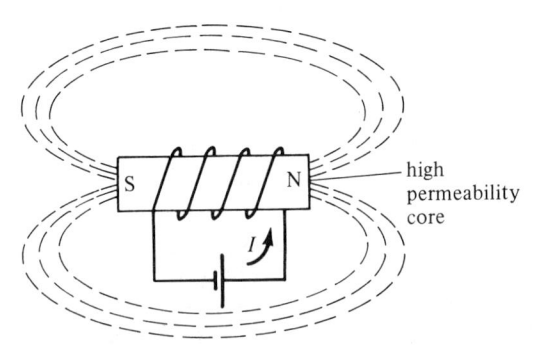

FIGURE 11.5 Magnetic Field Surrounding Solenoid.

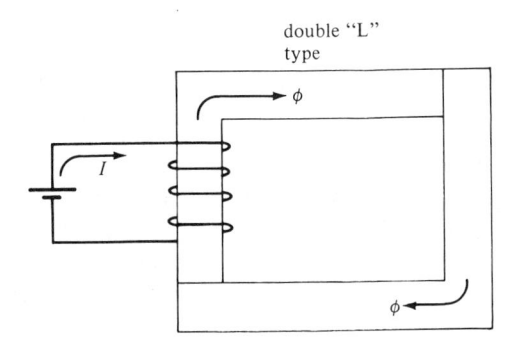

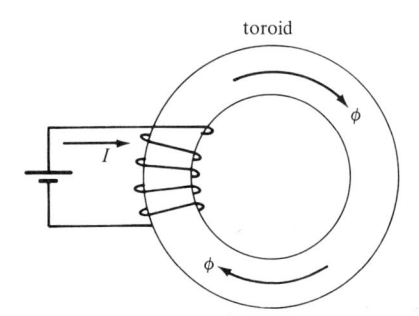

FIGURE 11.6 Continuous Cores.

field characteristics of bar magnets (Figure 11.5) with the added advantage that their magnetism can be controlled by changing the current.

The right-hand rule can be extended to the solenoid. Grasp the coil in the right hand, fingers wrapped in the direction of current, and the thumb points to the north pole.

Even greater concentration of flux is obtained by wrapping the solenoid around a bar or *core* made of high-relative-permeability material such as soft iron (Figure 11.5).

To eliminate the long flux paths through air of the core in Figure 11.5 and achieve further flux concentration, most magnetic devices, such as those shown in Figure 11.6, are made with closed cores. The *toroid*, a "doughnut" of compressed magnetic particles, confines all the flux inside it. The "double-L" style, on the other hand, has some *leakage* flux into the surrounding air.

11.3 THE MAGNETIC CIRCUIT

The coil–core arrangement in Figure 11.7 is called a *magnetic circuit* because it can be closely compared to the electric circuit alongside. Its "source" is current through the coil that develops a magnetic potential "rise." This rise is called *magnetomotive force* (*MMF*). It is represented by the letter $\mathscr{F}$ and defined by the equation

$$\mathscr{F} = NI \tag{11.3}$$

where $\mathscr{F}$ = magnetomotive force, in ampere-turns (AT)
N = number of turns
I = current through the N turns

11.3 THE MAGNETIC CIRCUIT

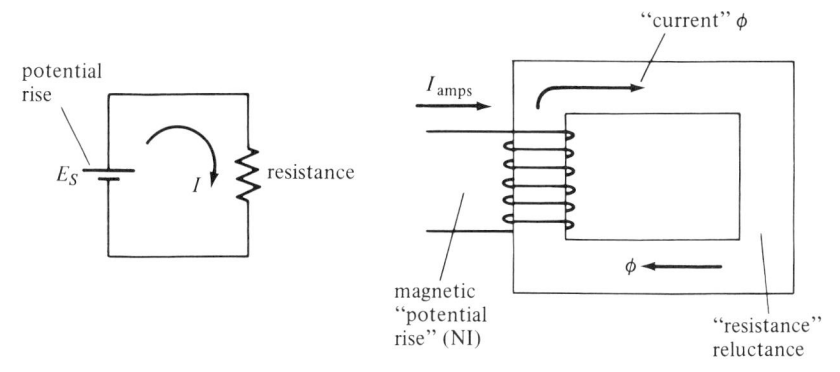

FIGURE 11.7 Comparing Magnetic and Electrical Circuits.

"Current" in the magnetic circuit is the flux developed by the MMF. The "resistance" to the development of flux is called *reluctance* (the core is reluctant to have flux established in it). Represented by the letter $\mathcal{R}$, reluctance is defined

$$\mathcal{R} = \frac{l}{\mu A} \frac{\text{amp-turns}}{\text{weber}} \tag{11.4}$$

where l = core length, in meters
  μ = permeability, in webers/amp-meter
  A = core cross-sectional area, in square meters

Magnetomotive force, flux, and reluctance are combined in an Ohm's law for magnetic circuits:

$$\phi = \frac{\mathcal{F}}{\mathcal{R}} \tag{11.5}$$

The reluctance in Equation 11.5 contains permeability, which has a nonlinear relationship with magnetomotive force; it can even have more than one value, as we shall see. As a result, Ohm's law for magnetic circuits is worthless for analyzing magnetic circuits. Its only value lies in its ability to show relationships between MMF, flux, and core properties. Actual analysis of magnetic circuits must be done by graphical techniques explained in sections to follow.

EXAMPLE 11.3

What is the total MMF developing flux in this toroidal core?

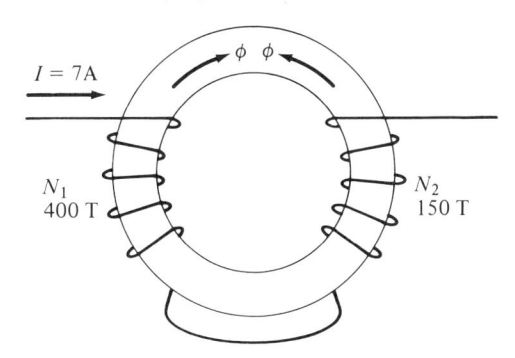

$I = 7\text{A}$
N_1 400 T
N_2 150 T

The left-hand coil creates clockwise flux (as indicated by the right-hand rule); the right-hand coil creates counterclockwise flux. Coils are in series; therefore, $I = 7$ A in both.

$$\text{total MMF} = 400I - 150I$$
$$= 250I$$
$$= 250(7)$$
$$= 1750 \text{ AT} \qquad \blacksquare$$

11.4 MAGNETIZING FORCE

If Equation 11.4 for reluctance is substituted into Ohm's law for magnetic circuits (Equation 11.5), we get

$$\phi = \frac{\mathscr{F}}{\mathscr{R}} = \frac{\mathscr{F}}{l/\mu A} = \frac{\mathscr{F}\mu A}{l}$$

Substitute this into Equation 11.1 for flux density:

$$B = \frac{\phi}{A} = \frac{\mathscr{F}\mu A}{Al} = \frac{\mathscr{F}\mu}{l} = \frac{\mathscr{F}}{l}\mu \qquad (11.6)$$

Note that the fraction multiplying μ in Equation 11.6 is magnetomotive force, $\mathscr{F}$, divided by core length, l (not coil length). This important ratio is given the name *magnetizing force* and is represented by the letter H. It is defined

$$H = \frac{\mathscr{F}}{l} \qquad \text{AT/m} \qquad (11.7)$$

Substituting NI for $\mathscr{F}$ gives us

$$H = \frac{NI}{l} \text{ AT/m} \qquad (11.8)$$

The CGS unit for H is the *oersted*:

$$1.26 \times 10^{-2} \text{ oersted} = 1 \text{ AT/m}$$

If H is substituted into Equation 11.6, we obtain

$$B = \mu H \qquad (11.9)$$

which makes the important point:

For a given magnetizing force, the greater the permeability of core material, the greater the flux density.

It can be seen from Equation 11.9 that the CGS units of permeability are gauss/oersted.

11.5 THE *B–H* CURVE

The relation between magnetizing force, H, and the flux density, B, which it develops in a ferromagnetic material is extremely important in the design of magnetic devices. Unfortunately, it cannot be expressed in a convenient math expression such as

11.5 THE B–H CURVE

Equation 11.9 because permeability is very dependent on H. Therefore, it is described graphically by a *B–H curve*.

To develop the $B-H$ curve for a particular core, the core's magnetizing force, H, is varied. This is done by changing coil current. For each H value, flux density, B, is measured. Then the points are plotted on a set of $B-H$ axes (see Figure 11.8). The $B-H$ curve has some unusual features. Let us examine it in greater detail.

Because of the battery and ground connection in Figure 11.8, potentiometer midpoint O is at 0 V with respect to ground. With the pot wiper at point O, no coil current flows; therefore, no flux is established in the core (the core has no initial magnetization).

When the wiper is moved below the zero point, towards the X end, current enters coil terminal A and flows downward through the coil. This develops a magnetizing force which combines the fields of some spinning electrons in what we will define as the positive direction. Flux density increases from zero, causing the curve to move to point a.

As the wiper moves lower, resistance decreases, causing I, and therefore H, to increase. The curve climbs higher as more electron fields shift. At H_{max} (point b), all electron fields are in the same direction. The core is *saturated*. Further increase in H produces no increase in B. The curve levels out.

Suppose that we next return the wiper to point O. This reduces H to zero and we would expect the curve to retrace its path from b through a to zero. But it doesn't. It follows the curve b-c-d instead. Note carefully how these points lie above the path for increasing H.

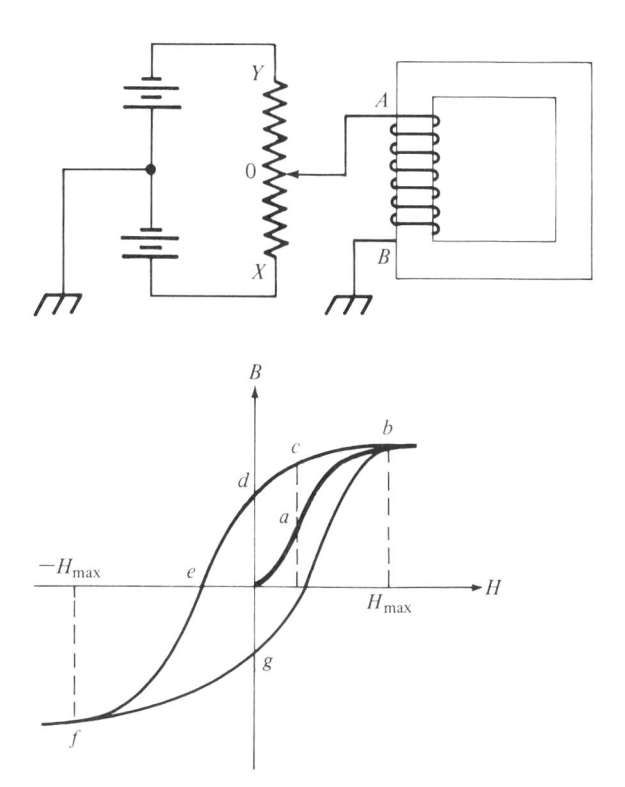

FIGURE 11.8 Developing a B-H Curve.

The curve at point *d* tells us two things:

- Even though $H = 0$, flux of density B_d remains in the core. This *remanent* magnetism occurs because some electron fields remain combined even though H is zero.
- Remanent magnetism makes permanent magnets possible.
- Originally, $B = 0$ when $H = 0$. Now, $B = B_d$ when $H = 0$. Reason—the core was saturated before reducing H to zero. In other words:

> The present state of core flux depends on the past state of core flux; *there is no one value of B for a given value of H.*

To reduce flux density to zero requires a negative magnetizing force, that is, one that combines electron fields in the negative direction. This occurs at some value of reverse coil current. It is obtained by advancing the wiper from *O* toward *Y*. This causes current to enter coil terminal *B* and flow upward through the coil.

H becomes more negative as the wiper approaches *Y*. Eventually, it reaches $-H_{max}$, causing core saturation in the negative direction (point *f*). All electron fields are now oriented opposite to what they were at point *b*.

Return the wiper to *O* and H is zero again while $B = B_g$. This has the same value as B_d, except that flux is in the opposite direction.

Advance the wiper toward *X* again and the curve returns to point *b*, closing the loop. Connect all points and the result is a *B–H* curve for the material of which the core is made.

Note that points *a* and *c* in the *B–H* curve have the same *H* value but B_c is greater than B_a. This tells us that flux density does not change as fast as magnetizing force does. In other words:

> Core material tends to oppose a change in flux density. This opposition is called *hysteresis* from a Greek word meaning "to lag behind."

For this reason, the *B–H* curve is also known as a *hysteresis loop*.

Energy needed to overcome a core's opposition to changing flux density is called *hysteresis loss*. It, together with I^2R heat lost in the wire (called *copper loss*), are

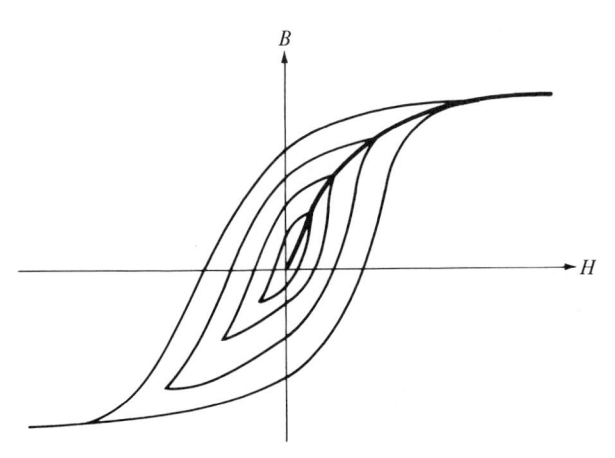

FIGURE 11.9 Normal Magnetization Curve.

11.6 AMPÈRE'S CIRCUITAL LAW

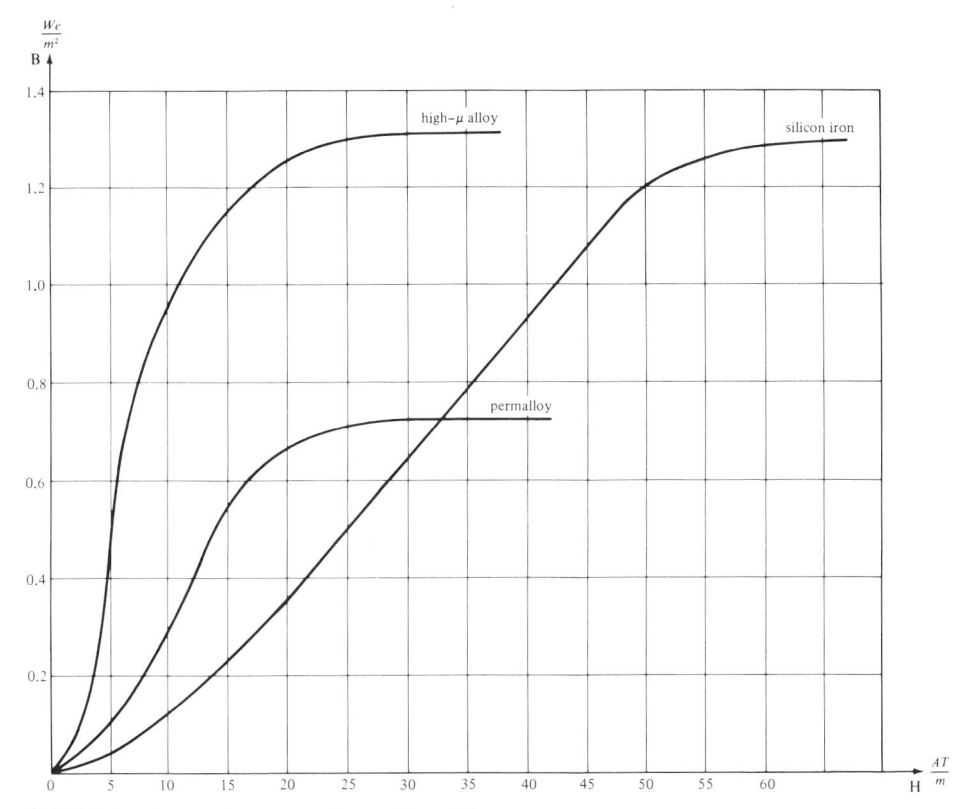

FIGURE 11.10 *B-H* Curves for Three Materials.

two reasons why magnetic devices get hot. It can be shown that hysteresis loss is proportional to the area inside the hysteresis loop.

If H is not carried to core saturation, smaller hysteresis loops result (Figure 11.9). The line connecting the tips of these curves is an average of all B values possible for each value of H. It is called the *normal magnetization curve*. Such curves are supplied to magnetic designers by the manufacturers of core materials. From now on when we speak of $B-H$ curves, we are referring to one of these. $B-H$ curves for three typical magnetic core materials are shown in Figure 11.10.

11.6 AMPÈRE'S CIRCUITAL LAW

Ampère's circuital law is to magnetic circuits as Kirchhoff's voltage law is to electric circuits. It says:

| Total applied MMF magnetizing a core equals the sum of the magnetic potential drops around the core.

In equation form,

$$\sum \mathscr{F} = \sum Hl \tag{11.10}$$

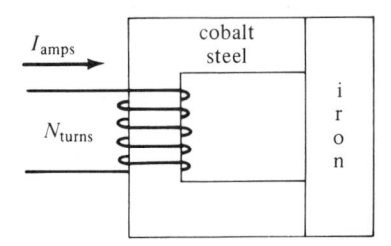

FIGURE 11.11 Magnetic Drops Around Closed Core.

Substituting Equation 11.3 into Equation 11.10 yields

$$\sum NI = \sum Hl \tag{11.11}$$

Applying Equation 11.11 to Figure 11.11, we have

$$NI = \underbrace{H_{\text{cobalt}} \times l_{\text{cobalt}}}_{\text{magnetic drops}} + \underbrace{H_{\text{iron}} \times l_{\text{iron}}}_{}$$

↑ magnetic rise

11.7 ANALYZING THE MAGNETIC CIRCUIT

There are two problems with which we will be concerned. First, given the flux, find the NI needed to produce it, and second, given the NI, find the resulting flux. Both require B–H curves and Ampère's circuital law for solution.

Here are examples of the first type.

EXAMPLE 11.4

Compute the current needed to produce a flux of 2.4×10^{-4} Wb in this permalloy core.

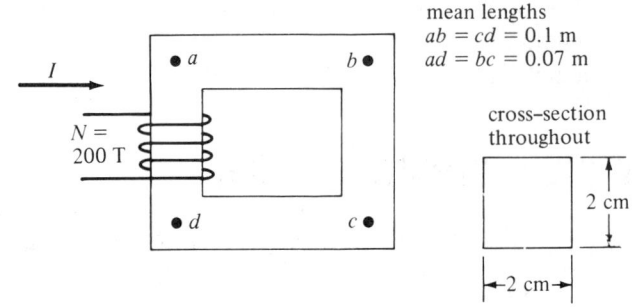

mean lengths
$ab = cd = 0.1$ m
$ad = bc = 0.07$ m

cross-section throughout

2 cm × 2 cm

$$\text{mean core length} = 2(0.1 + 0.07) \text{ m}$$
$$= 0.34 \text{ m}$$
$$\text{core area} = (2 \text{ cm})^2$$
$$= 4 \text{ cm}^2 \left(\frac{1 \text{ m}^2}{10^4 \text{ cm}^2} \right)$$
$$= 4 \times 10^{-4} \text{ m}^2$$

11.7 ANALYZING THE MAGNETIC CIRCUIT

$$B = \frac{\phi}{A}$$
$$= \frac{2.4 \times 10^{-4}}{4 \times 10^{-4}}$$
$$= 0.6 \text{ Wb/m}^2$$

From the B–H curve for permalloy in Figure 11.10,

$$H = 17 \text{ AT/m}$$

By Ampère's circuital law,

$$NI = Hl$$
$$I = \frac{Hl}{N}$$
$$= \frac{17(0.34)}{200}$$
$$= 28.9 \times 10^{-3} \text{ A}$$
$$= 28.9 \text{ mA}$$ ∎

EXAMPLE 11.5

The end of the core in Example 11.4 is changed to silicon iron and reduced to the dimensions shown. Find the new value of I needed to establish the same flux.

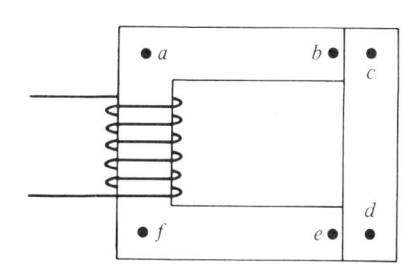

mean lengths
$ab = ef = 0.095$ m
$bc = ed = 0.0005$ m

cross-section for cd portion
1 cm
2 cm

area of silicon iron end $= (2 \times 10^{-2})(1 \times 10^{-2})$
$\qquad\qquad\qquad\qquad\quad = 2 \times 10^{-4} \text{ m}^2$
area of remaining core $= 4 \times 10^{-4} \text{ m}^2 \qquad$ from Example 11.4

$$B_{\text{permalloy}} = \frac{\phi}{A}$$
$$= \frac{2.4 \times 10^{-4}}{4 \times 10^{-4}}$$
$$= 0.6 \text{ Wb/m}^2$$

$H_{\text{permalloy}}$ from B–H curve $= 17$ AT/m
$Hl_{\text{permalloy}} = 17(ab + af + fe)$
$\qquad\qquad\; = 17(0.095 + 0.07 + 0.095)$
$\qquad\qquad\; = 17(0.26)$
$\qquad\qquad\; = 4.4$ AT

Both parts have the same flux; therefore,

$$B_{\text{silicon iron}} = \frac{\phi}{A} = \frac{2.4 \times 10^{-4}}{2 \times 10^{-4}}$$
$$= 1.2 \text{ Wb/m}^2$$

$H_{\text{silicon iron}}$ from B–H curve $= 50$ AT/m

$$\begin{aligned}Hl_{\text{silicon iron}} &= 50(bc + cd + de) \\ &= 50(0.005 + 0.07 + 0.005) \\ &= 50(0.08) \\ &= 4 \text{ AT}\end{aligned}$$

$$\begin{aligned}NI &= Hl_{\text{permalloy}} + Hl_{\text{silicon iron}} \\ &= \sum Hl = 4.4 + 4 = 8.4 \text{ AT}\end{aligned}$$

$$\begin{aligned}I &= \frac{Hl}{N} \\ &= \frac{8.4}{200} \\ &= 42 \text{ mA}\end{aligned}$$

Note carefully how the lower permeability and smaller dimensions of the silicon iron end in this example forced a current increase from 28.9 to 42 mA in order to establish the flux of Example 11.4. ∎

The next example is of the second type: Given NI, find the flux.

EXAMPLE 11.6

How much flux is produced in this toroidal core?

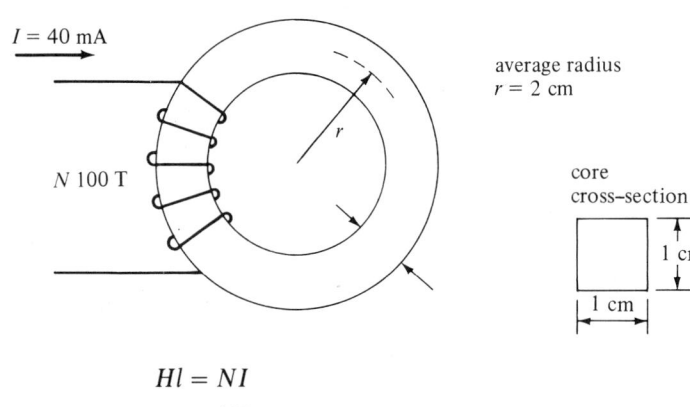

$I = 40$ mA

average radius $r = 2$ cm

N 100 T

core cross-section

1 cm × 1 cm

$$Hl = NI$$
$$H = \frac{NI}{l}$$

l = mean circumference of toroid
$$= 2\pi r_{\text{mean}}$$
$$= 2\pi(2 \times 10^{-2})$$
$$= 4\pi \times 10^{-2} \text{ m}$$

$$H = \frac{100(40 \times 10^{-3})}{4\pi \times 10^{-2}}$$
$$= 31.8 \text{ AT/m}$$

From $B-H$ curve for high-μ alloy:

$$B = 1.3 \text{ Wb/m}^2$$
$$\phi = BA$$
$$= 1.3(10^{-2})^2$$
$$= 1.3 \times 10^{-4} \text{ Wb}$$

11.8 AIR GAPS

A core's permeability is not constant. It varies according to the magnetizing force (see Figure 11.12), which, in turn, depends on coil current, temperature, and other factors. For applications in which such variations would cause problems, permeability can be stabilized by placing an *air gap* in the core. This reduces the permeability of a core material, μ_r, to a much lower figure, called design permeability, μ_d. If the gap is made wide enough, μ_d remains almost constant even though μ_r varies over a wide range. In other words, *permeability is traded for stability*.

For example, Figure 11.13 shows the effect on design permeability of a gap with a thickness 0.1% of core length. Note how design permeability stabilizes at 1000 for μ_r values beyond approximately 60,000. Another advantage of an air gap is that it prevents core saturation, which must be avoided in some applications. To pay for these gains, an enormous number of ampere-turns must be added to establish the desired flux in the high-reluctance, nonmagnetic gap.

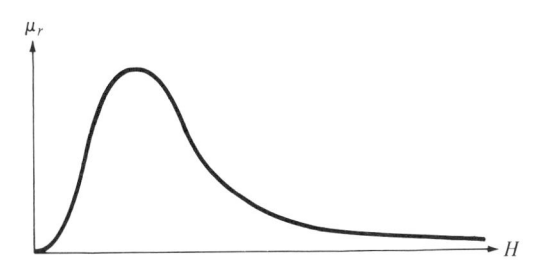

FIGURE 11.12 Permeability Variations with Magnetizing Force.

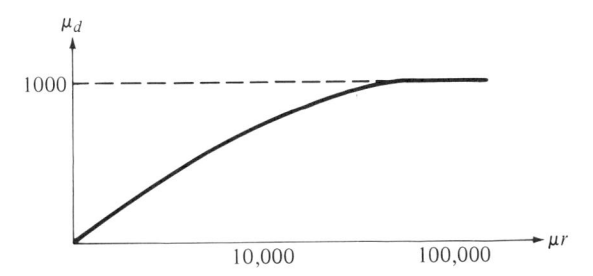

FIGURE 11.13 Effect of an Air Gap on Design Permeability.

A gap's "magnetic potential drop," $H_{gap}l_{gap}$, must be included in a core's Ampère's circuital law statement. But how do we get H_{gap}? Even though we know B_{gap} (it is equal to B_{core}), we can't get H_{gap} from a $B-H$ curve because there is none for air; air does not saturate. But we do know the permeability of free space (vacuum) or air; it is $4\pi \times 10^{-7}$ W/A-m. Substituting this into Equation 11.9, we have

$$H_{gap} = \frac{B_{gap}}{\mu_{air}} = \frac{B_{gap}}{4\pi \times 10^{-7}}$$
$$= (7.96 \times 10^5)B_{gap} \quad \text{AT/m} \quad (11.12)$$

EXAMPLE 11.7

Calculate the ampere-turns needed to set up a flux of 4.4×10^{-4} Wb in this high-μ alloy core:

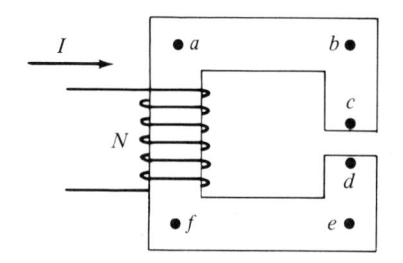

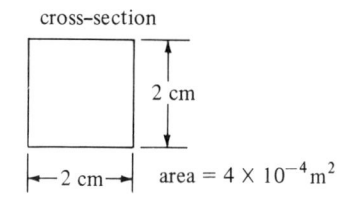

Mean lengths: $ab = ef = 5$ cm
$bc = de = 1.9$ cm
$af = 4$ cm
$cd = 2$ mm

$$B_{core} = \frac{\phi}{A} = \frac{4.4 \times 10^{-4}}{4 \times 10^{-4}} = 1.1 \text{ Wb/m}^2$$

From $B-H$ curves:

$$H_{core} = 14 \text{ AT/m}$$
$$H(2ab + af + 2bc) = 14[2(5 \times 10^{-2}) + (4 \times 10^{-2}) + 2(1.9 \times 10^{-2})]$$
$$= 14(10 + 4 + 3.8)10^{-2}$$
$$= 2.5 \text{ AT}$$
$$B_{gap} = B_{core}$$
$$H_{gap} = (7.96 \times 10^5)1.1$$
$$= 8.75 \times 10^5 \text{ AT/m}$$
$$H_{gap}l_{gap} = (8.75 \times 10^5)(2 \times 10^{-3})$$
$$= 1750 \text{ AT}$$
$$NI = H_{core}l_{core} + H_{gap}l_{gap}$$
$$= 2.5 + 1750$$
$$= 1752.5 \text{ AT}$$

Note how nearly all the ampere-turns were needed to establish flux in the nonmagnetic air gap. ∎

11.9 ELECTROMAGNETIC INDUCTION

As the conductor in Figure 11.14 is raised, it passes through, or *cuts*, flux lines of the magnet's field. This action develops a force inside the wire that pushes mobile charge carriers to one end. As a result, the wire becomes polarized, positively charged at one end, negative at the other. Cut the flux in the opposite direction by lowering the conductor, and carriers move in the opposite direction, reversing the polarization. Polarization is another way of saying that a voltage exists between the wire's ends.

Voltage developed across a conductor cutting flux lines is called an *induced voltage*. The process developing it is *electromagnetic induction*. If the conductor is part of a closed circuit, induced voltage causes current, as does any voltage.

Induced voltage magnitude is described by the formula

$$E = BlV \text{ volts} \tag{11.13}$$

where B = flux density, in webers/square meter
l = conductor length, in meters
V = conductor velocity perpendicular to flux lines, in meters/second

Velocity, V, means relative velocity between wire and flux; that is, the field may stand still and the conductor move, or vice versa.

Equation 11.13 says that induced voltage is increased by moving the conductor faster through a stronger field. Winding the conductor into a coil also raises potential by effectively increasing l. Each turn of the coil develops a voltage which is in series-aiding with the others. Note that no voltage is induced if the conductor moves sideways or stands still because no lines are being cut.

Electromagnetic induction, as described by Equation 11.13, is the basis of electrical generators, from a car's alternator to huge units that power a city. As we shall see in Chapter 13, generators consist of a coil of many turns rotating in a strong magnetic field.

Another form of electromagnetic induction occurs between stationary coils, such as those in Figure 11.15. Note how some of the flux developed by coil A passes through coil B. This flux is said to *link* the coils much like the center link shown in Figure 11.15, part (b), joins the outer links of a chain.

If the current in coil A is increased, the coil's flux increases, causing more flux linkages with coil B. As the linkages increase, voltage is induced and current flows in B. A reduction in coil A's current causes a linkage decrease. The result is a reversal of induced voltage polarity and current direction.

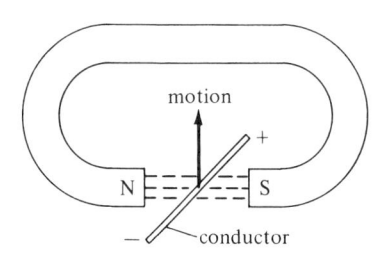

FIGURE 11.14 Conductor Cutting Flux Lines.

334   MAGNETISM

(a) magnetic flux linkages

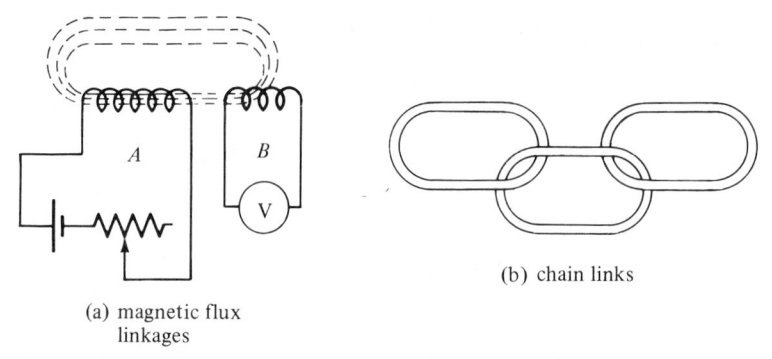

(b) chain links

FIGURE 11.15 Flux Linkages Between Two Stationary Coils.

Further experiments with the coils and flux linkages show that:

- No voltage is induced when current in coil A is constant, which tells us the number of linkages must be changing for induction to occur.
- The faster the current through coil A is increased or decreased, the greater the induced voltage in coil B. Induced voltage is therefore proportional to the rate at which linkages change, that is, the number of lines per second.
- If coil B is wound with N turns and each flux line from coil A links with all N of them, induced voltage is multiplied by N.

You may have noticed that two coils joined by changing linkages is the basis of the transformer, which we study in Chapter 21.

Michael Faraday, whom we met earlier, studied the development of voltage by changing flux linkages. From his work we have *Faraday's law of induced voltage*:

$$e = N\frac{d\phi}{dt} \qquad (11.14)$$

where e = induced voltage
N = number of turns
$d\phi/dt$ = rate at which flux linkages are changed per unit time, in webers/second

EXAMPLE 11.8

a. 65% of the 10^{-4} Wb developed by coil A pass through the 45 turns of coil B. If the current through A rises, causing a flux increase to 1.8×10^{-4} Wb in 27 μs, what is the voltage induced in B?

$$e_B = N\frac{d\phi}{dt} = 45\frac{0.65(1.8 - 1)10^{-4}}{(27 - 0)10^{-6}}$$

$$= \frac{45(0.65)(1.8 \times 10^{-4})}{27 \times 10^{-6}}$$

$$= 195 \text{ V}$$

11.9 ELECTROMAGNETIC INDUCTION

b. Current in *A* remains constant. What is the voltage induced in *B*?

Zero because $d\phi/dt = 0$ Wb/s.

c. Coil *A*'s current drops, causing flux to fall to 1.43×10^{-4} Wb in 41.8 μs. What is e_B during this time?

$$e_B = -N\frac{d\phi}{dt} = -45\frac{0.65(1.8 - 1.43)10^{-4}}{41.8 \times 10^{-6}}$$

$$= -\frac{45(0.65)(0.37 \times 10^{-4})}{41.8 \times 10^{-6}} = -25.9 \text{ V}$$

Note the change in polarity caused by a drop in flux through *B*.

If coil *B* in Example 11.8 was cut and a resistor wired between the ends, energy would be transferred to the resistor whenever voltage was induced in the coil. By the conservation of energy rule, this energy had to be supplied by some means; it cannot be created. Where did it come from?

When current flowed in the coil because of induced voltage, it developed its own flux field, as every current does. The Russian physicist Emil Lenz (1804–1865) found that the poles of this field were always in the direction to oppose a change in flux linkages. For example, in Figure 11.16, part (a), flux linkages increase as the magnet moves toward the coil. The resulting induced current has a field whose north pole repels the magnet's north pole. Because of this repulsion, energy (force times distance) must be expended to push the magnet toward the coil.

When the magnet is pulled away from the coil [Figure 11.16, part (b)], linkages decrease, causing current flow in the opposite direction. The coil's right end switches to become the south pole and the magnet must be pulled against the natural attraction of opposite poles.

Newton's law—for every action there is an equal and opposite reaction—explains this interaction between poles. *Lenz's law* says the same thing in electrical terms:

An induced current is always in a direction to oppose the motion producing it.

Lenz's law explains why it takes energy to turn electric generators. To get this energy, huge quantities of coal and oil are transported around the globe and burned to satisfy world demands for electricity. The resulting damage to our environment in the form of air pollution, acid rain, and oil spills is only beginning to be understood.

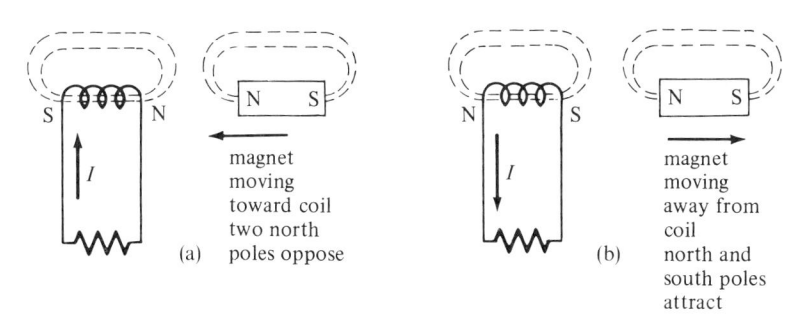

(a) magnet moving toward coil two north poles oppose

(b) magnet moving away from coil north and south poles attract

FIGURE 11.16 Lenz's Law

11.10 ELECTROMAGNETIC DEVICES

Magnetic recording tape for video (TV) and audio (sound) cassettes consists of a plastic film coated with ferrite oxide particles. Each of these acts as a tiny bar magnet. During recording, current, varying according to the signal being recorded, passes through a coil on the *recording head* [Figure 11.17, part (a)]. The resulting flux, also changing with the signal, bulges at the head gap, causing lines to dip into the ferrite layer. These magnetize the particles passing under the gap to a degree depending on the signal.

On playback, the tape passes under the gap in the *pickup head* [Figure 11.17, part (b)]. The flux of the magnetized particles develops a field within the toroidal head that varies according to the signal recorded. This induces a signal voltage in the coil which is then amplified.

Computer data are stored on and recovered from magnetic disks in much the same manner as illustrated in Figure 11.17.

A current-carrying conductor placed in a magnetic field (other than its own) experiences a force proportional to its length, the flux density of the field, and the current. This force is the basis of the electric motor, the D'Arsonval meter movement, and the *PM (permanent-magnet) loudspeaker*.

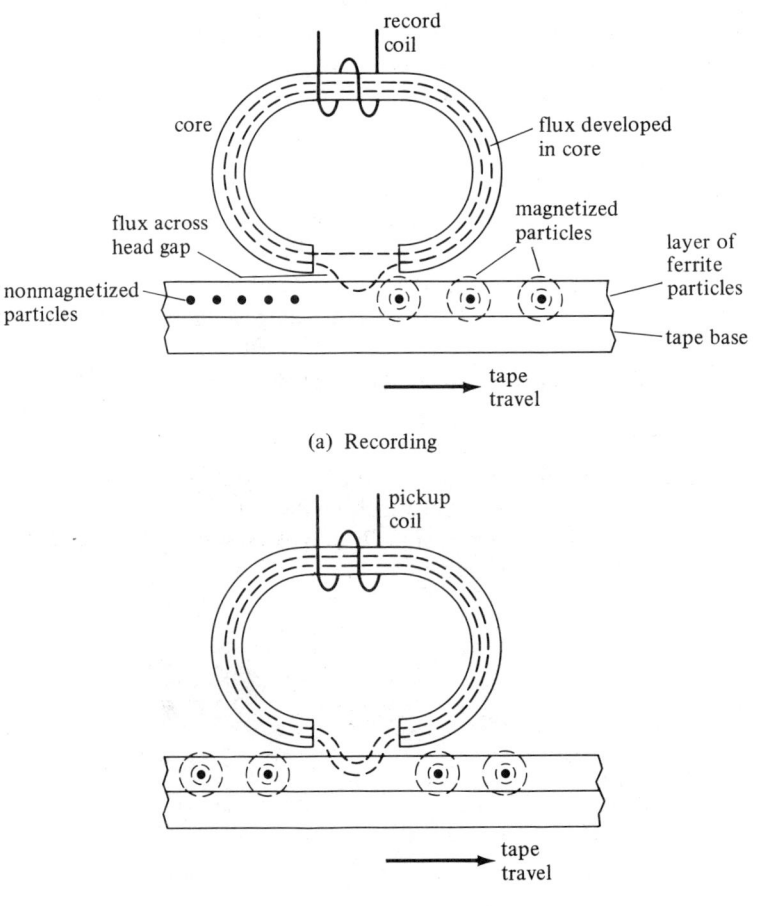

FIGURE 11.17 Magnetic Recording.

11.10 ELECTROMAGNETIC DEVICES

A field of high flux density is established in the PM speaker (Figure 11.18) by a permanent magnet made of ferrite materials or *Alnico*, an alloy of aluminum, nickel, cobalt, and iron. Located in this field is a *voice coil* of fine wire which is free to move. Attached to it is the *speaker cone*, a funnel-shaped piece of treated paper or cloth.

When current, varying according to an audio signal, flows through the voice coil, the forces developed by the field on the coil also vary. This causes the cone to vibrate, generating air pressure waves which our ears perceive as sound.

PM speakers work in reverse as *microphones*. Pressure waves from sound vibrate the cone and voice coil. Because the coil moves in a magnetic field, linkages change, developing in the coil a voltage analog of the sound. A PM speaker thus does double duty in applications such as *handy-talkies* and *intercom systems*.

Figure 11.19 shows a semiconductor in a magnetic field. Current through the material is in a direction perpendicular to the flux lines. The field bends the paths of the charge carriers, causing a higher concentration of them on one side of the

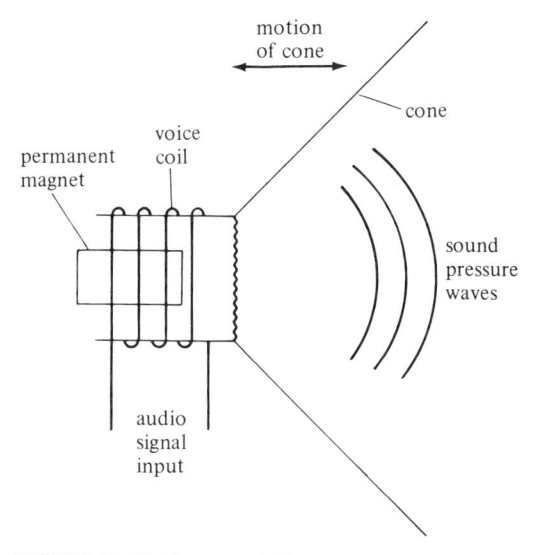

FIGURE 11.18 Permanent-Magnet Loudspeaker.

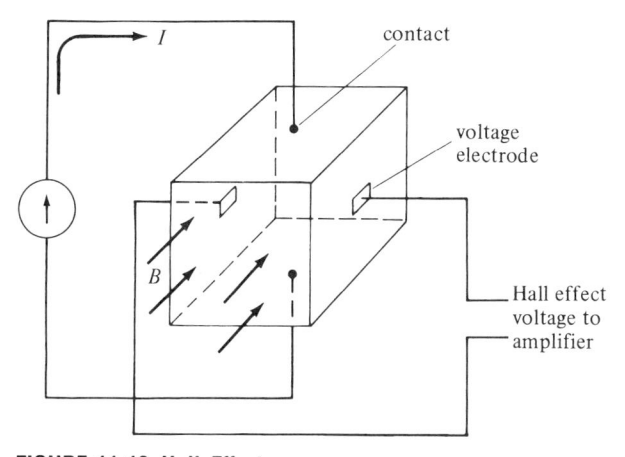

FIGURE 11.19 Hall Effect.

MAGNETISM

semiconductor. Because of the concentration difference, a voltage exists between the sides which is proportional to flux density. This voltage is the result of the *Hall effect*. Hall-effect sensors are used in a variety of applications, such as keyboard switches and automotive distributors.

EXERCISES

Sections 11.1 and 11.2

1. A flux of 750 Wb passes through the core cross sections shown. Calculate the flux density for each: a. in gauss; b. in Wb/m².

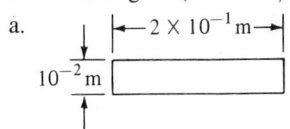

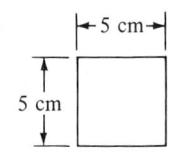

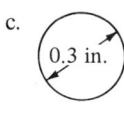

   a. 2×10^{-1} m, 10^{-2} m
   b. 5 cm × 5 cm
   c. 0.3 in.

2. Fill in the blanks.

	μ	μ_r
a.		350
b.	13.5×10^3	

3. Given: the following solenoid and magnetic field:

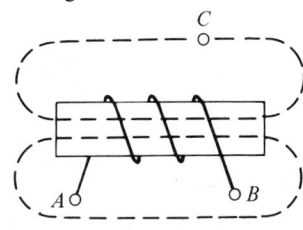

   What must the direction of I at terminal A be in order that:
   a. The solenoid's right end be the south pole?   b. Flux at point C points left?

4. What three things can be done to concentrate the magnetic field around a given current-carrying conductor?

Section 11.3

5. What must N_1 be in order that MMF total = 258 AT?

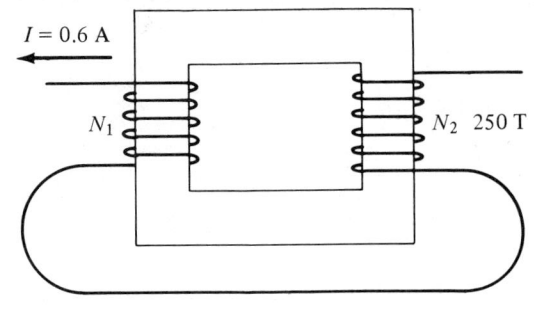

   $I = 0.6$ A, N_2 250 T

EXERCISES

6. A core has three coils wound on it: coil A: 2500 turns, 15 Ω resistance; coil B: 3800 turns, 30 Ω; coil C: 1200 turns, 10 Ω. What is the total MMF developed in each of the following cases? (Source voltage = 110 V dc.)
a. All coils connected in series electrically and aiding magnetically.
b. All coils in series, coil B opposing coils A and C magnetically.
c. Repeat parts a and b for the coils connected in parallel electrically.

Section 11.5

7. Why is it possible to have several values of B for a given H?

8. A core has two states of saturation. Explain.

9. How can we tell from a B–H curve when saturation occurs?

10. All other things being equal, which core will get hotter, one with a large area inside its hysteresis loop, or one with a small area? Explain.

Section 11.7

11. a. Compute the current needed to produce a flux of 4.7×10^{-4} Wb in this core of high-μ alloy.
b. What is the permeability of the core?

mean lengths
$ab = dc = 0.2$ m
$ad = bc = 0.05$ m

N
190 T

cross-section
2.5 cm
1.5 cm

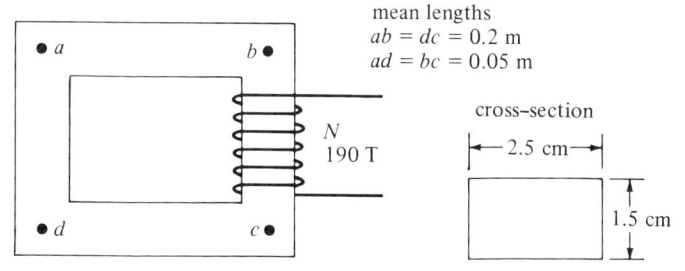

12. a. How many turns must be wrapped around the silicon–iron toroid to establish a flux of 1.6×10^{-4} Wb?
b. What is the permeability of the core?

I
480 mA

mean diameter = 7.5 cm

cross-section
$r = 0.75$ cm

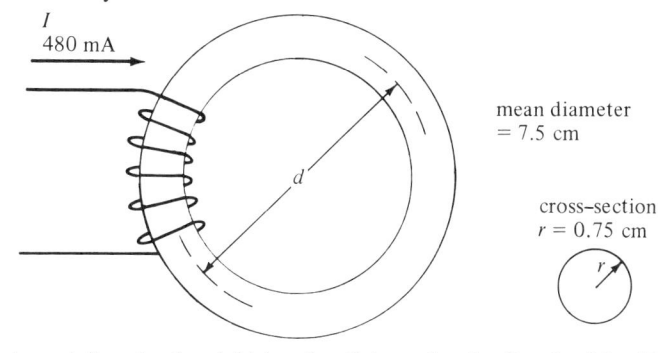

13. What current must flow in the right-hand coil to make the flux in this silicon–iron core be 1.86×10^{-3} Wb? Should the current enter or leave terminal A?

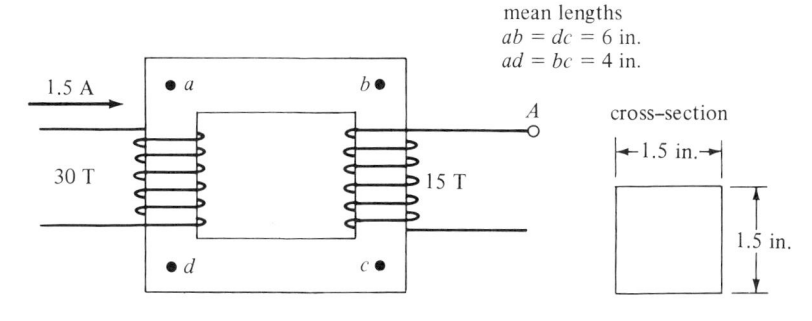

1.5 A
30 T
15 T

mean lengths
$ab = dc = 6$ in.
$ad = bc = 4$ in.

cross-section
1.5 in.
1.5 in.

14. What current must flow in this coil to establish a flux of 4×10^{-4} Wb in the toroid?

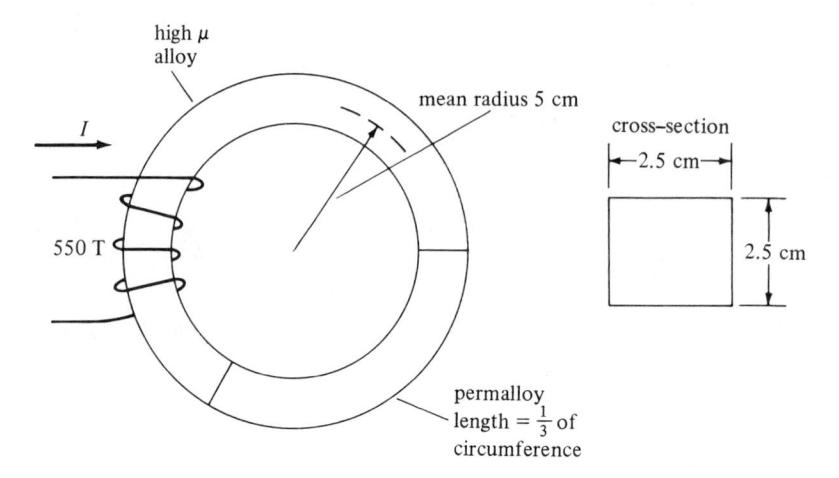

[Figure: Toroid with high μ alloy, mean radius 5 cm, 550 T coil carrying current I, permalloy section with length = $\frac{1}{3}$ of circumference. Cross-section: 2.5 cm × 2.5 cm.]

15. A flux of 6.25×10^{-4} Wb is to be established in this core. Find the NI necessary. *Hint:* Lengths indicated are not mean lengths.

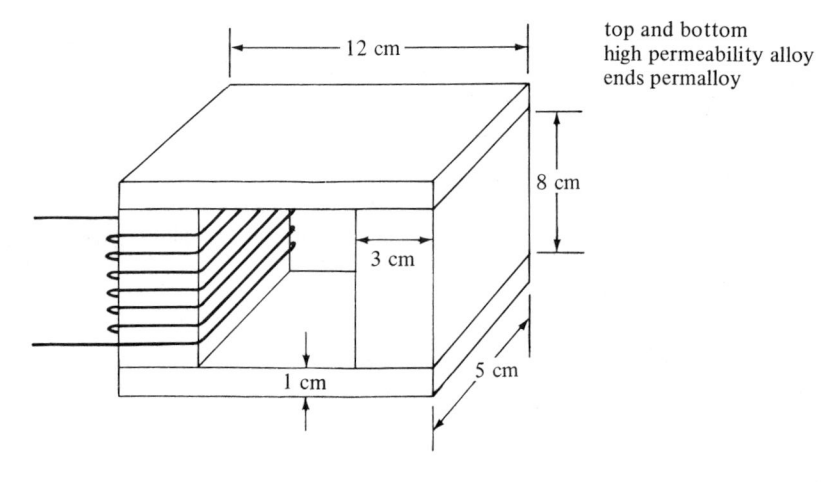

[Figure: Rectangular core, top and bottom high permeability alloy, ends permalloy. Dimensions: 12 cm wide, 8 cm tall, 5 cm deep, 3 cm inner opening height, 1 cm bottom thickness.]

16. What is the flux established in the core of Exercise 11 if 26 mA flows in its coil?

17. What flux is present in this toroidal core of silicon iron?

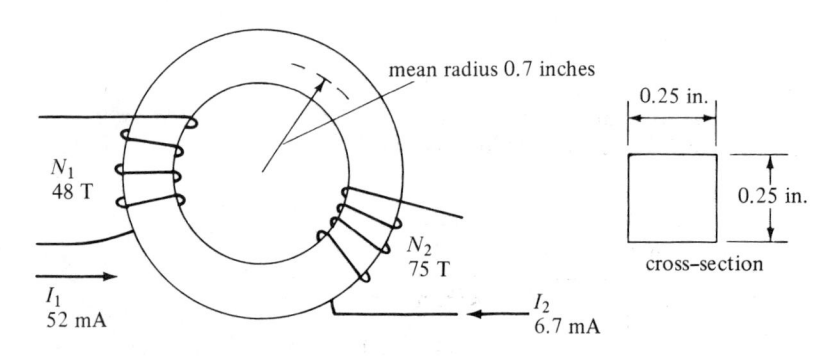

[Figure: Toroidal core, mean radius 0.7 inches. N_1 = 48 T with I_1 = 52 mA; N_2 = 75 T with I_2 = 6.7 mA. Cross-section: 0.25 in. × 0.25 in.]

18. The left-hand current for the core of Exercise 13 is reduced to 0.4 A and 0.67 A flows out of point A. What is the flux in the core?

EXERCISES

Sections 11.8

19. The core of Exercise 12 has a 2-mm gap cut into it. Calculate the turns needed to establish the same flux. Compare the answer to that of Exercise 12. Discuss differences.

20. Calculate the current needed to set up a flux of 3×10^{-4} Wb in the gap of this permalloy core.

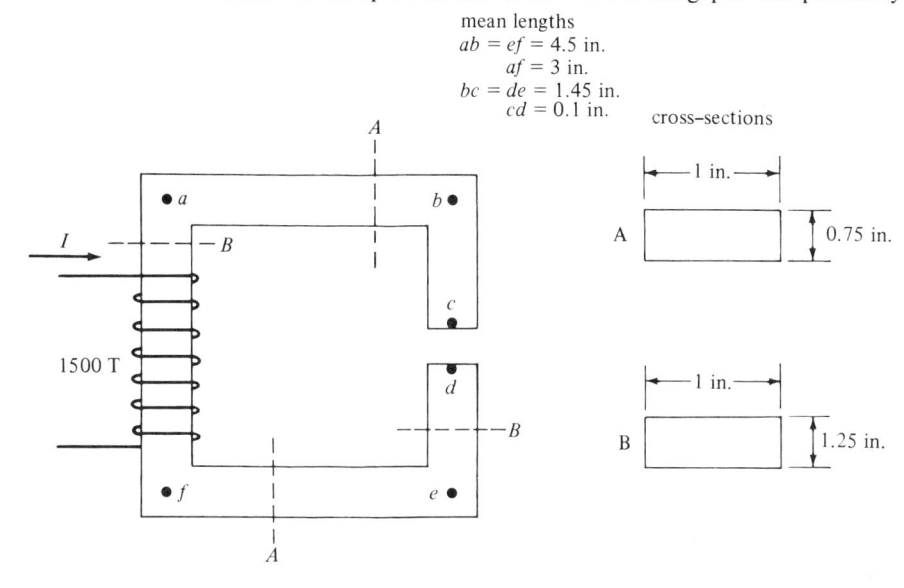

mean lengths
$ab = ef = 4.5$ in.
$af = 3$ in.
$bc = de = 1.45$ in.
$cd = 0.1$ in.

cross-sections

A: 1 in. × 0.75 in.
B: 1 in. × 1.25 in.

1500 T

Section 11.9

21. Will flux linkages between magnetic field and coil increase or decrease for each case?

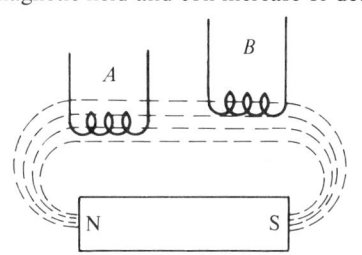

a. Coil A moved up and out of the field.   b. Coil B moved down.   c. Coil B moved sideways.

22. Explain in your own words what is meant by flux linkages.

23. Name three ways to increase induced voltage.

24. A coil of N turns has E volts induced in it when moved in a magnetic field of flux density B. What must be done to induce the same voltage when the coil is moved in a field of $4B$?

25. What voltage is induced when a flux of 3×10^{-2} Wb changes in a coil of 250 turns in 1.5 ms?

26. How many flux linkages must change in 2.7 s to induce 890 V in a coil of 330 turns?

27. How many turns must a coil have in order that 38 mV be induced in it when the linkages with it change at the rate of 1.9×10^{-4} Wb/s?

28. Which magnetic pole is the A end of this coil when: a. the coil moves left;   b. the magnet moves left;   c. both coil and magnet move right at the same speed;   d. both stand still?

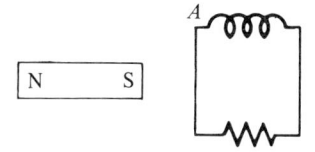

29. Explain why energy must be expended to change flux linkages with a current-carrying coil.
30. If the coil of Exercise 29 were opened, would energy still be required to change linkages? Explain.

Section 11.10

31. Explain how magnetic recording and playback are accomplished.
32. Explain how the PM speaker works as both a speaker and a microphone.

CHAPTER HIGHLIGHTS

Magnetism causes electricity and electricity causes magnetism.

Magnetic fields consist of flux lines that are continuous and have direction.

Any conductor carrying current has an electromagnetic field around it proportional to the current. The right-hand rule tells the orientation of that field.

Electromagnetic fields are concentrated by winding the conductor in the form of a coil around a closed core of high-permeability material.

Magnetic circuits are very nonlinear and must be analyzed by graphical means.

$B-H$ curves provide data on specific core materials and are vital to magnetic analysis and design.

Because of hysteresis, a core's reaction to a given MMF depends on its earlier state of magnetization.

Magnetic circuits are analyzed much as are electrical circuits; that is, the sum of the magnetic potential drops equals the sum of the rises.

An air gap is placed in a core to stabilize permeability and keep the core from saturating.

Flux linkages are magnetic flux lines encircling a closed conductor. When their number changes, voltage is induced in the conductor.

Lenz's law explains why energy is needed to turn an electric generator.

Formulas to Remember

$B = \dfrac{\phi}{A}$ flux density in Wb/m in SI, gauss in CGS

$\mu = \mu_0 \mu_r$ permeability of a material compared to that of free space

$\mathscr{F} = NI$ magnetomotive force

$\mathscr{R} = \dfrac{l}{\mu A}$ core reluctance

$\phi = \dfrac{\mathscr{F}}{\mathscr{R}}$ Ohm's law for magnetic circuits

$H = \dfrac{NI}{l}$ magnetizing force

$\sum NI = \sum Hl$ Ampère's circuital law

$H_g = 7.96 \times 10^5 \, B_g$ magnetizing force for an air gap

$e = N \dfrac{d\phi}{dt}$ Faraday's law of electromagnetic induction

CHAPTER 12

Inductance

In this chapter we build on the magnetic principles of Chapter 11 to study inductance, the third and final characteristic of components that limits current in a circuit. Inductance acts exactly the opposite of capacitance in almost all cases. Like capacitance (and resistance), we add it in some situations and work to get rid of it in others.

At the end of this chapter, you should be able to:

- Explain self-inductance.
- List the characteristics of a coil that determine its inductance.
- Explain the counter-EMF developed across an inductance.
- Compare the circuit characteristics of inductance, capacitance, and resistance.
- Calculate the energy stored in the electromagnetic field of an inductance.
- Describe the "inductive kick."
- List major inductor types.
- State what eddy currents are, why they are undesirable, and what is done to minimize them.
- Calculate total inductance of inductors in series, parallel, and series–parallel.
- Perform transient analysis on *RL* circuits.

12.1 SELF-INDUCTANCE

The circuit of Figure 12.1, part (a), has a high rheostat setting, hence low current and a weak magnetic field. Note carefully how flux lines of this field form linkages with the turns of the coil that produced them.

If rheostat resistance is reduced [Figure 12.1, part (b)], more current flows, causing more flux lines and increased linkages. As they increase, Faraday's law says that a voltage is induced across the coil. Because that voltage appears across the coil whose flux produced it, the voltage is said to be *self-induced*. Lenz's law tells us this voltage will be polarized to oppose (be counter to) the change causing it. Such a voltage is called a *counter-EMF*.

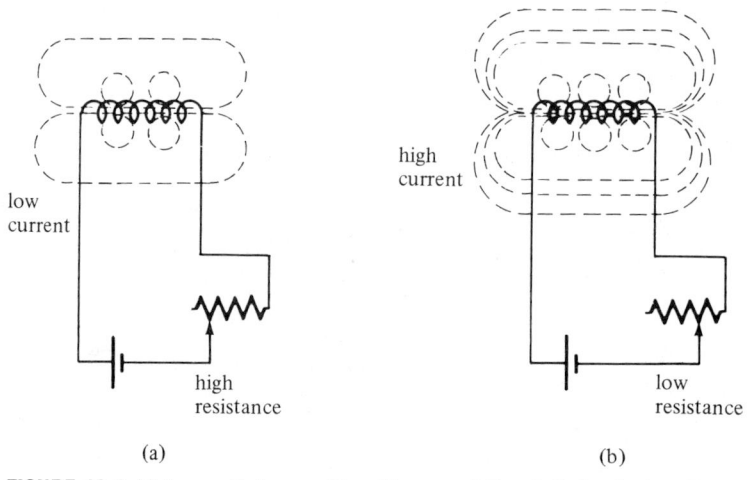

FIGURE 12.1 Linkages Between Flux Lines and the Coil Producing Them.

12.1 SELF-INDUCTANCE

To understand inductance, you must understand counter-EMFs. Therefore, let us examine them in greater detail. As we do, pay careful attention to the polarity of the induced counter-EMF.

In Figure 12.2, part (a), resistance decreases, causing current and flux linkages to increase (positive $d\phi/dt$). Note how the induced voltage is polarized so that it series-opposes the source voltage. It attempts to hold current at a low level by cancelling the voltage increase that is causing current to increase. In other words, *the counter-EMF opposes a current change.*

In Figure 12.2, part (b), resistance increases, so current and linkages decrease (negative $d\phi/dt$). Checking the induced voltage polarity again, we see that *the induced voltage now series-aids the source voltage.* It attempts to hold current at a high level by canceling the voltage decrease that is causing current to decrease. Again, the counter-EMF opposes a current change.

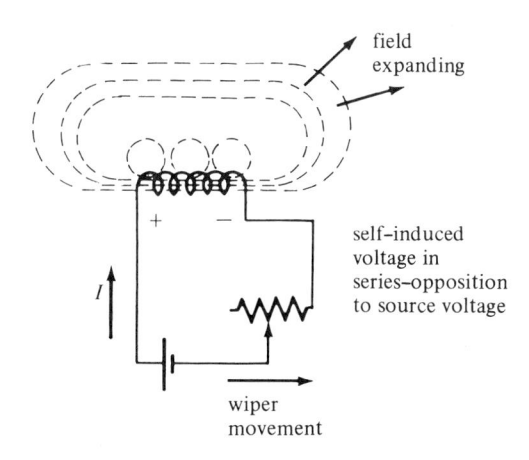

(a) decreasing resistance, increasing current and flux linkages

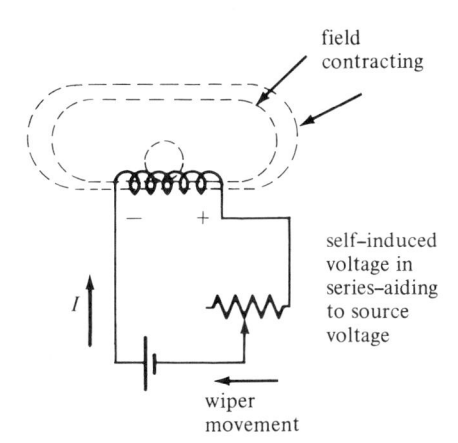

(b) increasing resistance, decreasing current and flux linkages

FIGURE 12.2. Counter-EMF Opposing Current Changes.

INDUCTANCE

The tendency of a coil to maintain current in its present state, to oppose current changes with an induced counter-EMF, is called *self-inductance*, or simply *inductance*. Measured in units of *henrys* (H) after an American, Joseph Henry (1797–1878), inductance is represented by the letter L.

The relation between coil inductance, turns, flux, and current is given in Equation 12.1:

$$L = N \frac{d\phi}{di} \quad \text{Henrys} \tag{12.1}$$

where N = number of turns on a coil
$d\phi$ = change in flux, in webers
di = change in current, in amperes

A formula for the counter-EMF developed across a coil, or *inductor*, is easily obtained by solving Equation 12.1 for N and substituting that value into Faraday's law (Equation 11.14):

From Equation 12.1: $\quad N = L \dfrac{di}{d\phi}$

Faraday's law: $\quad e = N \dfrac{d\phi}{dt}$

Substituting yields

$$e = L \frac{di}{d\phi} \frac{d\phi}{dt}$$

or

$$e_L = L \frac{di}{dt} \quad \text{volts} \tag{12.2}$$

The Henry can be defined on the basis of Equation 12.2:

> A coil has an inductance of 1 Henry if a current change of 1 ampere-per-second causes a counter-EMF of 1 volt.

EXAMPLE 12.1

What counter-EMF is developed by an 800-mH inductance to oppose a current change from 15 mA to 37 mA in 24 μs?

By equation 12.2,

$$\begin{aligned} e_L &= L \frac{di}{dt} \\ &= \frac{(800 \times 10^{-3})(37 \times 10^{-3} - 15 \times 10^{-3})}{24 \times 10^{-6}} \\ &= \frac{8(22) \times 10^{-4}}{24 \times 10^{-6}} \\ &= 733 \text{ V} \end{aligned}$$

∎

12.1 SELF-INDUCTANCE

Note that counter-EMF across an inductor is proportional to the rate at which inductor current changes (in contrast, you will recall that voltage across resistance is proportional to current value). The faster current changes, the greater the counter-EMF opposing that change, with the result that:

| Current through inductance cannot change instantaneously.

EXAMPLE 12.2

Current through a 3.7-H inductance changes according to this waveform. Determine the waveform of the counter-EMF.

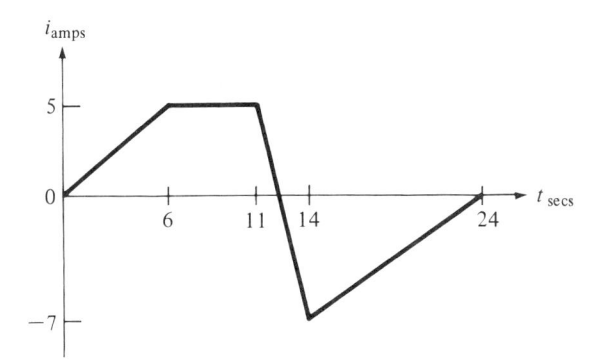

From $t = 0$ to $t = 6$ s

$$e_L = L\frac{di}{dt}$$

$$= \frac{3.7(5 - 0)}{6 - 0}$$

$$= 3.1 \text{ V}$$

From $t = 6$ s to $t = 11$ s:

$$e_L = 0 \text{ V} \quad \text{because} \quad \frac{di}{dt} = 0$$

From $t = 11$ s to $t = 14$ s:

$$e_L = \frac{3.7(-7 - 5)}{14 - 11}$$

$$= -14.8 \text{ V} \quad \text{note polarity shift because of negative slope}$$

From $t = 14$ s to $t = 24$ s:

$$e_L = \frac{3.7[0 - (-7)]}{24 - 14}$$

$$= 2.6 \text{ V}$$

Draw the voltage waveform:

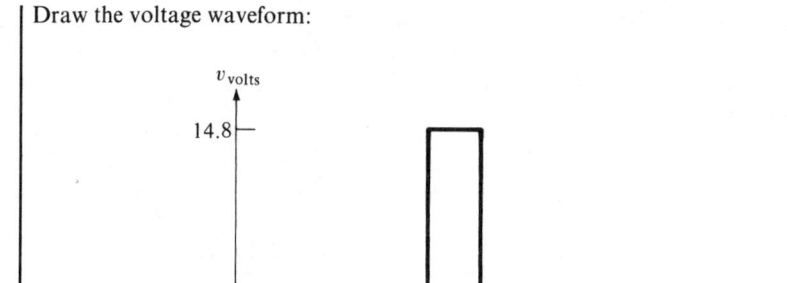

[Voltage waveform: v volts vs t secs. Values: 14.8 peak between t=11 and t=14; -2.6 between 0 and 6; level at 0 between 6 and 11; -3.1 between 14 and 24.]

Although current through an inductor cannot change instantaneously, this waveform shows that voltage can (note, for example, what happened at $t = 14$ seconds). ■

The effects of Equation 12.2 are seen in the waveforms of e_L and i_L when the switch is closed in the circuit of Figure 12.3 (note the symbol for the inductor). As the equation says, e_L is maximum when i_L is changing at its fastest rate. Later, when the current waveform has leveled off ($di/dt = 0$), e_L falls to zero.

Two important points are made by the waveforms of Figure 12.3:

- Since i_L reaches maximum value later than e_L, *inductive current is said to lag inductive voltage* (in contrast, you will recall that capacitive current leads capacitive voltage).
- An inductance is an open at the first instant of switch closure (voltage across but no current through), and a short later (current through but no voltage across. Again, capacitors are just the opposite. They can have voltage with no current.

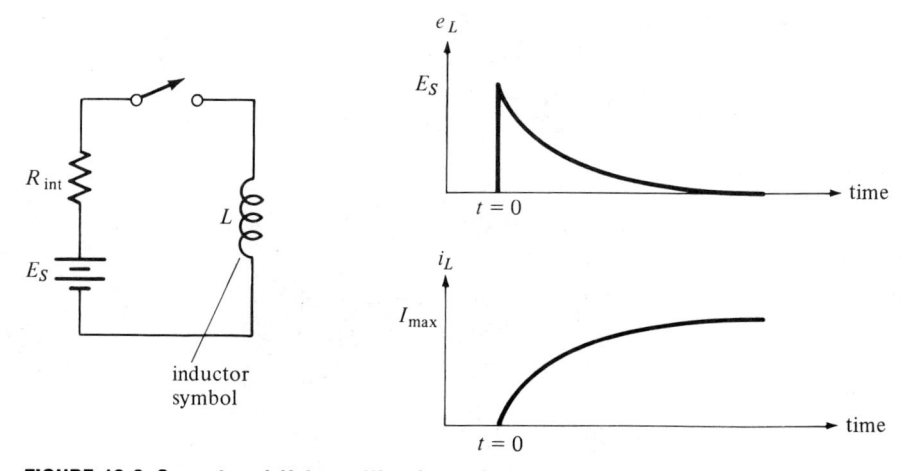

FIGURE 12.3 Current and Voltage Waveforms for Inductance.

12.1 SELF-INDUCTANCE

EXAMPLE 12.3

The switch in this circuit is closed at time $t = 0$.

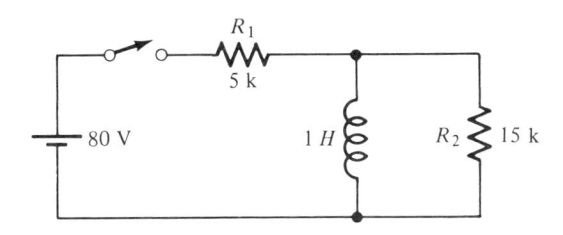

a. Find each resistor current at $t = 0^+$?

At $t = 0^+$, the inductance is an open; it has voltage across it but, as yet, no current through it. The circuit is now:

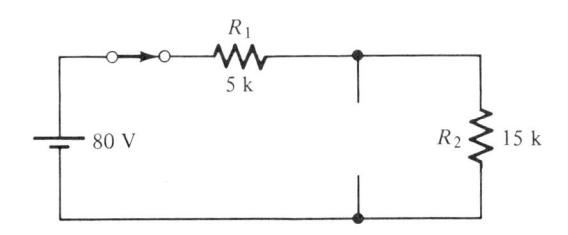

$$i_{R_1} = i_{R_2} = \frac{80}{5\text{ k} + 15\text{ k}}$$

$$= \frac{80}{20\text{ k}}$$

$$= 4\text{ mA}$$

b. What are the currents at $t = \infty$ (a long time after $t = 0$)?

At $t = \infty$, the inductance is a short (current through but no voltage across). The circuit is now

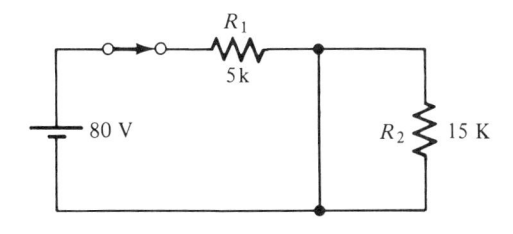

$$i_{R_1} = \frac{80}{5\text{ k}} = 16\text{ mA}$$

$$i_{R_2} = 0\text{ mA} \quad \text{since} \quad v_{R_2} = 0\text{ V}$$

■

12.2 STORED ENERGY

A source expends energy pushing current against an inductor's counter-EMF. This energy is stored in the inductor's electromagnetic field during times of current increase (flux field expanding), remains locked in potential form while current is constant, and is released when current decreases (field contracting). As we have seen, this released energy tries to maintain current when source voltage decreases.

By mathematics similar to those used to find the energy held by a charged capacitor, the energy stored by an inductance can be shown to be

$$W = \tfrac{1}{2}LI^2 \quad \text{joules} \tag{12.3}$$

When current to inductive devices such as motors, relays, and transformers is turned off, the flux field may collapse rapidly, causing a high rate of change of flux linkages ($d\phi/dt$). A surprisingly high voltage can thus be induced in even a modest inductance, enabling stored energy to be released in a high-temperature spark or arc called an *inductive kick*. In the worst cases, this arc melts switch contacts and destroys inductances by burning through insulation. Various protective measures can be taken to prevent this (see Exercise 12).

12.3 SOME COMPARISONS BETWEEN *R*, *L*, AND *C*

The following comparisons are made to reinforce your understanding of resistance, inductance, and capacitance:

> Only resistance dissipates energy. Inductance stores it in an electromagnetic field; $W = \tfrac{1}{2}LI^2$. Capacitance stores it in an electrostatic field; $W = \tfrac{1}{2}CE^2$.
>
> The voltage across and current through a resistance can both change instantaneously but inductive current and capacitive voltage cannot.
>
> Other comparisons are made in Figure 12.4.

12.4 PRACTICAL INDUCTORS

Inductors are often called *chokes* because of their tendency to "choke" out current changes with self-induced counter-EMFs. They consist of a coil of wire wrapped around a core of one material or another. For most types, the inductance can be approximated by the formula

$$L = \frac{N^2 \mu A}{l} \quad \text{henrys} \tag{12.4}$$

where N = number of turns
μ = core permeability
A = cross-sectional area of core, in square meters
l = average length of core, in meters

Substituting Equation 11.2 ($\mu = \mu_r \mu_0$) into Equation 12.4, we have

$$L = \mu_r \mu_0 \frac{N^2 A}{l} \tag{12.5}$$

12.4 PRACTICAL INDUCTORS   351

$v_L = L \dfrac{di}{dt}$

$i_C = C \dfrac{dv}{dt}$

$I = \dfrac{V}{R}$
$V = IR$

An inductor
is an open at $t = 0^+$
is a short at $t = \infty$
can have current
through with
no voltage
across.
i_L lags v_L

A capacitor
is a short at $t = 0^+$
is an open at $t = \infty$
can have voltage
across with no
current through.
i_C leads v_C

A resistance
has same
resistance
at all times.
Must have
voltage
across to
have current
through.
I_R and V_R in
time step
with each
other.

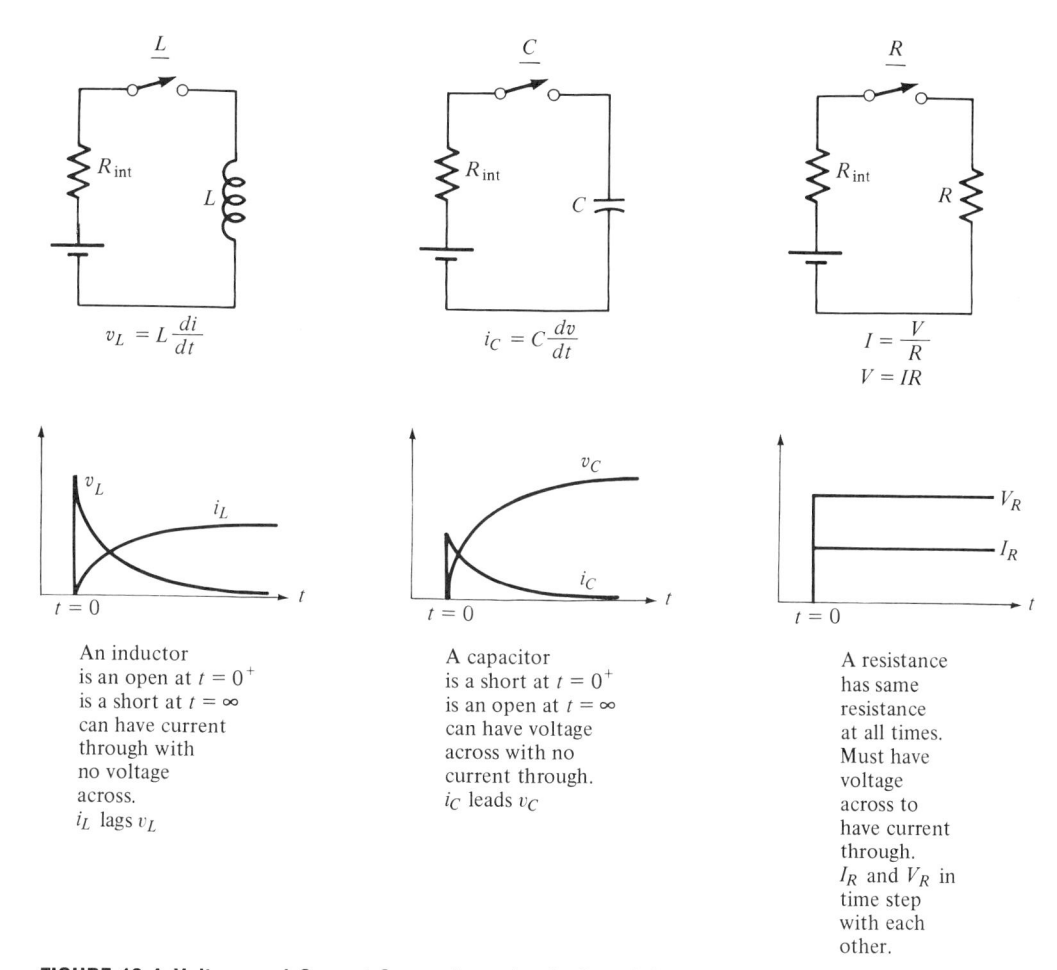

FIGURE 12.4 Voltage and Current Comparisons for L, C, and R.

But

$$\mu_0 \frac{N^2 A}{l}$$

is the inductance of the basic coil with an air core. Therefore, Equation 12.5 can be rewritten

$$L_{\text{with material core}} = \mu_r L_{\text{with air core}} \qquad (12.6)$$

which shows the value of a high-permeability core in increasing coil inductance.

N, A, and l in Equation 12.4 are all constants. The only variable is μ, which, as we saw in Chapter 11, varies widely. For those applications in which inductance must not change, the core is air gapped to stabilize μ. The price for this stability is a sharply reduced inductance from a given coil.

The two major inductor types are *iron core* and *air core*; there are many varieties of each. Their symbols are shown in Figure 12.5.

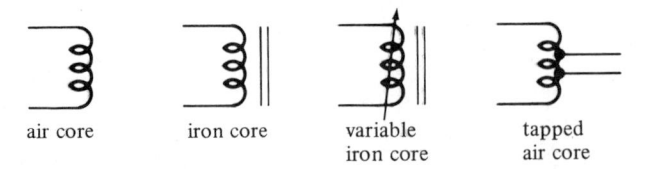

FIGURE 12.5 Inductor Symbols.

Air-core inductors are used for those applications in which iron cores are either too "lossy" or too unstable, that is, have characteristics that change with temperature, time, type of signal, and so on. They are found in communications equipment such as transmitters and receivers.

Iron-core inductors have a core made of high-permeability ferromagnetic material. For large values of inductance, perhaps 0.5 H and up, such a core will be *laminated*, meaning that it is made up of many layers, or *laminations*. To see why this is done, we must consider inductor energy losses.

Inductor losses come from three sources. We have already met two of them, hysteresis losses (energy needed to align atomic magnets) and copper losses (I^2R heat developed by winding resistance). The third loss occurs because the core itself acts as a closed conductor in its own flux field. As such, it has linkages with the field and when these change, voltage is induced and current flows in the core as it does in any conductor under the same circumstances. This current is called an *eddy current*. As it flows, I^2R heat is developed and the core gets hot.

To minimize eddy current losses, core resistance is increased. How? Core material and bulk are set by magnetic considerations and must not be changed. So instead of making the core a solid rod, it is stacked out of many paper-thin laminations, each insulated from its neighbors by varnish. Being thin, the laminations have small cross-sectional area and high resistance. When stacked, they produce a core of much higher resistance.

For maximum inductance, laminated cores are *closed* to confine flux lines to a high-permeability path with minimum leakage. Closed cores are made by sliding a prewound coil over the center post of a group of laminations shaped like the letter E or F, then closing the core with either an I-shaped end or another E- or F-shaped group (see Figure 12.6).

Small values of inductance, from 0.1 μH to several hundred millihenrys, look like carbon resistors. They consist of a coil wound around a rod, pressure formed of fine iron particles. The particles are varnish-insulated from each other to raise rod resistance and cut eddy current losses. The inductance of these units is indicated by

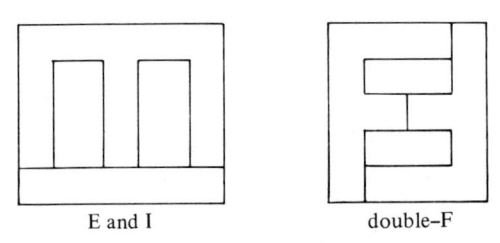

FIGURE 12.6 Typical Laminations Used In Closed Cores.

12.4 PRACTICAL INDUCTORS

colored bands using the EIA carbon-resistor color code (Table 3.1) with the following changes:

There are five bands. The first, a silver band twice the width of the other four, is located toward one end. It identifies the device as a choke. Bands 2, 3, and 4 indicate inductance in microhenrys. If band 2 or 3 is gold, it represents the decimal point for values less than 10; the other two bands represent significant figures. For values of 10 or more, bands 2 and 3 represent significant figures, with band 4 a multiplier. The fifth band indicates percent tolerance.

EXAMPLE 12.4

Determine the values of these two inductors:

a.

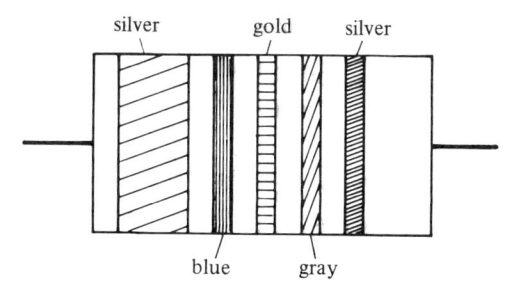

We see blue (first narrow band), decimal point (gold band), gray, and silver, indicating 10% tolerance. The value is 6.8 μH, plus or minus 10%.

b.

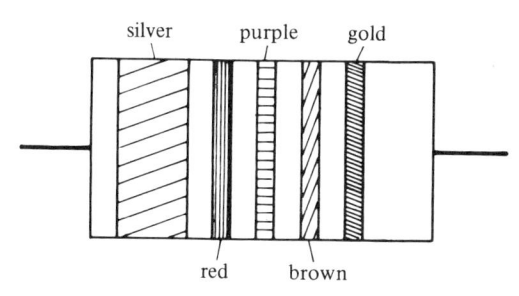

Neither the first nor the second band is gold. Therefore, we have red (2), purple (7), multiplier brown (times 10), and 5% tolerance. The final value is 270 μH, plus or minus 5%.

Variable inductors, sometimes called *slug-tuned* or *permeability-tuned* coils, consist of a nonmagnetic tube around which the coil is wound. Inside is a powdered iron slug moved by turning a threaded rod with a screwdriver. Turning the slug into the coil increases coil inductance by raising core permeability from that of air (slug out) to that of the iron (slug in). Slugs are also made of aluminum or brass. They reduce coil inductance when turned in. Fixed and variable inductors are see in Figure 12.7.

Iron-core inductors are suitable for low-frequency electronic signals. But eddy currents increase with frequency squared, which means that there is a limit to loss

FIGURE 12.7 Fixed and Variable Inductors. (Courtesy Coilcraft.)

reduction by making thinner laminations or finer iron particles. Therefore, inductors for higher frequencies are often made with a *ferrite* core.

Ferrite is a nonmetallic, ceramic, ferromagnetic material with resistivity so high that eddy current losses, even at high frequencies, are negligible. It has good permeability, low hysteresis loss, stability over time and temperature, and low cost.

Starting as a claylike mud, ferrite is pressure formed to the desired shape, then baked at high temperatures ($>2000°F$). Two of the most common shapes are:

- The *toroid:* a doughnut-shaped core around which the coil is wound (see Figure 12.8). Because its entire field is confined within the core, a toroid is *self-shielding*; that is, it has no external field to induce signals into nearby circuits.

FIGURE 12.8 Ferrite Cores. (Courtesy Coilcraft.)

12.5 INDUCTORS IN SERIES AND PARALLER

- The *pot core*: an excellent choice for stable, precision inductors. Coil wrapped around the inner circular shaft (top two inductors in Figure 12.8) is completely shielded by ferrite material. Note the hole on top, which is threaded for movement of the adjusting slug.

Inductors are often bulky, heavy, and expensive; in addition, their magnetic field induces unwanted signals in nearby circuits. Consequently, electronic designers avoid them when possible.

12.5 INDUCTORS IN SERIES AND PARALLEL

Assuming that they are far apart, so that flux from one coil does not form linkages with another, inductors in series and parallel combine like resistances in series and parallel.

For the series circuit of Figure 12.9, we have by KVL,

$$e_{\text{total}} = v_{L_1} + v_{L_2} + v_{L_3}$$

But Equation 12.2 says that

$$v_L = L\frac{di}{dt}$$

Substitute:

$$L_{\text{total}}\frac{di}{dt} = L_1\frac{di}{dt} + L_2\frac{di}{dt} + L_3\frac{di}{dt}$$

Divide both sides of the equation by (di/dt):

$$L_{\text{total}} = L_1 + L_2 + L_3$$

For N inductors in series:

$$L_{\text{total}} = L_1 + L_2 + \cdots + L_n \qquad (12.7)$$

For the three parallel-connected inductors in Figure 12.10, we have by KCL,

$$i_{\text{total}} = i_1 + i_2 + i_3$$

Therefore,

$$\frac{di_{\text{total}}}{dt} = \frac{di_1}{dt} + \frac{di_2}{dt} + \frac{di_3}{dt}$$

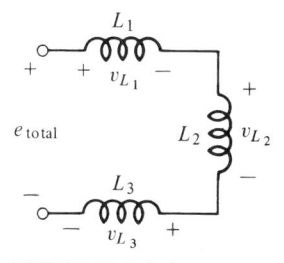

FIGURE 12.9 Inductors in Series.

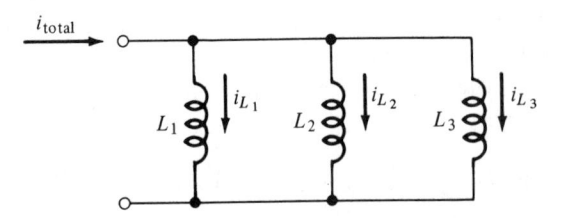

FIGURE 12.10 Inductors in Parallel.

But

$$\frac{di}{dt} = \frac{e}{L}$$

by Equation 12.2. Substitute:

$$\frac{e_{total}}{L_{total}} = \frac{e_{total}}{L_1} + \frac{e_{total}}{L_2} + \frac{e_{total}}{L_3}$$

Divide through by e_{total}:

$$\frac{1}{L_{total}} = \frac{1}{L_1} + \frac{1}{L_2} + \frac{1}{L_3}$$

or

$$L_{total} = \frac{1}{1/L_1 + 1/L_2 + 1/L_3}$$

For N inductors in parallel,

$$L_{total} = \frac{1}{1/L_1 + 1/L_2 + \cdots + 1/L_n} \quad (12.8)$$

All methods for calculating the total resistance of parallel-connected resistances apply for inductors in parallel. For example, the product-over-sum rule for two inductors in parallel yields

$$L_{total} = \frac{L_1 L_2}{L_1 + L_2} \quad (12.9)$$

Similarly, series–parallel inductor arrangements are analyzed with the techniques used on series–parallel resistive circuits.

EXAMPLE 12.5

What is the total inductance seen looking into terminals of the following circuit?

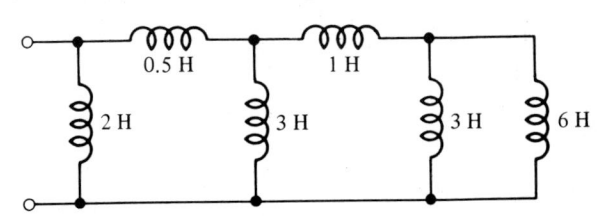

As with resistor circuits, we start reducing at the end farthest from the terminals. Here we see

$$\frac{6(3)}{6+3} = \frac{18}{9} = 2 \text{ H}$$

This combines with the series-connected 1 H to yield 3H. Two 3-H inductances combine to give $\frac{3}{2} = 1.5$ H, which is in series with the 0.5 H. The two add to provide $1.5 + 0.5 = 2$ H, which is in parallel with the 2 H across the terminals.
Final inductance = 2 H/2 = 1 H. ■

12.6 COMPONENTS IN THE REAL WORLD

Pure components contain only resistance, inductance, or capacitance. They do not exist in the real world, only in textbooks, where they simplify circuit analysis. But resistance, inductance, and capacitance are everywhere; real components contain portions of all three. A piece of wire $\frac{1}{4}$ in. long, for example has $\rho l/A$ resistance, an inductance of approximately 0.5 nH and capacitance with nearby conductors.

Figure 12.11 illustrates the unavoidable "impurities" present with each component. These impurities are called *parasitic* components.

For low-frequency signals we can relax, parasitic components are too small to cause problems; inductors really are inductors, a capacitor is a capacitor, and so on.

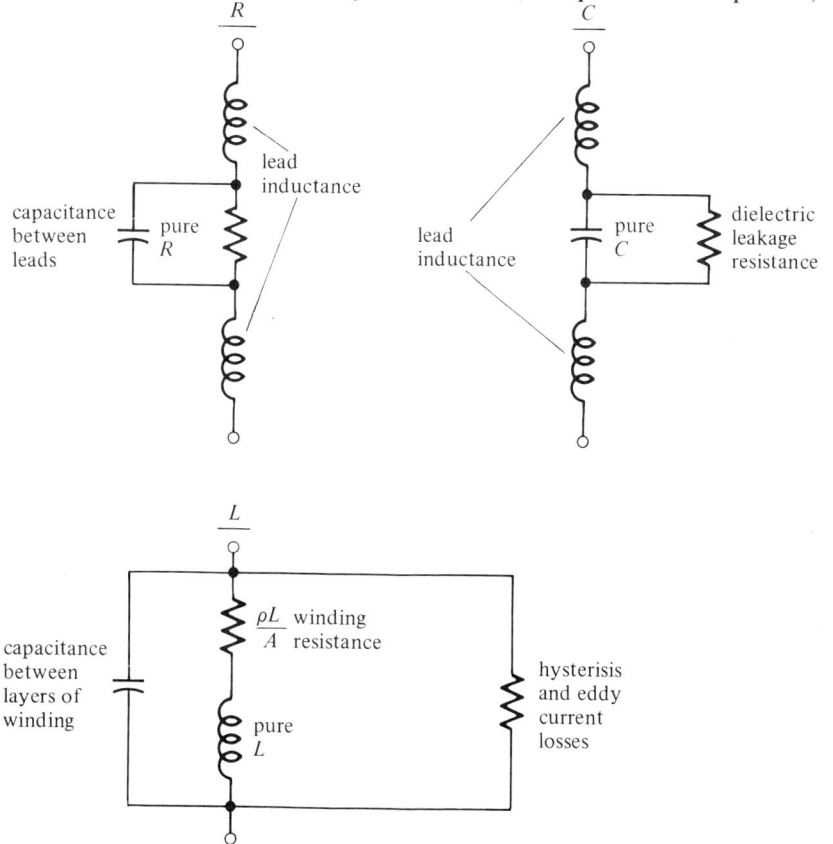

FIGURE 12.11 Parasitic Elements in Electronic Components.

With high-frequency signals, however, parasitic components have an enormous effect (we see why in later chapters) and can even become the principal component. An inductor, for instance, can change roles entirely and act as a capacitor. High-frequency equipment designers must be on guard against these undesirable effects.

Figure 12.12 illustrates an example of steps taken to reduce parasitic components. The construction of a high-dissipation resistor forces current flow in opposite directions. This cancels flux fields and reduces parasitic inductance.

Another problem with high-frequency signals is that they cause conductor resistance to increase. To see why, study the slice through a current-carrying conductor in Figure 12.13. Note how flux lines form concentric circles extending outward from the center. Because they do, current in the center is encircled or linked, by a greater number of lines than is current at the edge (point A). Equation 12.1 tells us that inductance is proportional to the rate of change of flux linkages. Since the center has more linkages, it follows that the center must have a higher rate of linkage change. Therefore, a conductor's center has greater inductance than its edge.

High-frequency currents flow more readily through low inductance (we will find out why later). Such currents therefore tend to flow along the low-inductance outer layer, or skin, of a conductor. The effective cross-sectional area of the conductor is thus decreased at high frequencies, resulting in a rise in resistance. This rise is called *skin effect*. Skin effect can be reduced with large-diameter silver-plated conductors.

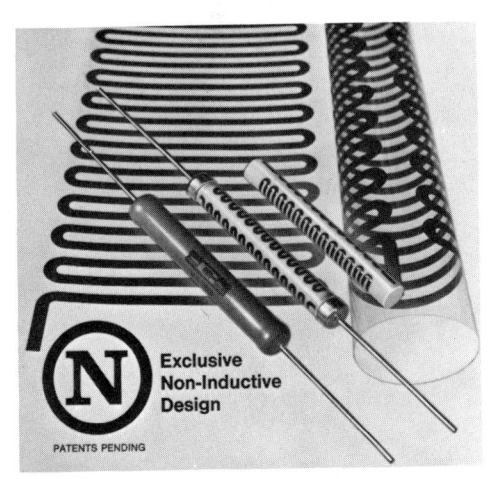

FIGURE 12.12 Non-inductive Power Resistor. (Courtesy Caddock Electronics, Inc.)

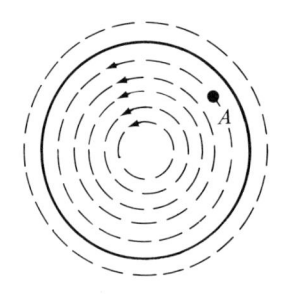

FIGURE 12.13 Flux Linkages in Conductor.

12.7 INDUCTIVE TRANSDUCERS

One type of inductive transducer depends on a change in coil inductance when ferromagnetic material passes by. For example, the relative permeability of a buried coil increases when a car passes over, causing a jump in inductance which is easily detected to measure traffic flow or activate traffic lights.

The same principle is used to measure pressure of gases or fluids. The pressure moves a diaphragm (see Figure 12.14), which, in turn, moves a slug within a coil. The resulting inductance changes causes current variations which are read on a meter as pressure changes.

Figure 12.15 illustrates an application of Faraday's law in a tachometer. A momentary voltage is induced in the coil each time the magnet passes (the same principle generates spark plug voltage in magnetos). These pulses are counted to show RPM. The pickup could also be a Hall-effect sensor.

12.8 TRANSIENT ANALYSIS

Because current and voltage are transients in RL circuits just as they were in RC circuits, they are also described by exponential equations and waveforms. These waveforms, illustrated in Figure 12.3, are repeated in Figure 12.16.

At t_B^-, the switch in the circuit of Figure 12.16 is in position A, no current flows, and v_L and v_R equal zero. At t_B^+ the switch has just arrived in position B. Current is starting

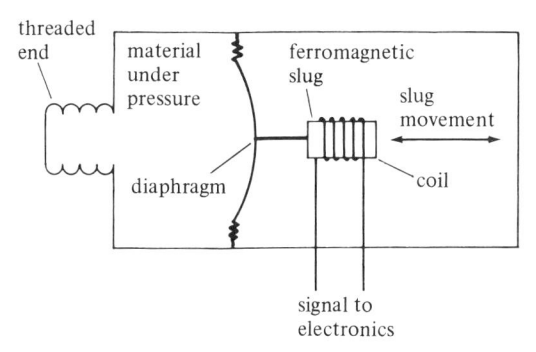

FIGURE 12.14 Inductive Transducer to Measure Pressure.

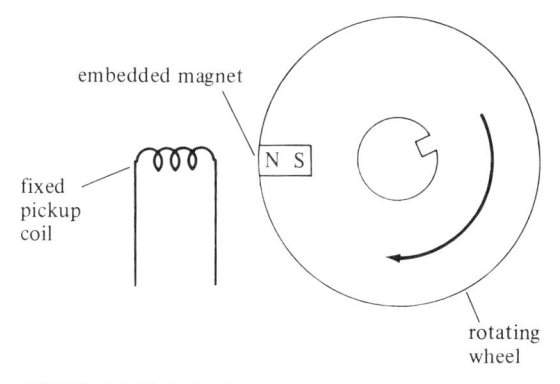

FIGURE 12.15 Inductive Tachometer.

360   INDUCTANCE

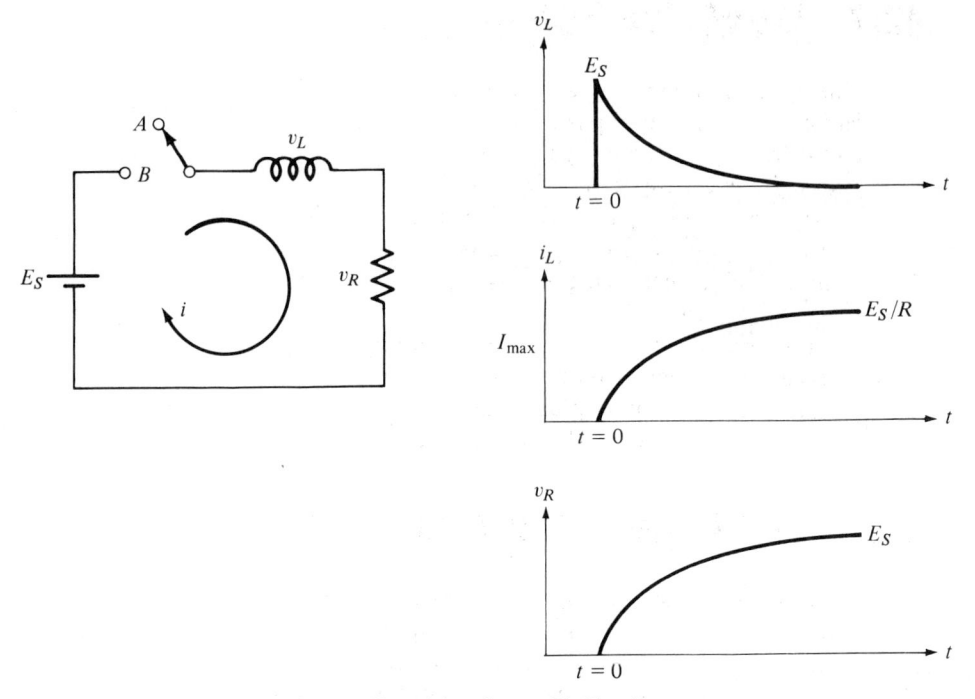

FIGURE 12.16 Voltage and Current Transients in an *RL* Circuit.

to rise from zero. But it cannot change instantaneously, so v_R is still zero. v_L, however, is maximum because di/dt is maximum; the current is changing at its highest rate.

As times passes, current increases causing v_R to increase. But the rate of current increase has decreased; the wave is not climbing as steeply. Therefore, v_L decreases. (Pay particular attention to the way v_R rises as v_L falls. Their sum always equals E_{source}.)

After sufficient time,

$$i = \frac{E_s}{R}$$

$$v_R = E_s$$

and

$$v_L = 0 \text{ V}$$

The transients are over.

At any instant,

$$v_R + v_L = E_s$$

by KVL. But

$$v_R = iR$$

by Ohm's law, and

$$v_L = L\frac{di}{dt}$$

12.8 TRANSIENT ANALYSIS

Substitution gives us

$$iR + L\frac{di}{dt} = E_s$$

This will be solved for i to give us an equation by which current can be determined at any instant. However, the presence of di/dt means that the equation must be solved by calculus, as was a similar equation for capacitive circuits. The result is

$$i_L = \frac{E_s}{R}(1 - e^{-t/\tau}) \text{ amperes} \tag{12.10}$$

where i_L = inductor current, in amperes
E_s = source voltage, in volts
R = circuit resistance, in ohms
e = base of natural logs, 2.718...
t = time, in seconds
τ = time constant, in seconds

Note carefully how this equation for transient inductor current has the same form as the equation for transient capacitor voltage (Equation 10.17). One major change between the two is the time constant, τ. For an RL circuit,

$$\tau = \frac{L}{R} \text{ seconds} \tag{12.11}$$

EXAMPLE 12.6

Answer the questions for the following circuit.

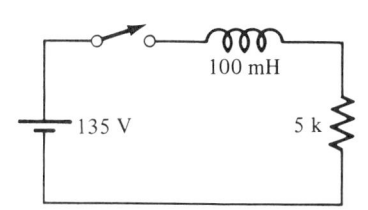

[Circuit diagram: 135 V source, switch, 100 mH inductor, and 5 k resistor in series]

a. Write the equation for the current.

$$\tau = \frac{L}{R} = \frac{100 \times 10^{-3}}{5 \times 10^3} = 20 \ \mu s$$

$$i = \frac{E_s}{R}(1 - e^{-t/\tau})$$

$$= \frac{135}{5 \text{ k}}(1 - e^{-t/20 \ \mu s})$$

$$= 27(1 - e^{-t/20 \ \mu s}) \text{ mA}$$

Instantaneous current can be determined at any time t by substituting that t into this equation.

b. What is the instantaneous current $t = 35$ μs after switch closure?

$$\begin{aligned} i &= 27(1 - e^{-(35 \times 10^{-6})/(20 \times 10^{-6})}) \\ &= 27(1 - e^{-1.75}) \\ &= 27(1 - 0.17) \\ &= 27(0.83) \\ &= 22.3 \text{ mA} \end{aligned}$$ ∎

Using calculus again, it can be shown that the transient voltage, v_L, across the inductor is

$$v_L = E_s e^{-t/\tau} \text{ volts} \tag{12.12}$$

EXAMPLE 12.7

For the circuit of Example 12.6.

a. Calculate the instantaneous voltage across the inductor 52 μs after switch closure.

This can be done two ways:

1. Use Equation 12.12:

$$\begin{aligned} v_L &= E_s e^{-t/\tau} \\ &= 135 e^{-(52 \times 10^{-6})/(20 \times 10^{-6})} \\ &= 135 e^{-2.6} \\ &= 135(0.07) \\ &= 10 \text{ V} \end{aligned}$$

2. Use Equation 12.10 to calculate i. Calculate v_R. Subtract it from E_s to get v_L.

b. How long will it take v_L to fall to 0 V?

As explained in the RC transient discussion (Chapter 10), v_L never arrives at 0. However, it will have dropped to $0.0067 E_s$ at the end of five time constants. We accept that as being zero for practical purposes.
Therefore,

$$\begin{aligned} 5\tau &= 5(20 \text{ μs}) \\ &= 100 \text{ μs needed for } v_L \text{ to } = 0 \text{ V} \end{aligned}$$

c. Draw the waveforms of i, v_L, and v_R for 5τ.

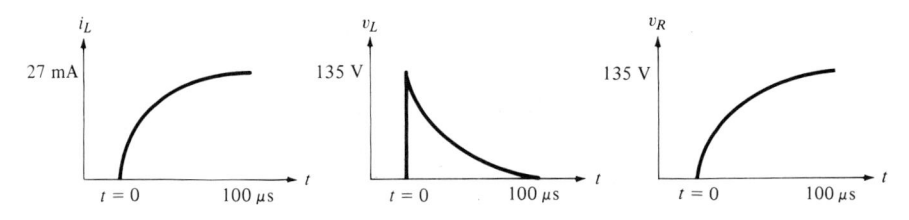

∎

EXAMPLE 12.8

Answer the questions for this circuit.

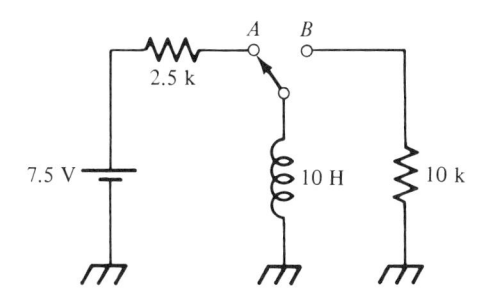

a. Write the equation for v_L while the switch is in position A.

$$\tau_A = \frac{L}{R} = \frac{10}{2.5 \text{ k}} = 4 \text{ ms}$$

$$v_L = 7.5e^{-t/4 \times 10^{-3}} \quad \text{volts}$$

b. What is i_L at the instant when the switch has been in position A for 5τ?

when $t = 5\tau$,

$$v_L = 0 \text{ V}$$

and

$$v_R = E_s = 7.5 \text{ V}$$

Therefore,

$$i_L = \frac{E_s}{R}$$

$$= \frac{7.5}{2.5 \text{ k}} = 3 \text{ mA}$$

c. The switch is moved from position A to position B in zero time. What is i_L at t_B^+?

Since current through an inductor cannot change instantaneously, inductor current at t_B^+ equals inductor current at $t_B^- = 3$ mA.

d. What is v_L at t_B^+?

i_L is still 3 mA. But it now passes through 10 k.
Therefore,

$$v_L = v_{10k} = iR$$
$$= 3 \text{ mA}(10 \text{ k})$$
$$= 30 \text{ V}$$

Note this result carefully remembering that the source voltage was just 7.5 V. Such voltage increases do not occur in RC circuits.

e. What is the equation for v_L while the switch is in position B?

Polarity is reversed from what it was in part a because flux is now collapsing into the coil.

$$\tau_B = \frac{10}{10 \text{ k}} = 1 \text{ ms}$$

$$v_L = -30e^{-t/10^{-3}} \quad \text{volts}$$

Note carefully that the time in this equation is zero at the instant this transient begins, that is, when the switch arrives in position B. It is not the same t as that in the equation for v_L when the switch is in A. That time ended when the switch was thrown to B.

f. Draw the waveforms of v_L for 5τ in switch postion A followed immediately by 5τ in position B.

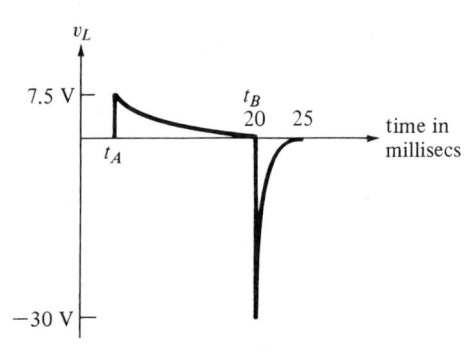

12.9 USING THÉVENIN'S THEOREM IN TRANSIENT ANALYSIS

When the circuit driving an inductor does not consist of a single source in series with a single resistance or cannot be reduced to that form, Thévenin's theorem must be used as it was in *RC* analysis.

EXAMPLE 12.9

Answer the questions for this circuit.

a. Determine the equations for v_L and i_L:

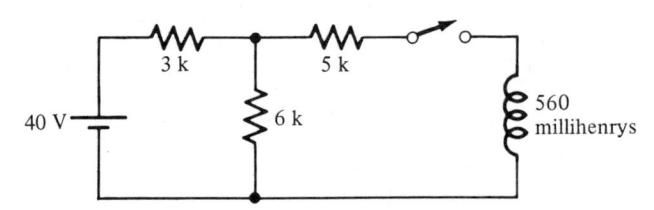

Because we cannot look back from the inductor and see a simple source/resistance circuit, the time constant and maximum coil voltage and current cannot be determined. The circuit driving the inductor must, therefore, be Théveninized. This job, you will recall, has two parts:

12.9 USING THÉVENIN'S THEOREM IN TRANSIENT ANALYSIS

1. Finding R_{int}: With the source turned off,
$$R_{int} = (3\text{ k} \| 6\text{ k}) + 5\text{ k} = 2\text{ k} + 5\text{k}$$
$$= 7\text{ k}$$

2. Finding E_{oc}:
$$E_{oc} = \text{voltage across the 6 k}$$
$$= \frac{40(6\text{ k})}{3\text{ k} + 6\text{ k}} = 26.7\text{ V}$$

Draw the new circuit:

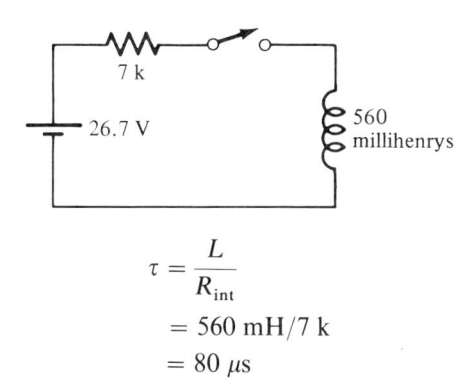

$$\tau = \frac{L}{R_{int}}$$
$$= 560\text{ mH}/7\text{ k}$$
$$= 80\ \mu s$$

Write the equations:

$$v_L = 26.7 e^{-t/80\ \mu s} \quad \text{volts}$$
$$i_L = \frac{E_{oc}}{R_{int}}(1 - e^{-t/\tau}) = \frac{26.7}{7\text{ k}}(1 - e^{-t/80\ \mu s})$$
$$= 3.8(1 - e^{-t/80 \times 10^{-6}}) \quad \text{mA}$$

b. Draw the waveforms of v_L and i_L:

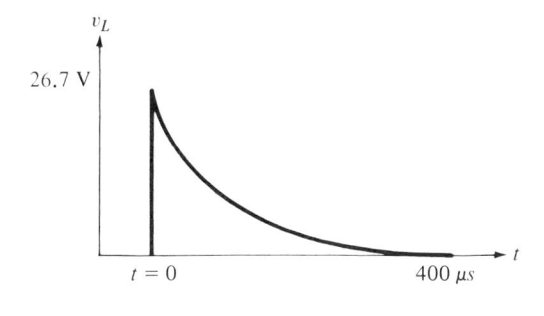

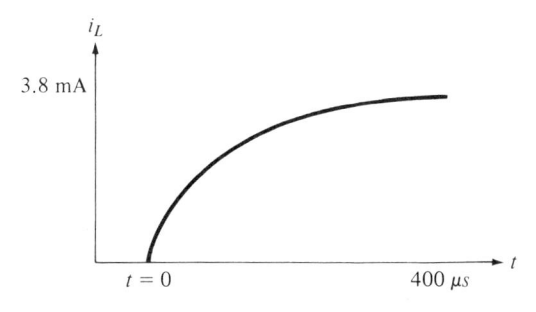

EXERCISES

Section 12.1

1. A rheostat is reduced slowly to increase the current through a coil and the number of linkages between the coil and its flux. A counter-EMF is developed. Discuss what happens to the counter-EMF if the rheostat is reduced quickly.

2. Why does inductive counter-EMF polarity reverse when current changes from increasing to decreasing?

3. Find the inductance of a 342-turn coil which experiences a flux linkage change of 0.27 mWb for a current change of 123 mA?

4. Find the induced counter-EMF across a 250-mH coil whose current changes 3.8 mA in 75 μs.

5. What inductance experiences a counter-EMF of 1.2 kV when its current changes 680 mA in 0.5 ms?

6. Current through a 1-H inductance increases from 3000 A to 3001 A in 1 h. Later, current through the same coil increases from 3 μA to 6 μA in 10 ns. Compare the two induced voltages. Explain differences.

7. Determine the waveform of induced voltage across a 2.4-H inductance when its current changes as shown.

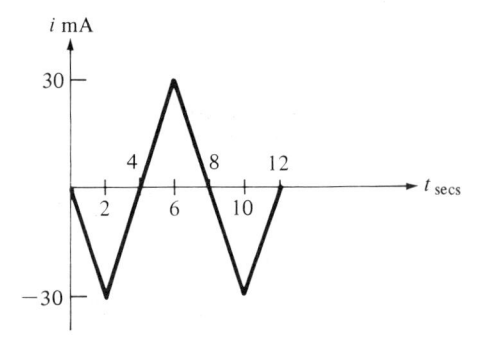

8. Repeat Exercise 7 for this current waveform:

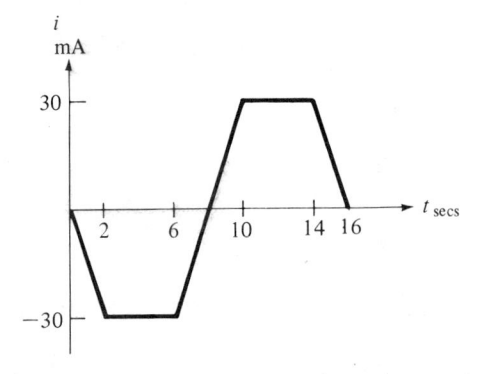

9. Calculate the current through each inductance at $t = 0^+$ and $t = \infty$ after switch closure.

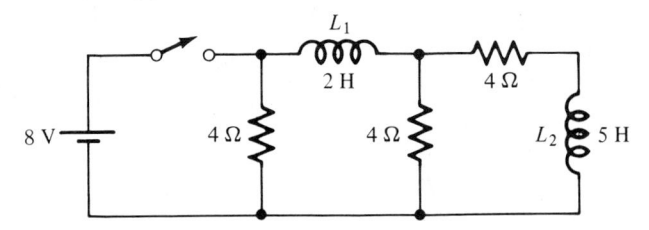

EXERCISES

10. Repeat Exercise 9 for this circuit:

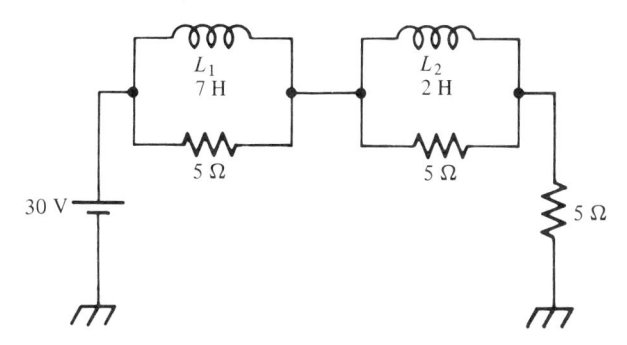

Section 12.2

11. What is the energy stored in the electromagnetic field of each inductance of Exercise 9 at $t = 0^+$ and $t = \infty$?

12. A semiconductor device called a *diode* is a short to current with the arrow (see symbol in schematic), and an open to current against the arrow. Keeping in mind the polarity of counter-EMF across inductance, explain how the diode protects this relay coil against inductive kick.

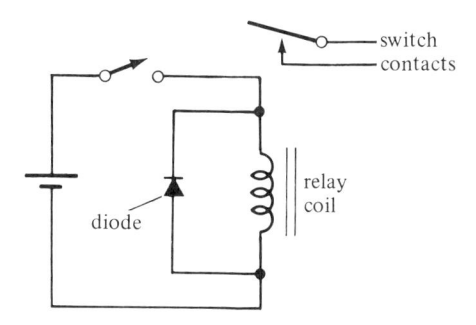

Section 12.3

13. Fill in the blanks with either R, L, or C, whichever best fits the situation.
a. v across leads i through _____.
b. v and i in step with each other _____.
c. i through reaches maximum before v across _____.
d. v across and i through reach their maximums simultaneously _____.
e. i through leads v across _____.
f. i through lags v across _____.

14. Compare the energy characteristics of resistance, capacitance, and inductance.

Section 12.4

15. Why are inductors often called chokes?

16. A 3-H choke consists of 420 turns on a core 0.15 m long and 4 cm^2 in cross-sectional area. What is the core's relative permeability?

17. How many turns must be wound on a toroidal core whose average diameter is 8 cm, and cross-sectional area is 2.25 cm^2, to get an inductance of 700 mH? $\mu_r = 1700$ Wb/A-m.

18. What are the three sources of losses in magnetic devices?

368 INDUCTANCE

19. What are eddy currents? Where are they developed and why? What is done to limit them?
20. What is meant by a closed core? What is the reason for using one?
21. Find the value of each inductor.

	Band color			
	Second	Third	Fourth	Fifth
a.	Red	Red	Brown	Silver
b.	Violet	Gold	Green	Gold
c.	Orange	White	Red	No color
d.	Gold	Brown	Blue	Gold

22. Explain how a variable inductor works.
23. Why are variable-inductor slugs sometimes made of nonmagnetic aluminum and brass?
24. Why does ferrite have such high resistivity? What benefit does this provide?

Section 12.5

25. What is the total inductance looking into the terminals of the following circuit? *Note:* There are no linkages between coils.

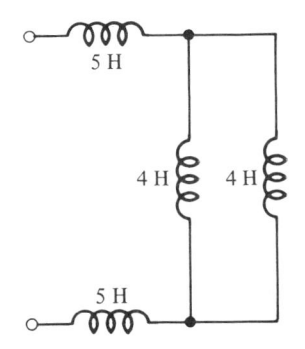

26. Repeat Exercise 25 for the circuit:

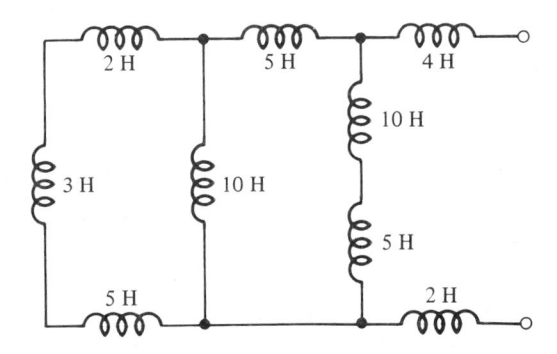

Section 12.6

27. What is meant by the term "parasitic" component? For what types of signals do parasitic components become important? For what types of signals can they usually be neglected?
28. Explain how skin effect raises conductor resistance to high-frequency signals.

EXERCISES

Section 12.7

29. Describe two types of inductive transducers.

30. *Extra Credit.* How could inductance be used to measure acceleration/deceleration? *Hint:* Study Fig. 12.14.

Section 12.8

31. A 4.5-Ω resistor and 9-H choke are series-connected across an 18-V source at $t = 0$ s.
 a. How long does the current take to reach maximum?
 b. What is the value of maximum current?
 c. The resistance and choke are disconnected and all voltages and currents allowed to decay to zero. Then the series combination is connected to a 40-kV source at $t = 0$. To what maximum does current climb, and how long will it take to do so?
 d. Based on the results of parts a and c, does the value of i_{max} have any bearing on the time needed to reach i_{max}?

32. Show that the units of τ in an RL circuit are seconds. *Hint:* Use Equation 12.2 for L and Ohm's law for R.

33. Answer the questions for this circuit:

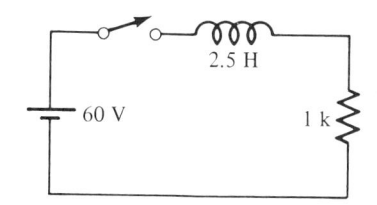

a. What is the time constant?
b. Write equations for v_L, v_R, and i for time after switch closure.
c. Calculate the current 4.8 ms after switch closure.
d. Without doing any calculations, estimate v_L for $t = 11.5$ ms. Then calculate the value. Compare the two.
e. What is v_R 15 ms after $t = 0$?
f. Sketch the v_L, v_R, and i waveforms from time $t = 0$ to $t > 5\tau$.

34. Repeat Exercise 33 for L increased to 200 H. Compare the results and explain differences.

35. Find the instantaneous energy stored in the magnetic field of the inductor of Exercise 33 when $t = 5.3$ ms.

36. Answer the questions for this circuit:

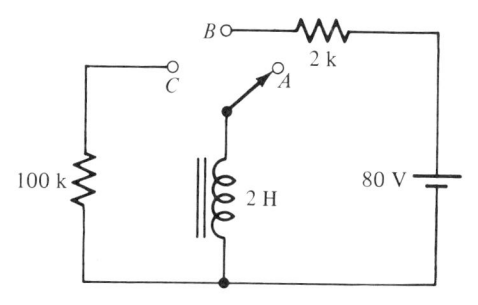

a. What are the equations for i_L, v_{2k}, and v_L when the switch is thrown to position B at time $t_B = 0$?
b. To what maximum will i_L climb?
c. The switch is thrown to position C in zero time. Write the equations for i_L, v_L, and v_{100k} for time $t > t_C$.
d. Draw the waveform of v_L assuming that the switch spends 5τ in B, is switched to C in zero time, and spends 5τ there.

37. Answer the questions for this circuit:

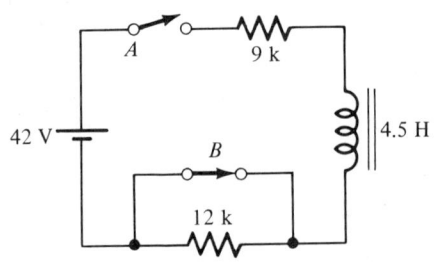

a. Switch A is closed at $t = 0$. What are v_L, i, and v_{9k} 5τ later at t_B^-?
b. Switch B is opened at t_B. What are i, v_L, v_{9k}, and v_{12k} at t_B^+?
c. Draw the waveform of v_L and i for the 5τ of part a followed by the next 5τ after switch B is opened. *Hint:* Think about the final values of v_L and i for each $t = 0$ and $\tau = 5\tau$.

38. *Extra Credit.* Write the equation for i for $t > t_B$ in Exercise 37.

Section 12.9

39. For the circuit:

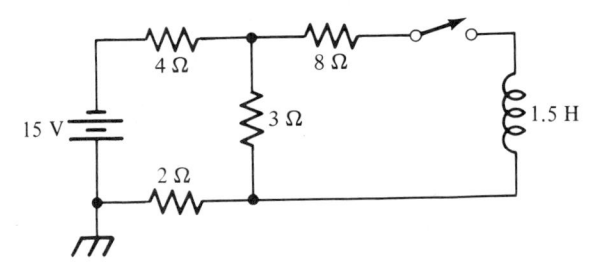

a. Find the equations for v_L and i_L for the time after switch closure.
b. Find the equation for $v_{8\Omega}$ for the time after switch closure.

40. For the following circuit, write the equations for v_L and i_L for time after switch closure.

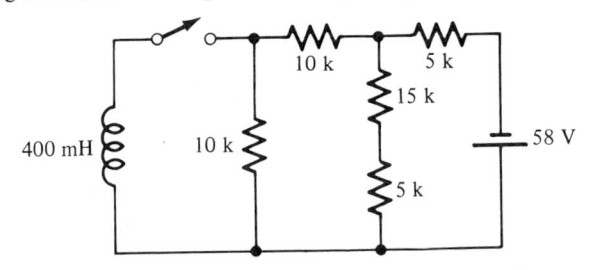

CHAPTER HIGHLIGHTS

A coil's flux forms linkages with its own turns. These induce a counter-EMF across the coil polarized to prevent current changes. This tendency of a coil is called inductance.
    Inductance tries to maintain current in its present state.
    Inductor current cannot change instantaneously.
    Current through inductance lags voltage across.
    Inductor is an open at switch closure (e but no i), a short much later (i but no e).
    Three inductor losses: eddy current, hysteresis, and copper.
    All components have some parasitic resistance, inductance, and capacitance, which can cause problems at high frequencies.

CHAPTER HIGHLIGHTS

RL transient circuits are analyzed with exponential equations just as *RC* circuits were.

Thévenin's theorem is used for those *RL* transient cases in which the inductor cannot see a single source/resistance combination.

Formulas to Remember

$L = N\dfrac{d\phi}{di}$	basic equation for inductance
$e = L\dfrac{di}{dt}$	voltage across an inductance
$W = \tfrac{1}{2}LI^2$	energy stored by an inductor's electromagnetic field
$L = \dfrac{N^2 \mu A}{l}$	formula for inductance
$L_{\text{total}} = L_1 + L_2 + \cdots + L_n$	total inductance of inductors in series
$L_{\text{total}} = \dfrac{1}{1/L_1 + 1/L_2 + \cdots + 1/L_n}$	total inductance of inductors in parallel
$i = \dfrac{E_s}{R}(1 - e^{-t/\tau})$	inductive transient current
$\tau = \dfrac{L}{R}$	time constant in *RL* circuits
$v_L = E_s e^{-t/\tau}$	inductive transient voltage

CHAPTER 13

Introduction to Alternating Current

So far, our work has concentrated on circuits drive by dc sources supplying constant voltage and current. With this chapter we reach an important turning point and begin studying circuits driven by ac (alternating current) sources.

An alternating voltage or current is one whose polarity changes or alternates, positive one portion of time, negative the next (dc, on the other hand, does not alternate; its polarity is constant).

Figure 13.1 shows several examples of alternating waveforms. Compare them to the dc waveform also shown. The sine wave is the most common and by far the most important, so much so that we will always take the term "alternating" to mean a sine wave unless told otherwise (the cosine wave is actually a shifted sine wave).

In this chapter we study the generation of sine waves and their important mathematical characteristics.

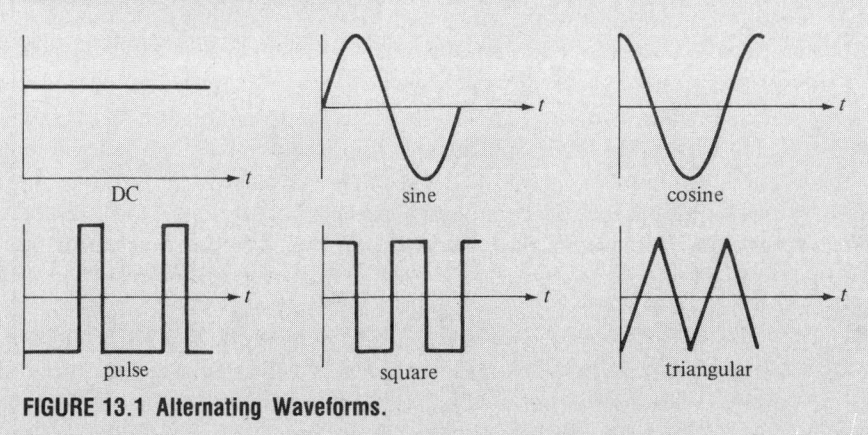

FIGURE 13.1 Alternating Waveforms.

At the end of this chapter, you should be able to:

- Tell the difference between scalar and vector quantities.
- Resolve a vector into its horizontal and vertical quantities with trigonometry.
- Discuss the generation of voltage and current sine waves by conductors rotating in magnetic fields.
- Show how sine waves can be described by rotating vectors called phasors.
- Work with angles in either radians or degrees.
- Explain how angular velocity differs from linear velocity.
- List the characteristics of sine waves.
- Calculate the instantaneous value of a sinusoidal voltage or current at any time t.
- Explain the effective or rms value of a current or voltage sine wave.

13.1 VECTORS

The parameters that we have worked with so far—volts, joules, watts, and so on—can all be described by an amount or *magnitude*, such as 10 Ω. These numbers are called *scalars*. Now, however, we are moving into a part of our studies in which some parameters have both magnitude and direction. These parameters are called *vectors*.

EXAMPLE 13.1

Which of the following are scalars and which are vectors?

1. Temperature of 32°C
2. Time of 104 ns
3. Force of 125 n down
4. Velocity of 40 kilometers per hour east

Answers: 1 and 2 indicate only magnitude and are scalars; 3 and 4, however, have both magnitude and direction and are vectors. ■

Vector V, shown as an arrow in Figure 13.2, represents conductor velocity through a horizontal flux field. The arrow indicates vector direction, its magnitude is indicated by arrow length In Figure 13.2, part (a), the vector is said to be in the *reference position*, that is, at 0°. This tells us that the conductor is moving horizontally, parallel to the flux; there is no *vertical component* to the motion. Figure 13.2, part (b), shows the conductor moving the same rate (same arrow length) at 90°. Conductor motion is now all vertical, with no *horizontal component*.

In Figure 13.2, part (c), the conductor moves through the field at an angle θ (the Greek lowercase letter theta, pronounced *thay-duh*). It thus moves both horizontally and vertically at the same time. Because of the shallow angle, the horizontal component, V_x, is longer than the vertical component, V_y. This tells us that the

13.1 VECTORS

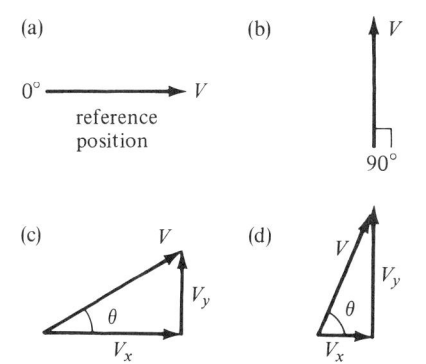

FIGURE 13.2 Velocity Vectors and Their Vertical and Horizontal Components.

horizontal motion is faster than the vertical. In contrast, Figure 13.2, part (d), shows a steeper angle; V_y is now much greater than V_x. Be aware, however, that conductor velocity is the same in all cases (as indicated by fixed length of V); only the respective components change.

Note how the three vectors in Figure 13.2, part (c) [or part (d)], form a right triangle. From trigonometry we know that

$$\sin \theta = \frac{V_y}{V} \quad (13.1)$$

Therefore,

$$V_y = V \sin \theta \quad (13.2)$$

Similarly,

$$\cos \theta = \frac{V_x}{V} \quad (13.3)$$

and

$$V_x = V \cos \theta \quad (13.4)$$

Equations 13.2 and 13.4 tell us if the velocity (magnitude) of V and θ are known, we can find the vertical and horizontal components of that velocity. This step is called *resolving* the vector into its X and Y components.

EXAMPLE 13.2

A conductor moves through a flux field at a velocity of 200 m/s. At what rate is the conductor moving horizontally and vertically through the field for each of the following angles?

a. 23°.

Use Equation 13.4 to resolve the horizontal component:

$$V_x = V \cos 23°$$
$$= 200(0.92)$$
$$= 184 \text{ m/s}$$

Use Equation 13.2 to resolve the vertical component:

$$V_y = V \sin 23°$$
$$= 200(0.4)$$
$$= 80 \text{ m/sec}$$

b. 75°.

$$V_x = V \cos 75°$$
$$= 200(0.26)$$
$$= 51.8 \text{ m/s}$$
$$V_y = V \sin 75°$$
$$= 200(0.97)$$
$$= 193.2 \text{ m/s}$$

Compare the velocities in parts a and b carefully.

c. 335°.

$$V_x = V \cos 335°$$
$$= 200(0.91)$$
$$= 181.3 \text{ m/s}$$
$$V_y = V \sin 335°$$
$$= 200(-0.42)$$
$$= -84.5 \text{ m/s}$$

Now the vertical component is negative (pointing down) because of the angle. ∎

13.2 GENERATING THE SINE WAVE

Electric generators are based on the principle described by Equation 11.13, $E = BlV$, which says that E volts are induced across a conductor of length l when it moves at a velocity V through a magnetic field of flux density B. They have two principal parts (see Figure 13.3), a *stator* and a *rotor* or *armature*. The stator (stationary) serves as both a supporting framework and a magnet to provide the B in the $E = BlV$ equation. Shown as a permanent magnet in the simplified diagram, the stator of a large generator is an electromagnet made by passing direct current through coils wound around the stator

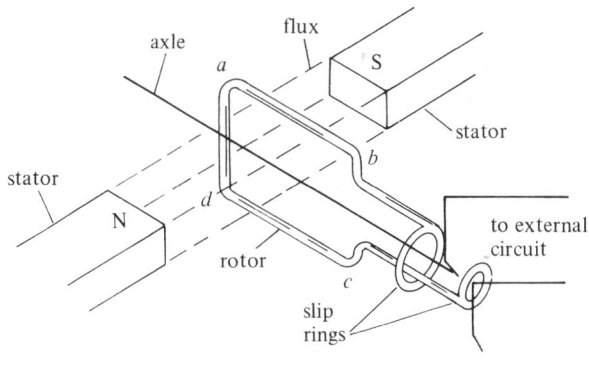

FIGURE 13.3 Simple Electric Generator.

13.2 GENERATING THE SINE WAVE

halves. The rotor, shown as a single turn, actually consists of many turns of wire (the l in BlV) wound around a form on an axle so that they can rotate (the velocity, V, in BlV) within the stator's magnetic field.

At one end of the rotor's axle is the means by which the input energy (usually mechanical) demanded by Lenz's law is coupled to the generator. This might be the pulley–fan belt arrangement of a car's alternator, a windmill, or a huge turbine turned by falling water. Mounted on the axle's other end are copper *slip rings* to which the ends of the rotor winding are connected. Pressing against these spinning rings are spring-loaded carbon rods called *brushes* that are fixed in position. This slip ring–brush arrangement allows the rotating armature to be connected to the nonrotating load.

Figure 13.4 shows our single-turn rotor in both end and top views "frozen" so that we can examine what is happening. Note the following: Side *a-b* moves up through the field while side *c-d* moves down. As a result, the polarities of their induced voltages are opposite. This is easily verified using the *right-hand rule of induced voltage*:

> Point the thumb of the right hand straight up, the first finger (forefinger) straight out, and the center finger to your left; the fingers are pointing in three mutually perpendicular directions. Holding the three in this relative position, turn the hand so that the thumb points in the direction of conductor motion and the forefinger in the direction of flux. The center finger then points in the direction of induced voltage and current. Remember: THumb = THrust, Forefinger = Flux, and Center finger = Current.

The induced polarities are series-aiding. Total loop voltage is therefore twice that developed in either side. No voltage is induced in sides *a-d* and *b-c* because they do not cut lines.

We assume that the loop is part of a closed circuit; therefore, current flows. Current into the page is indicated by the tail feathers of an arrow going away from the reader; current out of the page is a dot representing an arrow shaft coming at the reader. Note how current directions agree with polarities.

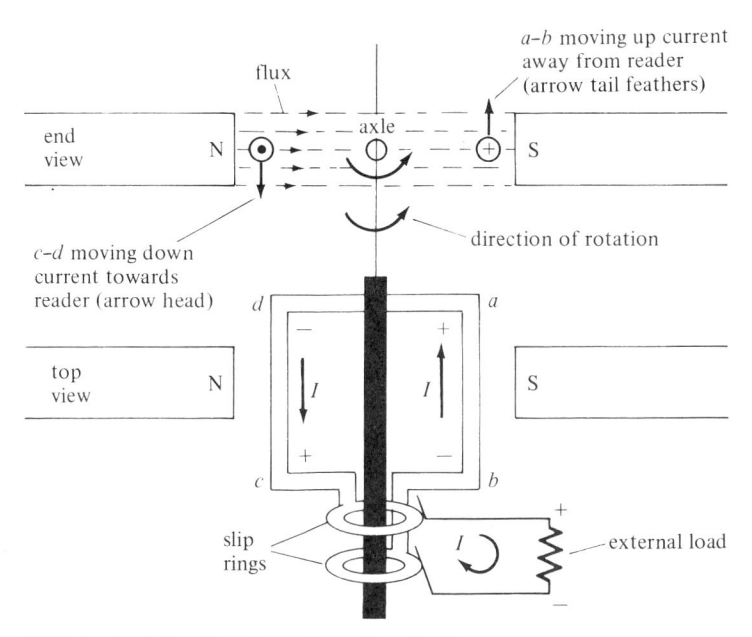

FIGURE 13.4 Rotating Loop in a Flux Field.

378   INTRODUCTION TO ALTERNATING CURRENT

Let us next turn the loop through a complete rotation. As we do, we will "freeze" the loop at various angles and consider the flux lines being cut (and hence the voltage being induced) at that instant. Pay close attention to the arrowhead and tail feathers at each position so that changing polarity is clearly understood.

Figure 13.5, part (a), shows the instant when the angle between flux and side-velocity vector, V_s, is zero. Since the vertical component of these velocities is zero, no lines are being cut. Induced voltage and current are zero.

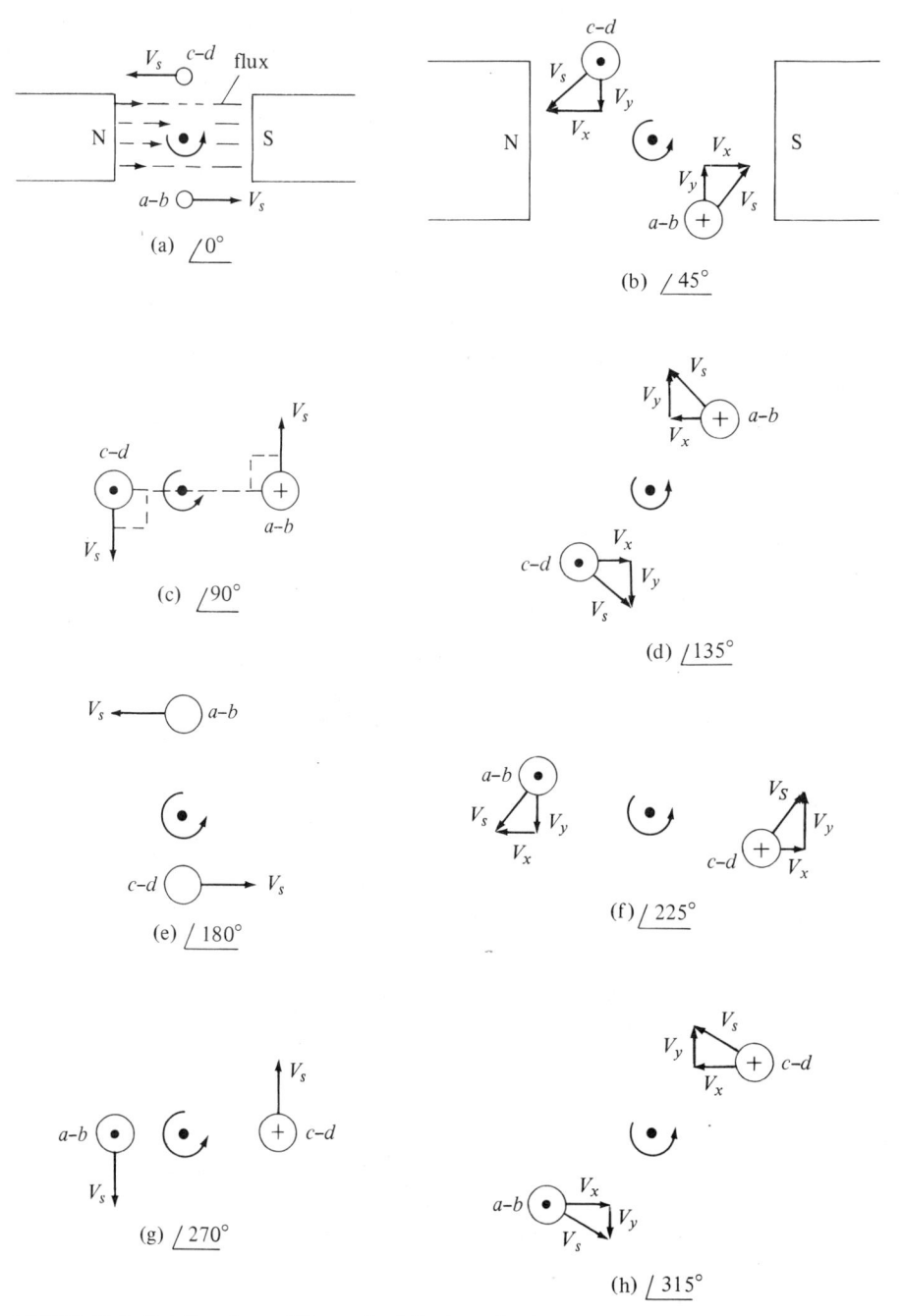

FIGURE 13.5 Rotating Loop "Frozen" at Various Angles.

13.2 GENERATING THE SINE WAVE

In Figure 13.5, part (b), the loop has rotated, putting the V_s vectors at a 45° angle. As always, V_x is parallel to the flux and cuts no lines. V_y however, does cut lines, causing induced voltage. Note that c-d current is out and a-b current is in.

Figure 13.5, part (c), again shows the loop at the instant when velocities and flux are perpendicular. V_x is now zero while V_y is maximum. Flux is being cut at the highest rate; therefore, induced voltage is at peak value. Current directions are unchanged.

The loop has rotated to 135° in Figure 13.5, part (d). As in part (b), V_y is less than maximum, so voltage is below peak value. In part (e) the angle has increased to 180°. Motion is again entirely horizontal; hence the voltage is zero.

When the loop rotates beyond the 180° position of Figure 13.5, part (e), the vertical components switch direction; side a-b now moves down through the field while side c-d moves up. By the right-hand rule, this direction change reverses the polarity of the induced voltage. Figure 13.5, part (f), for example, shows the loop at 225°. Induced voltage magnitude is the same as it was in part (d) but with opposite polarity. This polarity change is explained mathematically by the negative sine function for angles between 180 and 360°. Note carefully how the current directions in part (f) are opposite to those in part (d).

Figure 13.5, parts (g) and (h), are the same as parts (c) and (d), respectively, with the polarity change noted above. Another 45° beyond part (h) brings the loop back to the starting point part (a).

Each loop rotation repeats this pattern of voltage changes. That pattern is one cycle of a sine wave, or sinusoid (pronounced *sign-yuh-soid*). Three cycles of this important waveform are shown in Figure 13.6. Study the sine wave to become thoroughly familiar with its unique shape, which is neither circular nor elliptic. Note that it is symmetrical about the X axis; that is, the shape above the axis is the same as that below. Also, its peak or maximum value is the same in both positive and negative directions. Compare the amplitude for each angle on the X axis of Figure 13.6 with conductor position for that angle in Figure 13.5.

Why does a rotating conductor produce a sine wave?

As we have seen, induced voltage at any angle is a function of V_y, the vertical component of conductor velocity at that angle. And

$$V_y = V_s \sin \theta$$

by Equation 13.2. Substituting this velocity into Equation 11.13, we get

$$e = BlV$$
$$= BlV_s \sin \theta \qquad (13.5)$$

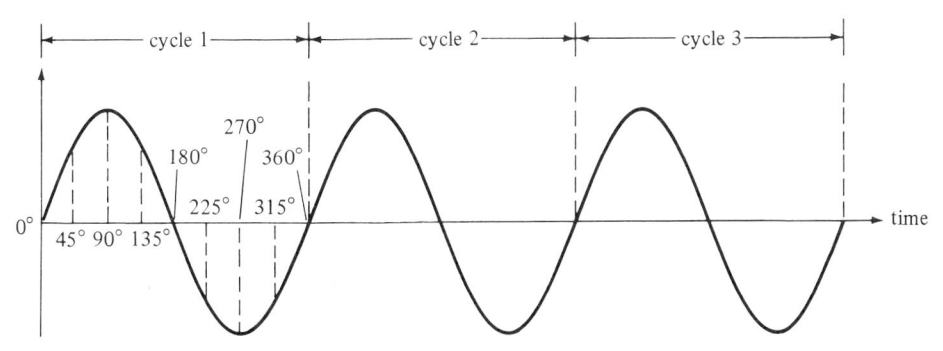

FIGURE 13.6 Three Cycles of a Sine Wave.

INTRODUCTION TO ALTERNATING CURRENT

Of the four factors on the right side of Equation 13.5, only sin θ varies. Consequently, induced voltage varies as the sine of an angle, that is *sinusoidally*. Since the sine function ranges between a minimum of -1 and a maximum of $+1$, induced voltage varies between $-BlV_s$ and $+BlV_s$.

A sine wave can also be developed by a rotating vector. Let us see how.

The vector M in Figure 13.7, rotating around the origin at a constant velocity (much like a spoke turns around a wheel's axle), is "frozen" at the instantaneous angle θ. If a light shines its rays parallel to the X axis, the vector casts a shadow like that seen in the figure. That shadow is called the vector's *vertical projection*. Its instantaneous height is described by the equation

$$\text{projection height} = M \sin \theta \tag{13.6}$$

where M = vector magnitude
θ = instantaneous angle made by M

In Figure 13.8, vector M rotates full circle and its instantaneous projection is plotted every 45°. Since projection height is a function of sin θ, a sine wave results when the points are connected. Thus we see that a sinusoid can be described by a rotating vector. Such a vector is called a *phasor*. In Chapter 14 we will find that phasor representation of sine waves provides a powerful means of analyzing ac circuits.

Since each conductor or phasor rotation takes the same time, sine waves, by our definition of Chapter 3, are *periodic*. They are also said to be *alternating* because their polarity changes or alternates between positive and negative. These opposites halves are called *alternations*. Current sine waves are called *alternating current*, or *ac*.

Because sinusoidal polarity changes, we must define when that polarity is positive, and when negative:

I Polarity is defined as positive when a sine wave is in the upper alternation (see Figure 13.9).

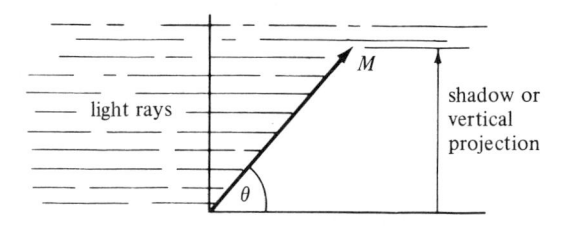

FIGURE 13.7 Vertical Projection of a Vector.

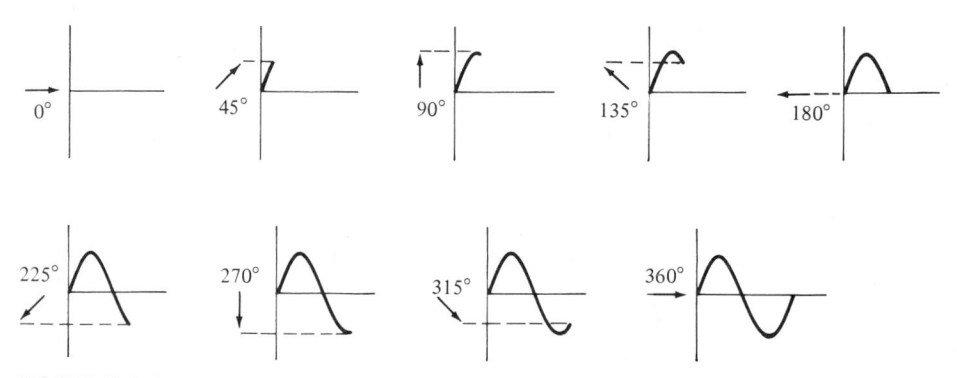

FIGURE 13.8 Phasor Development of Sine Wave.

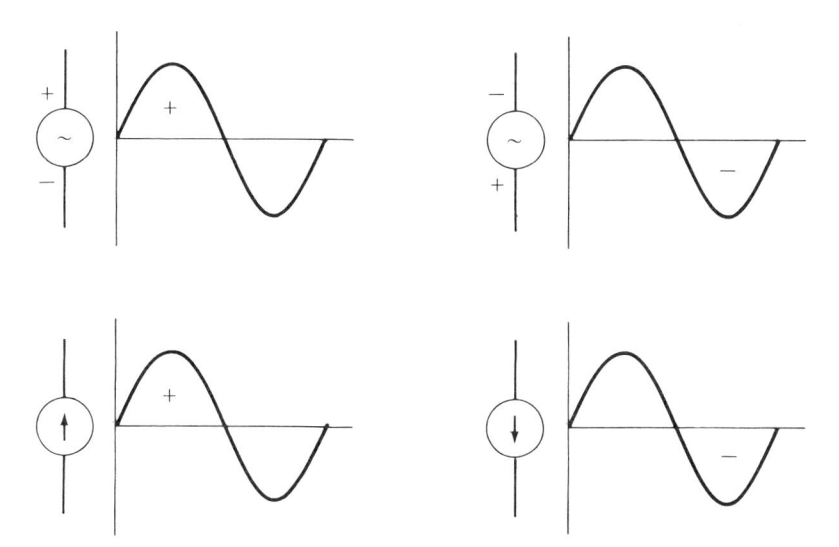

FIGURE 13.9 Polarity Definitions for Sinusoidal Voltage and Current Sources.

Sine waves have unique qualities. Any math operation done on them, such as adding, squaring, or multiplying, results in another sine wave. They are the only waveform unaffected by R, L, or C; that is, a voltage sinusoid causes a current sinusoid in all three. All periodic waves can be created, or synthesized, by combining sinusoids in special ways. This is discussed in Chapter 23.

Besides being induced in rotating conductors, sine waves are also generated by parallel LC circuits called *resonant* or *tuned* circuits. We study these in Chapter 19.

13.3 MEASURING THE SINE WAVE

Equation 13.5, $e = BlV_s \sin \theta$, is of little value because θ changes as the loop rotates. It can be put in a more useful form by first determining how θ varies with time. To do this, we must first discuss *angular velocity*.

A moving car has *linear* or straight-line *velocity* measured in distance traveled per unit time, that is, miles per hour, feet per second, and so on. Our generator's rotating loop, on the other hand, spins on its axle with an *angular velocity* measured in angle traveled per unit time. Represented by a Greek lowercase letter omega, ω, angular velocity is defined as

$$\omega = \frac{\text{angular distance, in radians}}{\text{time, in seconds}} \tag{13.7}$$

Note how ω is measured in the SI unit of angle, the *radian*, rather than degrees as might be expected. The conversion factor between radians and degrees is

$$2\pi \text{ radians} = 360° \tag{13.8}$$

Thus a conductor rotating through a complete circle in 1 second has an angular velocity of 360° per second or 2π radians per second. One spinning 50 rotations per second is going 18,000° per second or 100π radians per second; and so on.

Dividing 360° by 2π, we find that

$$1 \text{ rad} = 57.3° \tag{13.9}$$

* * * A Word of Caution * * *

Most calculators are in the degree mode when turned on and must be switched to work in radians. Study your instruction manual to see how this is done. Always be aware of the mode in which your calculator is operating when doing calculations with angles.

EXAMPLE 13.3

Convert each angle as indicated.

a. 180° to radians.

$$180° \left(\frac{2\pi \text{ radians}}{360°} \right) = \pi \text{ radians}$$

b. 90° to radians.

$$90° \left(\frac{\pi \text{ radians}}{180°} \right) = \frac{\pi}{2} \text{ radians}$$

c. 31.5° to radians.

$$31.5° \left(\frac{1 \text{ rad}}{57.3°} \right) = \frac{31.5}{57.3} = 0.55 \text{ rad}$$

d. $\pi/4$ radians to degrees.

$$\frac{\pi}{4} \text{ radians} \left(\frac{180°}{\pi \text{ radians}} \right) = \frac{180}{4} = 45°$$

Note how radian measurements are stated in terms of π as well as in numbers. ∎

The familiar formula $D = ST$ tells us the distance, D, traveled by a body moving S miles per hour for T hours. The same formula applies for angular distance covered by a rotating conductor. Distance, θ, is the angle through which the conductor turns while rotating ω radians per second for t seconds. In equation form, this is

$$\theta = \omega t \quad \text{radians} \tag{13.10}$$

Each time ωt increases by 2π, the conductor rotates a full circle.

Substitute Equation 13.10 in Equation 13.5 and we get:

$$e = BlV_s \sin \omega t \quad \text{volts}$$

This voltage reaches its maximum or peak value when $\omega t = \pi/2$ radians; that is,

$$E_{\text{peak}} = BlV_s \sin \frac{\pi}{2}$$
$$= BlV_s(1)$$
$$= BlV_s \tag{13.11}$$

(Note the uppercase E because the peak value is constant.)

Substituting Equation 13.11 into Equation 13.5, we have

$$e(t) = E_{\text{peak}} \sin \omega t \quad \text{volts} \tag{13.12}$$

13.3 MEASURING THE SINE WAVE

Because the induced voltage is now a function of time, we write it as $e(t)$, which is read "e of t." The statement does not mean e times t.

The current form of Equation 13.12 is

$$i(t) = I_{peak} \sin \omega t \quad \text{amperes} \quad (13.13)$$

Because Equations 13.12 and 13.13 allow us to determine voltage or current magnitude (or *amplitude*) at any instant, they are called *time functions*. When we do calculations with time functions, we are said to be working in the *time domain*.

EXAMPLE 13.4

Given: a voltage described by the time function.

$$e(t) = 43 \sin 2500t \quad \text{volts}$$

a. Find the peak amplitude.

$$E_{peak} = 43 \text{ V}$$

b. Find the angular velocity, ω.

$$\omega = 2500 \text{ rad/s}$$

c. The sine wave begins at 0 V at time $t = 0$. What is the instantaneous voltage 1.6 ms later?

$$e(t) = 43 \sin(2500)(1.6 \times 10^{-3})$$
$$= 43 \sin 4 \quad \text{(Note: This angle in radians.)}$$
$$= 43(-0.8)$$
$$= -34.6 \text{ V}$$

d. $e(t)$ is applied across a 2200-Ω resistance. Write the time function of the resulting current, $i(t)$.

By Ohm's law,

$$i(t) = \frac{E_{peak}}{R} \sin 2500t$$

$$= \frac{43}{2200} \sin 2500t = 19.5 \sin 2500t \quad \text{mA}$$

Note carefully that current through the resistance is sinusoidal and has the same angular velocity as the voltage.

e. What is $i(t)$ at $t = 0.48$ ms?

$$i(t) = 19.5 \sin 2500t \quad \text{mA}$$
$$= 19.5 \sin(2500)(0.48 \times 10^{-3})$$
$$= 19.5 \sin 1.2$$
$$= 19.5(0.93)$$
$$= 18.2 \text{ mA}$$

The number of cycles per second of a sine wave is called the wave's *frequency*. Represented by the letter f, frequency was once measured in cycles per second. Today, the term *hertz* is used, honoring the work of the German physicist Heinrich Hertz (1857–1894):

1 cycle per second is defined as a frequency of 1 hertz.

Since each loop rotation generates one cycle, it follows that high angular velocity is needed to develop high frequencies. The exact relation between ω and f can be derived:

$$\frac{1 \text{ loop rotation}}{\text{cycle}} \cdot \frac{2\pi \text{ radians}}{\text{rotation}} \cdot \frac{f \text{ cycles}}{\text{second}}$$
$$= 2\pi f \quad \text{rad/s}$$

But radians/second are the units of ω. Therefore,

$$\omega = 2\pi f \quad \text{rad/s} \tag{13.14}$$

One value of ω that should become familiar is 377, the angular velocity associated with the 60-Hz power frequency.

EXAMPLE 13.5

What is the frequency of the voltage sinusoid induced in a loop rotating 754×10^5 rad/s?

$$\omega = 2\pi f$$

Therefore,

$$f = \frac{\omega}{2\pi}$$
$$= \frac{754 \times 10^5}{2\pi}$$
$$= 12 \times 10^6 \text{ Hz}$$
$$= 12 \text{ MHz}$$

Equation 13.12 can be written in terms of f by substituting $2\pi f$ for ω:

$$e(t) = E_{\text{peak}} \sin 2\pi f t \quad \text{volts} \tag{13.15}$$

Similarly, Equation 13.13 is rewritten

$$i(t) = I_{\text{peak}} \sin 2\pi f t \quad \text{amperes} \tag{13.16}$$

Let's look at a few frequencies of interest starting with 0 Hz, or dc. At about 20 Hz we hit the bottom of the *audio range*, that is, the range of frequencies audible to most human beings. At 50 and 60 Hz we find the frequencies at which electric energy is transmitted in most countries, and at about 25 kHz, the high end of the audio range. Continuing upward, we arrive at 535 kHz, the low end of the standard broadcast AM band; the high end is at 1605 kHz. Just above the AM band the "short waves" begin. These extend to about 30 MHz and include frequencies occupied by "ham" (amateur)

13.3 MEASURING THE SINE WAVE

radio operators, international broadcasting, commercial and military communications, time signals, CB'ers, and much more. The VHF (very high frequency) TV band starts with channel 2 at 54 MHz and extends to channel 13 at 216 MHz; there is an 86-MHz-wide gap between channels 6 and 7 occupied by police and fire radios plus the FM allocation from 88 to 108 MHz. UHF (ultra high frequency) TV consists of 70 channels starting at 470 and ending at 890 MHz. Radar signals are found around 10 GHz (10^9 Hz), and visable light starts at about 10^{14} Hz.

If cycles per second are reciprocated, we get seconds per cycle, the time it takes one cycle to be completed. This time is called the *period* of the sine wave. It is represented by the letter T. Since period is understood to refer to one cycle, its units are given simply as seconds. From our discussion we see that

$$T = \frac{1}{f} \quad \text{seconds} \tag{13.17}$$

EXAMPLE 13.6

Given: a sine wave with a frequency of 18 kHz.

a. What is its period?

$$T = \frac{1}{f} = \frac{1}{18 \times 10^3} = 55.6 \ \mu s$$

b. What is the new period if the frequency is increased to 20 kHz?

$$T = \frac{1}{f} = \frac{1}{20 \times 10^3} = 50 \ \mu s$$

Note how the period decreased as the frequency increased.

Period can be determined from a waveform by measuring the time between a point on any wave and the same point on the next wave.

EXAMPLE 13.7

Determine the period and frequency of the following sinusoid.

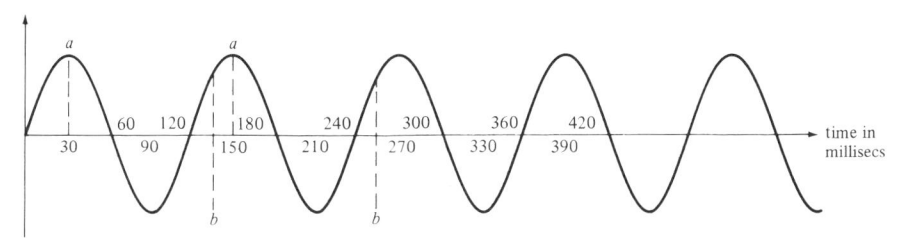

Note that the X axis or, abscissa, is in time rather than angle as before. The a dashed lines pass through two successive positive peaks; therefore, the time between them is the waveform's period:

$$T = (150 - 30) \text{ ms} = 120 \text{ ms}$$

$$f = \frac{1}{T} = \frac{1}{120 \times 10^{-3} \text{ s}} = 8.3 \text{ Hz}$$

Period could also be determined with the time between the *b* lines passing through the 30° point on two successive waves. ∎

EXAMPLE 13.8

Write the equation for the following waveform.

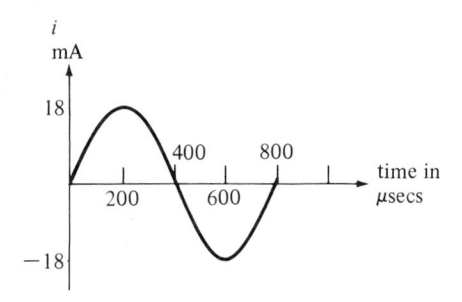

The shape of the waveform and the units on the *Y* axis tell us that this is a current sinusoid; therefore, the equation will be of the form

$$i(t) = I_{peak} \sin 2\pi ft \quad \text{amperes}$$

The period, measured on the *X* axis, is 800 μs. The frequency is, therefore,

$$f = \frac{1}{T} = \frac{1}{800 \times 10^{-6}} = 1.25 \times 10^3 \text{ Hz}$$

$$= 2\pi f = (6.28)(1.25 \times 10^3)$$
$$= 7.88 \times 10^3 \text{ rad/s}$$

Peak amplitude from the *Y* axis is 18 mA. Putting these items into the basic equation (Equation 13.13), we have

$$i(t) = 18 \sin 7880t \quad \text{mA}$$ ∎

The *peak-to-peak amplitude* of a sine wave is often of interest in electronics. It is simply twice the peak amplitude.

EXAMPLE 13.9

What is the peak-to-peak amplitude of the current sinusoid in Example 13.8?

The peak-to-peak amplitude of the sinusoid in Example 13.8 is 2 × 18 mA = 36 mA ∎

The cosine wave is a special form of sine wave. It could be developed by a second loop, *e–f*, *g–h*, on our generator rotor placed at right angles to the first [see Figure 13.10, part (a)]. As a result, this second conductor cuts maximum flux as the first

13.3 MEASURING THE SINE WAVE

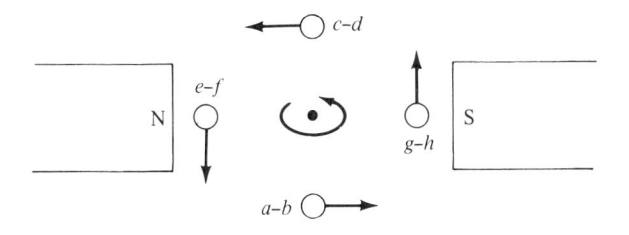

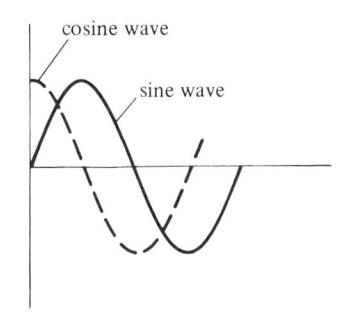

FIGURE 13.10 Generating the Cosine Wave.

travels parallel to the flux. The two waves are seen in Figure 13.10 part (b). Note how the cosine wave peaks $\pi/2$ radians ahead of the sinusoid.

Cosine and sine equations are almost identical. For example, a current can be described by a cosine equation as follows:

$$i(t) = I_{peak} \cos \omega t \quad \text{amperes} \tag{13.18}$$

or

$$i(t) = I_{peak} \cos 2\pi f t \quad \text{amperes} \tag{13.19}$$

When sketching a sine or cosine waveform, follow these steps:

1. Decide the units of the X axis. These can be degrees or radians, or divisions of time such as seconds, microseconds, and so on.
2. With the origin zero in all cases, label the axis in four equal portions as follows:

   in degrees: 90, 180, 270 and 360, or 0
   in radians: $\pi/2$, $2\pi/2$, $3\pi/2$, and $4\pi/2$
   in time: $T/4$, $2T/4$, $3T/4$, and $4T/4$

3. Mark the points of maximum and minimum amplitude on the Y axis.
4. Keep in mind the shape of the waves:

EXAMPLE 13.10

a. Sketch the waveform for the time function:

$$e(t) = 75 \cos 377 t \quad \text{volts}$$

With the X axis in milliseconds.

Drawing the waveform with time on the X axis: Period = $2\pi/\omega = 2\pi/377 = 16.7$ ms. Since this is a cosine wave, peak amplitude of 75 V occurs at $t = 0$ and 16.7 ms. The wave crosses the X axis at $T/4$ and $3T/4$, or 4.2 and 8.3 ms, respectively. Negative peak occurs at $2T/4 = 8.3$ ms. Draw the waveform:

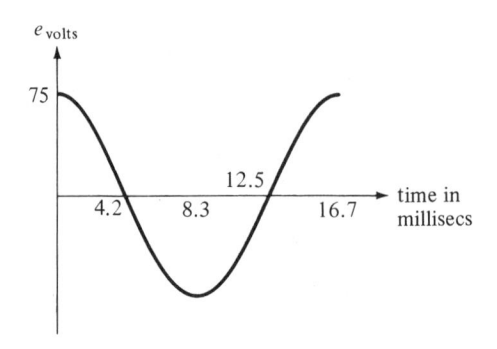

b. Repeat part a with the X axis in radians.

The wave crosses the X axis at $\pi/2$ and $3\pi/2$; the negative peak occurs at $2\pi/2$.

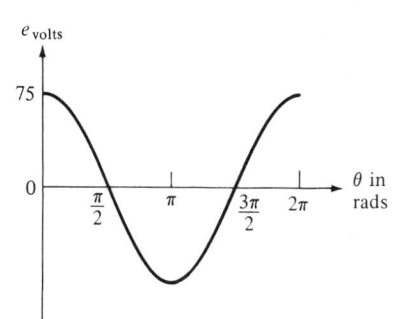

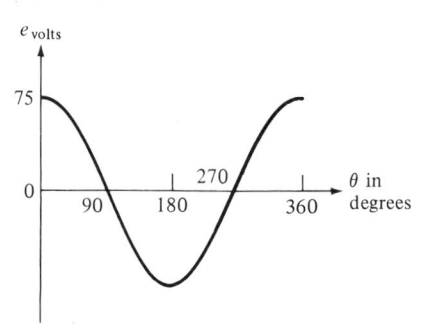

c. Repeat part a with the X axis in degrees.

The wave crosses the X axis is 90° and 270°; the negative peak occurs at 180°. The wave is drawn above. ∎

13.4 SINE WAVES ON DC LEVELS

Electronic circuits need power to operate. They usually get it in dc form. However, the signals being processed by the circuits are generally ac in nature. As a consequence, we often deal with *composite* signals, that is, signals with both ac and dc components. This is especially true in the study of transistor amplifiers.

A circuit that develops a composite signal is seen in Figure 13.11 part (a). It consists of a 3-V peak ac generator in series with a 5-V dc source. In part (b) the ac source has been "frozen" at +3 V. Note how the two polarities add, making the total instantaneous voltage 8 V. Part (c) shows the sine-wave source at its negative maximum. Polarities now cancel, so the total potential is 2 V. The result [part (d)] is a composite signal consisting of a 3-V peak sinusoid *superimposed* or *riding on* a 5-V dc level. Note how the sine wave, which is usually symmetrical about the X axis, now has a new "zero" line at 5 V. The equation for the waveform of Figure 13.11 is

$$e(t) = 5 + 3 \sin \omega t \quad \text{volts}$$

(The equation is actually an example of superposition.)

13.5 THE AVERAGE VALUE OF A SINE WAVE

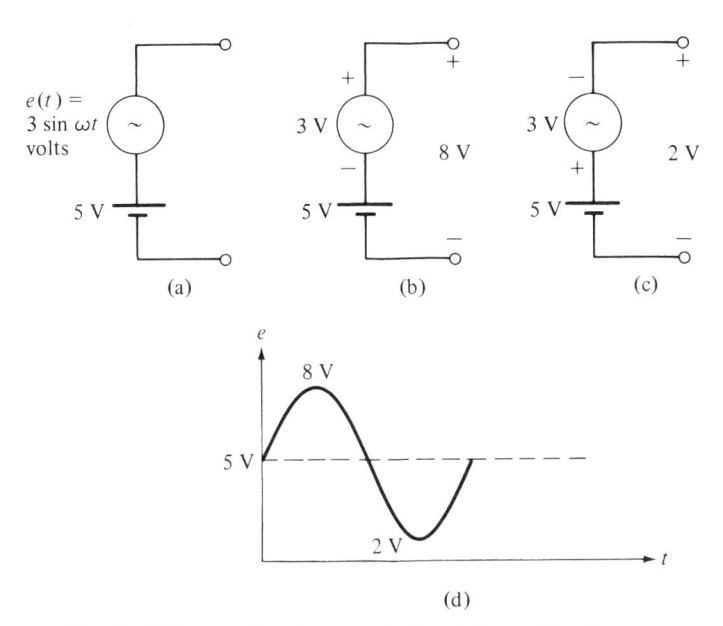

FIGURE 13.11 Composite Signal and Circuit Generating It.

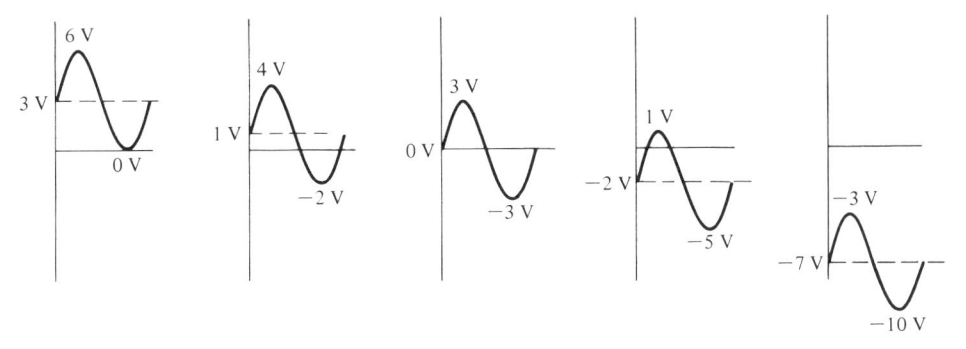

FIGURE 13.12 Changing the dc Component of a Composite Waveform.

Figure 13.12 illustrates the effect on the composite signal as the dc component is gradually reduced from +3 to −7 V. In each case, note how the sine wave is symmetrical about the dc level.

13.5 THE AVERAGE VALUE OF A SINE WAVE

We begin this section by reviewing the basic concept of averaging. Suppose that the weekly earnings of a salesperson during a 4-week period were $350, $350, $60, and $260. If asked to calculate the mean or average earnings, you would, no doubt, add the amounts, divide by the number of weeks in the period, and announce a figure of $255. But let's take a closer look at these calculations so that we can really understand what is happening.

Figure 13.13 shows income plotted versus time in weeks. The area of each of the plot's three rectangles is easily obtained by multiplying height times width. Since height is in units of dollars per week and width in weeks, their product, the units of these areas,

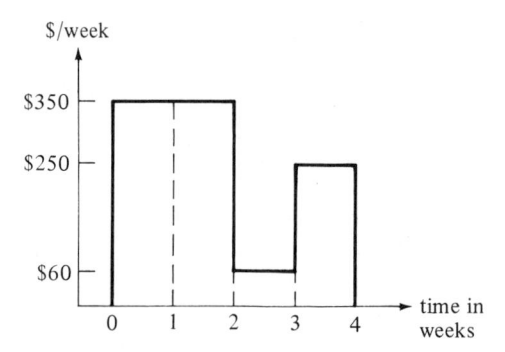

FIGURE 13.13 Plot for Averaging Income.

will be dollars. The three areas are:

Area 1: 350 dollars/week × 2 weeks = 700 dollars
Area 2: 60 dollars/week × 1 week = 60 dollars
Area 3: 260 dollars/week × 1 week = 260 dollars
Total area = (700 + 60 + 260) dollars
           = 1020 dollars

We next form a single rectangle having the same width (4 weeks) and area (1020 dollars) as above, but a height that is constant. This height is the average or *mean* value we are looking for.

Since area = width × height or, $A = WH$, we have

$$H = \frac{A}{W} = \frac{1020 \text{ dollars}}{4 \text{ weeks}}$$

$$= \frac{255 \text{ dollars}}{\text{week}}$$

the average value.

Suppose that the person did not work for three weeks. What is the new mean?

Since the additional weeks add zero area, total area remains at 1020 dollars. But the width has increased to 7 weeks, which means that the height must drop to maintain that area. The height of the new rectangle, the new average, is 1020/7 = $145.71/week.

From this problem we see that it takes two steps to find an average:

1. Find the total area under the *curve*. (*Note:* "Curve" is a mathematical term used to describe any graph, even a square wave.)
2. Divide that area by the width of the curve.

These steps can be put into a formula:

$$\text{mean value} = \frac{\text{area under curve}}{\text{width of curve}} \qquad (13.20)$$

Because our interest is in waveforms, the width of the curve is time on the X axis. The equation is thus rewritten

$$\text{average value} = \frac{\text{area under curve}}{\text{time of curve}} \qquad (13.21)$$

For a periodic waveform, the time of the curve is the period, T.

13.5 THE AVERAGE VALUE OF A SINE WAVE

Analog meter movements such as the D'Arsonval type cannot move fast enough to respond to rapid changes in voltage or current. Instead, they respond to the average value of a waveform.

EXAMPLE 13.11

A D'Arsonval meter is connected to a source developing each of the following waveforms. What will the meter read?

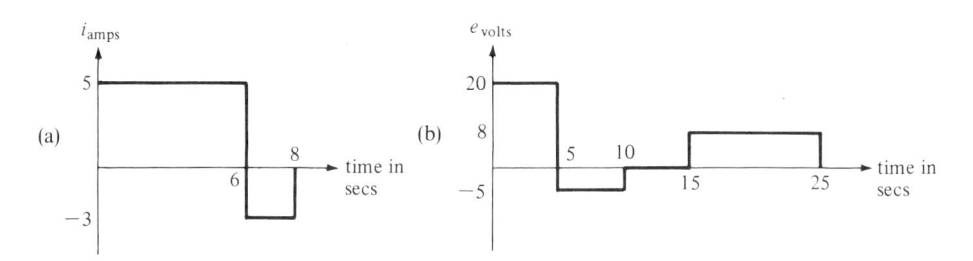

a. In calculating the mean current, pay close attention to signs.

$$\text{total area} = (5 \times 6) + (2 \times -3) = 30 - 6$$
$$= 24 \text{ Amp-secs}$$

$$I_{average} = \frac{\text{total area}}{\text{time}}$$

$$= \frac{24 \text{ Amp-secs}}{8 \text{ secs}}$$

$$= 3 \text{ A}$$

The meter reads 3 A.

b.

$$\text{total area} = (5 \times 20) + (5 \times -5) + (5 \times 0) + (10 \times 8)$$
$$= 100 - 25 + 0 + 80 = 155 \text{ V-s}$$

$$E_{average} = \frac{155 \text{ V-s}}{25 \text{ s}}$$

$$= 6.2 \text{ V}$$

The meter reads 6.2 V. ∎

Symmetrical waveforms have equal area above and below some horizontal line. The value of that line must, therefore, be the waveform's average value. Sine and cosine waves, for example, are symmetrical about the X axis; therefore:

| The average value of any sine or cosine wave centered on the X axis is zero.

In a composite waveform, however, the dc component is the mean value. (It might be said that a sine/cosine wave on the X axis is a composite signal with a dc component of zero.)

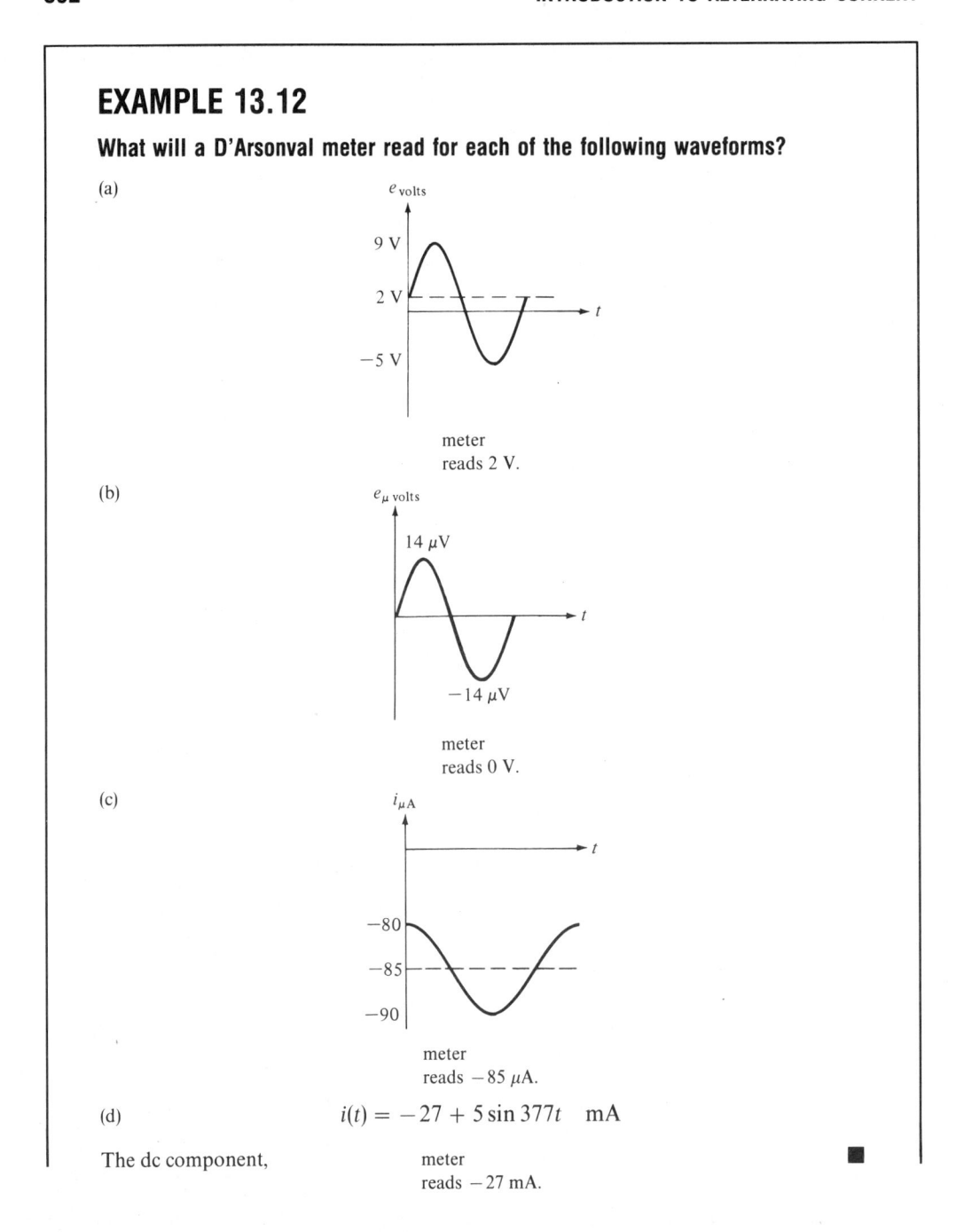

EXAMPLE 13.12

What will a D'Arsonval meter read for each of the following waveforms?

(a)

9 V
2 V
−5 V

meter reads 2 V.

(b)

14 μV
−14 μV

meter reads 0 V.

(c)

−80
−85
−90

meter reads −85 μA.

(d) $i(t) = -27 + 5\sin 377t$ mA

The dc component, meter reads −27 mA.

13.6 THE EFFECTIVE VALUE OF A SINE WAVE

Unlike dc with its constant value, ac has an infinite number of values between its negative and positive peaks. Which of these should be used in the familiar equation $p = i^2 R$ to calculate power? To answer that question, we must compare the energy transfer abilities of current in sine-wave and dc forms.

Consider the following experiment. The switch in Figure 13.14 is thrown left, connecting the dc source to the resistance immersed in water. Suppose that the energy

13.6 THE EFFECTIVE VALUE OF A SINE WAVE

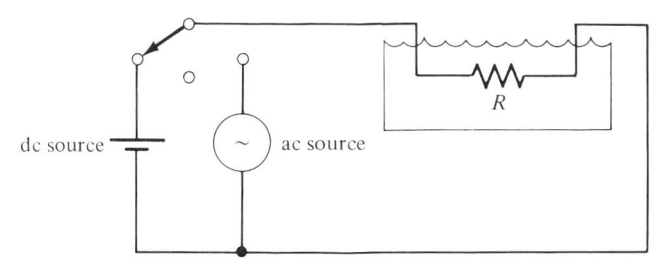

FIGURE 13.14 Energy Transfer by dc and ac Sources.

delivered from source to resistance raises the water temperature from 20°C to 60°C in some time *T*. The switch is then put in the middle position, allowing the water to cool to its original 20°C temperature. Finally, the ac source is connected to the resistance by throwing the switch to the right, and its current adjusted to make the water temperature rise again to 60°C in the same time *T*.

Since both sources raised the water temperature equally in equal time, the ac must have been as effective at delivering power as the dc. We say that the ac had an *effective value*, I_{eff}, equal to the dc. In other words:

$$I_{\text{dc}}^2 R = I_{\text{eff}}^2 R \tag{13.22}$$

The power developed in a resistance by a given number of dc amperes is the same as the power developed in that resistance by an equal number of effective ac amperes.

What is the effective value of an ac? Although peak value might be our immediate answer to that question, a moment's thought shows that this cannot be so, because sine waves peak for only an instant; they spend most of their time at lower levels. The effective value must therefore lie somewhere between zero and peak. Let us find it.

Consider the equation for instantaneous power, $p = i^2 R$, in which *i* is the current sinusoid:

$$i = I_{\text{peak}} \sin \omega t$$

Substitution gives us

$$p = (I_{\text{peak}} \sin \omega t)^2 R$$
$$= (I_{\text{peak}}^2 \sin^2 \omega t) R$$

The mean value of this power is what heats the water. It is easily found with the aid of the trigonometric identity

$$\sin^2 \omega t = \tfrac{1}{2} - \tfrac{1}{2} \cos 2\omega t$$

Substitute the identity into the power equation:

$$p = I_{\text{peak}}^2 R (\tfrac{1}{2} - \tfrac{1}{2} \cos 2\omega t)$$
$$= \frac{I_{\text{peak}}^2 R}{2} - \frac{I_{\text{peak}}^2 R}{2} \cos 2\omega t$$

Careful study of this expression shows it to be a composite waveform with a cosine wave riding on a dc level of $I_{\text{peak}}^2 R / 2$. This dc level is the power waveform's mean level that we are looking for. It delivers the same power to the water as $I_{eff}^2 R$ does. That is,

$$I_{eff}^2 R = \frac{I_{peak}^2 R}{2}$$

Solve for I_{eff}:

$$I_{eff}^2 \cancel{R} = \frac{I_{peak}^2 \cancel{R}}{2}$$

Take the square root of each side:

$$\sqrt{I_{eff}^2} = \sqrt{\frac{I_{peak}^2}{2}}$$

$$= \frac{\sqrt{I_{peak}^2}}{\sqrt{2}}$$

$$I_{eff} = \frac{I_{peak}}{\sqrt{2}}$$

$$= \frac{I_{peak}}{1.414}$$

Or

$$I_{eff} = 0.707 I_{peak} \tag{13.23}$$

In English, this equation says that the effective value of a sine wave is 70.7% of the peak value.

Let's review what we have just done. A current sinusoid was squared resulting in a composite waveform. The average of this squared wave, which we call the mean of the squares, was $I_{peak}^2/2$ (after division by R). The square root, or simply the root, of the mean of the squares was found, giving us the *root-mean-square* (*rms*) value of current. The rms value and the effective value are the same thing.

Rewrite Equation 13.23:

$$I_{eff} = I_{rms} = 0.707 I_{peak} \tag{13.24}$$

Solve Equation 13.23 for I_{peak}:

$$I_{peak} = \frac{I_{eff}}{0.707}$$

or

$$I_{peak} = 1.414 I_{eff} = \sqrt{2}\, I_{eff}$$
$$= 1.414 I_{rms} = \sqrt{2}\, I_{rms} \tag{13.25}$$

This tells us that the peak value of a sinusoid is 141.4% of the effective or rms value. Like Ohm's law, Equations 13.24 and 13.25 must be second nature to students of electronics.

Because the power equation, $p = e^2/R$, also involves the squaring of a sinusoid, the same expressions for voltage can be derived. They are

$$E_{eff} = E_{rms} = 0.707 E_{peak} \tag{13.26}$$

and

$$E_{peak} = \sqrt{2}\, E_{eff} = \sqrt{2}\, E_{rms} \tag{13.27}$$

13.6 THE EFFECTIVE VALUE OF A SINE WAVE

EXAMPLE 13.13

Given: $E_{eff} = 78$ V. Find the following.

a. The rms voltage.

$$E_{rms} = E_{eff} = 78 \text{ V}$$

b. The peak voltage.

$$\begin{aligned} E_{peak} &= \sqrt{2}\, E_{rms} \\ &= \sqrt{2}(78) \\ &= 110.3 \text{ V} \end{aligned}$$

c. The peak-to-peak voltage.

$$\begin{aligned} E_{p\text{-}to\text{-}p} &= 2E_{peak} \\ &= 2(110.3) \\ &= 220.6 \text{ V} \end{aligned}$$

d. The average voltage.

$$E_{ave} = 0 \text{ V}$$

EXAMPLE 13.14

Given: an ac with a peak value of 16.2 A.

a. What is the average power that it will supply to a 7.4-Ω load?

$$\begin{aligned} P = I_{eff}^2 R &= (0.707 I_{peak})^2 (7.4) \\ &= (11.5)^2 (7.4) = 132.3(7.4) \\ &= 978.7 \text{ W} \end{aligned}$$

b. What value of dc delivers the same power to the load?

$$\begin{aligned} I_{dc} = I_{eff} &= 0.707 I_{peak} \\ &= 0.707(16.2) = 11.5 \text{ A} \end{aligned}$$

Several important points about rms values must be thoroughly understood:

- Unless specifically indicated, any reference to an ac voltage or current is understood to be in rms form. For example, the 120-V potential in our homes is the rms value of a sinusoid whose peak value is $\sqrt{2}(120)$, or 170 V.
- Although rms is understood when referring to ac, peak value is always used in the time functions for instantaneous value, that is, Equation 13.12, $e(t) = E_{peak} \sin \omega t$, and equation 13.16, $i(t) = I_{peak} \sin 2\pi ft$.
- Like peak and average values of a given sinusoid, the rms value is a constant. Therefore, it is represented by an uppercase letter.
- The conversion, rms value = 0.707 (peak value), is for sine and cosine waves. Do not assume that it applies to other waveforms. Their rms value must be determined by finding the root of the mean of their squares (see the next examples).

EXAMPLE 13.15

Determine the rms value of this periodic square wave:

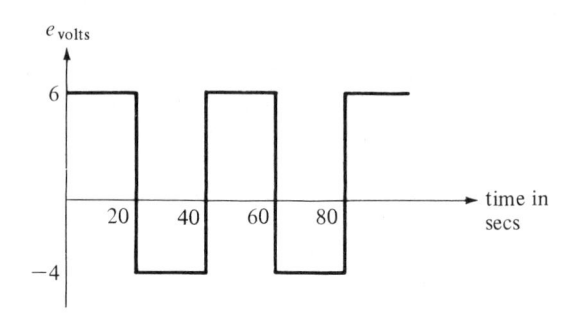

The plot of the squares is

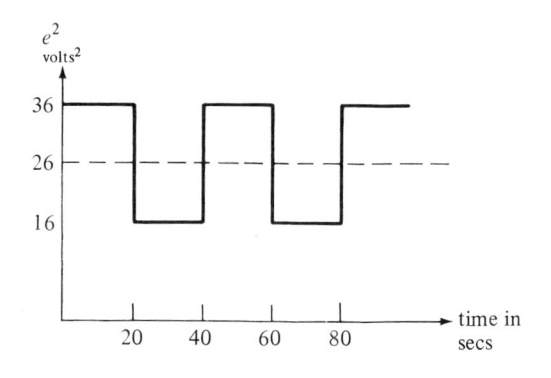

The mean of the squares is the value of the line dividing the squared curve into equal areas. This is 26 V^2. The rms value is

$$\sqrt{26 \text{ V}^2} = 5.1 \text{ V}$$

∎

EXAMPLE 13.16

Determine the rms value of this aperiodic waveform:

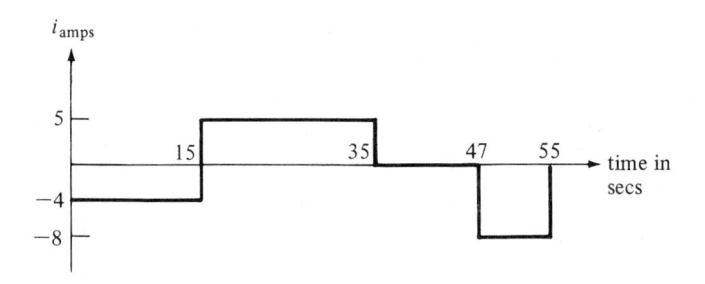

EXERCISES

The plot of the squares is

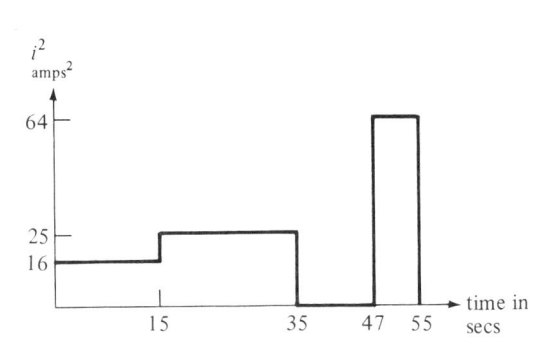

[Graph showing i² in amps² vs time in secs: value 16 from 0 to 15, value 25 from 15 to 35, value 0 from 35 to 47, value 64 from 47 to 55]

By Equation 13.21,

$$\text{Mean of the squares} = \frac{\text{total area under squares curve}}{\text{total time}}$$

$$= \frac{(16 \times 15) + (25 \times 20) + (0 \times 12) + (64 \times 8)}{55}$$

$$= \frac{240 + 500 + 0 + 512}{55}$$

$$= \frac{1252}{55}$$

$$= 22.7 \text{ A}^2$$

The rms value is $\sqrt{22.7 \text{ A}^2} = 4.8$ A ∎

EXERCISES

Section 13.1

1. Find the x and y components of each conductor velocity.
   a. 3 ft per second at 75°.
   b. 800 m/s at 240°.
2. The velocity of a conductor through a flux field is 50 m/s at an angle of 45°.
   a. What happens to the Y component of this velocity as the angle gets steeper?
   b. What is this component when the angle is 90°?
   c. What happens to the X component as the angle gets steeper?
   d. What is the X component when the angle equals 90°?
   e. What is the Y component when the angle is 0°?
   f. What is the X component when the angle is 0°?

Section 13.2

3. Answer the following questions about the voltage induced in a conductor rotating in a flux field.
   a. Why is there no voltage induced when θ, the angle between V_s and flux, is 0 and 180°?
   b. Why is the voltage maximum at $\theta = 90°$?
   c. Why is the induced voltage at any angle a function of sine θ?
   d. Why is the polarity for θ between 180 and 360° the opposite of what it is between 0 and 180°?

4. Assume that the conductor shown in the diagram is part of a closed circuit. What is the direction of induced current?

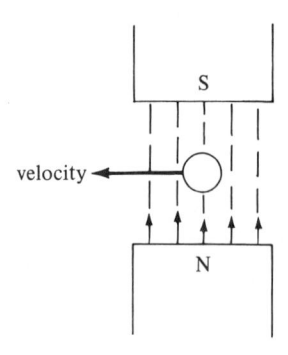

5. Use the right-hand rule to show that voltage induced in side *a-b* of the rotating loop in Figure 13.4 is in the opposite direction to that induced in side *c-d*. Show that the two voltages are series aiding.

6. Why is a sinusoid periodic?

Section 13.3

7. Convert as indicated.
a. To radians: (express answer in terms of π). (1) 180°; (2) 20°; (3) 60°; (4) 45°; (5) 15°.
b. To radians: (express answer in numbers). (1) 65°; (2) 18°; (3) 28°; (4) 82°.
c. To degrees. (1) 1.31 rad; (2) 0.192 rad; (3) 1.01 rad; (4) 0.663 rad.
d. To degrees. (1) $3\pi/12$ rad; (2) $4\pi/10$ rad; (3) $5\pi/4$ rad; (4) $7\pi/15$ rad.

8. Convert as indicated.
a. To radians: (express answer in terms of π). (1) 120°; (2) 75°; (3) 105°; (4) 135°.
b. To radians: (express answer in numbers). (1) 17°; (2) 143°; (3) 85.9°; (4) 37.2°.
c. To degrees. (1) 1.58 rad; (2) 0.47 rad; (3) 2.7 rad; (4) 4.8 rad.
d. To degrees. (1) $5\pi/6$ rad; (2) $7\pi/4$ rad; (3) $19\pi/12$ rad; (4) $31\pi/20$ rad.

9. A generator conductor starts rotating 10 rad/s at $t = 0$. For each of the times (1) 0.2 s, (2) 0.63 s, and (3) 850 ms:
a. Find the instantaneous angle between conductor velocity and generator flux field.
b. Find the sine of each angle.

10. A conductor rotates 10^4 rad/s for 250 μs.
a. What is the sine of the instantaneous angle between conductor velocity and flux.
b. What is the cosine of that angle?

11. A conductor rotating 1500 rad/s rotates through an angle of 78×10^5 rad.
a. What time was required to pass through this angle?
b. How many times did the conductor go completely around in this time?

12. Repeat Exercise 11 for a conductor rotating 650 rad/s that rotates through an angle of 210 rad.

13. The time function for a voltage sinusoid is written with three portions indicated by braces.

$$e(t) = 75 \sin 420t \text{ volts}$$

with 1 under "420t", 2 under "420t", and 3 under "sin 420t".

a. What is the maximum value of quantity 1?
b. What are the maximum and minimum values of quantity 2?
c. Repeat part b for quantity 3.
d. What are the units of quantities 1, 2, and 3?
e. Find a value of $t > 0$ for which $e(t) = 0$ V.

EXERCISES

f. Find a value of t for which $e(t) = 75$ V.
g. The sine-wave voltage is applied across a 25-Ω resistance. Write the equation for $i(t)$.
h. Draw the waveform of $e(t)$ and $i(t)$.

14. What determines the frequency of the sine wave induced in a conductor rotating in a magnetic field?

15. Given: a voltage described by the time function:

$$e(t) = 750 \sin 6720t \quad \text{V}$$

a. Find the peak amplitude.
b. Find the angular velocity.
c. Find the frequency.
d. Find $e(t)$ at $t = 285$ μs.
e. Find $e(t)$ at $t = 900$ μs.
f. Write the time-domain expression for the current that results when $e(t)$ is across a 250-k resistor.
g. Find the instantaneous current at $t = 500$ μs.

16. For the voltage

$$e(t) = 83 \sin(350 \times 10^3 \, t) \quad \text{mV}$$

a. Find the peak amplitude.
b. Find the angular velocity.
c. Find the frequency.
d. Find $e(t)$ at $t = 6.2$ μs.
e. Find $e(t)$ at $t = 9.5$ μs.
f. Write the time-domain expression that results when this voltage is across 2700 Ω.
g. Find the instantaneous current at $t = 2$ μs.

17. Fill in the blanks.

	rads/s	frequency (Hz)	T(s)
a.			5×10^{-7}
b.	4750		
c.		135 MHz	

18. Answer the questions for this sine wave:

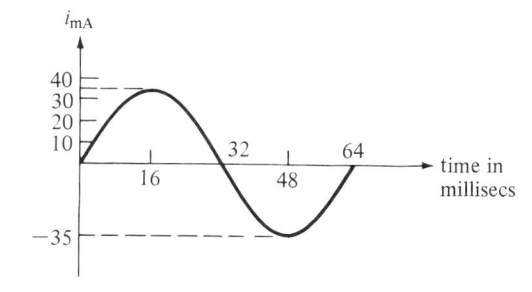

a. Calculate the period.
b. Calculate the frequency.
c. Calculate the peak amplitude.
d. Read the instantaneous value at $t = 26$ ms and $t = 52$ ms off the graph.
e. Write the equation for the current sinusoid.
f. With equation from part e, calculate the instantaneous current at the times of part d. Compare measured and calculated values.

19. Find the sinusoidal equation as indicated:
a. Find $i(t)$.

$e(t) = 17 \times 10^{-3} \sin 25000t$ volts

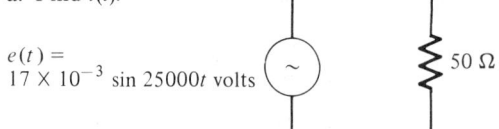

b. Find $e(t)$.

$i(t) = 7.7 \sin 377t$ amps

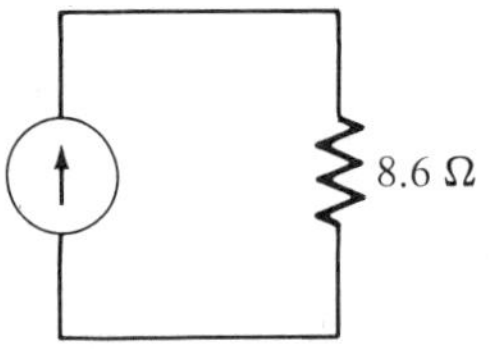

20. For the voltage in Exercise 19, part a, find: a. the period; b. the peak value; c. the peak-to-peak value; d. the instantaneous value at $t = 188$ μs.

21. For the current in Exercise 19, part b, find: a. the period; b. the peak value; c. the peak-to-peak value; d. the instantaneous value at $t = 2$ ms.

22. Write the time function for each waveform.

(a) $i_{\mu A}$

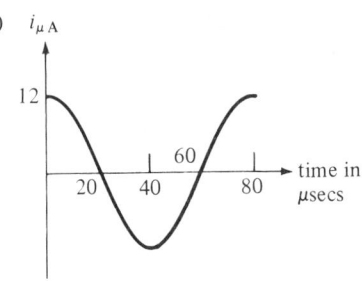

(b) e_{kV}

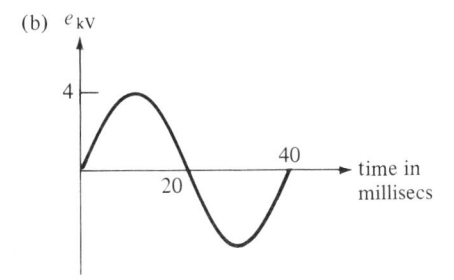

Section 13.4

23. A composite signal is described by the equation

$$i(t) = 5 + 4 \sin 377t \quad A$$

a. What is the magnitude of the dc component?
b. What is the peak amplitude of the ac component?
c. What is the peak amplitude of the composite?
d. Does the wave ever go negative?

24. Repeat Exercise 23 for

$$e(t) = 15 + 20 \cos 10,000t \quad V$$

25. A composite wave has a maximum value of 0.3 V and a minimum of -16.5 V. If the frequency is 3500 kHz, write the equation of the wave (the ac component is sinusoidal).

26. A composite signal has a dc component of 12.3 mA and a minimum value of -5.4 mA.
a. If the ac component is a cosine wave with a frequency of 144 MHz, what is the time function of the wave?
b. What is the wave's peak-to-peak amplitude?

Section 13.5

27. Find the average value of each waveform.

a.

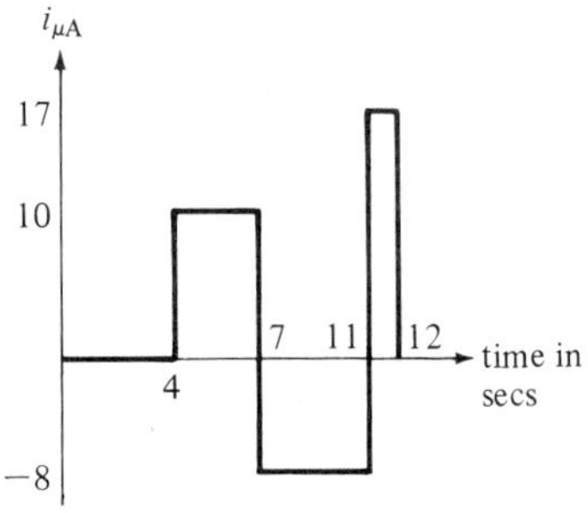

b.

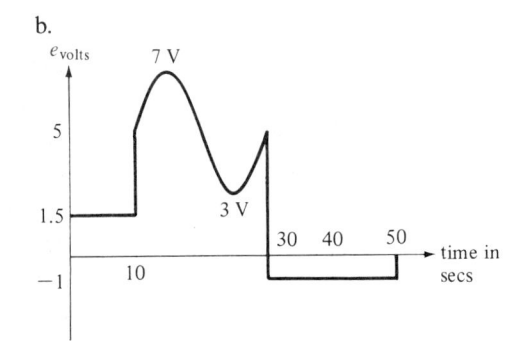

EXERCISES

28. What will a D'Arsonval meter read when driven by each of these waveforms?

a. 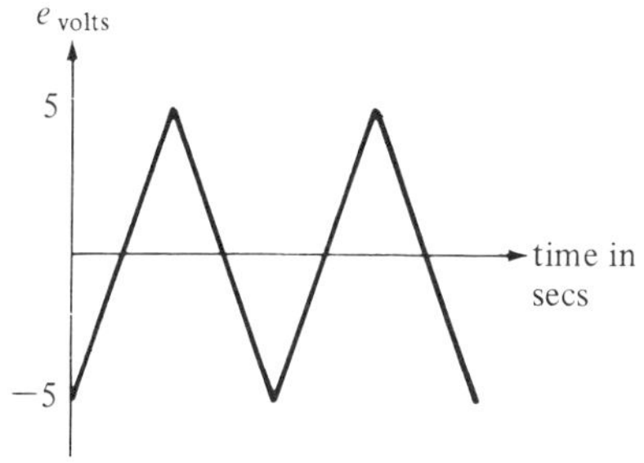 e volts — triangular wave between +5 and −5, time in secs

b. 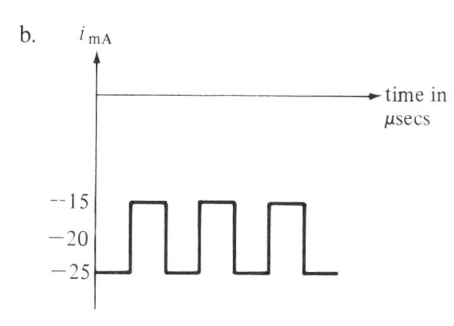 i mA — pulse train between −15 and −25, time in μsecs

c. 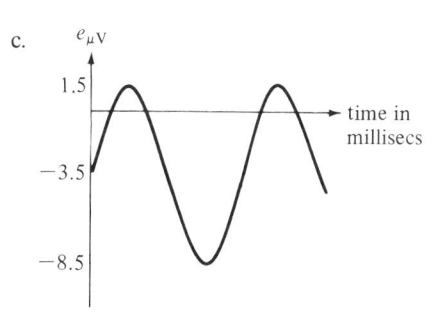 $e_{\mu V}$ — sine wave between 1.5 and −8.5 (with −3.5 axis), time in millisecs

d. 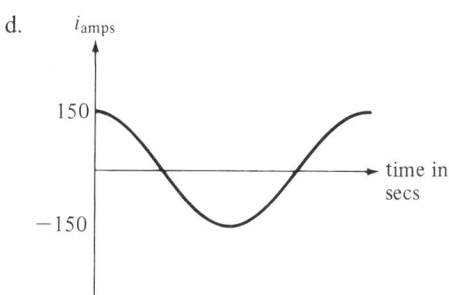 i_{amps} — sine wave between 150 and −150, time in secs

Section 13.6

29. Fill in the blanks.

	peak	peak-to-peak	rms	average
a.	15			
b.		74		
c.			20	
d.		150		
e.			3.8	

30. A 12-V car battery develops 97 W in a load. What is the rms value of ac voltage needed to develop the same power? What is the peak value of that ac?

31. What is the peak current through a light bulb that produces an average 60 W when connected to a 120-V line?

32. What value of dc current develops the same load power as the current shown in part a? What value of dc voltage develops the same load power as the voltage shown in part b?

a. 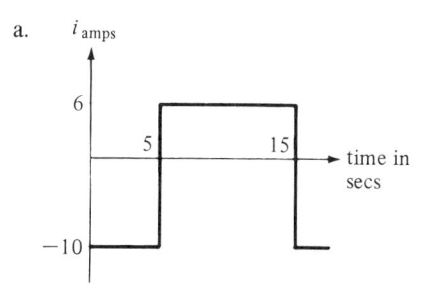 i amps — waveform: −10 from 0 to 5s, 6 from 5 to 15s, time in secs

b. 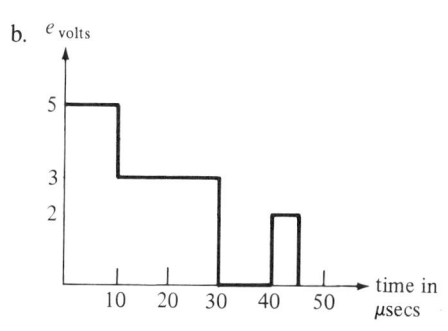 e volts — stepped waveform: 5 (0–10μs), 3 (10–30μs), 0 (30–40μs), 2 (40–50μs), time in μsecs

33. What is the average power dissipated by an 18-Ω resistance when the voltage $e(t) = 63\sin(377t - 28°)$ is applied across it?

34. The current through an 11-Ω resistance is

$$i(t) = 4.6\cos(1000t - 90°)$$

a. What is the peak power produced in the resistance?
b. What is the average power?

CHAPTER HIGHLIGHTS

Although they are not the only alternating waveform, sinusoids are the most common and the most important.

Scalars have magnitude only. Vectors have magnitude and direction.
The voltage induced in a conductor rotating in a magnetic field varies at a sinusoidal rate.
Sine waves can be described by rotating vectors called phasors.
Analog meters respond to a waveform's average value.
The effective or rms value of a sinusoid delivers the same energy to a load as a similar amount of dc.

Formulas to Remember

$$V_x = V\cos\theta \qquad V_y = V\sin\theta$$
$$360° = 2\pi \text{ rad}$$
$$1 \text{ rad} = 57.3°$$
$$\theta = \omega t$$
$$T = \frac{1}{f} \quad \text{period of sine wave}$$
$$\frac{\text{total area}}{\text{time}} \quad \text{equation for averaging}$$
$$I_{eff} = I_{rms} = 0.707 I_{peak}$$
$$E_{peak} = \sqrt{2}\, E_{rms}$$

Time functions for analysis done in the time domain:

$$e(t) = E_{peak}\sin\omega t = E_{peak}\sin 2\pi ft$$
$$i(t) = I_{peak}\sin\omega t = I_{peak}\sin 2\pi ft$$
$$e(t) = E_{peak}\cos\omega t = E_{peak}\cos 2\pi ft$$
$$i(t) = I_{peak}\cos\omega t = I_{peak}\cos 2\pi ft$$

CHAPTER 14

Phase and Phasors

Voltage and current sinusoids are usually out of step in ac circuits. That is, one reaches peak value before the other. This time separation, called a *phase difference* or *phase shift*, is one of the major ways in which dc and ac analysis differ.

In this chapter we learn to describe phase graphically and mathematically (the cause of phase shift is discussed in Chapter 15). We then study math techniques needed to analyze phase-shifted voltages and currents.

When finished with this chapter, you should be able to:

- Determine phase differences between sinusoids and phasors and identify which is *leading* and which is *lagging*.
- Express phase difference with diagrams and equations.
- Convert cosine expressions to sine expressions, and vice versa.
- Explain the role of the *j* operator.
- Distinguish among real, imaginary, and complex numbers.
- Define two forms of complex numbers and convert from one to the other.
- Perform math operations on complex numbers.
- Describe phasors with complex numbers.
- Transform sinusoids from *time domain* to *phasor domain*, and vice versa.
- Add and subtract out-of-phase voltages and currents.

14.1 PHASE RELATIONS

Figure 14.1 shows sine waves of voltage and current that might be found in a typical circuit containing inductance and resistance and driven by a sinusoidal source.

Note that the X axis is marked in units of time. We understand this to mean that early events lie near the origin at low values of time; later events fall farther to the right. With this in mind, we see that the peak of v_L arrives earlier (it is to the left of) the peak of either i or v_R. v_L is said to be leading i and v_R. On the other hand, it could be said that v_R and i lag behind v_L.

Because the waves do not reach their peak values at the same time, they are said to be *out of phase*. The amount by which they are out of phase is represented by the Greek lowercase letter ϕ and called the *phase difference*. Although phase difference could be described as a time, it is expressed as an angle, usually in degrees rather than radians.

Figure 14.2 shows three important phase relations. The signals in part (a) are in phase; that is, they reach peak value at the same time, their phase difference is 0°. Part (b) shows one signal at its positive peak at the same instant the other is at its negative peak. Since 180° separates the peaks of a waveform, the signals are said to be 180° out of phase.

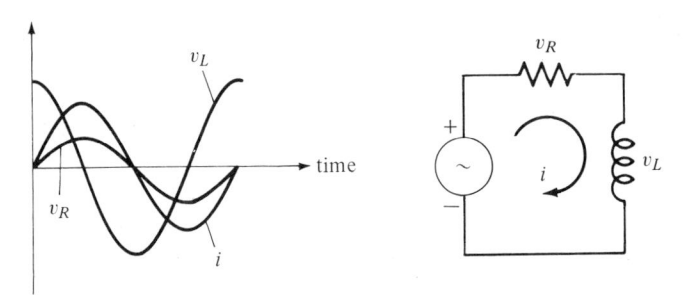

FIGURE 14.1 Phase Relations between Three Sinusoids.

14.1 PHASE RELATIONS

(a) in phase (0° out)

(b) 180° out

(c) 90° out (in quadrature)

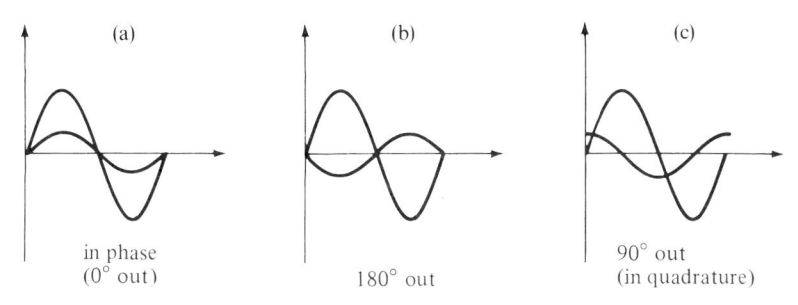

FIGURE 14.2 Three Important Phase Relations.

The signals in Figure 14.2, part (c), are 90° out of phase. Since 90° is a quarter rotation, the waves are also said to be *in quadrature*.

There are many applications for phase-shifted signals in electronics; for example:

- Right and left channels in FM stereo are separated by their phase difference.
- Antenna systems are made directional by driving them with phase-shifted signals.
- Color information is put on a TV signal with phase-shift methods.

Phase differences are important to such circuits as the phase-locked loop, the phase-shift oscillator, and the phase splitter. These differences are easily measured with an oscilloscope using methods discussed in Chapter 22.

There are two other ways to describe phase:

- As an angle in voltage and current time functions.
- With phasors.

In all methods:

| Phase is defined as lagging if current lags voltage and leading if current leads voltage.

Phase in Voltage and Current Time Functions

Consider a voltage $e(t) = 25\sin 100t$ volts, and current $i(t) = 2\sin(100t + 30°)$ amperes, from some circuit. Let us determine the effect of the 30° in the current equation.

e reaches its peak when the instantaneous value of its angle, $100t$ equals $\pi/2$ or 90°. Solving for t yields

$$t = \frac{\pi/2}{100} = \frac{\pi}{200}$$

i also peaks when its angle, $100t + 30°$, equals $\pi/2$ or 90°. However, the 30° means that $100t$ need only equal $\pi/3$, or 60°. Therefore,

$$100t = \frac{\pi}{3}$$

$$t = \frac{\pi}{300}$$

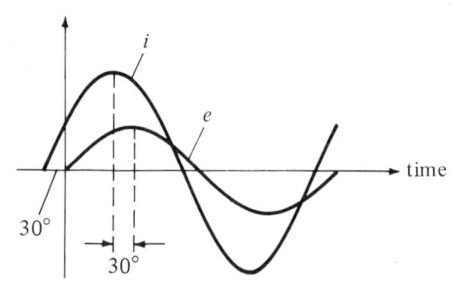

FIGURE 14.3 *i* leads *e* by 30°.

Since *t* is shorter for *i*, *i* must peak earlier than *e*. In other words, *i* leads *e*. The 30° has thus moved *i* ahead in time, giving *i* a headstart over *e* (see Figure 14.3). The phase difference, ϕ, is therefore 30°, leading.

By similar thinking,

$$i(t) = I_{peak} \sin(100t - 30°)$$

is moved back in time, causing *i* to lag *e* by 30°.
The conclusion:

- Adding angle to ωt makes a wave arrive earlier by moving it to the left.
- Subtracting angle from ωt delays a wave by moving it to the right.

When these extra angles are included in the general equations for *e(t)* and *i(t)*, we get

$$e(t) = E_{peak}(\sin \omega t \pm \alpha) \quad \text{volts} \quad (14.1)$$

and

$$i(t) = I_{peak} \sin(\omega t \pm \beta) \quad \text{ampere} \quad (14.2)$$

where α and β are the Greek letters alpha (*al-fuh*) and beta (*bay-duh*), respectively.

$$\text{phase angle, } \phi = \beta - \alpha \quad (14.3)$$

When ϕ is negative, phase is lagging. When ϕ is positive, phase is leading.

EXAMPLE 14.1

The voltage $e(t) = 115 \sin(1500t + 70°)$, causes the current, $i(t) = 3.7 \sin(1500t + 30°)$ amperes, to flow in a circuit.

a. What is the phase angle between *e* and *i*?

$$\phi = \beta - \alpha$$
$$= 30° - 70°$$
$$= -40°$$

b. Is this a leading or a lagging phase?

ϕ is negative; therefore, it is a lagging phase. Note that *e* reaches peak 40° before *i*; therefore, *i* lags *e*.

14.1 PHASE RELATIONS

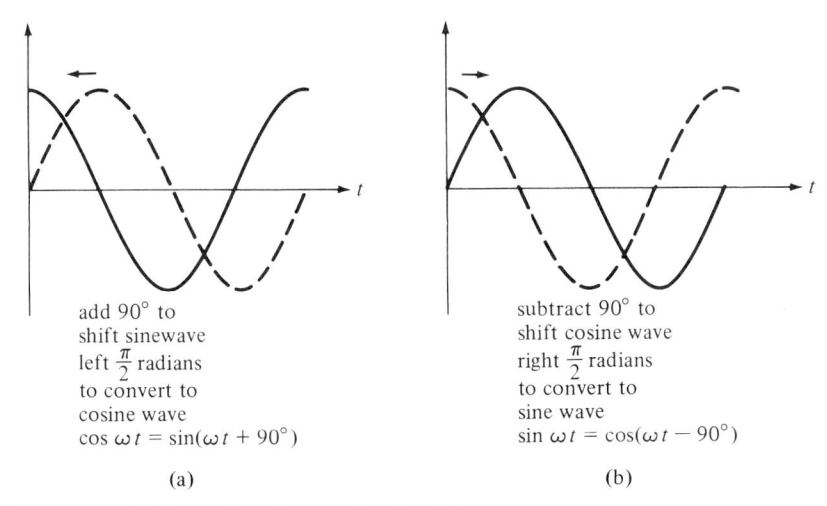

FIGURE 14.4 Converting Sine and Cosine Waves.

(a) add 90° to shift sinewave left $\frac{\pi}{2}$ radians to convert to cosine wave
$$\cos \omega t = \sin(\omega t + 90°)$$

(b) subtract 90° to shift cosine wave right $\frac{\pi}{2}$ radians to convert to sine wave
$$\sin \omega t = \cos(\omega t - 90°)$$

The following points should be made about Equations 14.1 and 14.2:

- The equations apply only to signals of the same frequency.
- Both functions should be in either sine or cosine form for easiest determination of ϕ. When they are not, use the following conversion formulas (see Figure 14.4):

A cosine wave is a sine wave shifted left 90°. Therefore, to change a cosine expression to sine form, add 90° to the angle and rewrite as a sine function.

$$\cos(\omega t + \alpha) = \sin(\omega t + \alpha + 90°) \tag{14.4}$$

A sine wave is a cosine wave shifted right 90°. Therefore, to change a sine expression to cosine form, subtract 90° from the angle and rewrite as a cosine function.

$$\sin(\omega t + \alpha) = \cos(\omega t + \alpha - 90°) \tag{14.5}$$

EXAMPLE 14.2

Convert each statement as indicated.

a. $15 \cos(4500t - 50°)$ to sine form.

$$15 \cos(4500t - 50°) = 15 \sin(4500t - 50° + 90°)$$
$$= 15 \sin(4500t + 40°)$$

b. $42 \sin(170t + 20°)$ to cosine form.

$$42 \sin(170t + 20°) = 42 \cos(170t + 20° - 90°)$$
$$= 42 \cos(170t - 70°)$$

EXAMPLE 14.3

What is the phase difference between the following pairs of sinusoids? Is this difference leading or lagging?

a. $e = 20 \sin 50t$ volts
$i = 7.3 \sin(50t - 20°)$ A

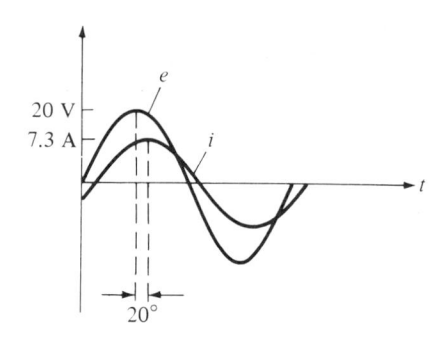

$\phi = -20° - 0° = -20°$.
i lags e by 20°.

b. $v = 79 \sin(420t - 60°)$ V
$i = 34 \sin(420t + 10°)$ A

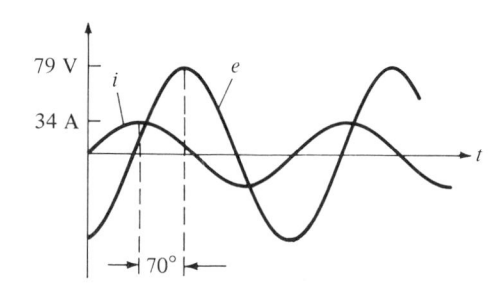

$\phi = 10° - (-60°) = 70°$.
i leads e by 70°.

c. $e = 20 \sin(50t - 20°)$ V
$i = 35 \cos(50t - 175°)$ A

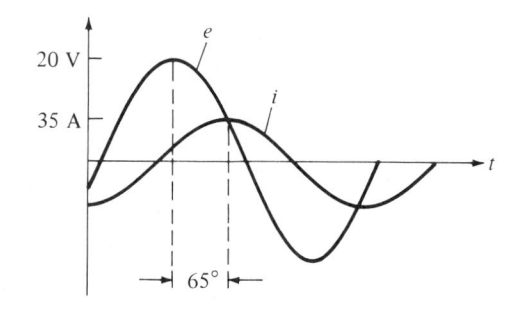

Converting the cosine wave:

$i = 35 \cos(50t - 175°) = 35 \sin(50t - 175° + 90°)$
$= 35 \sin(50t - 85°)$ A
$\phi = -85° - (-20°) = -65°$
i lags e by 65°. ∎

Showing Phase with Phasors

In Chapter 13 we learned that sine and cosine waves can by represented by rotating vectors called phasors. A sinusoid, $e(t) = 15 \sin 100t$ volts, for example, may be represented by a phasor spinning in the positive direction, counterclockwise, at an

14.1 PHASE RELATIONS

angular velocity of 100 rad/s. Phasor length could represent the sinusoid's peak value, but for convenience, it is set to the rms value (10.6 V in this case) because most ac calculations are done with effective values. When we represent sinusoids by phasors in this manner, we are said to be working in the *phasor domain*. To make phasors stand out, they are printed with an overbar, For example, $\bar{V}$.

Because all voltages and currents in a given circuit have the same frequency, the phasors representing them rotate at the same angular velocity. Some of these phasors, however, may be ahead of or behind the other because of phase differences (the effect is much like the fixed angle between revolving spokes). "Freeze"-action shots, called *phasor diagrams*, allow us to examine these differences. Figure 14.5 shows several

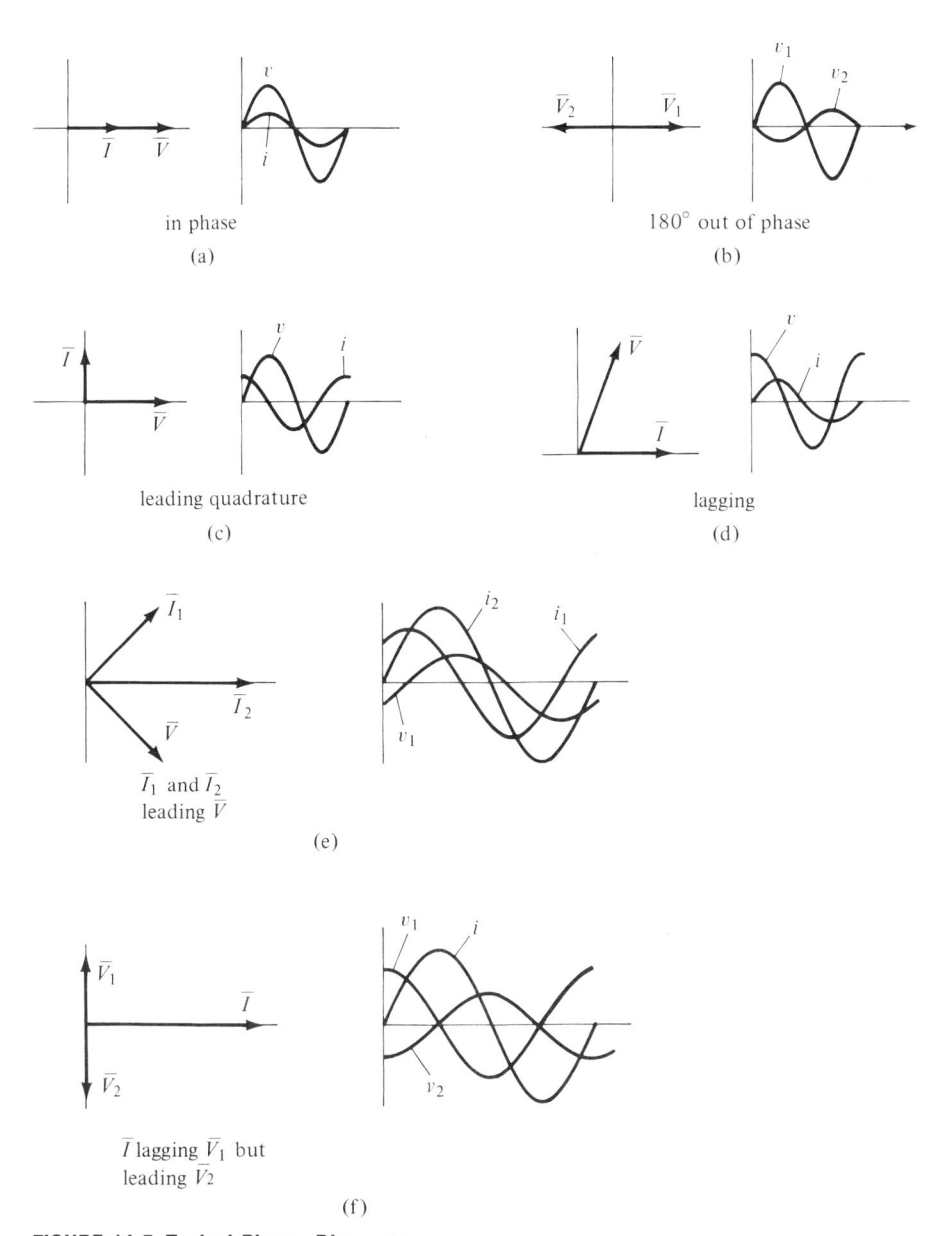

FIGURE 14.5 Typical Phasor Diagrams.

examples together with their equivalent sine-wave diagram for comparison. Pay close attention to the way the two types of diagrams say the same thing and how much easier it is to say it with phasors.

- Part (a): current and voltage in phase. The phasors are in the reference position since they lie on the $+X$ axis, the point from which angles are measured.
- Part (b): two voltage phasors 180° out of phase. $\bar{V}_1$, in reference position, has approximately twice the magnitude of $\bar{V}_2$.
- Part (c): current and voltage in leading quadrature.
- Part (d): lagging phase.
- Part (e): $\bar{I}_2$ in reference position is leading $\bar{V}$ and lagging $\bar{I}_1$. Magnitude of $\bar{I}_1$ approximately half that of $\bar{I}_2$.
- Part (f): $\bar{I}$ in leading quadrature with $\bar{V}_2$ and lagging quadrature with $\bar{V}_1$. $\bar{V}_1$ and $\bar{V}_2$ approximately equal in magnitude.

Before a phasor diagram can be drawn, all voltages and currents to be included in it must be in phasor form. This often means *converting* or *transforming sinusoids from the time domain to the phasor domain*. That transformation requires two pieces of information for each phasor:

- *Magnitude*. Take the rms value of the voltage or current being transformed.
- *Angle*. Take the phase shift from waveform diagrams or any angle added to or subtracted from ωt in an equation (first, convert any cosine waves to sinusoids using Equation 14.4).

EXAMPLE 14.4

Convert each statement from the time domain to the phasor domain.

Time domain   Phasor domain
a. $e = 100 \sin(50t + 30°)$ V $\Rightarrow \bar{E} = 70.7 \underline{/30°}$ V
                                       ↑   ↑
                                     rms angle
                              magnitude

Note that frequency information is not part of the phasor description of a sine wave; it can only be obtained from time-domain expressions.

b. $i = 35.4 \cos 1000t$ mA $\Rightarrow \bar{I} = 25 \underline{/90°}$ mA
c. $e = \sqrt{2}\, 65 \sin 377t$ kV $\Rightarrow \bar{E} = 65 \underline{/0°}$ kV ■

To convert in the opposite direction, from the phasor domain to time domain, frequency information must be supplied. Follow these steps:

- Convert rms magnitude to peak.
- Determine ω.
- Write the equation including phasor angle with ωt.

14.1 PHASE RELATIONS

EXAMPLE 14.5

Convert each statement from the phasor domain to the time domain at $\omega = 5$ krad/s.

Phasor domain	Time domain
a. $\bar{I} = 70.7\ \underline{/-30°}$ mA	$\Rightarrow i = 100 \sin(5000t - 30°)$ mA
b. $\bar{E} = 500\ \underline{/0°}$ V	$\Rightarrow e = 707 \sin 5000t$ V
c. $\bar{E} = 19\ \underline{/47°}\ \mu$V	$\Rightarrow e = 26.9 \sin(5000t + 47°)\ \mu$V

As stated earlier, phasor diagrams are actually "freeze" shots of phasors rotating ω radians per second. When we draw the diagram, we are choosing the position in which the phasors are frozen. That position can be anywhere; however, a convenient choice is to place the current phasor for a series circuit or the voltage phasor for a parallel circuit on the reference axis. Phasor diagrams need not be drawn with care; a simple sketch is sufficient.

EXAMPLE 14.6

Given: the following ac circuit:

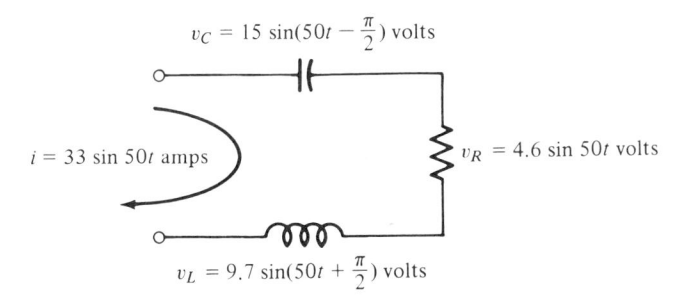

$v_C = 15 \sin(50t - \frac{\pi}{2})$ volts

$i = 33 \sin 50t$ amps

$v_R = 4.6 \sin 50t$ volts

$v_L = 9.7 \sin(50t + \frac{\pi}{2})$ volts

a. Convert each voltage and current to phasor form.

$$v_C = 15 \sin\left(50t - \frac{\pi}{2}\right) \Rightarrow 10.6\ \underline{/-\pi/2}\ \text{V}$$

$$v_R = 4.6 \sin 50t \Rightarrow 3.3\ \underline{/0°}\ \text{V}$$

$$v_L = 9.7 \sin\left(50t + \frac{\pi}{2}\right) \Rightarrow 6.9\ \underline{/\pi/2}\ \text{V}$$

$$i = 33 \sin 50t \Rightarrow 23.3\ \underline{/0°}\ \text{Am}$$

b. Sketch the phasor diagram with current in the reference position. For contrast, the sine-wave diagram is included.

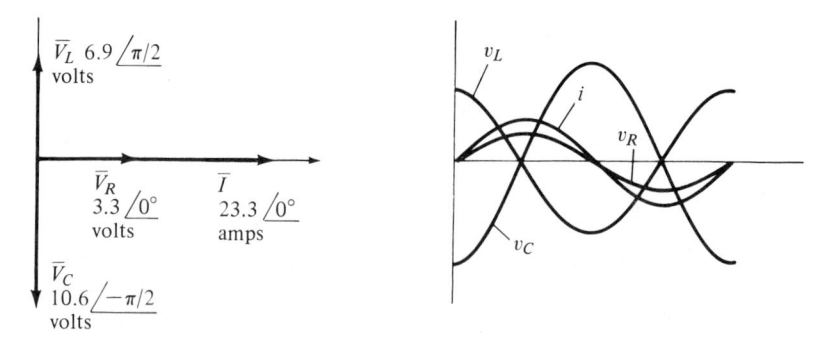

$\bar{V}_L$ 6.9 $\underline{/\pi/2}$ volts

$\bar{V}_R$ 3.3 $\underline{/0°}$ volts

$\bar{I}$ 23.3 $\underline{/0°}$ amps

$\bar{V}_C$ 10.6 $\underline{/-\pi/2}$ volts

c. Redraw the phasor diagram with V_c in the reference position.

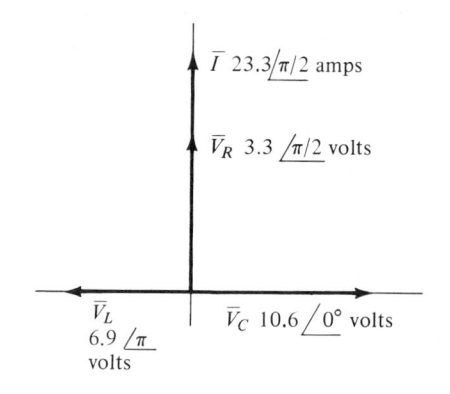

$\bar{I}$ 23.3 $\underline{/\pi/2}$ amps

$\bar{V}_R$ 3.3 $\underline{/\pi/2}$ volts

$\bar{V}_L$ 6.9 $\underline{/\pi}$ volts

$\bar{V}_C$ 10.6 $\underline{/0°}$ volts

Note how the phasors can be placed anywhere as long as the angles between them are unchanged. ∎

14.2 SINE-WAVE ADDITION: WHEN 2 + 2 DOES NOT EQUAL 4

The circuit in Figure 14.6 seems to show a violation of Kirchhoff's voltage law; the sum of the drops does not equal the source voltage. Why? What are we doing wrong?

Our error lies in adding the drops in the usual algebraic manner. Because of the phase difference, v_R peaks at the instant v_C is zero (see Figure 14.7, point c). Therefore, we must not add v_R peak to v_C peak and expect to get e_{source} peak.

The dashed-line sinusoid, e_{source} in Figure 14.7 shows one way to add the voltages on an instant-by-instant basis. For example, v_R is zero and v_C equals -2 at point a;

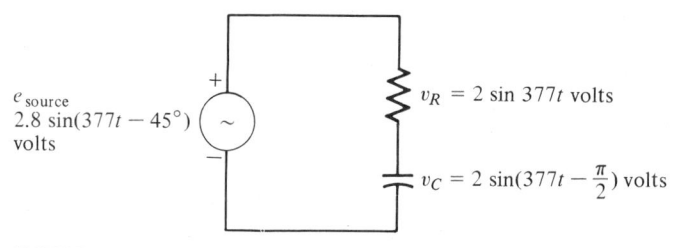

e_{source} 2.8 sin(377t − 45°) volts

$v_R = 2 \sin 377t$ volts

$v_C = 2 \sin(377t - \frac{\pi}{2})$ volts

FIGURE 14.6 Violation of KVL?

14.3 COMPLEX NUMBERS: THE j OPERATOR

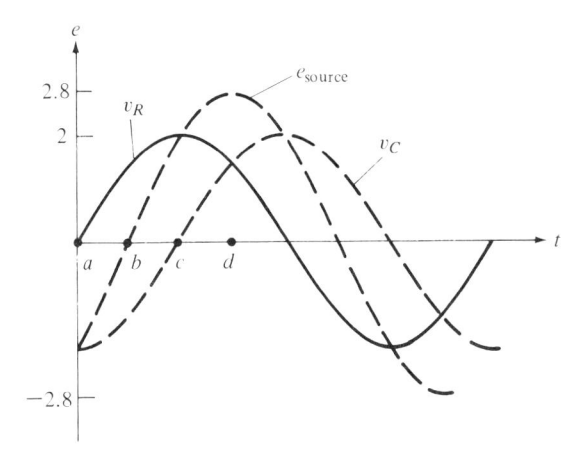

FIGURE 14.7 One Way to Add Sinusoids.

therefore, their total, e_{source}, equals -2. At b, the sinusoids have equal and opposite values, causing the total to be zero. And so on.

Instant d deserves special attention. It shows both voltages at 1.4. Their sum, 2.8, is the peak value of source voltage we did not get with algebraic addition.

Sinusoids may also be subtracted using the instant-by-instant method. But either operation is tricky and tedious, to say nothing of multiplying, dividing, squaring, and so on, this way.

At this point the question naturally arises: If the instant-by-instant approach is so difficult, how are we going to do the math required in ac analysis? The answer—sinusoids are first transformed to the phasor domain. Once that is done, ac analysis can be performed with mathematics only slightly more complicated than that used in dc analysis. When finished, we can, if we wish, transform our results back to the time domain or leave them in phasor form.

But first, we must learn to work with *complex numbers*.

14.3 COMPLEX NUMBERS: THE j OPERATOR

We know that $\sqrt{4} = \pm 2$, means that $(+2)^2$ or $(-2)^2$ equals 4. These numbers are called *real numbers*. They range from $-\infty$ through zero to $+\infty$. Should we want $\sqrt{-4}$, we run into a problem since no real number times itself equals a negative number. However, the rules of mathematics allow us to say:

$$\sqrt{-4} = \sqrt{-1 \times 4} = \sqrt{-1}\sqrt{4} = \pm\sqrt{-1} \times 2$$

Because this situation occurs often, the radical, $\sqrt{-1}$, is given a name:

$$\sqrt{-1} = j \tag{14.6}$$

Substituting, we have

$$\sqrt{-4} = \pm j2$$

The number $j2$ cannot be a real number. It is, instead, a member of the *imaginary number system*, which extends from $-j\infty$ through zero to $+j\infty$.

Do not be misled by the word "imaginary"; these numbers definitely exist. They just do not exist in the real number system.

| Any number preceded by j is an imaginary number.

Let us examine the role of the j operator more closely.

A vector, 10 units long, lies in the reference position in Figure 14.8 part (a). The same vector, multiplied by -1, points in the opposite direction in part (b). *Multiplication by -1 effectively rotated the vector $180°$ from the reference position to the negative X axis.*

If
$$-1 \times 10$$
rotates the vector $180°$, and
$$-1 = j \times j$$
we can say that
$$-1 \times 10 = j \times j \times 10$$

From this we conclude that if two multiplications by j rotate a vector $180°$, one multiplication must rotate it only $90°$ or $\pi/2$ radians:

| Multiplying a vector by j rotates that vector $90°$ in the positive, counterclockwise, direction.

It follows that multiplication by $-j$ rotates a vector in the negative, clockwise direction. Since j operates to rotate a vector while having no effect on the vector's magnitude, it is called the *j operator*. Figure 14.9 summarizes the j operations that rotate vector M around an entire circle.

Numbers that include both real and imaginary parts are called *complex numbers*. For example,
$$3 + j4$$
is a complex number with a *real part 3* and an *imaginary part j4*. A moment's thought shows that:

- Real numbers can be considered complex numbers with zero imaginary components.
- Imaginary numbers can be considered complex numbers with zero real components.
- Complex numbers are plotted on a complex plane defined by the usual vertical and horizontal axes (see Figure 14.10). However, the X axis is the *axis of the*

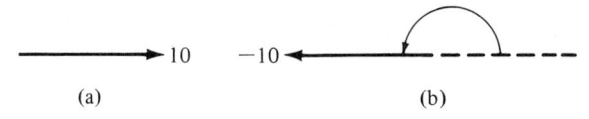

FIGURE 14.8 Multiplication by −1 Rotates a Vector π Radians.

14.4 COMPLEX NUMBERS AND PHASORS

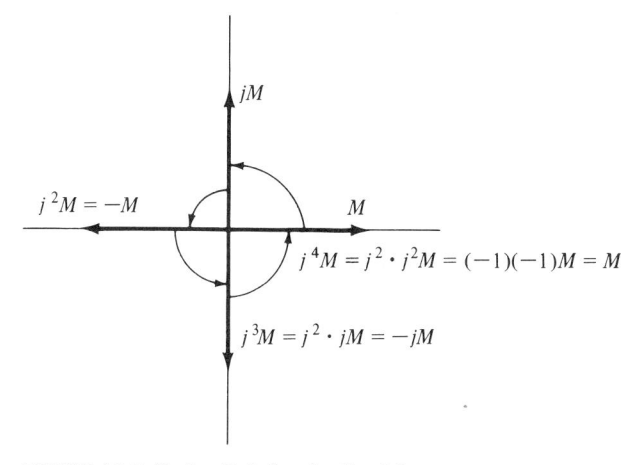

FIGURE 14.9 Vector Rotation by the *j* Operator.

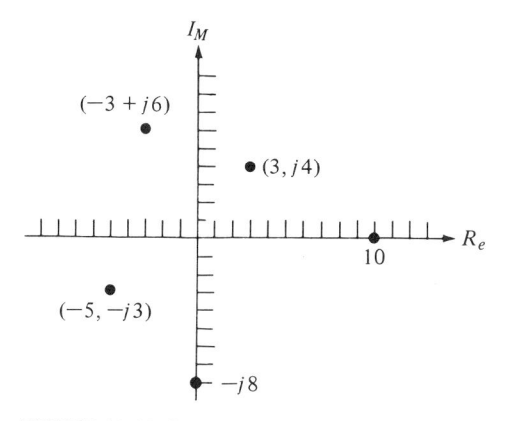

FIGURE 14.10 Complex Plane.

reals while the Y axis becomes the *axis of the imaginaries*. Plotted on the plane are the complex numbers $3 + j4$, $-j8$, $-3 + j6$, 10, and $-5 - j3$. Note carefully the component signs in each of the plane's four quadrants. To make complex numbers stand out, they are printed with an overbar, for example, $\bar{A}$.

14.4 COMPLEX NUMBERS AND PHASORS

The point $a + jb$ in Figure 14.11, part (a), is reached by going a units along the real axis, then up b units in the $+j$ direction. It can also be reached by a phasor M units long at angle θ [Figure 14.11, part (b)]. From this we see that

$$M \underline{/\theta} = a + jb \tag{14.7}$$

This tells us that *phasors can be described by complex numbers,* and vice versa.

The left side of Equation 14.7 represents the phasor $\bar{M} = |M|\underline{/\theta}$ in *polar form*. The overbar reminds us that a phasor is a complex number in polar form. $|M|$, a scalar, is phasor magnitude (the absolute value bars are omitted from now on, for convenience),

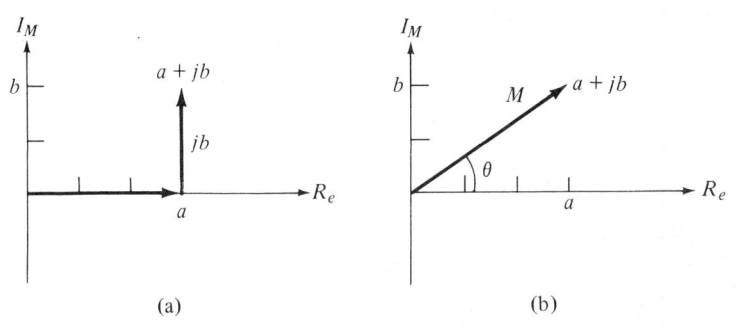

FIGURE 14.11 Phasor Description by Complex Number.

while θ is the angle measured from the reference position. The right side of equation 14.7, $a + jb$, represents the phasor in *rectangular* or *Cartesian* form (named after the French mathematician René Descartes).

Students of ac circuit analysis must be able to convert between polar and rectangular forms quickly and accurately.

Polar-to-Rectangular Conversion: $M \underline{/\Theta} \Rightarrow a + jb$

This is the same procedure as that used in Chapter 13 to resolve a vector into it X and Y components.

$$\text{real part, } a = M \cos \theta \tag{14.8}$$
$$\text{imaginary part, } jb = jM \sin \theta \tag{14.9}$$

Note: For convenience, store M and θ in memory for repeated use.

EXAMPLE 14.7

Convert each complex number from polar to rectangular form.

a. $5\underline{/53.1°}$.

$$\text{real part} = 5 \cos 53.1° = 5(0.6) = 3$$
$$\text{imaginary part} = 5 \sin 53.1° = 5(0.8) = 4$$
$$5\underline{/53.1°} \Rightarrow 3 + j4$$

b. $13\underline{/157.4°}$.

$$\text{real part} = 13 \cos 157.4° = 13(-0.92) = -12$$
$$\text{imaginary part} = 13 \sin 157.4° = 13(0.38) = 5$$
$$13\underline{/157.4°} \Rightarrow -12 + j5 \qquad \blacksquare$$

Rectangular-to-Polar Conversion: $a + jb \Rightarrow M\underline{/\Theta}$

Magnitude, M, is the hypotenuse of a right triangle. Its length is found by the Pythagorean theorem:

$$M = \sqrt{a^2 + b^2} \tag{14.10}$$

14.5 MATH OPERATIONS WITH COMPLEX NUMBERS

The angle is obtained from

$$\theta = \tan^{-1} \frac{b}{a} \tag{14.11}$$

Note: $\tan^{-1}$ is also called *arctan*.

EXAMPLE 14.8

Convert each complex number from Cartesian to polar form.

a. $7 - j4$.

$$M = \sqrt{7^2 + (-4)^2} = \sqrt{49 + 16} = \sqrt{65} = 8.1$$

$$\theta = \arctan -\frac{4}{7} = \tan^{-1}(-0.57) = -29.7°$$

$$7 - j4 \Rightarrow 8.1 \underline{/-29.7°}$$

b. $-4 - j3$.

$$M = \sqrt{(-4)^2 + (-3)^2} = \sqrt{16 + 9} = \sqrt{25} = 5$$

$$\phi = \tan^{-1} \frac{-3}{-4} = \arctan 0.75 = 36.9° + 180° = 216.9°$$

The 180° had to be added here so that the phasor lies in the third quadrant, where the component signs say it must. The calculator interprets arctan 0.75 as simply 36.9°.

$$-4 - j3 \Rightarrow 5 \underline{/216.9°} \quad ■$$

14.5 MATH OPERATIONS WITH COMPLEX NUMBERS

Before we can use complex numbers in circuit analysis, we must learn to do basic math operations with them.

Adding and Subtracting Complex Numbers

Polar Form

The only complex numbers in polar form that can be added/subtracted directly are those with the same angle, or angles differing by 180°. If they have different angles, convert to rectangular form and follow the procedure for rectangular numbers.

Rectangular Form

The general formats for addition and subtraction on $\bar{M}_1$ and $\bar{M}_2$, where

$$\bar{M}_1 = \pm A_1 \pm jB_1$$

and

$$\bar{M}_2 = \pm A_2 \pm jB_2$$

are:

For addition:

$$\bar{M}_1 + \bar{M}_2 = \underbrace{(\pm A_1 \pm A_2)}_{\text{summing reals}} + \underbrace{j(\pm B_1 \pm B_2)}_{\text{summing imaginaries}} \quad (14.12)$$

For subtraction:

$$\bar{M}_1 - \bar{M}_2 = \underbrace{[\pm A_1 - (\pm A_2)]}_{\text{subtracting reals}} + \underbrace{j[\pm B_1 - (\pm B_2)]}_{\text{subtracting imaginaries}} \quad (14.13)$$

EXAMPLE 14.9

Perform the indicated operations.

a. Add $5\,\underline{/35°}$ to $12\,\underline{/35°}$.

Polar forms have the same angle; therefore, the sum is $17\,\underline{/35°}$.

b. Add $9\,\underline{/50°}$ to $22\,\underline{/230°}$.

Angles differ by 180°. $22\,\underline{/230°}$ can be rewritten $22\,\underline{/50° + 180°}$. Recalling that rotation by 180° is the same as multiplication by -1, we have $9\,\underline{/50°} - 22\,\underline{/50°} = -13\,\underline{/50°}$ which can be converted to $13\,\underline{/230°}$, if desired.

c. Add: $(-12 + j20) + (5 + j13)$.

Add reals: $-12 + 5 = -7$
Add imaginaries: $20 + 13 = 33$
Result: $(-7 + j33)$

d. Add: $(5\,\underline{/40°}) + (-13\,\underline{/-20°})$.

Polar forms have different angles; therefore, convert to rectangular form:

$$5\,\underline{/40°} \Rightarrow 3.8 + j3.2$$
$$-13\,\underline{/-20°} \Rightarrow -12.2 + j4.5$$
$$\text{sum} = (3.8 - 12.2) + j(3.2 + 4.5)$$
$$= -8.4 + j7.7$$

This may be converted back to polar if desired.

e. Subtract: $(13 - j28) - (35\,\underline{/-65°})$.

Convert polar to Cartesian:

$$35\,\underline{/-65°} \Rightarrow 14.8 - j31.7$$
$$\text{difference} = (13 - 14.8) + j[-28 - (-31.7)]$$
$$= -1.8 + j3.7 \quad \blacksquare$$

Multiplying Complex Numbers

Complex numbers may be multiplied in either polar or rectangular form; however, the polar form is more convenient.

14.5 MATH OPERATIONS WITH COMPLEX NUMBERS

Polar Form

The general format for multiplying $\bar{M}_1$ and $\bar{M}_2$ in polar form is

$$(M_1 \underline{/\theta_1})(M_2 \underline{/\theta_2}) = M_1 M_2 \underline{/\theta_1 + \theta_2} \tag{14.14}$$

That is:

Multiply magnitudes and add angles.

Rectangular Form

We begin with

$$\bar{M}_1 \bar{M}_2 = (A_1 + jB_1) \cdot (A_2 + jB_2)$$

Applying the distributive law, we have

$$\bar{M}_1 \bar{M}_2 = A_1(A_2 + jB_2) + jB_1(A_2 + jB_2)$$
$$= A_1 A_2 + jA_1 B_2 + jB_1 A_2 + j^2 B_1 B_2$$

Combining reals and imaginaries, we get the general format for multiplication in rectangular form:

$$\bar{M}_1 \bar{M}_2 = (A_1 A_2 - B_1 B_2) + j(A_1 B_2 + A_2 B_1) \tag{14.15}$$

EXAMPLE 14.10

Multiply each of the following pairs of complex numbers.

a. $\bar{M}_1 = 5 \underline{/15°}$, $\bar{M}_2 = 12 \underline{/37°}$.

$$\bar{M}_1 \cdot \bar{M}_2 = 5(12) \underline{/15° + 37°}$$
$$= 60 \underline{/52°}$$

b. $\bar{P}_1 = 10 \underline{/22°}$, $\bar{P}_2 = -16 \underline{/-45°}$.

$$\bar{P}_1 \cdot \bar{P}_2 = (10 \underline{/22°})(-16 \underline{/-45°})$$
$$= [10(-16)] \underline{/22° + (-45°)}$$
$$= -160 \underline{/-23°}$$

c. $\bar{M}_1 = (3 + j4)$, $\bar{M}_2 = (2 + j11)$.

$$\bar{M}_1 \cdot \bar{M}_2 = (3 + j4)(2 + j11)$$
$$= 3(2 + j11) + j4(2 + j11)$$
$$= (6 + j33) + (j8 + j^2 44)$$

Combine reals and imaginaries:

$$= (6 - 44) + j(33 + 8)$$
$$= -38 + j41$$

d. $\bar{P}_1 = (6.3 \underline{/72°})$, $\bar{P}_2 = (-15 + j11)$.

One or the other must be changed to the opposite form. Let us change the polar to Cartesian:

$$6.3 \underline{/72°} = 6.3(\cos 72° + j \sin 72°) = 6.3(0.31 + j0.95)$$
$$= 2 + j6$$

Multiply:
$$(2 + j6)(-15 + j11) = 2(-15 + j11) + j6(-15 + j11)$$
$$= (-30 + j22) + (-j90 + j^2 66)$$
$$= (-30 - 66) + j(22 - 90)$$
$$= -96 - j68 \quad \blacksquare$$

Dividing Complex Numbers

Complex numbers can be divided in either polar or Cartesian form; however, as in multiplication, the polar form is more convenient

Polar Form

The general format for dividing $\bar{M}_1$ by $\bar{M}_2$ in polar form is

$$\frac{M_1 \underline{/\theta_1}}{M_2 \underline{/\theta_2}} = \frac{M_1}{M_2} \underline{/\theta_1 - \theta_2} \qquad (14.16)$$

That is, divide magnitudes and subtract the denominator angle from the numerator angle. (You will recall that the angles were added in multiplication.)

EXAMPLE 14.11

Divide each pair of phasors in polar form.

a. $\dfrac{15 \underline{/35°}}{5 \underline{/15°}} = \dfrac{15}{5} \underline{/35° - 15°} = 3 \underline{/20°}$

b. $\dfrac{-49 \underline{/-50°}}{11.5 \underline{/-28°}} = \dfrac{-49}{11.5} \underline{/(-50°) - (-28°)}$
$\qquad = -4.7 \underline{/-22°} \quad \blacksquare$

Rectangular Form

The rules of fractions tell us that

$$\frac{a + b}{c} = \frac{a}{c} + \frac{b}{c}$$

Similarly, a phasor in rectangular form, divided by a real number, c, is

$$\frac{a + jb}{c} = \frac{a}{c} + \frac{jb}{c} \qquad (14.17)$$

Division of complex numbers in rectangular form, such as

$$\frac{m + jn}{p + jq}$$

is done according to Equation 14.17. To do so, the denominator, $p + jq$, is first changed to a single real number by a process called *rationalizing the denominator*. How is this accomplished?

14.5 MATH OPERATIONS WITH COMPLEX NUMBERS

Of all the infinite number of complex numbers, just one can convert $p + jq$ to a real number. That special number is called the *complex conjugate* (pronounced *con-juh-git*) of $p + jq$. Represented by the symbol *, the conjugate is easily obtained; simply *switch the sign of the imaginary component*, that is:

$$\text{if } \bar{R} = p + jq, \quad \bar{R}^* = p - jq$$
$$\text{if } \bar{R} = p - jq, \quad \bar{R}^* = p + jq \quad (14.18)$$

For example, the conjugate of $3 + j10$ is $3 - j10$. For the conjugate of a number in polar form, reverse the sign of the angle; that is, if $\bar{R} = 5\,\underline{/30°}$, its conjugate is $\bar{R}^* = 5\,\underline{/-30°}$.

| A complex number becomes a real number when multiplied by its complex conjugate.

EXAMPLE 14.12

Divide the following complex numbers.

a. $\dfrac{3 + j4}{2 + j5}$.

The denominator is multiplied by its conjugate, $(2 - j5)$, to rationalize it, that is, to make it a real number. This means that the numerator must also be multiplied by $(2 - j5)$; otherwise, the value of the fraction is changed (in effect, we are multiplying the original problem by 1).

$$\frac{3 + j4}{2 + j5} \cdot \frac{2 - j5}{2 - j5} \quad \text{conjugate of denominator}$$

$$= \frac{6 - j15 + j8 - j^2 20}{4 - j10 + j10 + 25} \quad = +20$$

note that the imaginary terms drop out

$$= \frac{26 - j7}{29}$$

$$= \frac{26}{29} - j\frac{7}{29}$$

$$= 0.9 - j0.2$$

b. $\dfrac{-105 - j90}{-60 - j85}$.

$$\frac{-105 - j90}{-60 - j85} \cdot \frac{-60 + j85}{-60 + j85} \quad \text{conjugate of denominator}$$

$$= \frac{(6300 + 7650) + j(5400 - 8925)}{3600 - j5100 + j5100 + 7225}$$

$$= \frac{13{,}950 - j3525}{3600 + 7225}$$

$$= \frac{13{,}950}{10{,}825} - \frac{j3525}{10{,}825}$$

$$= 1.3 - j0.33 \qquad \blacksquare$$

We have now learned to transform voltage and current sinusoids to phasors and how to work with those phasors as complex numbers. Let us use these new skills to redo the sinusoidal addition problem of Figure 14.6 in Section 14.2. Notice how much easier the task is.

EXAMPLE 14.13

Find the source voltage in this RC series circuit redrawn from Figure 14.6.

Convert both sinusoids to phasors:

$$v_R = 2 \sin 377t \text{ volts}$$
$$v_C = 2 \sin(377t - \tfrac{\pi}{2}) \text{ volts}$$

$$v_R = 2\sin 377t \Rightarrow \bar{V}_r = 0.707(2\,\underline{/0°}) = 1.4\,\underline{/0°}\text{ V}$$
$$v_C = 2\sin(377t - 90°) \Rightarrow \bar{V}_c = 0.707(2\,\underline{/-90°}) = 1.4\,\underline{/-90°}\text{ V}$$

Note how phasors have rms magnitude and no frequency information. Converting phasors from polar to rectangular form

$$\bar{V}_R = 1.4 + j0$$
$$\bar{V}_C = 0 - j1.4 \text{ V}$$

By KVL,

$$\bar{E}_s = \bar{V}_R + \bar{V}_C$$
$$= (1.4 + j0) + (0 - j1.4)$$
$$= 1.4 - j1.4 \text{ V}$$

We could stop here, but the problem asks for source voltage in the time domain. Therefore, convert E_s to polar form:

$$\bar{E}_s = \sqrt{(1.4)^2 + (1.4)^2}\,\bigg/\,\tan^{-1}\frac{-1.4}{1.4}$$

$$2\,\underline{/-45°}\text{ V}$$

and then to the time domain:

$$e_s = \sqrt{2}\,2\sin(377t - 45°)$$

- convert to peak value ($\sqrt{2}$)
- ω obtained from v_r or v_c
- phase angle from polar form of $\bar{E}_s$

$$= 2.8\sin(377t - 45°)\quad\text{V}$$

∎

14.5 MATH OPERATIONS WITH COMPLEX NUMBERS

EXAMPLE 14.14

Find the current, i_L in this circuit:

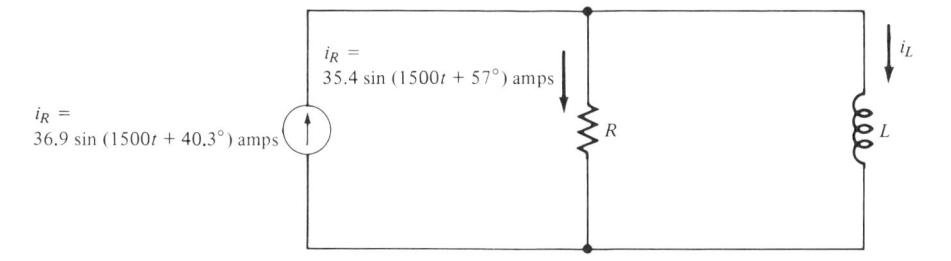

$i_R = 36.9 \sin(1500t + 40.3°)$ amps

$i_R = 35.4 \sin(1500t + 57°)$ amps

Convert i_S and i_R to the phasor domain:

$$i_S = 36.9 \sin(1500t + 40.3°) \Rightarrow \bar{I}_S = 26.1 \,\underline{/40.3°} \text{ A}$$
$$i_R = 35.4 \sin(1500t + 57°) \Rightarrow \bar{I}_R = 25 \,\underline{/57°} \text{ A}$$

By KCL,

$$\bar{I}_L = \bar{I}_S - \bar{I}_R = 26.1 \,\underline{/40.3°} - 25 \,\underline{/57°}$$
$$= (19.9 + j16.9) - (13.6 + j21) = 6.3 - j4.1$$
$$= 7.5 \,\underline{/-33°} \text{ A}$$

Transform to the time domain:

$$\bar{I}_L = 7.5 \,\underline{/-33°} \Rightarrow i_L = \sqrt{2} \, 7.5 \sin(1500t - 33°)$$
$$= 10.6 \sin(1500t - 33°) \text{ A} \quad \blacksquare$$

EXERCISES

Section 14.1

1. Answer the questions about this waveform.

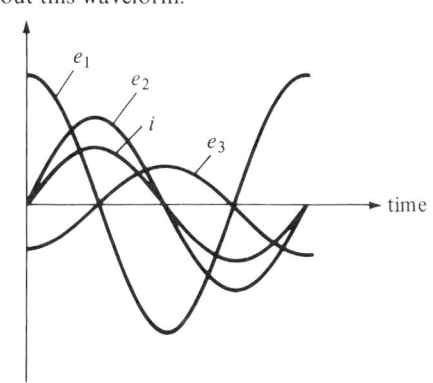

a. i is in phase with _____.
b. e_3 lags the voltage _____ by 90°.
c. e_1 leads the voltage _____ by $\pi/2$ rads.
d. i lags _____ by 90°.
e. e_3 and _____ are 180° out.
f. e_2 is in lagging quadrature with _____ and leading quadrature with _____.

2. Write the sinusoidal time functions of a 15-mA peak current and an 85-V peak voltage that is 15° ahead of the current. Angular velocity = 700 rad/s in both cases.

3. Write the time function for each waveform.

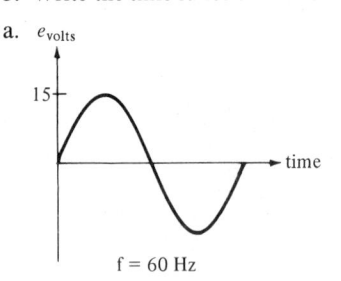

a. e_{volts}, peak 15, f = 60 Hz

b. $e_{millivolts}$, peak 13, shift $\frac{3\pi}{4}$, f = 25 Hz

4. Repeat Exercise 3 for these waveforms.

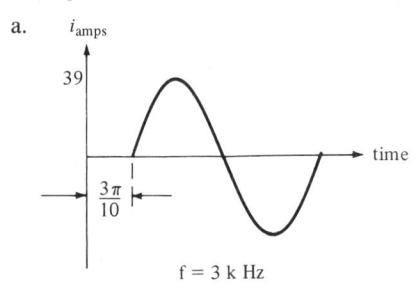

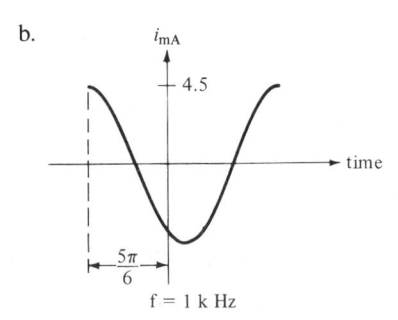

a. i_{amps}, peak 39, shift $\frac{3\pi}{10}$, f = 3 kHz

b. i_{mA}, peak 4.5, shift $\frac{5\pi}{6}$, f = 1 kHz

5. Find the phase angle for the following voltage–current combinations. State whether the phase is leading or lagging.
   a. $v = 78 \sin 500t$, $i = 31 \sin(500t - 30°)$
   b. $i = 3 \cos(377t + 30°)$, $e = 7 \cos(377t - 70°)$
   c. $e = 13 \sin(500t + 50°)$, $i = 9.7 \cos(500t - 25°)$
   d. $i = 70 \sin(1700t + 20°)$, $e = 175 \sin(540t - 83°)$

6. Repeat Exercise 5 for the following voltage–current combinations.
   a. $i = 6 \sin(700t + 30°)$ $e = 18 \sin(700t + 80°)$
   b. $e = 100 \cos(377t)$ $i = 10 \sin(377t - 20°)$
   c. $i = 28 \cos(1400t + 20°)$ $e = 14 \sin(1400t + 72°)$
   d. $e = 950 \sin(1700t + 10°)$ $i = 33 \cos(450t - 30°)$

7. Convert the time functions of Exercise 5 to the phasor domain.

8. Convert the time functions of Exercise 6 to the phasor domain.

9. Convert the phasors in each diagram to the time domain.

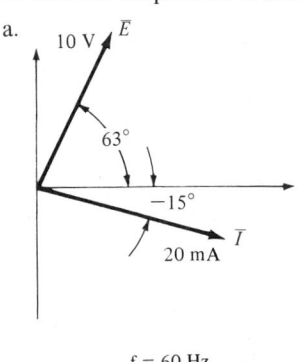

a. $\bar{E}$ = 10 V at 63°, $\bar{I}$ = 20 mA at −15°, f = 60 Hz

b. $\bar{E}_1$ = 33 V at 47°, $\bar{I}$ = 4 A at 20°, $\bar{E}_2$ = 50 V at −55°, f = 400 Hz

10. Describe the phase relation between the current and each voltage in the diagrams of Exercise 9.

EXERCISES

Section 14.3

11. What type of number is each of the following? a. 14; b. $j7$; c. $-j9$; d. -37; e. $4 - j7$.

12. Given: a vector, T, in the reference position. How much must it be rotated to move to each of these positions? a. $-jT$; b. $-T$; c. $j^3 T$; d. $j^4 T$.

Section 14.4

13. Convert the following complex numbers from polar to rectangular form: a. $12\angle 25°$; b. $7.6\angle -72°$; c. $15\angle -163°$; d. $9.2\angle 107°$.

14. Repeat Exercise 13 for these complex numbers: a. $15.8\angle 33°$; b. $25\angle -38°$; c. $12.6\angle 115°$; d. $-6.5\angle 129°$.

15. Convert the following complex numbers from rectangular to polar form: a. $5 + j12$; b. $-3 - j4$; c. $-2.5 + j6.1$; d. $11 - j7.9$.

16. Convert each of the following to polar form: a. $400 - j300$; b. $-1350 + j710$; c. $-12.7 - j4.7$; d. $240 + j185$.

Section 14.5

17. Add or subtract as indicated. Leave the answer in Cartesian form.
a. $(5 - j8.3) + (7 - j4.2)$
b. $(-3.8 + j14) - (-6.7 - j2.5)$
c. $(-11.5 - j13) + (-8.3 + j2.2)$
d. $(10 - j4.5) - (-9 + j9)$
e. $(9 - j14) + 12\angle 60°$
f. $33\angle 22° - 12\angle 47°$
g. $17\angle 85° + 5\angle 85°$
h. $-29\angle 18° - 7\angle -36°$
i. $-9.3\angle 27° + 5\angle 53.1° - 11.9\angle 24°$

18. Repeat Exercise 17 for these complex numbers.
a. $(12 + j17) + (8 + j3)$
b. $(-4 - j9) - (7 + j2)$
c. $(21 - j8) + (-3 - j11)$
d. $(5 + j17) - (13 - j8)$
e. $5\angle 45° + (7.3 + j10.2)$
f. $(-16 - j23) - 11\angle -28°$
g. $-7\angle -42° + 19\angle 72°$
h. $25\angle -65° + (3 - j9) - 17\angle -10°$
i. $4.5\angle 45° + 8.8\angle 55° - 3\angle 31° + 7.4\angle 67°$

19. Add or subtract these phase-shifted waves as indicated. Express the answer in the time domain.
a. $15\sin(100t + 40°) + 23\sin(100t - 29°)$
b. $37\sin(377t - 25°) - 12.5\cos(377t + 15°)$
c. $6\sin 50t + 14\sin(50t + 60°) - 19\sin(50t + 73°)$
d. $40\cos(1000t + 40°) - 27.3\sin(1000t - 10°) + 34.6\sin(1000t + 65°) - 23.5\cos(1000t - 12°)$

20. Repeat Exercise 19 for these phase-shifted waves.
a. $23\sin(2000t + 25°) + 30\sin(2000t + 57°)$
b. $11.5\sin 50t + 27.2\cos 50t$
c. $17.3\sin(90t + 15°) - 22\cos(90t - 30°) + 6.4\sin(90t + 72°)$
d. $31.4\sin 100t - 28.8\cos 100t - 19.7\sin(100t + 52°) + 83\cos(100t - 12°)$

21. Multiply the following using the polar form of complex number multiplication (Equation 14.14). Express each product in polar form with a positive magnitude.
a. $(5\angle 15°)(-2\angle 32°)$
b. $(2.5\angle 70°)(11.2\angle 37°)$
c. $(-12\angle -15°)(20\angle 37°)(3.7\angle -56°)$
d. $(-12\angle 24°)(-3 - j4)(4\angle 70°)$
e. $(2\angle 45°)(-17\angle 34°)(6\angle -10°)(-3\angle -53°)$
f. $j2.5(0.01\angle 5°)(1000\angle 40°)(535\angle 18°)$
g. $(5\angle -10°)(-j20)(15 - j10)(27\angle 63°)$
h. $(-2 + j7)(-13\angle 25°)(j9)(7)(20\angle 47°)$

22. Repeat Exercise 21 for the following.
a. $(7\angle 15°)(2\angle 50°)$
b. $(12.3\angle 27.5°)(8\angle -49°)$
c. $(42\angle -165°)(-j2.2)(7)$
d. $(3.6 + j2.4)(10)(17\angle 38°)$

e. $(25\underline{/25°})(-5-j7)(j3.7)$
f. $(-4.3-j11.2)(17\underline{/-22°})(3)(-8\underline{/61°})$
g. $(j4.7)(21+j7)(1.9\underline{/56°})(-7.6\underline{/21°})$
h. $(2+j6.5)(1.5\underline{/-75°})(-6.7-j21)(3.6\underline{/-18°})(j1.8)$

23. Multiply the following using the rectangular form of complex number multiplication (Equation 14.15). Express each product in rectangular form. ..
a. $(3+j4)(-2+j6)$
b. $(5+j11)(5-j11)$
c. $j8(-3-j6)$
d. $(-21-j13)(4+j7)$
e. $(-1+j2)(-3-j3)$
f. $(10+j5)(3-j6)$
g. $(-3-j4)(-3+j4)(5-j2)$
h. $(1+j2)(15+j1)(-2-j3)$

24. Repeat Exercise 23 for the following.
a. $(5-j8)(7-j4)$
b. $(-3.8+j1.4)(2-j1)$
c. $(-5-j4)(-5+j4)$
d. $(-3-j4)(5+j2.5)$
e. $(10-j4.5)(-9+j9)$
f. $j12(-1-j3)$
g. $(5+j2)(-6-j2)(-1+j4)$
h. $(2-j3)(4+j6)(-2-j7)(1+j5)$

25. Divide the following using the polar form of complex number division (Equation 14.16). Express each quotient in polar form with a positive magnitude.

a. $\dfrac{18\underline{/30°}}{6\underline{/12°}}$
b. $\dfrac{29\underline{/47°}}{13\underline{/-8°}}$
c. $\dfrac{(-9.3\underline{/37°})(2.6\underline{/10°})}{11.9\underline{/-8°}}$

d. $\dfrac{(-12\underline{/24°})(-17\underline{/34°})}{(-3\underline{/55°})(4\underline{/70°})}$
e. $\dfrac{25\underline{/82°}}{(5.6\underline{/-13°})(1.8\underline{/59°})}$
f. $\dfrac{(42\underline{/-165°})(-j2.2)}{7\underline{/35°}}$

g. $\dfrac{(2+j6.5)(4.5\underline{/-75°})}{(8\underline{/-49°})(2\underline{/55°})}$
h. $\dfrac{(j10)(15+j9)}{(22\underline{/35°})(-9\underline{/-33°})}$

26. Of all the infinite number of complex numbers, what is the only one that forms a real product when multiplying $5-j6$? Multiply by at least one other complex number to show that the product is not real.

27. Rationalize each denominator, then perform the division with numbers in rectangular form. Express each quotient in rectangular form.

a. $\dfrac{3+j15}{-2-j2}$
b. $\dfrac{5+j11}{-3-j6}$
c. $\dfrac{4-j7}{1-j1}$

d. $\dfrac{(-8.3+j2.2)(10+j4.5)}{5.7\underline{/31°}}$
e. $\dfrac{(16.8\underline{/72°})(-5.2-j8.3)}{1+j3.2}$
f. $\dfrac{5-j11}{(1+j2)(4+j7)}$

g. $\dfrac{(5+j3)(4\underline{/65°})}{(2-j2)(7-j1)}$
h. $\dfrac{(5.7\underline{/39°})(-j4)}{2(3+j7)}$

CHAPTER HIGHLIGHTS

Phase difference is a time difference between sinusoids. It is measured in units of angle.

- Lagging phase: current behind voltage.
- Leading phase: current ahead of voltage.

Summarizing ways to represent phase:
In sinusoidal form or time domain:
Mathematics:
The equations are

$$e(t) = E_{peak} \sin(\omega t \pm \alpha) \text{ volts}$$
$$i(t) = I_{peak} \sin(\omega t \pm \beta) \text{ amperes}$$

EXERCISES

Only form containing ω and frequency information. Adding angle to ωt shifts the sine wave left, to an earlier time; subtracting angle shifts the wave right, to a later time. If that angle is 90°, sine and cosine waves can be interchanged.

Get phase from

$$\phi = \beta - \alpha$$

Math operations are not done in this form.

Sine-wave diagrams:

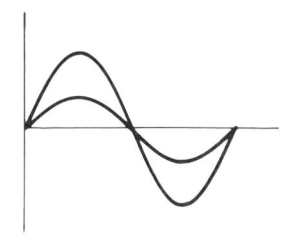

Difficult to sketch.
Phase information available but not easily obtained.

In phasor form or phasor domain:

Mathematics:

Phasors (rotating vectors) represent sinusoids by their vertical projection. n.

Magnitude given in rms value. Described by complex numbers in either of two forms:

$$\bar{E} = \overbrace{E\,\underline{/\alpha}}^{\text{polar form}} = \overbrace{E_{\text{real}} + jE_{\text{imaginary}}}^{\text{rectangular form}}$$
$$\bar{I} = I\,\underline{/\beta} = I_{\text{real}} + jI_{\text{imaginary}}$$

Form in which circuit analysis done.

Division and multiplication more conveniently done in polar form.
Addition and subtraction more conveniently done in rectangular form.

Phase easily determined.

$$\phi = \beta - \alpha$$

Phasor diagrams:

"Freeze"-action shots of phasors rotating counterclockwise, in radians per second:

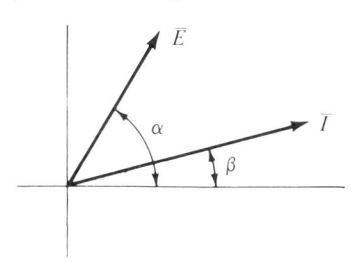

Much more easily drawn than waveforms.
Best for showing phase information.

Other Formulas to Remember

N is a real number. jN is an imaginary number.

Multiplication by j rotates a phasor counterclockwise (in the positive direction) 90°.
Multiplication by $-j$ rotates clockwise (in the negative direction) 90°.

If $\bar{A} = c + jd$, its complex conjugate is $\bar{A}^* = c - jd$.
Rectangular ⇔ polar conversions:

$$\bar{A} = |A|\underline{/\theta}$$
$$= \sqrt{c^2 + d^2}\,\underline{/\theta}$$
$$\theta = \arctan\frac{d}{c}$$
$$c = A\cos\theta$$
$$d = A\sin\theta$$

CHAPTER 15

Reactance and Impedance

Phase shift was no problem in dc circuits. Why now in ac circuits? What causes it?

This chapter answers these questions. It begins with a discussion of a math tool called the *derivative*. With it, we examine the voltage–current relations of inductance and capacitance to learn about *reactance*, a characteristic of L and C similar to resistance, yet totally different in important ways.

Resistance and reactance combine to form *impedance*, a measure of the total current-limiting ability of a circuit. As such, it replaces resistance in Ohm's law to give us Ohm's law for ac circuits.

The resistive portion of impedance dissipates energy as heat while the reactive part stores energy for later return to the source. As a result, power calculations involving impedance require new approaches that are covered in the last part of the chapter.

At the completion of this chapter, you should be able to:

- Explain what a derivative is.
- State the derivative of the sine function.
- Tell why pure inductance and capacitance cause phase shifts of 90°.
- Calculate inductive and capacitive reactance at any frequency.
- Define impedance and describe it with complex numbers.
- Use Ohm's law for ac circuits.
- Calculate the three powers—true, reactive and apparent—and sketch the power triangle for an ac circuit.
- Explain the importance of power factor.

15.1 THE DERIVATIVE OF A SINE WAVE

Voltage and current for pure inductance and capacitance are 90° out of phase (remember, "pure" means "containing no resistance"). Before we can understand why, we must study the *derivative* of a sine wave.

We have seen derivatives in earlier chapters but did not name them at the time. For example, in the equation for capacitive current from Chapter 10,

$$i = C\frac{dv}{dt}$$

dv/dt is the derivative of voltage with respect to time (the letter d, you will recall, means "small change in").

A waveform's derivative lets us determine how fast that waveform is changing at any point. High derivative values indicate rapid change, that is, waves climbing at steep angles with high slope [Figure 15.1, parts (a) and (e), for example]. Low derivative

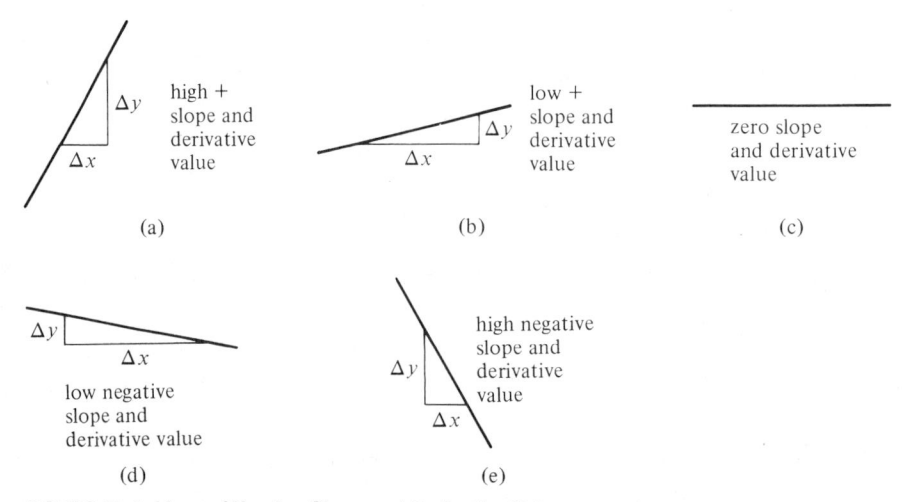

FIGURE 15.1 Lines of Varying Slope and Derivative Values.

15.1 THE DERIVATIVE OF A SINE WAVE

values indicate the slow change of waves climbing at shallow angles with low slope [parts (b) and (d)]. A derivative of zero indicates a line with zero slope and no vertical change, namely, a horizontal line [part (c)]. From this we see that derivative value and slope are equal at any point on a waveform. Let us consider next how a voltage sinusoid changes with time. By doing so, we will reveal the derivative of a sine wave.

At each of the points A through H on the sine wave of Figure 15.2, the rate at which the wave is rising or falling, the dv/dt, is measured. A dot (lowercase letter) indicates that rate by its distance above or below the X axis.

Point A, at the beginning and end of the wave, has the highest dv/dt, as indicated by the steepest positive slope. The a dots, therefore, are the highest dots above the X axis.

At point E the sinusoid has the same rate of change as at A. However, the slope is now negative; therefore, e lies as far below the axis as a lies above.

Points C and G have zero slope; hence dv/dt equals zero and c and g lie on the X axis.

At B and H we see positive slopes that are less than maximum but greater than zero. Their dots lie between the axis and the maximum position of a. D and F have the same rate of change as at B and H but in the negative direction. Their dots lie as far beneath the axis as b and h lie above it.

When all the points representing dv/dt values in Figure 15.2 are joined, the dashed cosine wave results. Its maximum and minimum values are $+\omega$ and $-\omega$, respectively; period and frequency are the same as those of the sine wave.

> Because it shows the instant-by-instant rates of change in a sine wave, a cosine wave is the derivative of a sine wave.

The exact statement for the derivative of a sine wave is

$$\frac{d[A \sin(\omega t + \theta)]}{dt} = \omega A \cos(\omega t + \theta) \tag{15.1}$$

where A is the peak amplitude.

The task of finding the derivative of a function is called *differentiation*. Derivatives and differentiation are studied in a branch of calculus called *differential calculus*.

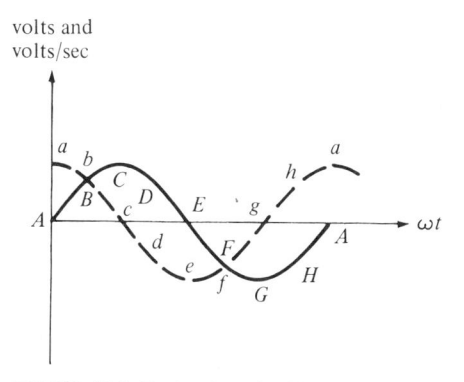

FIGURE 15.2 Derivative of a Sine Wave.

EXAMPLE 15.1

Given: the sinusoid:

$$e(t) = 75 \sin 100t \quad \text{V}$$

a. Find the derivative.

$$\frac{de}{dt} = 75(100) \cos 100t \quad \text{V/sec}$$

$$= 7500 \cos 100t \quad \text{V/sec}$$

Note the following about this derivative:

- Peak magnitude is ωE_{peak}.
- Frequency is the same as that of the sine wave; units are volts/second.

b. How fast is the sine wave changing at $t = 37$ms? Is the wave rising or falling?

Substituting the time into the derivative:

$$\frac{de}{dt} = 7500 \cos(100)(37 \times 10^{-3})$$

$$= 7500 \cos 3.7$$

$$= -6360.8 \quad \text{V}$$

Since the derivative is negative, the wave is falling.

c. Draw the waveforms of the voltage sinusoid and its derivative.

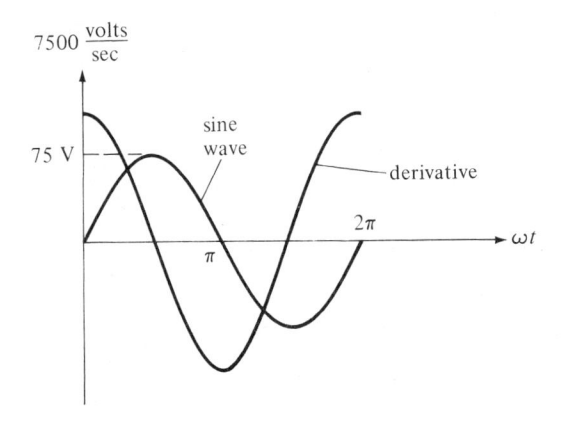

d. What is the derivative's new peak value if ω is increased to 275 rad/s?

The wave must change at a higher rate because of the increased radian frequency; therefore, de/dt must increase. The new peak value is

$$\omega(E_{peak}) = 275(75) = 20{,}625 \quad \text{V/s}$$

15.2 WHAT CAUSES PHASE SHIFT? REACTANCE

To understand the cause of phase shift, let us consider what happens when each of the pure circuit components, R, L, and C, is driven by sinusoidal voltage or current.

Resistance

A current

$$i_R = I_{peak} \sin \omega t \quad \text{amperes}$$

flowing through a resistance of R ohms, develops a voltage drop described by Ohm's law:

$$v_R = iR$$

across the resistance. Substitute for i:

$$v_R = I_{peak} R \sin \omega t \quad \text{volts}$$

But

$$I_{peak} R = V_{peak}$$

Therefore,

$$v_R = V_{peak} \sin \omega t \quad \text{volts}$$

Note carefully:

| There is no phase shift between the voltage across and current through resistance.

This is always the case.

Figure 15.3 shows the resistive phase relations.

If v_R and i_R are expressed as phasors and divided according to Ohm's law, we get

$$\frac{\bar{V}_R}{\bar{I}_R} = \frac{V\,\underline{/0°}}{I\,\underline{/0°}} = R\,\underline{/0°} = R \quad \text{ohms}$$

which shows resistance as a real number with magnitude R at an angle of $0°$ on the complex plane. By describing resistance this way, we gain an important advantage: The zero degree phase shift of resistance is handled automatically in our mathematics.

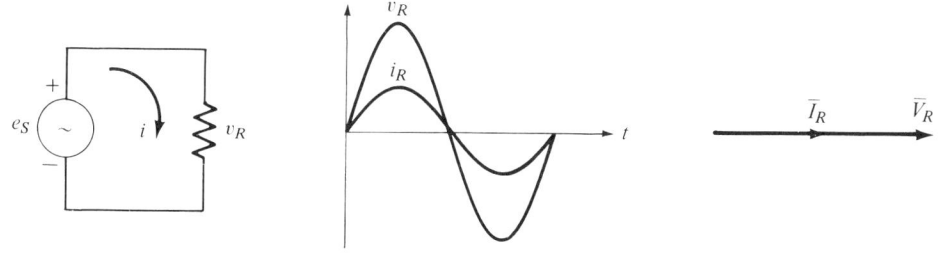

FIGURE 15.3 Resistive Phase Relations.

EXAMPLE 15.2

A current

$$i = 75 \sin 4000t \quad \text{A}$$

flows through a 2.5-Ω resistance.

a. Find the voltage drop across the resistance in phasor form.

Convert the current to a phasor:

$$i = 75 \sin \omega t \Rightarrow \bar{I} = 0.707(75 \underline{/0°})$$
$$\bar{I} = 53 \underline{/0°} \text{ A}$$
$$\bar{V}_R = \bar{I}R = 53 \underline{/0°} (2.5 \underline{/0°}) = 132.5 \underline{/0°} \text{ V}$$

Note how the resistive phase shift is handled automatically when calculations are done in phasor form.

b. Express the voltage drop as a time function.

$$V_{R_{\text{peak}}} = \sqrt{2}(132.5) = 185.5 \text{ V}$$
$$v_R = 185.5 \sin 4000t \quad \text{V}$$

c. Sketch the phasor and sine-wave diagrams of voltage and current.

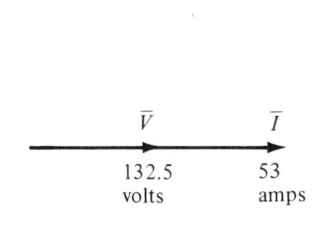

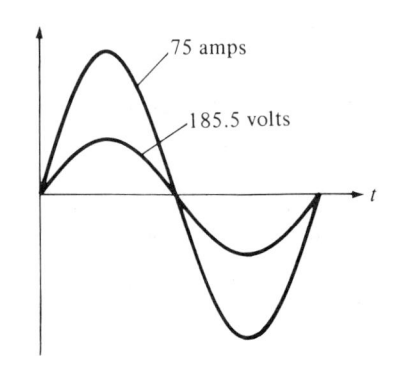

EXAMPLE 15.3

The voltage shown in the following phasor diagram is applied to a 15 k resistor.

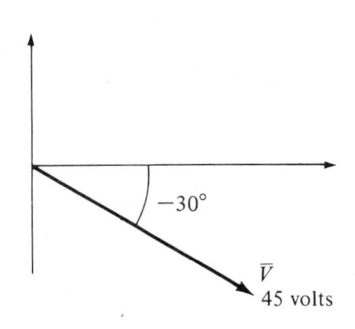

15.2 WHAT CAUSES PHASE SHIFT? REACTANCE

a. Find the current in phasor form.

$$\bar{I} = \frac{\bar{V}}{R} = \frac{45\,/\!-30°}{15\text{ k}\,/0°} = 3\,/\!-30°\text{ mA}$$

b. Sketch the phasor diagram of current and voltage.

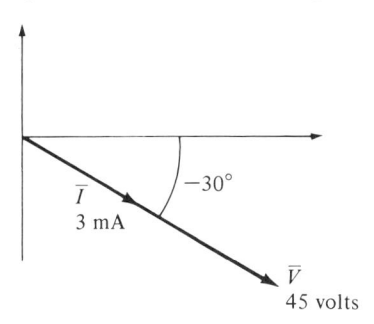

[Phasor diagram: $\bar{I}$ = 3 mA along horizontal axis pointing left-ish; $\bar{V}$ = 45 volts at $-30°$ below horizontal]

Inductance

Inductor phase shift was encountered during the discussion of *RL* transients in Chapter 12. We learned then that voltage across an inductance was maximum when current was minimum, and vice versa. Let us now investigate that phase shift for sinusoidal rather than exponential conditions.

The counter-EMF across a coil was described by Equation 12.2:

$$v_L = L\frac{di}{dt} \quad \text{volts} \tag{12.2}$$

We will now determine the nature of this voltage when the current is in sinusoidal form:

$$i_L = I_{\text{peak}} \sin \omega t \quad \text{amperes}$$

Substitute for *i* in Equation 12.2:

$$v_L = L\frac{di}{dt} = L\frac{d(I_{\text{peak}} \sin \omega t)}{dt}$$
$$= \omega L I_{\text{peak}} \cos \omega t \quad \text{volts} \tag{15.2}$$

But $\cos \omega t = \sin\left(\omega t + \frac{\pi}{2}\right)$; therefore,

$$v_L = \omega L I_{\text{peak}} \sin\left(\omega t + \frac{\pi}{2}\right) \quad \text{volts} \tag{15.3}$$

Note carefully that v_L is a sine wave and that it is shifted 90° ahead of i_L. In other words:

Current through pure inductance lags voltage across by $\pi/2$ radians or 90°. Inductance has a lagging phase angle.

Figure 15.4 shows inductive phase relations.
Let us consider Equation 15.3 on a units basis:

$$v(\text{volts}) = [\omega L(\text{unknown units})][I(\text{amperes})]\left[\sin\left(\omega t + \frac{\pi}{2}\right)(\text{no units})\right]$$

REACTANCE AND IMPEDANCE

$i_L = I_{peak} \sin \omega t$ amps

$v_L = \omega L \, I_{peak} \sin (\omega t + \frac{\pi}{2})$ amps

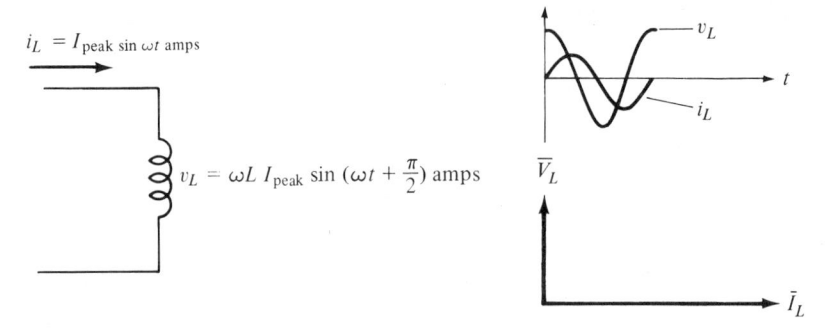

FIGURE 15.4 Inductive Phase Relations.

Solve for the units of ωL:

$$\omega L = \frac{\text{volts}}{\text{amperes}} \tag{15.4}$$

We know from Ohm's law that volts/amperes = ohms. Therefore, the units of ωL are ohms.

Do not make the mistake of thinking that ωL ohms are the same as ohms of resistance. They differ in several significant ways:

- ωL changes with frequency and causes phase shift; resistance does not.
- ωL ohms do not dissipate power as resistive ohms do. Energy is stored in an inductor's electromagnetic field during half of a sinusoid, and returned to the source during the other half. Power in inductance is studied later in this chapter.

Despite these important differences, ωL ohms can be used in Ohm's law to determine the reaction of inductance to voltage and current at any frequency. For this reason, ωL is called *inductive reactance*. It is represented by X_L (pronounced, "ex sub el"), and defined

$$X_L = \omega L = 2\pi f L \quad \text{ohms} \tag{15.5}$$

Note carefully:

| Inductive reactance, X_L, has units of ohms and is proportional to frequency.

EXAMPLE 15.4

Find the reactance of a 150-mH choke (inductor) at the following frequencies.

a. 60 Hz.

$$X_L = \omega L = 377(150 \times 10^{-3}) = 56.6 \, \Omega$$

b. 6 kHz.

$$X_L = 2\pi f L = 2\pi(6000)(150 \times 10^{-3}) = 5655 \, \Omega$$

15.2 WHAT CAUSES PHASE SHIFT? REACTANCE

c. 40 MHz.

$$X_L = 2\pi f L = 2\pi(40 \times 10^6)(150 \times 10^{-3})$$
$$= 37.7 \text{ M}$$

Pay careful attention to the increase in inductive reactance with frequency. ∎

EXAMPLE 15.5

What inductance is needed to provide an X_L of 750 Ω at f = 22 kHz?

$$X_L = 2\pi f L$$

Therefore,

$$L = \frac{X_L}{2\pi f} = \frac{750}{2\pi(22 \times 10^3)}$$
$$= 5.4 \text{ mH}$$ ∎

If v_L and i_L are expressed as phasors and divided according to Ohm's law, inductive reactance is seen to be represented by an imaginary number:

$$\frac{\bar{V}_L}{\bar{I}_L} = \frac{V_L \underline{/90°}}{I_L \underline{/0°}}$$
$$= X_L \underline{/90°} \qquad \text{polar form} \qquad (15.6)$$
$$= jX_L \qquad \text{rectangular form} \qquad (15.7)$$

By describing inductive reactance as an imaginary number, jX_L, we gain an important advantage: The 90° phase shift is handled automatically in our mathematics.

EXAMPLE 15.6

Given: an X_L = j6500 Ω. The current is

$$i_L = 17.3 \sin(10{,}400t - 45°) \quad \text{mA}$$

a. Find the inductor voltage in the phasor domain.

Convert i_L to $\bar{I}_L$:

$$\bar{I}_L = 0.707(17.3 \underline{/-45°}) \text{ mA}$$
$$= 12.2 \underline{/-45°} \text{ mA}$$
$$\bar{V}_L = \bar{I}_L X_L = (12.2 \underline{/-45°} \text{ mA})(6.5 \times 10^3 \underline{/90°})$$
$$= 79.3 \underline{/45°} \text{ V}$$

b. Find the inductor voltage in the time domain.

Convert to the time domain:

$$v_L = \sqrt{2}\,[79.3 \sin(10{,}400t + 45°)]$$
$$= 111 \sin(10{,}400t + 45°) \quad \text{V}$$

c. Draw the sine wave and phasor diagrams of voltage and current.

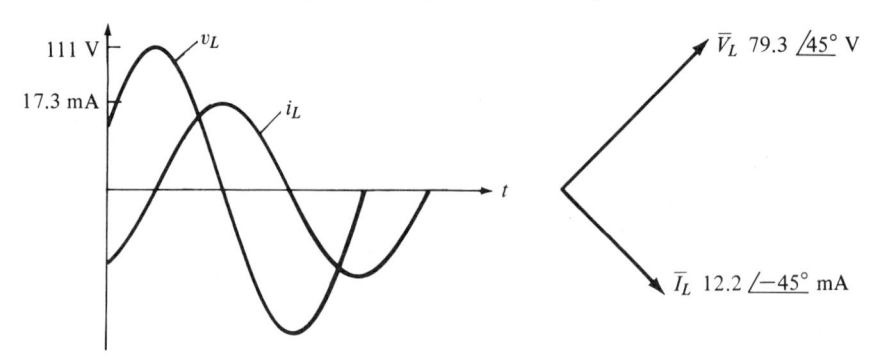

111 V, v_L
17.3 mA, i_L

$\bar{V}_L$ 79.3 $\underline{/45°}$ V

$\bar{I}_L$ 12.2 $\underline{/-45°}$ mA

| Note how both diagrams show current lagging voltage by 90°.

EXAMPLE 15.7

Find the inductance of this inductor:

$i_L = 2.8 \sin(20 \times 10^6 t - 50°)$ mA

$v_L = 37.5 \sin(20 \times 10^6 t + 40°)$ volts

Convert current and voltage to phasors:

$\bar{I}_L = 0.707(2.8 \underline{/-50°})$ mA $= 2 \underline{/-50°}$ mA
$\bar{V}_L = 0.707(37.5 \underline{/40°}) = 26.5 \underline{/40°}$ V

Solve for X_L:

$$X_L = \frac{\bar{V}_L}{\bar{I}_L} = \frac{26.5 \underline{/40°}}{2 \underline{/-50°} \text{ mA}}$$
$$= 13.3 \times 10^3 \underline{/40° - (-50°)}$$
$$= 13.3 \times 10^3 \underline{/90°}$$
$$= j13.3 \times 10^3 \, \Omega$$

Solve for inductance:

$$X_L = \omega L$$

Therefore,

$$L = \frac{X_L}{\omega} = \frac{13.3 \times 10^3}{20 \times 10^6}$$
$$= 665 \, \mu\text{H}$$

Capacitance

Capacitive phase shift was encountered in Chapter 10 during the discussion of *RC* transients. We learned then that voltage across a capacitor was minimum when the current was maximum, and vice versa. Let us now investigate that phase shift for sinusoidal rather than exponential conditions.

15.2 WHAT CAUSES PHASE SHIFT? REACTANCE

Capacitor current flows only during charge or discharge, that is, when the voltage across the capacitance is increasing or decreasing. The faster the voltage changes, the greater the current. This relation was shown by Equation 10.14:

$$i_C = C\frac{dv}{dt} \quad \text{amperes} \tag{10.14}$$

For

$$v_C = V_{peak} \sin \omega t$$

the equation becomes

$$i_C = C\frac{d(V_{peak} \sin \omega t)}{dt}$$
$$= \omega C V_{peak} \cos \omega t \quad \text{volts}$$

But $\cos \omega t = \sin(\omega t + \pi/2)$. Therefore,

$$i_C = \omega C V_{peak} \sin\left(\omega t + \frac{\pi}{2}\right) \quad \text{amperes} \tag{15.8}$$

Note carefully that i_C is a sine wave and that it is shifted $\pi/2$ radians ahead of v_C. In other words:

| Current through pure capacitance leads voltage across by $\pi/2$ radians or 90°. Capacitance has a leading phase angle.

Figure 15.5 shows capacitive phase relations.
Let us consider Equation 15.8 on a units basis:

$$i(\text{amperes}) = [\omega C(\text{unknown units})][V(\text{volts})]\left[\sin\left(\omega t + \frac{\pi}{2}\right)(\text{no units})\right]$$

Solve for the units of ωC:

$$\omega C = \frac{\text{amperes}}{\text{volts}}$$

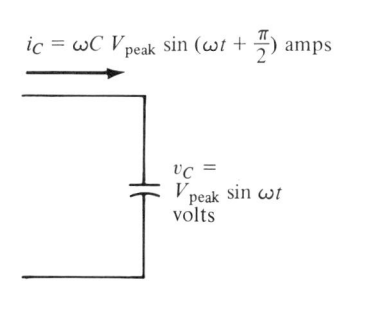

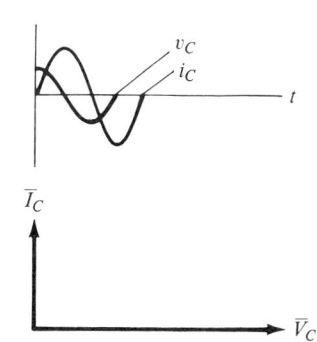

FIGURE 15.5 Capacitive Phase Relations.

Reciprocate both sides:

$$\frac{1}{\omega C} = \frac{\text{volts}}{\text{amperes}} = \text{ohms}$$

$1/\omega C$ is called *capacitive reactance*. It is represented by X_C ("ex sub c"):

$$X_C = \frac{1}{\omega C} = \frac{1}{2\pi f C} \quad \text{ohms} \qquad (15.9)$$

Note carefully:

Capacitive reactance, X_C, has units of ohms and is inversely proportional to frequency (the opposite of X_L).

Like those of X_L, the ohms of X_C differ from resistive ohms in several significant ways:

- $\frac{1}{\omega C}$ changes inversely with frequency and causes phase shift; resistance does not.
- $\frac{1}{\omega C}$ ohms do not dissipate power as resistive ohms do. Energy is stored in a capacitor's electrostatic field during half of a voltage sinusoid, and returned to the source during the other half. Power in capacitors is studied later in this chapter.

EXAMPLE 15.8

Find the reactance of a 0.015-μF capacitor at the following frequencies.

a. 60 Hz.

$$X_C = \frac{1}{\omega C} = \frac{1}{377(1.5 \times 10^{-8})}$$
$$= 177 \times 10^3 \; \Omega$$

b. 600 kHz.

$$X_C = \frac{1}{2\pi f C} = \frac{1}{2\pi(6 \times 10^5)1.5 \times 10^{-8}}$$
$$= 17.7 \; \Omega$$

c. 150 MHz. $X_C = \frac{1}{2\pi f C} = \frac{1}{6.28(1.5 \times 10^8)(1.5 \times 10^{-8})}$
$$= 71 \times 10^{-3} \; \Omega$$

Pay careful attention to the way that capacitive reactance drops with increasing frequency. Contrast this to inductive reactance. ■

15.2 WHAT CAUSES PHASE SHIFT? REACTANCE

EXAMPLE 15.9

What capacitance is needed to provide an X_c of 10,650 Ω at a frequency of 7.5 kHz? Express the answer in picofarads.

$$X_C = \frac{1}{2\pi f C}$$

Therefore,

$$C = \frac{1}{2\pi X_C f} = \frac{1}{6.28(10.65 \times 10^3)(7.5 \times 10^3)}$$

$$= 2 \times 10^{-9} \text{ farads} \frac{10^{12} \text{ pF}}{\text{farad}}$$

$$= 2000 \text{ pF}$$ ∎

If v_C and i_C are expressed as phasors and divided according to Ohm's law, capacitive reactance is seen to be represented by an imaginary number:

$$\frac{\bar{V}_C}{\bar{I}_C} = \frac{V_C \underline{/0°}}{I_C \underline{/90°}} = X_C \underline{/-90°} \qquad \text{polar form} \qquad (15.10)$$

$$= -jX_C \qquad \text{rectangular form} \qquad (15.11)$$

By describing capacitive reactance as an imaginary number, we gain an important advantage: the 90° phase shift is handled automatically in our mathematics.

EXAMPLE 15.10

A voltage,

$$v_C = 100 \sin(720t - 90°) \text{ V}$$

is applied across a 5-μF capacitor.

a. Find I_C in the time domain.

Find X_C:

$$X_C = -j\frac{1}{\omega C} = -j\frac{1}{720(5 \times 10^{-6})}$$

$$= -j278 \text{ Ω}$$

Write the time function:

$$i_C = \frac{v_C}{X_C} = \frac{100 \sin(720t - 90°)}{278 \underline{/-90°}}$$

$$= 360 \sin[720t - 90° - (-90°)] \text{ mA}$$

$$= 360 \sin 720t \quad \text{mA}$$

b. Find I_C in the phasor domain.

Convert to phasor form:
$$\bar{I}_C = 0.707(360\,\underline{/0°})$$
$$= 255\,\underline{/0°}\text{ mA}$$

c. Draw the sine wave and phasor diagrams of voltage and current.

Convert v_C to phasor form:
$$\bar{V}_C = 0.707(100\,\underline{/-90°})$$
$$= 70.7\,\underline{/-90°}\text{ V}$$

Draw the sine-wave and phasor diagrams:

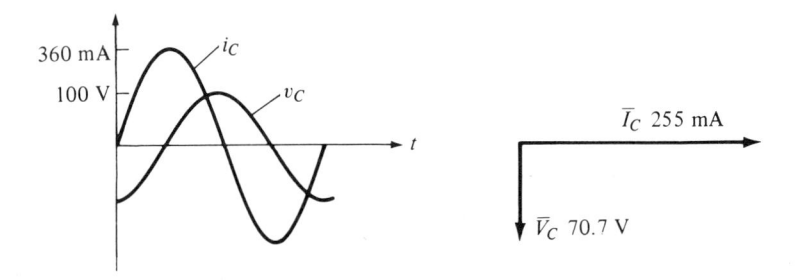

Note how both diagrams show current leading voltage by 90°.

EXAMPLE 15.11

A voltage,
$$\bar{V}_C = 131\,\underline{/27°}\text{ V}$$

causes a current,
$$\bar{I}_C = 239\,\underline{/117°}\text{ mA}$$

through a 0.22-μF capacitor. What is the frequency of the sinusoid?

$$\frac{\bar{V}_C}{\bar{I}_C} = \frac{131\,\underline{/27°}}{239\,\underline{/117°}\text{ mA}} = 548\,\underline{/-90°}\,\Omega$$

$$X_C = \frac{1}{2\pi f C}$$

Therefore,
$$f = \frac{1}{2\pi C X_C} = \frac{1}{6.28(22 \times 10^{-8})548}$$
$$= 1320\text{ Hz}$$

A good friend of electronics students for many years is ELI, the ICE man. He reminds us of inductive and capacitive phase relations by:

E across L leads I, I through C leads E.

Table 15.1 summarizes phase relations for R, X_C, and X_L.

15.3 IMPEDANCE

TABLE 15.1 Phase Summary for R, X_C, and X_L

Capacitive Reactance, X_c	Resistance, R	Inductive Reactance, X_L
$\bar{I}_C$ leads $\bar{V}_C$ by 90° or $\frac{\pi}{2}$ radians	$\bar{I}_R$ and $\bar{V}_R$ in phase	$\bar{I}_L$ lags $\bar{V}_L$ by 90° or $\frac{\pi}{2}$ radians
ICE leading quadrature		ELI lagging quadrature
$X_C = \|X_C\| \underline{/-90°} = -jX_C \, \Omega$	$R = R \underline{/0°} \, \Omega$	$X_L = \|X_L\| \underline{/90°} = jX_L$
X_C inversely proportional to frequency	R not affected by frequency	X_L directly proportional to frequency
$v_C = V_{peak} \sin \omega t$ V $i_C = I_{peak} \sin(\omega t + \frac{\pi}{2})$ A	$v_R = V_{peak} \sin \omega t$ V $i_R = I_{peak} \sin \omega t$ A	$v_L = V_{peak} \sin \omega t$ V $i_L = I_{peak} \sin(\omega t - \frac{\pi}{2})$ A

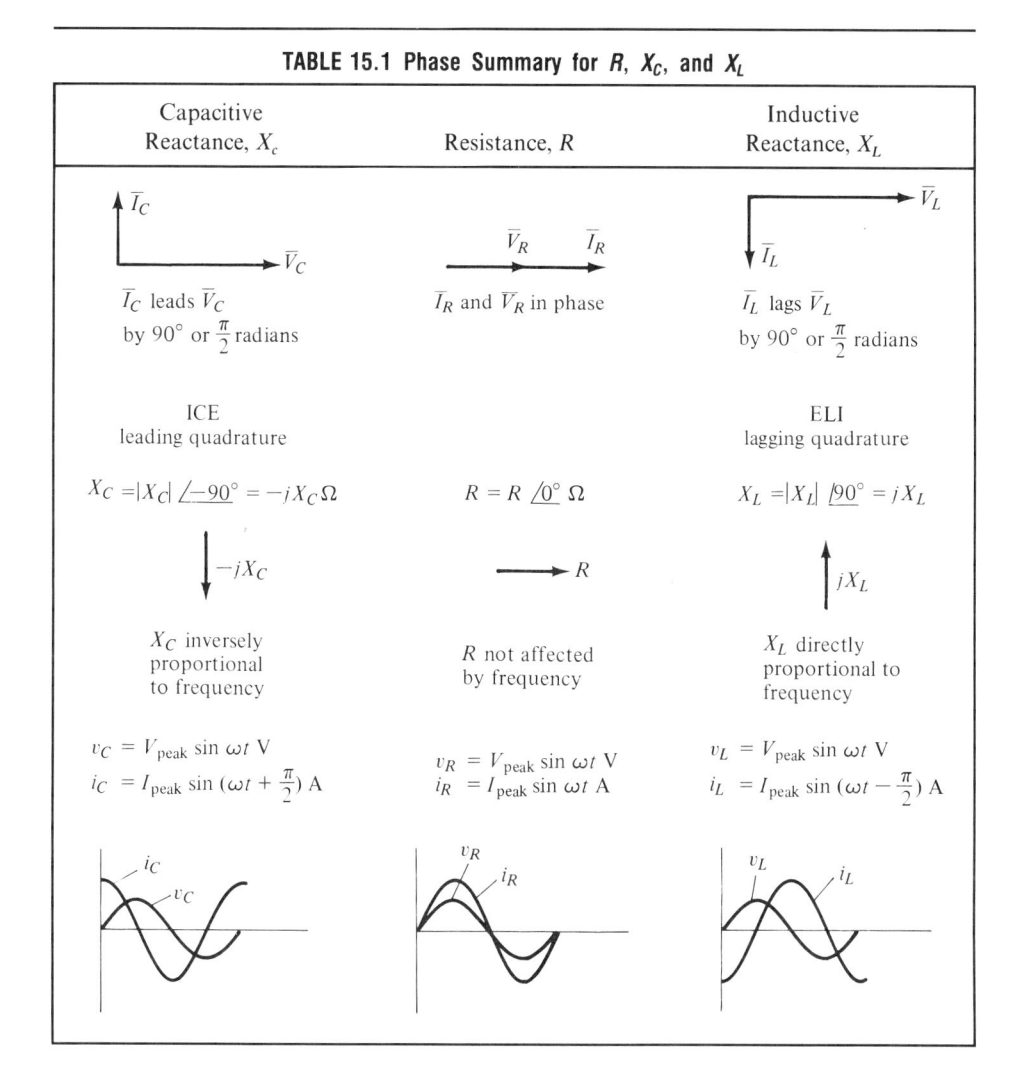

When resistance and reactance are connected, their phase characteristics combine to form a total circuit phase shift. An *RL* combination, for instance, has neither the in-phase characteristic of resistance nor the quadrature characteristic of inductance. Current through the combination lags applied voltage by an angle that depends on the relative magnitudes of R and X. If reactance is much greater than resistance, the phase angle falls nearer 90°. On the other hand, if resistance is the larger component, the angle lies closer to 0°.

Resistive and reactive ohms combine to form a circuit's total ability to limit or impede current flow. As we know, this ability cannot be found simply by adding R and X. They are represented by complex numbers and must be combined by the rules of complex algebra.

Figure 15.6 is a plot of the polar representations of resistance and inductive reactance for an *RL* circuit.

In Figure 15.7 the line representing X_L has been moved so that its tail is at the head (arrowhead) of the resistance. Note the third polar line added to the figure. It extends from the tail of R to the head of X_L. The three lines now form a right triangle with this new line as the hypotenuse. Figure 15.8 shows the equivalent triangle for an *RC* circuit.

The hypotenuse in the right triangles of Figures 15.7 and 15.8 is the complex sum of resistance and reactance. It is a measure of the total current-limiting ability of the resistor/reactance combination, and is called the circuit's *impedance*.

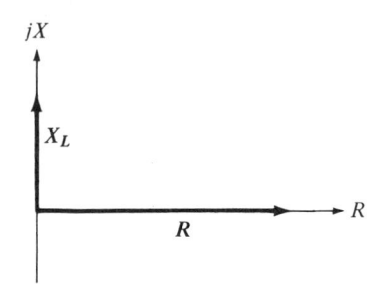

FIGURE 15.6 Impedance Diagram for an *RL* Circuit.

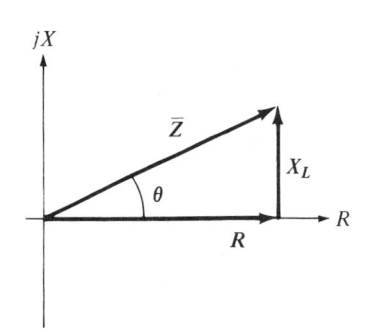

FIGURE 15.7 Reactance Moved to the Head of Resistance.

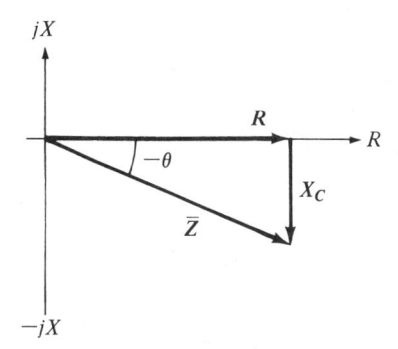

FIGURE 15.8 Impedance Diagram for an *RC* Circuit.

15.3 IMPEDANCE

TABLE 15.2 Impedance of Pure *R* and *X*

Resistance	Inductive reactance	Capacitive reactance
$\bar{Z} = R$	$\bar{Z} = X_L$	$\bar{Z} = X_C$
$= R\,\underline{/0°}$	$= X_L\,\underline{/90°}$	$= X_C\,\underline{/-90°}$
$= R + j0$	$= 0 + jX_L$	$= 0 - jX_C$

Represented by the letter Z, impedance is described by a complex number whose polar form is

$$\bar{Z} = Z\,\underline{/\pm \theta} \quad \text{ohms} \tag{15.12}$$

$$\text{where } Z = \sqrt{R^2 + X^2}$$
$$\theta = \tan^{-1}(\pm X/R)$$

Impedance in rectangular form is

$$\bar{Z} = R \pm jX \quad \text{ohms} \tag{15.13}$$

A moment's thought shows that resistance and the reactances are actually special cases of impedance in which one component is zero. Table 15.2 shows the three possibilities in polar and rectangular form.

EXAMPLE 15.12

Find the magnitude and angle of the impedance:

$$3 + j4\ \Omega$$

(In other words, convert to polar form.) Draw the impedance triangle.

$$Z = \sqrt{3^2 + 4^2} = \sqrt{9 + 16} = \sqrt{25} = 5\ \Omega$$

$$\theta = \arctan\frac{X}{R} = \arctan\frac{4}{3} = 53.1°$$

The sides of the impedance triangle are the three polar numbers representing impedance, resistance, and reactance.

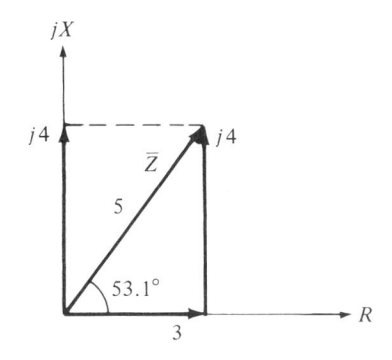

EXAMPLE 15.13

The frequency of the signal applied to the impedance of Example 15.12 is doubled. Find the new impedance in polar and rectangular form. Draw the new impedance triangle.

Resistance is unchanged but the reactive component doubles to $j8$ because this is X_L ($+j$ sign), which is directly proportional to frequency. The new impedance is $3 + j8$. Convert to polar form:

$$Z = \sqrt{3^2 + 8^2} = \sqrt{9 + 64} = \sqrt{73} = 8.5 \ \Omega$$

$$\theta = \tan^{-1}\frac{X}{R} = \tan^{-1}\frac{8}{3} = 69.4°$$

Note how the increased inductive reactance causes a longer Z vector at a steeper angle in the new impedance triangle.

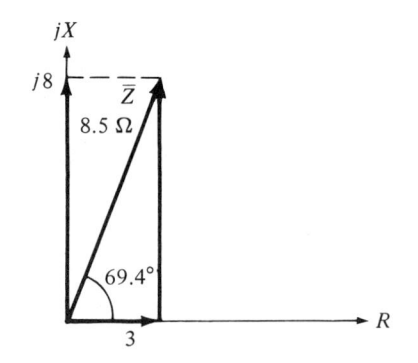

EXAMPLE 15.14

Given: the impedance

$$\bar{Z} = 13 \underline{/-22.6°} \ \Omega$$

Find the resistive and reactive components (convert to rectangular form). Draw the impedance triangle.

$$\bar{Z} = 13[\cos(-22.6°) + j\sin(-22.6°)]$$
$$= 13[0.92 - j0.38]$$
$$= 12 - j5 \ \Omega$$

The negative imaginary component indicates that the reactance is capacitive (in the language of electronics, we would say "the impedance looks capacitive" or "is capacitive"). The impedance triangle is

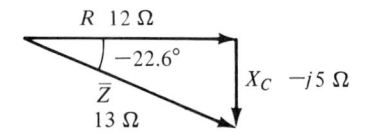

Note how the real and imaginary axes have been omitted to simplify this impedance triangle. We shall do this from now on.

15.3 IMPEDANCE

Because impedance completely describes the current-limiting ability of a circuit, it is substituted for resistance to give us an equation often called *Ohm's law for ac circuits*:

$$\bar{I} = \frac{\bar{E}}{\bar{Z}} \tag{15.14}$$

Due to the complex nature of all three elements in the equation, ac circuits can be analyzed using the techniques of complex numbers studied earlier.

EXAMPLE 15.15

A voltage

$$e = 255 \sin(1700t + 15°) \quad \text{V}$$

is applied across an impedance,

$$\bar{Z} = 12 - j20 \ \Omega$$

a. Find the current in phasor and time function forms.

Convert voltage to a phasor and impedance to polar form:

$$\bar{I} = \frac{\bar{E}}{\bar{Z}} = \frac{180 \ \underline{/15°} \ \text{V}}{23.3 \ \underline{/-59°} \ \Omega}$$
$$= 7.7 \ \underline{/15° - (-59°)}$$
$$= 7.7 \ \underline{/74°} \ \text{A}$$
$$i = \sqrt{2}(7.7) \sin(1700t + 74°)$$
$$= 10.9 \sin(1700t + 74°) \quad \text{A}$$

Notice how voltage was first converted to a phasor so that division could be done by the rules of complex numbers. The result was a current phasor which was transformed to the time domain.

b. Draw the phasor diagram.

Determine the phase angle:

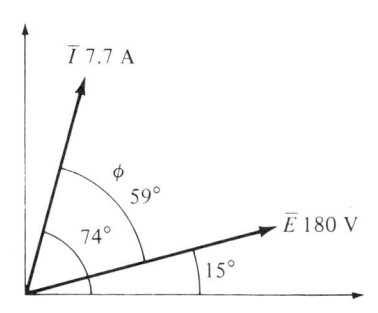

Current leads voltage by $74° - 15° = 59°$. The leading phase was to be expected because of the capacitive reactance of Z.

c. Compare the phase angle from part b to the angle of Z in part a.

The angle from part b, $59°$, is the inverse of the impedance angle, $-59°$, from part a. ∎

Example 15.15, part C, makes an important point:

The phase angle, ϕ, between a circuit's current and voltage, is the inverse of the circuit's impedance angle, θ.

Ohm's law shows why this is so.
For an inductive circuit:

$$\bar{Z} = Z\,\underline{/\theta} \quad (+ \text{ angle, inductive})$$

$$\bar{I} = \frac{E\,\underline{/0°}}{Z\,\underline{/\theta}} = I\,\underline{/0 - \theta} = I\,\underline{/-\theta}$$

$$\phi = \beta - \alpha = -\theta - 0 = -\theta, \quad \text{the inverse of } \theta$$

For a capacitive circuit:

$$\bar{Z} = Z\,\underline{/-\theta} \quad (- \text{ angle, capacitive})$$

$$\bar{I} = \frac{E\,\underline{/0°}}{Z\,\underline{/-\theta}} = I\,\underline{/0 - (-\theta)} = I\,\underline{/\theta}$$

$$\phi = \theta - 0 = \theta, \quad \text{the inverse of } -\theta$$

EXAMPLE 15.16

What is the phase angle between the voltage across and current through the impedance, $\bar{Z} = 87.3 + j107.2$?

$$\theta = \arctan \frac{107.2}{87.3} = 50.8°$$

Therefore, $\phi = -50.8°$, current lagging voltage. We can also tell that current is lagging by the positive (inductive) reactive component. ■

EXAMPLE 15.17

Draw the impedance triangle for the impedance that causes

$$i = 113 \sin(50 \times 10^6 t + 35°) \quad \text{mA}$$

for a voltage

$$e = 340 \sin(50 \times 10^6 t - 12°) \quad \text{V}$$

$$\bar{Z} = \frac{\bar{E}}{\bar{I}}$$

$$= \frac{(0.707)(340\,\underline{/-12°})}{0.707(113 \times 10^{-3}\,\underline{/35°})}$$

$$= 3000\,\underline{/-12° - 35°}$$

$$= 3000\,\underline{/-47°}\ \Omega$$

Convert to rectangular:

$$\bar{Z} = 3000[\cos(-47°) + j\sin(-47°)]$$

$$= 3000[0.68 - j0.73]$$

$$= 2046 - j2194\ \Omega$$

Draw the triangle:

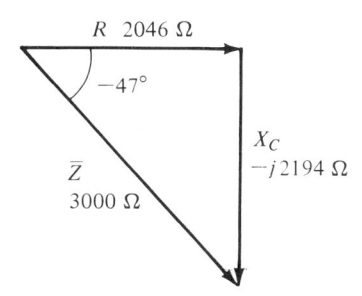

$$R\ 2046\ \Omega$$
$$-47°$$
$$\bar{Z}\ 3000\ \Omega$$
$$X_C\ -j2194\ \Omega$$

EXAMPLE 15.18

A current

$$\bar{I} = (50 - j92)\quad \text{mA}$$

flows through

$$\bar{Z} = (4.7 + j6.8) \times 10^3\ \Omega$$

Express the voltage across the impedance in the time domain for $\omega = 10^7$ rad/s.

Although the multiplication of I and Z could be done by first converting to polar numbers, they will be left in rectangular form for practice.

$$\begin{aligned}
\bar{E} &= \bar{I}(\bar{Z}) \\
&= [(50 - j92) \times 10^{-3}][(4.7 + j6.8) \times 10^3] \\
&= [50(4.7) + 92(6.8)] + j[50(6.8) - 92(4.7)] \\
&= (235 + 626) + j(340 - 432.4) \\
&= 861 - j92.4 \\
&= 865.9\ \underline{/-6.1°}\ \text{V}
\end{aligned}$$

Converting to time function:

$$\begin{aligned}
e &= \sqrt{2}(865.9)\sin(10^7 t - 6.1°) \\
&= 1224\sin(10^7 t - 6.1°)\quad \text{V}
\end{aligned}$$

15.4 AVERAGE POWER IN RESISTANCE AND REACTANCE

Besides causing phase shift and being sensitive to frequency, reactive ohms differ from resistive ohms in another important way. They cannot convert electrical energy into heat. Energy, you will recall, can only be "borrowed" from a source by reactance. It is stored in an electrostatic or electromagnetic field, and returned to the source later. As a result, average energy in and out is zero. Thus the power dissipation of reactance is also zero. These power characteristics can be shown with the aid of an equation which will now be derived. When

$$e = E_{\text{peak}} \sin \omega t \quad \text{volts}$$

and

$$i = I_{\text{peak}} \sin(\omega t + \phi) \quad \text{amperes}$$

are substituted in the power equation

$$p = ei$$

the product of two sinusoids occurs:

$$p = E_{peak}I_{peak} \sin \omega t \sin(\omega t + \phi)$$

This can be simplified using the trig identity

$$\sin A \sin B = \tfrac{1}{2}[\cos(A - B) - \cos(A + B)]$$

Substitute for A and B:

$$\begin{aligned}\sin \omega t \sin(\omega t + \phi) &= \tfrac{1}{2}[\cos\{\omega t - (\omega t + \phi)\} - \cos\{\omega t + (\omega t + \phi)\}] \\ &= \tfrac{1}{2}[\cos(-\phi) - \cos(2\omega t + \phi)] \\ &= \tfrac{1}{2}[\cos \phi - \cos(2\omega t + \phi)]\end{aligned}$$

since $\cos(-\phi) = \cos \phi$. Substitute into the power equation:

$$p = \frac{E_{peak}I_{peak}}{2}[\cos \phi - \cos(2\omega t + \phi)]$$

But

$$\frac{E_{peak}I_{peak}}{2} = \frac{E_{peak}}{\sqrt{2}}\frac{I_{peak}}{\sqrt{2}} = EI$$

where E and I are rms values. Therefore,

$$p = EI[\cos \phi - \cos(2\omega t + \phi)] \qquad (15.15)$$

(ϕ is, as usual, the phase angle between e and i, or the inverse angle of Z.)

Equation 15.15 describes a two-part composite waveform like those studied in Chapter 13. The parts are illustrated in Figure 15.9.

Two important points about this waveform were made during discussions on average and rms values of sinusoids in Section 13.5. They were:

■ The cosine portion contributes nothing to power transfer because its average is zero.
■ The waveform's average value is that of the constant on which the cosine rides.

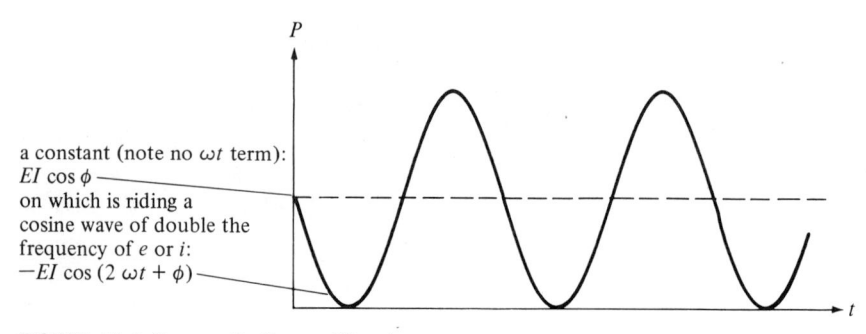

a constant (note no ωt term):
$EI \cos \phi$
on which is riding a
cosine wave of double the
frequency of e or i:
$-EI \cos(2\omega t + \phi)$

FIGURE 15.9 Composite Power Waveform.

15.4 AVERAGE POWER IN RESISTANCE AND REACTANCE

This average value is called *true power*. Measured in watts and represented by P_W, its magnitude tells a waveform's ability to deliver true power to a load:

$$P_W = EI \cos \phi \quad \text{watts} \tag{15.16}$$

Let us now use Equations 15.15 and 15.16 to examine the power dissipated by pure resistance and capacitive and inductive reactance.

Resistance

V_R and I_R in phase, $\phi = 0°$
Equation 15.15 becomes

$$\begin{aligned} p &= V_R I_R [\cos 0° - \cos(2\omega t + 0°)] \\ &= V_R I_R (1 - \cos 2\omega t) \quad \text{watts} \end{aligned}$$

This is the waveform shown in Figure 15.9. Note that the power curve does not drop below the X axis, which tells us that energy is completely absorbed by resistance; none is returned to the source.

The average value from Equation 15.16 is

$$P_W = V_R I_R \quad \text{watts} \tag{15.17}$$

By the usual applications of Ohm's law, this can also be represented by

$$P_W = I_R^2 R \quad \text{watts} \tag{15.18}$$

and

$$P_W = \frac{V_R^2}{R} \quad \text{watts} \tag{15.19}$$

Capacitance

V_C and I_C in quadrature, $\phi = 90°$
Equation 15.15 becomes

$$\begin{aligned} P &= V_C I_C [\cos 90° - \cos(2\omega t + 90°)] \\ &= -V_C I_C \cos(2\omega t + 90°) \end{aligned} \tag{15.20}$$

Note that the constant value in Equation 15.20 is zero ($\cos 90° = 0$), proving that average power to pure capacitance is zero.

From our work in Chapter 10, we know that capacitors store energy in an electrostatic field. They "borrow" that energy from the source during a rise in capacitor voltage and return it when the voltage falls. These energy transfers are seen in Figure 15.10, in which Equation 15.20 is plotted. Note how the power curve has equal areas above and below the X axis. Compare to the resistive curve, which is entirely above.

Inductance

V_L and I_L in quadrature, $\phi = -90°$.
Equation 15.15 becomes

$$\begin{aligned} p &= V_L I_L [\cos(-90°) - \cos(2\omega t - 90°)] \\ &= -V_L I_L \cos(2\omega t - 90°) \end{aligned} \tag{15.21}$$

Like Equation 15.20, the constant value of Equation 15.21 is zero ($\cos -90° = 0$); therefore, average power to pure inductance is zero.

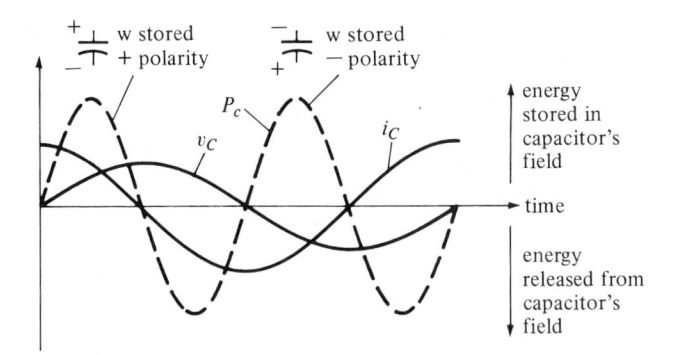

FIGURE 15.10 Sinusoidal Energy Transfer in Capacitance.

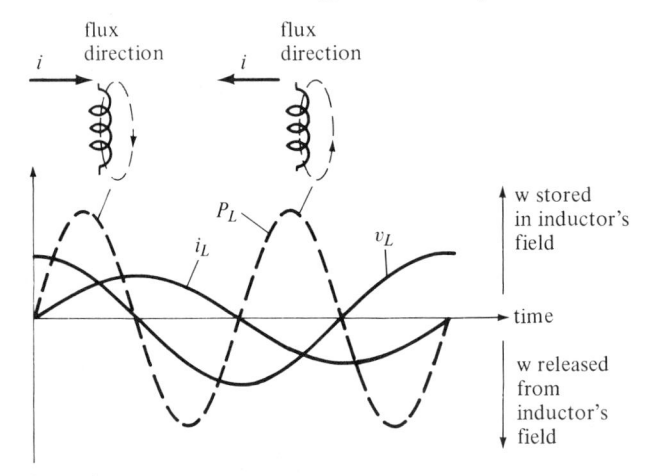

FIGURE 15.11 Sinusoidal Energy Transfer in Inductance.

From our work in Chapter 12, we know that energy is stored in an expanding magnetic flux field when inductor current increases. It returns to the source when that current decreases. These energy transfers are seen in Figure 15.11, in which Equation 15.21 is plotted. Compare this plot to that of the capacitor (Figure 15.10) and the resistor (Figure 15.9).

15.5 REACTIVE POWER

As we have seen, energy moves back and forth between the source and a reactance. Net energy transfer is zero; therefore, average power transfer is zero. However, reactance does draw current from a source and the product of this current and source voltage has units of joules/second, the same as watts. Yet we must not call this watts or refer to the product as true power. What do we call them?

The product is called *reactive power*. Its units are left just as they are: volts times amperes, or volt-amperes (no new parameter was invented). Since they are drawn by reactance, they are called *volt-amperes reactive*, which is abbreviated *var*.

Represented by P_Q, for quadrature, reactive power magnitude is determined by

$$P_Q = V_x I_x \quad \text{var} \quad (15.22)$$

where V_x = reactive voltage
 I_x = reactive current

15.6 APPARENT POWER

By applying Ohm's law, this equation may be written in two other familiar forms:

$$P_Q = I_x^2 X \quad \text{var} \tag{15.23}$$

and

$$P_Q = \frac{V_x^2}{X} \quad \text{var} \tag{15.24}$$

EXAMPLE 15.19

What is the reactive power drawn by 14 Ω of inductive reactance when placed across a 700 rms source?

$$P_Q = \frac{E^2}{X_L} = \frac{700^2}{14} = 35 \times 10^3 \text{ var}$$
$$= 35 \text{ kvar, inductive}$$

15.6 APPARENT POWER

If a source delivers true power, P_W, to resistance, and reactive power, P_Q, to reactance, it follows that it must deliver both to an impedance consisting of resistance and reactance. We consider next how P_W and P_Q are related in RC and RL circuits.

RC CIRCUIT

Figure 15.12 illustrates a source driving a capacitive impedance, and the phasor diagram of source rms voltage and current. Note the positive phase angle.

Current has been resolved into its real and imaginary components in Figure 15.13. The real, or in-phase, component, is due to the resistive portion of impedance and is

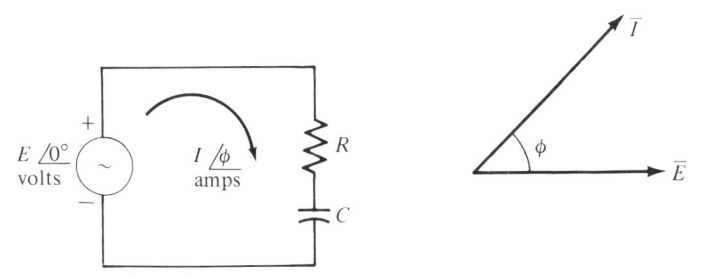

FIGURE 15.12 *RC* Circuit and Phasor Diagram.

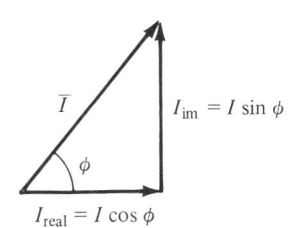

FIGURE 15.13 Real and Imaginary Components of *I*.

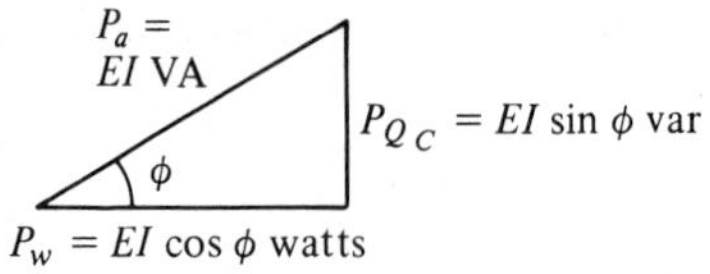

FIGURE 15.14 Power Triangle for an *RC* Circuit.

equal to

$$I_{real} = I \cos \phi \quad \text{amperes} \quad (15.25)$$

The imaginary, or quadrature, component, is due to the reactive portion of impedance and is equal to

$$I_{imag} = I \sin \phi \quad \text{amperes} \quad (15.26)$$

If each current in Figure 15.13 is multiplied by the source-voltage magnitude, E, we obtain the power triangle for an *RC* circuit (see Figure 15.14). Let us examine the triangle's three parts.

The power triangle's real component,

$$P_W = EI \cos \phi$$

is the average, or true, power dissipated by the resistive part of the impedance (see Equation 15.16). Its quadrature component,

$$P_Q = EI \sin \phi \quad (15.27)$$

is the reactive power drawn by the reactive part of the impedance.

The hypotenuse of the triangle is called the *apparent power* being supplied by the source. Represented by P_a and measured in *volt-amperes, or VA*, apparent power magnitude can be determined by the following equations:

$$P_A = EI \quad \text{VA} \quad (15.28)$$

or by the Pythagorean theorem,

$$P_A = \sqrt{P_W^2 + P_Q^2} \quad (15.29)$$

Power triangles are important aids to understanding circuit theory. They will be used frequently in work to come.

EXAMPLE 15.20

An impedance,

$$\bar{Z} = 12 - j5 \ \Omega$$

is connected to a source of

$$\bar{E} = 65 \underline{/0°} \ \text{V}$$

Determine the indicated values.
a. The circuit current in phasor form.

$$\bar{I} = \frac{\bar{E}}{\bar{Z}} = \frac{65 \underline{/0°}}{13 \underline{/-22.6°}} = 5 \underline{/22.6°} \ \text{A}$$

15.6 APPARENT POWER

b. The average power drawn.

$$P_W = EI \cos \phi = 65(5)(\cos 22.6°) = 300 \text{ W}$$

c. The reactive power drawn.

$$P_Q = EI \sin \phi = 65(5)(\sin 22.6°)$$
$$= 124.9 \text{ var, capacitive}$$

d. The apparent power.

By Equation 15.29,

$$P_A = \sqrt{P_W^2 + P_Q^2} = \sqrt{300^2 + 124.9^2} = 325 \text{ VA}$$

P_A could also have been determined by EI, $E^2/|Z|$, or by trigonometric means such as $P_Q/\sin \phi$ or $P_W/\cos \phi$.

e. Sketch the power triangle.

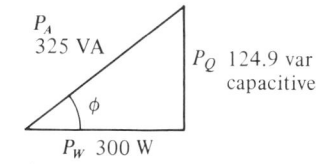

RL Circuit:

Figure 15.15 illustrates a source driving an inductive impedance, and the phasor diagram of source rms voltage and current. As usual, the phase angle is negative, indicating lagging current.

Using Equations 15.25 and 15.26, the current is resolved into its real and imaginary components (see Figure 15.16). When these components are multiplied by source

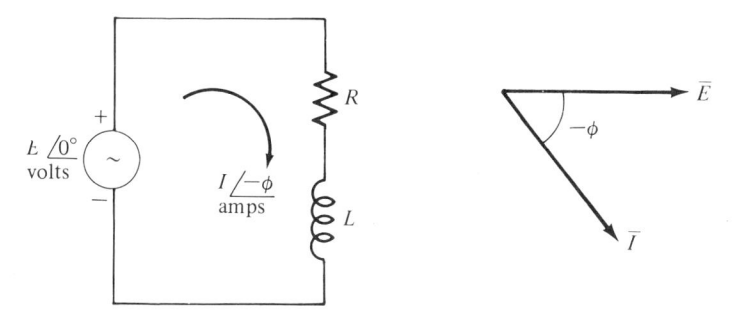

FIGURE 15.15 *RL* Circuit and Phasor Diagram.

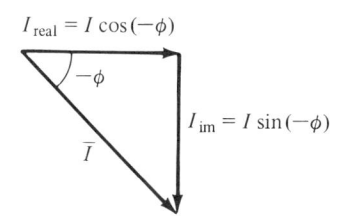

FIGURE 15.16 Real and Imaginary Components of *I*.

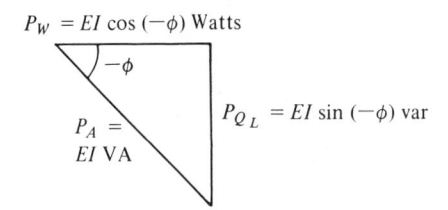

FIGURE 15.17 Power Triangle for *RL* Circuit.

voltage rms magnitude, E, the power triangle of Figure 15.17 results. The triangle's three components are obtained with the same formulas used for the *RC* circuit; they are summarized in Figure 15.17.

EXAMPLE 15.21

Find the resistive and reactive components of an impedance that draws 46 W and 38.6 inductive var from a 150-V source.

Draw the power triangle:

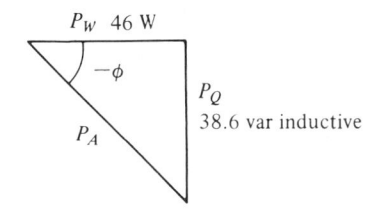

$$\phi = \tan^{-1} \frac{-38.6}{46} = \tan^{-1}(-0.84) = -40°$$

Therefore,

$$\theta = 40°$$

$Z = \dfrac{E}{I}$, but we don't know I. Find I:

$$P_W = EI \cos(-\phi) = 46 \text{ W}$$

Therefore,

$$I = \frac{46}{E \cos(-\phi)}$$

$$= \frac{46}{150 \cos(-40°)}$$

$$= \frac{46}{114.9} = 0.4 \text{ A}$$

$$\bar{Z} = \frac{E}{I}\underline{/40°} = \frac{150}{0.4}\underline{/40°} = 375\underline{/40°} \ \Omega$$

$$= 287.3 + j241 \ \Omega$$

15.7 POWER FACTOR

Here's another way to do the problem:

$$P_W = I^2 R \quad \text{by equation 15.23}$$

Therefore,

$$R = \frac{P_W}{I^2} = \frac{46}{(0.4)^2} = 287.5 \ \Omega$$

and

$$P_Q = I^2 X_L \quad \text{by Equation 15.28}$$

Therefore,

$$X_L = \frac{P_Q}{I^2} = \frac{38.6}{0.4^2} = 241.3 \text{ var} \qquad \blacksquare$$

Transformers and generators develop heat from I^2R and other losses. Design engineers must include provisions for removing this heat by air or liquid cooling. These provisions thus determine the maximum current a unit can supply without getting so hot that reliability suffers.

A purely resistive load may draw that maximum current and dissipate maximum power. On the other hand, an equal-valued reactive load draws the same current and develops the same losses, yet dissipates no power.

Therefore, manufacturers cannot rate large transformers and generators by power-transfer ability. If they did, a highly reactive load could draw safe power from a generator which is smoking from excessive reactive currents. Instead, high-power units are rated by their apparent-power capability, which is printed on the units in VA, kVA, or even MVA (megavolt-amperes).

15.7 POWER FACTOR

Factories using many high-power electric motors often draw currents with large inductive components. As we have seen, these currents cause I^2R losses within generators and transformers. They also increase line loss in the conductors between source and load. Such losses raise the cost of producing electricity by requiring increased burning of expensive fuel.

To offset these losses, power companies monitor a customer's demand and raise their rates if that demand becomes too reactive.

How is this monitoring done? Not by reading the factory's power meter. That is unaffected by reactive currents. Monitoring is done by observing the customer's *power factor*.

To understand power factor, consider a 10-V generator driving three different 5-Ω loads. In all cases, $I = 10/5 = 2$ A.

$\bar{Z} = 5 \underline{/0°} \ \Omega$ $\bar{Z} = 5 \underline{/-45°} \ \Omega$ $\bar{Z} = 5 \underline{/-90°} \ \Omega$
(pure R)   ($X = R$)   (pure X)
$P_W = EI \cos(0°)$ $P_w = EI \cos(-45°)$ $P_W = EI \cos(-90°)$
$\quad = 10(2)(\cos 0°)$ $\quad = 10(2)(\cos -45°)$ $\quad = 10(2)(\cos -90°)$
$\quad = 20(1)$ $\quad = 20(0.707)$ $\quad = 20(0)$
$\quad = 20$ W $\quad = 14.1$ W $\quad = 0$ W

$$PF = \cos\theta$$
$$= \frac{R}{Z} \quad \text{Equation 15.31}$$

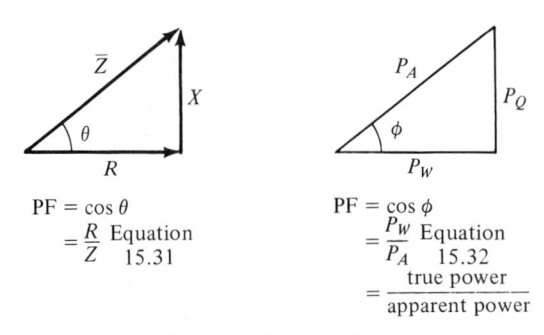

$$PF = \cos\phi$$
$$= \frac{P_W}{P_A} \quad \text{Equation 15.32}$$
$$= \frac{\text{true power}}{\text{apparent power}}$$

FIGURE 15.18 Power Factor Formulas.

A glance at these three shows that the only variable was cos ϕ, the cosine of the phase angle between voltage and current. What this tells us is:

> The closer cos ϕ is to 1, the more resistive the load and the higher the power dissipation.
> The closer cos ϕ is to 0, the more reactive the load and the lower the power dissipation.

Cos ϕ is very useful in power calculations and is called the load's *power factor*:

$$\text{power factor} = PF = \cos\phi \tag{15.30}$$

where ϕ is the phase angle between current and voltage. It is also the angle in the power triangle, and the inverse of the angle in the impedance triangle.

Since ϕ varies from -90 to $+90°$, PF is always positive and varies from 0 to 1. It is usually given as a percentage and said to be *lagging for inductive circuits* and *leading for capacitive circuits*.

In a similar manner, sin ϕ is used in the equation for reactive power. That equation is $P = EI \sin\phi$ var. As a result, sin ϕ is often called the *reactive factor*. Unlike the power factor, however, it can be negative as well as positive. Power factor may also be obtained from impedance and power triangles using the two formulas in Figure 15.18.

EXAMPLE 15.22

Given: a circuit with the voltage and current

$$e = 15\sin(1500t + 55°) \quad V$$
$$i = 3\sin(1500t - 20°) \quad A$$

a. What is the circuit power factor?

$$\phi = \beta - \alpha = (-20°) - 55° = -75°$$
$$PF = \cos\phi = \cos(-75°) = 0.259 \quad \text{or} \quad 25.9\%, \text{ lagging}$$

b. What is its reactive factor?

$$\text{reactive factor} = \sin\phi$$
$$= \sin(-75°)$$
$$= -0.966 \quad \text{or} \quad -96.6\% \text{ lagging}$$

EXAMPLE 15.23

A load having the impedance

$$\bar{Z} = 3000 - j7500$$

causes what power factor?

$$\theta = \tan^{-1}\frac{X}{R} = \tan^{-1}\frac{-7500}{3000}$$
$$= -68.2°$$

Therefore,

$$\phi = 68.2°$$
$$\text{PF} = \cos 68.2° = 0.37 \quad \text{or} \quad 37\%, \text{ leading}$$

EXAMPLE 15.24

A load draws 2850 kVa at a 13% PF.

a. Without doing any calculations, decide whether P_W should be closer to zero or 2850 W.

From the low power factor, we know that the load is highly reactive and P_W is close to 0.

b. Repeat part a for P_Q.

For the same reasons, P_Q will be high, close to 2850 kvar.

c. Calculate P_W and P_Q. Compare the results to predictions.

$$P_W = P_A \text{ PF} = 2850 \text{ kVA} (0.13) = 370.5 \text{ kW}$$

To find P_Q, we need ϕ.

$$\phi = \arccos 0.13 = 82.5°$$
$$P_Q = P_A \sin \phi = 2850 \text{ kVA} (\sin 82.5°)$$
$$= 2850(0.99) = 2826 \text{ kVA}$$

EXERCISES

Section 15.1

1. At which point would this wave have each of the following values of dv/dt? a. zero; b. medium + value; c. high − value; d. low + value.

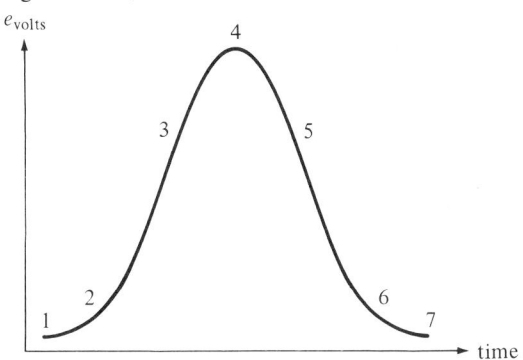

2. What is a sine wave doing, increasing or decreasing, when its derivative is a. +; b. −?

3. For what range of angles does a voltage sinusoid have a. $+dv/dt$; b. $-dv/dt$? Repeat for a cosine wave.

4. *Extra Credit.* Using the same technique with which $d(\sin \omega t)/dt$ was found, find $d(\cos \omega t)/dt$.

5. Determine the derivative of each of these sinusoids. Express the answer as a sinusoid.
a. $15 \sin 300t$; b. $307 \sin(40,000t + 41°)$; c. $0.07 \sin(11.5t - 15°)$.

6. Write the equation of the sine wave that has the derivative

$$75,000 \cos(1500t + 50°)$$

Section 15.2

7. For each frequency, find the inductive reactance of a 15-mH choke. Express answers in both polar and rectangular form. a. 100 Hz; b. 3.3 kHz; c. 500 kHz; d. 11.2 mHz.

8. Repeat Exercise 7 for a 0.015-μF capacitor. Compare changes in X_L and X_C in the two exercises as frequency increased.

9. What is the value of a 470-Ω resistance at each of the frequencies of Exercise 7? Discuss.

10. What will the value of inductive reactance do if: a. the voltage across an inductor is increased; b. the frequency of the applied signal is reduced; c. the peak value of the current through the inductor is decreased? Repeat the question for a capacitive reactance.

11. Find the current through each component. Express the answer in time and phasor domains.

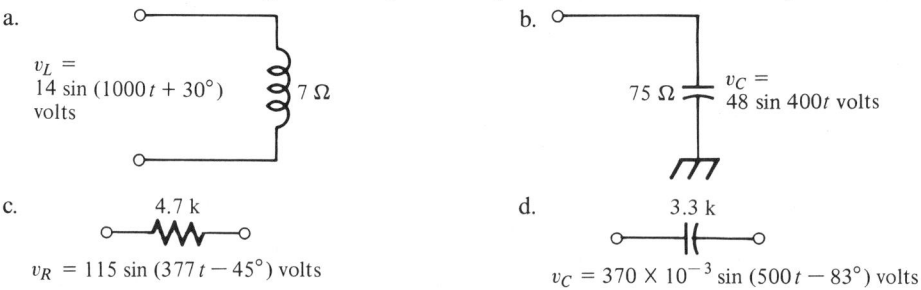

a. $v_L = 14 \sin(1000t + 30°)$ volts, 7 Ω

b. 75 Ω, $v_C = 48 \sin 400t$ volts

c. 4.7 k, $v_R = 115 \sin(377t - 45°)$ volts

d. 3.3 k, $v_C = 370 \times 10^{-3} \sin(500t - 83°)$ volts

12. Find the voltage across each component. Express answers in phasor form:

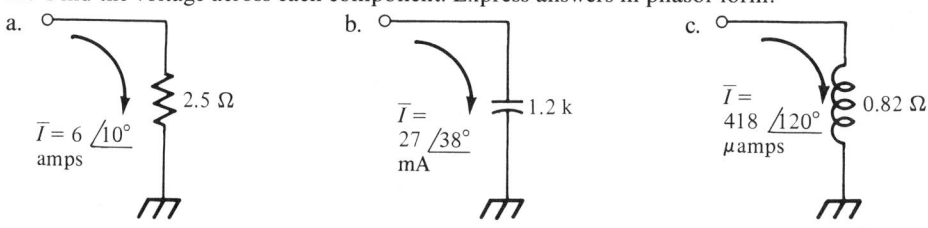

a. 2.5 Ω, $\bar{I} = 6 \angle 10°$ amps

b. 1.2 k, $\bar{I} = 27 \angle 38°$ mA

c. 0.82 Ω, $\bar{I} = 418 \angle 20°$ μamps

13. Which component, a resistance, inductor, or capacitor, would have each of the following voltage–current relations?

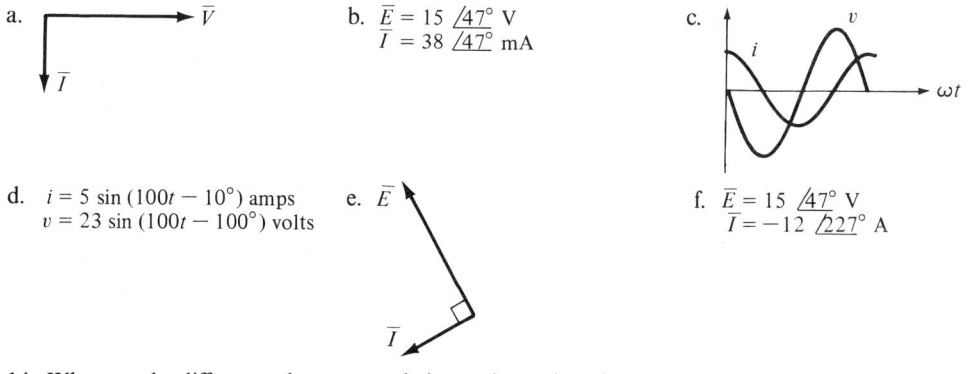

a. (phasor diagram with $\bar{V}$ horizontal and $\bar{I}$ downward)

b. $\bar{E} = 15 \angle 47°$ V
$\bar{I} = 38 \angle 47°$ mA

c. (waveform diagram showing v and i vs ωt)

d. $i = 5 \sin(100t - 10°)$ amps
$v = 23 \sin(100t - 100°)$ volts

e. (phasor diagram with $\bar{E}$ and $\bar{I}$ at right angle)

f. $\bar{E} = 15 \angle 47°$ V
$\bar{I} = -12 \angle 227°$ A

14. What are the differences between resistive and reactive ohms?

EXERCISES

Section 15.3

15. Find the resistive and reactive components of each impedance. Try to approximate values before doing calculations.   a. $2000\,\underline{/18°}\ \Omega$;   b. $3.7\,\underline{/-52°}\ \Omega$;   c. $830\,\underline{/-78°}\ \Omega$.

16. Current leads voltage in a particular circuit by 45°. What two things do we know about the circuit's impedance?

17. Convert each impedance to polar form. Indicate whether current through the impedance will lead or lag voltage across.   a. $1 - j3.2$;   b. $14 + j22$;   c. $3300 - j2700$;   d. $680\text{ k} + j500\text{ k}$.

18. Determine the magnitude of each impedance. Calculate the phase angle between the voltage across and current through each impedance.   a. $5\,\Omega$;   b. $35 - j20\,\Omega$;   c. $20 + j35\,\Omega$;   d. $-j5\,\Omega$;   e. $j5\,\Omega$

19. Given: the impedance, $\bar{Z} = 15 + j25$. Using the letters I, D, and NC, indicate what happens to each of the following when frequency increases.   a. R;   b. $|Z|$;   c. X_L;   d. θ;   e. L;   f. ω.

20. Repeat Exercise 19 holding the frequency constant but decreasing resistance.

21. Given: the impedance, $\bar{Z} = 3000 - j1700\,\Omega$. Using the letters, I, D, and NC, and M for becoming more negative and L for less negative, indicate what happens to each of the following when frequency decreases.   a. R;   b. $|Z|$;   c. X_C;   d. θ;   e. C;   f. ω.

22. Repeat Exercise 21 holding the frequency constant but increasing capacitance.

23. Given: an unknown resistance and an inductive reactance of $1000\,\Omega$. What must the resistance be in order that θ be.   a. 90°;   b. >45°;   c. ≈0°;   d. =45°;   e. ≈90°;   f. <45°.

24. Same resistance and reactance as Problem 23. What must the resistance be in order that Z magnitude, $|Z|$ be.   a. $=X_L$;   b. $\approx X_L$;   c. $\approx R$;   d. $=R$.

25. Given: a 1 k resistance connected to the reactance of a $0.03\text{-}\mu\text{F}$ capacitor. For what frequency or range of frequencies will θ be.   a. ≈ −90°;   b. < −45°;   c. = −45°;   d. > −45°;   e. ≈0°.

26. The voltage

$$e = 143\sin(1200t - 28°) \quad \text{volts}$$

causes the current,

$$i = 65\sin(1200t + 10°) \quad \text{mA}$$

to flow through an unknown impedance. Find the resistive and reactive components of that impedance and the capacitance.

27. A voltage

$$e = 78\sin(3400t + 25°) \quad \text{volts}$$

is applied to the impedance, $\bar{Z} = 35 - j20\,\Omega$.
a. Find the current in the phasor and time domains.
b. Determine the phase angle.
c. What is the new peak current value if the frequency is doubled?
d. The frequency is returned to its original value and the voltage doubled. What is the new peak current value?

28. A current of $(0.6 + j0.78)$ A flows through an impedance of $(11 + j11.8)\,\Omega$. Without doing any calculations:
a. State the approximate phase difference between the current and the voltage causing it.
b. State whether that voltage leads or lags the current.
c. Calculate the voltage in phasor form.

29. An RC combination has an impedance

$$\bar{Z} = 6220\,\underline{/-36.5°}\ \Omega$$

at 860 Hz. To what value must the frequency be changed in order that $\theta = -41.3°$?

REACTANCE AND IMPEDANCE

30. Suppose that the following question appeared on an exam:

| Find the value of resistance and inductance that has the impedance $10 + j15$ Ω.

Can the problem be done? Discuss.

Sections 15.4 and 15.5

31. Compute the reactive power in each case.

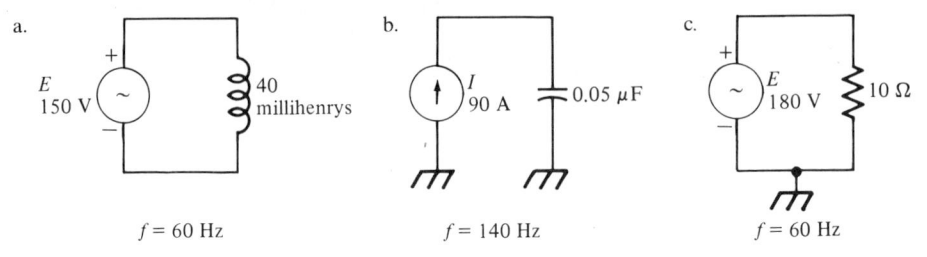

a.

$E = 150$ V, 40 millihenrys, $f = 60$ Hz

b.

$I = 90$ A, 0.05 μF, $f = 140$ Hz

c.

$E = 180$ V, 10 Ω, $f = 60$ Hz

32. In which circuit of Exercise 31 does each of the following conditions occur?
 a. $P_W = P_A$ b. $P_Q = P_A$ c. $\phi = 0°$ d. $\phi = 90°$

Section 15.6

33. A 4-kV generator is rated 2 MVA.
 a. What is the maximum current that it can supply?
 b. What is the minimum resistance that should be placed across it?
 c. What is the minimum reactance?
 d. What is the minimum magnitude of impedance?
 e. If 1.4 MW is drawn from the generator, what is the maximum reactive power that may be drawn at the same time?

34. A 200-W generator supplies 2 W to a load and overheats. What would be a probable value for each of the following?
 a. The load resistance relative to load reactance.
 b. The circuit phase angle.
 c. The power factor.
 d. The reactive factor.

Section 15.7

35. Fill in the blanks.

	E (V)	I (A)	ϕ (deg)	PF	P_A (VA)	P_Q (VAR)	P_W (W)
a.	100	4	60				
b.	35			lag	175		171.2
c.		7.3	−71				52.3
d.		5.7			141.4	100	
e.		1.6				15.8	2.2
f.	11.6			0.83		−37.5	

36. Each of the following circuit parameters is changed. Which ones affect circuit power factor?
 a. The peak voltage.   b. ω.   c. The resistance.   d. The peak current.
 e. The time.   f. The frequency.   g. The reactance.

EXERCISES

37. An impedance,

$$\bar{Z} = 20 + j53 \ \Omega$$

is connected to a source,

$$\bar{E} = 113 \ \underline{/0°} \ \text{V}$$

a. Find the circuit current in phasor form.
b. Find the apparent power.
c. Find the true power.
d. Find the reactive power.
e. Find the PF.
f. Draw the power triangle.

38. Repeat Exercise 37 for the impedance

$$\bar{Z} = 85 - j102 \ \Omega$$

connected to a $217 \ \underline{/0°}$-V source.

39. *Extra Credit.* Figure 15.9, 15.10, and 15.11, show the instantaneous power curve for pure R, L, and C, respectively. Sketch similar curves for the three impedances, $\bar{Z} = R - jX_C$, in which: a. $R > |X_C|$; b. $R = |X_C|$; c. $R < |X_C|$. Explain differences.

CHAPTER HIGHLIGHTS

A derivative tells the rate at which a variable changes.
The derivative of a sine wave is a cosine wave, or a sine wave shifted left 90°.
Resistance:

- Causes no phase shift.
- Dissipates nonreturnable energy as heat.
- Is not frequency dependent.
- Is represented by a vector on the real axis.

Reactance:

- Causes 90° phase shift.
- Stores and returns energy.
- Is frequency dependent.
- Is represented by quadrature vectors, either + (inductive), or − (capacitive).

ELI the ICE man
  Impedance is the total current-limiting ability of a circuit.
  Power factor indicates how reactive a load is. High PF—load resistive; low PF—load reactive.

Formulas to Remember

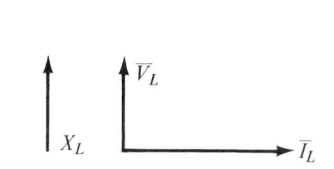

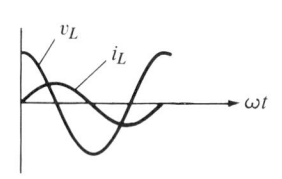

inductive reactance $X_L = j\omega L = j2\pi f L$ ohms

$$X_C = -j\frac{1}{\omega C} = -j\frac{1}{2\pi f C} \quad \text{ohms}$$

$$\bar{Z} = R \pm jX = \bar{Z}\underline{/\pm\theta} \quad \text{ohms}$$

Ohm's law for ac:
$$\bar{I} = \frac{\bar{E}}{\bar{\bar{Z}}}$$

Circuit phase angle, ϕ, is the inverse of the impedance angle, θ.
True power: $P_W = EI\cos\phi$ watts
Reactive power: $P_Q = EI\sin\phi$ var
Apparent power: $P_A = EI = \sqrt{P_W^2 + P_Q^2}$ VA

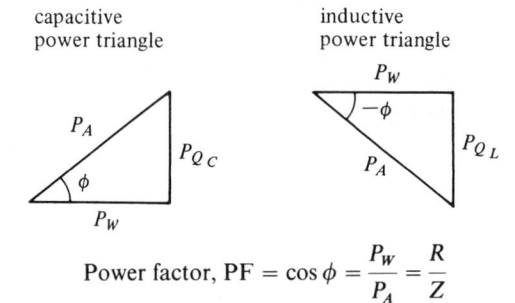

capacitive power triangle

inductive power triangle

$$\text{Power factor, PF} = \cos\phi = \frac{P_W}{P_A} = \frac{R}{Z}$$

CHAPTER 16

Analyzing Series and Parallel AC Circuits

In the preceding three chapters, we studied the basics of alternating current. We learned what sine waves are and various ways to represent them in diagrams and mathematics. We studied phase shift as well as its cause, namely, reactance and impedance. Ohm's law calculations have been done using complex numbers and phasor algebra. We also learned to calculate real and reactive power and the meaning of power factor.

The groundwork has been completed. We are ready to put these concepts to work and analyze ac circuits. Chapters 16 through 18 repeat the work done earlier on dc circuits in Chapters 4 through 7. All the concepts of series and parallel circuits, loop and nodal equations, Thévenin's and Norton's equivalents, and so on, will be rediscovered. As before, our goal is to find the voltage across, current through, and power dissipated by circuit elements. In addition, we will be concerned with phase differences between voltages and currents, and the effects of frequency changes on those differences. We also examine the power relationships in these circuits.

At the completion of this chapter, you should be able to:

- Find the total impedance of a group of impedances in series and parallel.
- Calculate the voltage across, current through, and real or reactive power for any component in series or parallel circuits.
- Determine phase difference between voltages and currents.
- Predict the results of a frequency change on circuit operation.
- Describe the frequency response characteristics of high-pass and low-pass filters.
- Analyze circuits in terms of conductance, susceptance, and admittance.
- Convert a series-RX circuit to its parallel equivalent, and vice versa.

16.1 Z_{TOTAL} OF SERIES IMPEDANCES

The resistive–reactive connection in Figure 16.1 is a series circuit because it has the same current through all elements.

By Kirchhoff's voltage law and the mathematics used to find total series resistance in Chapter 4, we can show that this circuit's total series impedance is

$$\bar{Z}_{total} = R_1 + R_2 - jX_{C_1} - jX_{C_2} + jX_{L_1} + jX_{L_2}$$

Using the rules of complex algebra, all real terms (resistances) are combined. Similarly, the imaginary terms (reactances) are combined. This gives us:

$$\begin{aligned}\bar{Z}_{total} &= R_{total} - jX_{C_{total}} + jX_{L_{total}} \\ &= (6 + 4) - j(3 + 5) + j(2 + 4) \\ &= 10 - j8 + j6 \\ &= 10 - j2 \ \Omega\end{aligned}$$

which is a series resistive–capacitive connection.

The discussion above makes two important points:

- Impedance in rectangular form describes a series connection of resistance and reactance.
- As resistances in series add to R_{total}, so, too, do reactances in series add to X_{total}, with proper regard to signs.

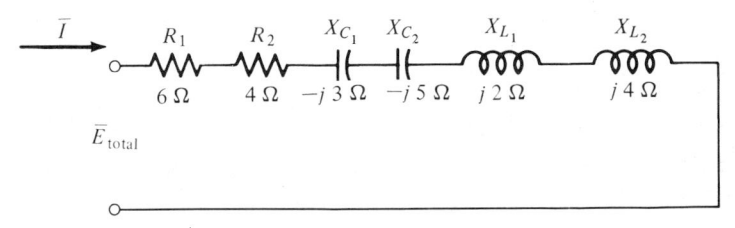

FIGURE 16.1 A Series AC Circuit.

16.2 ANALYZING SERIES AC CIRCUITS

The latter statement leads us to the following formula for the total impedance of any series circuit:

$$\bar{Z}_{total} = R_{total} + j(X_{L_{total}} - X_{C_{total}}) \qquad (16.1)$$

16.2 ANALYZING SERIES AC CIRCUITS

Series ac circuits obey the same rules as those for series dc circuits:

- Current is the same through all components.
- The voltage across a component is proportional to that component's impedance.
- Total voltage equals the sum of the individual voltages (KVL). Of course, in ac analysis, summing must be done according to the rules of complex algebra.
- True power may be calculated for an entire circuit with Equation 15.16:

$$P_W = EI \cos \phi \qquad \text{watts}$$

and reactive power with Equation 15.27:

$$P_Q = EI \sin \phi \qquad \text{vars}$$

- Power can also be determined for individual components using $P = I^2R$ or I^2X. Once voltages have been found by Ohm's law or some other means, the formulas

$$P_W = \frac{V_r^2}{R} \quad \text{and} \quad P_W = V_R I_R \qquad \text{watts}$$

may be used for true power, and

$$P_Q = \frac{V_X^2}{X} \quad \text{and} \quad P_Q = V_X I_X \qquad \text{vars}$$

for reactive power.

* * * * *But Remember the Earlier Warning:* * * * *

To determine voltage, current, impedance, and/or power for a component, use only the parameters associated with that specific component.

EXAMPLE 16.1

Perform the following operations for this series *RC* circuit.

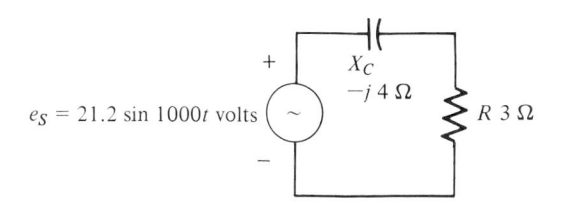

$e_S = 21.2 \sin 1000t$ volts

$X_C = -j\,4\,\Omega$

$R\,3\,\Omega$

a. Find Z_{total} and draw the impedance triangle.

$$\bar{Z}_{total} = R - jX_C$$
$$= 3 - j4 \ \Omega$$

In polar form:

$$\bar{Z}_{total} = 5 \ \underline{/-53.1°} \ \Omega$$

b. Find I.

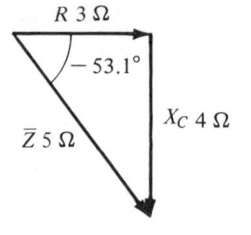

Convert E_s to a phasor function:

$$E_s = 0.707(21.2 \ \underline{/0°})$$
$$= 15 \ \underline{/0°} \ V$$

$$\bar{I} = \frac{\bar{E}}{\bar{Z}} = \frac{15 \ \underline{/0°}}{5 \ \underline{/-53.1°}}$$
$$= 3 \ \underline{/53.1°} \ A$$

Note that $\bar{I}$ leads $\bar{E}_s$, as it should.

c. Calculate $\bar{V}_R$ and $\bar{V}_C$ by Ohm's law.

$$\bar{V}_R = \bar{I}\bar{R}$$
$$= 3 \ \underline{/53.1°} \ (3 \ \underline{/0°})$$
$$= 9 \ \underline{/53.1°} \ V$$
$$\bar{V}_C = \bar{I}X_C$$
$$= 3 \ \underline{/53.1°} \ (4 \ \underline{/-90°})$$
$$= 12 \ \underline{/-36.9°} \ V$$

d. Draw the circuit phasor diagram.

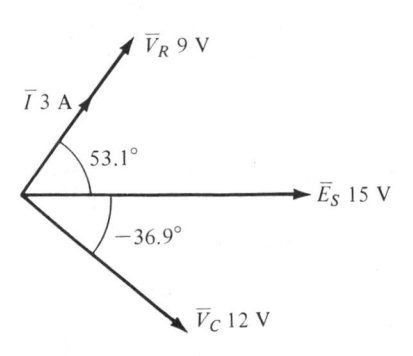

e. Express $\bar{I}$, $\bar{V}_R$, and $\bar{V}_C$ as time functions. Then draw the circuit's time-domain diagram.

$$I1 = 3 \ \underline{/53.1°} \Rightarrow 4.2 \sin(1000t + 53.1°) \quad A$$
$$\bar{V}_R = 9 \ \underline{/53.1°} \Rightarrow 12.7 \sin(1000t + 53.1°) \quad V$$
$$\bar{V}_C = 12 \ \underline{/-36.9°} \Rightarrow 17 \sin(1000t - 36.9°) \quad V$$

e_s given as $21.2 \sin 1000t$ V.

16.2 ANALYZING SERIES AC CIRCUITS

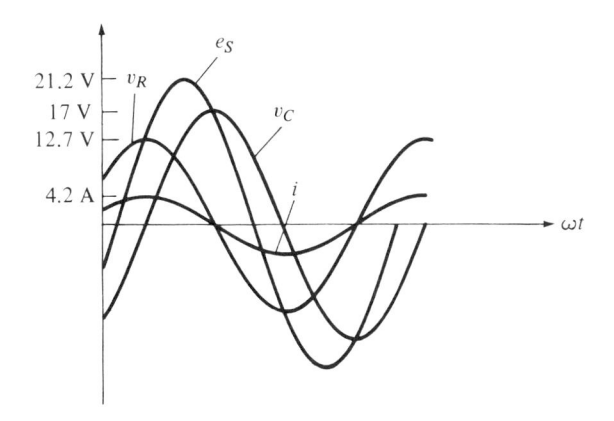

f. Calculate the power dissipated by the resistor.

$$P_W = \frac{V_R^2}{R} = \frac{9^2}{3} = 27 \text{ W}$$

g. Calculate the power drawn by entire circuit. Compare this power to that calculated in part f.

$$P_W = EI \cos \phi$$
$$= 15(3 \cos 53.1°)$$
$$= 27 \text{ W}$$

$P_{\text{resistor}} = P_{\text{circuit}}$ because the resistor is the only component dissipating power.

h. Calculate the reactive power drawn by the capacitor.

$$P_Q = I^2 X_C = 3^2(4) = 36 \text{ var, capacitive}$$

i. Calculate the apparent power.

$$P_A = E_s I_s$$
$$= 15(3)$$
$$= 45 \text{ VA}$$

j. Draw the circuit's power triangle.

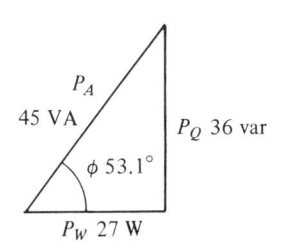

k. Determine the circuit power factor.

$$\text{PF} = \frac{P_W}{P_A} = \frac{27}{45}$$
$$= 0.6 \quad \text{or} \quad 60\% \text{ leading}$$

As a check:
$$PF = \cos 53.1° = 0.6$$
or
$$PF = \frac{R}{Z} = \frac{3}{5} = 0.6$$

∎

EXAMPLE 16.2

Perform the following steps for this circuit.

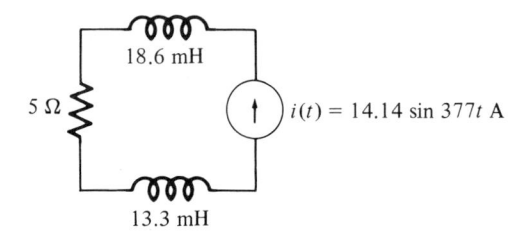

a. Find $\bar{Z}_{total}$.

X_{total} can be found two ways:

1. Find the reactance of each inductance and add them.
2. Find the total inductance. Then calculate its reactance.

Using method 1:
$$X_{L_1} = \omega L_1$$
$$= 377(18.6 \text{ mH})$$
$$= j7 \ \Omega$$
$$X_{L_2} = \omega L_2$$
$$= 377(13.3 \text{ mH})$$
$$= j5 \ \Omega$$
$$X_{L_T} = j7 + j5$$
$$= j12 \ \Omega$$
$$\bar{Z}_{total} = R + jX$$
$$= 5 + j12 \ \Omega$$

b. Calculate $\bar{V}_R$.

Convert current to phasor form:
$$\bar{I} = 10 \ \underline{/0°} \text{ A}$$
$$\bar{V}_R = \bar{I}R$$
$$= 10 \ \underline{/0°} \ (5 \ \underline{/0°})$$
$$= 50 \ \underline{/0°} \text{ V}$$

16.2 ANALYZING SERIES AC CIRCUITS

c. Calculate $\bar{V}_{X_{total}}$.

$$\bar{V}_{X_{total}} = \bar{I} X_{total}$$
$$= 10 \underline{/0°} \, (12 \underline{/90°})$$
$$= 120 \underline{/90°} \text{ V}$$

d. Find $\bar{E}_{source}$ by KVL; that is, add $\bar{V}_R$ and $\bar{V}_X$ as complex numbers.

$$\bar{E}_{source} = \bar{V}_R + \bar{V}_X$$
$$= 50 \underline{/0°} + 120 \underline{/90°} = 50 + j120$$
$$= 130 \underline{/67.4°} \text{ V}$$

e. Express source voltage and total reactive voltage as time functions.

$$e_s = \sqrt{2}(130)\sin(377t + 67.4°)$$
$$= 183.8 \sin(377t + 67.4°) \text{ V}$$
$$u_X = \sqrt{2}(120)\sin(377t + 90°)$$
$$= 169.7 \sin(377t + 90°) \text{ V}$$

EXAMPLE 16.3

Perform the following operations for this *RLC* series circuit.

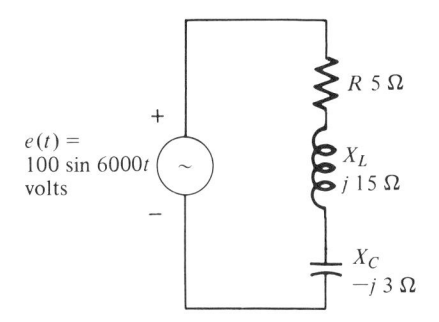

a. Draw the impedance triangle:

$$\bar{Z}_{total} = R + j(X_L - X_C)$$
$$= 5 + j(15 - 3)$$
$$= 5 + j12 \, \Omega$$
$$= 13 \underline{/67.4°} \, \Omega$$

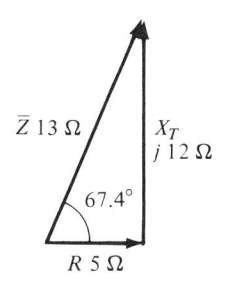

b. Calculate current as a phasor function.

Convert source voltage to a phasor:

$$\bar{E} = 70.7 \underline{/0°} \text{ V}$$

$$\bar{I} = \frac{\bar{E}}{\bar{Z}} = \frac{70.7 \underline{/0°}}{13 \underline{/67.4°}}$$

$$= 5.4 \underline{/-67.4°} \text{ A}$$

c. Calculate the three drops and draw the circuit phasor diagram.

$$\bar{V}_R = \bar{I}R$$
$$= (5.4 \underline{/-67.4°})(5)$$
$$= 27 \underline{/-67.4°} \text{ V}$$
$$\bar{V}_C = \bar{I}X_C$$
$$= (5.4 \underline{/-67.4°})(3 \underline{/-90°})$$
$$= 16.2 \underline{/-157.4°} \text{ V}$$
$$\bar{V}_L = \bar{I}X_L$$
$$= (5.4 \underline{/-67.4°})(15 \underline{/90°})$$
$$= 81 \underline{/22.6°} \text{ V}$$

Convert $\bar{E}_s$ to phasor form:

$$\bar{E}_s = 70.7 \underline{/0°} \text{ V}$$

Draw the phasor diagram:

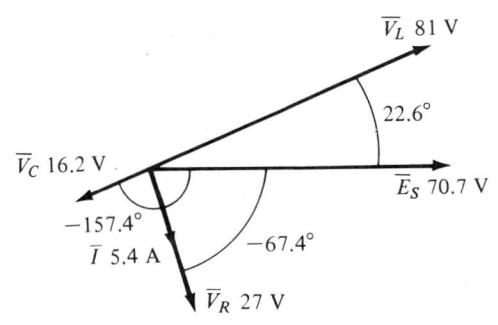

d. Calculate circuit power dissipation, P_W.

$$P_W = I^2 R$$
$$= (5.4)^2 5$$
$$= 145.8 \text{ W}$$

e. Calculate the two reactive powers. Find the total reactive power.

$$P_{Q_L} = \frac{V_L^2}{X_L} = \frac{81^2}{15}$$
$$= -437.4 \text{ var, inductive}$$

$$P_{Q_C} = V_C I$$
$$= (16.2)(5.4)$$
$$= 87.5 \text{ var, capacitive}$$

$$P_{Q_{\text{total}}} = 87.5 - 437.4$$
$$= -349.9 \text{ var, inductive}$$

f. Draw the circuit power triangle. Calculate P_A.

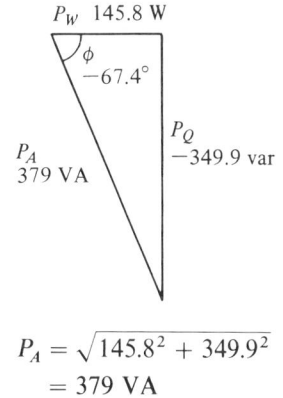

$$P_A = \sqrt{145.8^2 + 349.9^2}$$
$$= 379 \text{ VA}$$
$$\phi = \tan^{-1}\frac{-349.9}{145.8}$$
$$= \tan^{-1} -2.4$$
$$= -67.4°$$

16.3 THE VOLTAGE-DIVISION FORMULA

The voltage across any resistance in a series ac circuit may be obtained with the voltage-division formula just as it was in dc circuits:

$$\bar{V}_{R_N} = \bar{E}_{source}\frac{R_N}{\bar{Z}_{total}} \quad \text{volts} \tag{16.2}$$

This voltage is always in phase with the current through the resistance.

Voltage across any series reactance is obtained by

$$\bar{V}_{X_N} = \bar{E}_{source}\frac{X_N}{\bar{Z}_{total}} \quad \text{volts} \tag{16.3}$$

Of course, this voltage is always in leading quadrature with circuit current for inductance (ELI), and lagging quadrature with the current for capacitance (ICE). Since $\bar{V}_R$ is always in phase with $\bar{I}$, it follows that $\bar{V}_R$ will be in quadrature with any $\bar{V}_X$.

EXAMPLE 16.4

Using the voltage–division formula, find the voltage across the inductor in this circuit.

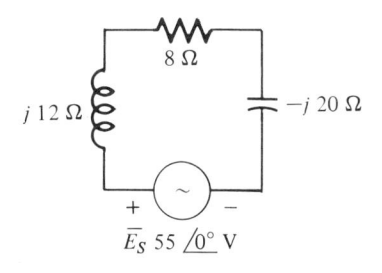

$$\bar{Z}_{total} = 8 + j12 - j20 = 8 - j8 \text{ or } 11.3 \underline{/-45°} \ \Omega$$

$$\bar{V}_L = \frac{\bar{E}_s X_L}{\bar{Z}_{total}}$$

$$= \frac{55 \underline{/0°} \ 12 \underline{/90°}}{11.3 \underline{/-45°}}$$

$$= 58.4 \underline{/135°} \text{ V} \qquad \blacksquare$$

16.4 RESPONSE OF *RC*, *RL*, AND *RLC* SERIES CIRCUITS TO CHANGING FREQUENCY

It isn't enough that the student of electronics be able to analyze series *RX* circuits. He or she must develop a "feel" for the characteristics of these circuits at low, medium, and high frequencies.

When operating at very low or very high frequencies, a circuit may have one reactance much greater than any of its other reactances or resistances. That reactance will dominate the circuit. It will have the majority of the source voltage dropped across it and will determine circuit characteristics such as phase angle and power factor.

On the other hand, reactance may fall to such a value that resistance dominates circuit action.

| A component having an impedance much greater than that of any other component in a series *RX* circuit determines the characteristics of that circuit.

With that statement in mind, let us examine first the series *RC* circuit. As we do, remember:

| Capacitors look like opens (high impedances) at low frequencies and shorts (low impedances) at high frequencies.

Figure 16.2 shows the changes in impedance magnitude of a series *RC* circuit as the frequency is increased from dc. Note carefully how the resistive component has constant length but the reactive component gets shorter as the frequency increases.

Students naturally ask at this point: "What is a high frequency? What is a low frequency?" There is no hard-and-fast answer to those questions because a great deal depends on the situation. What is high in one case may be low in another. With experience comes an awareness of the situations in which these circuit characteristics may or may not be important. For now, the important thing is to understand the characteristics thoroughly.

Turning our attention next to series *RL* circuits, remember:

| Inductors look like shorts at low frequencies and opens at high frequencies.

Figure 16.3 shows the changes in impedance magnitude of a series *RL* circuit as the frequency is increased from dc. Note again how resistance remains constant with frequency increases but the reactive component gets longer. Compare and contrast the *RL* characteristics with those of the *RC* circuit in Figure 16.2.

16.4 RESPONSE OF RC, RL, AND RLC SERIES CIRCUITS

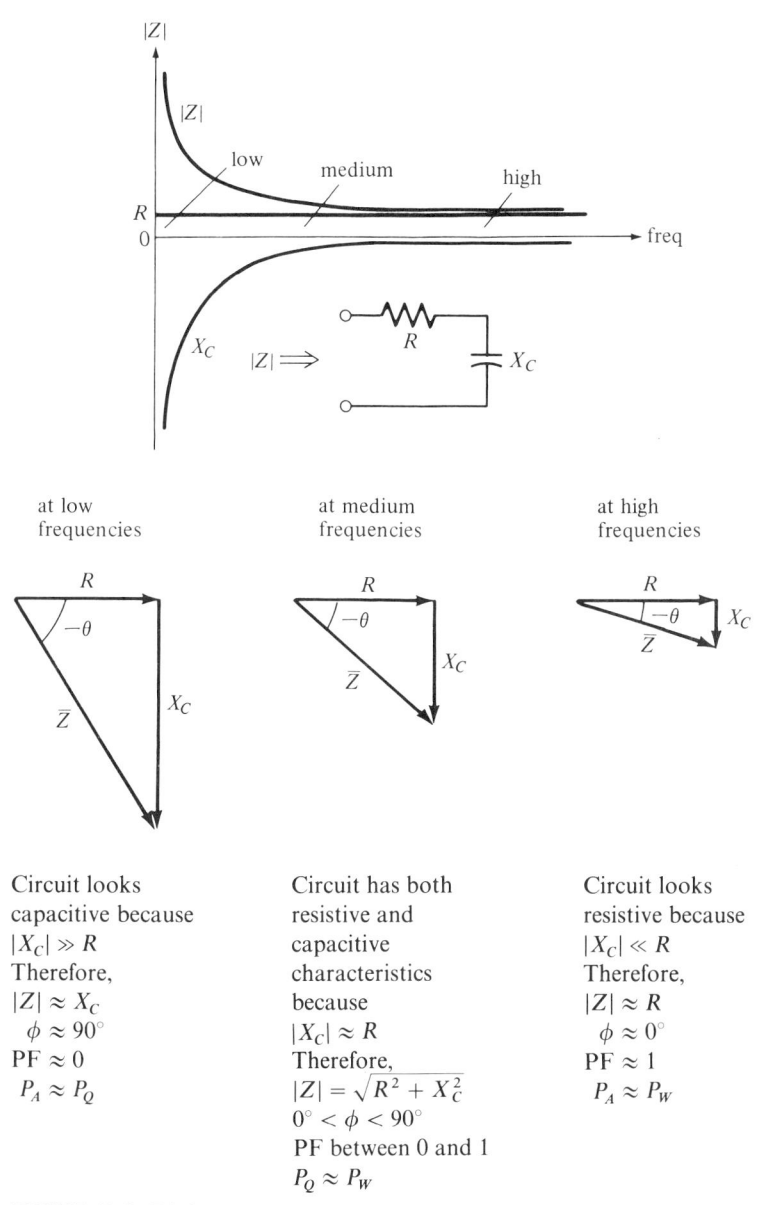

FIGURE 16.2 *RC* Circuit Action at Low, Medium, and High Frequencies.

at low frequencies

Circuit looks capacitive because
$|X_C| \gg R$
Therefore,
$|Z| \approx X_C$
$\phi \approx 90°$
PF ≈ 0
$P_A \approx P_Q$

at medium frequencies

Circuit has both resistive and capacitive characteristics because
$|X_C| \approx R$
Therefore,
$|Z| = \sqrt{R^2 + X_C^2}$
$0° < \phi < 90°$
PF between 0 and 1
$P_Q \approx P_W$

at high frequencies

Circuit looks resistive because
$|X_C| \ll R$
Therefore,
$|Z| \approx R$
$\phi \approx 0°$
PF ≈ 1
$P_A \approx P_W$

Something very special happens when we vary the frequency applied to a series circuit containing resistance and both capacitive and inductive reactance. At low frequencies, X_C is high and X_L low. At high frequencies, they reverse, X_C is low and X_L high. Somewhere between these extremes, a frequency is reached at which X_L and X_C are equal. Since they have opposite signs, they cancel, leaving only resistance in the circuit to limit current. The frequency at which this impedance minimum occurs is called the *resonant* frequency; the circuit is said to be in *resonance*.

Chapter 19 examines resonance in detail. For now, we simply investigate the action of a series *RLC* circuit for low, resonant, and high frequencies (Figure 16.4).

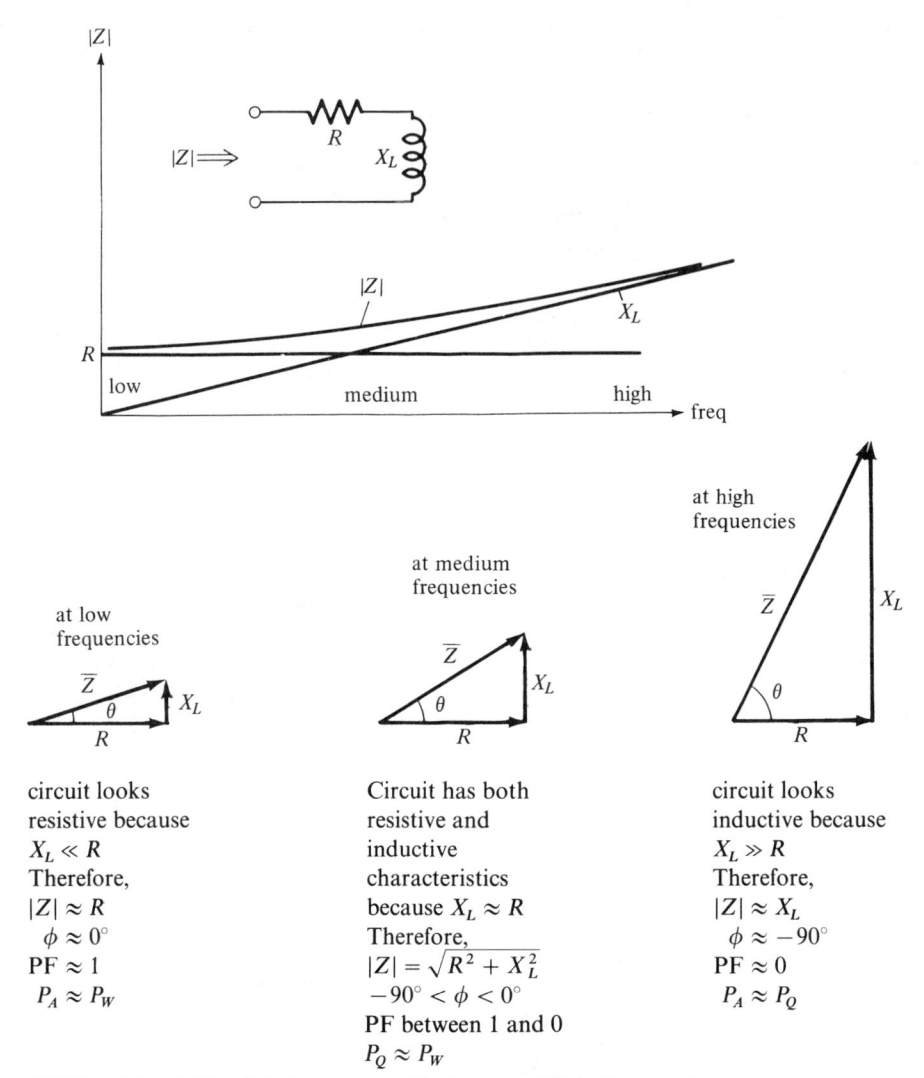

FIGURE 16.3 *RL* Circuit Action at Low, Medium, and High Frequencies.

at low frequencies

circuit looks resistive because
$X_L \ll R$
Therefore,
$|Z| \approx R$
$\phi \approx 0°$
PF ≈ 1
$P_A \approx P_W$

at medium frequencies

Circuit has both resistive and inductive characteristics because $X_L \approx R$
Therefore,
$|Z| = \sqrt{R^2 + X_L^2}$
$-90° < \phi < 0°$
PF between 1 and 0
$P_Q \approx P_W$

at high frequencies

circuit looks inductive because
$X_L \gg R$
Therefore,
$|Z| \approx X_L$
$\phi \approx -90°$
PF ≈ 0
$P_A \approx P_Q$

16.5   *RL* AND *RC* FILTER CIRCUITS

Consider the response of the *RL* voltage divider in Figure 16.5 as e_{in} varies over a wide range of frequencies.

At low frequencies, the inductor is a short, so $R \gg X_L$. Therefore, most of e_{in} is across *R* and the input of the driven circuit [Figure 16.6, part (a)]; $V_{out} \approx e_{in}$. (*Note:* The input impedance of the driven circuit is $\gg R$ at all frequencies and does not load the output of the *RL* circuit.)

At high frequencies, the inductor is an open, $X_L \gg R$. Most of e_{in} is dropped across this high-impedance, leaving little across *R* [Figure 16.6, part (b)]. Therefore, $v_{out} \approx 0$ V.

The *RL* circuit thus passes low frequencies and blocks, or filters out, high frequencies. As such, it is called a *low-pass filter*.

16.5 *RL* AND *RC* FILTER CIRCUITS

at low frequencies

$|X_C| \gg X_L$
and
$|X_C - X_L| \gg R$
Therefore,
$|Z| \approx |X_C|$
Circuit looks capacitive
$\phi \approx 90°$
PF ≈ 0 lead
$P_A \approx P_Q$

at the resonant frequency

$|X_C| = X_L$
Therefore,
$X_L - X_C = 0$
and
$|Z| = R$
Circuit looks resistive
$\theta = 0°$
PF $= 1$
$P_A = P_W$

at high frequencies

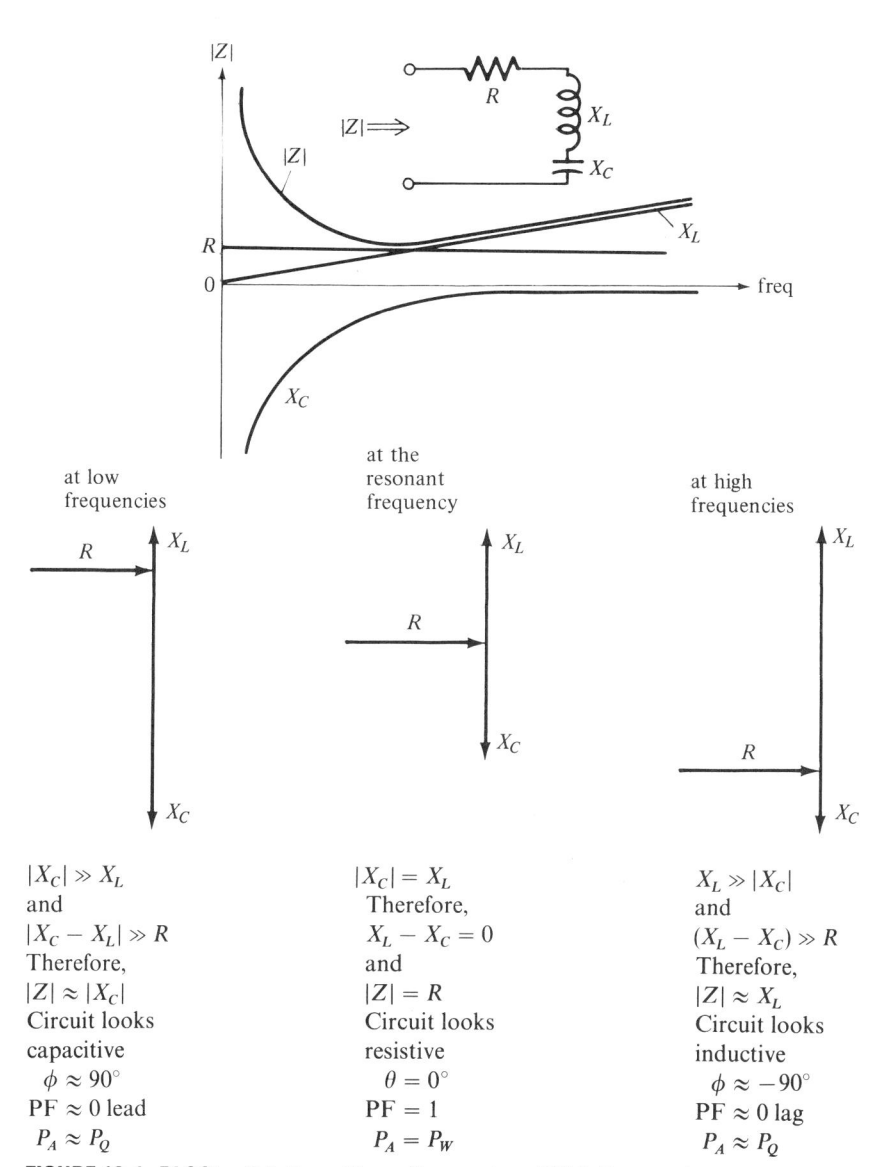

$X_L \gg |X_C|$
and
$(X_L - X_C) \gg R$
Therefore,
$|Z| \approx X_L$
Circuit looks inductive
$\phi \approx -90°$
PF ≈ 0 lag
$P_A \approx P_Q$

FIGURE 16.4 *RLC* Circuit Action at Low, Resonant, and High Frequencies.

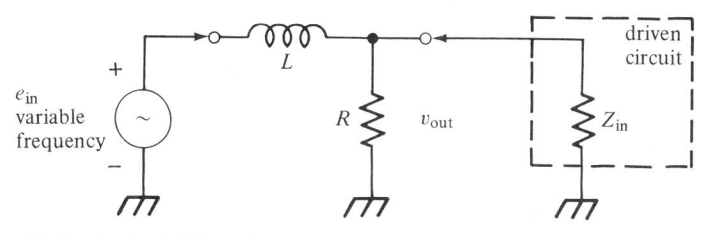

FIGURE 16.5 *RL* Filter Circuit.

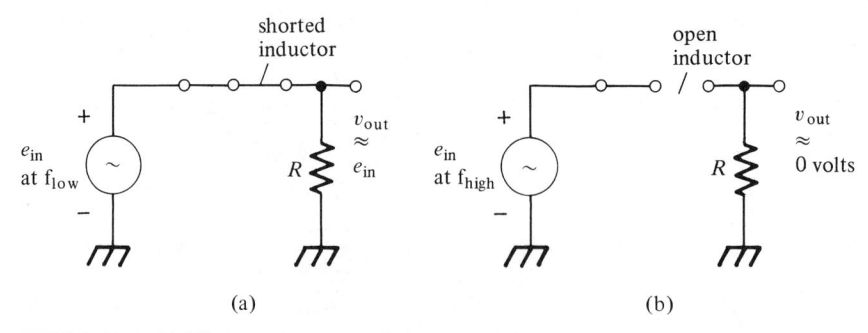

FIGURE 16.6 *RL* Filter at Low and High Frequencies.

Figure 16.7 shows an entire plot of v_{out} versus frequency for this filter. Such a plot is called a *frequency response*.

In the pass region of the plot, that is, at low frequencies, we have seen that $R \gg X_L$; therefore, $v_{out} \approx e_{in}$. As frequency increases, X_L increases, causing v_{out} to decrease. v_{out} eventually falls to 70.7% of e_{in} at a frequency called the *half-power frequency*, which is labeled, f_{HP}. Here X_L has risen to equal R.

To show that power drops to half when $X_L = R$, let us derive an equation for resistor power dissipation. At any frequency,

$$P_R = \frac{V_R^2}{R} \tag{16.4}$$

This power is maximum at low frequencies when $V_R \approx E_{in}$. Then

$$P_{R_{max}} \approx \frac{E_{in}^2}{R} \tag{16.5}$$

By voltage division,

$$V_R = \frac{E_{in} R}{R + jX_L} \tag{16.6}$$

Divide numerator and denominator by R:

$$= \frac{E_{in}}{1 + jX_L/R} \tag{16.7}$$

At f_{HP}, $X_L = R$; therefore,

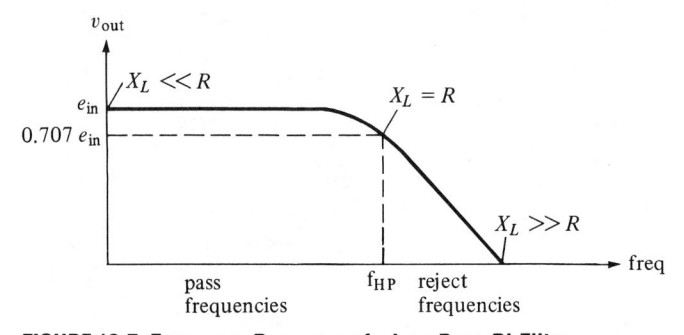

$$V_R = \frac{E_{in}}{1 + j1} = \frac{E_{in}}{1.414 \underline{/45°}} \tag{16.8}$$

FIGURE 16.7 Frequency Response of a Low-Pass *RL* Filter.

16.5 RL AND RC FILTER CIRCUITS

Taking just the magnitude:

$$V_R = \frac{E_{in}}{1.414}$$
$$= 0.707 E_{in} \quad (16.9)$$

Substitute in Equation 16.4:

$$P_R \text{ at } f_{HP} = \frac{(0.707 E_{in})^2}{R}$$
$$= 0.5 \frac{E_{in}^2}{R}$$

Substitute $P_{R_{max}}$ for E_{in}^2/R:

$$P_R \text{ at } f_{HP} = 0.5 P_{R_{max}}$$

The half-power frequency is chosen as the separation between a filter's pass and reject ranges.

A formula for f_{HP} is easily derived: At f_{HP},

$$X_L = R$$

or

$$2\pi f_{HP} = R$$

Solve for f_{HP}:

$$f_{HP} = \frac{R}{2\pi L} \quad (16.10)$$

For RC circuits, this will be

$$f_{HP} = \frac{1}{2\pi RC} \quad (16.11)$$

EXAMPLE 16.5

Calculate the half-power frequency for this circuit and draw the response curve.

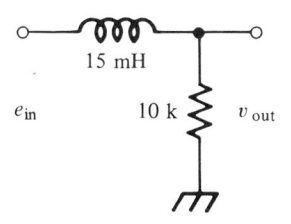

$$f_{HP} = \frac{R}{2\pi L}$$
$$= \frac{10,000}{2\pi(15 \times 10^{-3})}$$
$$= 106 \text{ kHz}$$

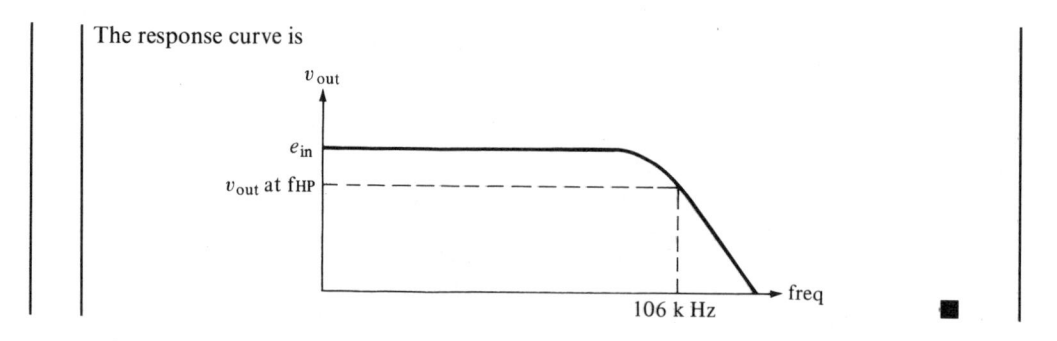

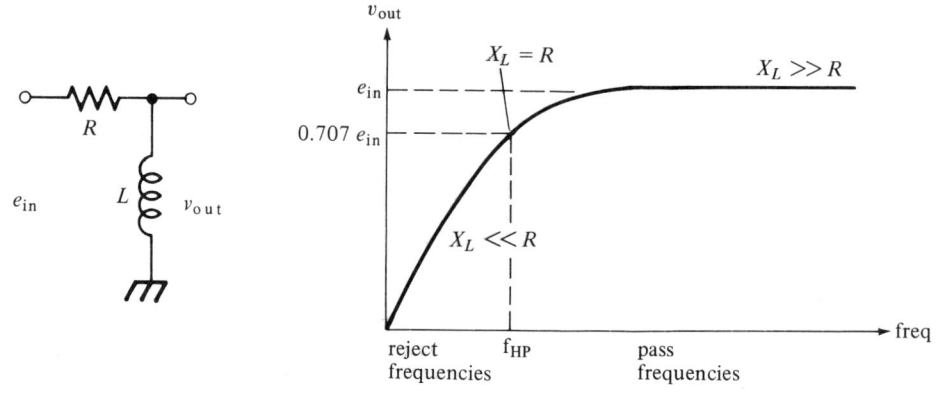

FIGURE 16.8 *RL* Circuit as a High-Pass Filter.

If the resistor and inductor swap positions so that v_{out} is taken across the inductor, the circuit becomes a high-pass filter. It is shown with its response curve in Figure 16.8.

It is left as an exercise for the student to decide where v_{out} should be taken from an *RC* circuit to get low- and high-pass characteristics.

16.6 Z_{TOTAL} OF PARALLEL IMPEDANCES: SUSCEPTANCE AND ADMITTANCE

The resistive–reactive connection in Figure 16.9 is a parallel circuit because all elements have the same voltage across them.

Using Kirchhoff's current law (KCL) and mathematics from Chapter 5, we can derive the familiar product-over-sum (POS) rule for Z_{total} of two parallel branches:

$$\bar{Z}_{total} = \frac{Z_1 Z_2}{Z_1 + Z_2} \tag{16.12}$$

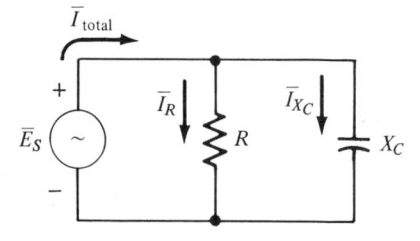

FIGURE 16.9 Parallel *RC* Circuit.

EXAMPLE 16.6

Find the total impedance looking into the terminals of this circuit.

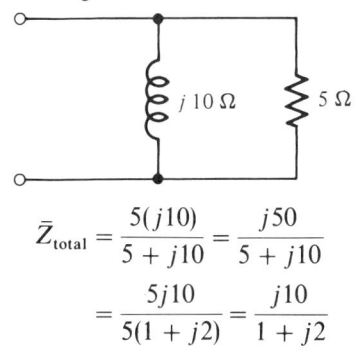

$$\bar{Z}_{total} = \frac{5(j10)}{5+j10} = \frac{j50}{5+j10}$$

$$= \frac{5j10}{5(1+j2)} = \frac{j10}{1+j2}$$

This division can be done in either of two ways:

- Convert numerator and denominator to polar form.
- Multiply numerator and denominator by complex conjugate of the denominator.

We will choose the second method:

$$\bar{Z}_{total} = \frac{j10}{1+j2} \cdot \frac{1-j2}{1-j2} = \frac{j10(1-j2)}{1^2+2^2}$$

$$= \frac{20}{5} + j\frac{10}{5} = 4 + j2 \; \Omega$$

(Note how the denominator is $R^2 + X_L^2$. Save time by remembering this pattern. It always results when a complex number is multiplied by its conjugate.)

As in dc analysis, three or more components in parallel may be combined by repeated use of the product-over-sum formula.

EXAMPLE 16.7

An $X_c = -j4 \; \Omega$ is placed in shunt with the circuit of example 16.6. Find the new $\bar{Z}_{total}$:

Original $\bar{Z}_{total}$ from Example 16.6 is $4 + j2$.

$$\text{new } \bar{Z}_{total} = \frac{(4+j2)(-j4)}{(4+j2)+(-j4)} = \frac{-j16+8}{4-j2} = \frac{8(1-j2)}{2(2-j1)}$$

$$= 4\frac{1-j2}{2-j1} = 4\frac{2.2\,\underline{/-63.4°}}{2.2\,\underline{/-26.6°}}$$

$$= 4\,\underline{/-36.8°}\; \Omega$$

Note how the additional parallel component caused the magnitude of $\bar{Z}_{total}$ to decrease and the phase to switch from lagging to leading.

The total parallel resistance of dc circuits was also found by reciprocating each resistance to get a conductance, and adding these to get G_{total}. This was then reciprocated to get R_{total}.

To use the same approach in ac analysis, both resistance and reactance are reciprocated and combined by complex algebra. The reciprocal of resistance is, as before, conductance, G, measured in siemens. The reciprocal of reactance is called *susceptance*.

Susceptance is a measure of how susceptible an inductor or capacitance is to the flow of current. Indicated by the letter B and also measured in siemens, *inductive susceptance* is

$$B_L = \frac{1}{X_L} = \frac{1}{j\omega L} = -j\frac{1}{\omega L} = \frac{1}{\omega L}\underline{/-90°} = \frac{1}{2\pi f L}\underline{/-90°} \quad \text{siemens} \quad (16.13)$$

and *capacitive susceptance* is

$$B_C = \frac{1}{X_C} = \frac{1}{-j(1/\omega C)} = j\omega C = \omega C\underline{/90°} = 2\pi f C\underline{/90°} \quad \text{siemens} \quad (16.14)$$

Note carefully the following points about susceptance:

1. Reciprocation causes a sign reversal.
   a. Positive inductive reactance becomes negative inductive susceptance.
   b. Negative capacitive reactance becomes positive capacitive susceptance.
2. Inductive susceptance is indirectly proportional to L and f.
3. Capacitive susceptance is directly proportional to C and f.

Like resistance and reactance, conductance and susceptance are plotted as real and imaginary numbers, respectively, on a complex plane.

EXAMPLE 16.8

What is the susceptance of a 167-mH inductance at a frequency of 4000 rad/s?

$$B_L = -j\frac{1}{\omega L} = -j\frac{1}{(4 \times 10^3)(167 \times 10^{-3})}$$
$$= -j1.5 \times 10^{-3}$$
$$= -j1.5 \text{ mS} \qquad \blacksquare$$

EXAMPLE 16.9

What is the susceptance of a 40-Ω capacitive reactance?

$$B_C = \frac{1}{X_C} = \frac{1}{40\underline{/-90°}} = 25 \times 10^{-3}\underline{/90°} \text{ S} \qquad \blacksquare$$

The complex combination of conductance and susceptance is called *admittance*. It indicates a circuit's total ability to admit current flow and is represented by the letter Y.

$$\bar{Y} = G \pm jB \qquad (16.15)$$

16.6 Z_{total} OF PARALLEL IMPEDANCES: SUSCEPTANCE AND ADMITTANCE

TABLE 16.1 Admittance of Pure Conductance and Susceptance

Conductance	Inductive Susceptance	Capacitive Susceptance
$\bar{Y} = G$	$\bar{Y} = B_L$	$\bar{Y} = B_C$
$= G\,\underline{/0°}$	$= B_L\,\underline{/-90°}$	$= B_C\,\underline{/90°}$
$= G + j0$	$= 0 - jB_L$	$= 0 + jB_C$

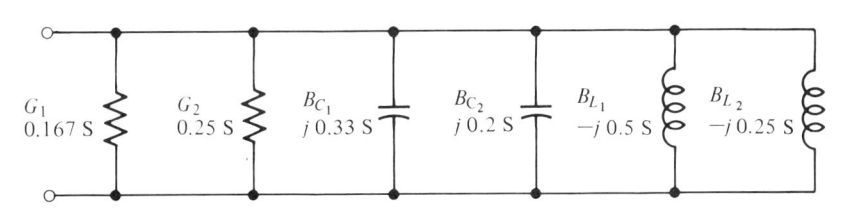

FIGURE 16.10 A Parallel G/B Circuit.

G_1 0.167 S, G_2 0.25 S, B_{C_1} j0.33 S, B_{C_2} j0.2 S, B_{L_1} $-j$0.5 S, B_{L_2} $-j$0.25 S

or, in polar form,

$$\bar{Y} = Y\,\underline{/\pm\Theta} \tag{16.16}$$

Conductance and susceptance are actually admittances with one component zero. Table 16.1 shows the three possibilities in polar and rectangular form together with their polar diagrams.

By KCL and the mathematics used to find total conductance in Chapter 5, we can show that the admittance of the circuit in Figure 16.10 is

$$\bar{Y}_{total} = G_1 + G_2 + jB_{C_1} + jB_{C_2} - jB_{L_1} - jB_{L_2}$$
$$= 0.167 + 0.25 + j0.33 + j0.2 - j0.5 - j0.25$$

Combine reals and imaginaries:

$$\bar{Y}_{total} = 0.417 + j0.53 - j0.75$$
$$= 0.417 - j0.22 \text{ S}$$

which, drawn as a circuit, is

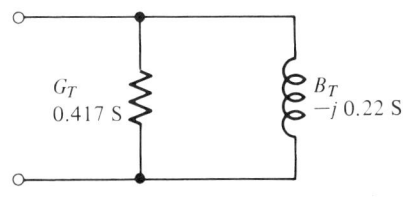

G_T 0.417 S, B_T $-j$0.22 S

FIGURE 16.11 Parallel Circuit Reduced to Two Elements.

The discussion above makes two important points:

- Admittance in rectangular form describes a parallel connection of G and B.
- As conductances in parallel add to G_{total}, so, too, do susceptances in parallel add to B_{total}. This leads to the following formula for the total admittance of a parallel circuit:

$$\bar{Y}_{total} = G_{total} + j(B_{C_{total}} - B_{L_{total}}) \qquad (16.17)$$

Total impedance is obtained by reciprocating Y_{total}, that is,

$$\bar{Z}_{total} = \frac{1}{\bar{Y}_{total}} \qquad (16.18)$$

EXAMPLE 16.10

Perform the following operations for this parallel circuit.

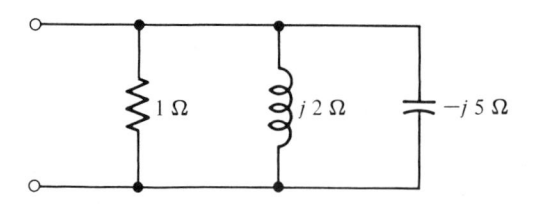

a. Calculate the admittance of each branch.

$$G = \frac{1}{1} = 1 \text{ S}$$

$$B_L = \frac{1}{X_L} = \frac{1}{j2} = -j0.5 \text{ S}$$

$$B_C = \frac{1}{X_C} = \frac{1}{-j5} = j0.2 \text{ S}$$

b. Draw the admittance triangle.

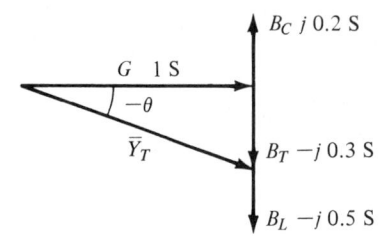

c. Calculate circuit input admittance.

$$\bar{Y}_{total} = G + j(B_C - B_L)$$
$$= 1 + j(0.2 - 0.5)$$
$$= 1 - j0.3 \text{ S}$$

16.6 Z_{total} OF PARALLEL IMPEDANCES: SUSCEPTANCE AND ADMITTANCE

d. Determine circuit input impedance.

$$\bar{Z}_{total} = \frac{1}{\bar{Y}}$$

$$= \frac{1}{1 - j0.3}$$

Convert denominator to polar form and divide:

$$\bar{Z}_{total} = \frac{1}{1 \underline{/-16.7°}}$$

$$= 1 \underline{/16.7°} \ \Omega$$

In the next example, impedances are converted to admittances. Note carefully the approach to the mathematics.

EXAMPLE 16.11

Convert each impedance to an admittance.

a. $\bar{Z} = 200 + j400 \ \Omega$.

$$\bar{Y} = \frac{1}{\bar{Z}}$$

$$= \frac{1}{200 + j400}$$

$$= \frac{1}{200(1 + j2)}$$

$$= \frac{5 \times 10^{-3}}{(1 + j2)} \cdot \frac{(1 - j2)}{(1 - j2)}$$

$$= \frac{5 \times 10^{-3}(1 - j2)}{1^2 + 2^2}$$

$$= 10^{-3}(1 - j2)$$

$$= (1 - j2) \ \text{mS}$$

b. $\bar{Z} = 50k - j120k$.

$$\bar{Y} = \frac{1}{\bar{Z}}$$

$$= \frac{1}{50k - j120k}$$

$$= \frac{1}{10^4(5 - j12)}$$

$$= \frac{10^{-4}}{13 \underline{/-67.4°}}$$

$$= 7.7 \times 10^{-6} \underline{/67.4°}$$

$$= (3 + j7.1) \ \mu\text{S}$$

16.7 ANALYZING PARALLEL AC CIRCUITS

General Comments
Parallel ac circuits obey the same rules as parallel dc circuits:

- Same voltage across all branches.
- Branch current inversely proportional to branch impedance, directly proportional to branch admittance.
- Total current equals the sum of the branch currents (Kirchhoff's current law, KCL). Of course, in ac analysis summing must be done according to the rules of complex algebra.

All the comments in Section 16.2 about calculating power in series circuits hold for parallel circuits with these additions:
Substituting G for R, we have additional equations for P_W:

$$P_W = \frac{I^2}{G} \quad \text{watt} \tag{16.19}$$

and

$$P_W = V^2 G \quad \text{watt} \tag{16.20}$$

Substituting B for X, we have additional equations for P_Q:

$$P_Q = \frac{I^2}{B} \quad \text{var} \tag{16.21}$$

and

$$P_Q = V^2 B \quad \text{var} \tag{16.22}$$

In admittance terms, the circuit power factor is

$$\text{PF} = \frac{G}{Y} \tag{16.23}$$

EXAMPLE 16.12

Perform the following steps for this circuit.

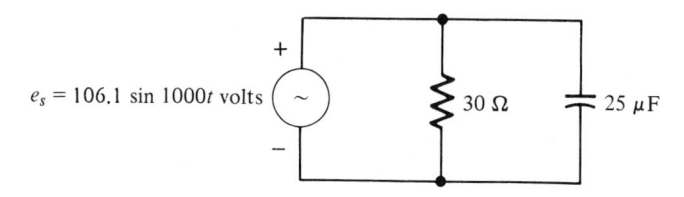

$e_s = 106.1 \sin 1000t$ volts, $30\ \Omega$, $25\ \mu F$

a. Find the admittance of each branch.

$$G = \frac{1}{R} = \frac{1}{30} = 33 \times 10^{-3}\ \text{S}$$

$$B_C = j\omega C = j10^3(25 \times 10^{-6})$$
$$= j25 \times 10^{-3}\ \text{S}$$

16.7 ANALYZING PARALLEL AC CIRCUITS

b. Find each branch current.

Convert source voltage to phasor form:

$$\bar{E}_s = 0.707(106.1) = 75 \underline{/0°} \text{ V}$$
$$\bar{I}_G = \bar{E}_s G = 75(33 \times 10^{-3}) = 2.5 \underline{/0°} \text{ A}$$
$$\bar{I}_{B_C} = \bar{E}_s B_C = 75(j25 \times 10^{-3}) = 1.9 \underline{/90°} \text{ A}$$

c. Add $\bar{I}_G$ and $\bar{I}_{B_C}$ to get $\bar{I}_{\text{total}}$.

$$\bar{I}_{\text{total}} = \bar{I}_G + \bar{I}_{B_C}$$
$$= (2.5 + j0) + (0 + j1.9) \text{ A}$$
$$= 2.5 + j1.9 \text{ or } 3.1 \underline{/37.2°} \text{ A}$$

d. Draw the phasor diagram.

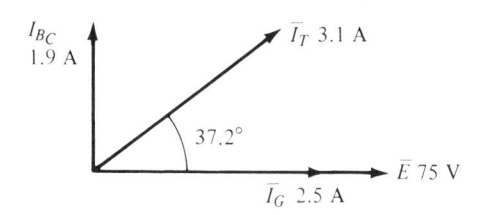

e. Calculate $\bar{Z}_{\text{total}}$ by Ohm's law.

$$\bar{Z}_{\text{total}} = \frac{\bar{E}_s}{\bar{I}_{\text{total}}}$$
$$= \frac{75 \underline{/0°}}{3.1 \underline{/37.2°}}$$
$$= 24.2 \underline{/-37.2°} \text{ or } 19.3 - j14.6 \text{ }\Omega$$

f. Calculate $\bar{Z}_{\text{total}}$ by POS. Compare results to those of part e.

$$X_C = \frac{1}{B_C} = \frac{1}{j25 \times 10^{-3}} = -j40 \text{ }\Omega$$

$$\bar{Z}_{\text{total}} = \frac{30(-j40)}{30 - j40} = \frac{-j1200}{30 - j40} = \frac{-j120}{3 - j4}$$
$$= \frac{-j120}{3 - j4} \cdot \frac{3 + j4}{3 + j4} = \frac{480 - j360}{25}$$
$$= 19.2 - j14.4 \text{ }\Omega$$

g. Calculate $\bar{Z}_{\text{total}}$ by reciprocating $\bar{Y}_{\text{total}}$. Compare the results with those of parts e and f.

$$\bar{Y}_{\text{total}} = G + jB = (33 + j25) \times 10^{-3} \text{ S}$$

$$\bar{Z}_{\text{total}} = \frac{1}{(33 + j25) \times 10^{-3}}$$
$$= \frac{1000}{41.4 \underline{/37.2°}}$$
$$= 24.2 \underline{/-37.2°} \text{ or } 19.3 - j14.6 \text{ }\Omega$$

488   ANALYZING SERIES AND PARALLEL AC CIRCUITS

h. Draw the admittance and impedance triangles.

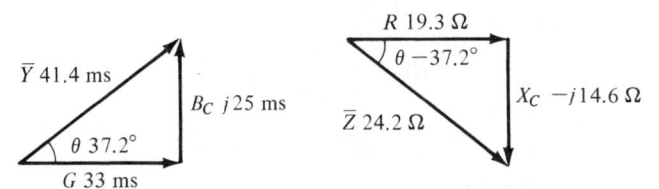

EXAMPLE 16.13

Perform the following operations for this *RL* circuit:

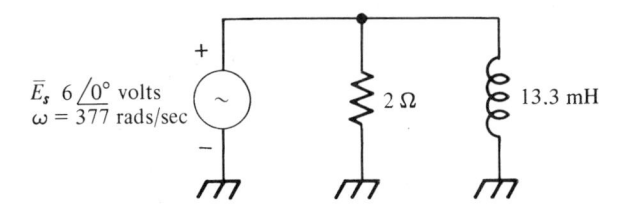

$\bar{E}_s$ 6 $\underline{/0°}$ volts
$\omega = 377$ rads/sec
$2\,\Omega$ $13.3\,\text{mH}$

a. Calculate the current through *R* and *L*.

$$\bar{I}_R = \frac{\bar{E}_s}{R} = \frac{6\,\underline{/0°}}{2\,\underline{/0°}}$$
$$= 3\,\underline{/0°}\;\text{A}$$
$$\bar{I}_L = \frac{\bar{E}_s}{X_L} = j\frac{6}{377(13.3 + 10^{-3})}$$
$$= \frac{6}{j5}$$
$$= 1.2\,\underline{/-90°}\;\text{A}$$

b. Using KCL at the upper node, determine $\bar{I}_{\text{total}}$.

By KCL,
$$\bar{I}_{\text{total}} = \bar{I}_R + \bar{I}_L$$
$$= 3\,\underline{/0°} + 1.2\,\underline{/-90°}$$
$$= 3 - j1.2 \quad \text{or} \quad 3.2\,\underline{/-21.8°}\;\text{A}$$

c. Determine $\bar{Y}_{\text{total}}$ by Ohm's law.

$$\bar{Y} = \frac{\bar{I}_s}{\bar{E}_s} = \frac{3.2\,\underline{/-21.8°}}{6\,\underline{/0°}}$$
$$= 0.53\,\underline{/-21.8°} \quad \text{or} \quad 0.5 - j0.2\;\text{S}$$

d. Using POS, find $\bar{Z}_{\text{total}}$.

From part a, $X_L = j5\,\Omega$.
$$\bar{Z}_{\text{total}} = \frac{R(jX_L)}{R + jX_L} = \frac{2(j5)}{2 + j5}$$
$$= \frac{j10}{2 + j5} \cdot \frac{2 - j5}{2 - j5} = \frac{50 + j20}{29}$$
$$= 1.7 + j0.7\,\Omega$$

16.7 ANALYZING PARALLEL AC CIRCUITS

e. Reciprocate $\bar{Z}_{total}$ of part d to obtain $\bar{Y}_{total}$. Compare with the results of part c.

$$\bar{Y}_{total} = \frac{1}{\bar{Z}_{total}}$$
$$= \frac{1}{1.7 + j0.7}$$
$$= \frac{1}{1.84 \underline{/22.4°}}$$
$$= 0.54 \underline{/-22.4°} \quad \text{or} \quad 0.5 - j0.2 \text{ S}$$

Results same as part c.

f. Calculate the resistor's dissipation.

$$P_R = \frac{V_R^2}{R} = \frac{6^2}{2} = 18 \text{ W}$$

g. Calculate the reactive power drawn by the inductor:

$$P_{Q_L} = I_L^2 X_L = (1.2^2)(5) = 7.2 \text{ var, inductive}$$

h. Determine circuit PF.

$$\text{PF} = \cos \phi$$

From part b, $\phi = -21.8°$.

$$\text{PF} = \cos(-21.8°) = 0.93 \quad \text{or} \quad 93\% \text{ lag}$$ ■

EXAMPLE 16.14

Perform the indicated operations for the following parallel *RLC* circuit.

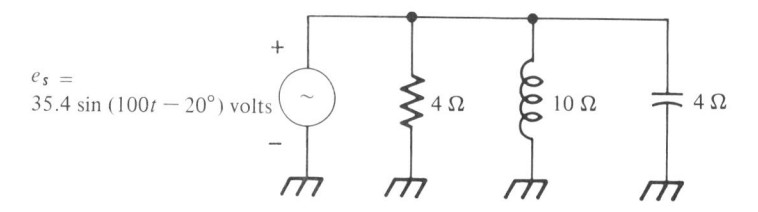

$e_s = 35.4 \sin(100t - 20°)$ volts; $4\,\Omega$, $10\,\Omega$, $4\,\Omega$

a. Find the admittance of each branch.

$$G = \frac{1}{R} = \frac{1}{4} = 0.25 \text{ S}$$

$$B_L = \frac{1}{X_L} = \frac{1}{j10} = -j0.1 \text{ S}$$

$$B_C = \frac{1}{X_C} = \frac{1}{-j4} = j0.25 \text{ S}$$

b. Find $\bar{Y}_{total}$ and $\bar{Z}_{total}$.

$$\bar{Y}_{total} = G + jB_C - jB_L$$
$$= 0.25 + j(0.25 - 0.1)$$
$$= 0.25 + j0.15 \quad \text{or} \quad 0.29 \underline{/31°} \text{ S}$$

$$\bar{Z}_{total} = \frac{1}{\bar{Y}_{total}}$$

$$= \frac{1}{0.25 + j0.15}$$

$$= \frac{1}{0.29 \underline{/31°}}$$

$$= 3.4 \underline{/-31°} \quad \text{or} \quad 2.9 - j1.8 \, \Omega$$

c. Draw the admittance triangle.

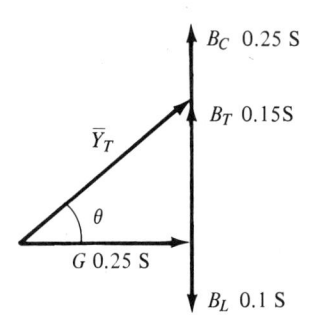

[Admittance triangle diagram showing B_C 0.25 S, B_T 0.15S, $\bar{Y}_T$, angle θ, G 0.25 S, B_L 0.1 S]

d. Calculate the three component currents. Draw the circuit's phasor diagram.

Convert source voltage to phasor form:

$$\bar{E}_s = 25 \underline{/-20°} \text{ V}$$

$$\bar{I}_R = \frac{\bar{E}_s}{R} = \frac{25 \underline{/-20°}}{4} = 6.25 \underline{/-20°} \text{ A}$$

$$\bar{I}_L = \frac{\bar{E}_s}{X_L} = \frac{25 \underline{/-20°}}{10 \underline{/90°}} = 2.5 \underline{/-110°} \text{ A}$$

$$\bar{I}_C = \frac{\bar{E}_s}{X_C} = \frac{25 \underline{/-20°}}{4 \underline{/-90°}} = 6.25 \underline{/70°} \text{ A}$$

Draw the phasor diagram:

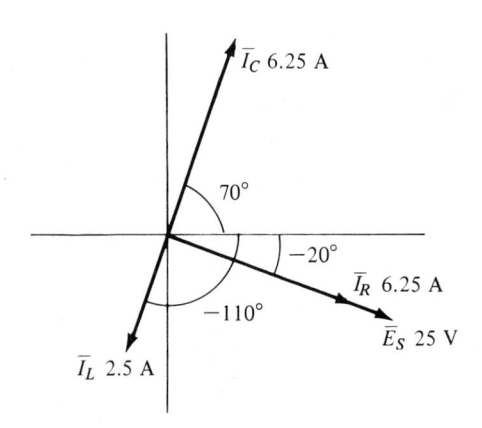

[Phasor diagram showing $\bar{I}_C$ 6.25 A at 70°, $\bar{I}_R$ 6.25 A at −20°, $\bar{E}_S$ 25 V, $\bar{I}_L$ 2.5 A at −110°]

16.7 ANALYZING PARALLEL AC CIRCUITS

e. Calculate apparent power, true power, and the two reactive powers.

Draw the power triangle:

$$P_W = E_s I_R = 25(6.25) = 156.3 \text{ W}$$
$$P_{Q_{ind}} = E_s I_L = 25(2.5) = -62.5 \text{ var}$$
$$P_{Q_{cap}} = E_s^2 B_C = 25^2(0.25) = 156.3 \text{ var}$$
$$P_{Q_{total}} = P_{Q_{cap}} - P_{Q_{ind}}$$
$$= 156.3 - 62.5$$
$$= 93.8 \text{ var}$$
$$P_A = E_s I_{total}$$
$$= 25(7.2) = 180 \text{ VA}$$

Draw the power triangle:

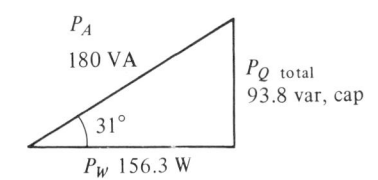

```
       P_A
       180 VA              P_Q total
                           93.8 var, cap
          31°
       P_W 156.3 W
```

f. Add the three component currents to find $\bar{I}_{total}$.

By KCL,

$$\bar{I}_{total} = \bar{I}_R + \bar{I}_L + \bar{I}_C$$
$$= 6.25 \underline{/-20°} + 2.5 \underline{/-110°} + 6.25 \underline{/70°}$$
$$= (5.9 - j2.1) + (-0.9 - j2.4) + (2.1 + j5.9)$$

Combine reals and imaginaries:

$$I_{total} = (5.9 - 0.9 + 2.1) + j(-2.1 - 2.4 + 5.9)$$
$$= 7.1 + j1.4 \quad \text{or} \quad 7.2 \underline{/11.2°} \text{ A}$$

EXAMPLE 16.15

A current, 17.9/56.3° A, enters the parallel combination of 50 Ω, j20 Ω, and −j5 Ω. What is the current through the capacitive branch?

$$\bar{Y}_{total} = \frac{1}{50} + \frac{1}{j20} + \frac{1}{-j5}$$
$$= (2 \times 10^{-2}) - (j5 \times 10^{-2}) + (j20 \times 10^{-2})$$
$$= (2 + j15) \times 10^{-2} \quad \text{or} \quad 15.1 \times 10^{-2} \underline{/82.4°} \text{ S}$$
$$\bar{E}_s = \frac{\bar{I}}{\bar{Y}} = \frac{17.9 \underline{/56.3°}}{15.1 \times 10^{-2} \underline{/82.4°}}$$
$$= 118.5 \underline{/-26.1°} \text{ V}$$
$$\bar{I}_C = \frac{\bar{E}_s}{\bar{X}_C} = \frac{118.5 \underline{/-26.1°}}{5 \underline{/-90°}}$$
$$= 23.7 \underline{/63.9°} \text{ A}$$

Note how this current leads the input voltage by 90°.

As in Chapter 5, current through any branch of a parallel circuit can be determined with the *current-division formula*:

$$\bar{I}_n = \frac{\bar{I}_{\text{total}}\bar{Z}_{\text{total}}}{\bar{Z}_n} \quad (16.24)$$

where I_n = current through branch n
Z_n = impedance of branch n
Z_{total} = total shunt impedance

For two parallel impedances, Z_1 and Z_2, the formula reduces to

$$\bar{I}_{Z_1} = \frac{\bar{I}_{\text{total}}\bar{Z}_2}{\bar{Z}_1 + \bar{Z}_2} \quad (16.25)$$

and

$$\bar{I}_{Z_2} = \frac{\bar{I}_{\text{total}}\bar{Z}_1}{\bar{Z}_1 + \bar{Z}_2} \quad (16.26)$$

Note again how component 2 is in the numerator for component 1 calculations, and vice versa.

EXAMPLE 16.16

Use the current-division formula to determine the current through each impedance in this circuit.

$\bar{I}_T$ 39 /0° A

$$\bar{I}_R = \frac{\bar{I}_{\text{total}}jX_L}{R + jX_L}$$

$$= \frac{39 \underline{/0°}(12 \underline{/90°})}{5 + j12}$$

$$= \frac{39(12 \underline{/90°})}{13 \underline{/67.4°}}$$

$$= 36 \underline{/22.6°} \text{ A}$$

$$\bar{I}_{X_L} = \frac{\bar{I}_{\text{total}}R}{(R + jX_L)}$$

$$= \frac{39 \underline{/0°}(5 \underline{/0°})}{5 + j12}$$

$$= \frac{39(5 \underline{/0°})}{13 \underline{/67.4°}}$$

$$= 15 \underline{/-67.4°} \text{ A} \qquad \blacksquare$$

16.8 RESPONSE OF PARALLEL *RC*, *RL*, AND *RLC* CIRCUITS TO CHANGING FREQUENCY

In Section 16.4 we learned to watch for an impedance much greater than any other impedance in a series RX circuit. We found that that impedance establishes the circuit's phase and power factor characteristics. It also determines whether the circuit appears resistive or reactive.

Parallel circuits are just the opposite. To develop a "feel" for them:

> Look for the impedance that is much less (or admittance that is much greater) than the others in a circuit at low, medium, or high frequencies. That impedance (or admittance) dominates the circuit. It carries the greatest current and establishes phase and power-factor characteristics.

Remember: Capacitors are opens at low frequencies, shorts at high frequencies. With that in mind, let us examine the characteristics of the parallel *RC* circuit in Figure 16.12 at low, medium, and high frequencies.

$\overline{Z} = R(-jX_C)/(R - jX_C)$
$\overline{Y} = G + jB_C$

at low frequencies

$|X_C| \gg R$
or
$B_C \ll G$

Therefore,
R/G dominates
$|Z| \approx R$
$|Y| \approx G$
$\phi \approx 0°$
PF ≈ 1
$P_A \approx P_W$

at medium frequencies

neither component dominates

Therefore,
$|Z|, |Y|, \theta$, PF, and P_A determined by formula
$0° < \phi < 90°$
$0 < $ PF $ < 1$ lead

at high frequencies

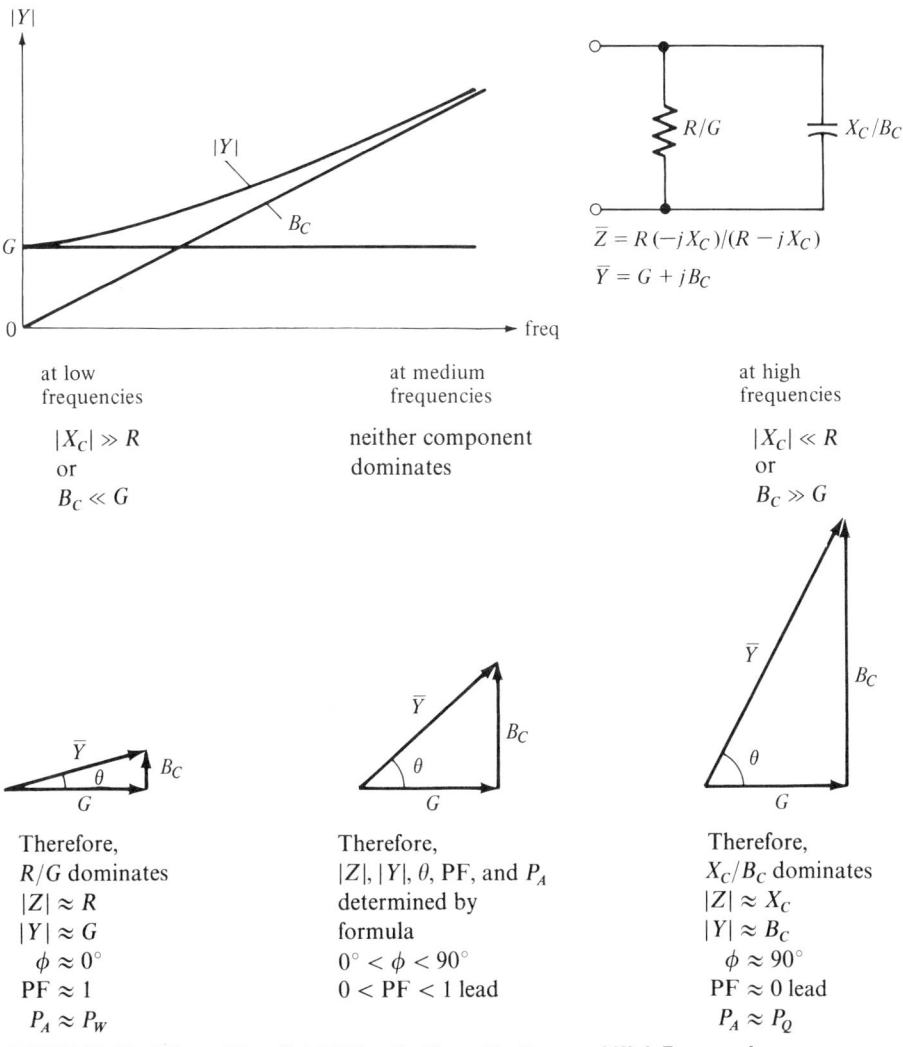

$|X_C| \ll R$
or
$B_C \gg G$

Therefore,
X_C/B_C dominates
$|Z| \approx X_C$
$|Y| \approx B_C$
$\phi \approx 90°$
PF ≈ 0 lead
$P_A \approx P_Q$

FIGURE 16.12 Action of Parallel *RC* Circuit at Low, Medium, and High Frequencies.

494   ANALYZING SERIES AND PARALLEL AC CIRCUITS

Note carefully how the susceptive magnitudes in Figure 16.12 increase with increasing frequency while conductive magnitudes remain constant.

Remembering that inductors are shorts at low frequencies and opens at high frequencies, let us next examine the characteristics of the parallel *RL* circuit in Figure 16.13 at low, medium, and high frequencies.

Consider next a circuit with all three components—resistance, capacitance and inductance—in parallel. The inductance shorts the circuit at low frequencies, the capacitor shorts it at high frequencies. Somewhere between these extremes is a frequency at which the reactive (susceptive) components are equal and opposite. Here

$$\overline{Z} = R(jX_L)/(R + jX_L)$$
$$\overline{Y} = G - jB_L$$

at low frequencies

$X_L \ll R$
or
$B_L \gg G$

Therefore,

X_L/B_L dominates

$|Z| \approx X_L$
$|Y| \approx B_L$
$\phi \approx -90°$
PF ≈ 0 lag
$P_A \approx P_Q$

at medium frequencies

neither component dominates

Therefore,

$|Z|, |Y|, \theta$, PF, and P_A

determined by formula
$-90° < \phi < 0°$
$0 <$ PF < 1 lag

at high frequencies

$X_L \gg R$
or
$B_L \ll G$

Therefore,

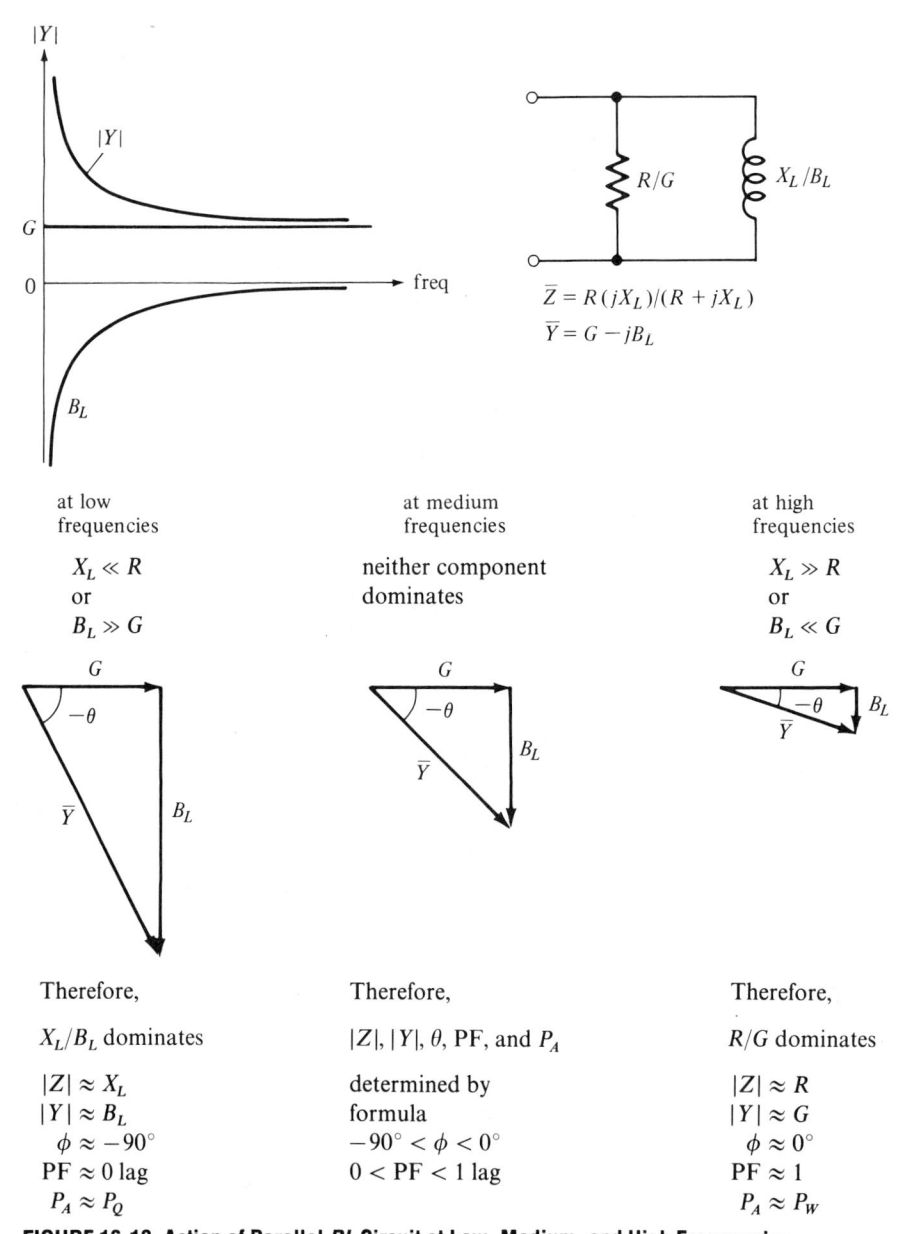

R/G dominates

$|Z| \approx R$
$|Y| \approx G$
$\phi \approx 0°$
PF ≈ 1
$P_A \approx P_W$

FIGURE 16.13 Action of Parallel *RL* Circuit at Low, Medium, and High Frequencies.

16.9 SERIES AND PARALLEL EQUIVALENT CIRCUITS   495

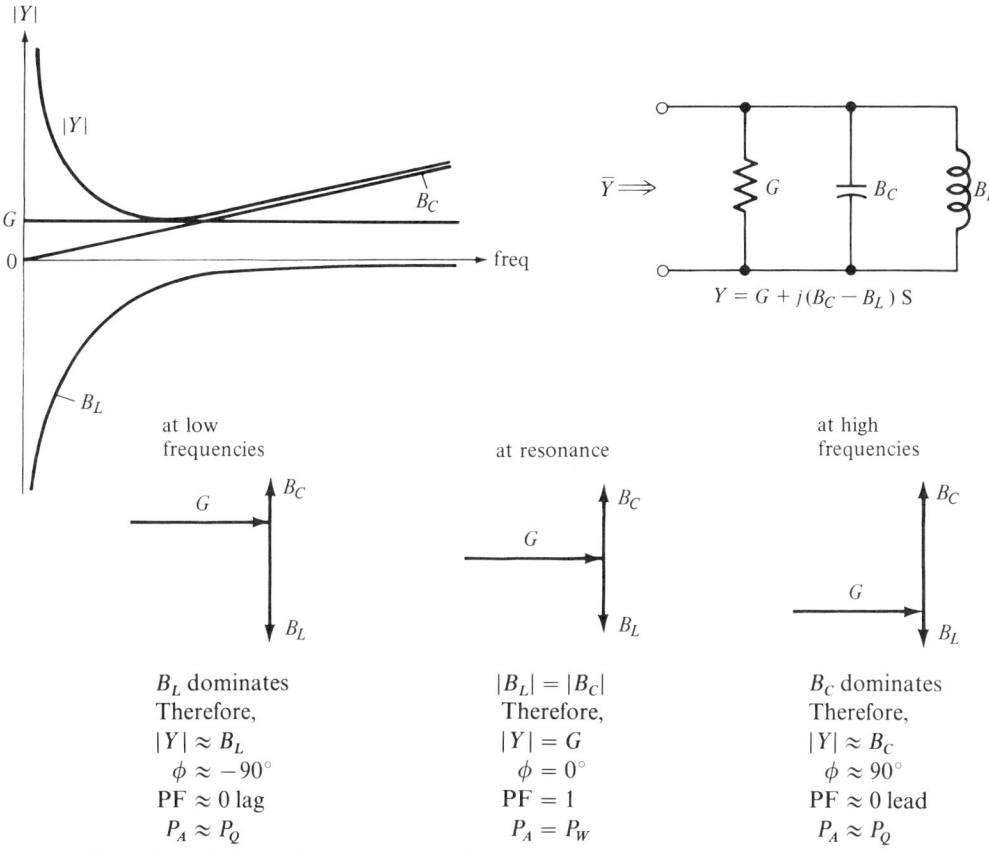

FIGURE 16.14 Action of Parallel *RLC* Circuit at Low, Resonant, and High Frequencies.

At low frequencies:
B_L dominates
Therefore,
$|Y| \approx B_L$
$\phi \approx -90°$
PF $\approx$ 0 lag
$P_A \approx P_Q$

At resonance:
$|B_L| = |B_C|$
Therefore,
$|Y| = G$
$\phi = 0°$
PF = 1
$P_A = P_W$

At high frequencies:
B_C dominates
Therefore,
$|Y| \approx B_C$
$\phi \approx 90°$
PF $\approx$ 0 lead
$P_A \approx P_Q$

they cancel, leaving only the resistance (conductance) in the circuit. The impedance is now maximum or, conversely, the admittance is minimum. The frequency at which this occurs is the *resonant* frequency.

As with series-resonant circuits, parallel resonance is discussed in Chapter 19. For now, let us investigate the action of a parallel *RLC* circuit at low, resonant, and high frequencies (see Figure 16.14).

16.9 SERIES AND PARALLEL EQUIVALENT CIRCUITS

A series ac circuit can be converted to an equivalent parallel circuit. This means that both circuits respond equally to a given source. Similarly, a parallel circuit can be converted to an equivalent series form.

Why convert to an equivalent circuit? Because some circuits are easier to analyze in one form than the other. Let us see how these conversions are done.

The impedance of the left-hand circuit in Figure 16.15 is

$$\bar{Z}_s = R_s + jX_{L_s}$$

Its admittance is

$$\bar{Y} = \frac{1}{\bar{Z}_s} = \frac{1}{R_s + jX_{L_s}}$$

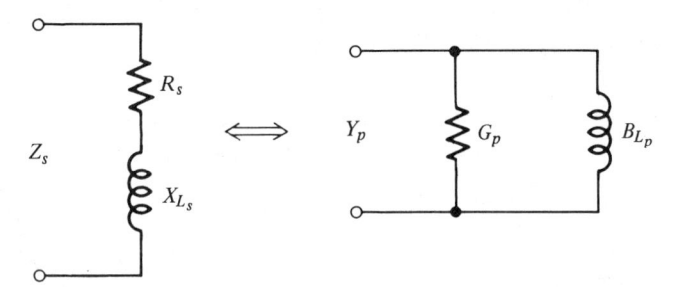

FIGURE 16.15 Converting Series-and Parallel-Connected Components.

Rationalizing the denominator:

$$\bar{Y} = \frac{1}{R_s + jX_{L_s}} \frac{R_s - jX_{L_s}}{R_s - jX_{L_s}} = \frac{R_s}{R_s^2 + X_{L_s}} - j\frac{X_{L_s}}{R_s^2 + (X_{L_s})^2} \quad (16.27)$$

Checking the units of both terms on the right side of this equation, we see that

$$\frac{\text{ohms}}{\text{ohms}^2} = \frac{1}{\text{ohms}} = \text{siemens}$$

Since the first term is real it must be conductance. Therefore,

$$G_p = \frac{R_s}{R_s^2 + (X_{L_s})^2} \quad (16.28)$$

Similarly, the second term is negative imaginary and must be inductive susceptance. Therefore,

$$B_{L_p} = \frac{X_{L_s}}{R_s^2 + (X_{L_s})^2} \quad (16.29)$$

Equation 16.27 thus describes the G/B circuit on the right of Figure 16.15, which is the parallel equivalent of the series circuit on the left.

By reciprocating both terms in Equation 16.27, we get the parallel equivalent components in terms of R_s and X_{L_s}:

$$R_p = \frac{R_s^2 + (X_{L_s})^2}{R_s} \quad (16.30)$$

$$X_{L_p} = \frac{R_s^2 + (X_{L_s})^2}{X_{L_s}} \quad (16.31)$$

Using the same approach, equations can be derived for the parallel equivalent of a series RC circuit:

$$R_p = \frac{R_s^2 + (X_{C_s})^2}{R_s} \quad (16.32)$$

$$X_{C_p} = \frac{R_s^2 + (X_{C_s})^2}{X_{C_s}} \quad (16.33)$$

Note the sum of two squares in these equations. Store it in your calculator memory for convenient recall.

16.9 SERIES AND PARALLEL EQUIVALENT CIRCUITS

EXAMPLE 16.17

Find the parallel equivalent of this *RL* circuit:

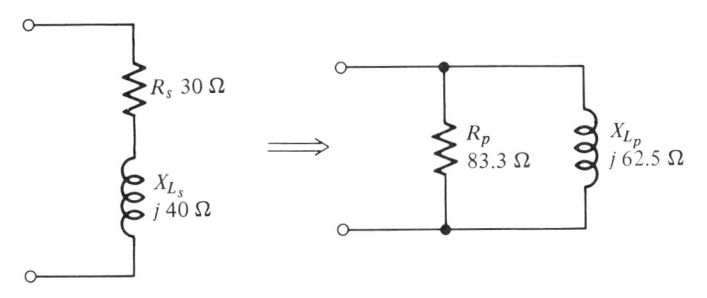

$$R_p = \frac{R_s^2 + (X_{L_s})^2}{R_s} = \frac{30^2 + 40^2}{30} = \frac{2500}{30} = 83.3 \ \Omega$$

$$X_{L_p} = \frac{R_s^2 + (X_{L_s})^2}{X_{L_s}} = \frac{2500}{40} = 62.5 \ \Omega$$

EXAMPLE 16.18

Convert this series *RX* circuit into its parallel equivalent.

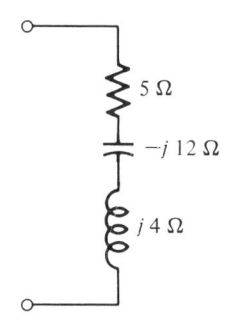

Combine the two reactances:

$$X_{\text{total}} = -j12 + j4 = -j8 \ \Omega$$

Calculate equivalent parallel components:

$$R_p = \frac{R_s^2 + X_s^2}{R_s} = \frac{5^2 + 8^2}{5} = \frac{25 + 64}{5} = 17.8 \ \Omega$$

$$X_p = \frac{R_s^2 + X_s^2}{X_s} = \frac{5^2 + 8^2}{-j8} = \frac{89}{-j8} = j11.1 \ \Omega$$

Note carefully the following important points from Examples 16.17 and 16.18:

- A component's parallel and series values are not the same.
- Series R, L, and C convert to parallel R, L, and C, respectively.
- Because a specific reactance occurs at only one frequency, conversions are valid only for the frequency at which reactance is determined.
- Both R_p and X_p are functions of X_s and are therefore frequency dependent.

We have derived equations that allow us to convert series-connected resistances and reactances to equivalent parallel circuits. Let us next do the opposite—derive equations for converting a parallel circuit to its series equivalent.

The input admittance of the right-hand circuit in Figure 16.15 is

$$\bar{Y}_p = G_p - jB_{L_p}$$

Its impedance is

$$\bar{Z}_p = \frac{1}{\bar{Y}_p}$$

$$= \frac{1}{G_p - jB_{L_p}}$$

Rationalize the denominator:

$$\bar{Z}_p = \frac{1}{G_p - jB_{L_p}} \frac{G_p + jB_{L_p}}{G_p + jB_{L_p}}$$

$$= \frac{G_p}{G_p^2 + (B_{L_p})^2} + j\frac{B_{L_p}}{G_p^2 + (B_{L_p})^2} \quad (16.34)$$

Checking the units of both terms on the right side of this equation, we get

$$\frac{\text{siemens}}{\text{siemens}^2} = \frac{1}{\text{siemens}} = \text{ohms}$$

Since the first term is real it must be resistance. Therefore,

$$R_s = \frac{G_p}{G_p^2 + (B_{L_p})^2} \quad (16.35)$$

Similarly, the second term is positive imaginary. It must be inductive reactance. Therefore,

$$X_{L_s} = \frac{B_{L_p}}{G_p^2 + (B_{L_p})^2} \quad (16.36)$$

Equation 16.34 thus describes the R_s/X_{L_s} circuit on the left of Figure 16.15 which is the series equivalent of the parallel circuit on the right.

Using the same approach, equations can be derived for the series equivalent of a parallel G/B_c circuit:

$$R_s = \frac{G_p}{G_p^2 + (B_{C_p})^2} \quad (16.37)$$

$$X_{C_s} = \frac{B_{C_p}}{G_p^2 + (B_{C_p})^2} \quad (16.38)$$

16.9 SERIES AND PARALLEL EQUIVALENT CIRCUITS

EXAMPLE 16.19

The *Q*, or *quality figure*, of an inductor is defined.

$$Q = \frac{\text{series reactance}}{\text{effective series resistance}}$$

$$= \frac{\omega L_s}{R_s}$$

What is the *Q* of this inductance?

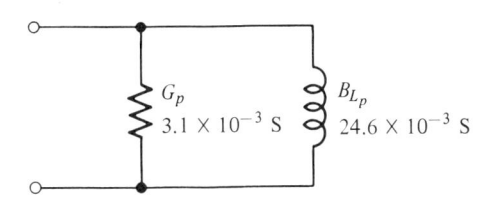

G_p = 3.1×10^{-3} S
B_{L_p} = 24.6×10^{-3} S

Converting the circuit to its series equivalent:

$$R_s = \frac{G_p}{G_p^2 + (B_{L_p})^2}$$

$$= \frac{3.1 \times 10^{-3}}{(3.1 \times 10^{-3})^2 + (24.6 \times 10^{-3})^2} = \frac{3.1 \times 10^{-3}}{615 \times 10^{-6}}$$

$$= 5 \, \Omega$$

$$X_s = \frac{B_{L_p}}{G_p^2 + (B_{L_p})^2}$$

$$= \frac{24.6 \times 10^{-3}}{615 \times 10^{-6}}$$

$$= 40 \, \Omega$$

$$Q = \frac{X_{L_s}}{R_s}$$

$$= \frac{40}{5}$$

$$= 8$$ ∎

EXAMPLE 16.20

A capacitor has the high resistance of its dielectric effectively in shunt with it. By the formulas above, this parallel *RC* circuit can be converted to a series *RC* circuit.

The ratio

$$\frac{R_s}{X_{c_s}}$$

expressed as a percentage, is a measure of capacitor quality called the *dissipation factor*. For most capacitors, its value is 1% or less. Find the dissipation factor of this capacitor:

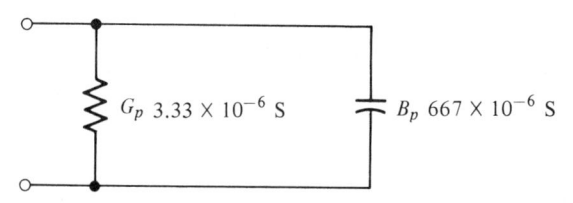

G_p 3.33 × 10⁻⁶ S B_p 667 × 10⁻⁶ S

Convert the circuit to its series equivalent:

$$Rs = \frac{G_p}{G_p^2 + B_p^2} \approx \frac{G_p}{B_p^2} = 7.5 \ \Omega \quad \text{because } B_p^2 \gg G_p^2.$$

$$X_{C_s} = \frac{B_{C_p}}{G_p^2 + (B_{C_p})^2} \approx \frac{B_{C_p}}{(B_{C_p})^2} = \frac{1}{B_{C_p}} = -j1500 \ \Omega$$

$$\%DF = \frac{Rs}{X_{C_s}} \times 100 = \frac{7.5}{1500} \times 100 = 0.5\%$$

■

When parallel components are given in impedance form rather than admittance, the POS formula yields

$$\bar{Z}_{\text{total}} = \frac{R_p(\pm jX_p)}{R_p \pm jX_p}$$

Rationalize the denominator:

$$Z_{\text{total}} = \frac{R_p(\pm jX_p)}{R_p \pm jX_p} \cdot \frac{R_p \mp jX_p}{R_p \mp jX_p}$$

$$= \frac{R_p X_p^2}{R_p^2 \pm X_p^2} \pm j\frac{R_p^2 X_p}{R_p^2 + X_p^2}$$

Note how this statement represents the impedance of a series RX circuit in which

$$R_s = \frac{R_p X_p^2}{R_p^2 + X_p^2} \quad (16.39)$$

$$X_s = \pm j\frac{R_p^2 X_p}{R_p^2 + X_p^2} \quad (16.40)$$

EXAMPLE 16.21

Perform the following steps for this circuit.

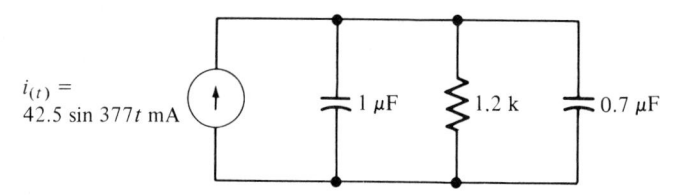

$i_{(t)} =$ 42.5 sin 377t mA   1 μF   1.2 k   0.7 μF

a. By current division, find $\bar{I}_R$.

16.9 SERIES AND PARALLEL EQUIVALENT CIRCUITS

Combine capacitors:
$$C_{total} = C_1 + C_2 = 1.7 \ \mu F$$

Calculate X_C.
$$X_C = \frac{1}{\omega C}$$
$$= \frac{1}{377(1.7 \times 10^{-6})}$$
$$= -j1560 \ \Omega$$

Convert source current to phasor form and apply current division:
$$\bar{I}_R = \frac{0.707(42.5 \ mA)(-j1560)}{1200 - j1560}$$
$$= \frac{46.8 \ \underline{/-90°}}{1968.1 \ \underline{/-52.4°}}$$
$$= 23.8 \ \underline{/-37.6°} \ mA$$

b. Find the circuit's power dissipation.
$$P_R = I^2 R$$
$$= (23.8 \times 10^{-3})^2 (1200)$$
$$= 679.7 \ mW$$

c. Convert the circuit to its series equivalent. (Watch how the 1000 is handled.)
$$R_s = \frac{R_p X_p^2}{R_p^2 + X_p^2}$$
$$= \frac{1.2(1.56)^2}{1.2^2 + 1.56^2}$$
$$= \frac{2.9}{3.9}$$
$$= 0.744 \ k$$
$$= 744 \ \Omega$$

Note how calculations are simplified by leaving the 1000 out until the end.
$$R_s = \frac{R_p X_p^2}{R_p^2 + X_p^2}$$
$$= 1.56 \frac{1.2^2}{3.9}$$
$$= -j580 \ \Omega$$

d. Calculate the circuit's power dissipation by $P_R = I^2 R$. Compare the results with those of part b.
$$P_R = I^2 R$$
$$= (30 \ mA)^2 (744)$$
$$= 670 \ mW$$

As it must, the series equivalent circuit dissipates the same power as the original parallel circuit. ∎

EXERCISES

Section 16.1

1. Determine component values for the simplest series circuit having each impedance at the indicated frequency.
 a. $\bar{Z} = 10 - j15\ \Omega$ at $\omega = 100$ rad/s.
 b. $\bar{Z} = 1.1\ \text{k} + j1.8\ \text{k}$ at $f = 2$ kHz.

2. Repeat Exercise 1 for these impedances.
 a. $\bar{Z} = 350\ \text{k} - j87\ \text{k}$ at $f = 183$ Hz.
 b. $\bar{Z} = 110\ \text{k} + j163\ \text{k}$ at $\omega = 13.6 \times 10^3$ rad/s.

3. Calculate the total impedance of each circuit.

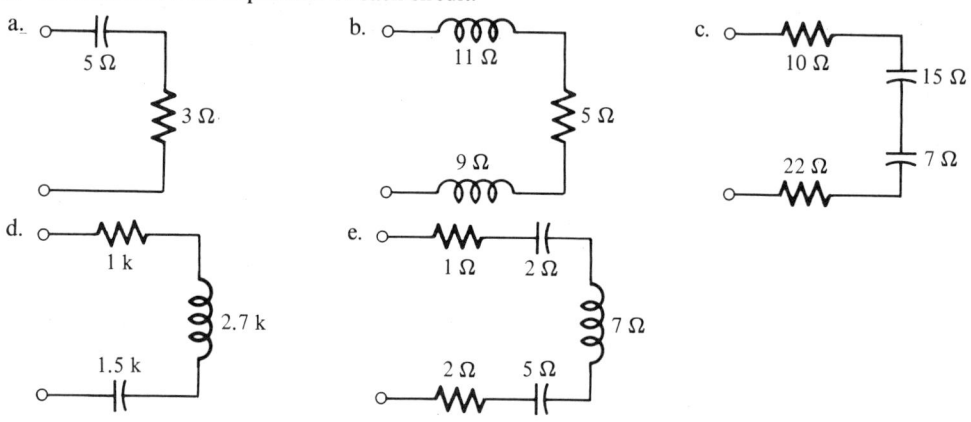

4. Calculate the new impedance of each of the circuits in Exercise 3 if the frequency of the applied signal is doubled.

Section 16.2

5. A 69-Ω resistance and a 0.6-μF capacitance are series-connected across a 190-V rms source, frequency = 1.5 kHz.
 a. Calculate $\bar{Z}_{\text{total}}$ and draw the impedance triangle.
 b. Calculate the circuit current.
 c. Calculate $\bar{V}_R$ and $\bar{V}_C$.
 d. Draw the phasor diagram.
 e. Express V_R, V_C, and I in the time domain.
 f. Calculate P_W, P_Q, and P_A and draw the power triangle.
 g. Calculate the circuit power factor.

6. A 470-Ω resistance and a 2200-pF capacitor are series-connected across a 185-V source, frequency = 181 kHz.
 a. Calculate $\bar{Z}_{\text{total}}$ and draw the impedance triangle.
 b. Calculate $\bar{V}_R$ and $\bar{V}_C$ by voltage division. Express in the phasor domain.
 c. Calculate the phase angle.
 d. Calculate P_W, P_Q, and P_A and draw the power triangle.
 e. Calculate the power factor.

7. A 1-k resistance is series-connected with a 3-H choke across a 166-V rms source at 60 Hz.
 a. Calculate X_L and $\bar{Z}_{\text{total}}$ and draw the impedance triangle.
 b. Calculate the current.
 c. Calculate $\bar{V}_R$ and $\bar{V}_L$. Express in the phasor domain.
 d. Draw the phasor diagram.
 e. Express $\bar{V}_R$, $\bar{V}_L$, and $\bar{I}$ in the time domain.
 f. Calculate P_W, P_Q, and P_A and draw the power triangle.
 g. Calculate the power factor.

EXERCISES

8. A relay coil draws 15 mA at an angle of 52° when connected to a 120-V, 60-Hz source. What are its inductance, resistance, and power factor?

9. Find the values of the two series-connected components that draw 40.8 W from a 235-V 50-kHz source. Power factor = 58%, leading.

10. What inductance must be connected in series with a 25-Ω resistance to make current lag the applied voltage by 57.5° at 25 Hz?

Section 16.3

11. An *RLC* series circuit consisting of $R = 20\ \Omega$, $L = 300\ \mu H$, and $C = 1\ \mu F$ is connected across a 50-V 12-kHz source.
   a. Find $\bar{Z}_{total}$.
   b. Calculate the current. Then determine the voltage across each element. Express in the phasor domain.
   c. Draw the phasor diagram.
   d. Calculate the three powers and draw the power triangle.

12. Repeat Exercise 11 for $R = 950\ \Omega$, $L = 140\ mH$, and $C = 0.01\ \mu F$. $E_{source} = 675$ V, frequency = 3540 Hz.

13. Answer the questions for the following circuit.

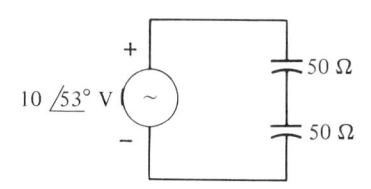

$10\ \underline{/53°}$ V, with two 50 Ω capacitors in parallel.

   a. Calculate the voltage across each capacitor.
   b. If the frequency doubles, what is the new reactance of each capacitor?
   c. What is the new voltage across each capacitor?

14. Answer the questions for the following circuit.

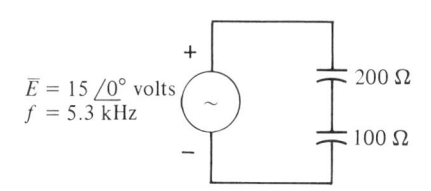

$\bar{E} = 15\ \underline{/0°}$ volts, $f = 5.3$ kHz, with 200 Ω and 100 Ω capacitors.

   a. What is the voltage across each capacitor?
   b. Calculate the value of each capacitance.
   c. Which capacitor has the greater voltage drop, the smaller capacitance or the larger? Compare this behavior to that of a resistive voltage divider.
   d. Which capacitor in a capacitive voltage divider must have the highest dielectric breakdown rating?

Section 16.4

15. Given: the following *RC* series circuit:

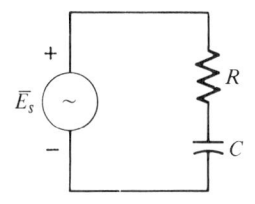

The source may be set to three frequencies:

1. f_1, a low frequency.
2. f_2, the frequency at which $|X_C| = R$.
3. f_3, a high frequency.

Indicate which frequency best fits the following statements by assigning a 1, 2, or 3 to each.
a. $|Z| \approx R$.
b. $\phi \approx 90°$.
c. $P_Q \approx 0$ var.
d. PF = 0.7.
e. E_{source} in phase with I.
f. PF ≈ 0 lead.
g. $P_A \approx P_W$.
h. Current in leading phase quadrature with E_{source}.
i. E_{source} lags I by 45°.
j. The circuit looks capacitive at this frequency.

16. Replace the capacitance in Exercise 15 with an inductor, then repeat the exercise for each of the following statements.
a. $|Z| \approx X_L$.
b. $\phi \approx 0°$.
c. P_Q and P_W have equal magnitudes.
d. E_{source} leads I by $\approx 90°$.
e. PF is low.
f. $P_A \approx P_Q$.
g. $\cos \phi \approx 1$.
h. $\phi \approx -45°$.
i. The circuit looks inductive at this frequency.

17. Repeat Exercise 15 for a series RLC circuit but use these frequencies:

1. f_1, a low frequency.
2. f_2, resonance $|X_L| = |X_C|$.
3. f_3, a high frequency.

a. $|Z| = R$
b. The circuit looks inductive.
c. I and E_{source} in phase.
d. $|X_C| \gg X_L$.
e. $P_A \approx P_{Q_{ind}}$.
f. I leads E_{source}.
g. PF = 1.
h. $P_A = P_W$.

18. Repeat Exercise 17 for each statement below.
a. The circuit looks capacitive.
b. $|X_L| \gg |X_C|$.
c. $P_A \approx P_{Q_{cap}}$.
d. The circuit looks resistive.
e. E_{source} leads I.
f. $P_Q = 0$.
g. $|X_L| = |X_C|$.

Section 16.5

19. Fill in the blanks for an RL filter.

	$R(\Omega)$	$L(H)$	$f(Hz)$
a.	12.6 k	0.5 H	
b.	3.3 k		2.1 mHz
c.		4 H	5 Hz

20. Music consists of all frequencies from about 20 Hz to 20 kHz. A small-diameter speaker called a tweeter reproduces the high frequencies best, while a large-diameter whoofer speaker reproduces lows best. Draw the schematic of a simple circuit that splits the output of a stereo into these ranges.

EXERCISES

21. Given: the RL filter circuit of Figure 16.5 with $E_{in} = 10$ V, $R = 10\,\Omega$, and $L = 10$ mH. Calculate V_{out} when the source frequency is: a. 15.9 Hz; b. 159 Hz; c. 1590 Hz.

22. Given: the RL filter circuit of Figure 16.8 with $E_{in} = 10$ V, $R = 10$ k, and $L = 159\,\mu$H. Calculate V_{out} when the source frequency is: a. 100 kHz; b. 10 MHz; c. 1 GHz.

23. Fill in the blanks for an RC filter:

	$R(\Omega)$	C(F)	f_{HP}(Hz)
a.		0.5 μF	28.9 Hz
b.	5 k		31.8 kHz
c.	750 Ω	220 pF	

24. Derive Equation 16.11.

25. Answer the following questions about RC filters.
   a. Across which element in an RC filter should the output be taken for low-pass response?
   b. Across which element should the output be taken for high-pass response?
   c. Given: a series RC filter with $R = 1$ k, $C = 0.159\,\mu$F, $E_{in} = 10$ V. Let V_{out} be taken across the capacitor. Calculate its value at 10 Hz, 1 kHz, and 1 MHz.
   d. Is the circuit high or low pass in part c? Compare results with your predictions of parts a and b.
   e. Repeat part c but take V_{out} across the resistor.
   f. Is the circuit high or low pass in part e? Compare results with your predictions of parts a and b.

26. Why is it important that $Z_{in} \gg R$ in Figure 16.5?

Section 16.6

27. Using POS, calculate the input impedance of each circuit.

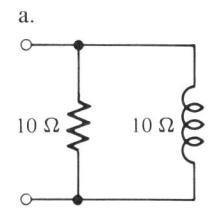

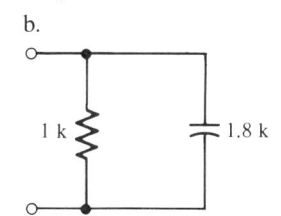

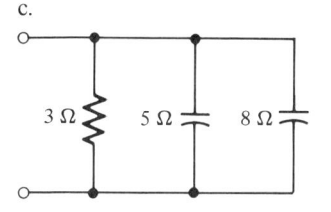

   a. [circuit: 10 Ω resistor in parallel with 10 Ω inductor]
   b. [circuit: 1 k resistor in parallel with 1.8 k capacitor]
   c. [circuit: 3 Ω resistor in parallel with 5 Ω capacitor in parallel with 8 Ω capacitor]

28. Double the frequency of the signal applied to each circuit in Exercise 27 and redo the exercise.

29. Fill in the blanks.

	L(H)	f(Hz)	$|B_L|$ S
a.	115 mH	5 kHz	
b.	3 μH		59
c.		60 Hz	0.1

30. Fill in the blanks.

	C(F)	ω(rad/s)	B_C(S)
a.	1 μF	377	377×10^{-6}
b.	20 μF		5×10^{-3}
c.		50×10^6	75×10^{-3}

31. Find the input admittance of each circuit ($\omega = 2$ krad/s).

a. 10 Ω

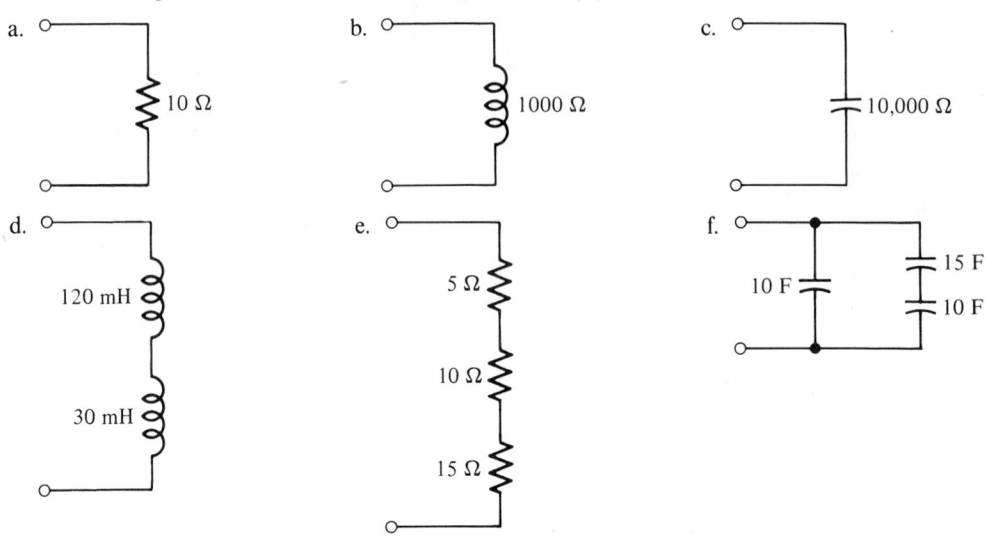

b. 1000 Ω

c. 10,000 Ω

d. 120 mH, 30 mH

e. 5 Ω, 10 Ω, 15 Ω

f. 10 F, 15 F, 10 F

32. Calculate the admittance of each component in Exercise 27, circuit a. Combine for total circuit admittance. Reciprocate this Y_{total} to get Z_{total}. Compare this Z_{total} to that of Exercise 27. Repeat for circuits b and c.

Section 16.7

33. A parallel RC combination of $R = 100$ Ω and $C = 1.3$ μF, is connected across a 130-V rms source at $f = 941$ Hz. Find: a. $\bar{Y}_{total}$; b. $\bar{I}_R$; c. $\bar{I}_C$; d. $\bar{I}_{total}$; e. ϕ; f. P_A; g. P_W; h. P_Q; i. PF.

34. Repeat Exercise 33 for $R = 4.7$ k, $C = 150$ pF, $E_{source} = 500$ mV and $f = 272$ kHz.

35. Perform the indicated operations for this circuit.

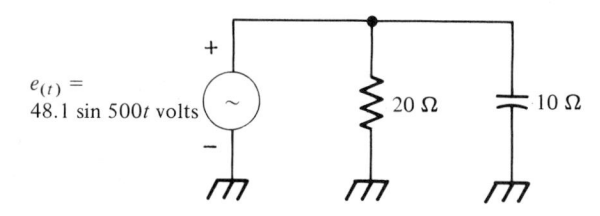

$e_{(t)} = 48.1 \sin 500t$ volts, 20 Ω, 10 Ω

a. Find the admittance of each component and $\bar{Y}_{total}$.
b. Draw the admittance triangle.
c. Determine total current and current through each component.
d. Draw the phasor diagram.
e. Verify KCL by showing that

$$\bar{I}_{total} = \bar{I}_R + \bar{I}_C$$

f. Calculate P_W, P_Q, and P_A and draw the power triangle.

36. A parallel RC combination has an admittance

$$\bar{Y} = 16 \times 10^{-5} \underline{/51.3°} \text{ S}$$

at 500 kHz. Find the values of R and C.

37. A parallel RL combination, $R = 680$ Ω, $L = 330$ mH, is connected across a 413-V rms source at a frequency of 251 Hz. Find: a. $\bar{Y}_{total}$; b. $\bar{I}_R$; c. $\bar{I}_L$; d. $\bar{I}_{total}$; e. ϕ; f. P_A; g. P_W; h. P_Q; i. PF.

38. Repeat Exercise 37 for $R = 45$ k, $L = 7$ mH, $E_{source} = 75.8$ V, and $f = 1.6$ MHz.

EXERCISES

39. A parallel RL combination has an admittance

$$\bar{Y}_{total} = 80.7 \times 10^{-3} \,\underline{/-34.3°} \text{ S}$$

at $f = 175$ kHz. Find the values of R and L.

40. A 90-V rms 60-Hz source delivers an apparent power of 90 kVA to a parallel RC load with a 45% leading power factor. What are the values of R and C?

41. Perform the indicated operations for this circuit:

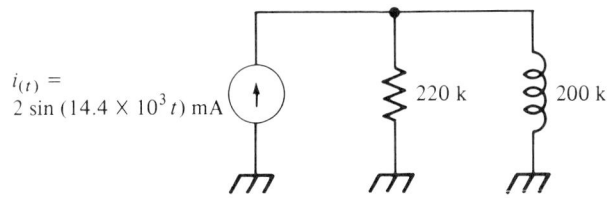

$i_{(t)} = 2 \sin(14.4 \times 10^3 t)$ mA ; 220 k ; 200 k

a. Find the admittance of each component and $\bar{Y}_{total}$ and draw the admittance triangle.
b. Find the source voltage. Express in the time domain.
c. Find the current through each component. Express the results in the time domain.
d. Draw the waveform diagram.

42. Given: a 4-mH choke shunted by an 85-Ω resistance. At what frequency will the total current into the combination lag the applied voltage by 74.1°?

43. The parallel RLC circuit of Example 16.16 has the following parameters:

$$E_{source} = 110 \text{ V}, R = 10 \,\Omega, X_L = j20 \,\Omega, X_C = -j5 \,\Omega$$

Perform the following operations.
a. Calculate $\bar{Y}_{total}$.
b. Calculate $\bar{I}_{total}$ by Ohm's law.
c. Determine the three component currents.
d. Draw the circuit's phasor diagram.
e. KCL says that the three currents of part c should add to I_{total}. Add them. Then compare the results to those of part b.
f. Determine the true or reactive power drawn by each component.
g. Find the apparent power and the power factor.

44. Repeat Exercise 43 for the following conditions:

$$E_{source} = 300 \text{ V} \quad R = 30 \text{ k} \quad X_L = 7.5 \text{ k} \quad X_C = 45 \text{ k}$$

45. A parallel RLC circuit has the total admittance

$$\bar{Y}_{total} = 38.2 \times 10^{-3} \,\underline{/-38.2°} \text{ S}$$

at $f = 15$ kHz. If the capacitance is 0.2 μF, find R and L.

46. In this circuit:

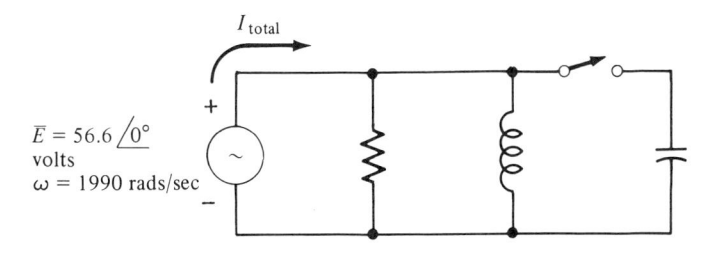

$\bar{E} = 56.6\,\underline{/0°}$ volts
$\omega = 1990$ rads/sec

$\bar{I}_{total} = 45.6\,\underline{/-34.3°}$ mA when the switch is open, and $40.4\,\underline{/-20.9°}$ mA when the switch is closed. What is the value of the capacitor?

Section 16.8

47. Using current division, find the current through each component.

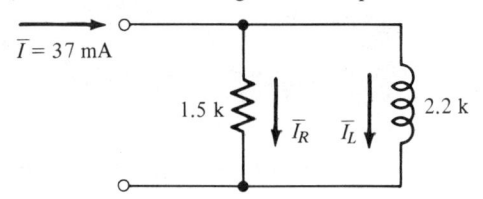

48. Repeat Exercise 47 for a parallel RC circuit of $R = 33$ k and $X_C = 47$ k. $I_{total} = 420$ mA.

Section 16.9

49. Given: the following parallel RC circuit:

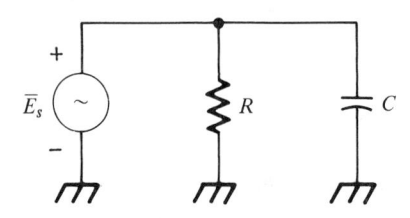

The source frequency may be set to:

1. f_1, low frequency.
2. f_2, the frequency at which $R = |X_C|$.
3. f_3, high frequency.

Indicate which frequency best fits the following statements by assigning a 1, 2, or 3 to each.
a. PF ≈ 1.
b. $\phi \approx 90°$.
c. The circuit looks capacitive.
d. E_{source} is in phase with I_{source}.
e. $P_A = \sqrt{2} P_W$.
f. $Y \approx jB$.
g. PF ≈ 0.7.
h. E_{source} lags I_{source} by ≈ 90°.
i. The circuit looks resistive.

50. Repeat Exercise 49 for a parallel RL circuit and the following statements.
a. The inductor dominates the circuit.
b. PF ≈ 1.
c. $|P_Q| \approx P_W$.
d. $\phi \approx -90°$.
e. The circuit looks inductive.
f. $R \gg X_L$.
g. $|X_L| = R$.
h. $P_Q \approx 0$.
i. The circuit looks resistive.

51. Repeat Exercise 49 for a parallel RLC circuit using f_1 as low frequency, f_2 as resonant frequency, and f_3 as high frequency.
a. The capacitor dominates the circuit.
b. $P_Q = 0$ var.
c. $|B_L| \ll B_C$.
d. The circuit looks inductive.
e. E_{source} and I_{source} are in phase.
f. The circuit looks capacitive.
g. $|B_L| = B_C$.
h. E_{source} leads I_{source}.
i. $P_W = P_A$.
j. $P_A \approx P_{Qc}$.

52. *Extra Credit.* By repeated use of POS, derive a formula for the input impedance of a parallel RLC circuit. Then set $X_L = X_C$, and show that $\bar{Z} = R$ at resonance.

53. Convert each series RX circuit to its parallel equivalent.
a. $R = 5 \, \Omega$, $X_L = 20 \, \Omega$.
b. $R = 32 \, \Omega$, $X_C = 22 \, \Omega$.

54. Repeat Exercise 53 for these RX combinations:
a. $R = 1100 \, \Omega$, $X_L = 1800 \, \Omega$.
b. $R = 3 \, \Omega$, $X_C = 500 \, \Omega$.

EXERCISES

55. Convert each parallel RX circuit to its series equivalent.
    a. $G = 33 \times 10^{-3}$ S, $B_C = 25 \times 10^{-3}$ S.
    b. $G = 0.5$ S, $B_L = 0.2$ S.
56. Repeat Exercise 55 for these combinations.
    a. $R = 10 \, \Omega$, $X_L = 10 \, \Omega$.
    b. $R = 1$ k, $X_C = 1.8$ k.
57. The frequency of the signal applied to Exercise 53, circuit a, is doubled.
    a. Determine the components of the parallel equivalent circuit now.
    b. Comment on changes caused by the frequency change. Did both impedances change or just the inductive reactance?
    c. Is an equivalent circuit equivalent at all frequencies?
58. Find the two-component series RX equivalent of this circuit:

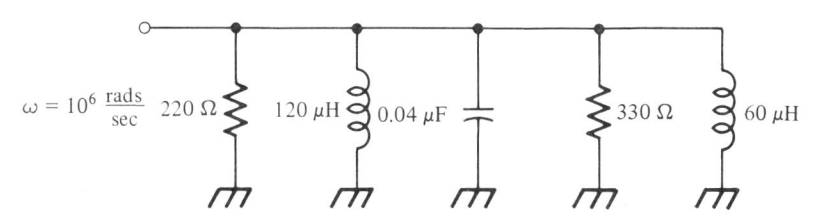

$\omega = 10^6 \frac{\text{rads}}{\text{sec}}$, 220 Ω, 120 μH, 0.04 μF, 330 Ω, 60 μH

59. Find the Q and PF of the inductor of Exercise 53, part a.
60. Find the dissipation factor (DF) and PF of the capacitor of Exercise 54, part b.

CHAPTER HIGHLIGHTS

Ac circuits are analyzed much like dc circuits. However, close attention must be paid to the rules of complex algebra.

Electronic personnel must have a "feel" for the way a circuit "looks" at frequency extremes:

- At low frequencies, capacitors are opens, inductors are shorts.
- At high frequencies, capacitors are shorts, inductors are opens.

Depending on the component across which the output voltage is taken, series RX circuits act as high- or low-pass filters.

Series RX circuits can be converted to equivalent parallel circuits, and vice versa. But such conversions hold only for the frequency at which equivalent reactances/susceptances were calculated.

Formulas to Remember

$\bar{Z}_{\text{total}} = R_{\text{total}} + j(X_{L_{\text{total}}} - X_{C_{\text{total}}})$ ohms — total impedance of a circuit

$\dfrac{\bar{E}_{\text{source}} \bar{Z}_N}{\bar{Z}_{\text{total}}}$ volts — voltage division

$f = \dfrac{1}{2\pi RC}$ and $\dfrac{R}{2\pi L}$ hertz — half-power frequencies for RC and RL filters

$\bar{Z}_{\text{total}} = \dfrac{\bar{Z}_1 \bar{Z}_2}{\bar{Z}_1 + \bar{Z}_2}$ ohms — product-over-sum for total impedance of two parallel impedances

$B_L = -j\dfrac{1}{\omega L} = -j\dfrac{1}{2\pi fL}$ ohms — inductive susceptance

$B_C = j\omega C = j2\pi fC$ ohms — capacitive susceptance

$\bar{Y} = G_{\text{total}} + j(B_{C_{\text{total}}} - B_{L_{\text{total}}})$ siemens — total admittance

$$P_W = \frac{I^2}{G} = V^2 G = EI \cos\phi \quad \text{watts}$$ true power in admittance terms

$$P_Q = \frac{I^2}{B} = V^2 B = EI \sin\phi \quad \text{vars}$$ reactive power in admittance terms

$$\bar{I}_{Z_1} = \frac{\bar{I}_{\text{total}} \bar{Z}_2}{\bar{Z}_1 + \bar{Z}_2} \quad \text{amperes}$$ current division

$$R_p = \frac{R_s^2 + X_s^2}{R_s} \quad X_p = \frac{R_s^2 + X_s^2}{X_s} \quad \text{ohms}$$ series RX-to-parallel RX conversion

$$R_s = \frac{G_p}{G_p^2 + B_p^2} \quad X_s = \frac{B_p}{G_p^2 + B_p^2} \quad \text{ohms}$$ parallel GB-to-series RX conversion (parallel components in admittance form)

$$R_s = \frac{R_p X_p^2}{R_p^2 + X_p^2} \quad X_s = \frac{X_p R_p^2}{R_p^2 + X_p^2} \quad \text{ohms}$$ parallel RX-to-series RX conversion (parallel components in impedance form)

CHAPTER 17

Analyzing Series–Parallel AC Circuits

As in Chapter 6, we move to a higher level of analysis, that of series–parallel, single-source circuits.

At the completion of this chapter, you should be able to:

- Analyze series–parallel ac circuits by the SPR (series–parallel reduction) method first discussed in Chapter 6.
- Calculate the capacitance needed to correct the power factor of an inductive circuit.
- Derive the balance equations for ac bridges.
- Transform Π circuits containing resistances and reactances to their T equivalents, and vice versa.
- Determine the action of multisource circuits by the method of superposition.
- Use the ladder method to analyze ac circuits.

17.1 ANALYSIS BY SERIES–PARALLEL REDUCTION

From Chapter 6 we recall that the series–parallel reduction (SPR) technique can only be used on series–parallel circuits that meet certain requirements. For an ac circuit, those requirements are:

- The circuit contains only one source or sources that can be combined into one, and
- All impedances are in series and parallel combinations that can be combined into one impedance.

Reviewing the steps used in SPR:

- Combine sources and source impedances.
- Calculate Z_{total} by combining impedances farthest from the source and working toward the source.
- Find I_{total} by Ohm's law.
- Work away from the source finding voltage drops by Ohm's law, voltage division, and KVL. Find currents by Ohm's law, current division, and KCL.
- Calculate powers.

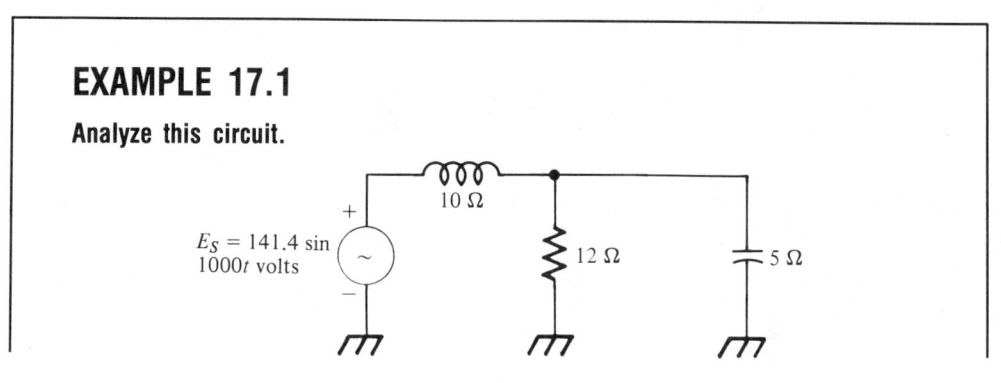

EXAMPLE 17.1

Analyze this circuit.

$E_S = 141.4 \sin 1000t$ volts, 10 Ω, 12 Ω, 5 Ω

17.1 ANALYSIS BY SERIES–PARALLEL REDUCTION 513

a. Find Z_{total}.

$$\bar{Z}_{total} = X_L + R \| X_C = j10 + 12 \| -j5$$

Use POS on the parallel RX combination:

$$\bar{Z}_{total} = \frac{12(-j5)}{12 - j5}$$
$$= 4.6 \underline{/-67.4°} \quad \text{or} \quad 1.8 - j4.3 \ \Omega$$

This is now a series RX arrangement in series with X_L.
Combine to get Z_{total}:

$$\bar{Z}_{total} = j10 + 1.8 - j4.3$$
$$= 1.8 + j5.7 \quad \text{or} \quad 6 \underline{/72.5°} \ \Omega$$

b. Find $\bar{I}_{total}$ by Ohm's law.

$$\bar{I}_{total} = \frac{\bar{E}}{\bar{Z}_{total}} = \frac{0.707(141.4)}{6 \underline{/72.5°}}$$
$$= \frac{100}{6 \underline{/72.5°}}$$
$$= 16.7 \underline{/-72.5°} \ \text{A}$$

c. Find the drop across the inductor by Ohm's law.

$$\bar{V}_L = \bar{I} X_L$$
$$= 16.7 \underline{/-72.5°} \ (j10)$$
$$= 167 \underline{/17.5°} \ \text{V}$$

Note how $\bar{V}_L$ is considerably greater than the source voltage. This often occurs in LC circuits. It does not cause a violation of Kirchhoff's voltage law, however; the sum of the potential drops around the source–inductor–capacitor loop continues to equal the source voltage.

d. Find $\bar{V}_R$ and $\bar{V}_C$ by KVL.

$$\bar{V}_R = \bar{V}_C = \bar{E}_{source} - \bar{V}_L = (100 + j0) - (160 + j50)$$
$$= -60 - j50 \ \text{V}$$

e. Find $\bar{I}_R$ by Ohm's law.

$$\bar{I}_R = \frac{\bar{V}_R}{R} = \frac{-60 - j50}{12}$$
$$= -5 - j4.2 \ \text{A} = 6.5 \underline{/-140°} \ \text{A}$$

f. Find $\bar{I}_C$ by KCL at the upper node.

$$\bar{I}_C = \bar{I}_{total} - \bar{I}_R = (5 - j16) - (-5 - j4.2)$$

Combine reals and imaginaries:

$$\bar{I}_C = (5 + 5) - j(16 - 4.2)$$
$$= 10 - j11.8 \ \text{A}$$

g. Calculate the powers and draw the power triangle.

$$P_W = I_R^2 R = 6.5^2(12) = 507.6 \text{ W}$$

$$P_{Q_C} = \frac{V_C^2}{X_C} = \frac{78.1^2}{5} = 1220 \text{ var}$$

$$P_{Q_L} = V_L I_L = 167.6(16.7) = -2799 \text{ var}$$

$$P_{Q\text{total}} = -2799 + 1220$$
$$= -1579 \text{ var}$$

$$P_A = \sqrt{507.6^2 + (-1579)^2}$$
$$= 1658.6 \text{ VA}$$

$$\phi = -72.7° \text{ from step b}$$

$$\text{PF} = \cos(-72.7°) = 0.297 \text{ or } 29.7\% \text{ lag}$$

Draw the power triangle:

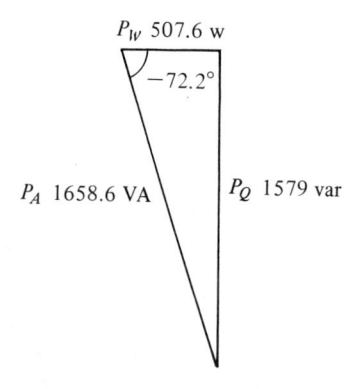

P_W 507.6 w
$-72.2°$
P_A 1658.6 VA
P_Q 1579 var

EXAMPLE 17.2

Perform the following operations for this circuit.

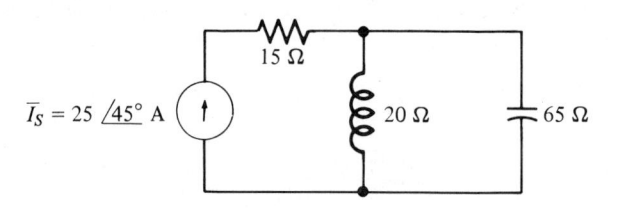

$\bar{I}_S = 25 \underline{/45°} \text{ A}$, 15 Ω, 20 Ω, 65 Ω

a. Without doing any calculations, tell whether the circuit is capacitive or inductive.

A greater percentage of I_{total} flows through X_L because its magnitude is less than that of X_C. Therefore, I_{source} lags E_{source}; the circuit is inductive.

b. Determine the time-domain equation for the inductor current at $\omega = 5$ krad/s.

By current division,

$$\bar{I}_L = \frac{\bar{I}_S X_C}{X_L + X_C} = \frac{(25 \underline{/45°})(65 \underline{/-90°})}{j20 - j65} = \frac{1625 \underline{/-45°}}{45 \underline{/-90°}}$$
$$= 36.1 \underline{/45°} \text{ A}$$

17.1 ANALYSIS BY SERIES–PARALLEL REDUCTION

Transform to the time domain:

$$i_L = (\sqrt{2})36.1 \sin(5000t + 45°)$$
$$= 51.1 \sin(5000t + 45°) \quad \text{A}$$

c. Find the source voltage.

We must first find $\bar{Z}_{\text{total}}$:

$$\bar{Z}_{\text{total}} = R + X_L \| X_C = 15 + \frac{1}{1/j20 + 1/-j65}$$

$$= 15 + \frac{1}{-j0.05 + j0.015} = 15 + \frac{1}{-j0.035}$$

$$= 15 + j28.9 \quad \text{or} \quad 32.6\,\underline{/62.6°}\ \Omega$$

The positive reactive component agrees with our decision of part a.

$$\bar{E}_{\text{source}} = \bar{I}_{\text{source}} \bar{Z}_{\text{total}} = 25\,\underline{/45°}\,(32.6\,\underline{/62.6°})$$
$$= 814\,\underline{/107.6°}\ \text{V}$$

EXAMPLE 17.3

The impedances indicated in this schematic diagram apply at medium frequencies.

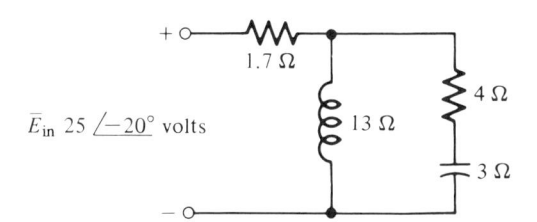

a. Calculate $\bar{I}_{\text{total}}$, ϕ, P_W, and P_Q at medium frequencies.

Converting series RX leg to equivalent parallel form:

$$R_P = \frac{R_s^2 + X_s^2}{R_s} \quad \text{(by Equation 16.30)}$$

$$= \frac{4^2 + 3^2}{4}$$

$$= \frac{25}{4} = 6.3\ \Omega$$

$$X_P = \frac{R_s^2 + X_s^2}{X_s} \quad \text{(by Equation 16.31)}$$

$$= \frac{4^2 + 3^2}{3}$$

$$= \frac{25}{3} = -j8.3\ \Omega$$

Redraw the impedances:

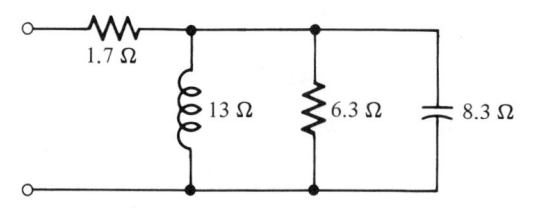

[Circuit diagram: 1.7 Ω resistor in series with parallel combination of 13 Ω inductor, 6.3 Ω resistor, and 8.3 Ω capacitor]

Combine parallel reactances by POS:

$$X_{total} = \frac{j13(-j8.3)}{j13 - j8.3}$$

$$= \frac{107.9}{j4.7}$$

$$= -j23 \text{ Ω}$$

The circuit is now:

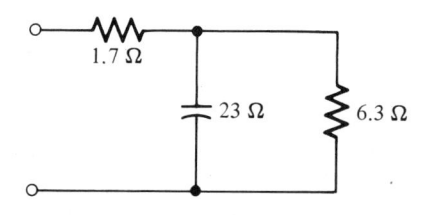

[Circuit diagram: 1.7 Ω resistor in series with parallel combination of 23 Ω capacitor and 6.3 Ω resistor]

Convert $R \parallel X_C$ to series equivalent:

$$R_s = \frac{R_P X_P^2}{R_P^2 + X_P^2} \quad \text{(by Equation 16.39)}$$

$$= \frac{6.3(23)^2}{6.3^2 + 23^2}$$

$$= \frac{6.3(529)}{568.7}$$

$$= 5.9 \text{ Ω}$$

$$X_s = \frac{X_P R_P^2}{R_P^2 + X_P^2} \quad \text{(by Equation 16.40)}$$

$$= \frac{23(6.3)^2}{6.3^2 + 23^2}$$

$$= \frac{23(39.7)}{568.7}$$

$$= -j1.6 \text{ Ω}$$

The impedances are now:

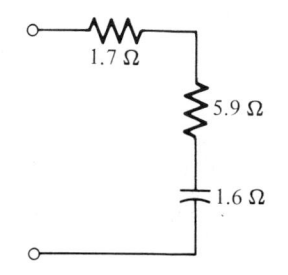

[Circuit diagram: 1.7 Ω resistor in series with 5.9 Ω resistor in series with 1.6 Ω capacitor]

17.1 ANALYSIS BY SERIES–PARALLEL REDUCTION

$$\bar{Z}_{total} = (1.7 + 5.9) - j1.6 = 7.6 - j1.6 \quad \text{or} \quad 7.8 \underline{/-11.9°} \ \Omega$$

$$\bar{I} = \frac{\bar{E}}{\bar{Z}_{total}} = \frac{25 \underline{/-20°}}{7.8 \underline{/-11.9°}} = 3.2 \underline{/-8.1°} \ A$$

$$\phi = (-8.1°) - (-20°) = 11.9° \quad \text{circuit is capacitive}$$

$$P_W = I^2 R = (3.2)^2 7.6 = 77.8 \ W$$

$$P_Q = I^2 X_C = (3.2)^2 1.6 = 16.4 \ \text{var, capacitive}$$

b. Repeat part a for approximate values at low frequencies, $f_{low} \ll f_{medium}$:

The approximate circuit at low frequencies is

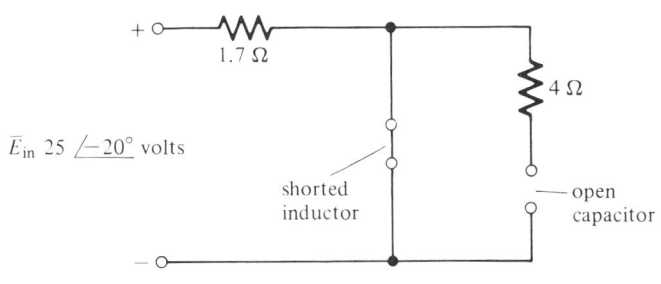

$$\bar{Z}_{total} \approx 1.7 \ \Omega$$

$$\bar{I} \approx \frac{\bar{E}_{in}}{\bar{Z}_{total}} = \frac{25 \underline{/-20°}}{1.7} = 14.7 \underline{/-20°} \ A$$

$$\phi \approx -20° - (-20°) = 0° \quad \text{circuit is resistive}$$

$$P_W \approx \frac{E_{in}^2}{R} = \frac{25^2}{1.7} = 367.7 \ W$$

$$P_Q \approx 0 \ \text{var}$$

c. Repeat part b for high frequencies, $f_{high} \gg f_{medium}$.

The approximate circuit at high frequencies is

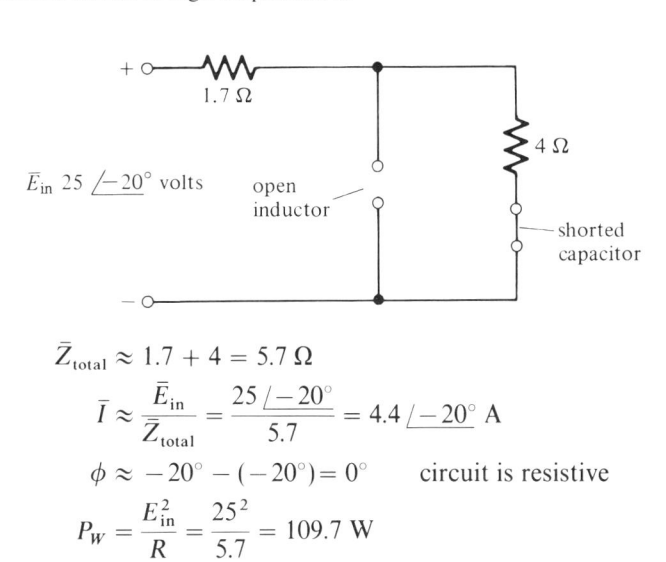

$$\bar{Z}_{total} \approx 1.7 + 4 = 5.7 \ \Omega$$

$$\bar{I} \approx \frac{\bar{E}_{in}}{\bar{Z}_{total}} = \frac{25 \underline{/-20°}}{5.7} = 4.4 \underline{/-20°} \ A$$

$$\phi \approx -20° - (-20°) = 0° \quad \text{circuit is resistive}$$

$$P_W = \frac{E_{in}^2}{R} = \frac{25^2}{5.7} = 109.7 \ W$$

$$P_Q \approx 0 \ \text{var}$$

■

EXAMPLE 17.4

Perform the indicated operations for this circuit.

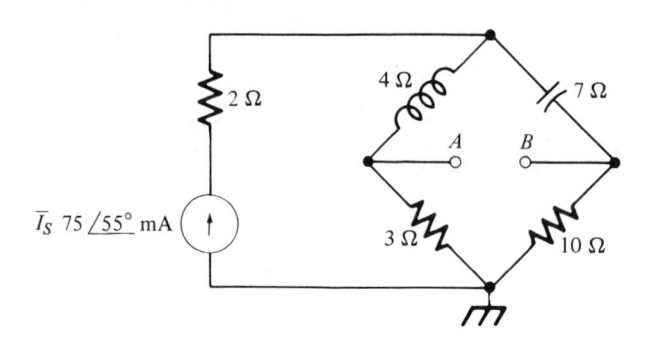

a. Determine the current through each leg.

Analysis is often simplified by redrawing the circuit in this manner:

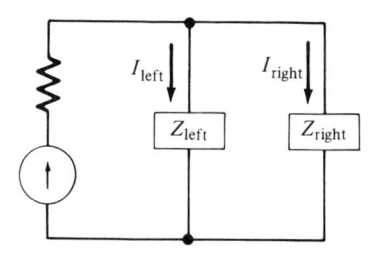

$\bar{I}_{\text{left}}$ by current division:

$$\bar{I}_{\text{left}} = \frac{\bar{I}_{\text{total}} \bar{Z}_{\text{right}}}{\bar{Z}_{\text{left}} + \bar{Z}_{\text{right}}}$$

$$= \frac{75 \underline{/55°} \text{ mA}(10 - j7)}{(3 + j4) + (10 - j7)}$$

$$= \frac{915.5 \underline{/20°} \text{ mA}}{13.3 \underline{/-13°}}$$

$$= 68.6 \underline{/33°} \text{ mA}$$

$\bar{I}_{\text{right}}$ by KCL:

$$\begin{aligned} I_{\text{right}} &= \bar{I}_{\text{total}} - \bar{I}_{\text{left}} \\ &= 75 \underline{/55°} \text{ mA} - 68.6 \underline{/33°} \\ &= (43 + j61.4) - (57.5 + j37.4) \\ &= -14.5 + j24 \quad \text{or} \quad 28 \underline{/121.1°} \text{ mA} \end{aligned}$$

b. Determine the voltage between points a and b.

Voltage at terminal A with respect to ground = voltage across 3-Ω resistance:

$$\begin{aligned} \text{voltage} &= \bar{I}_{\text{left}}(3 \text{ Ω}) \\ &= 3(68.6 \underline{/33°} \text{ mA}) = 205.8 \underline{/33°} \text{ mV} \end{aligned}$$

17.1 ANALYSIS BY SERIES–PARALLEL REDUCTION

Voltage at terminal B with respect to ground = voltage across 10-Ω resistance:

$$\text{voltage} = \bar{I}_{\text{right}}(10)$$
$$= 10(28\,\underline{/121.1°}) = 280\,\underline{/121.1°}\text{ mV}$$

Draw a loop containing V_{ab} and assume a polarity:

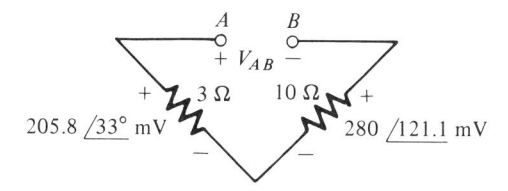

Write a loop equation and solve for V_{ab}:

$$\bar{V}_{ab} + 280\,\underline{/121.1°}\text{ mV} - 205.8\,\underline{/33°}\text{ mV} = 0$$
$$V_{ab} = (205.8\,\underline{/33°} - 280\,\underline{/121.1°})\text{ mV}$$
$$= (172.6 + j112.1) - (-144.6 + j239.8)\text{ mV}$$
$$= 317.2 - j127.7 \quad \text{or} \quad 341.9\,\underline{/-21.9°}\text{ mV}$$

EXAMPLE 17.5

Calculate the output of this voltage divider for the following conditions:

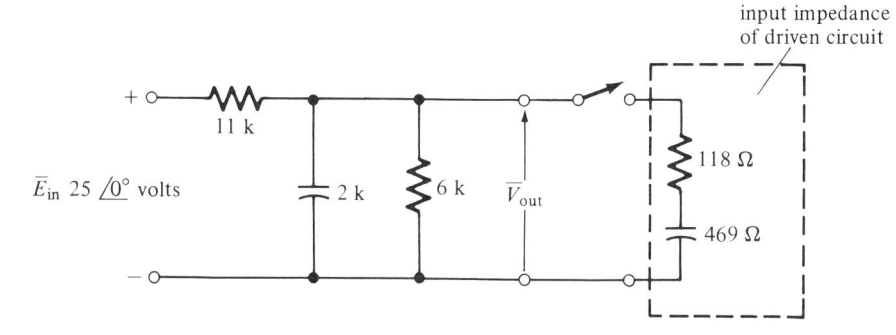

a. Unloaded conditions (switch open).

With the switch open, the circuit is

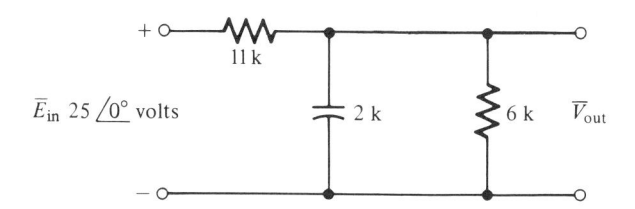

Impedance of parallel RX combination by POS:

$$Z = \frac{6\text{ k}(-j2\text{ k})}{6\text{ k} - j2\text{ k}} = \frac{-j12\text{ k}^2}{6.3\text{ k}\,\underline{/-18.4°}}$$
$$= 1.9\,\underline{/-71.6°}\text{ k}$$

V_{out} by voltage division:

$$V_{out} = \frac{25(1.9\text{ k}\,\underline{/-71.6°})}{(11\text{ k} + 1.9\text{ k}\,\underline{/-71.6°})}$$

$$= \frac{47.4\text{ k}\,\underline{/-71.6°}}{11.7\text{ k}\,\underline{/-8.8°}}$$

$$= 4.1\,\underline{/-62.8°}\text{ V}$$

b. Loaded conditions (switch closed)

With the switch closed, the circuit is

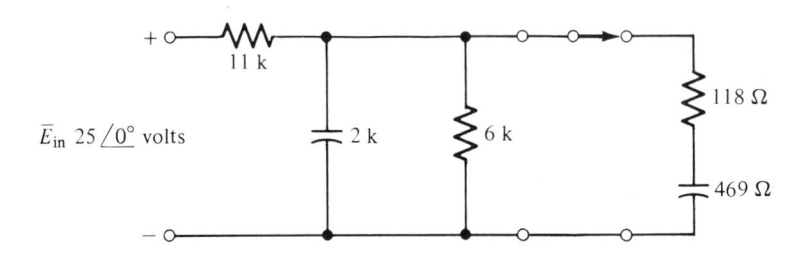

[Circuit diagram: $\bar{E}_{in}$ 25 /0° volts source, 11 k series resistor, parallel branches of 2 k capacitor, 6 k resistor, and series combination of 118 Ω resistor with 469 Ω capacitor]

Convert series RC branch to parallel equivalent:

$$R_P = \frac{(R_s^2 + X_s^2)}{R_s} = \frac{118^2 + 469^2}{118}$$

$$= \frac{234 \times 10^3}{118}$$

$$= 2\text{ k}$$

$$X_P = \frac{R_s^2 + X_s^2}{X_s} = \frac{234 \times 10^3}{469}$$

$$= -j500\text{ Ω}$$

The circuit is now

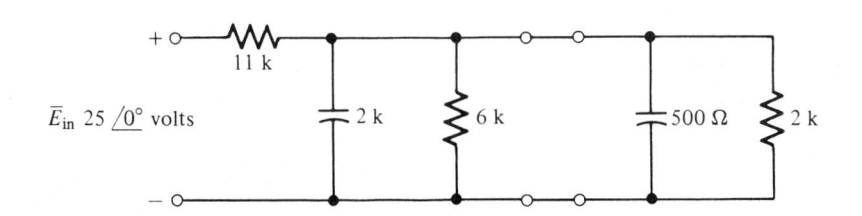

[Circuit diagram: $\bar{E}_{in}$ 25 /0° volts source, 11 k series resistor, parallel branches of 2 k capacitor, 6 k resistor, 500 Ω capacitor, and 2 k resistor]

Combine parallel elements by POS:

$$X_C = \frac{2000(500)}{2000 + 500}$$

$$= -j400\text{ Ω}$$

$$R = \frac{6\text{ k}(2\text{ k})}{6\text{ k} + 2\text{ k}}$$

$$= 1.5\text{ k}$$

17.1 ANALYSIS BY SERIES–PARALLEL REDUCTION

The circuit is now

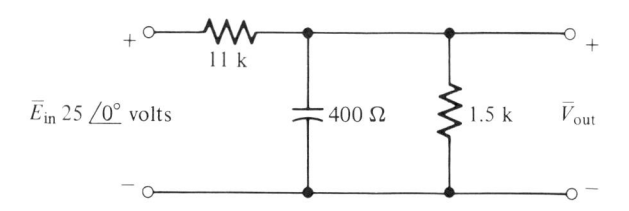

[Circuit diagram: $\bar{E}_{in}$ 25 /0° volts source, 11 k resistor in series, 400 Ω capacitor and 1.5 k resistor in parallel, $\bar{V}_{out}$ across the parallel combination]

Impedance of parallel combination by POS:

$$\bar{Z} = \frac{1500(-j400)}{1500 - j400}$$

$$= 387 \,/\!-75.1° \; \Omega$$

$\bar{V}_{out}$ by voltage division:

$$V_{out} = \frac{25(387\,/\!-75.1°)}{11\text{ k} + 387\,/\!-75.1°} = \frac{9660\,/\!-75.1°}{11.1\text{ k}\,/\!-1.9°}$$

$$= 870\,/\!-73.2° \text{ mV}$$

EXAMPLE 17.6

Determine the components of this circuit's power triangle:

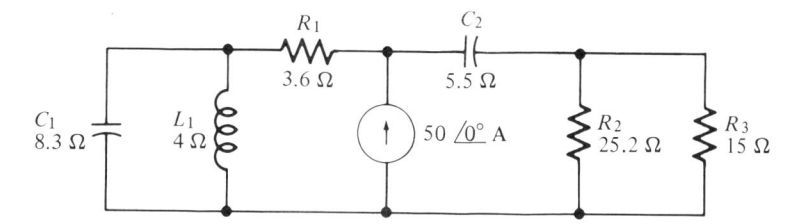

[Circuit diagram with $C_1 = 8.3\,\Omega$, $L_1 = 4\,\Omega$, $R_1 = 3.6\,\Omega$, $C_2 = 5.5\,\Omega$, current source 50 /0° A, $R_2 = 25.2\,\Omega$, $R_3 = 15\,\Omega$]

The circuit is simplified by redrawing:

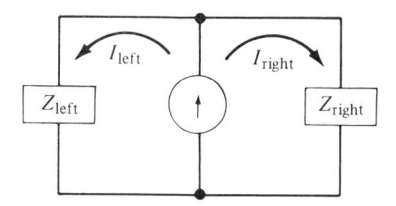

[Simplified circuit showing Z_{left} with I_{left}, current source, Z_{right} with I_{right}]

To find $\bar{I}_{left}$ and $\bar{I}_{right}$ by current division, we need $\bar{Z}_{left}$ and $\bar{Z}_{right}$:

$$\bar{Z}_{left} = R_1 + X_{C_1} \| X_{L_1}$$

$$= 3.6 + \frac{(-j8.3)(j4)}{j4 - j8.3}$$

$$= 3.6 + j7.7 \quad \text{or} \quad 8.5\,/64.9° \; \Omega$$

$$\bar{Z}_{right} = X_{C_2} + R_2 \| R_3$$

$$= -j5.5 + \frac{25.2(15)}{25.2 + 15}$$

$$= 9.4 - j5.5 \quad \text{or} \quad 10.9\,/\!-30.3° \; \Omega$$

The circuit has now been reduced to

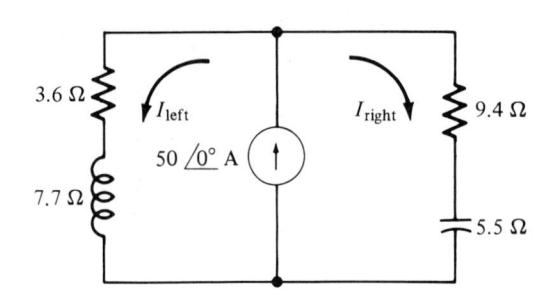

[Circuit diagram: 3.6 Ω resistor and 7.7 Ω inductor in series on left branch with current I_{left}; 50 /0° A current source in middle; 9.4 Ω resistor and 5.5 Ω capacitor in series on right branch with current I_{right}.]

Find $\bar{I}_{\text{left}}$ and $\bar{I}_{\text{right}}$:

$\bar{I}_{\text{left}}$ by current division:

$$I_{\text{left}} = \frac{\bar{I}_{\text{total}} \bar{Z}_{\text{right}}}{\bar{Z}_{\text{left}} + \bar{Z}_{\text{right}}}$$

$$= \frac{50(10.9 \,/\!\!-30.3°)}{(9.4 - j5.5) + (3.6 + j7.7)}$$

$$= \frac{545 \,/\!\!-30.3°}{13.2 \,/9.6°}$$

$$= 41.3 \,/\!\!-39.9° \text{ A}$$

$\bar{I}_{\text{right}}$ by KCL $= \bar{I}_{\text{total}} - \bar{I}_{\text{left}}$:

$$\bar{I}_{\text{right}} = 50 - 41.3 \,/\!\!-39.9°$$
$$= 50 - (31.7 - j26.5)$$
$$= 18.3 + j26.5 \quad \text{or} \quad 32.2 \,/55.3° \text{ A}$$

$$P_{W \text{ total}} = P_{W \text{ left}} + P_{W \text{ right}}$$
$$= 41.3^2(3.6) + 32.2^2(9.4)$$
$$= 6140 + 9746$$
$$= 15{,}886 \text{ W}$$

$$P_{Q \text{ total}} = P_{Q \text{ left}} + P_{Q \text{ right}}$$
$$= 41.3^2(7.7) \text{ var inductive} + 32.2^2(5.5) \text{ var capacitive}$$
$$= -13{,}133.8 + 5702.6 \text{ var}$$
$$= -7431 \text{ var}$$

Determine P_A by Pythagorean theorem:

$$P_A = \sqrt{P_W^2 + P_Q^2}$$
$$= \sqrt{15{,}886^2 + (-7431)^2}$$
$$= 17{,}538 \text{ VA}$$

$$\text{power factor} = \frac{P_W}{P_A}$$
$$= \frac{15{,}886}{17{,}538}$$
$$= 91\% \text{ lag}$$

17.1 ANALYSIS BY SERIES–PARALLEL REDUCTION

Draw the power triangle:

P_W 15886 W, $-25.1°$, P_Q 7431 var, P_A 17538 VA

The problem could also have been done this way:

Find $\bar{Z}_{total}$:

$$\bar{Z}_{total} = \frac{1}{\bar{Y}_{total}}$$

$$\bar{Y}_{total} = \frac{1}{\bar{Z}_{left}} + \frac{1}{\bar{Z}_{right}}$$

$$= \frac{1}{10.9\,\underline{/-30.3°}} + \frac{1}{8.5\,\underline{/64.9°}}$$

$$= 0.092\,\underline{/30.3°} + 0.0118\,\underline{/-64.9°}$$

$$= (0.079 + j0.046) + (0.05 - j0.107)$$

$$= 0.129 - j0.061 \text{ or } 0.143\,\underline{/-25.3°}\ \text{S}$$

$$\bar{Z}_{total} = \frac{1}{\bar{Y}_{total}}$$

$$= \frac{1}{0.143\,\underline{/-25.3°}}$$

$$= 7\,\underline{/25.3°} \quad \text{or} \quad 6.3 + j3\ \Omega$$

$$P_{W\,total} = I^2 R_{total}$$
$$= 50^2(6.3)$$
$$= 15{,}750\ \text{W}$$

$$P_{Q\,total} = I^2 X_{total}$$
$$= 50^2(3)$$
$$= 7500\ \text{var, inductive} \qquad \blacksquare$$

In Chapter 15 we learned that factories with assembly lines powered by large motors present highly inductive loads. These draw excessive line currents at low power factors. However, the inductive components are easily reduced or canceled by adding capacitive reactance to the circuit. For this purpose, banks of power-factor-correcting capacitors are often located near factories and other users of bulk power.

EXAMPLE 17.7

A factory using many electric motors draws 800 A from a 4-kV 60-Hz line with an 80% lagging power factor. Answer the following questions.

a. What capacitance muct be placed across the factory lines to increase the power factor to 94%?

Find the original power triangle before capacitance is added:

$$P_A = EI = 4\ \text{kV}(800) = 3200\ \text{kVA}$$
$$\text{PF} = 0.8$$

Therefore, $\phi = \arccos 0.8 = 36.9°$

$$\frac{P_Q}{P_A} = \sin \phi$$

Therefore,

$$P_Q = P_A \sin \phi = 3200 \sin 36.9°$$
$$= 3200(0.6) = 1920 \text{ kvar, inductive}$$

$$\frac{P_W}{P_A} = \cos \phi$$

Therefore,

$$P_W = P_A \cos \phi = 3200 \cos 36.9°$$
$$= 3200(0.8) = 2559 \text{ kW}$$

Draw the original power triangle:

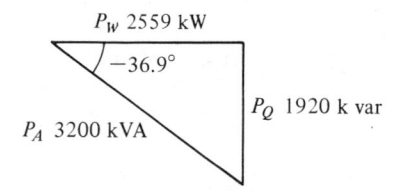

P_W 2559 kW
$-36.9°$
P_A 3200 kVA P_Q 1920 k var

We know two things about the new power triangle after capacitor addition:

- The power-factor-correcting capacitance does not change the load resistance; therefore P_w remains at 2559 kW.
- Since the new power factor is 94%, the angle, ϕ, between P_W and P_A equals $\arccos 0.94 = 20°$.

Draw the new power triangle:

P_W 2559 kW
$-20°$
P_A new P_Q new

$$\frac{P_{Q_{\text{new}}}}{P_{W_{\text{original}}}} = \tan 20°$$

Therefore,

$$P_{Q_{\text{new}}} = P_W \tan 20° = 2559(0.36)$$
$$= 931.4 \text{ kvar}$$

$$P_{Q_{\text{new}}} = P_{Q_{\text{original}}} - P_{Q_{\text{correcting capacitor}}}$$

Therefore,

$$P_{Q_{\text{cap}}} = P_{Q_{\text{original}}} - P_{Q_{\text{new}}}$$
$$= (1920 - 931.4) \text{ kvar}$$
$$= 988.6 \text{ kvar}$$

$$P_{Q_{\text{cap}}} = \frac{V_C^2}{X_C} = V_C^2 \omega C$$

17.1 ANALYSIS BY SERIES–PARALLEL REDUCTION

Therefore,

$$C = \frac{P_Q}{\omega V_C^2} = \frac{988.6 \times 10^3}{377(16 \times 10^6)}$$
$$= 164 \ \mu\text{F}$$

b. What is the drop in line current caused by the addition of the capacitor?

$$P_{A_{\text{new}}} = EI_{\text{new}} = \sqrt{(P_{W_{\text{original}}})^2 + (P_{Q_{\text{new}}})^2}$$

Therefore,

$$I_{\text{new}} = \frac{\sqrt{(P_{W_{\text{original}}})^2 + (P_{Q_{\text{new}}})^2}}{E}$$
$$= \frac{\sqrt{(2559\text{k})^2 + (931.4\text{k})^2}}{4000}$$
$$= \frac{2720 \text{ kVA}}{4 \text{ kV}}$$
$$= 681 \text{ A}$$

The line current has dropped 14.9%, from 800 to 681 A.

c. What is the new current drawn by the factory?

The factory current remains at 800 A since its impedance has not changed.

d. What is the current through the capacitor?

$$P_{Q_{\text{cap}}} = V_{\text{cap}} I_{\text{cap}}$$

Therefore,

$$I_C = \frac{P_Q}{V_C} = \frac{988.6 \text{ kvar}}{4 \text{ kV}}$$
$$= 247 \text{ A}$$

EXAMPLE 17.8

What capacitance must be placed in shunt with a $44\underline{/38°}$-Ω impedance load so that total circuit power factor is 95% lagging?

The load admittance is

$$\bar{Y} = \frac{1}{\bar{Z}} = \frac{1}{44\underline{/38°}}$$
$$= 22.7 \times 10^{-3} \underline{/-38°} \text{ S}$$
$$= (17.9 - j14) \times 10^{-3} \text{ S}$$

This represents a parallel $G-jB$ circuit whose admittance triangle is

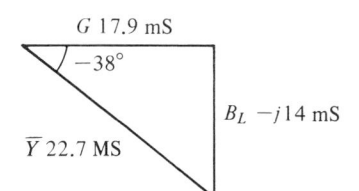

G 17.9 mS
$-38°$
B_L $-j14$ mS
$\bar{Y}$ 22.7 MS

With the capacitance added, the new phase angle will be arccos 0.95 = −18.2°. Draw the new admittance triangle.

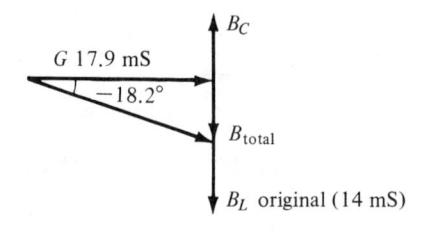

$$\frac{B_{\text{total}}}{G} = \tan(-18.2°)$$

Therefore,

$$B_{\text{total}} = G\tan(-18.2°) = 17.9 \times 10^{-3}(-0.33)$$
$$= -5.9 \times 10^{-3} \text{ S}$$
$$-B_{\text{total}} = -B_{L_{\text{original}}} + B_{C_{\text{added}}}$$

Therefore,

$$B_{C_{\text{added}}} = B_{L_{\text{original}}} - B_{\text{total}}$$
$$= (14 - 5.9) \times 10^{-3}$$
$$= 8.1 \times 10^{-3} \text{ S}$$
$$B_C = \omega C$$

Therefore,

$$C = \frac{B_C}{\omega} = \frac{8.1 \times 10^{-3}}{377}$$
$$= 21.5 \text{ μF}$$ ∎

17.2 AC BRIDGES

Figure 17.1 shows the schematic diagram of an ac bridge used for measuring impedances. Note that its source is sinusoidal and the four legs are impedances rather than resistances.

Like the dc bridge discussed in Chapter 6, one leg of this ac bridge is the unknown being measured. Another leg is variable. It is adjusted until balance is achieved as

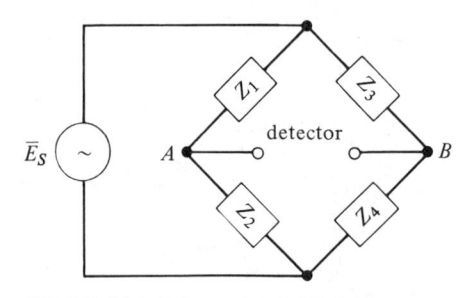

FIGURE 17.1 Schematic of AC Bridge.

17.2 AC BRIDGES

indicated by a minimum or *null reading* on the *detector* connected between points A and B.

For accurate readings, the null signal can be amplified and sent to a variety of electronic measuring devices such as the oscilloscope. When "ballpark" accuracy is sufficient and the source frequency is in the audio range, the detector may be headphones to which the operator listens while adjusting the bridge for minimum volume.

Balance in ac bridges means the same thing it did in the dc case, namely, the voltage between points A and B is zero. As a result, no current flows through the detector. This occurs when

$$E_{z_2} = E_{z_4}$$

By voltage division,

$$E_{z_2} = \frac{E_s Z_2}{Z_1 + Z_2}$$

and

$$E_{z_4} = \frac{E_s Z_4}{Z_3 + Z_4}$$

Setting the right sides equal and repeating the algebra steps used in the derivation of Equation 6.1, we obtain the balance equation for ac bridges:

$$\frac{Z_1}{Z_2} = \frac{Z_3}{Z_4} \tag{17.1}$$

By cross-multiplying, a more convenient form of Equation 17.1 is obtained:

$$Z_1 Z_4 = Z_2 Z_3 \tag{17.2}$$

Balance conditions for ac bridges are not as simple as they are for dc bridges, however, because at least two branch impedances include reactance. When they do, they become complex numbers, and we know that:

> For complex numbers to be equal, their real parts must be equal and their imaginary parts must be equal.

What this is saying is that there are two conditions that must be met to balance an ac bridge. This means that two controls must be adjusted for detector null, in contrast to one on a dc bridge.

Here are the steps to follow when deriving the balance equations for one of the many types of ac bridges:

- Substitute bridge branch impedances into Equation 17.2.
- Perform the algebra needed to break this equation into real and imaginary parts.
- Equate real components and solve for one of the unknowns, either R_x or X_x.
- Equate imaginary components and solve for the other unknown.
- Perform the algebra needed to get a statement for R_x without any X_x in it.
- Repeat the step to obtain a statement for X_x without any R_x in it.
- If either statement contains ω, balance is frequency dependent.

Keep these comments in mind as we derive the balance conditions for the *Maxwell bridge*. This circuit contains a precision capacitor, C_s, and is especially suited for measuring the unknown R_x and L_x components of inductances (see Figure 17.2).

Substituting the four branch impedances of the Maxwell bridge into Equation 17.2:

$$\frac{R_1(-jX_{C_s})}{R_1 - jX_{C_s}}(R_x + jX_x) = R_2 R_3$$

or

$$-jR_1 X_{C_s}(R_x + jX_x) = R_2 R_3 (R_1 - jX_{C_s})$$
$$-jR_1 X_{C_s} R_x + R_1 X_{C_s} X_x = R_1 R_2 R_3 - jX_{C_s} R_2 R_3$$

Equate reals:

$$R_1 X_{C_s} X_x = R_1 R_2 R_3$$

Solve for X_x:

$$X_x = \frac{R_2 R_3}{X_{C_s}}$$
$$\omega L_x = R_2 R_3 \omega C_s$$

or

$$L_x = R_2 R_3 C_s \qquad (17.3)$$

Equate imaginaries:

$$-jR_1 X_{C_s} R_x = -jR_2 R_3 X_{C_s}$$

Solve for R_x:

$$R_x = \frac{R_2 R_3}{R_1} \qquad (17.4)$$

Note that there are two balance conditions, both of which are independent of frequency.

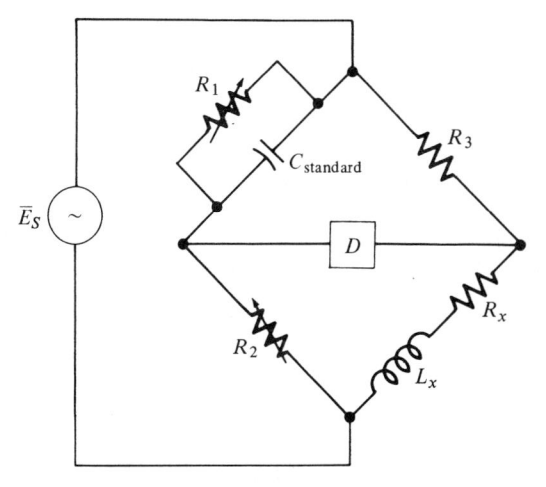

FIGURE 17.2 Maxwell Bridge

17.2 AC BRIDGES

EXAMPLE 17.9

R_3 and C_s of a Maxwell bridge are 600 Ω and 205 pF, respectively. When an unknown inductor is connected to leg 4 and the brigde adjusted for null, R_1 = 1330 Ω and R_2 = 600 Ω. What are the equivalent series R and L components of the unknown?

$$R_x = \frac{R_2 R_3}{R_1} = \frac{600(600)}{1330} = 271 \text{ Ω}$$

$$L_x = R_2 R_3 C_s = 600(600)(205 \times 10^{-12}) = 73.8 \text{ μH}$$ ∎

The *capacitance bridge*, seen in Figure 17.3 together with its balance conditions, is used to measure capacitances. The derivations are left as an exercise for the reader.

Another common bridge, frequently used in test equipment, is the *Wien bridge* (pronounced *veen*; however, most North Americans say *ween*). It is shown in Figure 17.4.

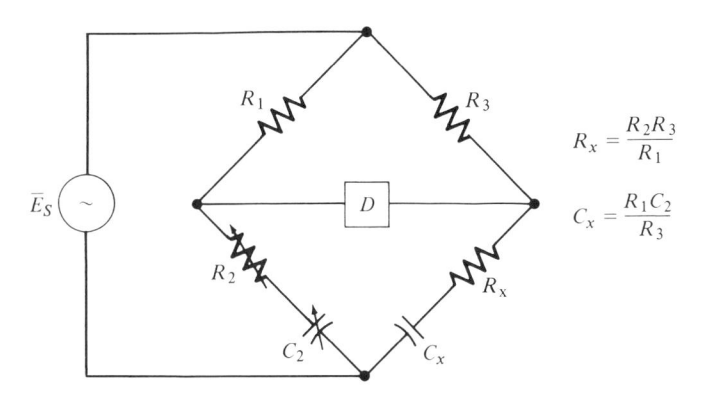

FIGURE 17.3 Capacitance Bridge and Balance Conditions.

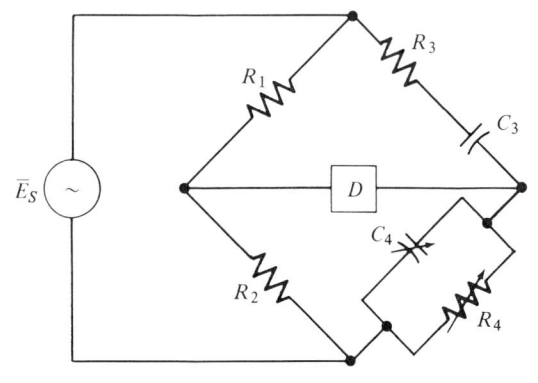

FIGURE 17.4 Wien Bridge.

Like other ac bridges, the Wien circuit measures impedances. But replace the unknown branch with known components having certain specific values and the bridge measures frequency. Let us determine those values.

Substitute the branch impedances into Equation 17.2:

$$R_1 \frac{R_4(-jX_4)}{R_4 - jX_4} = R_2(R_3 - jX_3)$$

$$-jR_1R_4X_4 = (R_4 - jX_4)(R_2R_3 - jR_2X_3)$$
$$= R_2R_3R_4 - R_2X_3X_4 - jR_2R_3X_4 - jR_2R_4X_3$$

Equate reals:

$$R_2R_3R_4 - R_2X_3X_4 = 0$$

Therefore,

$$R_2R_3R_4 = R_2X_3X_4$$

$$R_3R_4 = \frac{1}{(2\pi f)^2 C_3 C_4}$$

or

$$f^2 = \frac{1}{(2\pi)^2 R_3 R_4 C_3 C_4}$$

If we make $R_3 = R_4$ and $C_3 = C_4$, then

$$f = \frac{1}{2\pi R_4 C_4} \tag{17.5}$$

Equate imaginaries:

$$R_1 R_4 X_4 = R_2 R_3 X_4 + R_2 R_4 X_3$$

or

$$\frac{R_1 R_4}{2\pi f C_4} = \frac{R_2 R_3}{2\pi f C_4} + \frac{R_2 R_4}{2\pi f C_3}$$

$$\frac{R_1 R_4 - R_2 R_3}{C_4} = \frac{R_2 R_4}{C_3}$$

Cross-multiply:

$$C_3(R_1 R_4 - R_2 R_3) = C_4 R_2 R_4$$

Letting $R_3 = R_4$ and $C_3 = C_4$, as before:

$$R_1 R_4 = R_2 R_4 + R_2 R_4$$

or

$$R_1 = 2R_2 \tag{17.6}$$

A Wien bridge, made according to the specifications above ($R_3 = R_4$, $C_3 = C_4$, and $R_1 = 2R_2$) will thus measure unknown frequencies. To do so, the circuit is nulled by adjusting C_4 and R_4, reading C_4 and R_4 values from dials, and substituting those values into Equation 17.5.

17.3 T-Π TRANSFORMATIONS

In Chapter 6 we learned that Π and T (or Δ–Y) networks, so named because of their characteristic shape (see Figure 17.5), are often found in electronics. We learned to change, or *transform*, one into an equivalent form of the other. By doing so, a circuit is altered to allow SPR analysis, yet the circuit as a whole operates with no change.

The formulas by which Π to T and T to Π transformations are made were derived for purely resistive circuits in Appendix A. They are the same for ac circuits except that impedances are substituted for resistances. The formulas are:

For Π-to-T conversion

The denominator is the same in all three equations namely, the sum of the three Π elements. Therefore, let us define:

$$D = Z_x + Z_y + Z_z$$

Then

$$Z_a = \frac{Z_x Z_z}{D} \quad (17.7)$$

$$Z_b = \frac{Z_x Z_y}{D} \quad (17.8)$$

$$Z_c = \frac{Z_y Z_z}{D} \quad (17.9)$$

Note how each numerator is the product of the two Π impedances in Figure 17.5 that meet at one end of the T impedance in that formula.

For T-to-Π conversion

The numerator is the same in all three equations, namely, the sum of the three T products. Therefore, let us define:

$$N = Z_a Z_b + Z_b Z_c + Z_c Z_a$$

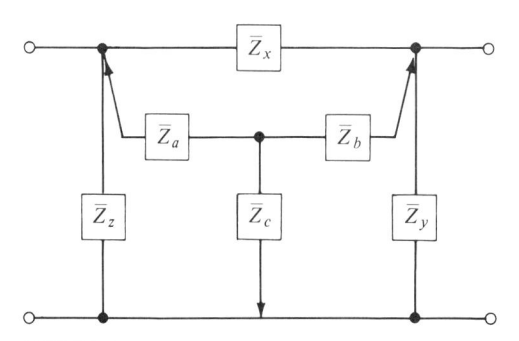

FIGURE 17.5 Π and T Networks of Three Impedances.

Then

$$Z_x = \frac{N}{Z_c} \tag{17.10}$$

$$Z_y = \frac{N}{Z_a} \tag{17.11}$$

$$Z_z = \frac{N}{Z_b} \tag{17.12}$$

Note how each denominator is a single T impedance. To find which one, draw an imaginary line from the Π impedance in Figure 17.5 through the center node joining the three T impedances to the farthest T impedance. That T element is the Π conversion denominator.

As in the dc case, the formulas are simplified when the three impedances to be converted are the same. But a word of caution: These are complex impedances. For them to be equal, both magnitude and angle must be the same. When this is the case, Π-to-T formulas become

$$Z_T = \frac{Z_\Pi}{3} \tag{17.13}$$

T-to-Π formulas become

$$Z_\Pi = 3Z_T \tag{17.14}$$

Two final points about Π–T transformations:

- Be very careful to place the equivalent circuit between the right points in the original circuit.
- If information is needed about a component, do not transform that component out of existence.

EXAMPLE 17.10

Because the elements in this circuit are not in series or in parallel, Z_{total} cannot be determined by combining components. Therefore, convert the Π circuit (circled) to its T equivalent and calculate Z_{total}:

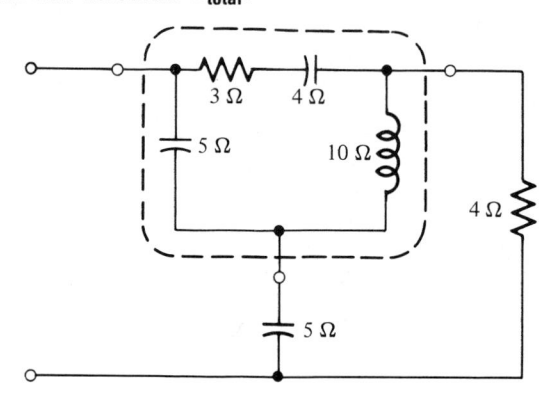

17.3 Π–T TRANSFORMATIONS

The denominator in these Π-to-T conversions is

$$D = (3 - j4 + j10 - j5) = 3 + j1 \quad \text{or} \quad 3.2\,\underline{/18.4°}$$

$$\bar{Z}_a = \frac{(3 - j4)(-j5)}{D}$$

$$= \frac{25\,\underline{/-143.1°}}{3.2\,\underline{/18.4°}}$$

$$= 7.8\,\underline{/-161.5°} \quad \text{or} \quad -7.4 - j2.5\ \Omega$$

$$\bar{Z}_b = \frac{(3 - j4)(j10)}{D}$$

$$= \frac{50\,\underline{/36.9°}}{3.2\,\underline{/18.4°}}$$

$$= 15.6\,\underline{/18.5°} \quad \text{or} \quad 14.8 + j5\ \Omega$$

$$\bar{Z}_c = \frac{(-j5)(j10)}{D}$$

$$= \frac{50}{3.2\,\underline{/18.4°}}$$

$$= 15.6\,\underline{/-18.4°} \quad \text{or} \quad 14.8 - j4.9\ \Omega$$

Draw the transformed circuit:

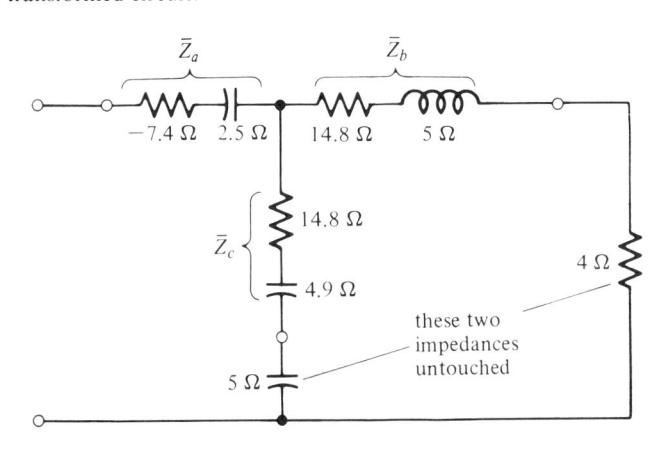

Represent the circuit in simpler form:

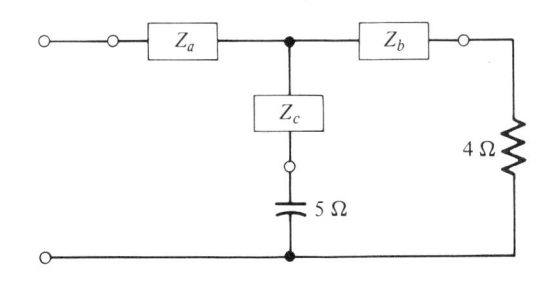

Find Z_{total}:

$$\bar{Z}_{\text{total}} = \bar{Z}_a + (\bar{Z}_c - j5) \| (\bar{Z}_b + 4)$$

Define $\bar{Z}_1 = \bar{Z}_c - j5$:

$$\bar{Z}_1 = 14.8 - j4.9 - j5$$
$$= 14.8 - j9.9 \quad \text{or} \quad 17.8 \underline{/-33.8°} \; \Omega$$

Define $\bar{Z}_2 = \bar{Z}_b + 4$:

$$\bar{Z}_2 = 14.8 + j5 + 4$$
$$= 18.8 + j5 \quad \text{or} \quad 19.5 \underline{/14.9°} \; \Omega$$

The parallel combination of these two is defined as $\bar{Z}_3$. It is

$$\bar{Z}_3 = \frac{\bar{Z}_1 \bar{Z}_2}{\bar{Z}_1 + \bar{Z}_2}$$
$$= \frac{17.8 \underline{/-33.8°}(19.5 \underline{/14.9°})}{14.8 - j9.9 + 18.8 + j5}$$
$$= \frac{347.1 \underline{/-18.9°}}{34 \underline{/-8.3°}}$$
$$= 10.2 \underline{/-10.6°} \quad \text{or} \quad 10 - j1.9 \; \Omega$$

$$\bar{Z}_{\text{total}} = \bar{Z}_a + \bar{Z}_3$$
$$= -7.4 - j2.5 + 10 - j1.9$$
$$= 2.6 - j4.4 \quad \text{or} \quad 5.1 \underline{/-59.4°} \; \Omega \quad \blacksquare$$

EXAMPLE 17.11

The circuit of Example 17.10 is redrawn to point out the T (circled) consisting of the $j10$-Ω and the two $-j5$-Ω reactances. Convert this T to its Π equivalent. Calculate Z_{total} looking into the terminals. Compare the results to those of Example 17.10.

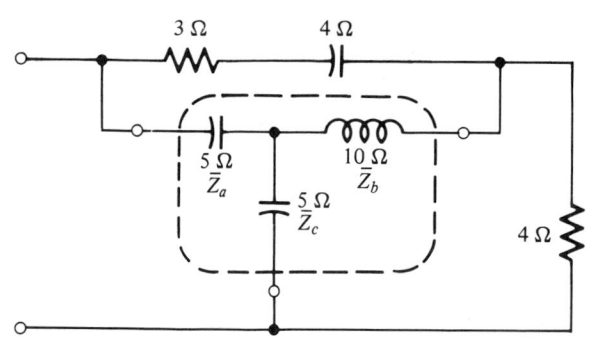

The numerator in these T-to-Π conversions is

$$N = (-j5)(j10) + (j10)(-j5) + (-j5)(-j5)$$
$$= 75$$

$$\bar{Z}_x = \frac{N}{\bar{Z}_c} = \frac{75}{-j5} = j15 \quad \text{or} \quad 15 \underline{/90°} \; \Omega$$

$$\bar{Z}_y = \frac{N}{\bar{Z}_a} = \frac{75}{-j5} = j15 \quad \text{or} \quad 15 \underline{/90°} \; \Omega$$

$$\bar{Z}_z = \frac{N}{\bar{Z}_b} = \frac{75}{j10} = -j7.5 \quad \text{or} \quad 7.5 \underline{/-90°} \; \Omega$$

17.4 THÉVENIN AND NORTON AC SOURCES

Draw the transformed circuit:

Define $\bar{Z}_1 = (3 - j4) \| j15$:

$$\bar{Z}_1 = \frac{(3 - j4)(j15)}{3 - j4 + j15}$$

$$= \frac{75 \, \underline{/36.9°}}{11.4 \, \underline{/74.7°}}$$

$$= 6.6 \, \underline{/-37.8°} \quad \text{or} \quad 5.2 - j4 \, \Omega$$

Define $\bar{Z}_2 = j15 \| 4$:

$$\bar{Z}_2 = \frac{4(j15)}{4 + j15}$$

$$= \frac{60 \, \underline{/90°}}{15.5 \, \underline{/75.1°}}$$

$$= 3.9 \, \underline{/14.9°} \quad \text{or} \quad 3.7 + j1$$

Define $\bar{Z}_3 = \bar{Z}_1 + \bar{Z}_2$:

$$\bar{Z}_3 = (5.2 - j4) + (3.7 + j1)$$

$$= 8.9 - j3 \quad \text{or} \quad 9.4 \, \underline{/-18.6°}$$

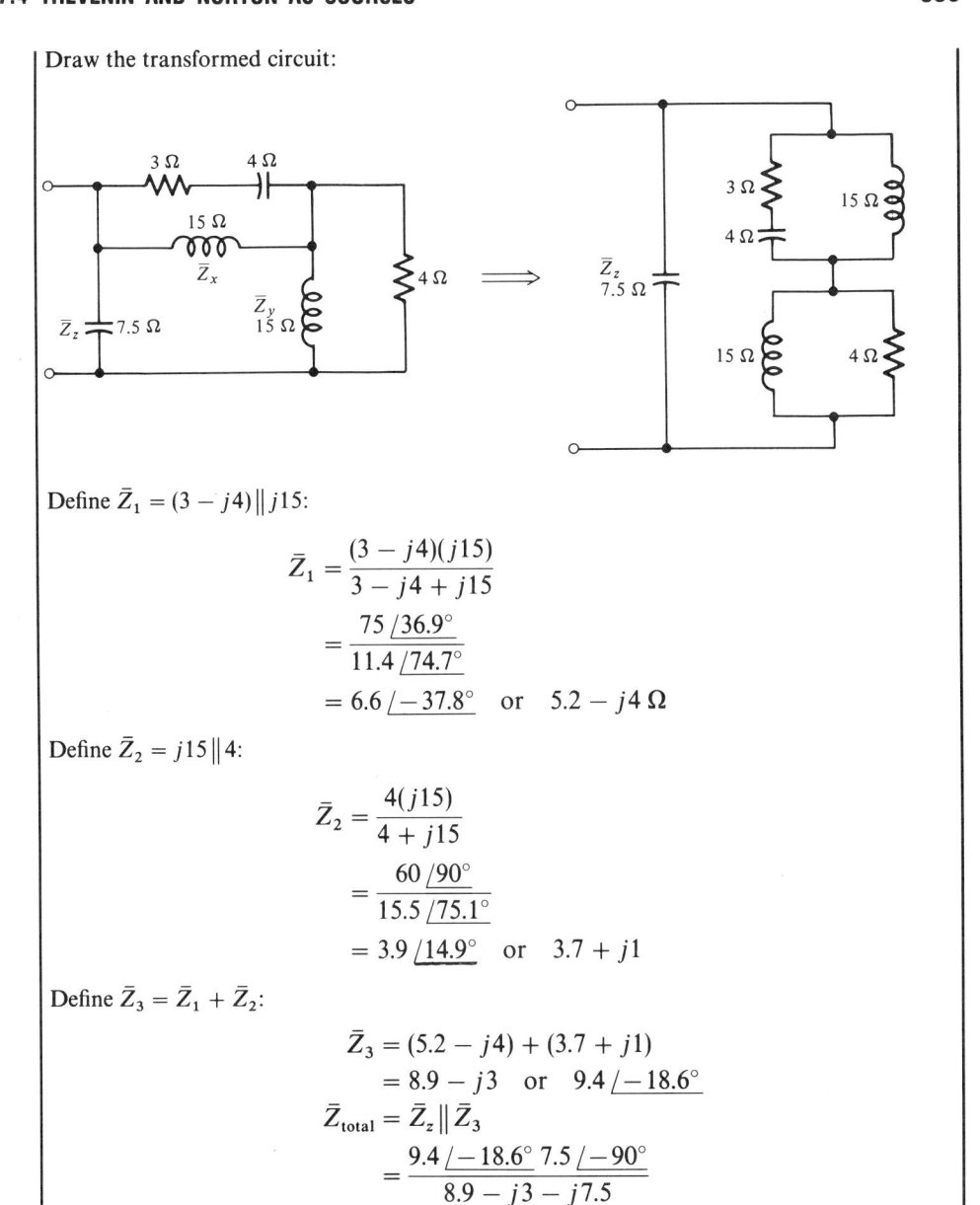

$$\bar{Z}_{\text{total}} = \bar{Z}_z \| \bar{Z}_3$$

$$= \frac{9.4 \, \underline{/-18.6°} \; 7.5 \, \underline{/-90°}}{8.9 - j3 - j7.5}$$

$$= \frac{70.4 \, \underline{/-108.6°}}{13.8 \, \underline{/-49.7°}}$$

$$= 5.1 \, \underline{/-58.9°} \, \Omega$$

The result from Example 17.10 was $5.1 \, \underline{/-59.4°} \, \Omega$. ■

17.4 THÉVENIN AND NORTON AC SOURCES

The *Thévenin* and *Norton models* of ac sources are much like their dc counterparts except that they have an *internal impedance*, Z_{int}, rather than an internal resistance and the ideal source is sinusoidal. The models are seen in Figure 17.6.

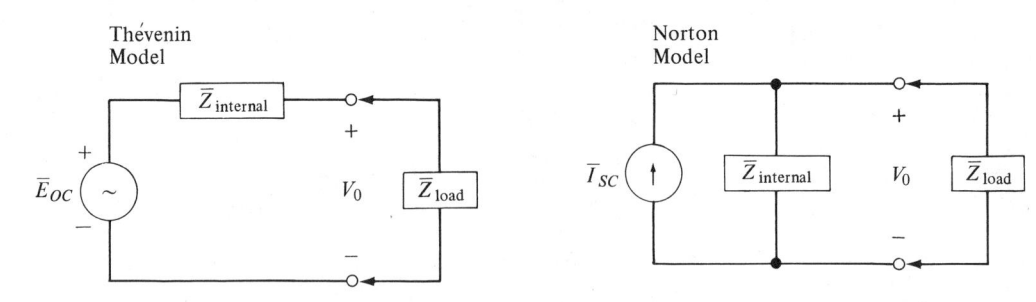

FIGURE 17.6 Thévenin and Norton models of AC sources.

Let us review the characteristics of these sources:

Output Voltage and Current Characteristics

The Thévenin Model

The output voltage,

$$\bar{V}_o = \bar{E}_{oc} - \bar{I}_{load}\bar{Z}_{int} \quad \text{by KVL} \tag{17.15}$$

or

$$\bar{V}_o = \frac{\bar{E}_{oc}\bar{Z}_{load}}{\bar{Z}_{load} + \bar{Z}_{internal}} \quad \text{by voltage division} \tag{17.16}$$

The load current,

$$\bar{I}_{load} = \frac{\bar{E}_{oc}}{\bar{Z}_{load} + \bar{Z}_{internal}} \tag{17.17}$$

For open-circuit conditions ($\bar{Z}_{load} = \infty$):

$$\bar{V}_o = \bar{E}_{oc}, \text{ its highest value}$$
$$\bar{I}_{load} = 0 \text{ A}$$

For short-circuit conditions ($\bar{Z}_{load} = 0\,\Omega$):

$$\bar{V}_o = 0 \text{ V}$$
$$\bar{I}_{load} = \bar{I}_{sc} = \frac{\bar{E}_{oc}}{\bar{Z}_{internal}}, \text{ its highest value}$$

The Norton Model

The output voltage,

$$\bar{V}_o = \bar{I}_{sc}(\bar{Z}_{int} \| \bar{Z}_{load}) \tag{17.18}$$

The load current,

$$\bar{I}_{load} = \frac{\bar{I}_{sc}\bar{Z}_{int}}{\bar{Z}_{int} + \bar{Z}_{load}} \quad \text{by current division} \tag{17.19}$$

For open-circuit conditions ($\bar{Z}_{load} = \infty$ ohms):

$$\bar{V}_o = \bar{I}_{sc}\bar{Z}_{internal}, \text{ its highest value}$$
$$\bar{I}_{load} = 0 \text{ A}$$

17.4 THÉVENIN AND NORTON AC SOURCES

For short-circuit conditions ($\bar{Z}_{load} = 0\ \Omega$):

$$\bar{V}_o = 0\text{ V}$$
$$\bar{I}_{load} = \bar{I}_{sc}, \text{ its highest value}$$

Maximum Power Transfer: Matched Conditions

For both sources the maximum power transfer theorem says:

> Maximum power is delivered to a load impedance when that impedance equals the complex conjugate of the internal impedance of the source; that is,
>
> $$\bar{Z}_{load} = \bar{Z}_{internal}{}^* \qquad (17.20)$$
>
> In rectangular form, this means that
>
> $$R_{load} = R_{internal}$$
>
> and
>
> $$\pm jX_{load} = \mp jX_{internal} \qquad (17.21)$$
>
> In polar form, Equation 17.20 means that
>
> $$|Z_{load}| = |Z_{internal}|$$
>
> and
>
> $$\theta_{load} = -\theta_{internal} \qquad (17.22)$$
>
> When $\bar{Z}_{load}$ is set equal to $\bar{Z}^*_{internal}$ for maximum power transfer, the reactive components cancel. A moment's thought shows that this is the same condition as before in the dc case; that is, $R_{load} = R_{internal}$. It should also be noted that this is the series-resonance condition discussed in Chapter 16.
> A circuit operating under conditions of maximum power transfer is said to be operating under "matched conditions"; the load "matches" the source internal impedance.

EXAMPLE 17.12

Answer the questions for this Thévenin source:

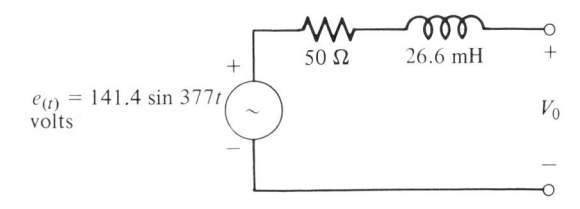

$e_{(t)} = 141.4 \sin 377t$ volts; 50 Ω; 26.6 mH; V_0

a. What are the maximum and minimum rms values of $\bar{V}_o$?

$$\bar{V}_{o_{max}} = E_{oc_{rms}} = 0.707(141.4) = 100\text{ V}$$
$$\bar{V}_{o_{min}} = E_{sc_{rms}} = 0\text{ V}$$

b. What are the maximum and minmium rms values of $\bar{I}_{load}$?

Calculate X_L of the internal inductance:

$$X_L = \omega L$$
$$= 377(26.5 \times 10^{-3})$$
$$= j10\ \Omega$$
$$\bar{Z}_{internal} = 50 + j10\ \Omega$$
$$\bar{I}_{load\,max} = \bar{I}_{sc} = \frac{\bar{E}_{oc}}{\bar{Z}_{internal}}$$
$$= \frac{100}{50 + j10} = \frac{100}{51\,\underline{/11.3°}}$$
$$= 2\,\underline{/-11.3°}\ A$$
$$\bar{I}_{load\,min} = \bar{I}_{load\,open\,circuit} = 0\ A$$

c. For what load impedance does the circuit operate under matched conditions?

$$\bar{Z}_{matched} = \bar{Z}_{load}\ \text{for maximum power transfer}$$
$$= \bar{Z}^*_{internal}$$

Therefore,

$$\bar{Z}_{load} = 50 - j10\ \Omega$$

d. What is this maximum power?

For matched conditions, the total circuit impedance is

$$\bar{Z}_{total} = \bar{Z}_{internal} + \bar{Z}_{load}$$
$$= 50 + j10 + 50 - j10$$
$$= 100\ \Omega$$
$$\bar{I} = \frac{\bar{E}}{\bar{Z}_{total}} = \frac{100}{100}$$
$$= 1\ A$$
$$P_{load} = I^2 R_{load}$$
$$= 1^2(50)$$
$$= 50\ W$$

EXAMPLE 17.13

Given: the following Norton source:

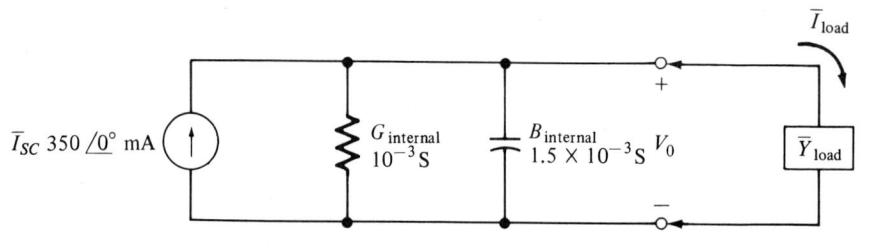

$\bar{I}_{SC}\ 350\,\underline{/0°}$ mA, $G_{internal}\ 10^{-3}$ S, $B_{internal}\ 1.5 \times 10^{-3}$ S, V_0, $\bar{Y}_{load}$, $\bar{I}_{load}$

a. What is the maximum value of $\bar{I}_{load}$?

$$\bar{I}_{sc}$$

17.4 THÉVENIN AND NORTON AC SOURCES

b. What is the maximum value of $\bar{V}_o$?

$$\bar{V}_o = \bar{E}_{oc} = \frac{\bar{I}_{sc}}{\bar{Y}_{int}} \quad (\bar{Y}_{load} = 0 \text{ S})$$

$$= \frac{350 \times 10^{-3}}{(1 + j1.5) \times 10^{-3}} = \frac{350}{1.8 \underline{/56.3°}}$$

$$= 194 \underline{/-56.3°} \text{ V}$$

c. What is the value of $\bar{Y}_{load}$ for maximum power transfer?

$\bar{Y}_{load}$ must match (equal) $\bar{Y}^*_{internal}$. Therefore,

$$\bar{Y}_{load} = G - jB_L$$
$$= (1 - j1.5) \times 10^{-3} \text{ S}$$

Converting from One Source Model to the Other

The steps for conversion are the same as used for dc sources. Let us review them:

To go from Thévenin to Norton

$\bar{I}_{sc}$ must be determined. Therefore, short-circuit the Thévenin source and solve for $\bar{I}_{sc}$.

$$\bar{I}_{sc} = \frac{\bar{E}_{oc}}{\bar{Z}_{internal}}$$

Make $\bar{I}_{sc}$ the output of an ideal current source connected in parallel with $\bar{Z}_{internal}$. For convenience, $\bar{Z}_{internal}$ may be converted to $\bar{Y}_{internal}$.

To go from Norton to Thévenin

$\bar{E}_{oc}$ must be determined:

$$\bar{E}_{oc} = \frac{\bar{I}_{sc}}{\bar{Y}_{internal}} \quad \text{or} \quad \bar{I}_{sc} \bar{Z}_{internal}$$

Make this the output of an ideal voltage source in series with $\bar{Z}_{internal}$. If Norton source has parallel $\bar{Y}_{internal}$, convert to equivalent series impedance.

EXAMPLE 17.14

Convert each source model to the opposite form.

a. Thévenin ⇒ Norton.

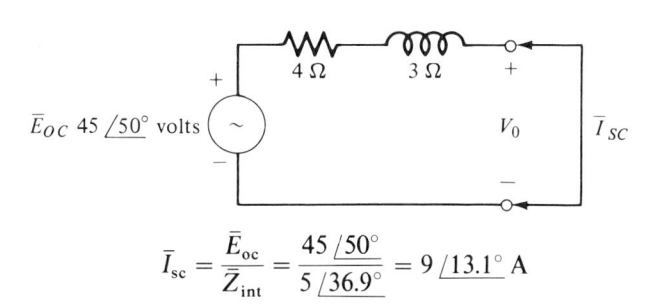

$$\bar{I}_{sc} = \frac{\bar{E}_{oc}}{\bar{Z}_{int}} = \frac{45 \underline{/50°}}{5 \underline{/36.9°}} = 9 \underline{/13.1°} \text{ A}$$

Convert $\bar{Z}_{internal}$ to $Y_{internal}$ using Equations 16.28 and 16.29:

$$G_p = \frac{R_s}{R_s^2 + X_s^2}$$

$$= \frac{4}{4^2 + 3^2}$$

$$= \frac{4}{25}$$

$$= 160 \times 10^{-3} \text{ S}$$

$$B_p = \frac{X_s}{R_s^2 + X_s^2}$$

$$= \frac{3}{4^2 + 3^3}$$

$$= \frac{3}{25}$$

$$= -j120 \times 10^{-3} \text{ S}$$

Draw the Norton equivalent:

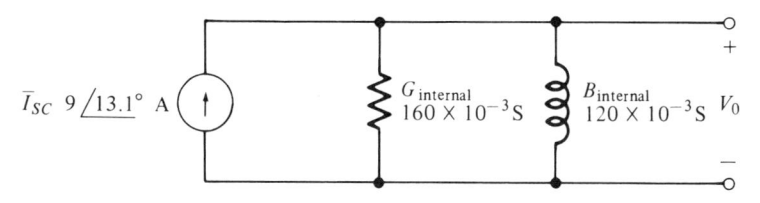

[Circuit diagram: Norton equivalent with $\bar{I}_{SC}\; 9\;\underline{/13.1°}$ A current source, $G_{internal}\; 160 \times 10^{-3}$ S, $B_{internal}\; 120 \times 10^{-3}$ S, output V_0]

b. Norton ⇒ Thévenin.

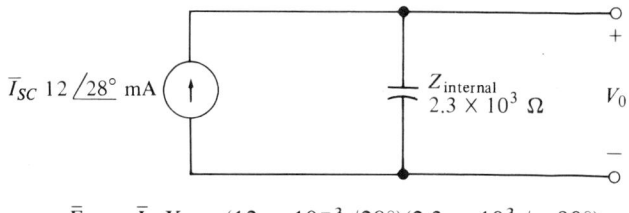

[Circuit diagram: $\bar{I}_{SC}\; 12\;\underline{/28°}$ mA current source in parallel with $Z_{internal}\; 2.3 \times 10^3\; \Omega$, output V_0]

$$\bar{E}_{oc} = \bar{I}_{sc} X_C = (12 \times 10^{-3}\;\underline{/28°})(2.3 \times 10^3\;\underline{/-90°})$$
$$= 27.6\;\underline{/-62°} \text{ V}$$

Draw the Thévenin equivalent:

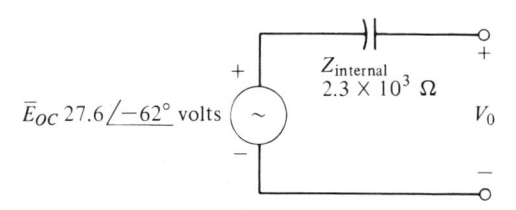

[Circuit diagram: Thévenin equivalent with $\bar{E}_{OC}\; 27.6\;\underline{/-62°}$ volts source in series with $Z_{internal}\; 2.3 \times 10^3\; \Omega$, output V_0] ∎

Combining Sources

1. *For ideal sources:*
    a. Ideal current sources must not be connected in series unless they supply the same short-circuit current.

17.4 THÉVENIN AND NORTON AC SOURCES

    b. Ideal voltage sources must not be connected in parallel unless they supply the same open-circuit voltage.
2. *For Thévenin sources in series:*
    a. Add internal impedances.
    b. Add source potentials in phasor form.
3. *For Norton sources in series:* Convert to Thévenin form first, then combine according to the instructions above. Convert the total Thévenin circuit back to Norton if desired.
4. *For Norton sources in parallel:* By Millman's theorem, Norton sources in parallel are combined by summing current sources in phasor form. Then sum $Z_{internals}$ or $Y_{internals}$.
5. *For Thévenin sources in parallel:* Convert to Norton form first, then combine according to the instructions above. Convert total Norton circuit back to Thévenin if desired.

EXAMPLE 17.15

Reduce this circuit to a single Norton equivalent:

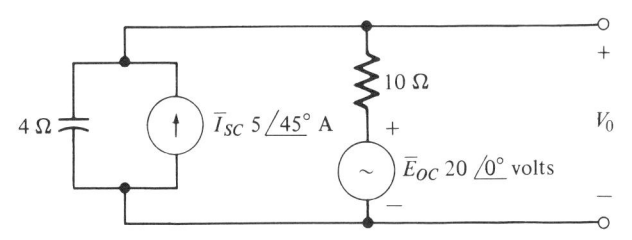

Convert the Thévenin to a Norton:

$$\bar{I}_{sc} = \frac{\bar{E}_{oc}}{\bar{R}_{int}} = \frac{20\,\underline{/0°}}{10} = 2\,\underline{/0°}\ \text{A}$$

$$G_{int} = \frac{1}{R_{int}} = \frac{1}{10} = 0.1\ \text{S}$$

Combine current sources:

$$\bar{I}_{total} = 5\,\underline{/45°} + 2\,\underline{/0°} = 3.5 + j3.5 + 2$$
$$= 5.5 + j3.5 \quad \text{or} \quad 6.5\,\underline{/32.5°}\ \text{A}$$

Combine internal admittances:

$$\bar{Y}_{total} = G + jB = 0.1 + \frac{1}{-j4}$$
$$= 0.1 + j0.25\ \text{S}$$

Draw the final circuit:

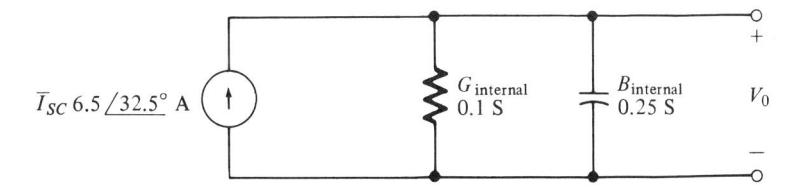

17.5 SUPERPOSITION

Superposition, you will recall from Chapter 6, is a powerful method for analyzing circuits with two or more sources. Let us review the procedure:

1. Turn off all sources but one. This is done so that source internal impedance or admittance remains in the circuit; that is, ideal voltage sources are short circuited, and ideal current sources are open circuited.
2. The circuit can now be analyzed by the SPR (series–parallel reduction) approach reviewed in Section 17.1. Find the parameter (current or voltage) being sought.
3. Turn off the source left on in step 1. Turn on another. Repeat step 2.
4. Repeat step 3 until the contribution of every source has been determined.
5. The final value equals the sum of the individual contributions.

EXAMPLE 17.16

Find the current through the inductor by superposition.

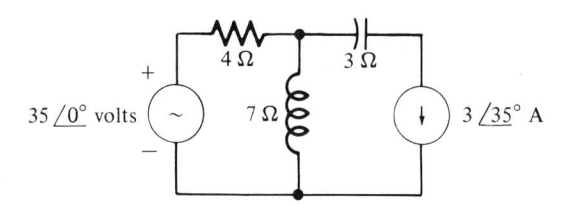

Finding the contribution from the voltage source with current source opened: The circuit is now:

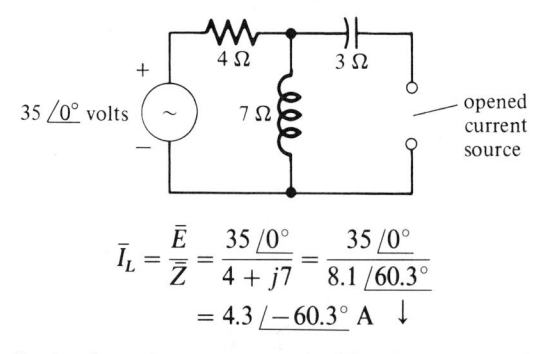

$$\bar{I}_L = \frac{\bar{E}}{\bar{Z}} = \frac{35\ \underline{/0°}}{4 + j7} = \frac{35\ \underline{/0°}}{8.1\ \underline{/60.3°}}$$
$$= 4.3\ \underline{/-60.3°}\ \text{A} \downarrow$$

Finding the contribution from the current source with voltage source shorted: The circuit is now

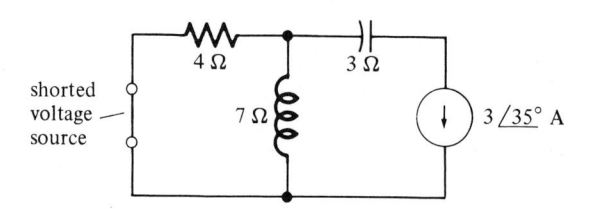

17.5 SUPERPOSITION

By current division,

$$\bar{I}_L = \frac{\bar{I}_{\text{source}}(4)}{4 + j7}$$

$$= \frac{3\,\underline{/35°}\,(4)}{4 + j7}$$

$$= \frac{12\,\underline{/35°}}{8.1\,\underline{/60.3°}}$$

$$= 1.5\,\underline{/-25.3°}\ \text{A}\ \uparrow$$

The currents through the inductor are

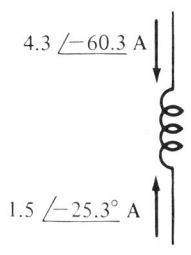

$$4.3\,\underline{/-60.3}\ \text{A}\ \downarrow$$

$$1.5\,\underline{/-25.3°}\ \text{A}\ \uparrow$$

Combine to get I_{total}:

$$\bar{I}_{\text{total}} = (4.3\,\underline{/-60.3°}) - (1.5\,\underline{/-25.3°})$$
$$= (2.1 - j3.7) - (1.4 - j0.6)$$
$$= 0.7 - j3.1\ \ \text{or}\ \ 3.2\,\underline{/-77.3°}\ \text{mA}\ \blacksquare$$

Because superposition works for linear relationships only, it cannot be used to determine total power by summing individual powers. If power is needed, it may be obtained by finding total voltage across or current through a component by superposition, then using V^2/R or I^2R. This also holds for reactive power.

EXAMPLE 17.17

Using superposition, determine the reactive power drawn by the inductance in this circuit.

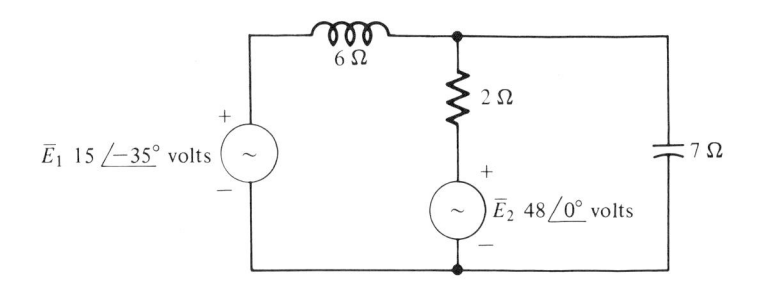

$\bar{E}_1\ 15\,\underline{/-35°}$ volts

6 Ω

2 Ω

7 Ω

$\bar{E}_2\ 48\,\underline{/0°}$ volts

This reactive power can be determined by either V^2/X or I^2X. We will use the voltage form. Turning E_2 off (by shorting the source), the circuit becomes

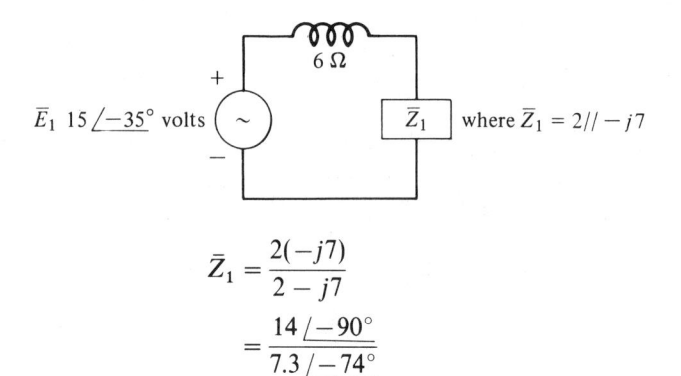

$\bar{E}_1$ $15\,\underline{/-35°}$ volts, $6\,\Omega$ inductor, $\bar{Z}_1$ where $\bar{Z}_1 = 2 // -j7$

$$\bar{Z}_1 = \frac{2(-j7)}{2 - j7}$$

$$= \frac{14\,\underline{/-90°}}{7.3\,\underline{/-74°}}$$

$$= 1.9\,\underline{/-16°}\ \Omega$$

Find $\bar{V}_{L_1}$ by voltage division:

$$\bar{V}_{L_1} = \frac{\bar{E}_1 X_L}{X_L + \bar{Z}_1}$$

$$= \frac{15\,\underline{/-35°}\,(6\,\underline{/90°})}{6\,\underline{/90°} + 1.9\,\underline{/-16°}}$$

$$= \frac{90\,\underline{/55°}}{j6 + 1.8 - j0.5}$$

$$= \frac{90\,\underline{/55°}}{1.8 + j5.5}$$

$$= \frac{90\,\underline{/55°}}{5.8\,\underline{/71.9°}}$$

$$= 15.6\,\underline{/-16.9°}\ \text{V}$$

Turning E_1 off and E_2 on, the circuit becomes

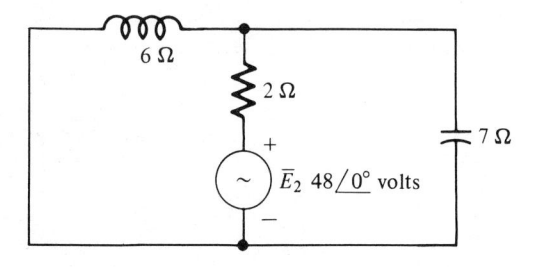

$6\,\Omega$ inductor, $2\,\Omega$ resistor, $\bar{E}_2$ $48\,\underline{/0°}$ volts, $7\,\Omega$ capacitor

The capacitor and inductor are in parallel. Their total impedance, $\bar{Z}_2$, by POS is

$$\bar{Z}_2 = \frac{j6(-j7)}{j6 - j7}$$

$$= \frac{42}{-j1}$$

$$= j42\ \Omega$$

$\bar{V}_{L_2}$ by voltage division:

$$\bar{V}_{L_2} = \frac{\bar{E}_s X_L}{2 + \bar{Z}_2}$$

$$= \frac{48(42\,/90°)}{(2 + j42)}$$

$$= \frac{2016\,/90°}{42\,/90°}$$

$$= 48\,/0°\ \text{V}$$

$$\bar{V}_{\text{total}} = \bar{V}_{L_1} - \bar{V}_{L_2}$$
$$= 48 - (15.6\,/-16.9°)$$
$$= 48 - (14.9 - j4.5)$$
$$= 33.1 + j4.5$$
$$= 33.4\,/7.7°\ \text{V}$$

V_{L_1} 15.6 /−16.9° volts

V_{L_2} 48 /0° volts

Finally,

$$P_Q = \frac{V_t^2}{X} = \frac{33.4^2}{6} = 185.9\ \text{var, inductive} \qquad \blacksquare$$

17.6 THE LADDER METHOD

The *ladder method*, or assumed current method, was first discussed in Section 6.7. In it, you will recall, a current, I, is assumed to flow through the branch farthest from the source. Working back to the source, all voltages and other currents are determined by KCL, KVL, and Ohm's law, and are written in terms of I. Finally, the actual value of I is calculated using the known source current or voltage. Once I is known, all voltages or currents throughout the circuit are easily determined.

When reactance is present in a circuit, ladder calculations can become tricky and tedious. For this reason, the method is not recommended for large circuits.

EXAMPLE 17.18

Perform the following operations for this circuit:

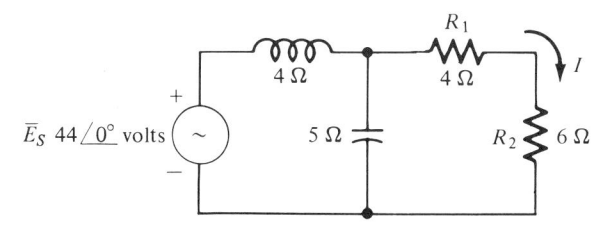

a. Using the ladder method, find the current, I.

Voltage across R_1 and $R_2 = (6 + 4)I = 10I$
This is also the value of $\bar{V}_C$. Therefore,

$$\bar{I}_C = \frac{\bar{V}_C}{X_C} = \frac{10I}{5\,/-90°} = 2I\,/90°\ \text{A}$$

$$\bar{I}_L = \bar{I}_R + \bar{I}_C = I + 2I\,\underline{/90°}$$
$$= I(1 + j2)$$
$$= 2.2I\,\underline{/63.4°}\text{ A}$$
$$\bar{V}_L = \bar{I}_L X_L = 2.2I\,\underline{/63.4°}\,(4\,\underline{/90°})$$
$$= 8.8I\,\underline{/153.4°}\text{ V}$$

By KVL,

$$\bar{E}_s = \bar{V}_L + \bar{V}_C$$
$$= 8.8I\,\underline{/153.4°} + 10I$$
$$= I(8.8\,\underline{/153.4°} + 10)$$
$$= I(-7.9 + j3.9 + 10)$$
$$44\,\underline{/0°} = I(2.1 + j3.9)$$
$$= 4.4I\,\underline{/61.7°}\text{ V}$$

Therefore,

$$\bar{I} = \frac{44}{4.4\,\underline{/61.7°}}$$
$$= 10\,\underline{/-61.7°}\text{ A}$$

Note that every current and voltage equation was written in terms of I.

b. What is the voltage across the capacitor?

From the diagram,

$$\bar{V}_C = 10I = 10(10\,\underline{/-61.7°})$$
$$= 100\,\underline{/-61.7°}\text{ V}$$

c. What is $\bar{V}_L$?

$$\bar{V}_L = 8.8I\,\underline{/153.4°}$$
$$= 8.8\,\underline{/153.4°}\,(10\,\underline{/-61.7°})$$
$$= 88\,\underline{/91.7°}\text{ V}$$

d. Verify KVL around the loop: $\bar{E}_s, \bar{V}_L, \bar{V}_C$:

$$\bar{E}_s - (\bar{V}_L + \bar{V}_C) = 0$$
$$44 - (88\,\underline{/91.7°} + 100\,\underline{/-61.7°}) = 0$$
$$44 - (-2.6 + j88 + 47.4 - j88) = 44 - 44.8 \approx 0$$

e. What would the new value of $\bar{I}_C$ be if $\bar{E}_s$ was changed to $39.2\,\underline{/20°}$ V?

$$\bar{E}_s = 4.4I\,\underline{/61.7°} \Rightarrow 39.2\,\underline{/20°}\text{ V}$$

Therefore,

$$I = \frac{39.2\,\underline{/20°}}{4.4\,\underline{/61.7°}}$$
$$= 8.9\,\underline{/-41.7°}\text{ A}$$
$$\bar{I}_C = 2I\,\underline{/90°}$$
$$= 2(8.9\,\underline{/-41.7°})\underline{/90°}$$
$$= 17.8\,\underline{/48.3°}\text{ A}\qquad\blacksquare$$

EXERCISES

Section 17.1

1. For each of the following sets of conditions, find: a. $\bar{Z}_{total}$; b. $\bar{I}_{total}$; c. phase between source voltage and current; d. $\bar{I}_R$; e. $\bar{I}_L$.

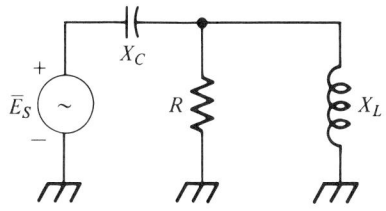

1. $\bar{E}_{source} = 25 \underline{/0°}$ V, $X_C = -j4.4\ \Omega$, $X_L = j4\ \Omega$, $R = 3\ \Omega$.
2. $\bar{E}_{source} = 75 \underline{/-20°}$ V, $X_C = -j10\ \Omega$, $X_L = j8\ \Omega$, $R = 5\ \Omega$.
3. $\bar{E}_{source} = 120 \underline{/0°}$ V, $X_C = -j1.5\ \Omega$, $X_L = j20\ \Omega$, $R = 10\ \Omega$.

2. For each of the following sets of conditions, find: a. $\bar{Z}_{total}$; b. $\bar{E}_{source}$; c. ϕ; d. $\bar{V}_R$; e. $\bar{V}_L$.

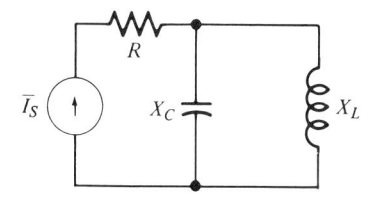

1. $\bar{I}_{source} = 2 \underline{/0°}$ A, $R = 15\ \Omega$, $X_C = 10\ \Omega$, $X_L = 8\ \Omega$
2. $\bar{I}_{source} = 34 \underline{/-45°}$ A, $R = 3\ \Omega$, $X_C = 6\ \Omega$, $X_L = 2\ \Omega$
3. $\bar{I}_{source} = 7 \underline{/16°}$ mA, $R = 10$ k, $X_C = 7.5$ k, $X_L = 4.8$ k

3. For the following circuit, find: a. $\bar{I}_{total}$; b. $\bar{I}_{R_1}$; c. $\bar{V}_{R_2}$; d. the phase angle between $\bar{E}_s$ and $\bar{V}_C$.

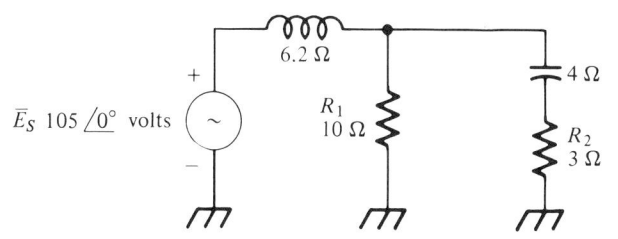

4. The impedances in Exercise 3 were calculated at medium frequencies. Find the approximate value of V_{R_2} at: a. low frequencies; b. high frequencies. Assume $f_{low} \leq f_{medium}/10$ and $f_{high} \geq 10 f_{medium}$.

5. For this circuit:

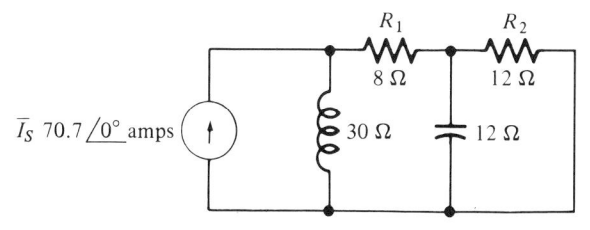

Find: a. P_{R_1}; b. $\bar{I}_C$.

6. The impedances in Exercise 5 were calculated at medium frequencies. Calculate the approximate values of P_{R_1} and $\bar{I}_C$ at: a. very low frequencies; b. very high frequencies.

548   ANALYZING SERIES–PARALLEL AC CIRCUITS

7. Find the simplest two-component series circuit hidden in this box.

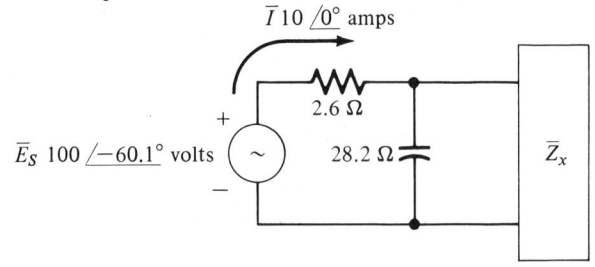

$\bar{I}$ 10 $\underline{/0°}$ amps

$\bar{E}_S$ 100 $\underline{/-60.1°}$ volts, 2.6 Ω, 28.2 Ω, $\bar{Z}_x$

8. For this circuit:

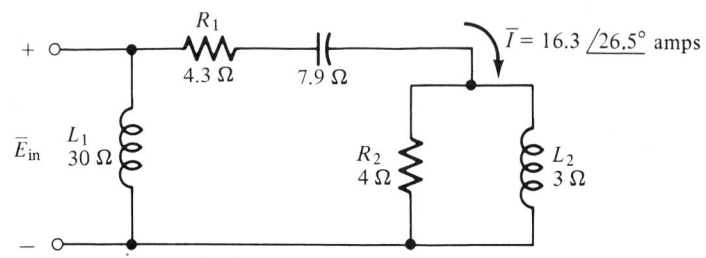

R_1 4.3 Ω, 7.9 Ω, $\bar{I} = 16.3 \underline{/26.5°}$ amps

$\bar{E}_{in}$, L_1 30 Ω, R_2 4 Ω, L_2 3 Ω

Determine: a. the input voltage; b. the components of the power triangle.

9. Perform the following operations for their circuit.

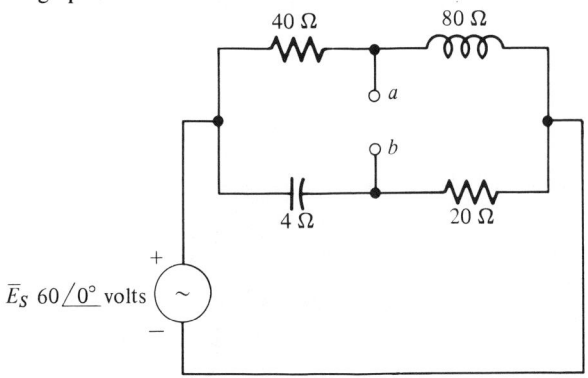

40 Ω, 80 Ω, 4 Ω, 20 Ω, $\bar{E}_S$ 60 $\underline{/0°}$ volts

a. Find the voltage, $\bar{V}_{ab}$.
b. Assume the opposite polarity of $\bar{V}_{ab}$ from that chosen in part a. Recalculate $\bar{V}_{ab}$. Comment on differences.
c. The frequency of the signal applied to the circuit drops to one-tenth its original value. Calculate the new value of $\bar{V}_{ab}$.

10. For this circuit:

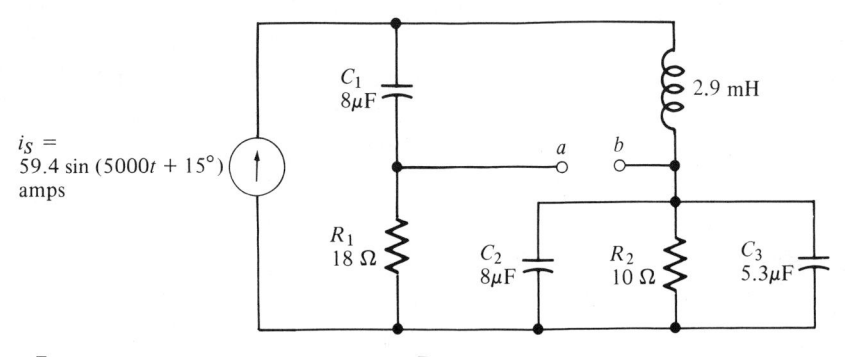

C_1 8μF, 2.9 mH, $i_S = 59.4 \sin(5000t + 15°)$ amps, R_1 18 Ω, C_2 8μF, R_2 10 Ω, C_3 5.3μF

Find: a. $\bar{Y}_{total}$ seen by the source; b. the voltage, $\bar{V}_{ab}$.

EXERCISES

11. An induction-type electric motor places a 62.9 /63°-Ω load across a 220-V 60-Hz line.
   a. What is the circuit power factor?
   b. A 73.7-μF capacitor is wired in series with the motor. What is the new power factor?
   c. What is the voltage across the motor with the power-factor-correcting capacitor in series? What is wrong?
   d. What capacitance must be placed in shunt with the motor to provide the power factor of part b?
   e. What is the voltage across the motor with the correcting capacitor in shunt? What conclusion do you arrive at regarding the wiring of power-factor-correcting capacitors?

12. An impedance, $\bar{Z} = 34.2 + j94$ Ω, is connected across a 400-V 60-Hz line. What capacitance must be shunted across it to provide a power factor of 89.1% lagging?

13. Repeat Exercise 12 finding the capacitance needed to correct the circuit to unity power factor, that is, PF = 1.

14. Verify that the impedance seen by the 400-V line has unity PF when the capacitance of Exercise 13 is connected.

Section 17.2

15. Given: the Maxwell bridge in Figure 17.2 and the values $R_2 = 2$ k $R_3 = 6$ k, $C = 2200$ pF, $R_1 = 14.1$ k, $L_x = 26.4$ mH, and $R_x = 1.1$ k. Is the bridge balanced?

16. Derive the balance equations for the capacitance bridge in Figure 17.3.

17. The Wien bridge circuit of Figure 17.4 has the following values: source frequency = 5 kHz, $C_4 = 420$ pF, $R_4 = 75.7$ k, and $R_1 = 2R_2 = 10$ k. The engineer nulls the bridge. Does she see a reading on the detector?

18. Repeat Exercise 17 for a source frequency of 2800 Hz, $C_4 = 345$ pF, $R_1 = 14$ k $= 2R_2$, and $R_4 = 112$ k.

19. Derive the balance conditions for each of the following bridges. Which of the bridges have frequency-dependent balance conditions?

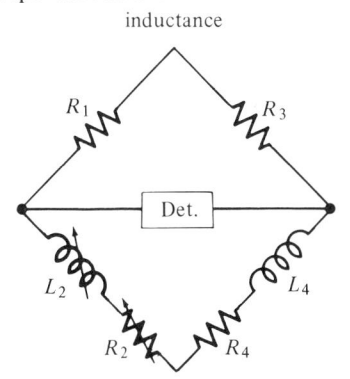

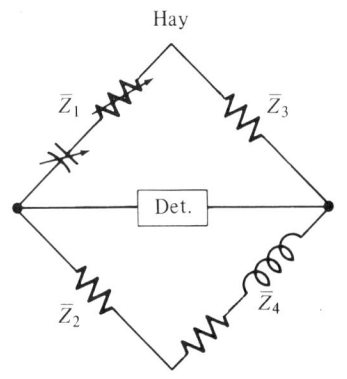

inductance

Hay

20. Repeat Exercise 19 for each of these bridges.

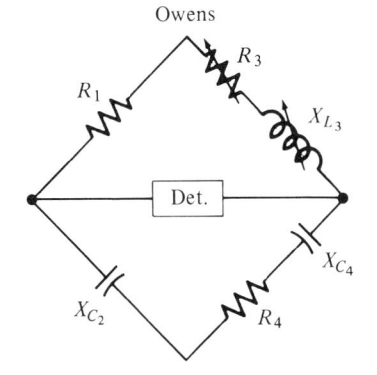

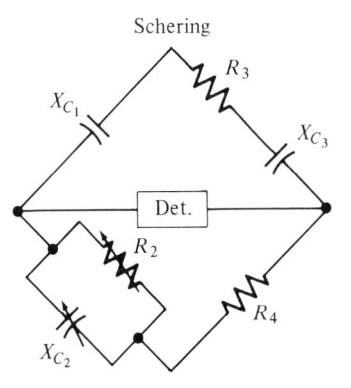

Owens

Schering

21. Let the bridge in Figure 17.1 be balanced. Swap the detector and source. Then show that balance is unaffected by this change.

22. *Extra Credit.* Besides measuring frequency, the Wien bridge of Figure 17.4 will measure impedance by placing an unknown in the series RX branch and adjusting the parallel RX circuit for null. Consider the series $R_3 X_3$ branch to be $R_x X_x$, and derive the balance equations.

Section 17.3

23. Perform the following operations. For this network:

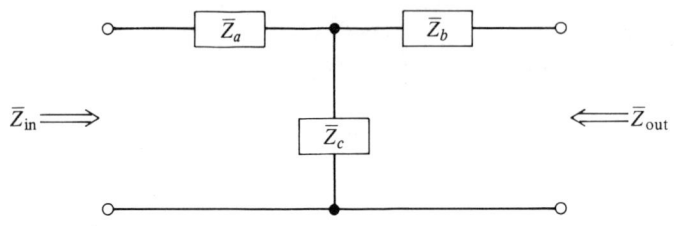

a. Find $\bar{Z}_{in}$ (output open).
b. Find $\bar{Z}_{out}$ (input open).
c. Transform the T to its Π equivalent.
d. Repeat parts a and b on the transformed circuit of part c. Are the two networks equivalent for $\bar{Z}_{in}$ and $\bar{Z}_{out}$?

24. Repeat Exercise 23 for $\bar{Z}_a = 5 + j12$, $\bar{Z}_b = 8\,\underline{/-55°}$, and $\bar{Z}_c = 5\,\Omega$.

25. A Π network has the following impedances: $\bar{Z}_x = 8\,\Omega$, $\bar{Z}_y = 4 + j3\,\Omega$, and $\bar{Z}_z = 15\,\underline{/30°}\,\Omega$. Calculate the T-network equivalent.

26. Repeat Exercise 25 for $\bar{Z}_x = \bar{Z}_y = \bar{Z}_z = 72 + j54\,\Omega$.

27. Solve for the source current in this circuit by converting the Π, consisting of R_2, C, and L_2, to its equivalent T.

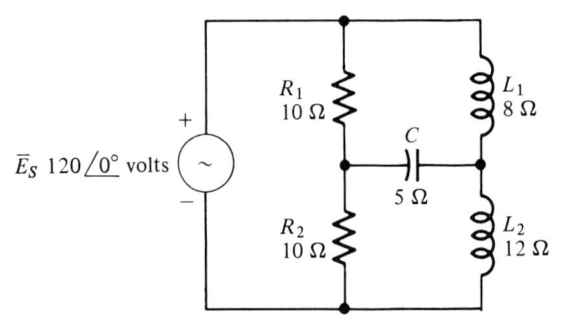

28. Redo Exercise 27 by converting the T, consisting of R_1, R_2, and C, to its Π equivalent. Compare the results.

Section 17.4

29. Given: the Thévenin source:

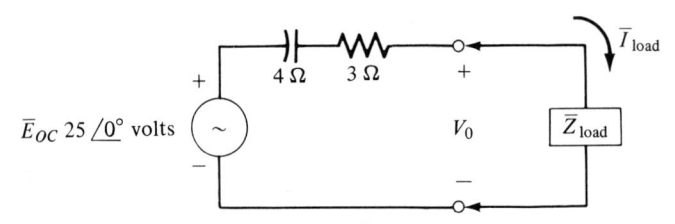

a. What are the maximum and minimum values of $\bar{V}_o$?
b. What are the maximum and minimum values of $\bar{I}_{load}$?

EXERCISES

c. What are $\bar{V}_o$ and $\bar{I}_{load}$ when the load impedance is $10 \underline{/60°}\ \Omega$?
d. Express $\bar{Z}_{load}$ for maximum power transfer in polar form.
e. What is this value of maximum power?

30. Repeat Exercise 29 for this Norton source:

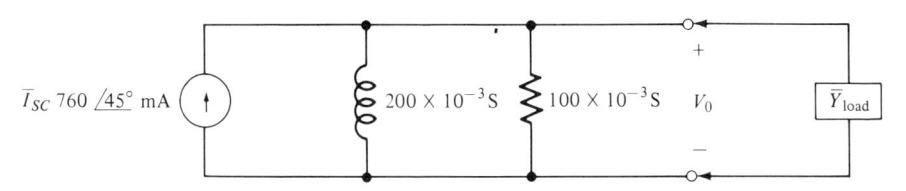

$\bar{I}_{SC}\ 760\ \underline{/45°}$ mA, 200×10^{-3} S, 100×10^{-3} S, V_0, $\bar{Y}_{load}$

31. Convert each source model to the opposite form. Use $\bar{Y}_{internal}$ for the Norton circuit and $\bar{Z}_{internal}$ for the Thévenin.

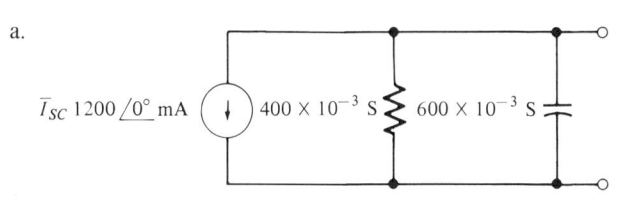

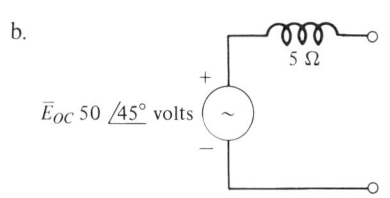

a. $\bar{I}_{SC}\ 1200\ \underline{/0°}$ mA, 400×10^{-3} S, 600×10^{-3} S

b. $\bar{E}_{OC}\ 50\ \underline{/45°}$ volts, $5\ \Omega$

32. What is the power factor of a circuit operating under matched conditions?

33. Combine into one Norton source:

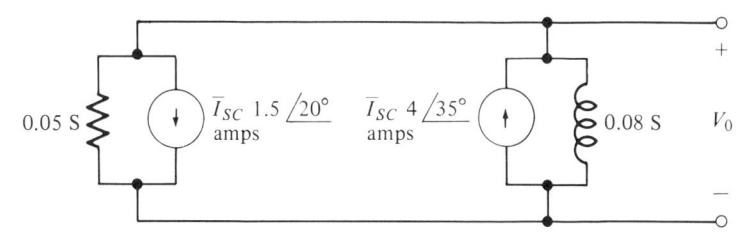

0.05 S, $\bar{I}_{SC}\ 1.5\ \underline{/20°}$ amps, $\bar{I}_{SC}\ 4\ \underline{/35°}$ amps, 0.08 S, V_0

34. A signal generator with an open-circuit output of 2.5 V rms and a $\bar{Z}_{internal}$ of $50\ \underline{/-10°}\ \Omega$ drives a parallel RL circuit under matched conditions at a frequency of 450 kHz. What is the value of the load resistance and inductance? What is the average power delivered to the load?

35. Combine into one Thévenin source:

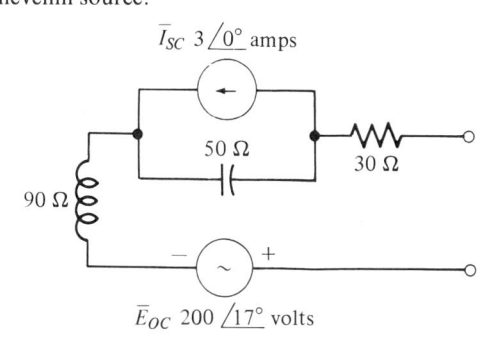

$\bar{I}_{SC}\ 3\ \underline{/0°}$ amps, $50\ \Omega$, $30\ \Omega$, $90\ \Omega$, $\bar{E}_{OC}\ 200\ \underline{/17°}$ volts

36. For this circuit:

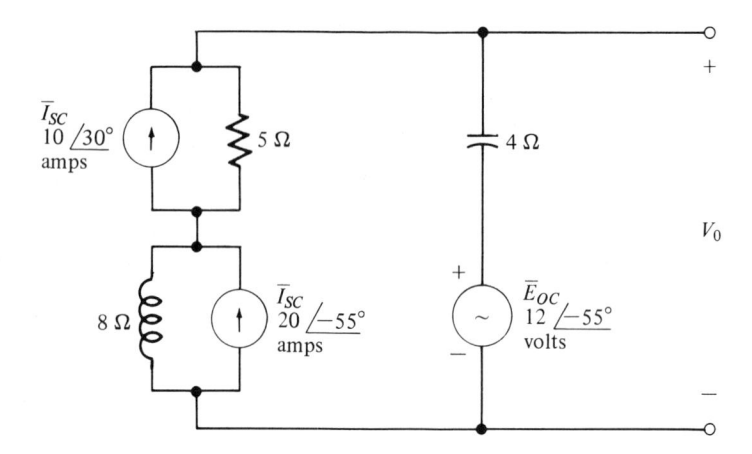

a. Determine the total Norton circuit.
b. Determine the load admittance for matched conditions.

Section 17.5

37. Determine the current, I, in each circuit by superposition.

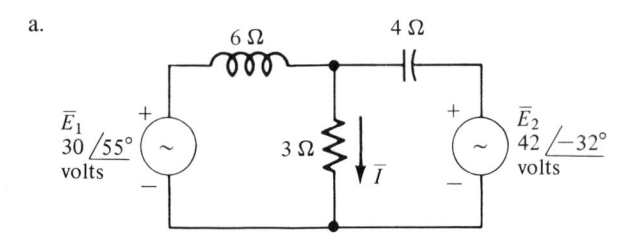

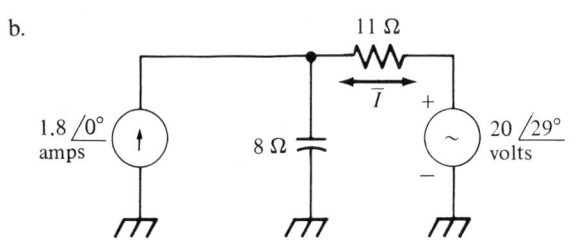

a.

b.

38. Using superposition, determine the reactive power drawn by the capacitor in each circuit.

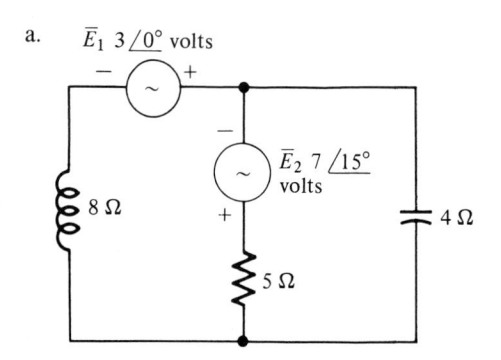

a. $\bar{E}_1\ 3\ \underline{/0°}$ volts

EXERCISES

b.

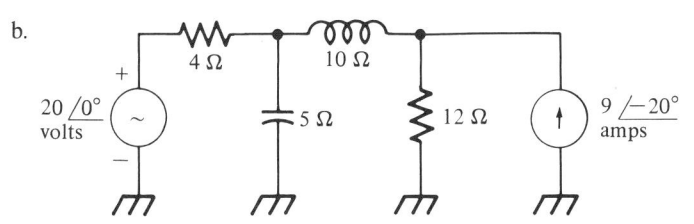

39. Use superposition to find the current through the 10-Ω resistance in this circuit.

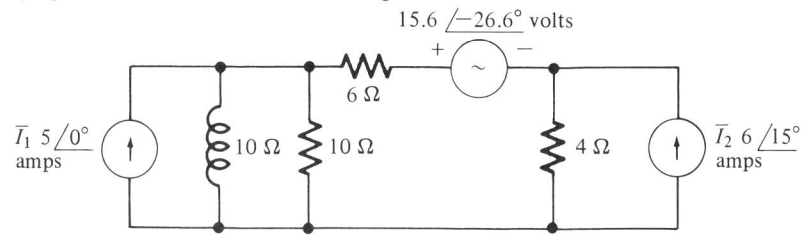

40. Use superposition to find the reactive power drawn by the inductor in this circuit.

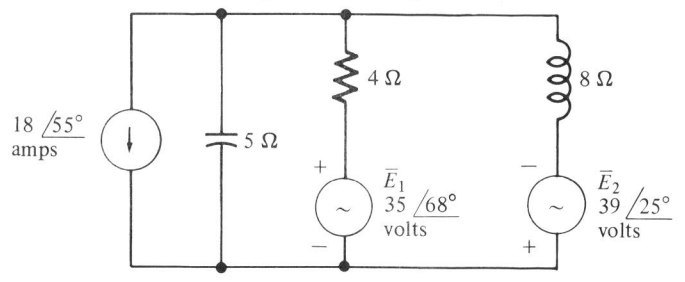

Section 17.6

41. For this circuit:

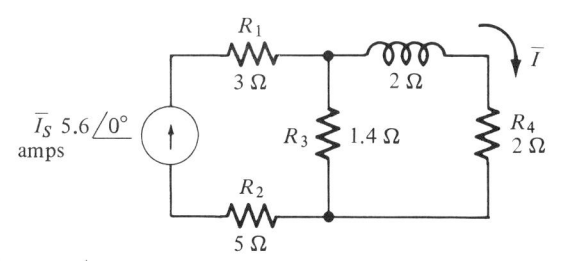

a. Find the current, $\bar{I}$, by the ladder method.
b. Find the source voltage.
c. Check the results of part a by redoing the exercise with SPR techniques.

42. Determine the following for this circuit:

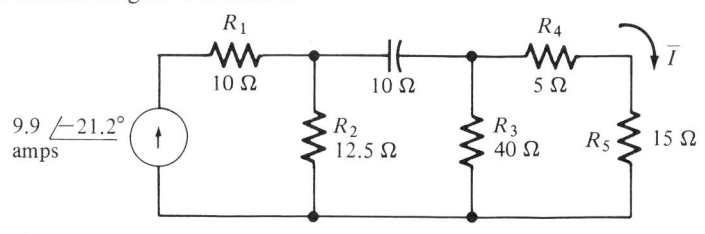

a. The current, $\bar{I}$, by the ladder method.
b. Using the equations derived in part a, determine the power dissipated by R_3.

c. Determine the voltage across R_2 by using the equations derived in part a.
d. What must the source current be in order that 960.4 var be drawn by the capacitor?
e. Redo part a by SPR. Compare the results.

43. For this circuit, determine the following values.

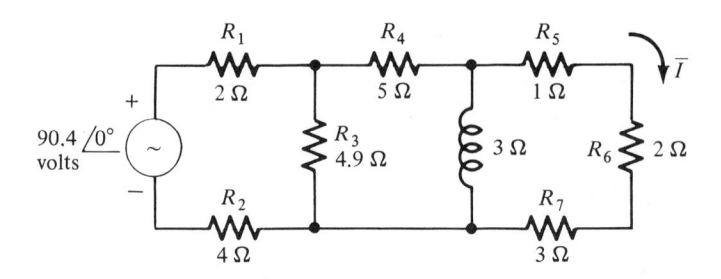

a. The current, $\bar{I}$, by the ladder method.
b. The average power dissipated by R_4.
c. The voltage across R_3.
d. The inductor current if $\bar{E}_{source}$ is increased to 127.5 $\underline{/0°}$ V.

CHAPTER HIGHLIGHTS

Ac circuits have different characteristics at low, medium, and high frequencies.
    Capacitors shunt a factory's inductive load to improve circuit power factor.
    There are many ac bridges, each named for its developer and each specially capable of measuring certain ac parameters.
    Because we are dealing with complex numbers, ac bridge balance equations must be equal in both magnitude and angle.
    Π-to-T and T-to-Π conversions can be made in ac circuits as they were in dc circuits, but again, care must be taken when handling complex numbers.
    AC Thévenin and Norton source models are much like their dc counterparts except that they have ideal sinusoidal sources and an internal impedance.
    Matched refers to circuits whose impedances match for maximum power transfer.
    Thévenin and Norton sources can be converted and combined as they were in the dc case.
    To find total current or voltage by superposition, let each source operate by itself, then add the individual contributions.
    The ladder method is especially useful for finding voltages and currents when changes are to be made in a circuit.

Formulas to Remember

$\bar{Z}_1 \bar{Z}_4 = \bar{Z}_2 \bar{Z}_3$ ac bridge balance

$\bar{Z}_a = \dfrac{\bar{Z}_x \bar{Z}_y}{\bar{Z}_x + \bar{Z}_y + \bar{Z}_z}$ Π-to-T conversions

$\bar{Z}_x = \dfrac{\bar{Z}_a \bar{Z}_b + \bar{Z}_b \bar{Z}_c + \bar{Z}_c \bar{Z}_a}{\bar{Z}_a}$ T-to-Π conversions

$\bar{Z}_T = \dfrac{\bar{Z}_\Pi}{3}$ Π-to-T conversions when all branches the same

$\bar{Z}_\Pi = 3\bar{Z}_T$ T-to-Π conversions when all branches the same.

$\bar{Z}_{load} = \bar{Z}^*_{internal}$ maximum power or matched conditions

CHAPTER 18

Advanced AC Circuit Analysis

In this chapter we combine our new skills in phasor algebra with the techniques of advanced dc analysis and apply them to ac circuits. These techniques include loop (or mesh) analysis based on KVL, nodal analysis based on KCL, and the Thévenin and Norton theorems by which a complicated circuit may be reduced to a source and internal impedance. (Note: Much of the present material on loop, nodal, Thévenin, and Norton techniques was covered in greater detail in Chapter 7; a thorough review of that chapter might prove helpful.)

At the conclusion of this chapter, you should be able to:

- Analyze multisource ac circuits with loop and nodal analysis.
- Find the Thévenin or Norton equivalent of the ac circuit driving any component.

18.1 LOOP ANALYSIS OF AC CIRCUITS

Loop analysis, a powerful technique for analyzing circuits based on Kirchhoff's voltage law, was first discussed in Chapter 7. In it, a number of loop (or mesh) equations are written, each summing voltages around a closed loop. The resulting system of equations is then solved for the unknown currents.

Let us review the steps:

a. Count the circuit's branches (B) and nodes (N). Substitute these into the $B - N + 1$ rule to determine how many equations are needed to solve the problem (this will be two or three in our case).
b. Choose loops so that desired information can be obtained with the least mathematics; that is, have only one current through the component being investigated. Make sure that every branch is included in at least one equation. Complicated circuits can sometimes be done more easily by first redrawing them with generalized impedances (Z_1, Z_2, etc.). After the equations are written and solved with these impedances, specific values are substituted in the solution for the final answer.
c. Draw a loop current around each loop. Identify each current and indicate its direction with an arrowhead.
d. Go entirely around a loop summing all voltages in a loop equation. Pay close attention to polarity and j-operator signs. Repeat for remaining loops.
e. Solve the resulting set of equations for desired current. Check results by KVL around one or more loops or by doing the analysis another way.

EXAMPLE 18.1

The resistor current in this circuit was calculated to be $-5 - j\,4.2$ A by SPR in Example 17.1. Verify that value by loop analysis.

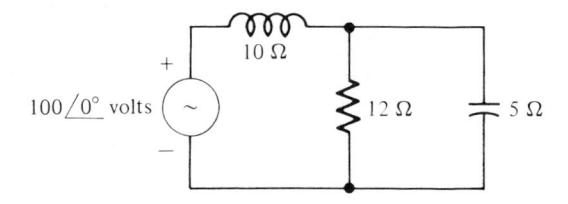

The solution requires two loop equations. They could be drawn around the inner loops but that places two currents in the resistor. Choosing one inner loop and the outer loop requires less work.

18.1 LOOP ANALYSIS OF AC CIRCUITS

Draw the loops:

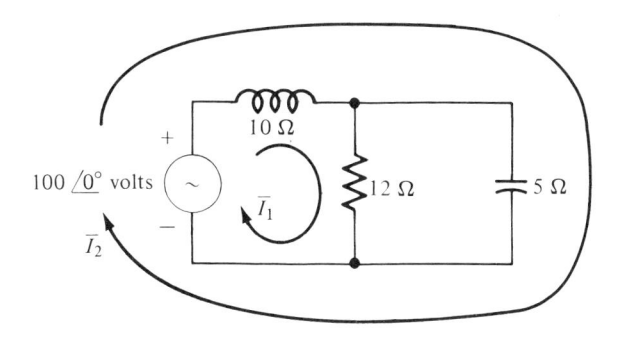

Write the equations:

$$(12 + j10)\bar{I}_1 + j10\bar{I}_2 = 100$$
$$j10\bar{I}_1 + (j10 - j5)\bar{I}_2 = 100$$

Using the loop equation check learned in dc circuits, we note that the contribution of I_2 in loop 1 is the same as the contribution of I_1 in loop 2. Solve for I_1 by Cramer's rule:

$$\bar{I}_1 = \frac{\begin{vmatrix} 100 & j10 \\ 100 & j5 \end{vmatrix}}{\begin{vmatrix} 12 + j10 & j10 \\ j10 & j5 \end{vmatrix}}$$

$$= \frac{(100(j5) - (j10)100)}{j5(12 + j10) - (j10)^2}$$

$$= \frac{j500 - j1000}{-50 + j60 + 100}$$

$$= \frac{500 \,/\!-90°}{50 + j60}$$

$$= 6.4 \,/\!-140.2° \quad \text{or} \quad -4.9 - j4.1 \text{ A}$$

These results compare favorably with those obtained by SPR.

EXAMPLE 18.2

Find the reactive power drawn by the capacitor in this circuit:

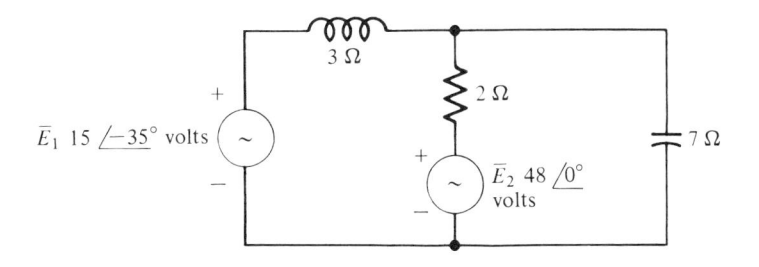

I_{cap} will be determined by loop analysis.

Write loop equations around the inner loops as indicated:

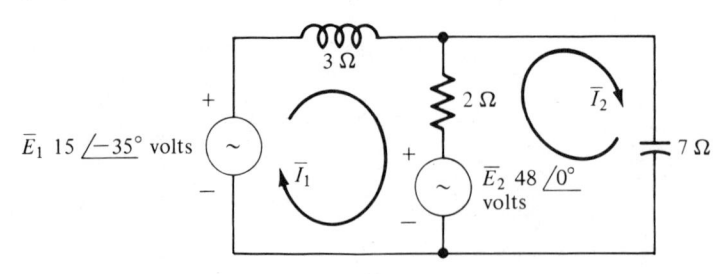

$$-15\underline{/-35°} + (2 + j3)\bar{I}_1 - 2\bar{I}_2 + 48 = 0$$
$$-48 - 2\bar{I}_1 + (2 - j7)\bar{I}_2 = 0$$

or

$$(2 + j3)\bar{I}_1 - 2\bar{I}_2 = 15\underline{/-35°} - 48 = 36.7\underline{/-166.5°}$$
$$-2\bar{I}_1 + (2 - j7)I_2 = 48$$

Solve for I_2 by Cramer's rule:

$$\bar{I}_2 = \frac{\begin{vmatrix} 2 + j3 & 36.7\underline{/-166.5°} \\ -2 & 48 \end{vmatrix}}{\begin{vmatrix} 2 + j3 & -2 \\ -2 & 2 - j7 \end{vmatrix}}$$

$$= \frac{96 + j144 + 73.4\underline{/-166.5°}}{4 + 21 + j6 - j14 - 4} = \frac{96 + j144 - 71.4 - j17.1}{21 - j8}$$

$$= \frac{24.6 + j126.9}{21 - j8} = \frac{129.3\underline{/79°}}{22.5\underline{/-20.9°}}$$

$$= 5.7\underline{/99.9°}\text{ A}$$

$$P_Q = I_2^2 X_C = 5.7^2(7)$$
$$= 227.4 \text{ var, capacitive}$$

EXAMPLE 18.3

Derive the bridge balance equation using loop equations.

Draw the bridge with generalized impedances and three loop currents:

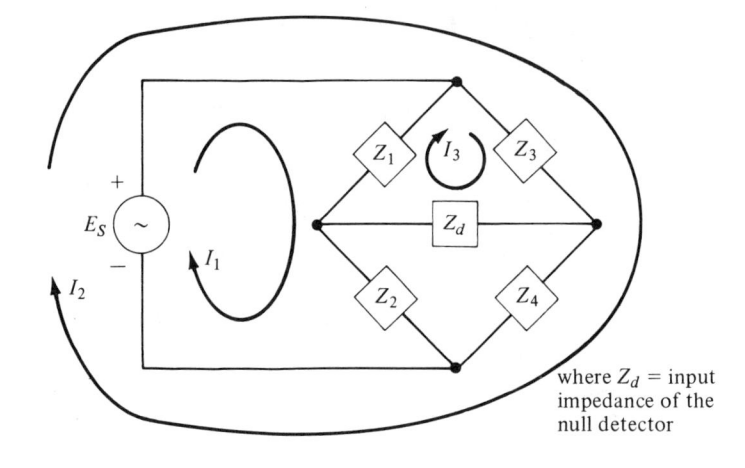

where Z_d = input impedance of the null detector

18.2 NODAL ANALYSIS OF AC CIRCUITS

Write the loop equations:

$$(Z_1 + Z_2)I_1 + 0I_2 - Z_1 I_3 = E_s$$
$$0I_1 + (Z_3 + Z_4)I_2 + Z_3 I_3 = E_s$$
$$-Z_1 I_1 + Z_3 I_2 + (Z_1 + Z_3 + Z_d)I_3 = 0$$

When the bridge is balanced, I_3 will be zero. The solution of I_3 by Cramer's rule requires the division of a numerator determinant (N) by a denominator determinant (D). Setting this ratio equal to zero, we have

$$\frac{N}{D} = 0$$

Multiply both sides of this equation by D:

$$D\frac{N}{D} = D(0)$$

or

$$N = 0$$

In other words, we need only expand the numerator determinant to determine balance conditions (of course, this applies only when the denominator determinant does not equal zero, division by zero being undefined).

Write and expand the numerator determinant for I_3:

$$\begin{vmatrix} Z_1 + Z_2 & 0 & E_s \\ 0 & Z_3 + Z_4 & E_s \\ -Z_1 & Z_3 & 0 \end{vmatrix}$$
$$= [(Z_1 + Z_2)(Z_3 + Z_4)0 + (0(E_s)(-Z_1) + (E_s(0)Z_3)]$$
$$- [E_s(Z_3 + Z_4)(-Z_1) + (Z_1 + Z_2)E_s(Z_3) + (0)(0)(0)]$$
$$= [0] - [E_s(Z_3 + Z_4)(-Z_1) + E_s Z_3(Z_1 + Z_2)]$$

Set this equal to zero and factor E_s:

$$-E_s[(Z_3 + Z_4)(-Z_1) + Z_3(Z_1 + Z_2)] = 0$$

or

$$Z_3(Z_1 + Z_2) = Z_1(Z_3 + Z_4)$$

or

$$Z_1 Z_3 + Z_2 Z_3 = Z_1 Z_3 + Z_1 Z_4$$

Therefore,

$$Z_1 Z_4 = Z_2 Z_3$$

These are the same balance conditions as those derived in Equation 17.2. It is interesting to note that detector input impedance does not appear in the statement. This is as it should be. Since Z_d has no current through it at balance, there is no potential difference across it that affects the equations. ∎

18.2 NODAL ANALYSIS OF AC CIRCUITS

Nodal analysis is the method discussed in Chapter 7 in which one node is chosen as a reference point (for convenience, that node is identified by a ground symbol). A Kirchhoff's current law equation is written summing all currents at each of the $N - 1$

remaining nodes. The resulting system of equations is then solved for the desired unknown node voltages.

Let us review the steps:

a. Choose the reference node (usually the bottom one). All node voltages will be considered positive with respect to this point.
b. Label the voltage at each of the remaining nodes.
c. Write the nodal equations. This is often done more easily by working with generalized impedances (Z_1, Z_2, etc.) and admittances (Y_1, Y_2, etc.) rather than with specific values. For convenience, treat all currents as leaving nodes. Make sure to include all currents associated with a node. Observe j-operator signs carefully.
d. Solve the system of equations for the unknown node voltages.
e. Check results by KVL, KCL, or doing the problem with another technique.

EXAMPLE 18.4

The resistor current in this circuit was found to be $-5 - j4.2$ A in Example 17.1 by SPR, and Example 18.1 by loop analysis. Verify that value with nodal analysis.

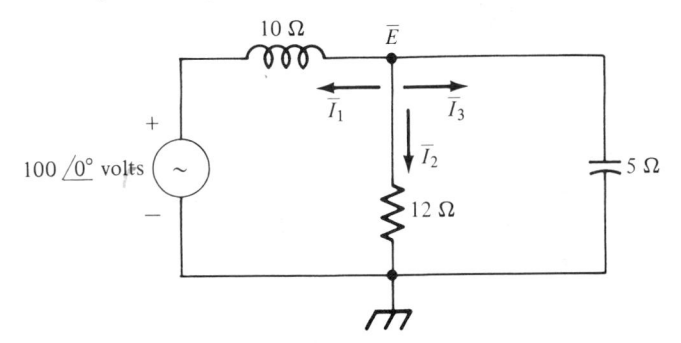

There are two nodes in this circuit. The bottom one is grounded, signifying its role as a reference point. The upper node is assumed to have a potential of E volts positive with respect to the reference node. Writing a single nodal equation for the three currents that combine at the upper node, we have

$$\bar{I}_1 + \bar{I}_2 + \bar{I}_3 = 0$$

Let us consider each of these currents.

Note that the right end of the inductor is E volts above reference, while the left end is 100 V above reference. For I_1 to leave the node as shown, the node voltage, E, must be greater than 100 V. The difference between the two, $\bar{E} - 100\,\underline{/0°}$, is the potential difference across the inductor. Dividing this by the reactance, $j10\,\Omega$, gives I_1; that is,

$$\bar{I}_1 = \frac{\bar{E} - 100}{j10}$$

I_2 and I_3 are easily obtained by Ohm's law since the entire E volts is across the 12 Ω and $-j5$ Ω, respectively. Combine the three in a nodal equation:

$$\frac{\bar{E} - 100}{j10} + \frac{\bar{E}}{12} + \frac{\bar{E}}{-j5} = 0$$

18.2 NODAL ANALYSIS OF AC CIRCUITS

Multiplying each fraction so that all denominators equal, CD, the common denominator, we have

$$\frac{(12)(-j5)(E-100)}{CD} + \frac{(j10)(-j5)E}{CD} + \frac{(j10)(12)E}{CD} = 0$$

where $CD = (12)(j10)(-j5)$. Multiply both sides by CD to eliminate denominators:

$$-j60(\bar{E} - 100) + 50\bar{E} + j120\bar{E} = 0$$
$$\bar{E}(-j60 + 50 + j120) + j6000 = 0$$
$$\bar{E} = \frac{6000\,\underline{/-90°}}{50 + j60} = \frac{6000\,\underline{/-90°}}{78.1\,\underline{/50.2°}}$$
$$= 76.8\,\underline{/-140.2°}\ \text{V}$$

Solve for I_R:

$$\bar{I}_r = \frac{\bar{E}}{R} = \frac{76.8\,\underline{/-140.2°}}{12}$$
$$= 6.4\,\underline{/-140.2°} \quad \text{or} \quad -4.9 - j4.1$$

These results compare favorably with those obtained by SPR and loop analysis. ∎

Nodal mathematics can be simplified by converting impedances to admittances. This changes current calculations from E/Z to EY, thereby removing denominators and the extra work they demand.

EXAMPLE 18.5

Find the inductor current in this circuit:

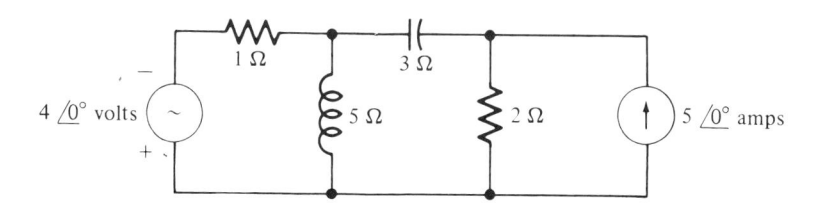

Redraw the circuit with node voltages and reference node indicated and impedances converted to admittances:

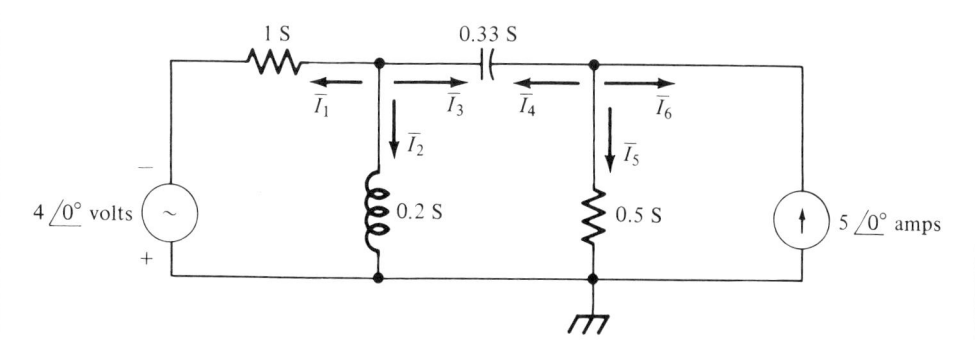

The right end of the 1-S conductance is E_a volts above reference. The left end is 4 V below reference. Therefore, the potential difference across the conductance is $E_a + 4$ V, and

$$I_1 = (E_a + 4)1 \quad \text{amperes}$$

For I_3 to flow as shown, E_a must be greater than E_b. The difference between the two is the voltage across the capacitor. Therefore,

$$I_3 = (E_a - E_b)j0.33$$

Write the nodal equation at node A:

$$(E_a + 4)1 + E_a(-j0.2) + (E_a - E_b)(j0.33) = 0$$

Node B shows I_4 through the capacitor in the opposite direction to I_3. This, of course, cannot really happen, but as we saw in dc circuits, the mathematics will tell the true direction of capacitor current. For I_4 to flow as shown, E_b must be greater than E_a. Therefore,

$$I_4 = (E_b - E_a)j0.33$$

Write the equation at node B:

$$(E_b - E_a)(j0.33) + E_b(0.5) - 5 = 0$$

Note the 5 A is negative because it is flowing into the node, opposite to our definition of positive current. Rewrite the two nodal equations:

$$E_a(1 - j0.2 + j0.33) - E_b(j0.33) = -4$$
$$-E_a(j0.33) + E_b(0.5 + j0.33) = 5$$

Finally,

$$E_a(1 + j0.13) - E_b(j0.33) = -4$$
$$-E_a(j0.33) + E_b(0.5 + j0.33) = 5$$

Set up the solution for E_a by Cramer's rule:

$$\bar{E}_a = \frac{\begin{vmatrix} -4 & -j0.33 \\ 5 & 0.5 + j0.33 \end{vmatrix}}{\begin{vmatrix} 1 + j0.13 & -j0.33 \\ -j0.33 & 0.5 + j0.33 \end{vmatrix}}$$

$$= \frac{-4(0.5 + j0.33) - 5(-j0.33)}{(1 + j0.13)(0.5 + j0.33) - (-j0.33)(-j0.33)}$$

$$= \frac{-2 - j1.3 + j1.7}{0.46 + j0.4 + 0.1}$$

$$= \frac{-2 + j0.4}{0.56 + j0.4}$$

$$= \frac{2 \,\underline{/168.7°}}{0.68 \,\underline{/35.5°}}$$

$$= 2.9 \,\underline{/133.2°} \text{ V}$$

$$\bar{I}_{\text{inductor}} = \frac{\bar{E}_a}{X_L} = \frac{2.9 \,\underline{/133.2°}}{5 \,\underline{/90°}}$$

$$= 0.6 \,\underline{/43.2°} \text{ A} \qquad \blacksquare$$

18.2 NODAL ANALYSIS OF AC CIRCUITS

EXAMPLE 18.6

Determine the reactive power drawn by the inductor in this circuit.

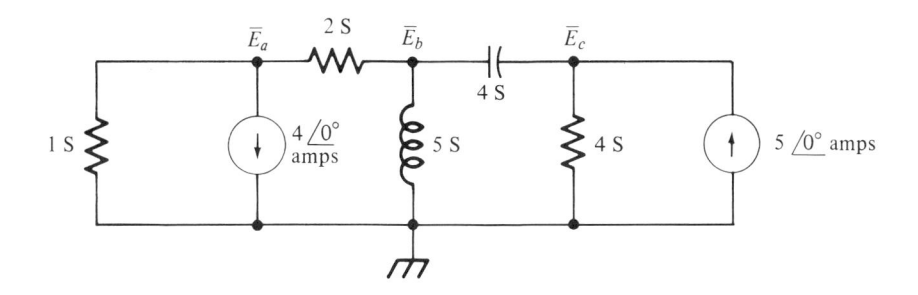

Nodal analysis will be used to determine the potential difference across the inductor. Note that the components are given in admittance form.

At node A:

$$E_a(1) + 4 + (E_a - E_b)2 = 0$$

At node B:

$$(E_b - E_a)2 + E_b(-j5) + (E_b - E_c)(j4) = 0$$

At node C:

$$(E_c - E_b)(j4) + E_c(4) - 5 = 0$$

After combining and rewriting we get the system of equations:

1. $3E_a - 2E_b = -4$
2. $-2E_a + (2 - j1)E_b - j4E_c = 0$
3. $-j4E_b + (4 + j4)E_c = 5$

Solving for E_b by substitution: Solve equation 1 for E_a:

4. $E_a = \dfrac{-4 + 2E_b}{3}$

Solve equation 3 for E_c:

5. $E_c = \dfrac{5 + j4E_b}{4 + j4}$

Substitute equations 4 and 5 into 2:

$$-2\left(\frac{-4 + 2E_b}{3}\right) + (2 - j1)E_b - j4\left(\frac{5 + j4E_b}{4 + j4}\right) = 0$$

Solve for E_b:

$$\frac{8}{3} - \frac{4}{3}E_b + E_b(2-j1) - j\left(\frac{5+j4E_b}{1.4\,\underline{/45°}}\right) = 0$$

$$E_b\left(-\frac{4}{3} + 2 - j1 + \frac{4}{1.4\,\underline{/45°}}\right) = -\frac{8}{3} + j\frac{5}{1.4\,\underline{/45°}}$$

$$\bar{E}_b = \frac{-2.7 + 3.6\,\underline{/45°}}{-1.3 + 2 - j1 + 2.9\,\underline{/-45°}}$$

$$= \frac{-2.7 + 2.6 + j2.6}{0.7 - j1 + 2 - j2}$$

$$= \frac{-0.1 + j2.6}{2.7 - j3}$$

$$= \frac{2.6\,\underline{/92.2°}}{4\,\underline{/-48°}}$$

$$= 0.64\,\underline{/140.2°}\text{ V}$$

$$P_q = E_b^2 Y_L = (0.64)^2(5) = 2.1 \text{ var, inductive} \qquad \blacksquare$$

18.3 THÉVENIN ANALYSIS OF AC CIRCUITS

The idea that a complicated circuit of many sources and linear, bilateral impedances can be represented by an ideal voltage source and series impedance is one of the most important in electronics. Let us review the steps by which this Thévenin equivalent circuit is determined:

1. Remove the component or components being investigated.
2. Look into the pair of terminals created by this removal and determine:
   a. *Open-circuit voltage*, E_{oc}: Finding E_{oc} is a circuit analysis problem; therefore, use any of the analytical tools discussed so far.
   b. *Internal impedance*: Find with all sources turned off, that is, voltage sources are shorted, and current sources are opened.
3. Construct the Thévenin equivalent circuit (see Figure 18.1).

Since reactance changes with frequency, a particular Thévenin circuit is only equivalent for the frequency at which reactances were determined.

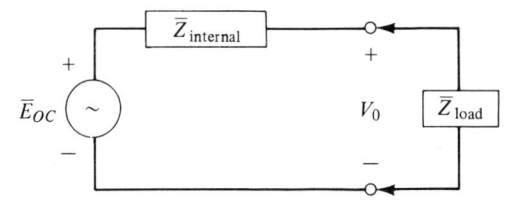

FIGURE 18.1 Generalized Thevenin Equivalent Circuit

18.3 THÉVENIN ANALYSIS OF AC CIRCUITS

EXAMPLE 18.7

The resistive current in this circuit has been found to be approximately $6.5/\!-\!140°$ A by a variety of techniques. Verify the value by first Théveninizing the circuit driving the resistor, then using Ohm's law:

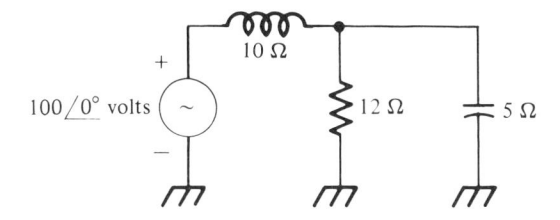

Remove the 12-Ω resistance. Calculate the open-circuit voltage across the hole created by this removal. Note that this is the same as the capacitor voltage.

$$\bar{E}_{oc} = \bar{V}_{cap}$$
$$= \frac{100(-j5)}{j10 - j5} \quad \text{by voltage division}$$
$$= \frac{500/\!-\!90°}{5/90°}$$
$$= 100/\!-\!180° \text{ V}$$

Short the voltage source. Look into the terminals and calculate $\bar{Z}_{internal}$:

$$\bar{Z}_{int} = -j5 \| j10$$
$$= \frac{-j5(j10)}{-j5 + j10} = \frac{50}{j5}$$
$$= 10/\!-\!90° \text{ Ω}$$

Draw the Thévenin equivalent circuit and reconnect the 12-Ω resistance:

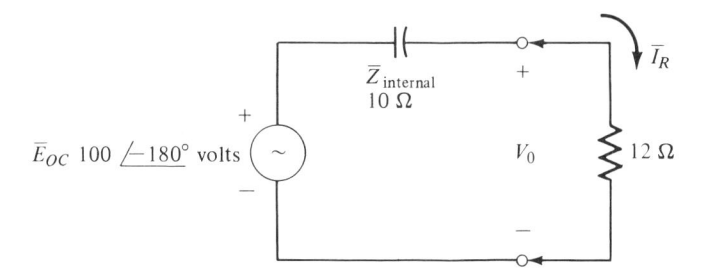

Calculate $\bar{I}_R$:

$$\bar{I}_R = \frac{100/\!-\!180°}{12 - j10}$$
$$= \frac{100/\!-\!180°}{15.6/\!-\!39.8°}$$
$$= 6.4/\!-\!140.2° \text{ A}$$

EXAMPLE 18.8

The reactive power drawn by the capacitor in this circuit was originally calculated to be approximately 227 var by loop analysis in Example 18.2. Verify that figure by Thévenin analysis.

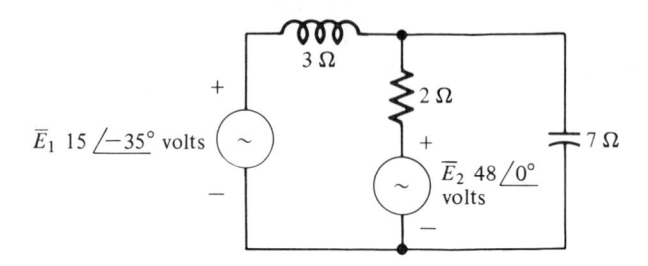

After removing the capacitor, the circuit becomes

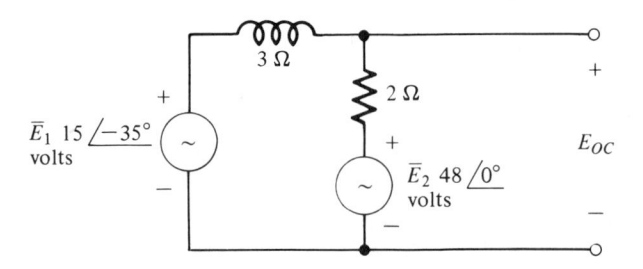

With the instantaneous source polarities shown, current flows CCW because of the greater E_2 potential. Here are two ways E_{oc} could be determined:

$$E_{oc} = E_2 - \text{the drop across the 2-}\Omega \text{ resistance}$$

or

$$E_{oc} = E_1 + \text{the drop across } j3 \, \Omega$$

We will use the second approach.
Find the potential across the inductor by voltage division:

$$\bar{V}_{ind} = \frac{(48 - 15 \underline{/-35°})(j3)}{2 + j3}$$

$$= \frac{(35.7 + j8.6)(j3)}{3.6 \underline{/56.3°}}$$

$$= \frac{-25.8 + j107.1}{3.6 \underline{/56.3°}}$$

$$= \frac{110.2 \underline{/103.5°}}{3.6 \underline{/56.3°}}$$

$$= 30.6 \underline{/47.2°} \text{ V}$$

18.3 THÉVENIN ANALYSIS OF AC CIRCUITS

Draw a circuit to help us determine E_{oc}:

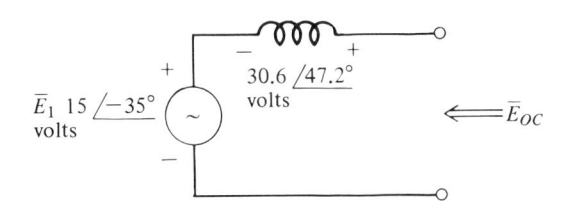

$$\bar{E}_{oc} = \bar{V}_{inductor} + \bar{E}_1$$
$$= 30.6 \underline{/47.2°} + 15 \underline{/-35°}$$
$$= 20.8 + j22.5 + 12.3 - j8.6$$
$$= 33.1 + j13.9 \quad \text{or} \quad 35.9 \underline{/22.8°} \text{ V}$$

When the two sources are shorted for determing $Z_{internal}$, the reactance and resistance are in parallel. Therefore,

$$\bar{Z}_{internal} = 2 \| j3$$
$$= \frac{2(j3)}{2 + j3}$$
$$= \frac{j6}{3.6 \underline{/56.3°}}$$
$$= 1.7 \underline{/33.7°} \quad \text{or} \quad 1.4 + j0.9 \text{ }\Omega$$

Draw the Thévenin equivalent circuit:

$$\bar{I} = \frac{\bar{E}_{oc}}{\bar{Z}_{total}}$$
$$= \frac{\bar{E}_{oc}}{\bar{Z}_{internal} + \bar{Z}_{load}}$$
$$= \frac{35.9 \underline{/22.8°}}{1.4 + j0.9 - j7}$$
$$= \frac{35.9 \underline{/22.8°}}{6.3 \underline{/-77.1°}}$$
$$= 5.7 \underline{/99.9°} \text{ A}$$
$$P_Q = I^2 X_C$$
$$= (5.7)^2(7)$$
$$= 230.3 \text{ var, capacitive}$$

The error between this result and that obtained by loop analysis is approximately 1.3%.

EXAMPLE 18.9

The capacitance in Example 18.8 is doubled cutting capacitive reactance to 3.5 Ω. What is the new value of reactive power drawn?

If this problem was being done by loop equations, another complete analysis would be necessary. But having determined the Thévenin equivalent in Example 18.8, we need only change the capacitance and do a simple series analysis:

$$\bar{Z}_{total} = \bar{Z}_{internal} + \bar{Z}_{load}$$
$$= (1.4 + j0.9) - j3.5$$
$$= 1.4 - j2.6 \quad \text{or} \quad 3 \underline{/-61.7°} \; \Omega$$
$$\bar{I} = \frac{\bar{E}_{oc}}{\bar{Z}_{total}} = \frac{35.9 \underline{/22.8°}}{3 \underline{/-61.7°}}$$
$$= 12.2 \underline{/84.5°} \; A$$
$$P_Q = I^2 X_C$$
$$= (12.2)^2(3.5)$$
$$= 517.3 \text{ var}$$

EXAMPLE 18.10

What value of Z_{load} draws maximum power from the Thévenin equivalent circuit of Example 18.9?

For maximum power transfer:

$$\bar{Z}_{load} = \bar{Z}^*_{internal}$$
$$= (1.4 + j0.9)^*$$
$$= (1.4 - j0.9) \; \Omega$$

18.4 NORTON ANALYSIS OF AC CIRCUITS

Norton's theorem, you will recall, is much like Thévenin's. It allows a complicated circuit of sources and linear, bilateral components to be reduced to an ideal current source shunted by an internal impedance. Let us review the steps for determining this important circuit:

a. Remove the component or components being investigated.
b. Short the terminals created by this removal.
c. Determine the current through this short. This is an analysis problem; therefore, use any of the analytical techniques discussed so far.
d. Remove the short. Look into the terminals and determine $Z_{internal}$ with all sources off.
e. Construct the Norton equivalent circuit (see Figure 18.2).

Like the Thévenin equivalent, a Norton circuit is only good for the frequency at which reactances were calculated.

18.4 NORTON ANALYSIS OF AC CIRCUITS

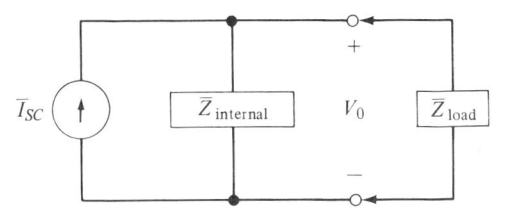

FIGURE 18.2 Generalized Norton Equivalent Circuit

EXAMPLE 18.11

The resistor current in this circuit has been found to be approximately 6.5 /−140° A by a variety of techniques. Confirm this value by finding the Norton equivalent of the circuit driving the resistance then using Ohm's law.

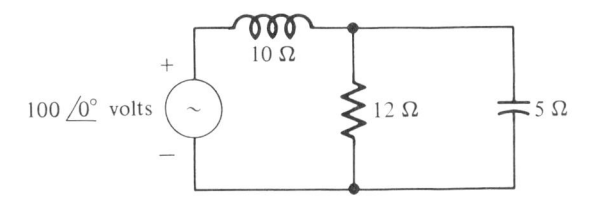

Remove the resistance. Place a short across the gap caused by this removal and calculate I_{sc}. (*Note:* The capacitor has no effect on this current because of the short.)

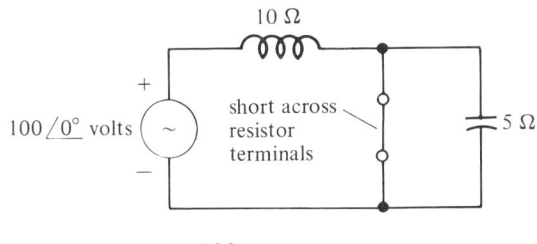

$$\bar{I}_{sc} = \frac{100}{j10} = 10\,/\!-90°\text{ A}$$

Remove the short. Look into the terminals with the source shorted. Calculate Z_{internal} (as we have seen, this is done just as it was for Thévenin analysis):

$$\bar{Z}_{\text{internal}} = -j5 \,\|\, j10 = \frac{-j5(j10)}{-j5 + j10}$$

$$= 10\,/\!-90°\ \Omega$$

Draw the Norton equivalent circuit:

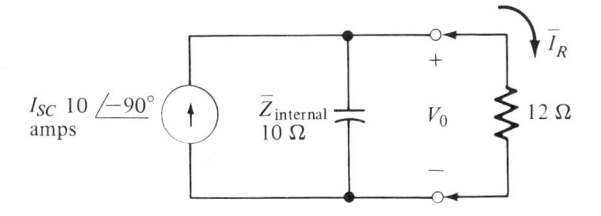

Calculate $\bar{I}_R$ by current division:

$$\bar{I}_R = \frac{10\underline{/-90°}(10\underline{/-90°})}{12 - j10}$$

$$= \frac{100\underline{/-180°}}{15.6\underline{/-39.8°}}$$

$$= 6.4\underline{/-140.2°} \text{ A}$$ ∎

EXAMPLE 18.12

The reactive power drawn by the capacitor in this circuit has been calculated to be approximately 227 var by loop and Thévenin techniques. Verify that value by finding the Norton equivalent of the circuit driving the capacitor.

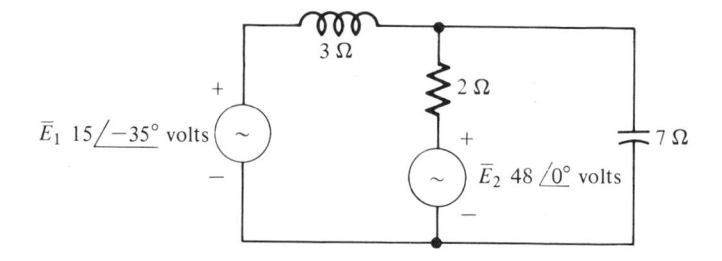

Remove the capacitor and short the resulting terminals:

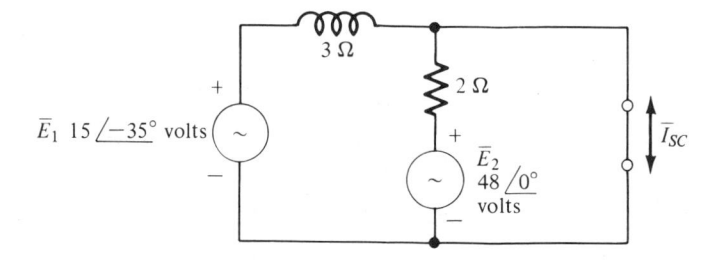

Convert both Thévenin sources into Norton sources:

$$\bar{I}_{\text{short circuit-left}} = \frac{15\underline{/-35°}}{3\underline{/90°}}$$

$$= 5\underline{/-125°} \text{ A}$$

$$\bar{I}_{\text{short circuit-right}} = \frac{48}{2}$$

$$= 24\underline{/0°} \text{ A}$$

We now have

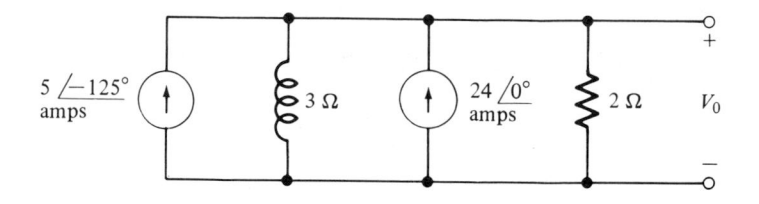

Combine the two Norton sources by Millman's theorem provides a single Norton supplying $I_{\text{short circuit}}$ for the final circuit:

$$\bar{I}_{sc} = 5 \underline{/-125°} + 24 \underline{/0°} = 21.1 - j4.1 = 21.5 \underline{/-11°} \text{ A}$$

(This current could also have been determined by loop analysis or superposition.)

Z_{internal}, obtained with both sources off, is

$$\bar{Z}_{\text{internal}} = \frac{2(j3)}{2 + j3}$$

$$= \frac{6 \underline{/90°}}{3.6 \underline{/56.3°}}$$

$$= 1.7 \underline{/33.7°} \text{ }\Omega$$

Draw the Norton equivalent with the 7-Ω capacitor reconnected:

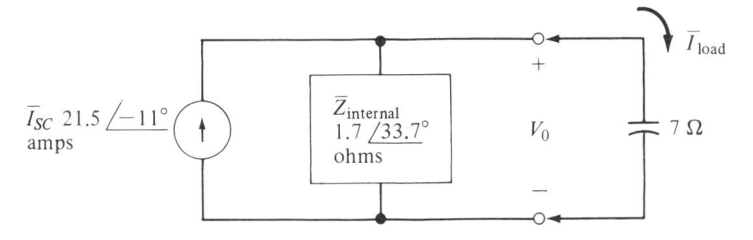

Finding I_{load} by current division:

$$\bar{I}_{\text{load}} = \frac{21.5 \underline{/-11°}(1.7 \underline{/33.7°})}{1.7 \underline{/33.7°} + 7 \underline{/-90°}}$$

$$= \frac{35.7 \underline{/22.7°}}{6.2 \underline{/-76.9°}}$$

$$= 5.7 \underline{/99.6°} \text{ A}$$

Calculate reactive power:

$$P_q = I_L^2 X_C = (5.7)^2(7) = 227.4 \text{ var}$$ ■

EXERCISES

Note: In order to strengthen their analytical skills, students are strongly urged to check the following exercises by doing them a different way.

Section 18.1

1. Find the inductor current by loop analysis using the loop currents shown.

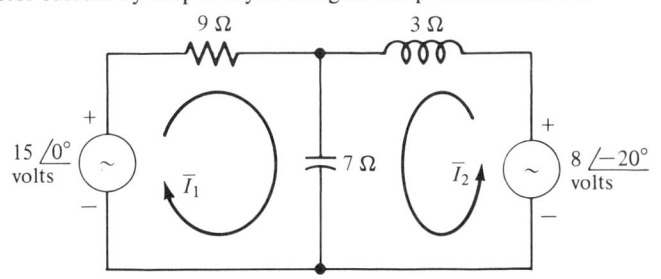

2. Repeat Exercise 1 but reverse the direction of I_1 and I_2. Discuss the difference in results.

3. Using loop analysis to find the current, determine the reactive power drawn by the inductor in this circuit. Use loop currents I_1 and I_2.

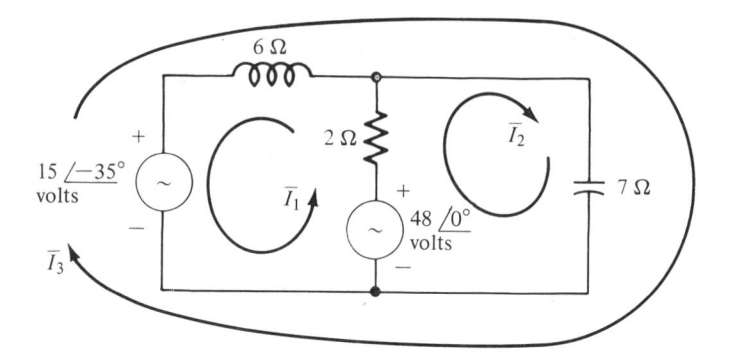

4. Repeat Exercise 3 with loop currents I_2 and I_3. Compare the results.

5. a. Using loop analysis with currents I_1 and I_2, find the source current in the following circuit.
b. Repeat the analysis using currents I_2 and I_3. Compare the results.

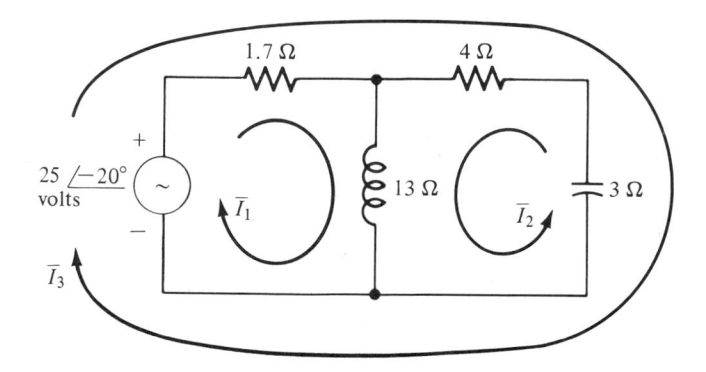

c. Repeat part b with currents I_1 and I_3.

6. a. Using currents I_1 and I_3, determine the resistor current by loop analysis.

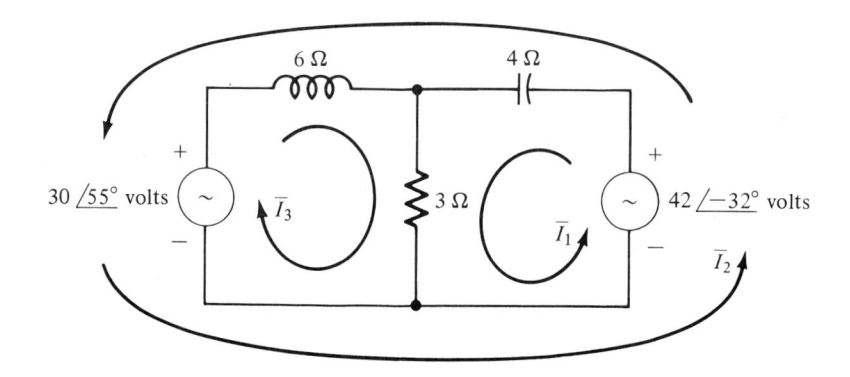

b. Repeat the analysis using currents I_1 and I_2. Compare the results.
c. Repeat part b with currents I_2 and I_3.

EXERCISES

7. Find the capacitor current by loop analysis.

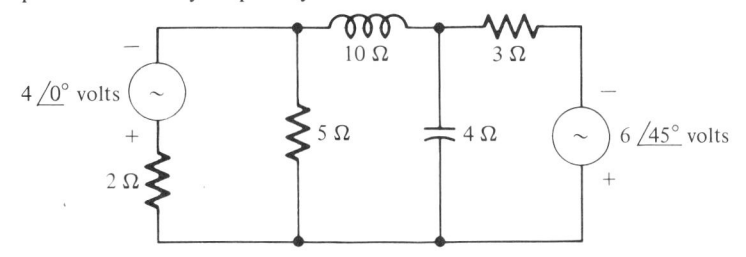

8. Find the capacitor current by loop analysis. Draw your currents so that only one passes through the capacitor.

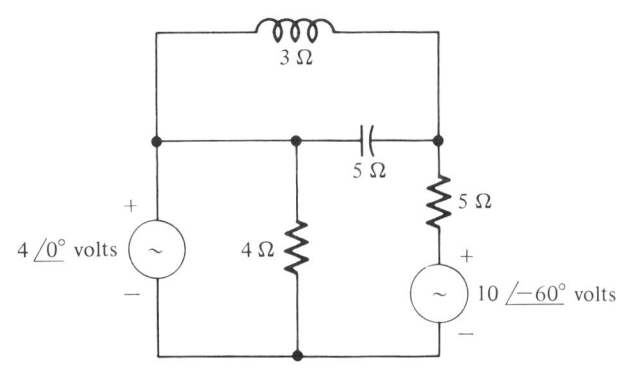

9. Find the current through R_3 by loop analysis.

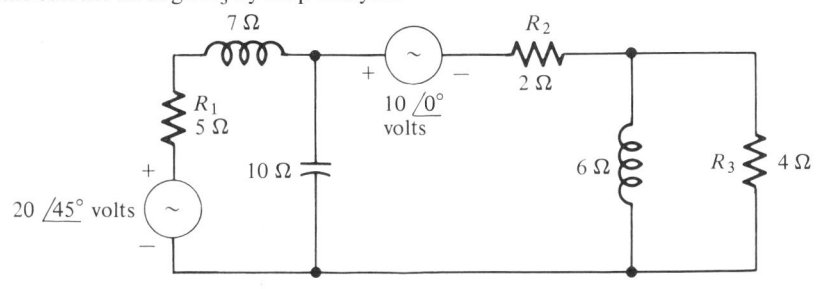

10. Write equations around the three loops indicated on this inductance bridge diagram. Set up the solution for I_3 by Cramer's rule. Set this equal to zero. Solve the real part for the value of R_4 needed for balance. Solve the imaginary part for the value of L_4 needed for balance. Z_d = null detector input impedance. *Hint:* Remember, only the numerator determinant need be expanded.

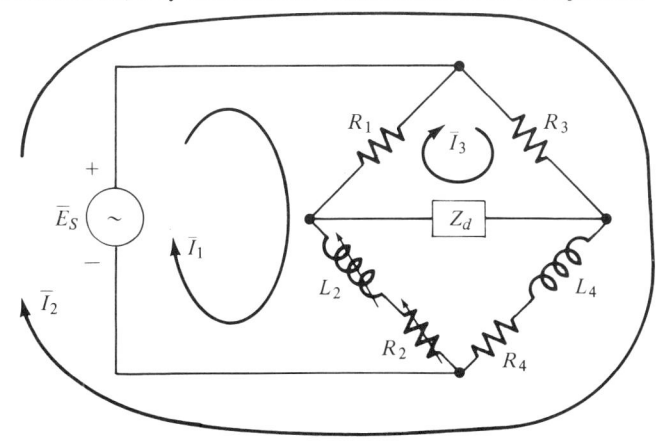

Section 18.2

11. Using nodal equations, determine the voltage at point A with respect to ground for each circuit.

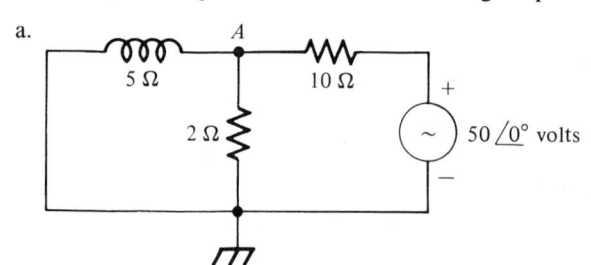

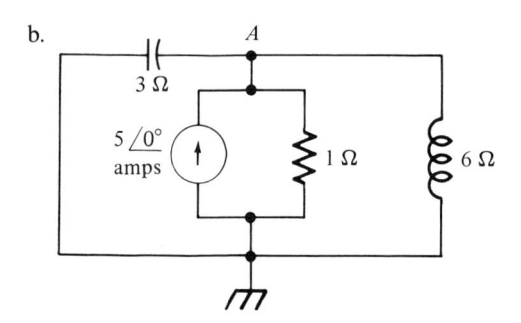

a.

b.

12. Find the inductor current in this circuit by nodal analysis:

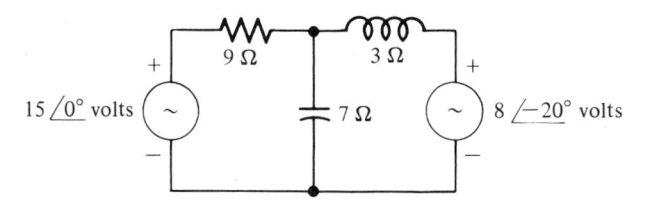

13. Without changing the circuit in any way, find the resistor current by nodal analysis.

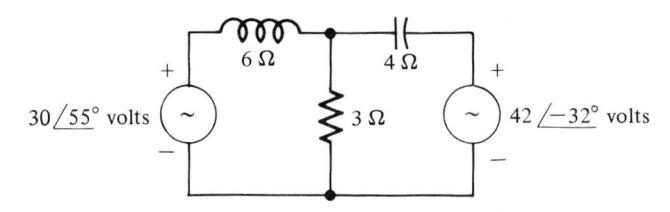

14. Repeat Exercise 13, but first convert the two Thévenin sources into their Norton equivalents. Compare nodal equations from the two exercises.

15. Using nodal equations, determine the reactive power drawn by the capacitor in each circuit.

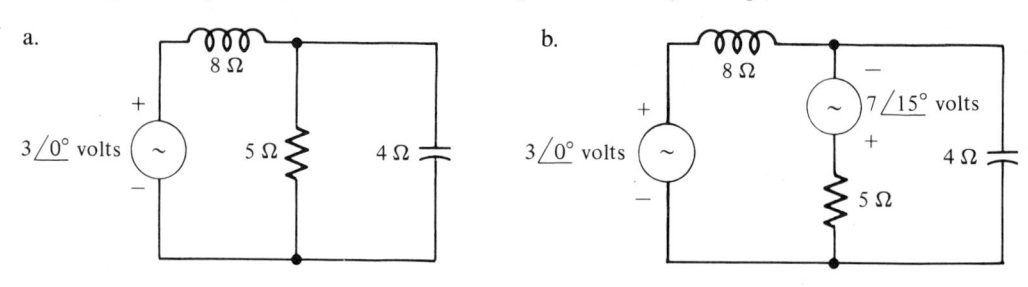

a.

b.

EXERCISES

c.

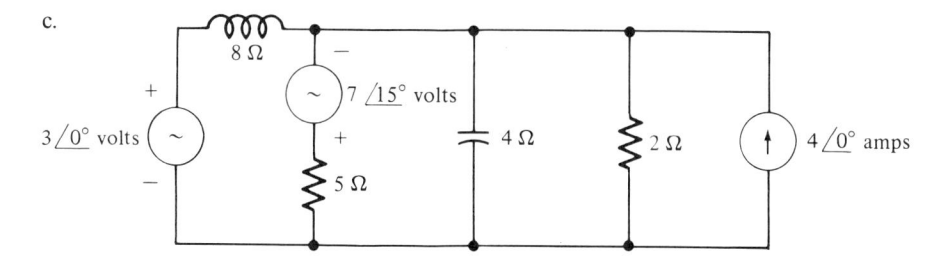

16. Using nodal equations, solve for the capacitor current.

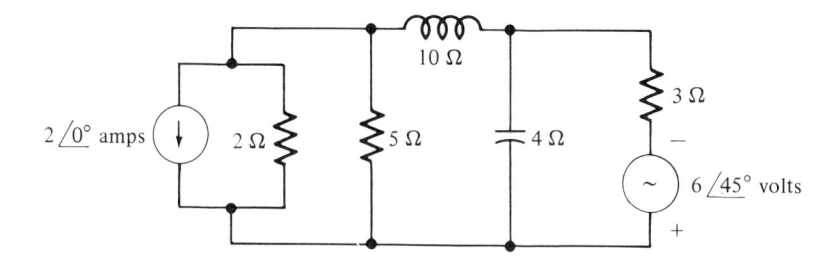

17. Find the inductor current by nodal analysis.

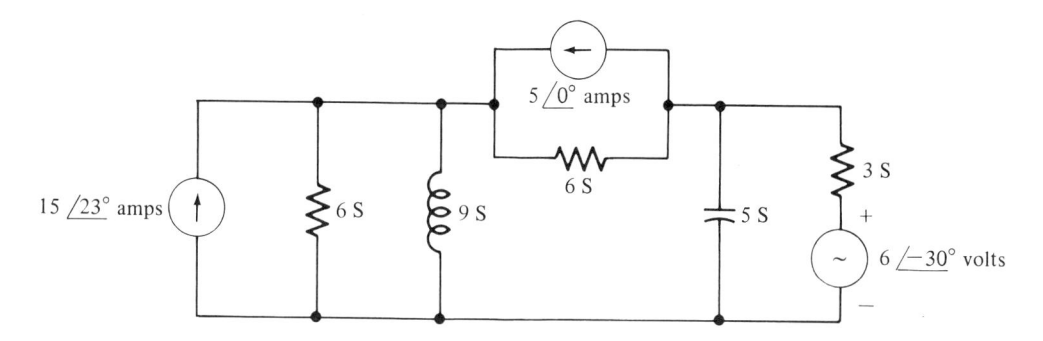

18. a. Write the nodal equations for this ac bridge. Use generalized admittances.

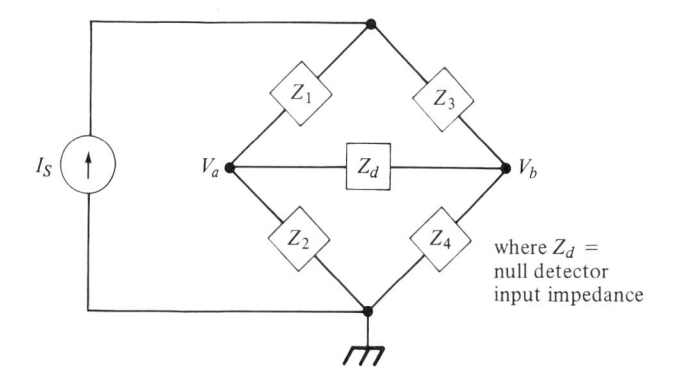

  b. Expand and solve the numerator determinant for E_a with respect to ground. Repeat for E_b.
  c. Using the condition $E_a - E_b = 0$ v, for balance, solve for the familiar balance equation.
  d. Why could the denominator determinants be disregarded in part c?

Section 18.3

19. For each part, find the Thévenin equivalent of the circuit driving the capacitor.

a.

b.

20. Find the Thévenin equivalent for the circuit driving the inductor in Exercise 19, part b.

21. In each part, find the Thévenin equivalent of the circuit driving the inductor.

a.

b.

22. Repeat Exercise 15 by finding the Thévenin equivalent circuit driving the capacitor in each case. Then calculate reactive power drawn. Compare the results.

23. What value of Z_{load} draws maximum power from each Thévenin circuit in Exercise 21?

24. Théveninize the circuit driving the 10-V source and internal impedance connected between points A and B. Then determine the current through this source.

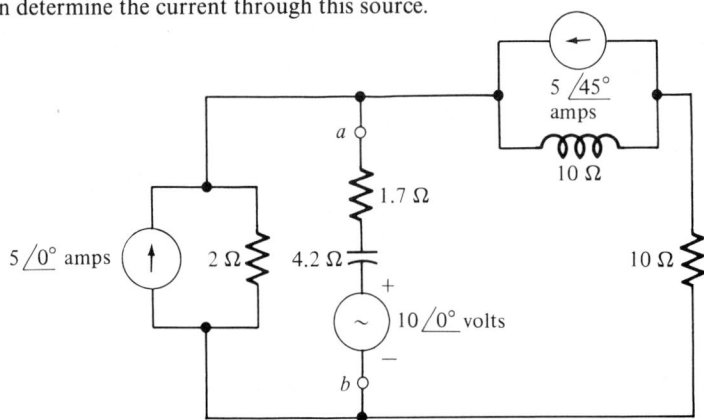

EXERCISES

Section 18.4

25. For each part, determine the Norton equivalent of the circuit driving the capacitor. Then calculate capacitor current.

a.

b.

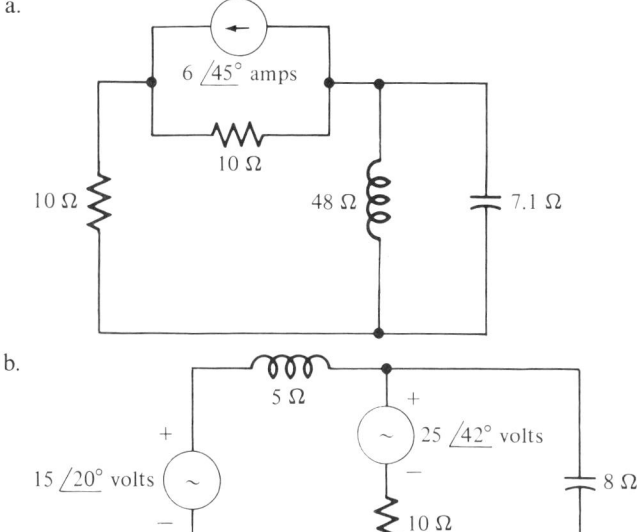

26. Find the Norton equivalent of the circuit driving the inductor in Exercise 25, part b.

27. Repeat Exercise 21 finding the Norton circuit driving each inductor.

28. a. For the following circuit, determine the Norton equivalent circuit seen by the resistor. Using this equivalent circuit, determine $I_{resistor}$.

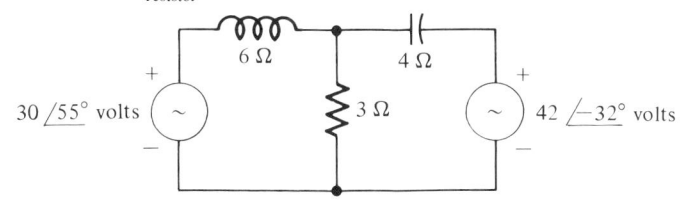

b. Repeat part a for the capacitor.
c. Repeat part a for the inductor.
d. Using I_R, I_C, I_L from parts a, b, and c, respectively, confirm Kirchhoff's current law at the upper node of the original circuit.

29. Repeat Exercise 24 finding the Norton equivalent of the circuit driving the 10-V source and its internal impedance.

30. Find the Norton equivalent of the circuit driving the capacitor in Exercise 16.

31. Find the Norton equivalent of the circuit driving the inductor in Exercise 16.

CHAPTER 19

Resonant Circuits

Pushing a child on a swing raises the child and stores potential energy in the child–swing system. Release the swing and it swings, or *oscillates*, at a rate called its *natural frequency*. This frequency is solely determined by rope length (the basis of the pendulum clock).

The swing's oscillations follow a waveform called a *damped* or *decaying* sinusoid (see Figure 19.1) in which peak amplitudes decreases. This decay is the result of energy loss from windage and rope heating during each cycle. Obviously, the greater the loss per cycle, the faster the oscillations die out.

The swing, like any pendulum, has only one natural frequency. Other oscillating bodies, such as a taut guiter string, often have two or more. A body will oscillate continuously at any of these frequencies if energy losses are replaced by properly timed pulses of external energy. When this is done, the body and pulses are said to be *in resonance*. The frequency at which this occurs is the *resonant frequency*.

Minimum energy is required to maintain resonant oscillations. Any excess causes oscillation amplitude increases which can create problems. For example, a car sometimes hits a series of bumps at just the right speed so that a wheel's

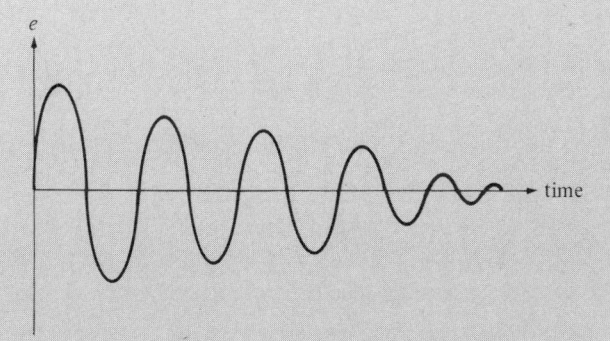

FIGURE 19.1 Damped or Decaying Sinusoid.

spring–shock absorber suspension system receives impulses at, or close to, its resonant frequency. This causes the bounce of one bump to be reinforced by that from the next. The resulting rough ride could be made much smoother by a small change in speed. And troops break step rather than march across bridges. If they didn't, the impact of their pounding boots might have a rhythm close to a natural bridge frequency. As a result, dangerous vibration amplitudes could develop, possibly causing bridge collapse.

Oscillations occur in electric circuits, too. As we saw in earlier chapters, inductance and capacitance store energy in their fields. When these components are series or parallel connected, that energy surges back and forth between the two. It does so at a natural resonant frequency determined by L and C. Unless replaced, the energy is gradually dissipated by circuit resistance.

Oscillating circuits of inductance and capacitance are called *resonant*, or *tuned* circuits. They are also called *resonators*. Tuned circuits have the ability to pass or select signals of one frequency and block others; hence they are vital to electronics and modern life. We are using one every time we change TV channels or tune a radio. Although a few comments were made on the subject of resonance in Chapter 16, it is studied in detail in this chapter.

At the completion of this chapter, you should be able to:

- Define and calculate the resonant frequency, Q, half-power frequencies, and bandwidth for series- and parallel-resonant *RLC* circuits.
- Describe the generation of sine waves by parallel *RLC* "tank" circuits.
- Explain and design *L*-network impedance–matching circuits.
- Discuss the frequency response characteristics of band-pass and band-reject filters.

19.1 THE SERIES RESONANT CIRCUIT

Figure 19.2 contains plots showing how pure R, X_C, and X_L change with frequency. Note carefully how they illustrate the characteristics that we have come to know so well in past chapters:

- R: [part (a)] constant with changing frequency (we disregard skin effect here); therefore, the plot is a horizontal line.
- X_C: [part (b)] inversely proportional to frequency. An open ($-$ infinite ohms) at dc, its magnitude decreases (value becomes less negative) toward zero; it is effectively a short at high frequencies.
- X_L: [part (c)] directly proportional to frequency. A short at dc, it increases linearly, becoming effectively an open at high frequencies.

The plots of Figure 19.2 are redrawn in Figure 19.3. They are added to form a fourth plot marked $|Z_{\text{total}}|$, representing the magnitude of the total impedance of a series *RLC* combination. Beneath the diagram are three polar plots repeated from Figure 16.4. Each describes the state of the circuit at one of the frequencies, f_a, f_s, and f_b, marked on the X axis.

Let us examine circuit operation as the frequency increases from f_a through f_s to f_b:

- At f_a (very low frequency). $|X_C| \gg |X_L|$. Therefore, the circuit is capacitive, R has little effect and $|Z| \approx |X_{\text{total}}| \approx |X_C|$. Leading PF ≈ 0. $P_A \approx P_Q$.

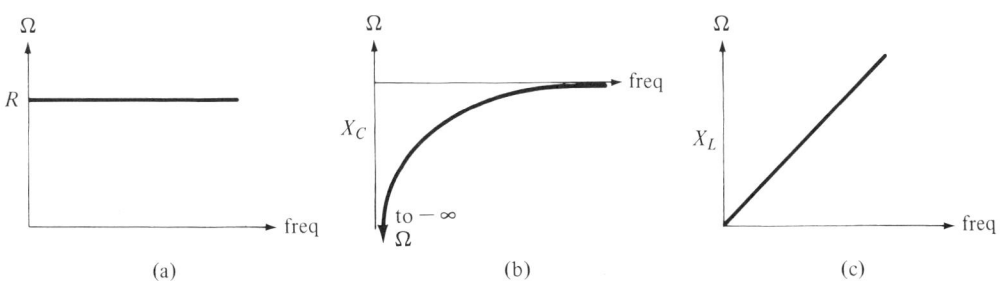

FIGURE 19.2 Variation in R, X_C and X_L with Frequency.

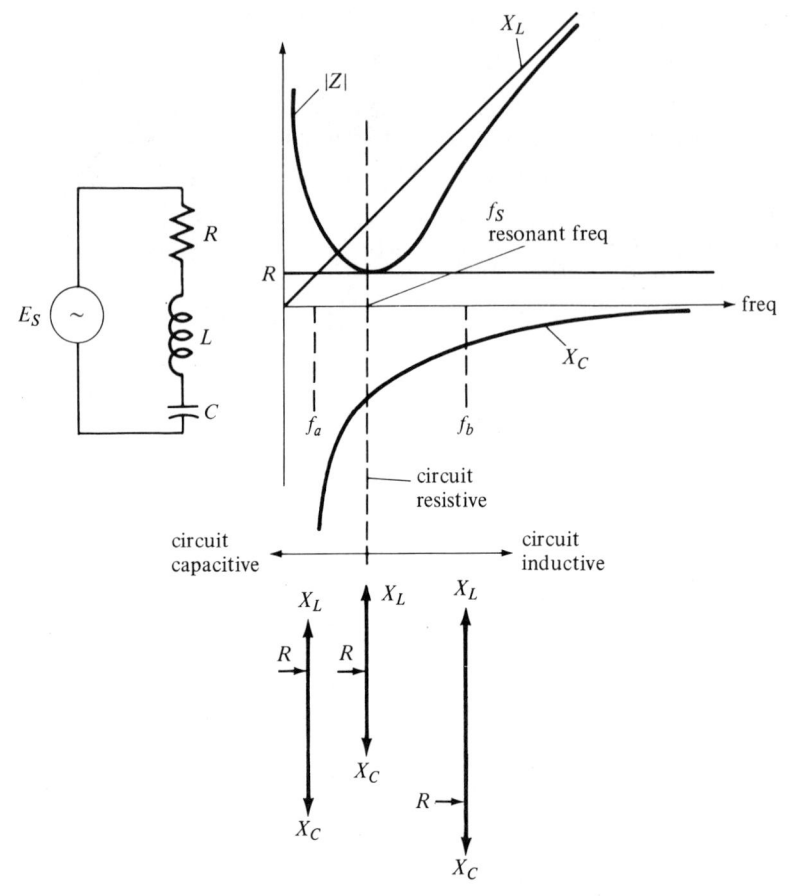

FIGURE 19.3 Frequency Characteristics of Series *RLC* Circuit.

- As frequency increases above f_a, X_C decreases as X_L increases. Impedance becomes less capacitive as X_{total} decreases. R has greater effect. Frequency finally reaches f_s (series-resonant frequency):
  $|X_L| = |X_C|$; therefore, $X_{total} = 0$;
  therefore, $|Z| = R$, the circuit is resistive. PF unity (100%). $P_A = P_W$.
- As frequency increases above f_s, X_L begins to dominate until at.
  f_b (very high frequency), $|X_L| \gg |X_C|$.
  Therefore, the circuit is inductive, R has little effect and $|Z| \approx |X_{total}| \approx |X_L|$. Lagging PF ≈ 0. $P_A \approx P_{Q_L}$.

Figure 19.4 shows the Z-magnitude graph removed from Figure 19.3 for clarity. Another plot illustrating the current resulting from this impedance has been added on the same set of axes. Note that the curves are not symmetrical about the dashed f_s line.

A separate graph showing how the phase angle changes as frequency passes through f_s is included in Figure 19.4. Study carefully the relations between plots.

Knowing that X_{total} of a series *RLC* circuit is zero at resonance provides the key that allows us to derive a formula for f_s. Since

$$X_{total} = 0$$

19.1 THE SERIES RESONANT CIRCUIT

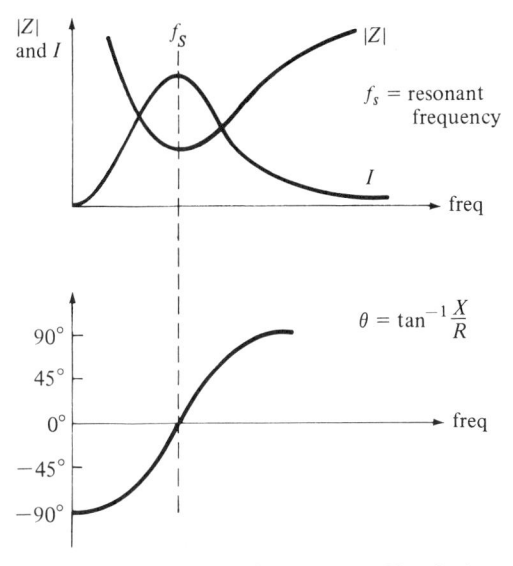

FIGURE 19.4 Series *RLC* Impedance Magnitude and Phase Changes versus Frequency.

we have

$$X_{ind} - X_{cap} = 0$$

Therefore,

$$X_{ind} = X_{cap}$$

or

$$2\pi f_s L = \frac{1}{2\pi f_s C}$$

Solve for f_s:

$$f_s^2 = \frac{1}{(2\pi)^2 LC}$$

Therefore,

$$f_s = \frac{1}{2\pi\sqrt{LC}} \qquad (19.1)$$

or

$$\omega_s = \frac{1}{\sqrt{LC}} \qquad (19.2)$$

where f, L, C and ω are all in basic units of hertz, henries, farads, and radians per second, respectively. Note that the frequency of resonance is not a function of resistance.

Equations 19.1 and 19.2 tell us that there are infinite combinations of inductance and capacitance that resonate at any one frequency. Is it necessary that we use any one combination? That question is answered in following sections.

EXAMPLE 19.1

A 120-μH inductor with a series resistance of 42Ω is series connected to a 0.002-μF capacitor. $E_{source} = 100$ V.

a. What is the resonant frequency?

$$f_s = \frac{1}{2\pi\sqrt{LC}}$$

$$= \frac{1}{2\pi[(120 \times 10^{-6})(2 \times 10^{-9})]^{1/2}}$$

$$= 325 \text{ kHz}$$

b. What is the maximum current flowing in the circuit?

Maximum current flows when impedance is minimum, that is, at resonance when $|Z| = R$. Therefore,

$$i_{max} = \frac{100}{42}$$

$$= 2.4 \text{ A}$$

c. Repeat parts a and b for R increased to 65 Ω.

f_s remains at 325 kHz because it is not a function of resistance.

$$i_{max} = \frac{100}{65}$$

$$= 1.5 \text{ A}$$

d. The capacitance is reduced to 1100 pF. What is the new f_s?

$$f_s = \frac{1}{2\pi\sqrt{LC}}$$

$$= \frac{1}{2\pi[(120 \times 10^{-6})(1100 \times 10^{-12})]^{1/2}}$$

$$= 438 \text{ kHz}$$

Note that a decrease in capacitance caused an increase in resonant frequency. ∎

Example 19.1 makes an important point that should not go unnoticed. By varying L and/or C, we can select one frequency, f_s, for which a series RLC circuit has minimum impedance and therefore maximum current. All other signals encounter higher impedance, causing them to pass lower current.

A deeper understanding of resonance can be gained by a brief consideration of energy movement in the RLC circuit. Energy is traded back and forth between the inductor's electromagnetic field and the capacitor's electrostatic field. At frequencies below f_s, $V_{capacitor}$ is greater than $V_{inductor}$. The capacitor thus stores more energy than the inductor can supply. Because this extra energy must come from the source, the circuit appears capacitive to the source. At frequencies above f_s, the inductor has the

19.2 THE QUALITY FACTOR, Q

greater voltage and draws source energy; the circuit appears inductive. At f_s, the reactances have equal potential differences therefore equal energy storage. Energy released by one reactance is completely stored by the other. The source now supplies only that energy being dissipated by circuit resistance. The load thus appears resistive; its impedance has no reactive component.

19.2 THE QUALITY FACTOR, Q

Let us return to our child–swing system for a moment. Because of losses, each swing oscillation begins with less stored energy than the preceding one. If these losses are not replaced by pushing, stored energy gradually dissipates and swinging stops. The faster the energy is dissipated, the higher the decay rate and the sooner the swinging ends.

The relation between stored energy and energy lost per cycle for any oscillating body is expressed mathematically by a parameter called Q, for *quality factor*. It is defined

$$Q = 2\pi \frac{W_{stored}}{W_{dissipated\ per\ cycle}} \quad (19.3)$$

Note Q is dimensionless.

Q provides a number by which oscillating systems can be compared. The higher its value, the longer oscillations continue for a given energy input. Tuning forks, for example, have Q values of about 2000, while the vibrating medium in certain atomic clocks have figures around 10^9.

Figure 19.5, part (a) illustrates the oscillating waveform of a high-Q system. Note the nearly constant amplitude. The increasingly faster decay times of Figure 19.5, parts (b) and (c), display medium- and low-Q systems, respectively.

Since LC circuits are systems of oscillating energy, they can be described by their Q. In fact, Q and resonant frequency are the two most important parameters of tuned circuits.

Let us derive formulas for Q_s, the Q of series resonant circuits.

We learned earlier that a resonant circuit has all its energy stored in the inductor at instants of peak current. By Equation 12.3, that energy is

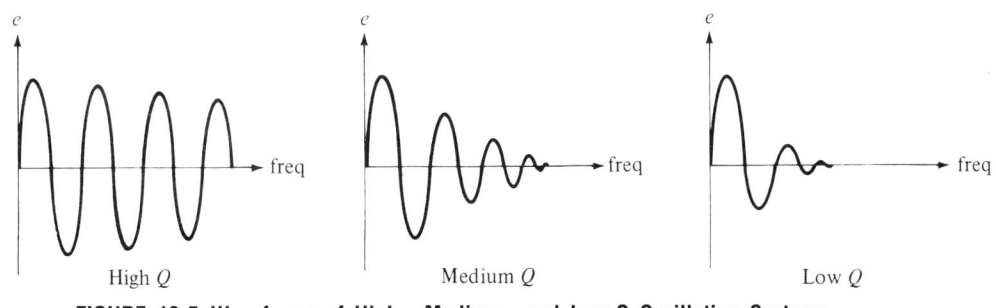

$$W_{ind} = \tfrac{1}{2} L I_{peak}^2 \quad (19.4)$$

FIGURE 19.5 Waveforms of High-, Medium-, and Low-Q Oscillating Systems.

or
$$W_{ind} = LI_{rms}^2 \tag{19.5}$$

where W_{ind} = energy stored in inductor
I_{peak} = peak inductor current
I_{rms} = rms inductor current

The energy dissipated by the resistor per cycle is
$$W_{res} = P_{res}T$$
$$= I_{rms}^2 RT$$

where W_{res} = energy dissipated by resistance
P_{res} = power dissipated by resistance
T = period of oscillation

Substitute $1/f_s$ for T:
$$W_{res} = \frac{I_{rms}^2 R}{f_s} \tag{19.6}$$

Substitute Equations 19.5 and 19.6 into Equation 19.3:
$$Q_s = 2\pi \frac{LI_{rms}^2}{\frac{I_{rms}^2 R}{f_s}}$$
$$= \frac{2\pi f_s L}{R} \tag{19.7}$$
$$= \frac{\omega_s L}{R} \tag{19.8}$$

or
$$Q_s = \frac{X_L}{R} \tag{19.9}$$

Since $X_L = X_C$ at resonance, we can also write these equations as
$$Q_s = \frac{X_C}{R} \tag{19.10}$$
$$= \frac{1}{2\pi f RC} \tag{19.11}$$
$$= \frac{1}{\omega RC} \tag{19.12}$$

Another equation for Q_s is easily derived. Since Q equals either circuit reactance divided by resistance, we can say that

$$Q = \frac{X}{R}$$

19.2 THE QUALITY FACTOR, Q

Multiply numerator and denominator by I_{rms}^2:

$$= \frac{I_{rms}^2 X}{I_{rms}^2 R}$$

This we recognize as the ratio of reactive power to average power dissipated by circuit resistance. Thus

$$Q_s = \frac{P_{reactive}}{P_{true}} \qquad (19.13)$$

Typical Q values found in electronics range from perhaps 10 to a few hundred with figures around a thousand on occasion; speciality devices go to about 10,000. Students are urged, however, not to think of Q as money in the bank: the more, the better. Both high-Q and low-Q circuits are needed in electronics.

EXAMPLE 19.2

What is the Q_s of the circuit in Example 19.1?

$$Q_s = \frac{2\pi f L}{R}$$

$$= \frac{2\pi(325 \times 10^3)(120 \times 10^{-6})}{42}$$

$$= 5.8 \qquad \blacksquare$$

Let us next consider the voltage across the inductor in the series RLC circuit of Figure 19.3. By voltage division,

$$V_{ind} = \frac{E_s X_L}{Z}$$

where V_{ind} = rms inductor voltage
E_s = rms source voltage

But, at f_s,

$$Z = R$$

Therefore,

$$V_{ind} = \frac{E_s X_L}{R}$$

But

$$Q_s = \frac{X_L}{R}$$

Substitute:

$$V_{ind} = Q_s E_s \qquad (19.14)$$

By a similar derivation,

$$V_{cap} = Q_s E_s \tag{19.15}$$

These equations can be modified to give two other definitions of Q_s:

$$Q_s = \frac{V_{ind}}{E_s} \tag{19.16}$$

and

$$Q_s = \frac{V_{cap}}{E_s} \tag{19.17}$$

Note carefully what Equations 19.14 and 19.15 are saying: The voltage across a reactive component in a series resonant circuit is Q times the source voltage. In high-Q circuits, this effect develops unexpectedly high potentials which deliver nasty shocks and cause serious problems with insulation breakdowns and component failures for the unwary circuit designer. It also explains the voltage magnification seen earlier in analysis problems.

We will see Q multiplication again in parallel-resonant circuits.

EXAMPLE 19.3

What is the peak source voltage that may be applied to this circuit?

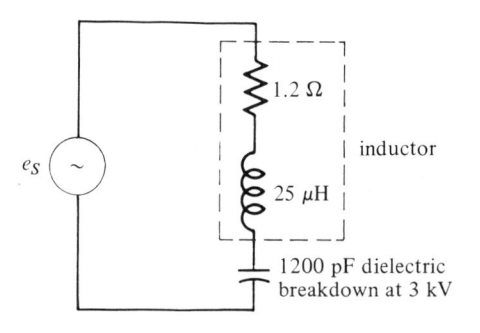

$$f_s = \frac{1}{2\pi\sqrt{LC}}$$

$$= \frac{1}{2\pi\sqrt{(1200 \times 10^{-12})(25 \times 10^{-6})}}$$

$$= 919 \text{ kHz}$$

$$X_C = \frac{1}{2\pi f C}$$

$$= \frac{1}{2\pi(919 \times 10^3)(1200 \times 10^{-12})}$$

$$= 144 \ \Omega$$

Therefore,

$$Q_s = \frac{X_C}{R}$$
$$= \frac{144}{1.2}$$
$$= 120$$
$$Q_s = \frac{V_C}{e_{s_{\text{peak}}}}$$

$$e_{s_{\text{peak}}} = \frac{V_C}{Q_s}$$
$$= \frac{3000}{120}$$
$$= 25 \text{ V} \qquad \blacksquare$$

Because of their unavoidable "impurity" resistance, inductors and capacitors are *lossy*, that is, they lose energy. The ratio of a reactor's reactance to its resistance is the component's Q:

$$Q_{\text{component}} = \frac{X}{R} \qquad (19.18)$$

The parameter can be considered a measure of reactance quality, high Q indicating a high ratio of reactance (energy storage) to resistance (energy dissipation).

Tuned circuit Q is usually determined by the Q of the most lossy component. Capacitor Q's are so high, typically in the thousands, that those components are considered lossless at most frequencies. Inductors, on the other hand, usually contribute most, if not all, of the resistance in a resonant circuit. Consequently, Q_{circuit} is usually Q_{ind}. The R in most of our problems will be inductor resistance, R_{ind}.

19.3 THE IMPORTANCE OF Q

Q plays a very important role in resonant circuits that was not brought out by the equations of Section 19.2. To examine that role, let us derive an expression showing how resistance, inductance, and capacitance combine to determine Q.

Equation 19.8 says that

$$Q_s = \frac{\omega_s L}{R}$$

Substitute Equation 19.2 for ω_s:

$$Q_s = \frac{1}{\sqrt{LC}} \frac{L}{R}$$
$$= \frac{1}{R} \frac{\sqrt{L^2}}{\sqrt{LC}}$$
$$= \frac{1}{R} \sqrt{\frac{L}{C}} \qquad (19.19)$$

where L and C are the inductance and capacitance tuned to resonance (henrys and farads, respectively), and R is total circuit resistance.

Equation 19.19 tells us:

- To get high Q, make the resistance low and the ratio of inductance to capacitance high (from now on, we shall refer to this as the *LC ratio*).
- To get low Q, make the resistance high and the *LC* ratio low.

Let us examine how Q varies with changes in resistance and the *LC* ratio.

With the *LC* ratio fixed the frequency applied to a series *RLC* circuit with R ohms is varied through resonance. The result is the lowest curve in Figure 19.6. Still holding the *LC* ratio constant, R is decreased to $R/2$, then $R/4$, and finally to $R/10$. As it should, current increases with each resistance decrease.

Just as important as the current increases, however, is the changing shape of the curves. Note carefully how it goes from broad and low for high resistance (R ohms), to narrow and high for low resistance ($R/10$ Ω). The reason for this is seen in the effect of resistance on impedance magnitude as described by the familiar equation, $|Z| = \sqrt{R^2 + X_{total}^2}$. When R is high, it dominates this magnitude; and R is not frequency dependent. As a result, the frequency must change a great deal in the vicinity of f_s before X_{total} increases enough to have much effect on the magnitude. Consequency, the low peak current changes gradually. On the other hand, when resistance is low, its influence on impedance magnitude is minimal. Now, small changes in frequency around f_s cause large changes in X_{total} and therefore in impedance magnitude. This causes the increasingly steeper drops from peak current above and below f_s in the higher-resistance curves.

Next, we will hold the resistance constant and increase the *LC* ratio to raise Q_s. The curves of Figure 19.7 result. Note that peak current remains constant (no change in resistance) and that the curves become narrower with steeper sides, or *skirts*, as the ratio increases. This occurs because the higher the *LC* ratio, the greater the reactance swing,

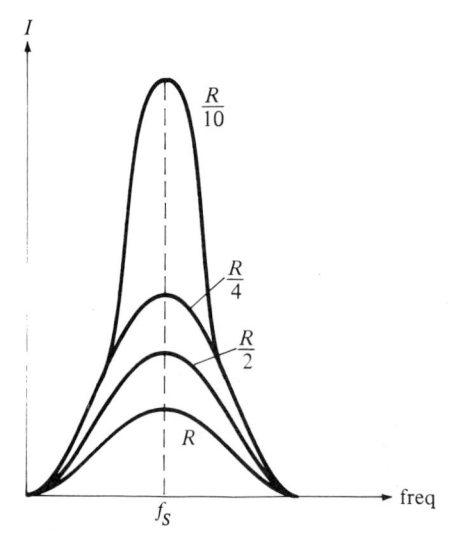

FIGURE 19.6 Changing Response Curves with Decreasing Resistance.

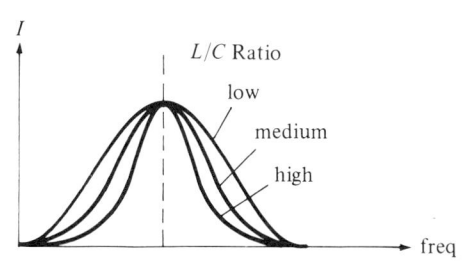

FIGURE 19.7 Changing Response Curves with Increasing *LC* Ratios.

and therefore the greater the impedance magnitude swing for a given frequency variation.

Note carefully in both Figures 19.6 and 19.7 that the high-*Q* cases have the narrowest curves. That is the important point:

| The higher the *Q* of a resonant circuit, the narrower its response curve.

But how do we know how narrow the curve should be? In other words, what *Q* value must be used when designing a resonant circuit? To answer that question, we must study bandwidth.

19.4 BANDWIDTH

When the reactances of a series *RLC* circuit cancel at resonance, $Z = R$ and current is maximum. Power delivered to the resistance must also be maximum. If the applied frequency is raised above f_s, X_L increases and X_C decreases until a point is reached (as it was for the filters in Section 16.5) when $X_{total} = R$. Then

$$|Z| = \sqrt{R^2 + X_{total}^2}$$
$$= \sqrt{R^2 + R^2}$$
$$= \sqrt{2}\,R$$

Current at this frequency drops to $0.707 I_{max}$, so power delivered falls to

$$P = I^2 R$$
$$= (0.707 I_{max})^2 R$$
$$= 0.5 I_{max}^2 R = 0.5 P_{max}$$

Since this *half-power frequency* is above f_s, it is called the *upper half-power frequency*, f_2. Decreasing the applied frequency below f_s causes the same *Z*-magnitude increase but with X_C dominating. Eventually, X_{total} again equals R at the *lower half-power frequency*, f_1. The two frequencies are shown in Figure 19.8.

Knowing that total reactance is equal to resistance at f_2 and f_1 enables us to derive formulas for these frequencies. At the upper frequency,

$$\omega_2 L - \frac{1}{\omega_2 C} = R$$

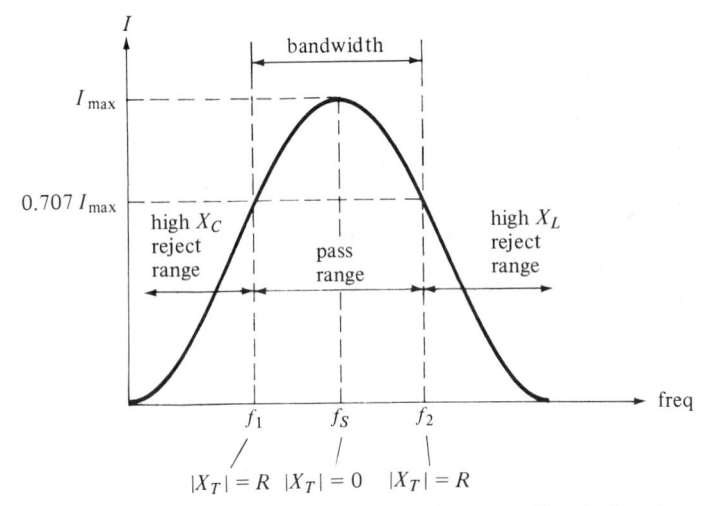

FIGURE 19.8 Response Curve of Series–Resonant Circuit Showing Half-Power Frequencies, Pass and Reject Regions, and Bandwidth.

Divide through by L and multiply by ω_2:

$$\omega_2^2 - \frac{1}{LC} = \frac{R\omega_2}{L}$$

Rearrange into a quadratic:

$$\omega_2^2 - \frac{R}{L}\omega_2 - \frac{1}{LC} = 0$$

When solved by the quadratic formula,

$$x = \frac{-b \pm \sqrt{b^2 - 4ac}}{2a}$$

we get

$$\omega_2 = \frac{-(-R/L) \pm \sqrt{(-R/L)^2 - 4(-1/LC)}}{2}$$

which becomes

$$\omega_2 = \frac{R}{2L} \pm \sqrt{\left(\frac{R}{2L}\right)^2 + \frac{1}{LC}}$$

The quantity beneath the radical will always be greater than $R/2L$. Therefore, the negative possibility is discarded because it causes a negative frequency, which is meaningless. This leaves

$$\omega_2 = \frac{R}{2L} + \sqrt{\left(\frac{R}{2L}\right)^2 + \frac{1}{LC}} \qquad (19.20)$$

Because $f_2 = \omega_2/2\pi$, we have

$$f_2 = \frac{1}{2\pi}\left[\frac{R}{2L} + \sqrt{\left(\frac{R}{2L}\right)^2 + \frac{1}{LC}}\right] \qquad (19.21)$$

19.4 BANDWIDTH

By a similar derivation the lower half-power frequency is found to be

$$\omega_1 = -\frac{R}{2L} + \sqrt{\left(\frac{R}{2L}\right)^2 + \frac{1}{LC}} \tag{19.22}$$

and

$$f_1 = \frac{1}{2\pi}\left[-\frac{R}{2L} + \sqrt{\left(\frac{R}{2L}\right)^2 + \frac{1}{LC}}\right] \tag{19.23}$$

Between the half-power frequencies is the pass region of the resonance response curve (see Figure 19.8). Signals within this band are said to be *selected*, or *tuned*. Those outside the band are sharply reduced or *attenuated*; the farther the signal is from f_s, the greater the attenuation.

The width of the pass region is defined as the circuit's *bandwidth*:

$$\text{BW} = f_2 - f_1 \tag{19.24}$$

When Equations 19.21 and 19.23 are substituted into Equation 19.24, the results are:

$$\begin{aligned}\text{BW} &= f_2 - f_1 \\ &= \frac{1}{2\pi}\left[\frac{R}{2L} + \sqrt{\left(\frac{R}{2L}\right)^2 + \frac{1}{LC}} - \left(-\frac{R}{2L} + \sqrt{\left(\frac{R}{2L}\right)^2 + \frac{1}{LC}}\right)\right] \\ &= \frac{1}{2\pi}\left[\frac{2R}{2L}\right] \\ &= \frac{R}{2\pi L}\end{aligned} \tag{19.25}$$

Multiply numerator and denominator of Equation 19.25 by f_s:

$$\begin{aligned}\text{BW} &= \frac{Rf_s}{2\pi f_s L} \\ &= \frac{Rf_s}{X_L}\end{aligned}$$

But

$$\frac{R}{X_L} = \frac{1}{Q_s}$$

Therefore,

$$\text{BW} = \frac{f_s}{Q_s} \tag{19.26}$$

This equation makes an important point which agrees with the conclusions of Section 19.3, namely:

| Bandwidth and Q are indirectly related

- High-Q circuits have narrow bandwidths.
- Low-Q circuits have broad bandwidths.

EXAMPLE 19.4

An *RLC* series circuit has $R = 134\ \Omega$, $L = 1.4$ H AND $C = 14.8\ \mu$F.

a. Find f_s.

$$f_s = \frac{1}{2\pi\sqrt{LC}}$$
$$= \frac{1}{2\pi[1.4(14.8 \times 10^{-6})]^{1/2}}$$
$$= 35\text{ Hz}$$

b. Find Q_s.

$$Q_s = \frac{1}{R}\sqrt{\frac{L}{C}}$$
$$= \frac{1}{134}\left(\frac{1.4}{14.8 \times 10^{-6}}\right)^{1/2}$$
$$= 2.3$$

c. Find the upper and lower half-power frequencies.

$$f_2 = \frac{1}{2\pi}\left[\frac{R}{2L} + \sqrt{\left(\frac{R}{2L}\right)^2 + \frac{1}{LC}}\right]$$
$$= \frac{1}{2\pi}\left[\frac{134}{2(1.4)} + \sqrt{\left(\frac{134}{2(1.4)}\right)^2 + \frac{1}{1.4(14.8 \times 10^{-6})}}\right]$$
$$= 43.4\text{ Hz}$$

$$f_1 = \frac{1}{2\pi}\left[-\frac{R}{2L} + \sqrt{\left(\frac{R}{2L}\right)^2 + \frac{1}{LC}}\right]$$
$$= \frac{1}{2\pi}\left[-\frac{134}{2(1.4)} + \sqrt{\left(\frac{134}{2(1.4)}\right)^2 + \frac{1}{1.4(14.8 \times 10^{-6})}}\right]$$
$$= 28.2\text{ Hz}$$

d. Find BW.

$$\text{BW} = f_2 - f_1$$
$$= 43.4 - 28.2$$
$$= 15.2\text{ Hz}$$

∎

Since

$$\text{BW} = \frac{f_s}{Q_s}$$

and

$$Q_s = \frac{1}{R}\sqrt{\frac{L}{C}}$$

19.4 BANDWIDTH

it follows that

$$\text{BW} = \frac{Rf_s}{\sqrt{L/C}} \quad (19.27)$$

This equation tells us two things: first, low resistance gives narrow bandwidth (this agrees with the findings of Figure 19.6), and second, bandwidth can be adjusted by changing series resistance in an RLC resonant circuit.

EXAMPLE 19.5

What resistance must be added to the circuit of Example 19.4 to extend the bandwidth to 18 Hz?

$$\text{BW} = \frac{Rf_s}{\sqrt{L/C}}$$

Therefore,

$$R = \frac{\text{BW}\sqrt{L/C}}{f_2}$$

$$= \frac{18\left(\dfrac{1.4}{14.8 \times 10^{-6}}\right)^{1/2}}{35}$$

$$= 158.2 \, \Omega$$

$$\text{amount to be added} = 158.2 - 134$$

$$= 24.2 \, \Omega \quad \blacksquare$$

Even though it is reasonable to assume that f_s splits the response curve evenly with half the bandwidth on either side, Example 19.4 shows that this is not the case. The upper half-power frequency is 8.4 Hz above f_s, while the lower is only 6.8 Hz below. Half-power frequencies become more evenly spaced about f_s for higher Q circuits. For $Q > 10$ the small difference is neglected and we say that f_s splits the bandwidth so that the half-power frequencies are approximately $f_s \pm \text{BW}/2$. Specifically, for circuits with $Q > 10$,

$$f_2 \approx f_s + \frac{\text{BW}}{2} \quad (19.28)$$

and

$$f_1 \approx f_s - \frac{\text{BW}}{2} \quad (19.29)$$

Having discussed bandwidth, we are ready to answer the question about the choice of Q when designing a resonant circuit.

An electronic signal carrying intelligence in any form, such as music, digital pulses, and so on, is not a single sinusoid but an infinite number of frequencies clustered in a *band.* The difference between the highest and lowest of these frequencies is called the *signal bandwidth.* A radio signal on the standard broadcast AM band, for example, is 10 kHz wide, 5 kHz either side of the center frequency. For a radio to receive one of these signals, its tuned circuits must do two things. First, select a station by resonating to its frequency on the dial. Second, pass the entire signal by having a bandwidth as wide as that of the signal. If the bandwidth is too narrow, some frequencies will be lost, causing loudspeaker output to sound unnatural.

Thus we see that a tuned circuit's resonant frequency and bandwidth are determined by the nature of the signal being tuned by that circuit. With f_s and BW known, the choice of Q is made for us by Equation 19.26, $Q_s = f_s/\text{BW}$. The circuit's resistance, inductance, and capacitance are chosen to obtain this Q_s and f_s.

EXAMPLE 19.6

A tuned circuit is to be designed to pass a 200-kHz-wide FM signal at f_s = 95 MHz. On hand is a 2.5-pF capacitor and wire of negligible resistance from which the coil will be wound. Design the circuit.

By equation 19.26,

$$Q_s = \frac{f_s}{\text{BW}}$$

$$= \frac{95 \times 10^6}{200 \times 10^3}$$

$$= 475$$

$$f_s = \frac{1}{2\pi\sqrt{LC}}$$

Therefore,

$$L = \frac{1}{(2\pi f_s)^2 C}$$

$$= \frac{1}{[2\pi(95 \times 10^6)]^2 (2.5 \times 10^{-12})}$$

$$= 1.1 \ \mu\text{H}$$

By Equation 19.19,

$$Q_s = \frac{1}{R}\sqrt{\frac{L}{C}}$$

Therefore,

$$R = \frac{1}{Q_s}\sqrt{\frac{L}{C}}$$

$$= \frac{1}{475}\left(\frac{1.1 \times 10^{-6}}{2.5 \times 10^{-12}}\right)^{1/2}$$

$$= 1.4 \ \Omega$$

All of this must be added since wire resistance is negligible. Final circuit:

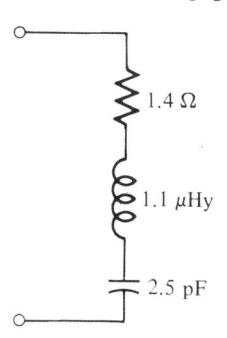

```
    1.4 Ω
    1.1 μHy
    2.5 pF
```

Had inductor resistance been greater than 1.4 Ω, another coil choice would have been necessary.

19.5 THE PARALLEL-RESONANT CIRCUIT

Current and impedance relationships of parallel *RLC*-resonant circuits are exactly opposite those of their series-resonant counterparts. When parallel reactive components cancel at resonance, impedance rises to its maximum. For this reason, parallel resonance is sometimes referred to as *antiresonance*.

The majority of tuned circuits in electronics are of the parallel type. Often called *tank circuits* because of their energy storage capabilities, their basic configuration is seen in Figure 19.9. R_{ind}, the resistance of the inductor wire, is usually the circuit's only resistance. As we shall see, however, resistance can be added to adjust bandwidth as it was in the series case.

To ease analysis of the circuit, the series *RL* branch is converted to its parallel equivalent using the formulas derived in Section 16.9.

$$R_p = \frac{R_s^2 + X_s^2}{R_s} \tag{19.30}$$

$$X_p = \frac{R_s^2 + X_s^2}{X_s} \tag{19.31}$$

This gives us the circuit shown in Figure 19.10.

X_L and X_C are shorts at low and high frequencies, respectively, causing low Z_{total} and high I_{line}. As in the series-resonant case, increasing frequency causes X_{L_p} to increase

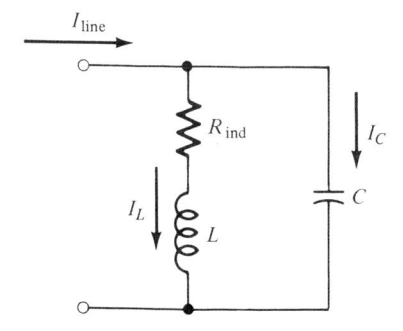

FIGURE 19.9 Basic Parallel *RLC* Resonant Circuit.

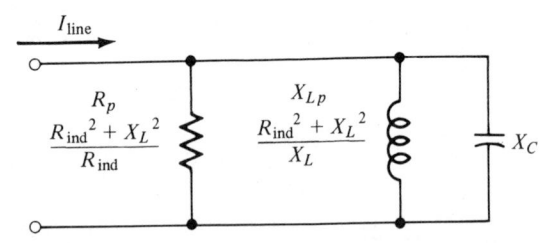

FIGURE 19.10 Parallel Circuit in Converted Form.

and X_C to decrease. However, in a parallel circuit, these changes cause Z_{total} to increase, not decrease as it did before. This, of course, causes I_{line} to decrease.

The impedance reaches peak value of R_p and line current nulls when the reactances cancel at the parallel-resonant frequency, f_p. These changes are illustrated in Figure 19.11. Note carefully how they are opposite to those of the series-resonant circuit.

Let us derive the formula for f_p. At resonance,

$$X_{L_p} = X_C$$

or

$$\frac{R_{ind}^2 + X_L^2}{X_L} = X_C$$

Therefore,

$$R_{ind}^2 + X_L^2 = X_L X_C$$

or

$$R_{ind}^2 + (\omega_p L)^2 = \frac{L}{C}$$

$$(\omega_p L)^2 = \frac{L}{C} - R_{ind}^2 = \frac{L - R_{ind}^2 C}{C}$$

$$\omega_p^2 = \frac{L - R_{ind}^2 C}{L^2 C}$$

$$= \frac{1}{LC} - \frac{R_{ind}^2 \frac{C}{L}}{LC}$$

$$= \frac{1}{LC}\left(1 - \frac{R_{ind}^2 C}{L}\right)$$

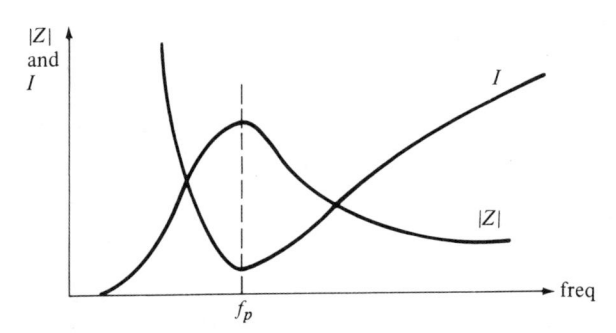

FIGURE 19.11 Frequency Characteristics of Z Magnitude and Current in Parallel *RLC* Circuit.

19.5 THE PARALLEL-RESONANT CIRCUIT

Therefore,

$$\omega_p = \frac{1}{\sqrt{LC}} \sqrt{1 - \frac{R_{ind}^2 C}{L}} \quad \text{rad/s} \quad (19.32)$$

The first factor on the right of this equation is immediately recognized as the formula for series resonance. Therefore,

$$\omega_p = \omega_s \sqrt{1 - \frac{R_{ind}^2 C}{L}} \quad \text{rad/s} \quad (19.33)$$

In the usual manner, these equations are rewritten:

$$f_p = \frac{1}{2\pi \sqrt{LC}} \sqrt{1 - \frac{R_{ind}^2 C}{L}} \quad \text{Hz} \quad (19.34)$$

and

$$f_p = f_s \sqrt{1 - R_{ind}^2 C/L} \quad \text{Hz} \quad (19.35)$$

Note that the parallel resonant frequency is a function of inductor resistance. However, in most cases, the term $R_{ind}^2 C/L$ is very small and can be neglected. When it is,

$$f_p \approx f_s = \frac{1}{2\pi \sqrt{LC}} \quad (19.36)$$

A convenient formula is easily derived for maximum impedance magnitude in terms of capacitance, series R_{ind} and inductance. At resonance, nance,

$$Z_p = R_p$$

By Equation 19.30,

$$R_p = \frac{R_{ind}^2 + X_L^2}{R_{ind}}$$

Therefore,

$$Z_p = \frac{R_{ind}^2 + X_L^2}{R_{ind}} \quad (19.37)$$

Also at resonance,

$$X_{L_p} = X_C$$

or

$$\frac{R_{ind}^2 + X_L^2}{X_L} = X_C \quad \text{by Equation 19.31.}$$

Therefore,

$$R_{ind}^2 + X_L^2 = X_L X_C$$

Substitute for $R_{ind}^2 + X_L^2$ in Equation 19.37:

$$Z_p = \frac{X_L X_C}{R_{ind}} = \frac{\omega L}{R_{ind} \omega C}$$

$$= \frac{L}{R_{ind} C} \quad (19.38)$$

EXAMPLE 19.7

Given: the parallel *RLC* circuit shown:

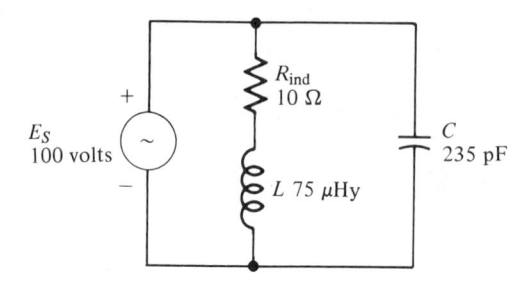

E_S 100 volts; R_{ind} 10 Ω; L 75 μHy; C 235 pF

Determine each of the following values.

a. f_p.

$$f_p = \frac{1}{2\pi\sqrt{LC}}\sqrt{1 - \frac{R_{ind}^2 C}{L}}$$

$$= \frac{1}{2\pi[(75 \times 10^{-6})(235 \times 10^{-12})]^{1/2}}\left[1 - \frac{10^2(235 \times 10^{-12})}{75 \times 10^{-6}}\right]^{1/2}$$

$$= 1.2 \text{ MHz}[1 - 0.000313]^{1/2}$$

$$\approx 1.2 \text{ MHz}$$

b. Z_p at resonance.

$$Z_p = \frac{L}{R_{ind} C}$$

$$= \frac{75 \times 10^{-6}}{10(235 \times 10^{-12})}$$

$$= 31.9 \text{ k}$$

c. I_{line} at resonance.

$$I_{line} = \frac{E_s}{Z}$$

$$= \frac{100}{31.9 \text{ k}}$$

$$= 3.1 \text{ mA}$$

d. I_{cap} at resonance.

$$I_{cap} = \frac{E_s}{X_C}$$

$$= \frac{E_s}{1/2\pi fC}$$

$$= 2\pi E_s fC$$

$$= 2\pi(100)(1.2 \times 10^6)(235 \times 10^{-12})$$

$$= 177 \text{ mA}$$

The reason $I_{cap} \gg I_{line}$ is explained below. ∎

19.6 *Q* OF THE PARALLEL-RESONANT CIRCUIT

As it did in the series case, energy surges back and forth in the parallel-resonant circuit, being stored by the reactances and dissipated by the resistance. In fact, the *R*, *L*, and *C* are in series as far as oscillating energy is concerned. Hence the *Q* formula from series circuits:

$$Q_s = \frac{X_s}{R_s}$$

applies to parallel circuits. However, most tank circuit analysis is done with parallel components, so it is convenient to have an equation for parallel Q_p in terms of R_p and X_p. Let us derive that equation.

The other conversion formulas derived in Section 16.9 were

$$R_s = \frac{R_p X_p^2}{R_p^2 + X_p^2}$$

and

$$X_s = \frac{X_p R_p^2}{R_p^2 + X_p^2}$$

Substitue these into the basic *Q* equation:

$$Q_s = \frac{X_s}{R_s}$$
$$= \frac{R_p^2 X_p/(R_p^2 + X_p^2)}{R_p X_p^2/(R_p^2 + X_p^2)}$$
$$Q_p = \frac{R_p}{X_p} \tag{19.39}$$

Note that Equation 19.39 refers to the circuit itself with no additional resistances such as from source or load. Under these conditions,

$$Q_p = Q_s \tag{19.40}$$

Pay careful attention to the requirement that R_p be high for high *Q*. This makes sense when we consider that high resistance draws low power.

Let us derive next two approximations that simplify series $R_s X_s$ to parallel $R_p X_p$ conversions. From Equation 19.30,

$$R_p = \frac{R_{\text{ind}}^2 + X_L^2}{R_{\text{ind}}}$$
$$= R_{\text{ind}} + \frac{X_L^2}{R_{\text{ind}}} \cdot \frac{R_{\text{ind}}}{R_{\text{ind}}}$$
$$= R_{\text{ind}} + \frac{X_L^2 R_{\text{ind}}}{R_{\text{ind}}^2}$$
$$= R_{\text{ind}}(1 + Q_s^2) \tag{19.41}$$

For $Q_s > 10$,

$$R_p \approx Q_s^2 R_{\text{ind}} \tag{19.42}$$

From Equation 19.31,

$$X_p = \frac{R_{ind}^2 + X_L^2}{X_L}$$

$$= \frac{R_{ind}^2}{X_L} + \frac{X_L^2}{X_L}$$

$$= \frac{R_{ind}^2}{X_L} + X_L$$

$$= \frac{X_L}{X_L}\frac{R_{ind}^2}{X_L} + X_L = \frac{R_{ind}^2 X_L}{X_L^2} + X_L$$

$$= X_L\left(1 + \frac{1}{Q_s^2}\right) \qquad (19.43)$$

For $Q_s > 10$,

$$X_p \approx X_L \qquad (19.44)$$

When other resistances are connected across a tank circuit, additional losses occur that reduce Q. For example, Figure 19.12, part (a), shows a circuit shunted by R_{int} and R_1. A new formula must be derived that includes their effects on Q.

The series RL branch is converted to its parallel equivalent, giving us the new configuration in Figure 19.12, part (b). The shunt resistances are totaled [part (c)] into a single resistance labeled R_t. It is defined

$$R_t = R_p \text{ combined with all resistances in shunt with it} \qquad (19.45)$$

R_t is the resistance to use when calculating the Q of circuit c. This Q must not be called Q_s or Q_p, however, because it involves components outside the actual resonant circuit. We will label it Q_t. Substitute R_t for R_p in Equation 19.39:

$$Q_t = \frac{R_t}{X_p} \qquad (19.46)$$

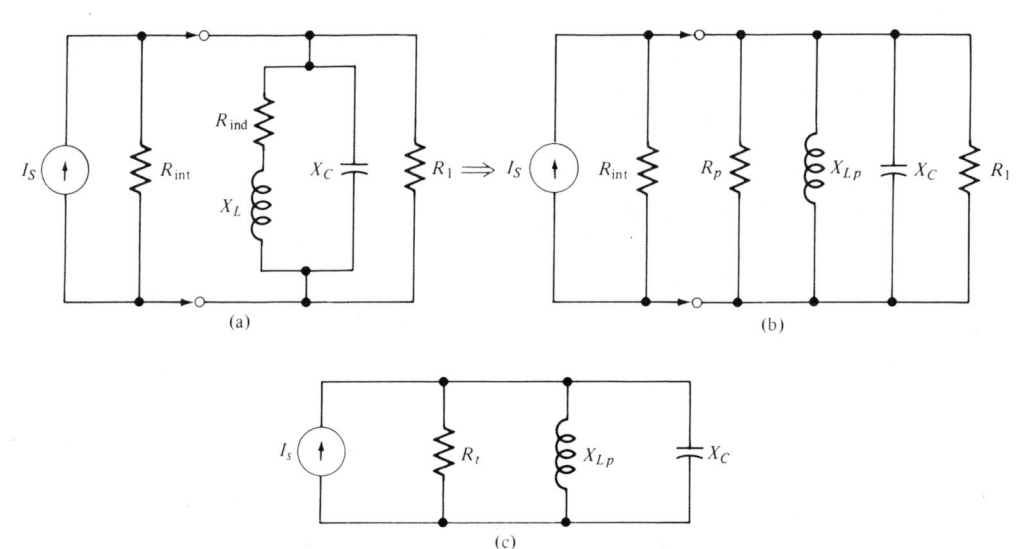

where $X_p = $ either X_{L_p} or X_C.

FIGURE 19.12 Effects of External Resistances on Parallel Circuit Q.

19.6 Q OF THE PARALLEL-RESONANT CIRCUIT

EXAMPLE 19.8

The parallel tuned circuit of Example 19.7 is redrawn. Perform the following steps.

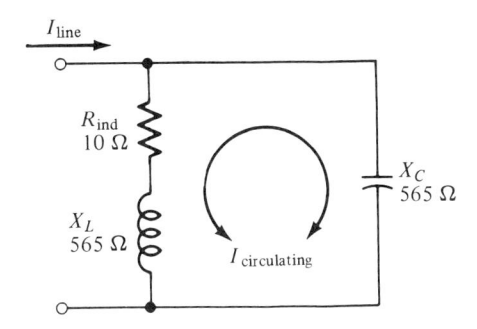

a. Calculate Q_p.

Since there are no additional shunt resistances,

$$Q_p = Q_s$$
$$= \frac{X_L}{R_{ind}} = \frac{565}{10}$$
$$= 56.5$$

b. Convert the tank circuit to its parallel equivalent.

$Q_s > 10$; therefore, approximations hold.

$$R_p \approx Q_s^2 R_{ind}$$
$$= (56.5)^2 10$$
$$= 31.9 \text{ k}$$
$$X_{L_p} \approx X_L$$
$$= 565 \text{ }\Omega$$

Draw the circuit:

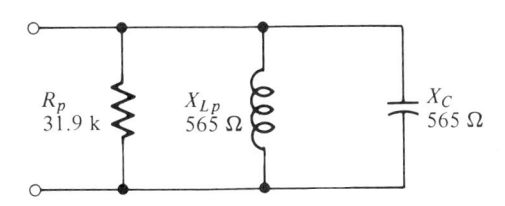

c. A current source with $R_{int} = 100$ k drives this circuit. What is the new Q?

The circuit is now

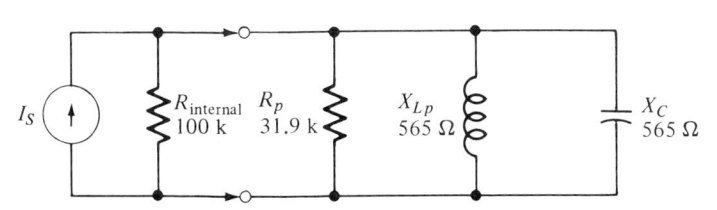

Source resistance shunts R_p; therefore, it affects Q_p:

$$R_t = R_p \| R_{int}$$
$$= 24.2 \text{ k}$$
$$Q_t = \frac{R_t}{X_p}$$
$$= \frac{24.2 \times 10^3}{565}$$
$$= 42.8 \qquad \blacksquare$$

Since tank circuit impedance is maximum at resonance, it follows that line current, I_{line}, should be minimum. At the same time, however, the current carrying the energy oscillating between reactive fields is maximum. This is called *circulating current* and labeled I_{circ} (the currents are seen in the circuit of Example 19.8). Let us derive an equation describing the relation between I_{line} and I_{circ}:

$$I_{circ} = I_{cap}$$
$$= \frac{E_s}{X_C}$$
$$= I_{line}(Z \text{ at resonance})\omega C$$

Substitute Z at resonance from Equation 19.38:

$$I_{circ} = \frac{I_{line} L \omega C}{R_{ind} C} = \frac{I_{line} \omega L}{R_{ind}}$$

But

$$Q_s = \frac{\omega L}{R_{ind}}$$

Therefore,

$$I_{circ} = Q_s I_{line} \qquad (19.47)$$

EXAMPLE 19.9

I_{cap} for the tank of Example 19.7 is actually the circuit's circulating current. It was calculated to be 177 mA by Ohm's law in part d of the example. Verify the value using Equation 19.47.

$$I_{circ} = Q_s I_{line}$$
$$Q_s = \frac{X_s}{R_{ind}} = \frac{565}{10}$$
$$= 56.5$$
$$I_{line} = 3.1 \text{ mA from Example 19.7, part c}$$
$$Q_s I_{line} = 56.5(3.1 \text{ mA})$$
$$= 177 \text{ mA}$$

Results are the same as in Example 19.7, part d. $\qquad \blacksquare$

19.7 BANDWIDTH OF THE PARALLEL-RESONANT CIRCUIT

Current multiplication in parallel RLC circuits was encountered in earlier chapters. Note its similarity to voltage multiplication in series-resonant circuits. It, too, causes component failures, but from overheating due to high currents.

Like its series counterpart, the parallel-resonant circuit draws peak power at resonance and has half-power frequencies above and below f_p. The formulas for these frequencies, derived using techniques similar to those for series circuits, are

$$\omega_2 = \frac{1}{2R_tC} + \sqrt{\left(\frac{1}{2R_tC}\right)^2 + \frac{1}{LC}} \tag{19.48}$$

Therefore,

$$f_2 = \frac{1}{2\pi}\left[\frac{1}{2R_tC} + \sqrt{\left(\frac{1}{2R_tC}\right)^2 + \frac{1}{LC}}\right] \tag{19.49}$$

and

$$\omega_1 = \frac{-1}{2R_tC} + \sqrt{\left(\frac{1}{2R_tC}\right)^2 + \frac{1}{LC}} \tag{19.50}$$

Therefore,

$$f_1 = \frac{1}{2\pi}\left[\frac{-1}{2R_tC} + \sqrt{\left(\frac{1}{2R_tC}\right)^2 + \frac{1}{LC}}\right] \tag{19.51}$$

For cases when $Q_t > 10$, the half-power frequencies are approximated by $f_p \pm \text{BW}/2$. As before with series-resonant circuits,

$$\text{BW} = f_2 - f_1$$

Substract Equation 19.51 from 19.49:

$$\text{BW} = \frac{1}{2\pi R_tC} \tag{19.52}$$

Note that high R_t results in narrow bandwidth.

Multiply numerator and denominator of Equation 19.52 by f_p:

$$\text{BW} = \frac{f_p}{f_p 2\pi R_tC}$$
$$= \frac{f_p X_C}{R_t}$$
$$= \frac{f_p}{Q_t} \tag{19.53}$$

As in the series case, high Q causes narrow bandwidth, low Q causes broad bandwidth.

Remember: Bandwidth is affected by total circuit resistance, not just R_p. Any source, load, and/or bandwidth adjusting resistances must be combined with R_p to get R_t before computing f_2, f_1, and BW.

EXAMPLE 19.10

For the following circuit, find the indicated quantities.

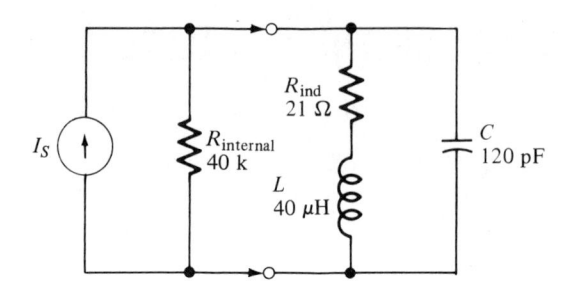

a. Find f_p.

The quantity $R_{ind}^2 C/L$ in the resonance formula is $\ll 1$ and can be neglected. Therefore,

$$f_p \approx \frac{1}{2\pi\sqrt{LC}}$$

$$= \frac{1}{2\pi[(40 \times 10^6)(120 \times 10^{-12})]^{1/2}}$$

$$= 2.3 \text{ MHz}$$

b. Find Q_s.

$$Q_s = \frac{X_L}{R_{ind}} = \frac{2\pi f L}{R}$$

$$= \frac{2\pi(2.3 \times 10^6)(40 \times 10^{-6})}{21}$$

$$= 27.5$$

$Q_s > 10$; therefore, approximations may be used.

c. Find R_p.

$$R_p \approx Q_s^2 R_{ind}$$
$$= (27.5)^2 21$$
$$= 15.9 \text{ k}$$

d. Find BW.

R_t must be used in BW calculations.

$$R_t = R_{int} \| R_p$$
$$= 40 \text{ k} \| 15.9 \text{ k}$$
$$= 11.4 \text{ k}$$

$$BW = \frac{1}{2\pi R_t C} \quad \text{by Equation 19.52}$$

$$= \frac{1}{2\pi(11.4 \times 10^3)(120 \times 10^{-12})}$$

$$= 116.4 \text{ kHz.}$$

19.7 BANDWIDTH OF THE PARALLEL-RESONANT CIRCUIT 607

e. Find the approximate half-power frequencies.

$$f_2 = f_p + \frac{BW}{2}$$

$$= 2.3 \times 10^6 + \frac{116.4 \times 10^3}{2}$$

$$= 2358 \text{ kHz}$$

$$f_1 = f_p - \frac{BW}{2}$$

$$= 2.3 \times 10^6 - \frac{116.4 \times 10^3}{2}$$

$$= 2242 \text{ kHz}$$

f. Find the shunt resistance needed to widen the BW to 130 kHz.

$$BW = \frac{1}{2\pi R_t C}$$

Solving for R_t:

$$R_t = \frac{1}{2\pi BW C}$$

$$= \frac{1}{2\pi(130 \times 10^3)(120 \times 10^{-12})}$$

$$= 10.2 \text{ k}$$

What R must shunt 11.4 k to produce 10.2 k?
By the product-over-sum,

$$10.2 \text{ k} = \frac{11.4 \text{ kR}}{11.4 \text{ k} + R}$$

Solve for R:

$$11.4 \text{ k} + R = \frac{11.4 \text{ kR}}{10.2 \text{ k}} = 1.12R$$

$$R = \frac{11.4 \text{ k}}{0.12}$$

$$= 95 \text{ k}$$

g. Draw the final circuit:

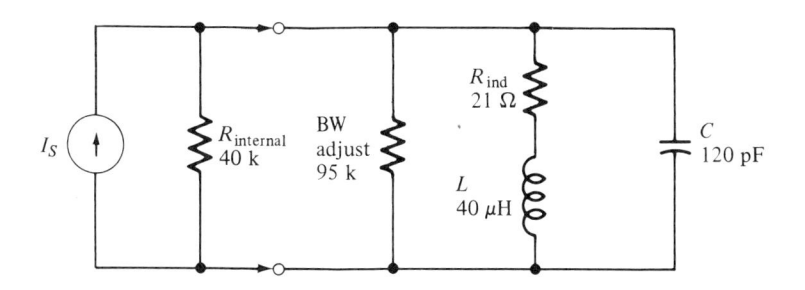

19.8 THE TANK CIRCUIT AS AN AC GENERATOR

Let us close the switch in the circuit of Figure 19.13, part (a), for an instant and follow the resulting energy transfers:

Part (a): instant of switch closure. Energy being stored in both fields as current flows down through L and C.
Part (b): first instant after switch opening. Inductor current continues to flow since it cannot change instantaneously; therefore, the capacitor discharges to supply this current. Energy transfers from C to L.
Part (c): capacitor voltage (and energy) fall to zero, and there is maximum flux around the coil.

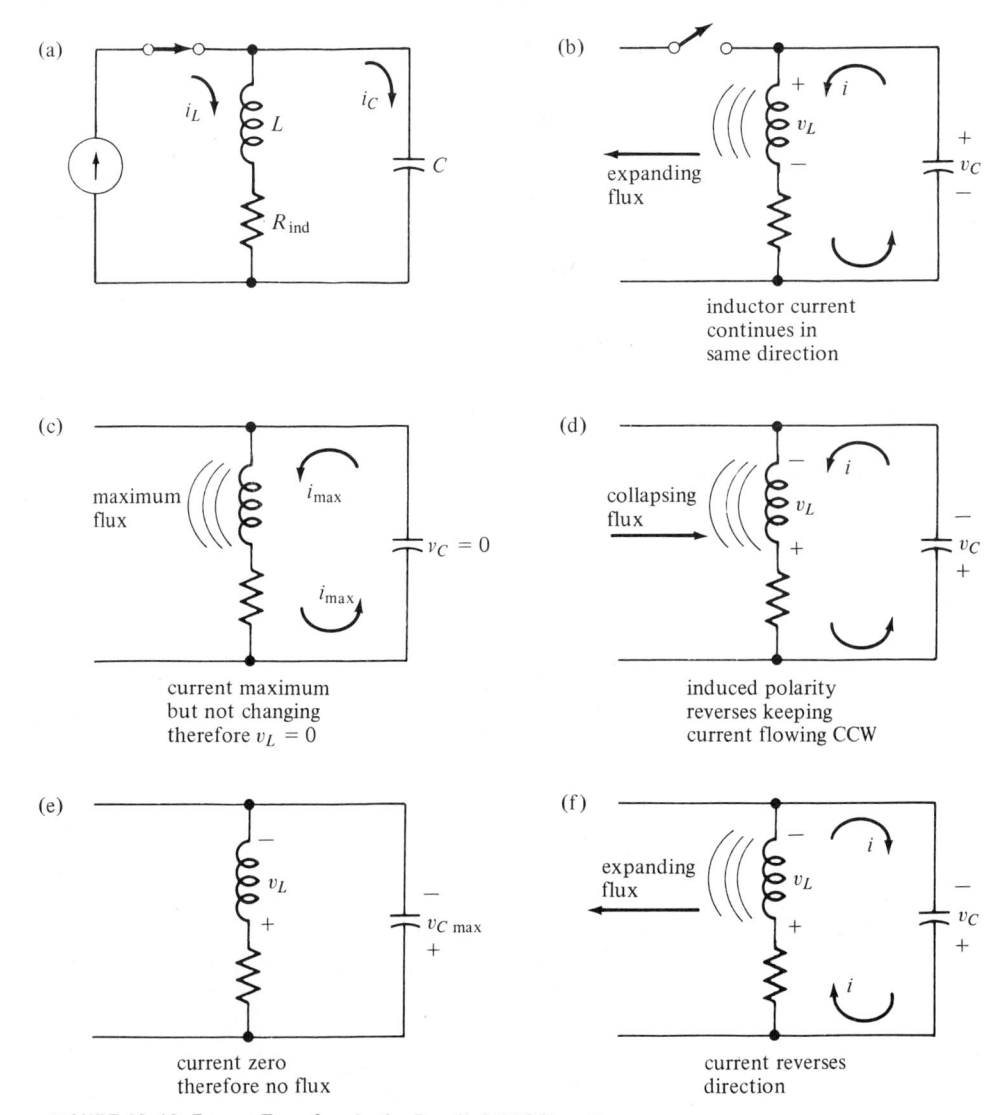

FIGURE 19.13 Energy Transfers in the Parallel RLC Circuit.

Part (d): flux now collapses inward inducing a voltage of opposite polarity across the inductor. This maintains current in the CCW direction transferring energy from L into C but in a direction opposite to that during switch closure. Part (e): flux has collapsed to zero; therefore, inductor energy is zero. All energy has been transferred to the capacitor. Current is zero but voltage is maximum. Part (f): capacitor now discharges sending current in the CW direction. This restores inductor energy but in the opposite direction than before. Eventually, the capacitor is discharged and the flux field collapses again, inducing voltage that causes current to continue. Energy transfers from L to C as the capacitor is charged in the opposite direction and the cycle is ready to be repeated.

Tank circuit oscillations of charge and energy occur at the natural LC resonant frequency at which one reactance releases energy as fast as the other can accept it. They are called *ringing*, because of their similarity to a bell's mechanical vibrations. The coasting action by which they continue after completion of the initial energy input is called the *flywheel effect*.

Energy loss in resistance causes oscillation decay. But replace that energy by pulsing the switch at the right intervals (like pushing the swing) and amplitude remains constant. Such high-speed switching is done by transistors.

Something simple yet very important has happened in our discussion above that should not go unnoticed. The tank circuit received energy from a dc source and converted that energy into sinusoidal form at a frequency determined by L and C. In other words, the circuit converts dc to ac at any frequency we choose.

Parallel-resonant LC combinations determine the operating frequency of radio transmitters and receivers as well as many types of test equipment; there are numerous other applications.

19.9 MATCHING IMPEDANCE WITH RESONANT CIRCUITS: THE *L* NETWORK

By the maximum power transfer theorem, we know that a load receives maximum power from a source when $Z_{\text{load}} = Z^*_{\text{source internal}}$. A source-load system operating in this state is said to be matched.

If the impedances cannot be changed (usually the case), a special circuit inserted between source and load transforms the smaller impedance to look like the larger. The circuit thus ensures matched conditions and is called an *impedance matcher*.

Maximum power transfer applications aren't the only times when it's necessary to change impedance levels, however. Some circuits, notably filters, must drive a specific impedance to work to design specifications. If the input impedance of the filter's load is not what the filter's output wants to 'see," it must be transformed to the required value.

Although there is much to be said about impedance matching, we will limit our discussion to the *L network*, which is based on principles discussed earlier in this chapter. For convenience, inductor resistance will be considered zero.

Equation 19.30, $R_p = (X_s^2 + R_s^2)/R_s$, converts the resistance in a series $R_s X_s$ circuit to its equivalent in a parallel $R_p X_p$ connection. The equation can be rewritten

$$R_p = R_s(1 + Q_s^2) \tag{19.54}$$

(see the derivation of Equation 19.41). Note how the equation multiplies R_s by $(1 + Q_s^2)$ to obtain a higher R_p. That multiplication is the basis of the impedance transforming capability of the L network.

Solving Equation 19.54 for Q_s, we get

$$Q_s = \sqrt{\frac{R_p}{R_s} - 1} \qquad (19.55)$$

This equation, together with the Q relations, $Q_s = X_s/R_s$ and $Q_p = R_p/X_p$, are all that is needed to design L networks.

EXAMPLE 19.11

Design an L network to match a 50-Ω load to a 2500-Ω source.

Connect a reactance in series with the 50-Ω load to form a series $R_s X_s$ circuit. How much X_s?
Substitute R_p and R_s in Equation 19.55:

$$\begin{aligned} Q_s &= \sqrt{\frac{R_p}{R_s} - 1} \\ &= \sqrt{\frac{2500}{50} - 1} \\ &= \sqrt{49} \\ &= 7 \\ Q_s &= \frac{X_s}{R_s} \end{aligned}$$

Solve for X_s:

$$\begin{aligned} X_s &= Q_s R_s \\ &= 7(50) \\ &= 350 \ \Omega \end{aligned}$$

The reactance may be of either type. Let it be inductive. Our $R_s X_s$ circuit is now

$$\begin{array}{c} X_S \\ j350 \ \Omega \end{array} \qquad \begin{array}{c} R_S \\ R_{\text{load}} \\ 50 \ \Omega \end{array}$$

This circuit has an impedance of $50 + j350 \ \Omega$. It also has a parallel $R_p X_p$ equivalent. We know that $R_p = 2500 \ \Omega$. Let us solve for X_p.

Since the series and parallel circuits are equivalent,

$$Q_s = Q_p$$

and

$$Q_p = \frac{R_p}{X_p}$$

19.9 MATCHING IMPEDANCE WITH RESONANT CIRCUITS: THE L NETWORK

Solve for X_p:

$$X_p = \frac{R_p}{Q_p}$$
$$= \frac{2500}{7}$$
$$= 357 \, \Omega$$

The equivalent parallel circuit is

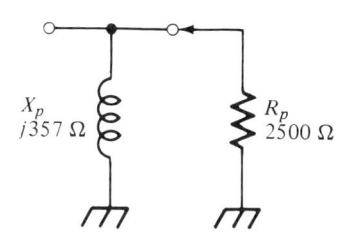

Its impedance is also $50 + j350 \, \Omega$ (as an exercise, prove this using POS).

Note that the 50-Ω load is now the desired 2500 Ω in the parallel circuit. However, it has an undesired X_p in shunt. This is removed by tuning it out with an equal and opposite reactance, namely, an X_C of $-j357 \, \Omega$. Only R_p remains and it matches the source impedance. The final design is

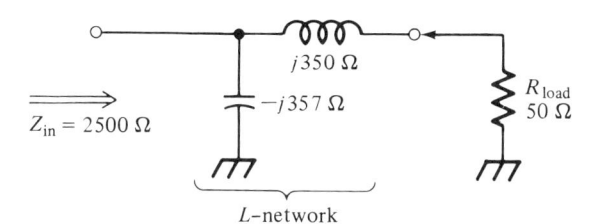

It is important to understand that the $R_s X_s$ circuit only acts like 2500 Ω in shunt with $j357 \, \Omega$; the $R_p X_p$ circuit is never actually built. The network gets it names from the L-shaped X_C–X_L connection. Specific inductive and capacitive values are determined for any frequency with the usual X_C or X_L equation. ■

Some final comments:

- Because the smaller impedance is multiplied, it must be the one in series with X_s.
- Either X_C or X_L may be used for X_s, although other circuit considerations may require one or the other. For example, X_L would be necessary if a dc path between input and load was required.
- A particular network only matches at its design frequency.

We have seen how a series-to-parallel RX conversion multiplies R_s by $(1 + Q_s^2)$ to obtain a higher R_p. It follows that a parallel-to-series conversion divides R_p by the same quantity to obtain a lower R_s.

EXAMPLE 19.12

Transform a 5100-Ω load to match a 300-Ω source at 1.8 MHz. Dc must be blocked from the load.

Since the 5100 Ω is to be stepped down, it must be R_p; the 300-Ω impedance is R_s. X_s must be capacitive to block dc from the load. X_p is therefore inductive. By Equation 19.55,

$$Q_p = \sqrt{\frac{R_p}{R_s} - 1}$$

$$= \sqrt{\frac{5100}{300} - 1}$$

$$= 4$$

$$X_p = \frac{R_p}{Q_p}$$

$$= \frac{5100}{4}$$

$$= 1275 \ \Omega$$

Solve for the inductance having this reactance at 1.8 MHz:

$$L = \frac{X_L}{2\pi f}$$

$$= \frac{1275}{2\pi(1.8 \times 10^6)}$$

$$= 113 \ \mu\text{H}$$

When this is connected across the 5100 Ω R_p, we have

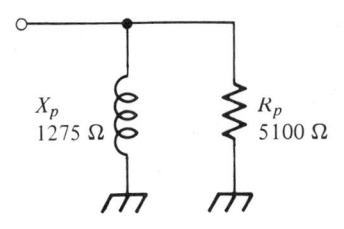

X_p 1275 Ω || R_p 5100 Ω

This has a series equivalent of

X_S 1200 Ω — R_S 300 Ω

19.9 MATCHING IMPEDANCE WITH RESONANT CIRCUITS: THE L NETWORK

in which

$$X_s = Q_s R_s$$
$$= 4(300)$$
$$= 1200 \ \Omega$$

X_s is tuned out by a series-connected X_C of equal value. Solve for C at 1.8 MHz:

$$C = \frac{1}{2\pi f X_C} = \frac{1}{2\pi(1.8 \times 10^6)(1200)}$$
$$= 73.7 \text{ pF}$$

The final circuit is

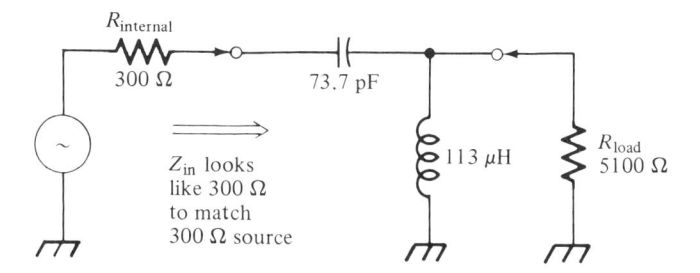

Source and load impedances are usually complex rather than the pure resistances of our examples. Let us consider one of many possibilities, a load of resistance and shunt inductive reactance. The approach is to tune out the reactance, then proceed with a resistance-to-resistance design as above.

EXAMPLE 19.13

A circuit's input impedance is 600-Ω shunted by 5.4 μH. Design an L-network to match this load to a 72-Ω resistive source at 14.1 MHz. Dc must pass between the two.

Solve for the capacitance needed to tune out the 5.4 μH:

$$C = \frac{1}{(2\pi f)^2 L}$$
$$= \frac{1}{[2\pi(14.1 \times 10^6)]^2 (5.4 \times 10^{-6})}$$
$$= 23.6 \text{ pF}$$

With the load's reactance canceled, proceed with the design:

$$Q_s = \sqrt{\frac{R_p}{R_s} - 1}$$
$$= \sqrt{\frac{600}{72} - 1}$$
$$= 2.7$$

The 600 Ω must be stepped down; therefore,

$$X_p = \frac{R_p}{Q_p} = \frac{600}{2.7}$$
$$= 222 \ \Omega$$

This must be capacitive, so X_s can be inductive to pass dc. Therefore,

$$C_p = \frac{1}{2\pi f X_p}$$
$$= \frac{1}{2\pi(14.1 \times 10^6)222}$$
$$= 51 \text{ pF}$$
$$X_s = Q_s R_s$$
$$= 2.7(72)$$
$$= 195 \ \Omega$$

Solve for L_s:

$$L_s = \frac{X_s}{2\pi f}$$
$$= \frac{195}{2\pi(14.1 \times 10^6)}$$
$$= 2.2 \ \mu\text{H}$$

C_p is in parallel with the 23.6 pF used to cancel load reactance. Combine the two:

$$C_{\text{total}} = (51 + 23.6) \text{ pF} = 74.6 \text{ pF}$$

The final design is

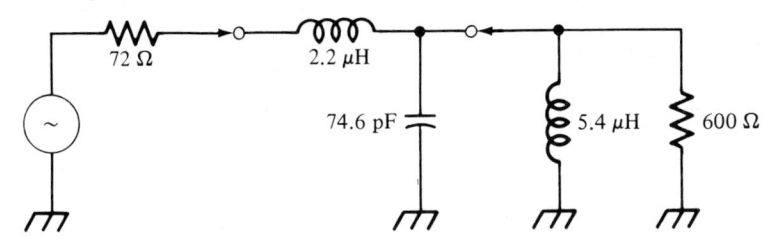

19.10 RESONANT FILTER CIRCUITS

The *RC* and *RL* filters studied in Section 16.5 were either low pass or high pass. They rejected, or *attenuated*, all signals on one side of their half-power frequency and passed all those on the other. Filters containing resonant circuits, on the other hand, pass or reject a relatively narrow band of frequencies and are called *band-pass* or *band-reject* filters, respectively.

The filter's pass or reject region lies between the half-power frequencies of its tuned circuit. It is therefore centered on an f_s or f_p, which can be any frequency. The circuits are of limited application because the attenuation of signals close to f_s/f_p changes enormously depending on skirt steepness. This, in turn, is a function of circuit Q.

Let us consider each type.

19.10 RESONANT FILTER CIRCUITS

Band-Pass Filters

Band-pass frequency characteristics are seen in Figure 19.14, part (a). Circuit (b) in the figure has these characteristics because of its series-resonant RLC branch in series with the signal path. For $f < f_1$, $X_c \gg R_{out}$. Similarly, for $f > f_2$, $X_L \gg R_{out}$. In these regions, therefore, the bulk of E_{in} is dropped across a reactance and the output is heavily attenuated. At resonance, however, the branch looks like R_{ind} and the circuit reduces to an ordinary voltage divider. v_{out} maximum is then determined by the usual voltage-division formula:

$$v_{out} = \frac{e_{in} R_{out} \| R_{load}}{R_{ind} + R_{out} \| R_{load}} \tag{19.56}$$

Of course, if $R_{load} \gg R_{out}$, it can be disregarded. If desired, the half-power frequencies may be determined with the help of Equations 19.21 and 19.23, or, if $Q > 10$, by approximations 19.28 and 19.29.

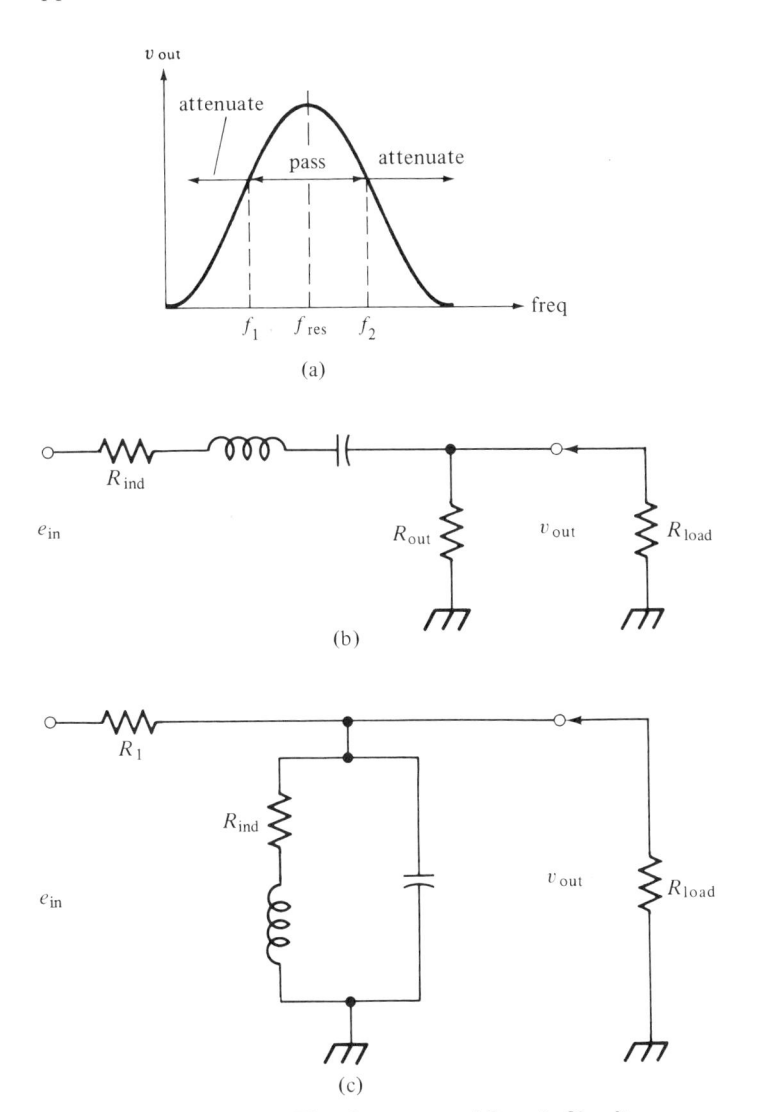

FIGURE 19.14 Band-Pass Filter Response and Sample Circuits.

EXAMPLE 19.14

Perform the indicated operations for this circuit:

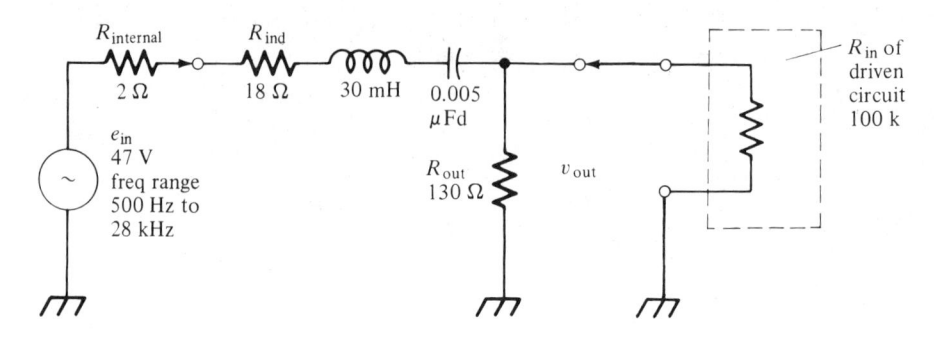

[Circuit diagram: $R_{internal}$ = 2 Ω, R_{ind} = 18 Ω, L = 30 mH, C = 0.005 μFd, R_{out} = 130 Ω, e_{in} = 47 V, freq range 500 Hz to 28 kHz; R_{in} of driven circuit = 100 k; v_{out} across R_{out}.]

a. Determine the upper and lower pass-band limits if the pass region is centered on 13 kHz.

Q is needed to find the half-power frequencies. Total circuit resistance must be used in this calculation; however, R_{in} of driven circuit $\gg R_{out}$ and can be disregarded.

$$Q = \frac{2\pi f L}{R_{total}}$$

$$= \frac{2\pi(13 \times 10^3)(30 \times 10^{-3})}{2 + 18 + 130}$$

$$= 16.3$$

$$BW = \frac{f_{center}}{Q}$$

$$= \frac{13{,}000}{16.3}$$

$$= 800 \text{ Hz}$$

$Q > 10$, therefore, half-power frequencies by approximation are

$$f_2 = 13 \text{ kHz} + \frac{800 \text{ Hz}}{2}$$

$$= 13.4 \text{ kHz}$$

$$f_1 = 13 \text{ kHz} - \frac{800 \text{ Hz}}{2}$$

$$= 12.6 \text{ kHz}$$

b. Determine the output voltage at center and limit frequencies.

$$v_{out} \text{ at } f_s \approx \frac{e_{in} R_{out}}{R_{int} + R_{ind} + R_{out}}$$

$$= \frac{47(130)}{2 + 18 + 130}$$

$$= 40.7 \text{ V}$$

$$v_{out} \text{ at limit frequencies} = 0.707 \, v_{out_{max}}$$

$$= 0.707(40.7) = 28.7 \text{ V}$$

19.10 RESONANT FILTER CIRCUITS

c. Determine v_{out} at 0.1 and 10 times f_{center}.

$$0.1 f_{center} = 1.3 \text{ kHz}$$
$$X_L = 2\pi(1.3 \times 10^3)(30 \times 10^{-3})$$
$$= j245 \text{ }\Omega$$
$$X_C = \frac{1}{2\pi(1.3 \times 10^3)(5 \times 10^{-9})}$$
$$= -j24.5 \text{ k}$$

Then $Z \approx X_C$.

$$v_{out} \approx \frac{e_{in} R_{out}}{X_C}$$
$$= \frac{47(130)}{24.5 \times 10^3 \underline{/-90°}}$$
$$= 249 \underline{/-90°} \text{ mV}$$

At $10 f_{center}$, X_L dominates and $v_{out} \approx 249 \underline{/90°}$ mV. Note the severe attenuation outside the pass-band.

d. Draw the response curve.

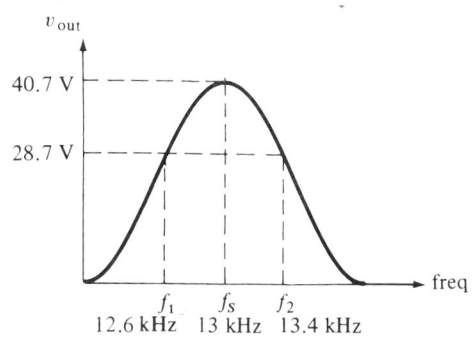

Figure 19.14, part (c), illustrates how band-pass response is obtained by shunting the filter's output with a parallel-resonant circuit. Signals above and below f_p are shorted by the low reactances of C and L, respectively. At resonance, however, the tank is a high impedance. If designed so that $Q_s^2 R_{ind} \gg R_1$ and if $R_{load} \gg Q_s^2 R_{ind}$, then $v_{out} \approx e_{in}$ for frequencies within the pass-band.

Band-Reject Filters

Band-reject frequency characteristics are seen in Figure 19.15, part (a). Circuit (b) in the figure obtains these characteristics by shorting the reject signals with a series-resonant *RLC* branch across its output. The circuit is designed to make $R_1 \ll X_C$ for $f < f_1$, and $R_1 \ll X_L$ for $f > f_2$. This places the bulk of e_{in} across a reactance and, as a result, at the output, for frequencies outside the resonant bandwidth. For best rejection, R_{ind} should be much less than R_1. R_{load} should be much greater than both R_{ind} and R_1 at all frequencies.

Figure 19.15, circuit (c), includes a tank circuit in series with the signal path. Tank reactances are low for frequencies outside the resonant bandwidth and $v_{out} \approx e_{in}$. At resonance, however, $Z_{tank} \gg R_{out}$, the output drops and input signals are heavily

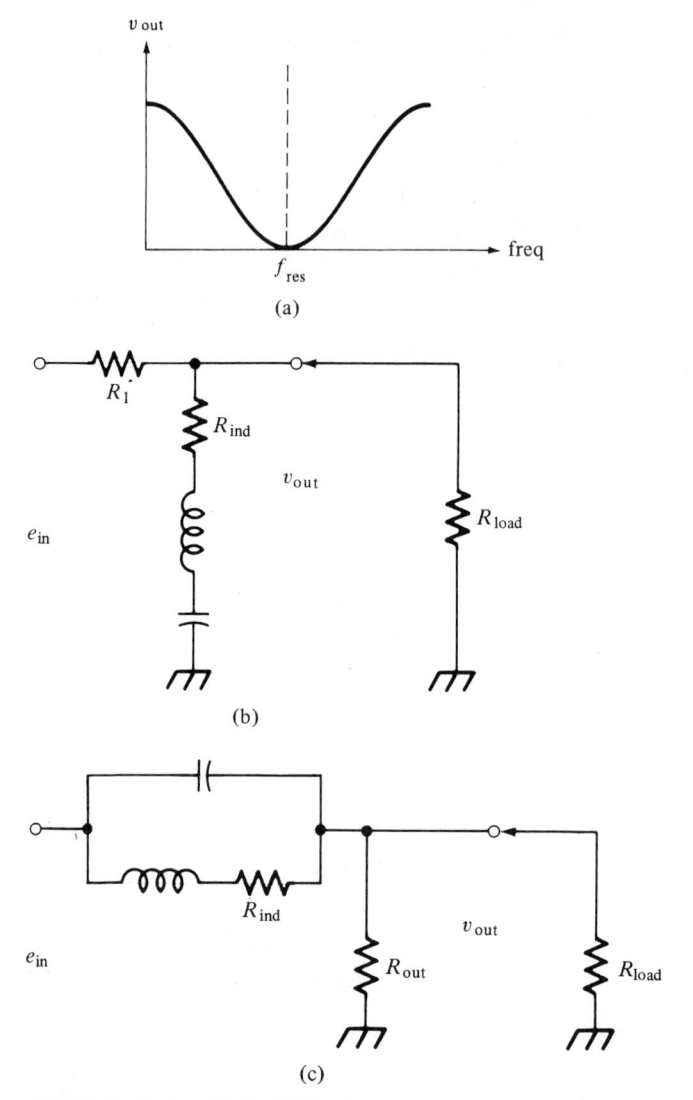

FIGURE 19.15 Band-Reject Filter Response.

attenuated. For best results, R_{load} must be very much greater than R_{out} and the circuit should be driven by a high-impedance source.

Double-Tuned Filters

Figure 19.16 illustrates double-tuned circuits, each of which has both band-pass and band-reject characteristics. In circuit (a), the antiresonant combination of C and L_p forms a reject filter by presenting a high impedance in series with R_{load}. At frequencies above f_p, the $C-L_p$ combination becomes capacitive and eventually resonates with L_s. However, this is series resonance. Therefore, the entire circuit has minimum impedance, causing maximum voltage across R_{load}; the circuit has become band-pass. By changing L_s to C_s, the pass region is moved below f_p. This happens because the tank is inductive at frequencies below f_p. The next example describes the design of this type.

19.10 RESONANT FILTER CIRCUITS

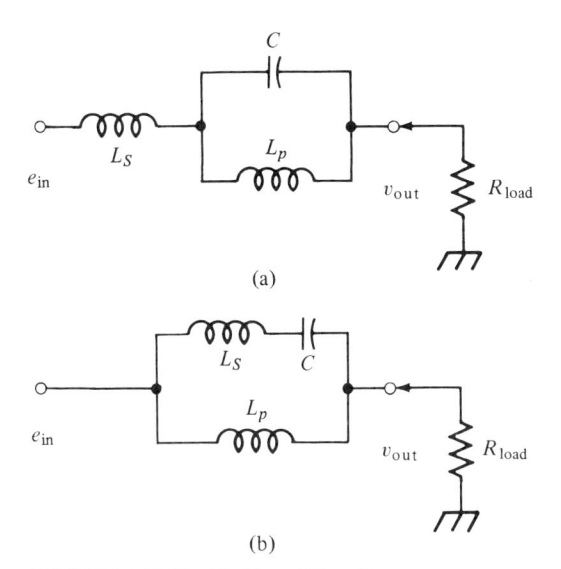

FIGURE 19.16 Double-Tuned Circuits.

EXAMPLE 19.15

Design a double-tuned filter of the type in Figure 19.16, part (a), but with one inductor and two capacitors. It is to have a reject region centered at 185 kHz and a pass region at 70 kHz. If $L = 74.1$ μH, find the values of the capacitors.

L and C_p are antiresonant at 185 kHz. Therefore,

$$C_p = \frac{1}{(2\pi f)^2 L}$$

$$= \frac{1}{[2\pi(185 \times 10^3)]^2 (74.1 \times 10^{-6})}$$

$$= 0.01 \; \mu F$$

L and C_p are in parallel. We must know their X_{total} at 70 kHz in order to find C_s. This is done by adding susceptances for 70 kHz and inverting the total:

$$B_C = j2\pi f C$$
$$= j2\pi(7 \times 10^4)(10^{-8})$$
$$= j44 \times 10^{-4} \; S$$

$$B_L = -j\frac{1}{2\pi f L}$$
$$= -j\frac{1}{2\pi(7 \times 10^4)(74.1 \times 10^{-6})}$$
$$= -j307 \times 10^{-4} \; S$$

$$B_{total} = B_C - B_L$$
$$= (j44 - j307) \times 10^{-4}$$
$$= -j263 \times 10^{-4} \; S$$

By the negative sign we know this to be an inductive susceptance.

$$X_{total} = \frac{1}{B_{total}}$$

$$= \frac{1}{-j263 \times 10^{-4}}$$

$$= j38 \; \Omega$$

C_s must tune out this X_L. Solve for C_s:

$$C_s = \frac{1}{2\pi f X_C}$$

$$= \frac{1}{2\pi(7 \times 10^4)(38)}$$

$$= 0.06 \; \mu F$$

Draw the final circuit:

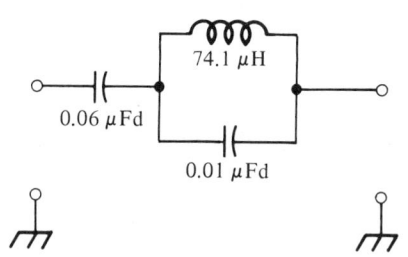

[Circuit diagram: 0.06 μFd capacitor in series with a parallel combination of 74.1 μH inductor and 0.01 μFd capacitor]

■

Figure 19.16, circuit (b), operates in much the same manner as circuit (a). Its passband is determined by series resonance between C and L_s. At frequencies below the pass band, the series circuit looks capacitive. This effective capacitance parallel-resonates with L_p at one of these lower frequencies to establish the reject band. By substituting a capacitor for inductor, L_p, the reject band may be placed above the pass band. This is due to the inductive character of the series RLC branch at frequencies above resonance.

19.11 OTHER RESONANT DEVICES

Certain crystals, notably quartz, have a special quality called the *piezoelectric* (pronounced *pee-zo* or *pee-ay-zo electric*) effect, which causes the crystal to vibrate when a voltage is placed across it. Vibration frequency is determined by crystal thickness and other factors.

Vibrating piezoelectric crystals have the electrical characteristics of the high-Q, double-tuned LC-circuit illustrated, together with the crystal symbol, in figure 19.17. C_o in that circuit is capacitance external to the quartz, that is, from the holder and stray capacitances. C_i, L, and R are internal characteristics of the mineral slab itself. As might be expected, L_i and C_i series resonate at one frequency, while C_o and L_i are antiresonant at another. Crystals are used in advanced filters and for setting frequencies in radio transmitters and receivers.

Another piezoelectric device having the same equivalent circuit as a quartz crystal, but made of zirconate titanate, is the *ceramic resonator*. Shown in Figure 19.18, these

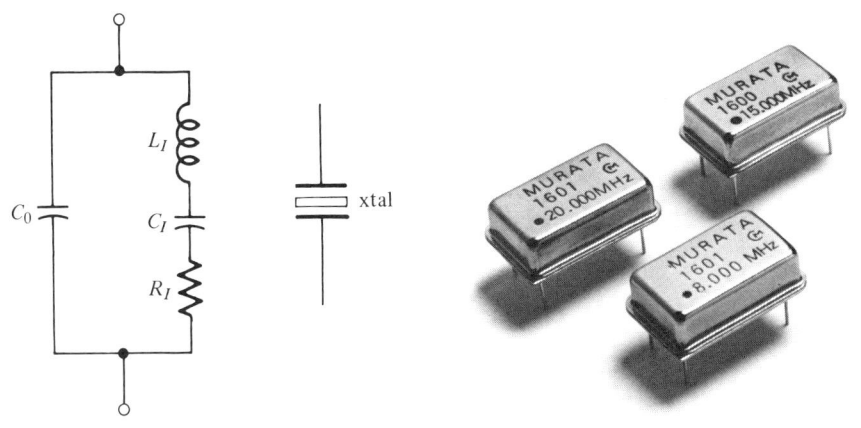

FIGURE 19.17 The Quartz Crystal, Equivalent Diagram, and Schematic Symbol. (Courtesy Murata Erie North America, Inc.)

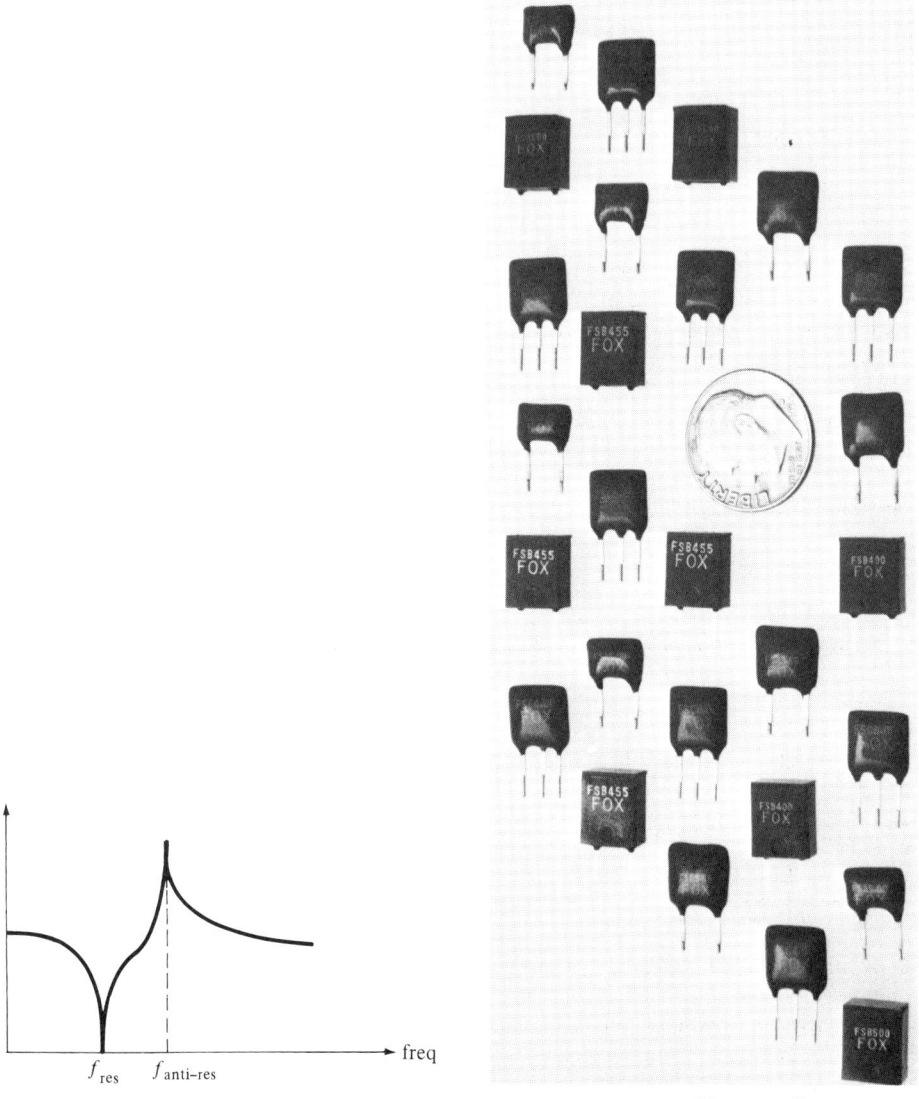

FIGURE 19.18 Ceramic Resonators and Typical Frequency Response. (Courtesy Fox Electronics.)

small units are excellent for PC board applications and are rapidly replacing LC tuned circuits in transistor radios and similar small-scale electronics. Typical frequency characteristics have been included in the figure.

EXERCISES

Section 19.1

1. Fill in the blanks.

	f(Hz)	L(H)	C(F)
a.		500 μH	20 pF
b.	48 kHz		0.001 μF
c.	6.3 MHz	4.5 μH	
d.		750 μH	0.4 μF
e.	577 kHz		2000 pF
f.	35 Hz	1.4 H	

2. A series LC circuit has maximum current at $f = 1$ kHz. If $C = 0.05\ \mu F$, what is the inductance?

3. For each of the following conditions, state whether the frequency of the signal applied to a series RLC circuit is: a. less than f_s; b. equal to f_s; or c. greater than f_s.
1. $|X_C| > |X_L|$ 2. $\theta +$ 3. I_{max} 4. $|X_L| > |X_C|$
5. $|X_L| = |X_C|$ 6. $|Z_t| \approx X_L$ 7. PF lagging   8. $\phi +$

4. A variable capacitor has minimum and maximum capacitances of 35 and 365 pF, respectively. Will it tune the entire AM broadcast band from 540 to 1600 kHz with a 250-μH coil?

Section 19.2 through 19.4

5. Fill in the blanks.

	f	L or C	Q	R
a.	1.6 MHz	500 μH	61.3	
b.	48 kHz	11 mH		226
c.	6.3 MHz	L	41.4	4.3
d.		1.4 H	2.3	134
e.	577 kHz	2000 pF		0.06
f.	9.2 kHz	0.4 μF	14,400	

6. By supercooling LC circuits, resistance decreases so that a nearly ideal oscillating system (one with no losses) is obtained.
a. Such a system would oscillate how long on a single energy input?
b. What is the value of Q in the system?

7. At resonance, a series RLC circuit draws 5 A from a 30-V source. If $X_C = -80\ \Omega$, find Q and V_{cap} at resonance.

EXERCISES

8. Why should high-Q series-resonant circuits be driven by low-impedance sources, that is, sources with low internal resistance?

9. Indicate the reading on a voltmeter connected between each of the following sets of points (circuit is resonant, its $Q = 200$). a. V_{ad}; b. V_{ab}; c. V_{bc}; d. V_{cd}; e. V_{bd}.

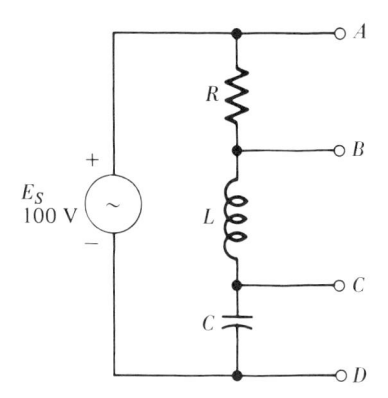

10. A circuit resonates at 500 kHz. It is driven by a zero-output impedance source, $E_{rms} = 70.7$ V, and has an inductance of 100 μH. If circuit resistance is 26.2 Ω, what is the maximum voltage that the capacitor's dielectric must stand?

11. Fill in the blanks.

	Q	R	$L(H)$	$C(F)$
a.		108	15 mH	0.02 μF
b.	20.4		75 μH	20 pF
c.	52.7	163		6500 pF
d.	195	1.3	640 μH	

12. What is the minimum resistance needed in this circuit?

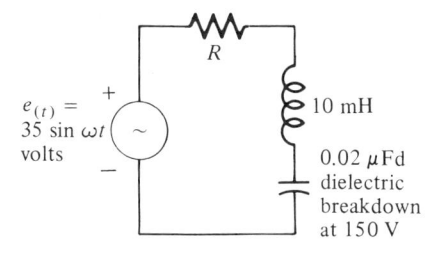

13. Indicate for each of the following whether the statement more likely describes a high-Q or low-Q circuit (use H or L).
a. Steep skirts.
b. Low R.
c. Low LC ratio.
d. Broad response curve.
e. High maximum current.
f. Narrow BW.
g. Half-power frequencies close together.

h. Circuit impedance magnitude changes slowly with changing frequency.
i. $V_C = 3E_s$ at f_s.
j. Narrow pass region.
k. High-energy storage, low-energy loss.

14. Design a series-resonant circuit to pass a 24.5-kHz-wide signal centered on 4050 kHz. On hand is a 28-μH coil of 1.8-Ω resistance and a capacitor variable from 20 to 80 pF.

15. Fill in the blanks (use approximations of Equations 19.28 and 19.29).

	f_s	f_2	f_1	BW(Hz)	Q
a.	1000 kHz				10
b.		14.35 MHz		100 kHz	
c.	237 Hz	238.3 Hz			
d.	71.8 kHz				15.6

16. *Extra Credit.* By letting $\omega = \omega_2$ (see Equation 19.20), prove the statement $|X_{total}| = |X_L - X_C| = R$ at ω_2.

17. Given: an RLC series circuit with $R = 115.2\ \Omega$, $L = 1146\ \mu H$, and $C = 150$ pF.
Find: a. f_s; b. Q_s; c. BW; d. the half-power frequencies by approximation;
e. Repeat part d by exact formula. Compare the results.

18. Repeat Exercise 17 for $R = 10\ \Omega$, $L = 1623\ \mu H$, and $C = 0.4\ \mu F$.

Sections 19.5 through 19.7

19. A 130-μH inductance having a resistance of 20 Ω is shunted by 950 pF.
a. Will the value of the radical in Equation 19.34 have significant effect on f_p of this circuit?
b. Calculate the approximate value of f_p.
c. Determine I_{source} at resonance for $E_{source} = 85$ V.
d. Determine I_{cap} at resonance.

20. Will the value of the radical in Equation 19.34 have significant effect on the f_p of a 15-μH coil having a resistance of 85 Ω and shunted by 394 pF?

21. A 500-μH ($R = 250\ \Omega$) coil parallel resonates with a 20-pF capacitor at 1.6 MHz.
a. What is Q_s?
b. Calculate Z at resonance.
c. Convert the circuit to its parallel equivalent.
d. Compare R_p from part c to Z from part b.
e. A 10-mA current source with a 100-k internal resistance drives the tank. What is I_{line}?
f. Calculate $I_{circulating}$ by Ohm's law and $Q_s I_{line}$. Compare results.
g. What is Q_t for the circuit?

22. Find the L and C in this circuit:

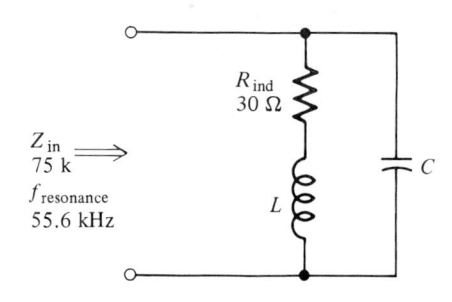

EXERCISES

23. An unwary technician connects an oscilloscope across the tank of Exercise 21 to observe the voltage waveform. The lead to the scope is coaxial cable 3 ft long with a capacitance of 35 pF/ft. What happens?

24. Given: a 0.001-μF capacitor shunted by an 11-mH coil with an $R_{ind} = 73.7$ Ω. Find: a. f_p;   b. Q_s;   c. R_p and X_p;   d. BW. The circuit is then loaded by the 200-k input resistance of the next circuit. Find: e. R_t;   f. Q_t;   g. the new BW.   h. Using the R_p-to-R_s conversion formula, convert R_t to R_s. Then calculate Q_s by X_s/R_s. Compare to value from part f.

25. For this circuit: determine: a. f_p,   b. Q_s,   c. BW,   d. f_2 and f_1.

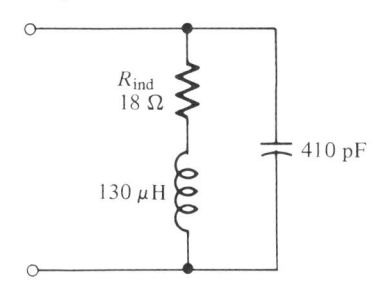

e. What is the minimum internal resistance a current source driving this circuit may have without causing the circuit's BW to exceed 26.6 kHz?

26. Design a tank circuit to resonate at 7.5 MHz with a 100-kHz BW. On hand is a 15-μH coil ($R_{ind} = 5$ Ω) and a variety of other components.

27. Design a parallel tuned circuit to resonate at the standard international distress frequency, 500 kHz, with a 10-kHz BW. On hand is an inductor, variable from 580 to 990 μH ($R_{ind} = 30$ Ω), a 130-pF capacitor, and a variety of resistors.

28. The capacitor and variable inductor from Exercise 27 form a parallel-resonant circuit.
a. Calculate the minimum and maximum frequencies to which the circuit will tune.
b. Calculate the BW at each of these extremes.

Section 19.8

29. Explain how a tank circuit provides an entire sine wave from an energy pulse of much shorter duration.

30. How is lost energy replaced in tank circuits?

Section 19.9

31. Design an L network to match a 600-Ω load to a 50-Ω source at 2.8 MHz. Let X_s be capacitive.

32. Repeat Exercise 31 to match a 1250-Ω load to a 385-Ω source at 750 kHz, X_s inductive.

33. Design an L network to match a 50-Ω load to a 1-k source at 7MHz. Dc must pass between them.

34. Repeat Exercise 33 to block dc between a 5.3-k source and a 150-Ω load at 30 MHz.

35. A load, consisting of 360 Ω shunted by 400 pF, is to be matched to a 50-Ω source at 1.1 MHz. Design an L network that blocks dc.

36. Match a load, consisting of an 80-Ω resistance shunted by 1.8 μH to a 4200-Ω source at 10 MHz. Provide a dc path between source and load.

Section 19.10

37. The circuit of Figure 19.14, part (b), has an X_C of 1375 Ω at its center frequency of 100 kHz. If the BW is 7.6 kHz and $R_{out} = 100$ Ω, find: a. the pass-band limit frequencies;   b. the inductor resistance, R_{ind} (R_{load} is open);   c. L and C;   d. BW if load = 500Ω.

38. The circuit of Figure 19.14, part (b), has an inductive reactance of 1300 Ω at its frequency of maximum v_{out}, 5 MHz. If $Q_s = 12.5$ and $R_{ind} = 4$ Ω, find: a. the limit frequencies;   b. R_{out} (R_{load} is open);   c. the value of R_{load} that sets BW to 375 kHz.

39. Sketch the frequency characteristics of this circuit. Show minimum and maximum V_{out}, f_{center}, and the limit frequencies. Indicate the pass and reject regions.

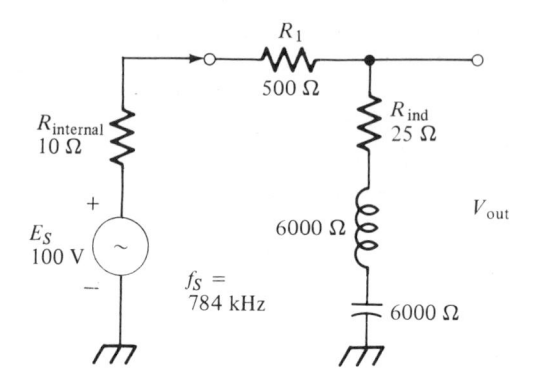

40. R_1 in Exercise 39 is doubled to 1 k. Find the new limit frequencies and BW.
41. Design a double-tuned filter of the type in Figure 19.16, part (a), which has its pass region centered on 100 kHz, its reject region on 20 kHz. The filter includes a 0.1-μF capacitor.
42. Repeat Exercise 41 for a filter with $f_{reject} = 225$ kHz and $f_{pass} = 49.1$ kHz. The circuit includes a 100-μH inductor.
43. Design a double-tuned filter of the type in Figure 19.16, part (b), which has its pass region centered on 380 kHz, its reject region on 100 kHz. The filter includes a 1590-pF capacitor.
44. Repeat Exercise 43 for a filter with $f_{reject} = 520$ kHz and $f_{pass} = 150$ kHz. The circuit includes a 60-μH coil.
45. *Extra Credit.* Derive a formula for the pass frequency of the circuit of Exercise 42. Then test your derivation by using the component values of Exercise 42 in it to see if $f_{pass} = 49.1$ kHz.

CHAPTER HIGHLIGHTS

The resonant frequencies of series *RLC* and high-*Q* (*Q* > 10) parallel *RLC* circuits are not affected by resistance.

High-*Q* circuits have narrow-bandwidth, low-*Q* circuits have broad bandwidth.
To design a tuned circuit, we must know its resonant frequency and either its *Q* or bandwidth.
Upper and lower half-power frequencies are defined as the limits of the tuned response curve.
The flywheel effect causes tank circuits to ring and generate sinusoids.
Transforming impedance levels is an important function of reactive circuits.
Band-pass filters attenuate all frequencies outside their pass band and pass all those within. Band-reject filters pass all frequencies outside their reject band and attenuate all those within.
The piezoelectric effect enables certain minerals to act as high-*Q* resonant circuits.

Formulas to Remember

$$f = \frac{1}{2\pi LC}$$ series and parallel resonant (for $Q > 10$) frequency

$$Q = (2\pi)\frac{W_{stored}}{W_{dissipated}}$$ definition of Q

$$Q_s = \frac{X_s}{R_s} = \frac{1}{R}\sqrt{\frac{L}{C}}$$ Q of series *RLC* circuit

CHAPTER HIGHLIGHTS

$V_C = V_L = Q_s E_s$ — reactive voltage at series resonance

$\text{BW} = f_2 - f_1$ — bandwidth

$\text{BW} = \dfrac{f_s}{Q_s}$ — bandwidth of series RLC circuit

$\text{BW} = \dfrac{f_p}{Q_t}$ — bandwidth of parallel RLC circuit

$Q_t = \dfrac{R_{\text{total}}}{X_p}$ — Q of loaded parallel RLC circuit

$I_{\text{circ}} = Q_p I_{\text{line}}$ — circulating current in resonant parallel RLC circuit

CHAPTER 20

Polyphase Circuits

Bulk electric energy is shipped as ac because ac allows transmission at potentials as high as 700 kV. Such voltages mean low currents for the same power. As a result, lineloss, the I^2R loss in conductor resistance, is significantly lower.

The ac frequency in North America is 60 Hz. It was chosen as a compromise between engineering and economic considerations; as frequency increases, transmission-line inductive reactance increases but equipment costs decrease. Fifty hertz is commonly found in other parts of the world while 400 Hz is used aboard aircraft to cut the weight of inductors, transformers, and generators.

Bulk ac power is carried by high-tension circuits called *three-phase (3-ϕ) transmission systems*. There are economic and engineering reasons for this:

- A three-phase system carries energy more efficiently, and therefore at lower cost, than single-phase circuits.
- Three-phase motors run with less vibration, higher torque, and better starting characteristics.
- Some three-phase equipment has lower cost than comparable single-phase pieces.

We study 3-ϕ systems in this chapter. However, since this book is geared toward electronics, not power transmission (a whole field in itself), we will only be introduced to the subject.

At the completion of this chapter, you should be able to:

- Explain the term "phase" and describe single-, two-, and three-phase transmission systems.
- Describe the 3-ϕ generator in both wye- and delta-connected three- and four-wire forms.
- Analyze 3-ϕ systems using wye- or delta-connected loads, either balanced or unbalanced.

20.1 SINGLE AND TWO-PHASE CIRCUITS

The ac circuits discussed so far have had a single sinusoidal source supplying energy to a load via two conductors. Such circuits are called *two-wire single-phase systems* [see Figure 20.1, part (a)].

Figure 20.1, part (b) shows another single-phase system having three wires and two sources connected in series aiding (these are actually two halves of a transformer winding with the center point grounded). This is the technique used to bring electricity into our homes. It provides two separate 120-V circuits, one between each outer wire and the center or, *neutral*, conductor, plus an additional 240-V supply between outer conductors for high-power appliances such as stoves, water heaters, and so on. The neutral is normally grounded.

The generation of single-phase sinusoids was discussed in Chapter 13. We learned, then, that a coil, rotating in a magnetic field, has a voltage induced in it. We also found that the voltage varied sinusoidally because its instantaneous magnitude was a function of the sine of the angle between the flux lines and the coil's instantaneous velocity vector.

If two coils are mounted on the same axle, two separate voltages, called *phases*, will be generated. Their phase difference will be the same as the angular spacing of the rotating windings. Figure 20.2, part (a), for example, shows two coils spaced 90°. The generator thus becomes a two-phase source supplying two voltages in quadrature. Figure 20.2, parts (b) and (c), show these voltages in sine-wave and phasor diagrams.

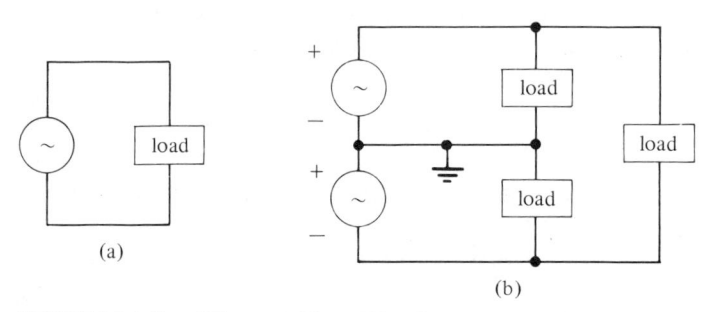

FIGURE 20.1 Two-Wire and Three-Wire Single-Phase Systems.

20.2 THREE-PHASE POWER

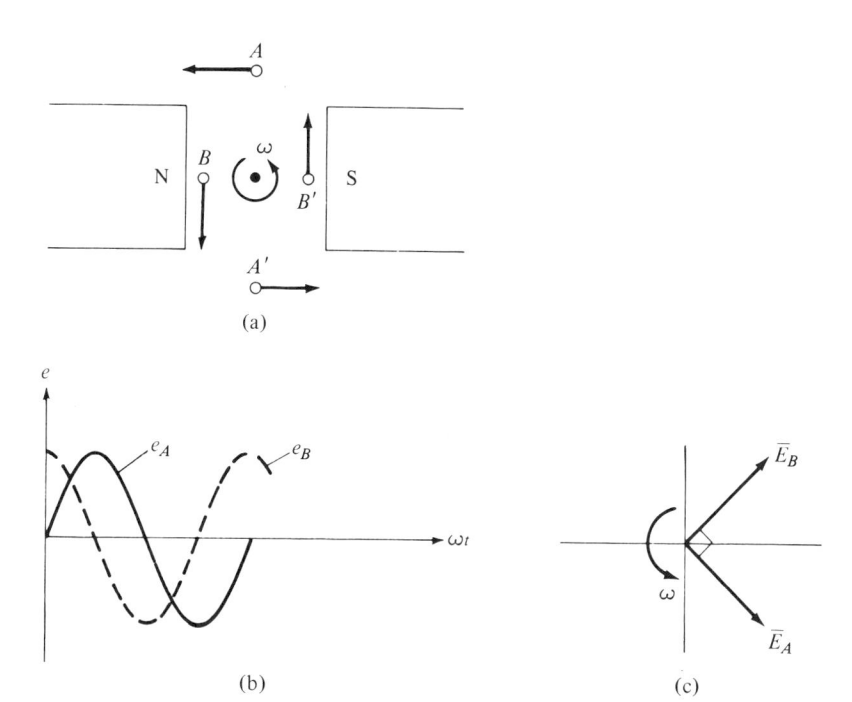

FIGURE 20.2 Two Phase Generator and Its Output.

20.2 THREE-PHASE POWER

Most electric energy is generated in 3-ϕ form by generators having three rotating coils spaced 120°. The voltages induced in these coils are of equal magnitude but spaced 120°. They are shown in a sine-wave diagram in Figure 20.3, part (a), together with the appropriate time-domain equations. Note carefully the angular relations. The equivalent phasor information is found in part (b) of the figure.

The individual coils in a 3-ϕ generator do not feed separate single-phase loads through three independent pairs of conductors. Instead, they are connected within the generator to form a single 3-ϕ system. Figure 20.4 shows the two ways this is done.

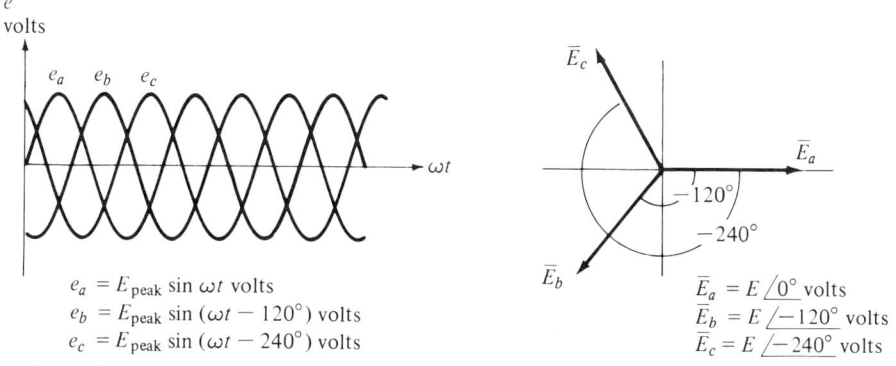

$e_a = E_{\text{peak}} \sin \omega t$ volts
$e_b = E_{\text{peak}} \sin (\omega t - 120°)$ volts
$e_c = E_{\text{peak}} \sin (\omega t - 240°)$ volts

$\bar{E}_a = E \,\underline{/0°}\,$ volts
$\bar{E}_b = E \,\underline{/-120°}\,$ volts
$\bar{E}_c = E \,\underline{/-240°}\,$ volts

FIGURE 20.3 Three-Phase Voltage.

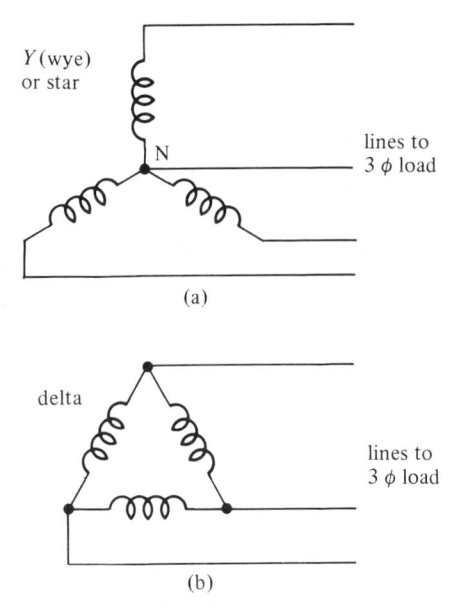

FIGURE 20.4 Coil Connections in 3-ϕ Generators.

In the *wye* or *star* connection [Figure 20.4, part (a)] the inner end of each coil joins the others to form a fourth generator connection called the *neutral* and designated by the letter *n*. Part (b) shows the *delta* connection, which has no neutral.

Note that both connections carry energy over three or four lines, rather than six, a savings in conductors as well as the line loss accompanying them. This reduced line loss explains the higher efficiency of 3-ϕ power transmission systems.

Another advantage of these systems can be seen by comparing the instantaneous power characteristics of single- and three-phase methods.

In Chapter 15 we learned how single-phase power swings from a maximum value one instant to zero the next (see, for example, Figure 15.9). Figure 20.5, on the other hand, is a plot of instantaneous power in each phase of a 3-ϕ system; note the angular separation of 120°. Like single-phase power, each phase power varies between a minimum and maximum, but *the sum of the three is constant*. Its value equals three

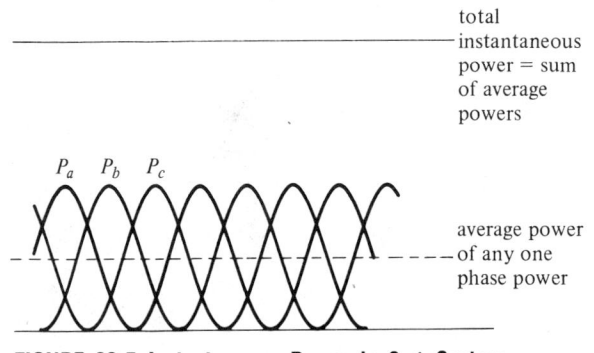

FIGURE 20.5 Instantaneous Power in 3-ϕ System.

20.3 THE WYE-CONNECTED GENERATOR

The three coils of a wye (or star)-connected 3-ϕ generator are shown in greater detail in Figure 20.6. The dashed line to the neutral point signifies that this connection may or may not be made to the load. If it is, the generator is called a *four-wire wye-connected source*. If the neutral floats, that is, is not connected to anything, the generator is a *three-wire, wye-connected source*.

The potentials, $\bar{E}_{an}$, $\bar{E}_{bn}$, and $\bar{E}_{cn}$, found across the A, B, and C source windings, respectively, in Figure 20.6, are phase voltages. Of equal magnitude, they are easily measured between neutral and line terminals. Since we use this magnitude regularly in analysis, it will be called E_p for convenience.

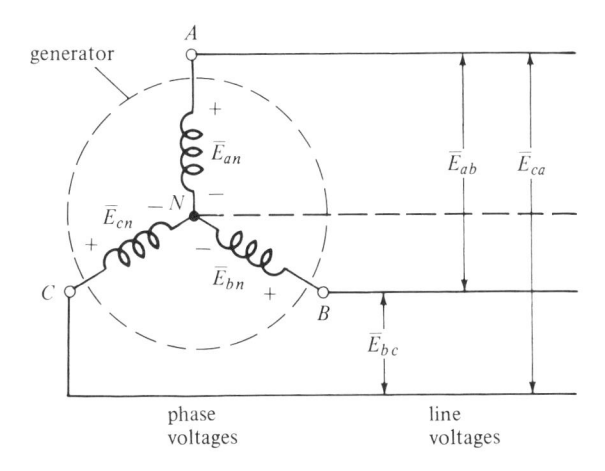

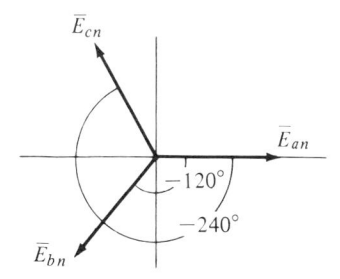

FIGURE 20.6 Wye-Connected 3-ϕ Source and Phasor Diagram.

The three phase voltages in complex form, with E_{an} in the reference position, are

$$\bar{E}_{an} = E_p \underline{/0°} \text{ V}$$
$$= E_p(1 + j0) \qquad (20.1)$$
$$\bar{E}_{bn} = E_p \underline{/-120°} \text{ V}$$
$$= E_p(-0.5 - j0.866) \qquad (20.2)$$
$$\bar{E}_{cn} = E_p \underline{/-240°} \text{ V}$$
$$= E_p(-0.5 + j0.866) \qquad (20.3)$$

From line to line we have the line voltages $\bar{E}_{ab}$, $\bar{E}_{bc}$, and $\bar{E}_{ca}$. Their value, together with that of the phase voltages, is important when working with 3-ϕ systems. Let us derive an equation that describes the relation between the two.

Consider just the two phase voltages, $\bar{E}_{an}$ and $\bar{E}_{bn}$, with instantaneous polarities shown in Figure 20.7.

Write a loop equation:

$$\bar{E}_{ab} + \bar{E}_{bn} - \bar{E}_{an} = 0$$

or

$$\bar{E}_{ab} = \bar{E}_{an} - \bar{E}_{bn}$$

Substitute Equations 20.1 and 20.2:

$$\bar{E}_{ab} = \bar{E}_p(1 + j0) - \bar{E}_p(-0.5 - j0.866)$$
$$= \bar{E}_p(1.5 + j0.866)$$
$$= \bar{E}_p(1.732) \underline{/30°} \text{ volts}$$

But $1.732 = \sqrt{3}$; therefore,

$$\bar{E}_{ab} = \sqrt{3}\, E_p \underline{/30°} \text{ V} \qquad (20.4)$$

By similar derivations:

$$\bar{E}_{bc} = \sqrt{3}\, E_p \underline{/-90°} \text{ V} \qquad (20.5)$$

and

$$\bar{E}_{ca} = \sqrt{3}\, E_p \underline{/150°} \text{ V} \qquad (20.6)$$

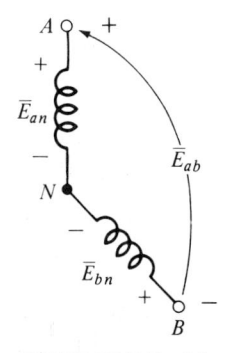

FIGURE 20.7 Deriving Line-Voltage Magnitude.

20.3 THE WYE–CONNECTED GENERATOR

These derivations tell us:

The line voltage of a wye-connected 3-ϕ generator is $\sqrt{3}$ times the phase voltage:

$$E_{line} = \sqrt{3}\, E_p \tag{20.7}$$

EXAMPLE 20.1

One of the most common phase voltages for three-phase systems feeding factories is 277 V. What is the line voltage? Draw the diagram of a four-wire system using a source supplying this voltage, lines, and load. Show all line and phase voltages.

$$\bar{E}_{line} = \sqrt{3}\,\bar{E}_p = 3(277)$$
$$= 479.8 \text{ V}$$

This is commonly referred to as a 277/480 circuit.
Draw the circuit:

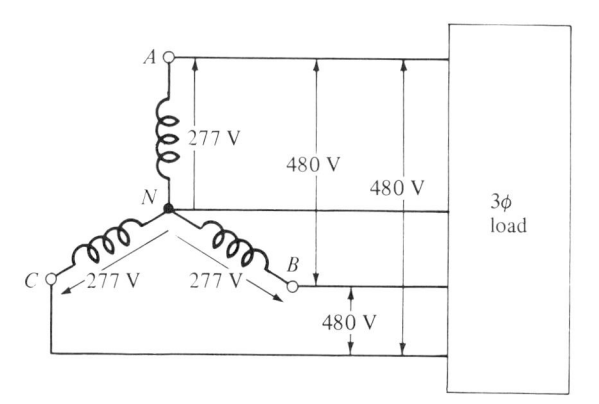

The phase and line voltages associated with a wye-connected 3-ϕ generator are seen in the phasor diagram of Figure 20.8. Note the following:

- Like the phase voltages, the line voltages are 120° apart.
- Each line voltage is 30° ahead of the nearest phase voltage.
- Line-voltage magnitude is $\sqrt{3}$ times phase voltage magnitude.

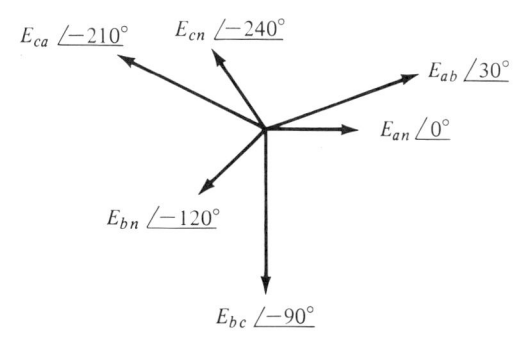

FIGURE 20.8 Phasor Diagram of Line and Phase Potentials for a Wye-Connected Source.

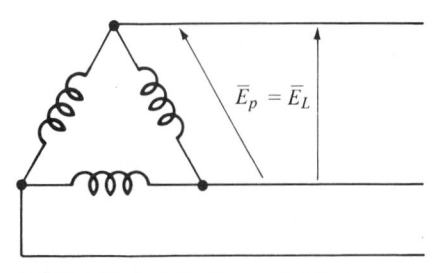

FIGURE 20.9 Delta-Connected 3-ϕ Generator.

The coils of a 3-ϕ generator can also be delta connected, as shown in Figure 20.9. Note there is no neutral connection and that phase voltage and line voltage are the same.

Delta-connected generators are seldom used, for the following reasons: Delta-connected line voltage equals the phase voltage, while wye-connected line voltage equals $\sqrt{3}$ times the phase voltage. Therefore, a delta-connected source needs 1.732 times the armature windings to develop the same line voltage as a wye-connected source. This raises costs. The wye connection offers a fourth, or neutral, terminal which is not available with the delta connection. This makes possible a four-wire system with grounded neutral, which is desirable from a safety standpoint.

We will not discuss delta-connected generators any further.

20.4 LOADING THE WYE-CONNECTED 3-ϕ GENERATOR: THE BALANCED WYE-CONNECTED LOAD

Like generators, 3-ϕ loads can be wye or delta connected. They are also said to be *balanced* when all three impedances are the same, and *unbalanced* when impedances are not the same.

Figure 20.10 shows a *four-wire grounded-neutral system* consisting of a wye-connected source feeding a wye-connected balanced load. Although the diagram might look complicated at first, a moment's study shows that the neutral actually splits the

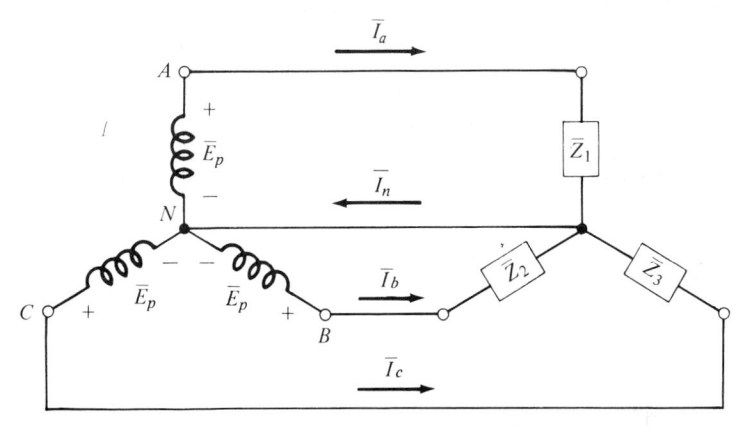

FIGURE 20.10 Four-Wire Y-Y System: Balanced Load.

20.4 LOADING THE WYE-CONNECTED 3-φ GENERATOR

system into three ordinary source/load circuits. Let us analyze one of them. In most cases, its figures can be multiplied by 3 to get total results.

The top phase of Figure 20.10, consisting of source phase A and load Z_1, has been redrawn in Figure 20.11. The potential across the impedance is V_{p_a}, the phase voltage. Generator phase current, signified by $\bar{I}_{p_{gen}}$, equals the line current, $\bar{I}_a$, as well as the phase current in the load $\bar{I}_{p_{load}}$.

The analysis uses familiar series-circuit methods. Remembering that we are speaking of just one phase impedance in a 3-φ load, we have:

Current:
$$\bar{I}_{p_{load}} = \frac{\bar{E}_p}{\bar{Z}_1}$$

True power:
$$P_{W_{z1}} = I_a^2 R_1 = \frac{E_p^2}{R_1} = E_p I_a \cos\phi \quad \text{watts} \quad \text{where } \phi = \text{angle between } V_p \text{ and } I$$

Reactive power:
$$P_{Q_{z1}} = I_a^2 X_1 = \frac{E_p^2}{X_1} = E_p I_a \sin\phi \quad \text{var}$$

Apparent power:
$$P_A = E_p I_a \quad \text{VA}$$

Power factor:
$$PF = \cos\phi = \frac{R_1}{Z_1} = \frac{P_{W_1}}{P_{A_1}}$$

As stated earlier, a balanced load has equal impedances. Since phase voltages are also equal, it follows that total power to a balanced load is

$$P_{W_{total}} = 3(\text{single-phase watts}) \qquad (20.8)$$

total reactive power is

$$P_{Q_{total}} = 3(\text{single-phase var}) \qquad (20.9)$$

and total apparent power is

$$P_{Q_{total}} = 3(\text{single-phase VA}) \qquad (20.10)$$

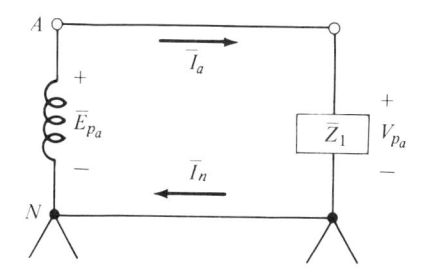

FIGURE 20.11 Analyzing the 3-φ System.

Something interesting happens when we investigate the magnitude of current on the neutral line. A check of Figure 20.10 shows this current to be the complex sum of the three line currents; that is, by KCL;

$$\bar{I}_n = \bar{I}_a + \bar{I}_b + \bar{I}_c \tag{20.11}$$

By Ohm's law and our earlier choice of phase angles, this becomes

$$\bar{I}_n = \frac{E_{pa}\underline{/0°}}{\bar{Z}_1} + \frac{E_{pb}\underline{/-120°}}{\bar{Z}_2} + \frac{E_{pc}\underline{/-240°}}{\bar{Z}_3}$$

assuming purely resistive impedances and remembering that $\bar{Z}_1 = \bar{Z}_2 = \bar{Z}_3$. Convert to rectangular form and factor:

$$\bar{I}_n = \frac{E_p[1 + j0 + (-0.5 - j0.866) + (-0.5 + j0.866)]}{\bar{Z}_1}$$

$$= \frac{E_p[1 - 1 - j0.866 + j0.866]}{\bar{Z}_1} = \frac{E_p(0)}{\bar{Z}_1}$$

$$= 0 \text{ A}$$

Since neutral line current is zero, the conductor can be eliminated to make a three-wire system. The point should be emphasized, however, that this only applies for a balanced load such as a 3-ϕ motor.

Engineers try to keep loads balanced but only an approximation is possible because of customer's constantly changing requirements. As a consequence, the neutral line must be included. However, it usually carries lower current that the other three lines and can therefore be a higher-gauge (thinner), lower-cost conductor.

EXAMPLE 20.2

A 3-ϕ wye-connected source, having a line voltage of 480 V, feeds a 3-ϕ motor. Each phase of the motor has the impedance $Z = 5 + j12 \; \Omega$. Perform the following steps.

a. Find the phase voltage magnitude.

$$\bar{E}_p = \frac{\bar{E}_{line}}{\sqrt{3}}$$

$$= \frac{480}{\sqrt{3}}$$

$$= 277 \text{ V}$$

b. Find phase current.

$$\bar{I}_p = \frac{E_p}{Z_p}$$

$$= \frac{277}{5 + j12}$$

$$= \frac{277}{13\underline{/67.4°}}$$

$$= 21.3\underline{/-67.4°} \text{ A}$$

20.5 THE UNBALANCED WYE-CONNECTED LOAD

c. If phase voltages A, B, and C, are at angles $0°$, $-120°$, and $+120°$, respectively, state the three line currents in phasor form.

Each line current lags its associated phase voltage by $67.4°$. Therefore,

$$\bar{I}_a = 21.3 \underline{/-67.4°} \text{ A}$$
$$\bar{I}_b = 21.3 \underline{/-187.4°} \text{ A}$$
$$\bar{I}_c = 21.3 \underline{/52.6°} \text{ A}$$

d. What are the true, reactive, and apparent powers drawn by each phase?

From part b, $\phi = -67.4°$:

$$P_W = E_p I_p \cos \phi$$
$$= 277(21.3)\cos(-67.4°)$$
$$= 2267.4 \text{ W}$$
$$P_Q = E_p I_p \sin \phi$$
$$= 277(21.3)\sin(-67.4°)$$
$$= -5447 \text{ var}$$
$$P_A = E_p I_p$$
$$= 277(21.3)$$
$$= 5900 \text{ VA}$$

e. What is the total true power drawn by the motor? Repeat for reactive and apparent power.

$$P_{W_{total}} = 3P_{W_{phase}}$$
$$= 3(2267.4)$$
$$= 6.8 \text{ kW}$$
$$P_{Q_{total}} = 3P_{Q_{phase}}$$
$$= 3(5447)$$
$$= 16.3 \text{ kvar}$$
$$P_{A_{total}} = 3P_{A_{phase}}$$
$$= 3(5900)$$
$$= 17.7 \text{ kVA}$$

f. Verify that $\bar{I}_n = 0$ A.

By Equation 20.11,

$$\bar{I}_n = \bar{I}_a + \bar{I}_b + \bar{I}_c$$
$$= 21.3 \underline{/67.4°} + 21.3 \underline{/-187.4°} + 21.3 \underline{/52.6°}$$
$$= 21.3[(0.38 - j0.92) + (-0.99 + j0.13) + (0.6 + j0.8)]$$
$$= 21.3[(0.38 - 0.99 + 0.6) + j(-0.92 + 0.13 + 0.8)]$$
$$\approx 0 \text{ A}$$

20.5 THE UNBALANCED WYE-CONNECTED LOAD

With the exception of motors and certain other loads such as large industrial ovens, most 3-ϕ loads are unbalanced. As a consequence, the neutral line carries current, meaning that the four-line system must be used.

The analysis of such a system simply repeats the work done for the balanced load except that it is done three times, once for each phase impedance. Total powers are obtained by complex addition of the phase powers.

EXAMPLE 20.3

The unbalanced load shown below is fed by a 3-ϕ wye-connected source having a phase voltage of 120 V rms. Phase voltage angles are $-10°$, $-130°$, and $+110°$ for *A*, *B*, and *C*, respectively.

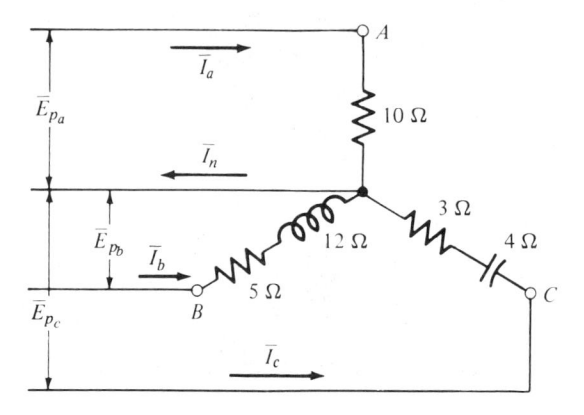

a. Calculate $\bar{I}_{line}$, P_W, P_Q and P_A for each phase.

For phase *A*:

$$\bar{I}_{line} = \frac{\bar{E}_{p_a}}{\bar{Z}_a}$$

$$= \frac{120 \underline{/-10°}}{10}$$

$$= 12 \underline{/-10°} \text{ A}$$

$$P_W = \frac{(E_{p_a})^2}{Z}$$

$$= \frac{120^2}{10}$$

$$= 1440 \text{ W}$$

$$P_Q = 0 \text{ var}$$
$$P_A = 1440 \text{ VA}$$

For phase *B*:

$$\bar{I}_{line} = \frac{\bar{E}_{p_b}}{\bar{Z}_b}$$

$$= \frac{120 \underline{/-130°}}{5 + j12}$$

$$= \frac{120 \underline{/-130°}}{13 \underline{/64.7°}}$$

$$= 9.2 \underline{/-194.7°} \text{ A}$$

20.5 THE UNBALANCED WYE-CONNECTED LOAD

$$P_W = E_p I_b \cos \phi$$
$$= 120(9.2)\cos(-64.7°)$$
$$= 471.8 \text{ W}$$
$$P_Q = E_p I_b \sin \phi$$
$$= 120(9.2)\sin(-67.4°)$$
$$= -998 \text{ var, inductive}$$
$$P_A = E_p I_b$$
$$= 120(9.2)$$
$$= 1104 \text{ VA}$$

For phase C:

$$\bar{I}_{\text{line}} = \frac{\bar{E}_p}{\bar{Z}_c}$$
$$= \frac{120 \underline{/110°}}{3 - j4}$$
$$= \frac{120 \underline{/110°}}{5 \underline{/-53.1°}}$$
$$= 24 \underline{/163.1°} \text{ A}$$
$$P_W = E_p I_c \cos \phi$$
$$= 120(24)\cos(53.1°)$$
$$= 1728 \text{ W}$$
$$P_Q = E_p I_c \sin \phi$$
$$= 120(24)\sin(53.1°)$$
$$= 2304 \text{ var, capacitive}$$
$$P_A = E_p I_c$$
$$= 120(24)$$
$$= 2880 \text{ VA}$$

b. Calculate the total true power

$$\text{Total power} = \sum \text{phase powers}$$
$$= 1440 + 471.8 + 1728$$
$$= 3640 \text{ W}$$

c. Calculate the total reactive power.

$$\text{Total vars} = \sum \text{phase vars}$$
$$= 0 - 998 + 2304$$
$$= 1306 \text{ var, capacitive}$$

d. Draw the system power triangle and determine total system apparent power and power factor.

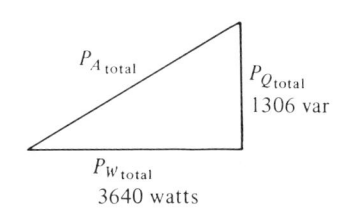

$P_{A\text{total}}$
$P_{Q\text{total}}$ 1306 var
$P_{W\text{total}}$ 3640 watts

Because each P_A is at a different angle, we cannot add them to get total apparent power; Pythagorean theorem most be used with total power and total var.

$$P_{A_{total}} = \sqrt{P_{W_{total}}{}^2 + P_{Q_{total}}{}^2}$$
$$= \sqrt{(3640)^2 + (1306)^2}$$
$$= 3867 \text{ VA}$$
$$\text{PF} = \frac{P_W}{P_A} = \frac{3640}{3867}$$
$$= 94.1\% \text{ leading}$$

e. Calculate the current in the neutral line; I_n.

$$\bar{I}_n = \bar{I}_a + \bar{I}_b + \bar{I}_c$$
$$= 12\underline{/-10°} + 9.2\underline{/-194.7°} + 24\underline{/163.1°}$$
$$= (11.8 - j2.1) + (-8.9 + j2.3) + (-23 + j7)$$
$$= (11.8 - 8.9 - 23) + j(-2.1 + 2.3 + 7)$$
$$= -20.1 + j7.2 \text{ or } 21.4\underline{/160.3°} \text{ A}$$

20.6 THE DELTA-CONNECTED LOAD

Let us examine next the phase and line voltages and currents for the delta-connected load in the three-wire system of Figure 20.12. We consider a balanced system first.

The voltage across each delta impedance is called the *load phase voltage*. $V_{p_{load}}$. A check of Figure 20.12 shows these voltages to be the line voltages (see, for example, V_{c_a}). Therefore,

$$V_{p_{load}} = \sqrt{3}\, E_{p_{gen}} \qquad (20.12)$$

Using the angular relations derived earlier (see Figure 20.8), these load phase voltages are:

$$\bar{V}_{ab} = \sqrt{3}\, E_p \underline{/30°} \qquad (20.13)$$
$$\bar{V}_{bc} = \sqrt{3}\, E_p \underline{/-90°} \qquad (20.14)$$
$$\bar{V}_{ca} = \sqrt{3}\, E_p \underline{/150°} \qquad (20.15)$$

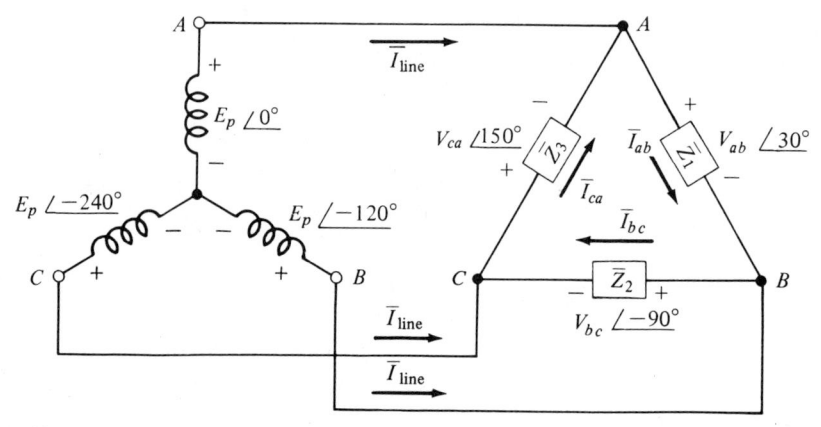

FIGURE 20.12 Delta-Connected Load.

20.6 THE DELTA–CONNECTED LOAD

Phase currents, by Ohm's law, are

$$\bar{I}_{phase} = \frac{\bar{V}_{load\ phase}}{\bar{Z}_{phase}} \tag{20.16}$$

As usual, phase currents are equal in a balanced system.

Line current magnitude can be determined by writing a KCL equation at any of the load nodes. For example, at node A,

$$\bar{I}_{line} = \bar{I}_{ab} - \bar{I}_{ca}$$

By the same mathematics used to derive Equation 20.7, it can be shown that

$$\bar{I}_{line} = \sqrt{3}\,\bar{I}_{p\,load} \tag{20.17}$$

for the delta-connected load.

Power computations for the delta load are done in the usual manner. As with the balanced-wye load, total powers are obtained by multiplying phase powers by 3.

EXAMPLE 20.4

A wye-connected generator having a phase voltage of 12 kV feeds a balanced delta-connected load each leg of which has an impedance, $20 + j28\ \Omega$. Find the magnitudes of each of the following.

a. $V_{p\,load}$.

$$\bar{V}_{p\,load} = \sqrt{3}\,E_{p\,gen}$$
$$= \sqrt{3}(12\ \text{kV})$$
$$= 20.8\ \text{kV}$$

b. $I_{p\,load}$.

$$\bar{I}_{p\,load} = \frac{\bar{E}_{p\,load}}{|Z|}$$
$$= \frac{20.8 \times 10^3}{\sqrt{20^2 + 28^2}} = \frac{20.8 \times 10^3}{34.4}$$
$$= 604\ \text{A}$$

c. I_{line}.

$$\bar{I}_{line} = \sqrt{3}\,\bar{I}_{p\,load}$$
$$= 3(604)$$
$$= 1046\ \text{A}$$

d. The total true power drawn.

$$\text{single-phase power} = E_p I_p \cos \phi$$
$$= (20.8 \times 10^3)(604)\cos\left(\tan^{-1}\frac{28}{20}\right)$$
$$= (12{,}563 \times 10^3)(0.58)$$
$$= 7302\ \text{kW}$$
$$\text{total power} = 3(7302)$$
$$= 21{,}907\ \text{kW} \quad \text{or} \quad 21.9\ \text{MW}$$

644 POLYPHASE CIRCUITS

Like the unbalanced wye connection, the unbalanced delta connection is analyzed one phase at a time. Line currents for these circuits cannot be determined by Equation 20.17 but must be found by nodal equations.

EXAMPLE 20.5

The three unbalanced loads of Example 20.3 are rewired in the delta connection shown below, and fed by a 3-ϕ source with a line voltage of 480 V.

a. Find the three phase currents in phasor form.

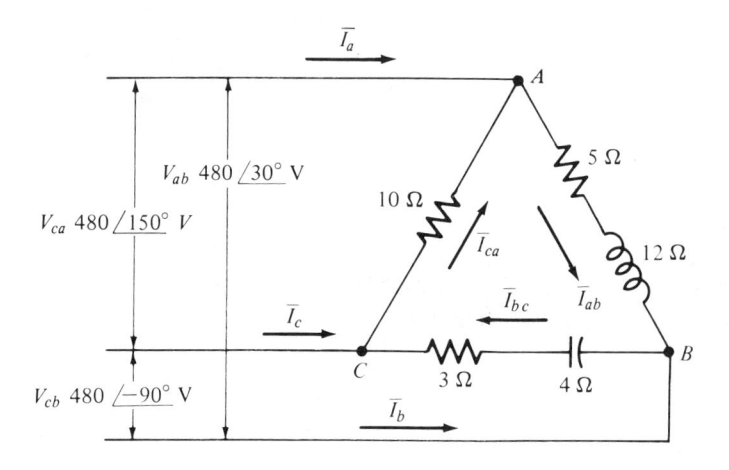

$$\bar{I}_{ab} = \frac{\bar{V}_{ab}}{\bar{Z}_1}$$

$$= \frac{480\,\underline{/30°}}{5 + j12}$$

$$= \frac{480\,\underline{/30°}}{13\,\underline{/67.4°}}$$

$$= 36.9\,\underline{/-37.4°} \quad \text{or} \quad 29.3 - j22.4 \text{ A}$$

$$\bar{I}_{bc} = \frac{\bar{V}_{bc}}{\bar{Z}_2}$$

$$= \frac{480\,\underline{/-90°}}{3 - j4}$$

$$= \frac{480\,\underline{/-90°}}{5\,\underline{/-53.1°}}$$

$$= 96\,\underline{/-36.9°} \quad \text{or} \quad 76.8 - j57.6 \text{ A}$$

$$\bar{I}_{ca} = \frac{\bar{V}_{ca}}{\bar{Z}_3}$$

$$= \frac{480\,\underline{/150°}}{10}$$

$$= 48\,\underline{/150°} \quad \text{or} \quad -41.6 + j24 \text{ A}$$

EXERCISES

b. Find the three line currents.

$$\bar{I}_a = \bar{I}_{ab} - \bar{I}_{ca}$$
$$= (29.3 - j22.4) - (-41.6 + j24)$$
$$= 70.9 - j46.4 \quad \text{or} \quad 84.7 \underline{/-33.2°} \text{ A}$$
$$\bar{I}_b = \bar{I}_{bc} - \bar{I}_{ab}$$
$$= (76.8 - j57.6) - (29.3 - j22.4)$$
$$= 47.5 - j35.2 \quad \text{or} \quad 59.1 \underline{/-36.5°} \text{ A}$$
$$\bar{I}_c = \bar{I}_{ca} - \bar{I}_{bc}$$
$$= (-41.6 + j24) - (76.8 - j57.6)$$
$$= -118.4 + j81.6 \quad \text{or} \quad 143.8 \underline{/145.4°} \text{ A}$$

An Assignment:

Take time to study the next high-tension transmission line you pass. Watch for groups of three heavy (1/2 in. or more) conductors, one for each system on the towers. These are balanced, three-wire systems with ground as a return for unbalanced, neutral currents. Note insulator size and try to estimate line voltage. If there is more than one system, do they appear to be at the same potentials? How can you tell? Look for a thin, lightning protection line running between tower tops. Note that it is not insulated from the tower. Follow it down to a ground rod at the tower base.

EXERCISES

Section 20.1 and 20.2

1. Explain how electric lines to homes provide three circuits at two different voltages with just three wires.

2. Each of five evenly spaced coils around the armature of a 60-Hz generator has 260-V peak induced across it. Write the equations for the voltages in the time and phasor domains, placing the first at 0°.

Section 20.3

3. Fill in the blanks for a wye-connected 3-ϕ generator.

	E_p	E_L
a.	1750 V	
b.		540 kV

4. Given: a 3-ϕ wye-connected generator with the following phase voltages.
a. $\bar{E}_{an} = 920 \underline{/35°}$ V
b. $\bar{E}_{bn} = 920 \underline{/-85°}$ V
c. $\bar{E}_{cn} = 920 \underline{/155°}$ V

Write the statements for the line voltages in phasor form.

Section 20.4

5. A balanced, wye-connected load has an impedance of 10 Ω in each leg. The generator feeding it has a phase voltage of 317 V.
a. Find the magnitude of the line voltage.

b. Find the magnitude of the load phase voltage.
c. Find the magnitude of the load phase current.
d. Find the magnitude of the line current.

6. Repeat exercise 5 changing each load impedance to $10 + j10\ \Omega$.

7. A 3-ϕ wye-connected source has a line voltage of 1732 V and feeds a balanced load of $4 - j10\ \Omega$ per leg over a four-wire system. Its phase voltages, A, B, and C, are at $0°, -120°$, and $-240°$, respectively.
a. Find the phase currents in phasor form.
b. Show that neutral current is zero.
c. Draw the system's power triangle.

8. Repeat Exercise 7 for a load impedance of $12 + j20\ \Omega$ per leg.

9. A 12-hp motor is wye-connected to a 3-ϕ system having a line voltage of 480 V. Assume that the motor is 100% efficient and that the PF of each phase is 94%, lagging.
a. Determine the line current drawn by the motor.
b. Determine the total power drawn. (Remember: 1 hp = 746 W.)
c. Repeat parts a and b if the motor is only 73% efficient.

10. The diagram shows a 3-ϕ source feeding a load over lines each of which has an impedance of $3 + j4\ \Omega$. Find the line voltage at the source. *Hint:* The source cannot tell the line impedance from the load phase impedance.

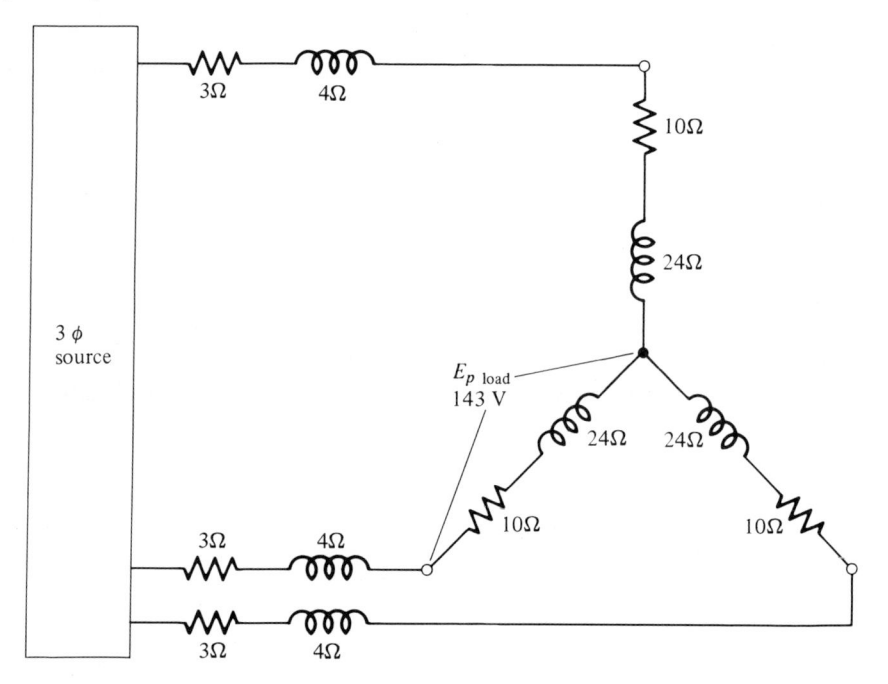

11. The total power drawn by a 3-ϕ wye-connected balanced load is 12 kW at 100% PF. If system line voltage is 4 kV, find the impedance of each load phase in rectangular form.

12. Repeat Exercise 11 for a balanced delta-connected load drawing a $P_{Q_{total}}$ of 1.32 megavar, inductive, at a 95.8% lagging PF.

Section 20.5

13. The phase voltages of a 3-ϕ generator have a magnitude of 277 V and are at $0°, -120°$, and $-240°$ for E_a, E_b, and E_c, respectively. Over a four-wire system, this source feeds a load with phase impedances 10, 15, and 22 Ω for Z_a, Z_b, and Z_c, respectively.
a. Find the line currents.
b. Find the total P_W.

EXERCISES

   c. Find the total P_Q.
   d. Draw the system power triangle and determine total P_A and PF.
   e. Find I_n.

14. Repeat Exercise 13 for $\bar{Z}_a = 10$, $\bar{Z}_b = 3 + j4$, and $\bar{Z}_c = 12 - j5$.
15. A 3-ϕ source has phase voltages $\bar{E}_{an}$, $\bar{E}_{bn}$, and $\bar{E}_{cn}$, of 120 V, at $-35°$, $-155°$, and $85°$, respectively. Determine $\bar{I}_n$ and the phase currents for $\bar{Z}_a = 10\,\underline{/45°}\ \Omega$, $\bar{Z}_b = 15\,\underline{/-25°}\ \Omega$, and $\bar{Z}_c = 25\,\underline{/0°}\ \Omega$, in phasor form.
16. *Extra Credit.* Given: an unbalanced wye-connected load with phase voltages $\bar{E}_{an} = 120\,\underline{/15°}$ V, $\bar{E}_{bn} = 120\,\underline{/-105°}$ V, and $\bar{E}_{cn} = 120\,\underline{/135°}$ V. Total $P_a = 7818$ VA at a 91.1% leading PF. If $\bar{Z}_a = 1 - j2\ \Omega$ and $\bar{Z}_b = \bar{Z}_c$, find $\bar{Z}_b$ and $\bar{Z}_c$.

Section 20.6

17. A wye-connected 3-ϕ generator having an $\bar{E}_p = 277$ V feeds a balanced delta-connected load each leg of which is a 27-Ω resistance. Determine the magnitude of: a. $\bar{E}_{p_{\text{load}}}$; b. $\bar{I}_{p_{\text{load}}}$; c. $\bar{I}_{\text{line}}$; d. $P_{W_{\text{total}}}$.
18. Repeat Exercise 17 for phase impedance $= 9 - j14\ \Omega$.
19. The line voltages feeding a balanced delta-connected load are

$$\bar{E}_{ab} = 2420\,\underline{/0°}\ \text{V}$$
$$\bar{E}_{bc} = 2420\,\underline{/-120°}\ \text{V}$$
$$\bar{E}_{ca} = 2420\,\underline{/120°}\ \text{V}$$

Each load impedance is $6 + j10.4\ \Omega$.
   a. Calculate the phase currents $\bar{I}_{ab}$, $\bar{I}_{bc}$, and $\bar{I}_{ca}$, in phasor form.
   b. Calculate the line current magnitude.
   c. Draw the load's power triangle and determine total PF.
20. A 3-ϕ source with $\bar{E}_p = 600$ V feeds a balanced delta-connected load each leg of which has an impedance $\bar{Z}_p = 12\,\underline{/60°}\ \Omega$.
   a. Determine the line currents, load power triangle, and power factor.

From our work in Π–T transformations, we know that a balanced T (the wye connection of this chapter) is equivalent to a balanced Π (the delta here) if impedances are divided by 3. Therefore:

   b. Repeat part a for a balanced-wye connected load, each leg of which is $\bar{Z}_{\text{delta}}/3$ or, in this case, $4\,\underline{/60°}\ \Omega$.
   c. Compare results of parts a and b. Are the two loads equivalent as far as the source is concerned?
21. For this load calculate: a. phase currents; b. line currents; c. load power triangle and PF.

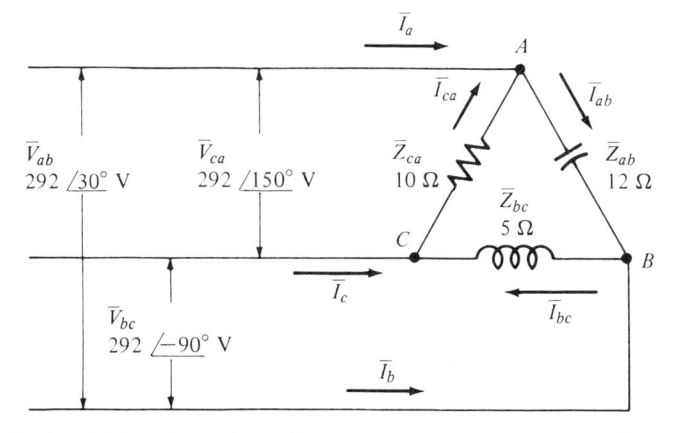

$\bar{V}_{ab}\ 292\,\underline{/30°}$ V $\bar{V}_{ca}\ 292\,\underline{/150°}$ V $\bar{Z}_{ca}\ 10\ \Omega$ $\bar{Z}_{ab}\ 12\ \Omega$ $\bar{Z}_{bc}\ 5\ \Omega$

$\bar{V}_{bc}\ 292\,\underline{/-90°}$ V

22. Repeat Exercise 21 with load impedance $\bar{Z}_{ab}$ changed to $13\,\underline{/-67.4°}\ \Omega$ and $\bar{Z}_{bc}$ changed to $7.1\,\underline{/45°}\ \Omega$; $\bar{Z}_{ca}$ unchanged.

CHAPTER HIGHLIGHTS

Factories and other high-demand loads get their electric energy over 3-ϕ transmission systems for reasons of economy and improved machine performance.

3-ϕ systems have three voltages spaced 120° between adjacent phases.

Most 3-ϕ generators are wye connected in a three-wire system for balanced loads (all load impedances equal) or four-wire for unbalanced loads (load impedances not equal). The delta connection is possible but seldom used.

Loads, too, can by wye or delta connected.

Formulas to Remember

For wye-connected load:

$$E_{\text{phase}} = \frac{E_{\text{line}}}{\sqrt{3}}$$

$$I_{\text{line}} = I_{\text{phase}}$$

For delta-connected load:

$$E_{\text{phase}} = E_{\text{line}}$$

$$I_{\text{phase}} = \frac{I_{\text{line}}}{\sqrt{3}}$$

For balanced load:

line currents equal,
phase currents equal,
$I_{\text{neutral}} = 0$,
phase powers equal,
total P (either P_W, P_Q, or P_A) $= 3P_{\text{phase}}$

For unbalanced load:

Find individual currents and powers by analysis.

$\bar{I}_n = \bar{I}_a + \bar{I}_b + \bar{I}_c$

$P_{W_{\text{total}}} = \sum P_{W_{\text{phase}}}$

$P_{Q_{\text{total}}} = \sum P_{Q_{\text{phase}}}$

$P_{A_{\text{total}}} = \sqrt{(P_{W_{\text{total}}})^2 + (P_{Q_{\text{total}}})^2}$

CHAPTER 21

Transformers

Electric energy is transmitted at high potentials as ac because ac voltage can be stepped up and down by one of the most efficient machines ever devised, the *transformer*.

Transformers are simple devices consisting of two or more coils of wire linked, or *coupled*, by magnetic flux. Depending on the number of turns in these coils, transformers step voltage, current, and impedance up or down, and because they contain no moving parts, they have no friction and are highly reliable.

Before starting this chapter, the student is strongly advised to review sections 11.2, 11.3, 11.9, 12.1, and 12.4.

At the completion of this chapter, you should be able to:

- Explain transformer action, both ideal and actual.
- Calculate changes in voltage, current, and impedance from step-up and step-down transformers.
- Analyze transformer circuits by reflecting impedances.
- Use the dot convention to obtain flux, current direction, and polarity information.
- Determine total inductance of series-connected coils with mutual inductance.
- Describe the iron-core transformer equivalent circuit and explain the frequency response curve obtained from it.
- Compare and contrast the operation of tightly coupled versus loosely coupled transformers.
- Identify tapped, variable, and autotransformers on schematics.

21.1 IRON-CORE TRANSFORMER THEORY

The left-hand winding of the basic transformer (Figure 21.1) receives source energy and is called the *primary*. The right-hand coil supplies energy to a load and is called the *secondary*. When the coils are wound on a high-permeability closed core of laminations, the unit is called an *iron-core transformer*.

Figure 21.2 illustrates a typical power transformer (abbreviated xfmr) used regularly in electronics. Its 115V, 50/60 Hz primary is connected by the black leads in the photo. The secondary provides 36V split into two 18V halves by a center tap. Maximum secondary current is 0.75A.

Our study of fundamentals is simplified by considering the iron-core transformer to be an *ideal transformer*. That is:

a. Winding resistances are negligible.
b. Core losses are negligible.
c. All flux is confined to the core; that is, there is no leakage.

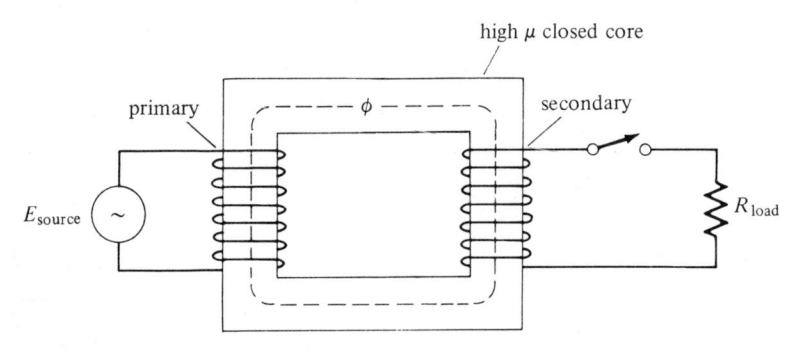

FIGURE 21.1 Basic Transformer.

21.1 IRON-CORE TRANSFORMER THEORY

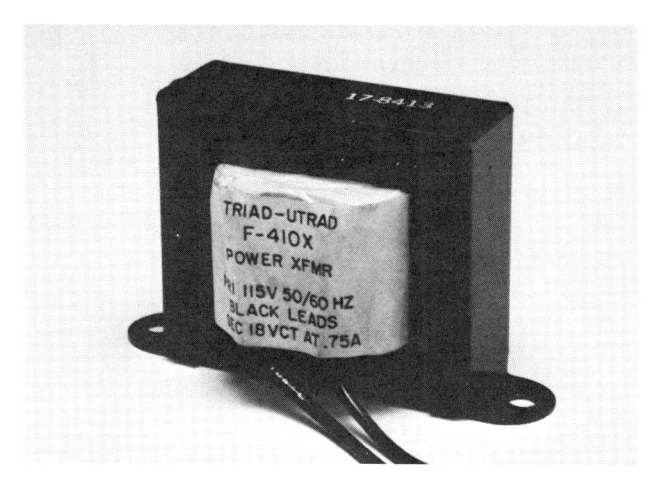

FIGURE 21.2 Power Transformer. (Courtesy of Triad-Utrad.)

    d. Core permeability is so high that negligible primary current is required to develop core flux.

Quality transformers, made with modern materials, come close to these requirements and are almost ideal.

Having defined the ideal transformer, our study proceeds in three stages: (1) the primary alone, (2) the transformer with unloaded secondary, and (3) the transformer with loaded secondary.

The Primary Alone

When source current flows in the primary windings, a magnetomotive force, $\mathscr{F}_p = N_p I_p$ ampere-turns, causes a flux of ϕ webers to be developed in the core. Since the current changes sinusoidally, ϕ must do likewise; that is

$$\phi = \phi_{\max} \sin 2\pi f t \qquad \text{webers} \tag{21.1}$$

By Faraday's law,

$$\begin{aligned} e_p &= N_p \frac{d\phi}{dt} \\ &= N_p \frac{d[\phi_{\max} \sin 2\pi f t]}{dt} \end{aligned} \tag{21.2}$$

Substitute the derivative of $\sin 2\pi f t$ (see Equation 15.1):

$$e_p = N_p 2\pi f \phi_{\max} \cos 2\pi f t \qquad \text{volts}$$

The effective value of this is

$$\begin{aligned} E_p &= \frac{2\pi N_p f \phi_{\max}}{\sqrt{2}} \\ &= 4.4 N_p f \phi_{\max} \end{aligned} \tag{21.3}$$

This is the induced voltage across the primary. By KVL, it must equal the source voltage (by the first ideal transformer requirement, $I_p R_p \approx 0$ V).

Transformer with Unloaded Secondary

The flux developed by the primary's magnetomotive force passes through, or *links*, the secondary. The coils are said to be *coupled*. Since the flux varies sinusoidally, the voltage induced in the secondary will also. That is, by Faraday's law,

$$e_s = N_s \frac{d\phi}{dt} \tag{21.4}$$

Equation 21.4 makes an important point which must be emphasized before we move on:

> Changing core flux is necessary to induce secondary voltage. If flux is constant, no matter what its intensity, $e_s = 0$ V. This is the reason dc cannot be transformed. Except for the instants of turn-on and turn-off, there is no $d\phi/dt$ with dc.

By the steps used to derive Equation 21.3, we have

$$E_s = 4.4 N_s f \phi_{max} \tag{21.5}$$

Solve Equations 21.3 and 21.5 for $4.4 f \phi_{max}$:

$$\frac{E_p}{N_p} = 4.4 f \phi_{max} = \frac{E_s}{N_s}$$

or

$$\frac{E_p}{N_p} = \frac{E_s}{N_s}$$

or

$$\frac{E_p}{E_s} = \frac{N_p}{N_s} \tag{21.6}$$

The ratio N_p/N_s in Equation 21.6 is called the *turns ratio* (some books refer to it as the transformation ratio). It is usually given in the form 1:10, or 130:85, rather than divided into a whole number. For convenience, the ratio is represented by the letter a, which is defined

$$a = \frac{N_p}{N_s} \tag{21.7}$$

For

$a > 1$: $N_p > N_s$. Therefore, $E_p > E_s$.
  This is called a *step-down* transformer.
$a < 1$: $N_p < N_s$. Therefore, $E_p < E_s$.
  This is called a *step-up* transformer.

Note the transformer symbol introduced in the next example. Since circuit functions generally move from left to right, the primary is placed on the left. Vertical

21.1 IRON-CORE TRANSFORMER THEORY

lines between coils indicate a laminated iron core; their absence would signal an air core.

EXAMPLE 21.1

Tell whether each transformer is step-up or step-down and find the missing quantities.

$$N_p = 400\text{ T} \qquad N_s = 100\text{ T}$$
$$E_p = 800\text{ V} \qquad E_s$$

a. Find N_p/N_s, a, and E_s.

$$\frac{N_p}{N_s} = \frac{400}{100}$$
$$= 4:1$$
$$a = 4$$
$$E_s = \frac{E_p}{a}$$
$$= \frac{800}{4}$$
$$= 200 \text{ V}$$

Since $E_s < E_p$, this is a step-down transformer.

b. Find N_p/N_s and a.

$$E_p = 120\text{ V} \qquad E_s = 4\text{ kV}$$

$E_s > E_p$, this is a step-up.

$$\frac{N_p}{N_s} = \frac{120}{4000}$$
$$= 1:33$$

Notice that the actual primary or secondary turns are not needed in order to express the turns ratio.

$$a = \frac{1}{33}$$
$$= 0.03$$

c. Find a, N_s, and $\dfrac{N_p}{N_s}$.

N_p
1285 T

E_p
14 kV

E_s
220 V

$$a = \dfrac{E_p}{E_s}$$

$$= \dfrac{14 \text{ kV}}{220 \text{ V}}$$

$$= 63.6$$

Unit is a step-down.

$$N_s = \dfrac{N_p}{a}$$

$$= \dfrac{1285}{63.6}$$

$$= 20.2 \text{ T}$$

$$\dfrac{N_p}{N_s} = 63.6:1$$ ∎

Transformer with Loaded Secondary

Closing the switch in the secondary circuit of Figure 21.1 causes the induced voltage, E_s, to transfer energy to the load. Because this energy must come from the source, an interesting question arises: How does the secondary pass its energy demands to the source when the only link between the two is a flux field?

When secondary current flows, it causes a magnetomotive force, $\mathscr{F}_s = N_s I_s$, that creates new flux in the core. Because of Lenz's law, this flux opposes the original flux that created it. At first, it would seem that this opposition would reduce or cancel total flux. However, this does not happen. Equation 21.3 says that flux cannot decrease without E_p decreasing, which it cannot because $E_p = E_{source}$ and E_{source} is constant. Therefore, ϕ must be constant.

For ϕ to remain constant when the secondary is loaded, the primary magnetomotive force, $\mathscr{F}_p = N_p I_p$, must increase to offset $N_s I_s$. Since N_p cannot change, I_p must increase, which, of course, draws more source energy. *The primary is thus self-regulating.* For $I_s = 0$, $I_p \approx 0$. But let I_s increase and I_p increases to neutralize secondary magnetomotive force and keep core flux constant.

From the discussion above we see that

$$\mathscr{F}_p = N_p I_p = \mathscr{F}_s = N_s I_s$$

Solve for the turns ratio in terms of current:

$$\dfrac{N_p}{N_s} = \dfrac{I_s}{I_p} \tag{21.8}$$

Note carefully that the turns ratio for currents is the inverse of what it was for voltages: What is step-up for voltage is step-down for current. Conservation of energy explains

21.1 IRON-CORE TRANSFORMER THEORY

why this must be so. For an ideal transformer,

$$P_{\text{apparent out}} = P_{\text{apparent in}}$$

But

$$P_{\text{apparent out}} = E_s I_s$$

which, from Equations 21.6 and 21.8, is

$$P_{\text{apparent out}} = \frac{N_s}{N_p} E_p \frac{N_p}{N_s} I_p$$

$$= E_p I_p$$

which is $P_{\text{apparent in}}$. In other words, if $E_s > E_p$, $I_s < I_p$, and vice versa.

Finally, let us determine how the transformer transforms impedance between windings. By Ohms's law,

$$Z_s = \frac{E_s}{I_s}$$

$$= \frac{E_p/a}{aI_p}$$

$$= \frac{(1/a^2)E_p}{I_p}$$

But

$$\frac{E_p}{I_p} = Z_p$$

Therefore,

$$Z_p = a^2 Z_s \tag{21.9}$$

This equation tells us that a source sees a transformer's secondary load impedance as a^2 times that impedance across the primary. In other words, transformers transform impedances by the square of the turns ratio.

We studied reactive matching circuits and their ability to shift impedance levels in Chapter 19. As pointed out by Equation 21.9, transformers also perform this function and are used in a variety of matching applications.

EXAMPLE 21.2

A radio signal begins at low power in a piezoelectric crystal and is increased many times by a series of amplifiers to a level sufficient to drive the final amplifier, the last circuit driving the antenna. High-power finals, greater than a few kilowatts, are one place where tubes still reign over transistors. But tubes have two drawbacks: they have high Z_{out} compared to an antenna's Z_{in}, and they require high voltage.

Given: an AM broadcast final, operating under the following conditions:

$$P_{\text{out}} = 50 \text{ kW}$$
$$\eta = 60\%$$
$$Z_{\text{out}} = 4500 \text{ }\Omega$$
$$E_{\text{in}} = 5 \text{ kV}$$

The power transformer feeding the amplifier has a 440-V primary and can be considered ideal.

a. Find the turns ratio of the transformer needed to match this amplifier to a 72-Ω antenna.

$$Z_p = a^2 Z_s$$

Therefore,

$$a = \left(\frac{Z_p}{Z_s}\right)^{1/2}$$

$$= \left(\frac{4500}{72}\right)^{1/2}$$

$$= 7.9$$

The 72-Ω antenna impedance is multiplied by the matching transformer to 4500 Ω for maximum power transfer from the 4500-Ω source.

b. Find the power transformer primary current and turns ratio.

$$P_{in} \text{ to amplifier} = \frac{P_{out}}{\eta}$$

$$= \frac{50 \text{ kW}}{0.6}$$

$$= 83.3 \text{ kW}$$

Amplifier P_{in} equals power supply output power, $E_{supply} I_{supply}$. Therefore,

$$I_{supply} = \frac{P_{in}}{E_{supply}}$$

$$= \frac{83.3 \text{ kW}}{5 \text{ kV}}$$

$$= 16.7 \text{ A}$$

Since the power transformer is ideal,

$$P_{in \text{ to amplifier}} = P_{pri} = E_p I_p$$

Therefore,

$$I_p = \frac{P_{in \text{ to amplifier}}}{E_p}$$

$$= \frac{83.3 \text{ kW}}{440}$$

$$= 189.3 \text{ A}$$

The turns ratio is

$$\frac{N_p}{N_s} = \frac{E_p}{E_s} = \frac{440}{5000}$$

$$= 1:11.4$$

which means that the secondary has 11.4 times the turns of the primary. ■

21.2 THE DOT SYSTEM

To summarize the operations of the idealized iron-core transformer:

- Voltages are transformed directly with the turns ratio.
- Currents are transformed inversely with the turns ratio.
- Impedances are transformed directly with the square of the turns ratio.
- $P_p = P_s \qquad VA_p = VA_s$
- Frequency remains the same.

21.2 THE DOT SYSTEM

Transformers windings are interconnected to form wye and delta hookups in 3-ϕ systems and in series and parallel to increase voltage and current capabilities. To make these connections, attention must be paid to the phase of each winding so that voltages combine as desired. Such phase information is indicated on transformers as well as schematic diagrams by large dots placed at one terminal of each winding. These dotted terminals indicate points having the same instantaneous polarity. That is, if one of a transformer's dotted terminals is negatively polarized at some instant, all the transformer's dotted terminals are negatively polarized at that instant. A half-cycle later all become positive.

Figure 21.3 illustrates the use of the dot system in determining polarities. Study each part carefully.

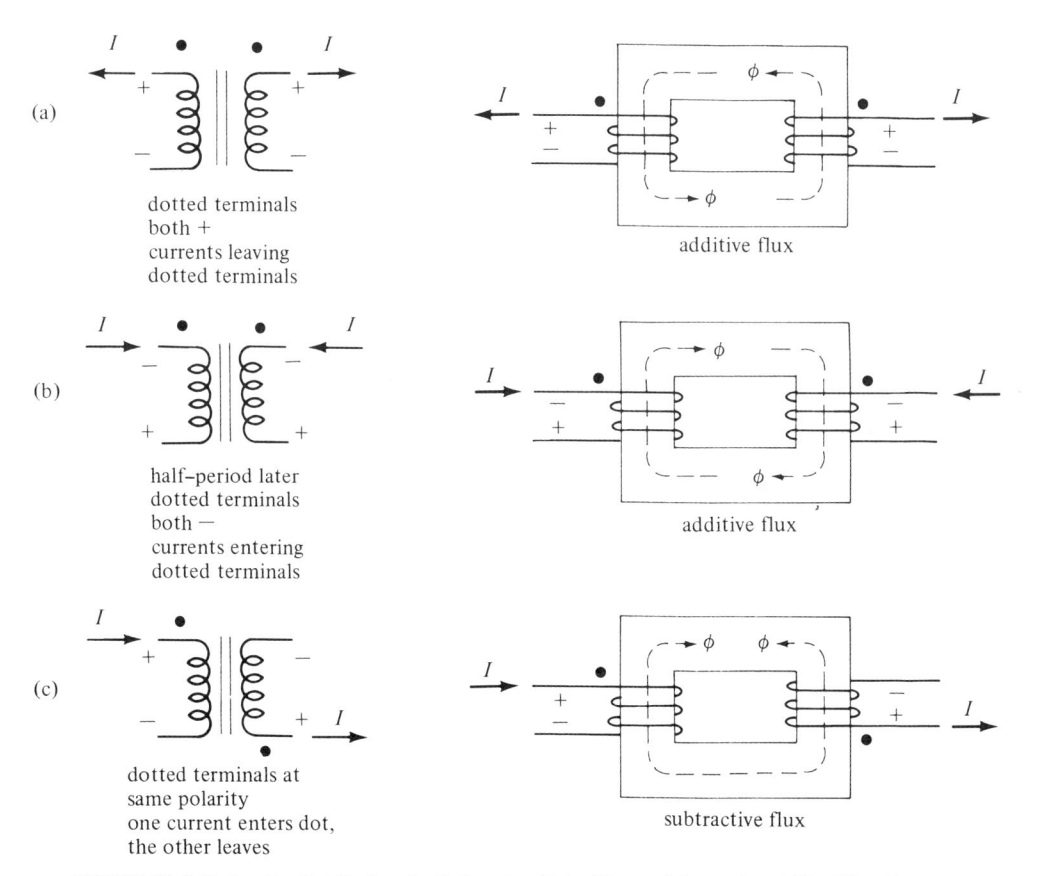

(a) dotted terminals both +
currents leaving dotted terminals

additive flux

(b) half-period later dotted terminals both −
currents entering dotted terminals

additive flux

(c) dotted terminals at same polarity
one current enters dot, the other leaves

subtractive flux

FIGURE 21.3 Using the Dot System to Determine Polarities and Current and Flux Directions.

The dot system also provides information on core flux directions. These, you will recall, are easily determined with the right-hand rule of Section 11.2 (if the right-hand fingers encircle coil in current direction, the thumb points in the flux direction). But coil diagrams are not included on schematics. They are not needed because the dots tell what we want to know:

> Currents flowing the same direction with respect to dotted terminals produce flux in the same direction.

Test this point on the three cases in Figure 21.3. Each allows flux direction to be determined by both right-hand rule and dots.

Figure 21.4, part (a) shows the dot convention on the diagram of a dual-secondary 25-kVA 4160/240-V pole transformer used by power companies. (The units are usually

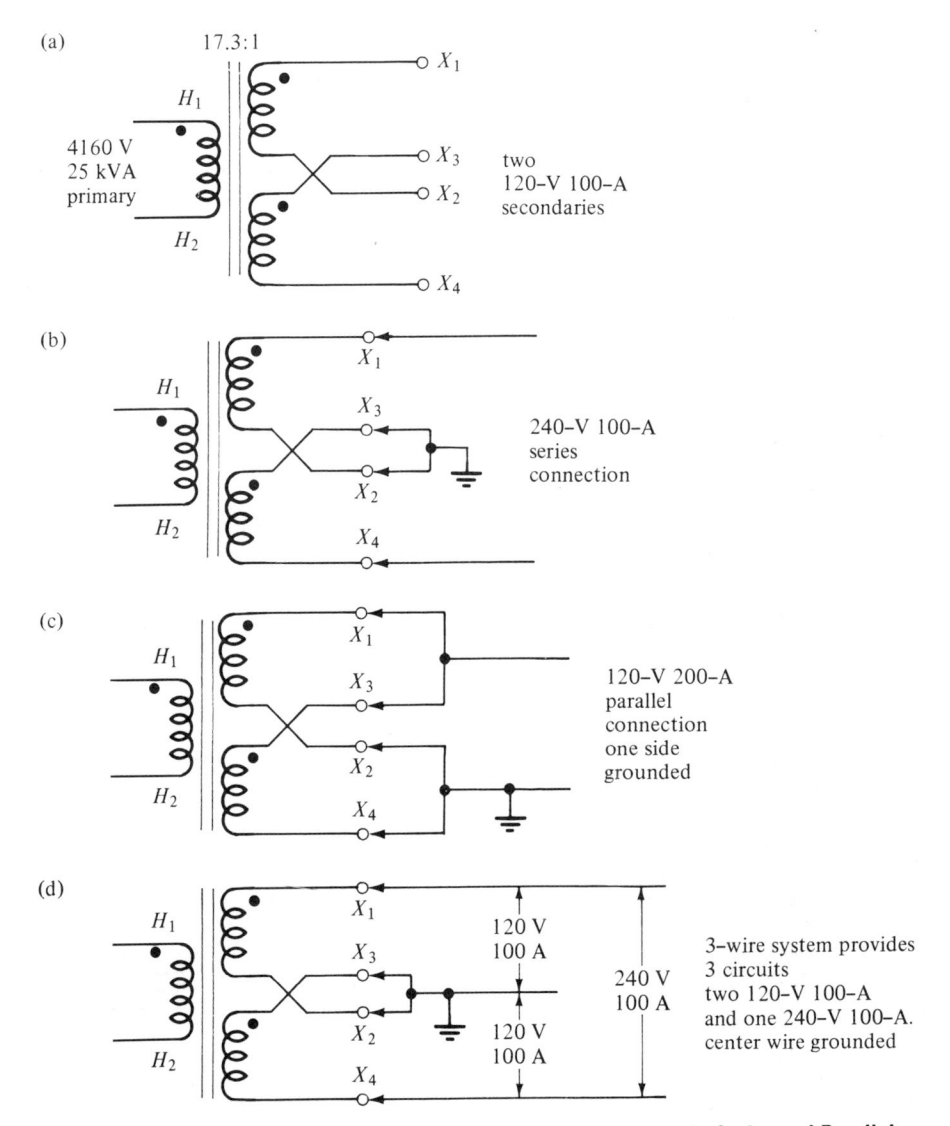

FIGURE 21.4 Use of the Dot Convention When Connecting Windings in Series and Parallel.

called "cans" in the industry, although "tubs" and "pole pigs" have been heard around North America; there are undoubtedly other names.) The lettering and subscripts are industry standards. Secondary terminals X_3 and X_2 are interchanged to allow convenient connections between adjacent terminals for a variety of wiring combinations. The turns ratio (17.3:1) is shown as it is usually indicated on schematics.

Figure 21.4 parts b, c, and d, show secondary connections, each satisfying a particular customer's needs. Study all three, paying close attention to the interconnection of dotted terminals. Be sure to understand why each obtains its indicated voltage and current capabilities. Method d might be recognized as the hookup used to supply homes with one 240-V and two 120-V circuits.

21.3 ANALYZING CIRCUITS CONTAINING IDEAL TRANSFORMERS

Let us first analyze a circuit containing an ideal transformer using well-known techniques from earlier chapters. Then we repeat the analysis introducing a new approach, which reduces the math work considerably.

EXAMPLE 21.3

Find the primary current, I_p, by loop analysis.

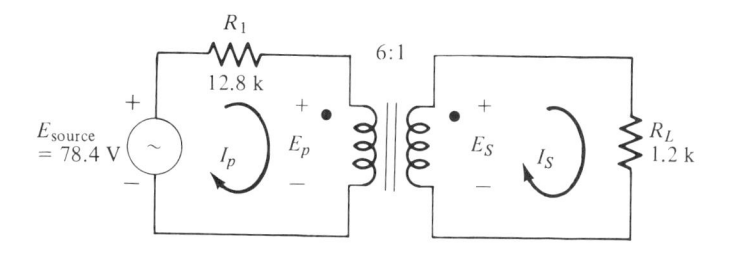

Writing an equation around each loop:

$$78.4 = 12.8 \text{ k}I_p + E_p$$
$$E_s = 1.2 \text{ k}I_s$$

E_s and I_s can be eliminated by use of the transformer equations and the turns ratio:

$$I_s = 6I_p \quad \text{and} \quad E_s = \frac{E_p}{6}$$

Substitute:

$$78.4 = 12.8 \text{ k}I_p + E_p$$
$$\frac{E_p}{6} = 6I_p(1.2 \text{ k})$$

or

$$E_p = 43.2 \text{ k}I_p$$

Substitute for E_p in first equation:

$$78.4 = 12.8 \text{ k}I_p + 43.2 \text{ k}I_p$$
$$= 56 \text{ k}I_p$$

Therefore,

$$I_p = \frac{78.4}{56 \text{ k}}$$
$$= 1.4 \text{ mA} \qquad \blacksquare$$

Our second analysis uses a concept called *reflected impedance*. Based on Equation 21.9, this concept shows that a secondary impedance is transferred, or *reflected*, into the primary circuit if it is multiplied by a^2.

Let us redo Example 21.3 by this method.

EXAMPLE 21.4

Find I_p in the circuit of Example 21.3 by reflecting R_{load} into the primary.

The equivalent circuit after reflecting R_{load} is

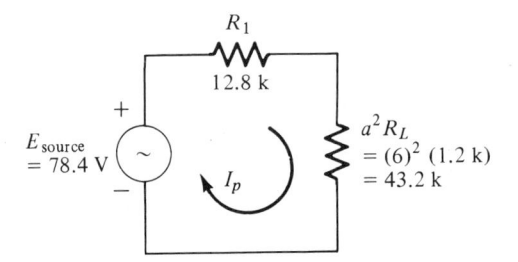

$E_{\text{source}} = 78.4$ V

$R_1 = 12.8$ k

$a^2 R_L = (6)^2 (1.2 \text{ k}) = 43.2$ k

By Ohm's law,

$$I_p = \frac{E_{\text{source}}}{R_1 + R'_{\text{load}}}$$

where R'_{load} is the reflected load resistance, $a^2 R_L$.

$$I_p = \frac{78.4}{12.8 \text{ k} + 43.2 \text{ k}}$$
$$= 1.4 \text{ mA} \qquad \blacksquare$$

Had we wanted I_{load}, the earlier loop equations could be solved for it. Let us get it instead by reflecting the entire primary circuit into the secondary. As this is done, note carefully the use of the turns ratio, a; impedance reflections require a^2, currents and voltages only a.

21.3 ANALYZING CIRCUITS CONTAINING IDEAL TRANSFORMERS

EXAMPLE 21.5

Find I_{load} in the circuit of Example 21.3 by reflecting the primary circuit into the secondary.

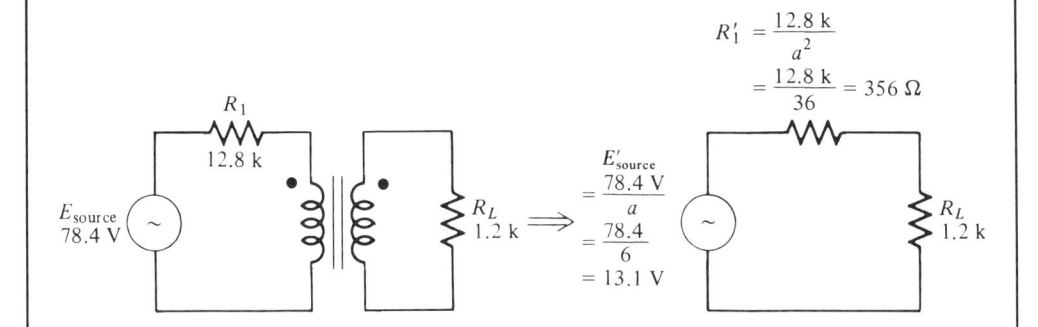

$$R_1' = \frac{12.8\text{ k}}{a^2}$$
$$= \frac{12.8\text{ k}}{36} = 356\ \Omega$$

$$E'_{source} = \frac{78.4\text{ V}}{a}$$
$$= \frac{78.4}{6}$$
$$= 13.1\text{ V}$$

$$I_{load} = \frac{E'_{source}}{R_1' + R_{load}}$$

where E'_{source} = reflected source voltage
R_1' = reflected R_1

$$= \frac{356}{356 + 1200}$$
$$= 8.4\text{ mA}$$

The dots in Example 21.5 were such that the source was reflected with its original polarity. If they are at opposite ends of the windings, the reflected polarity must be reversed. This point is seen in the next example which also illustrates the need for a higher level of analysis, even after reflections.

EXAMPLE 21.6

Find the load current by reflecting the primary circuit into the secondary.

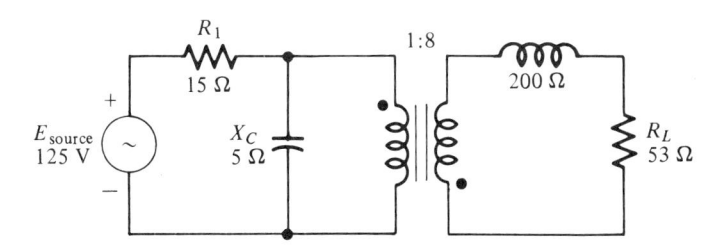

Notice that this is a step-up transformer. Primary impedances are still divided by a^2 but a is 1/8 in this case, meaning that they are actually multiplied by 64. The equivalent circuit

with reflected components is

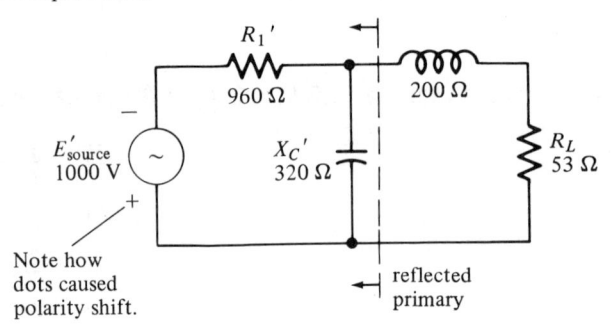

Note how dots caused polarity shift.

This problem could be done by:

- Getting Thévenin or Norton equivalent of reflected primary.
- Loop or nodal equations.
- Series–parallel reduction.

Let us Thévenize the reflected primary.

$$\bar{Z}_{\text{Thev}} = \frac{(960)(-j320)}{960 - j320} = 96 - j288 \; \Omega$$

$$\bar{E}_{\text{Thev}} = \frac{1000(320\underline{/-90°})}{960 - j320} = 316.2\underline{/-71.6°} \; \text{V}$$

Draw the Thévenin equivalent circuit and calculate I_{load}:

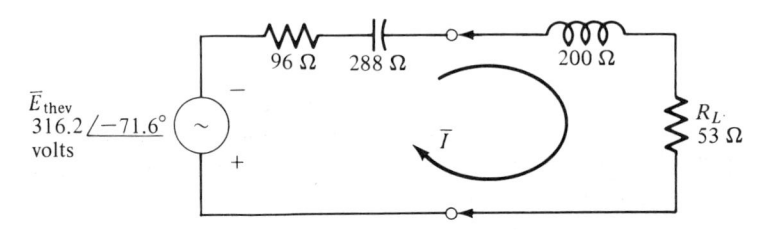

$$\bar{I}_{\text{load}} = \frac{\bar{E}_{\text{Thev}}}{\bar{Z}_{\text{total}}}$$

$$= \frac{-316.2\underline{/-71.6°}}{(96 + 53) + j(200 - 288)} \quad \text{(minus sign required because } I_{\text{load}} \text{ against source polarity)}$$

$$= \frac{-316.2\underline{/-71.6°}}{149 - j88}$$

$$= -1.8\underline{/-41°}$$

$$= 1.8\underline{/-221°} \; \text{A}$$

21.4 FLUX IN THE NONIDEAL CORE

Having covered the basics with an idealized transformer, let us next turn our attention to the theory of actual, nonideal units.

Primary magnetomotive force, $\mathscr{F}_p = N_p I_p$, develops flux, ϕ_p, in a nonideal core. That flux divides into two parts (see Figure 21.5). The larger part remains within the

21.4 FLUX IN THE NONIDEAL CORE

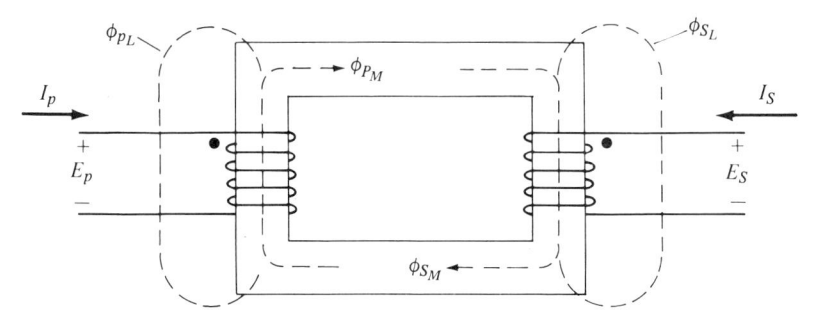

FIGURE 21.5 Leakage and Mutual Fluxes.

core and links every winding on the core. Because it passes through all windings, it is called *mutual flux* and labeled ϕ_{p_M}. The smaller portion leaves the core and passes through the surrounding air even though air permeability is much lower than that of the iron core. This part links only the winding that produces it, namely, the primary. It is called *leakage flux* and labeled ϕ_{p_L}.

In equation form,

$$\phi_p = \phi_{p_M} + \phi_{p_L} \tag{21.10}$$

Since a given number of ampere-turns develops a certain flux, the lower the leakage, the higher the mutual flux. A parameter providing a measure of this relationship is the *coefficient of coupling*. Represented by the letter k, it is defined as the ratio of the mutual flux produced by a coil to the total flux produced by that coil:

$$k = \frac{\phi_{\text{mutual}}}{\phi_{\text{total}}} \tag{21.11}$$

Substituting the relationship of Equation 21.10 gives

$$k = \frac{\phi_{\text{mutual}}}{\phi_{\text{mutual}} + \phi_{\text{leakage}}} \tag{21.12}$$

A check of the definition shows that k can never be greater than 1.

The ideal transformer investigated earlier was assumed to have no leakage. Its coefficient of coupling is therefore 1. The k values of power transformers designed for peak efficiency at 60 Hz are close to this ideal. They are obtained by using high-permeability cores and winding the coils on top of each other. Tuned transformers used in communications work, on the other hand, are designed for Q and bandwidth considerations, not efficiency. Their k values may be as low as 0.01.

Flux developed by the secondary MMF also splits into mutual and leakage components (see Figure 21.5). It is described by the equation

$$\phi_s = \phi_{s_M} + \phi_{s_L} \tag{21.13}$$

By the dot system or right-hand rule, we see that both primary and secondary mutual fluxes in Figure 21.5 are clockwise in the core. Therefore,

$$\phi_{m_{\text{total}}} = \phi_{p_M} + \phi_{s_M} \tag{21.14}$$

If either current or one of the coils was reversed, the fluxes would be subtractive, giving us

$$\phi_{m_{\text{total}}} = \phi_{p_M} - \phi_{s_M} \tag{21.15}$$

664   TRANSFORMERS

Because secondary mutual flux, ϕ_{s_M}, passes through the primary, total primary flux is

$$\phi_{P_{\text{total}}} = \underset{\substack{\nearrow \\ \text{developed by} \\ \text{primary MMF}}}{\phi_p} + \underset{\substack{\nwarrow \\ \text{developed by} \\ \text{secondary MMF}}}{\phi_{s_M}}$$

Substitute Equation 21.10 for ϕ_p:

$$\phi_{P_{\text{total}}} = \phi_{p_M} + \phi_{p_L} + \phi_{s_M} \quad (21.16)$$

Any change in $\phi_{P_{\text{total}}}$ induces voltage across the primary according to Faraday's law. Another component of primary voltage is the IR drop across winding resistance. Combining these components in an equation, we have

$$E_p = I_p R_p + N_p \frac{d\phi_{P_{\text{total}}}}{dt} \quad (21.17)$$

Substitute Equation 21.16 for $\phi_{P_{\text{total}}}$:

$$E_p = I_p R_p + N_p \frac{d[\phi_{p_M} + \phi_{p_L} + \phi_{s_M}]}{dt}$$

or

$$E_p = I_p R_p + N_p \frac{d[\phi_{p_M} + \phi_{p_L}]}{dt} + N_p \frac{d\phi_{s_M}}{dt} \quad (21.18)$$

It is important that we understand each of the voltages in Equation 21.18:

- $I_p R_p$: voltage drop across winding resistance. Considered zero in ideal model.
- $N_p \frac{d(\phi_{p_M} + \phi_{p_L})}{dt}$: a function of changing I_p. Normal voltage across a coil that occurs with changing current. The I_p needed to establish these two flux components with $I_s = 0$ (secondary open) is called *magnetizing*, or *excitation*, current.
- $N_p \frac{d\phi_{s_M}}{dt}$: a function of changing I_s. Present whether I_p is flowing or not.

A similar equation can be written for the secondary:

$$E_s = I_s R_s + N_s \frac{d[\phi_{s_M} + \phi_{s_L}]}{dt} + N_s \frac{d\phi_{p_M}}{dt} \quad (21.19)$$

EXAMPLE 21.7

A transformer with primary and secondary having negligible resistance and 35 and 92 turns, respectively, develops a primary flux of 28 W. The flux changes 1.7 W/s and 5.6 W link only the primary.

a. Find the leakage flux.

$$\phi_L = \text{flux linking only the primary} = 5.6 \text{ Wb}$$

21.5 MUTUAL INDUCTANCE

b. Find ϕ_m.

$$\phi_m = \phi_p - \phi_{pL}$$
$$= 28 - 5.6$$
$$= 22.4 \text{ Wb}$$

c. Find k.

$$k = \frac{\phi_{pM}}{\phi_p}$$
$$= \frac{22.4}{28}$$
$$= 0.8$$

d. Find the voltage induced across the primary and open-circuited secondary.

$$E_p = N_p \frac{d\phi_p}{dt}$$
$$= \frac{35(1.7)}{1}$$
$$= 59.5 \text{ V}$$

$$E_s = N_s \frac{d\phi_{pM}}{dt}$$
$$= N_s k \frac{d\phi_p}{dt}$$
$$= \frac{92(0.8)1.7}{1}$$
$$= 125.1 \text{ V} \quad \blacksquare$$

21.5 MUTUAL INDUCTANCE

Equations 21.18 and 21.19 of the preceding section can be put in a more convenient form by rewriting the Faraday law portions in terms of inductance. This is done with the equality

$$e = N \frac{d\phi}{dt}$$
$$= L \frac{di}{dt}$$

(see the derivation of Equation 12.2.)

When this substitution is made on the second term of Equation 21.18, $N_p[d(\phi_{pM} + \phi_{pL})]/dt$ becomes $L_p(di_p/dt)$, where L_p is the primary inductance. As seen earlier, this voltage is the normal potential difference developed across an inductance by a changing current.

The third term of Equation 21.18, $N_p[d\phi_{s_M}/dt]$, is also rewritten in terms of inductance but because it depends on mutual flux, it is called *mutual inductance* and represented by an *M* (rather than an *L*) to make it stand out.

When these changes are substituted into Equation 21.18, we get

$$E_p = I_p R_p + L_p \frac{dI_p}{dt} + M \frac{dI_s}{dt} \qquad (21.20)$$

Similar substitutions in Equation 21.19 yield

$$E_s = I_s R_s + L_s \frac{dI_s}{dt} + M \frac{dI_p}{dt} \qquad (21.21)$$

for the secondary. Note that M is the same in both equations; there is no M_p or M_s: a transformer has a single value of M.

Our discussions so far have centered on the coils of Figure 21.5, in which currents entered dotted terminals creating additive fluxes. This caused positive mutual flux and therefore positive mutual inductance. If, however, either coil was wound in the opposite direction or either current left a dot, the fluxes, as we saw earlier, would be subtractive. Mutual inductance is now negative. Remember:

> Currents flowing the same direction with respect to dotted terminals produce positive mutual inductance.

Figure 21.3, parts (a) and (b), have $+M$, while part (c), with subtractive fluxes, has $-M$.

Negative mutual inductance decreases coil voltages rather than increases them as before. As a result, Equations 21.20 and 21.21 become

$$E_p = I_p R_p + L_p \frac{dI_p}{dt} - M \frac{dI_s}{dt} \qquad (21.22)$$

and

$$E_s = I_s R_s + L_s \frac{dI_s}{dt} - M \frac{dI_p}{dt} \qquad (21.23)$$

Let us consider next two examples of the effects of positive and negative mutual inductance.

Positive and Negative M Between Series-Connected Coils

The total inductance of series-connected coils was shown in Chapter 12 to be $L_t = L_1 + L_2 + \cdots + L_n$. That formula applies when there are no flux linkages between coils [see Figure 21.6, part (a)]. Let us see what happens when there are linkages.

The M with arrows above the series-connected coils of Figure 21.6, part (b), indicates mutual inductance between them. Dot position shows that any current, regardless of direction, must travel the same direction with respect to the dotted terminals. M is therefore positive.

From Equations 21.20 and 21.21, we have

$$e_1 = L_1 \frac{di_1}{dt} + M \frac{di_2}{dt}$$

and

$$e_2 = L_2 \frac{di_2}{dt} + M \frac{di_1}{dt}$$

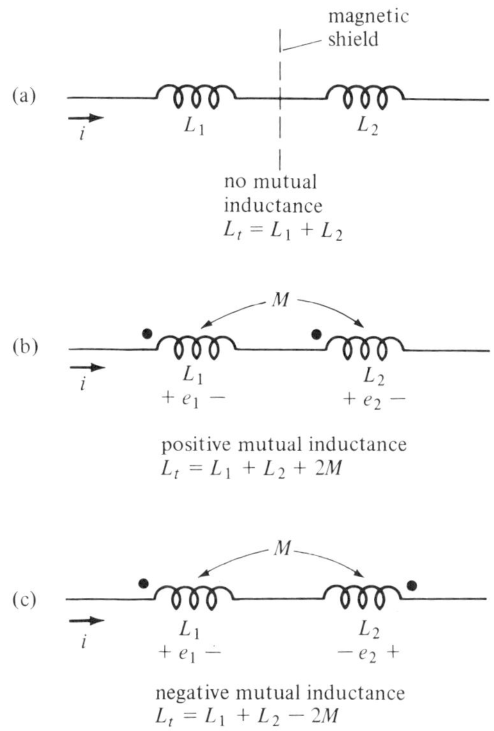

21.5 MUTUAL INDUCTANCE

FIGURE 21.6 Series-Connected Coils with Mutual Inductance.

if iR drops are disregarded. But

$$i_2 = i_1 = i$$

and

$$e_{total} = e_1 + e_2$$

Therefore,

$$e_{total} = L_1 \frac{di}{dt} + M \frac{di}{dt} + L_2 \frac{di}{dt} + M \frac{di}{dt}$$

$$= \frac{di}{dt}(L_1 + L_2 + 2M)$$

$$= \frac{di}{dt} L_{total}$$

Therefore, total inductance is

$$L_{total+M} = L_1 + L_2 + 2M \qquad (21.24)$$

in which the (+) sign indicates positive mutual inductance.

Figure 21.6, part (c), illustrates series-connected coils with negative M. By a similar derivation, the total inductance is shown to be

$$L_{total-M} = L_1 + L_2 - 2M \qquad (21.25)$$

EXAMPLE 21.8

Find total inductance in each case.

a.

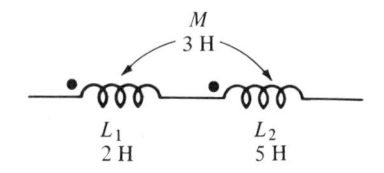

$$L_1 = 2\text{ H}, \quad L_2 = 5\text{ H}, \quad M = 3\text{ H}$$

The dot position indicates $+M$. Therefore,

$$L_{\text{total}} = L_1 + L_2 + 2M$$
$$= 2 + 5 + 2(3)$$
$$= 13\text{ H}$$

b.

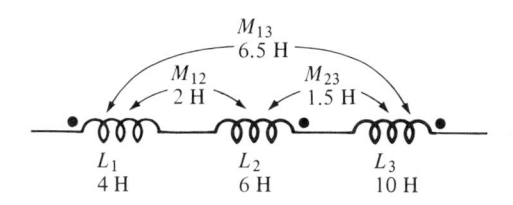

$$L_1 = 4\text{ H}, \quad L_2 = 6\text{ H}, \quad L_3 = 10\text{ H}$$
$$M_{12} = 2\text{ H}, \quad M_{23} = 1.5\text{ H}, \quad M_{13} = 6.5\text{ H}$$

By the dots we see that M_{12} and M_{13} are negative while M_{23} is positive.

$$L_{\text{total}} = L_1 + L_2 + L_3 - 2M_{12} + 2M_{23} - 2M_{13}$$
$$= 4 + 6 + 10 - 2(2) + 2(1.5) - 2(6.5)$$
$$= 20 - 4 + 3 - 13$$
$$= 6\text{ H}$$

■

A transformer's mutual inductance is easily determined with an inductance bridge by measuring total inductance with the windings in series. One winding's connections are then reversed and the measurement repeated. The larger of the two readings occurs for $+M$ and equals $L_p + L_s + 2M$. The smaller, for $-M$, is $L_p + L_s - 2M$. Subtracting the smaller from the larger gives $L_{\text{total}_{\text{max}}} - L_{\text{total}_{\text{min}}} = 4M$. Solve for M. L_p and L_s are easily measured on the bridge with the opposite winding open-circuited.

Negative *M* in a Transformer with Loaded Secondary

As always, the dotted terminals on the transformer of Figure 21.7 have the same polarity. However, current direction tells us that the primary is a load taking energy from the source. The secondary, on the other hand, is a source. Voltage induced in it forces current that delivers energy to the load.

A check of the currents in Figure 21.7 shows I_p entering a dot, I_s leaving. M is therefore negative, which means that the mutual flux, ϕ_{s_M}, developed by $N_s I_s$, subtracts

21.5 MUTUAL INDUCTANCE

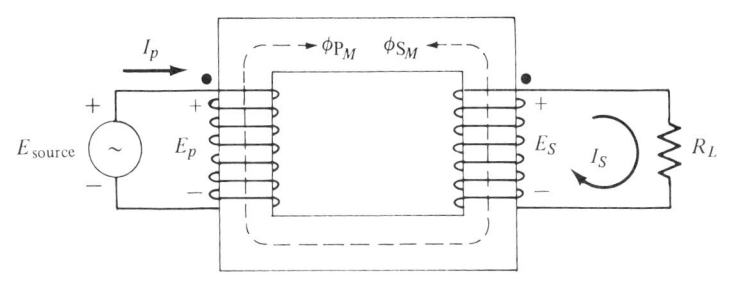

FIGURE 21.7 Loaded Transformer has Negative Mutual Inductance.

from that due to I_p, namely, ϕ_{p_M}. But as we have seen, ϕ_{total} remains constant, which means that I_p increases to develop greater primary magnetomotive force. This is the self-regulating feature discussed earlier.

That regulating feature is easily seen with the aid of an equation for total magnetomotive force developed by the primary:

$$\text{MMF}_{pri_{total}} = N_p I_{exciting} + N_p \frac{I_s}{a} \tag{21.26}$$

The first term (considered zero in the ideal) accounts for leakage and the establishing of minimal core flux. It requires about 3 to 6% of the primary ampere turns of a well-designed transformer and is nearly constant no matter what I_s. The second term varies directly with I_s, causing total primary MMF to do likewise.

Finally, let us rewrite the equation for the coefficient of coupling, k, as a function of inductance (it was originally defined in flux terms). By a series of steps similar to that used to rewrite Equations 21.18 and 21.19, k is redefined in the more convenient form

$$k = \frac{M}{\sqrt{L_p L_s}} \tag{21.27}$$

EXAMPLE 21.9

One way to obtain variable inductance is to use mutually coupled series-connected coils arranged mechanically so that one coil can be moved relative to the other. This changes the degree of coupling, and thus k. If $L_1 = 16$ mH and $L_2 = 9$ mH, over what range must k vary in order that total inductance of the series combination vary from 27 to 44 mH?

By Equation 21.24,

$$L_{total} = L_1 + L_2 + 2M$$

Substitute for M from Equation 21.27:

$$L_{total} = L_1 + L_2 + 2k\sqrt{L_1 L_2}$$
$$= 16 \text{ mH} + 9 \text{ mH} + 2k\sqrt{(16 \text{ mH})(9 \text{ mH})}$$
$$= 25 \text{ mH} + 2k(12 \text{ mH})$$
$$L_{min} = 27 \text{ mH} = 25 \text{ mH} + 24 \text{ mH}(k_{min})$$

Solve for k_{min}:

$$k_{min} = \frac{(27-25)\text{ mH}}{24\text{ mH}}$$

$$= 0.083 \quad \text{very light or loose coupling, coils far apart.}$$

$$L_{max} = 44\text{ mH} = 25\text{ mH} + 24\text{ mH}(k_{max})$$

$$k_{max} = \frac{(44-25)\text{ mH}}{24\text{ mH}}$$

$$= \frac{19}{24}$$

$$= 0.79 \quad \text{tight or close coupling, coils close together.} \quad \blacksquare$$

21.6 EQUIVALENT CIRCUIT OF THE IRON-CORE TRANSFORMER

If certain "impurity" components are combined with an ideal transformer, a *model* or *equivalent circuit* of an actual transformer results. It can be useful in achieving understanding of iron-core transformers, particularly those operating over a broad frequency range. Let us examine the reason each component is included and its effect on the circuit's frequency response.

The equivalent circuit is seen in Figure 21.8. R_p and R_s are the $\rho l/A$ resistances of the primary and series, respectively. Heat developed by wasted energy in these resistances is called *copper loss*. Series inductances, L_p and L_s, are the result of the leakage flux discussed earlier. Capacitors, C_p and C_s, are formed by two layers of windings separated by insulating paper placed between layers. C_{ps} also has paper insulation for a dielectric, but its conductors are the primary and secondary coils, not two layers of the same coil.

After passing through R_p and L_p, primary current, I_p, splits into I'_p and I_{excite}. I'_p flows to the ideal transformer primary to develop the ampere-turns needed to hold constant flux against negative mutual inductance. Part of I_{excite} flows through L_m, the magnetizing inductance, to establish the mutual flux always present in a real transformer (it was neglected in the ideal approximation). The remainder passing through R_c accounts for energy loss in the core.

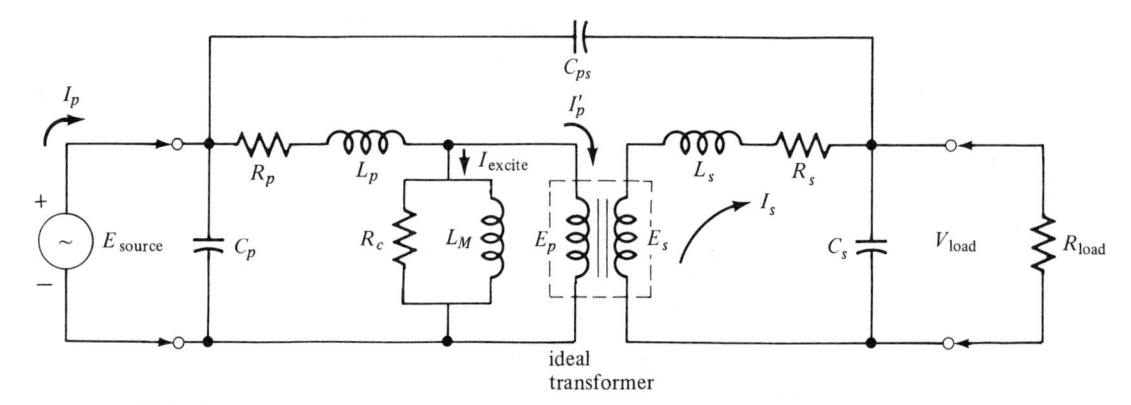

FIGURE 21.8 Equivalent Circuit of the Iron-Core Transformer.

21.7 LOOSELY-COUPLED TRANSFORMERS   671

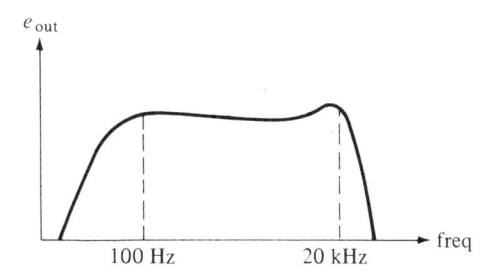

FIGURE 21.9 Frequency Response of Typical Audio Transformer.

Core loss, we recall from Chapters 11 and 12, is caused by eddy currents and hysterisis. Eddy currents are currents formed in the core because the core acts as a secondary in the primary's field. These currents waste energy as heat and are minimized by laminating the core to raise its resistance. In high-frequency applications, ferrite cores must be used as laminated cores become too lossy. Hysterisis, also resulting in heat loss, is the result of energy needed to shift magnetic orientation of core atoms.

Let us determine next the frequency response characteristics of a typical audio transformer, that is, one made to transform signals over the audio frequency range from about 20 Hz to 20 kHz. To do this, e_{source} is held constant and its frequency varied over the desired range. At selected frequencies, perhaps every 200 Hz, e_{out} is read and plotted with frequency on the X axis, e_{out} on the Y axis. Figure 21.9 shows a typical response curve.

At frequencies below approximately 100 Hz, e_{out} is low because the reactance of L_m shorts the primary. As the frequency climbs, L_m reactance increases, causing the output to do likewise. However, it is shunted by a constant R_c and is of no further importance for frequencies above approximately 100 Hz (the reactance of L_p and L_s is too low at these frequencies to have any affect). As a result, the curve begins the *flat region*, the group of frequencies for which transformer operation most nearly approximates the ideal. Audio transformer designers attempt to make this region perfectly flat over the entire audio range so that all frequencies are passed equally.

Just above the flat region a low-Q resonant voltage rise occurs when C_s series resonates with L_s. Above this point the response falls or drops off steeply as C_s shorts the output at high frequencies.

21.7 LOOSELY-COUPLED TRANSFORMERS

The idealized approximations developed in Section 21.1 applied for tightly coupled iron-core power transformers, those whose high k values could be assumed equal to 1. Air-core transformers used in communications at RF (radio frequencies), however, are loosely coupled for Q and bandwidth considerations. They have very low k values, perhaps only 0.01. Consequently, they require a different analytical approach that produces some interesting results.

Consider the equivalent circuit of a loosely coupled transformer with loaded secondary in Figure 21.10. Note the positive mutual inductance between windings and the lack of core elements, R_c and L_m, in the primary circuit.

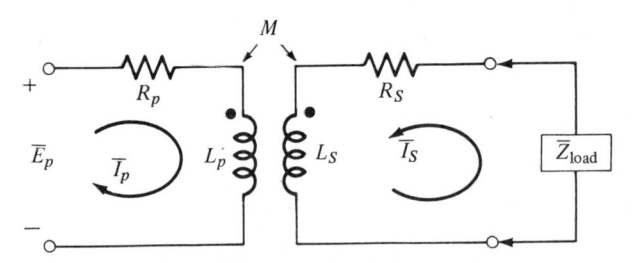

FIGURE 21.10 Equivalent Circuit of Loosely Coupled Transformer with Loaded Secondary.

Writing coupled-circuit loop equations around primary and secondary circuits:

$$\bar{E}_p = (R_p + j\omega L_p)\bar{I}_p + j\omega M \bar{I}_s \qquad (21.28)$$
$$= \bar{Z}_p \bar{I}_p + j\omega M \bar{I}_s$$
$$0 = j\omega M \bar{I}_p + (R_s + \bar{Z}_{\text{load}} + j\omega L_s)\bar{I}_s$$
$$= j\omega M \bar{I}_p + (\bar{Z}_s + \bar{Z}_{\text{load}})\bar{I}_s \qquad (21.29)$$

where

$$\bar{Z}_p = R_p + j\omega L_p$$
$$\bar{Z}_s = R_s + j\omega L_s$$

Solve equation 21.29 for I_s:

$$\bar{I}_s = \frac{-j\omega M \bar{I}_p}{\bar{Z}_s + \bar{Z}_{\text{load}}}$$

Substitute into equation 21.28:

$$\bar{E}_p = (R_p + j\omega L_p)\bar{I}_p + j\omega M \frac{-j\omega M}{\bar{Z}_s + \bar{Z}_{\text{load}}} \bar{I}_p$$

Factor out I_p:

$$= \bar{I}_p \left(R_p + j\omega L_p + \frac{(\omega M)^2}{\bar{Z}_s + \bar{Z}_{\text{load}}} \right)$$

Solve for $\bar{Z}_{\text{in}}$, the transformer input impedance:

$$\bar{Z}_{\text{in}} = \frac{\bar{E}_p}{\bar{I}_p} = R_p + j\omega L_p + \frac{(\omega M)^2}{\bar{Z}_s + \bar{Z}_{\text{load}}}$$
$$= \bar{Z}_p + \frac{(\omega M)^2}{\bar{Z}_s + \bar{Z}_{\text{load}}} \qquad (21.30)$$

The first term in Equation 21.30 is the primary circuit impedance, $R_p + j\omega L_p$. The second results from the coupling of secondary impedance into the primary and is called *coupled*, or *reflected*, impedance. It equals the mutual reactance squared divided by total secondary impedance. As a result of the squaring, Z_{in} is independent of the sign of M. In other words, *dots make no difference* here.

Note also that total secondary impedance is in the denominator of the right-hand term of Equation 21.30. Division moves the reactive part of this impedance to the numerator with the opposite sign. Therefore, if Z_{load} is resistive, $j\omega L_s$ becomes $-j\omega L_s$,

21.7 LOOSELY-COUPLED TRANSFORMERS

a capacitive reactance, in the primary circuit. This cancels all or part of $j\omega L_p$, thereby reducing Z_{in}.

One final point. The reflected secondary impedance in Section 21.3 replaced the ideal transformer primary. In our present, loosely coupled, nonideal case, secondary impedance comes over in series with $Z_{primary}$. The transformer is eliminated in both cases, however.

To summarize some significant differences between tightly and loosely coupled transformers:

	Tightly coupled	Loosely coupled
k	≈ 1	$\ll 1$
Reflected Z	$a^2 Z_s$	$\dfrac{(\omega M)^2}{Z_s}$
	Replaces primary	In series with primary
Dotted terminals	Necessary	Unnecessary
Sign of reflected X_s	Unchanged	Reversed

EXAMPLE 21.10

Couple the secondary impedance into the primary and find i_p.

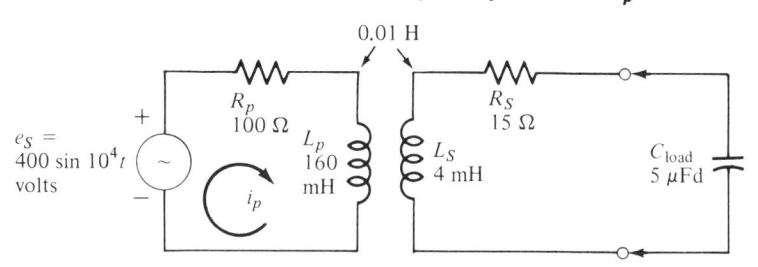

$e_s = 400 \sin 10^4 t$ volts; R_p 100 Ω; L_p 160 mH; $M = 0.01$ H; R_s 15 Ω; L_s 4 mH; C_{load} 5 μFd.

Find $\bar{Z}_{in}$:

$$\bar{Z}_{in} = \bar{Z}_p + \frac{(\omega M)^2}{\bar{Z}_s + \bar{Z}_{load}}$$

$$= R_p + j\omega L_p + \frac{(\omega M)^2}{R_s + j\omega L_s - j(1/\omega C_{load})}$$

$$= 100 + j10^4(160 \times 10^{-3}) + \frac{(10^4 \times 10^{-2})^2}{15 + j10^4(4 \times 10^{-3}) - j[1/10^4(5 \times 10^{-6})]}$$

$$= 100 + j1600 + 240 - j320$$

$$= 340 + j1280 \quad \text{or} \quad 1324.4 \,\underline{/75.1°}\, \Omega$$

Calculate i_p:

$$i_p = \frac{e_s}{\bar{Z}_{in}} = \frac{400}{1324.4 \,\underline{/75.1°}}$$

$$= 302 \sin(10^4 t - 75.1°) \text{ mA}$$

Secondary voltage and current can be obtained for loosely coupled circuits by first considering Equation 21.29 again:

$$0 = j\omega M \bar{I}_p + (\bar{Z}_s + \bar{Z}_{load})\bar{I}_s$$

or

$$-j\omega M \bar{I}_p = (\bar{Z}_s + \bar{Z}_{load})\bar{I}_s$$

Note the units of the left-hand term are volts (ohms times amps). This is the voltage induced in the secondary by changing primary current. Therefore, rewriting yields

$$\bar{E}_s = -j\omega M \bar{I}_p \tag{21.31}$$

which, of course, means

$$\bar{I}_s = \frac{\bar{E}_s}{\bar{Z}_s + \bar{Z}_{load}}$$

EXAMPLE 21.11

Find $\bar{E}_s$ and $\bar{I}_s$ for the circuit of Example 21.10.

Substitute the values from Example 21.10 into Equation 21.31:

$$\bar{E}_s = -j\omega M \bar{I}_p$$
$$= (10^4)(10^{-2})\underline{/-90°}(0.707)(302 \times 10^{-3})\underline{/-75.1°}$$
$$= 21.4 \underline{/-165.1°} \text{ V}$$

$$\bar{I}_s = \frac{\bar{E}_s}{\bar{Z}_s + \bar{Z}_{load}}$$

From Example 21.10:

$$\bar{Z}_s + \bar{Z}_{load} = 15 + j20$$

Therefore,

$$\bar{I}_s = \frac{21.4 \underline{/-165.1°}}{15 + j20}$$
$$= 0.86 \underline{/-218.2°} \text{ A}$$

21.8 OTHER TYPES OF TRANSFORMERS

There are numerous types of transformers designed for specific applications. Let us describe some of the more general kinds.

Audio Transformers

An audio transformer passes or, *couples*, signals between parts, or *stages*, of an audio amplifier, that is, an amplifier amplifying audio signals (≈ 20 to 20,000 Hz). Such a transformer must pass all these frequencies equally in both amplitude and phase; otherwise, distortion occurs. This requires careful design and high-quality core materials to minimize unwanted reactances and distortion from changing core characteristics.

21.8 OTHER TYPES OF TRANSFORMERS

Pulse Transformers

As we shall see in Chapter 23, pulses consist of an infinite number of superimposed sinusoids of increasing frequency. Pulse transformers, like audio transformers, must therefore pass a range of frequencies. For the pulse unit, however, that range extends into the megahertz. As a result, a pulse transformer must have an extra-high-permeability core, very low leakage ($k > 99\%$), and minimum interwinding capacitance.

Isolation Transformers.

One of the most important jobs of transformers is *isolation*, the passing of signals between circuits with no direct connection between those circuits. This may be done for noise reduction, to block dc, or to raise a system above ground. For example, Figure 21.11, part (a), illustrates a voltage measurement by test equipment having one input terminal grounded (as is usually the case). As a result, Z_1 is shorted; the test equipment has affected that which is being tested. Figure 21.11, part (b), shows how a 1:1 turns-ratio isolation transformer raises the Z_2 potential above ground to allow a valid reading. Of course, care must be taken to ensure that the isolation transformer's frequency characteristics do not affect the signal being passed.

Tapped Transformers

Transformer windings are often made with connections called *taps*, brought out from certain points in the coil. Center taps, identified by CT on schematics [see Figure 21.12, part (a)], are most common. They split primaries and/or secondaries into balanced halves to provide two signals 180° out of phase. Taps also provide different output voltages or a choice of impedance-matching ratios [see Figure 21.12, part (b)].

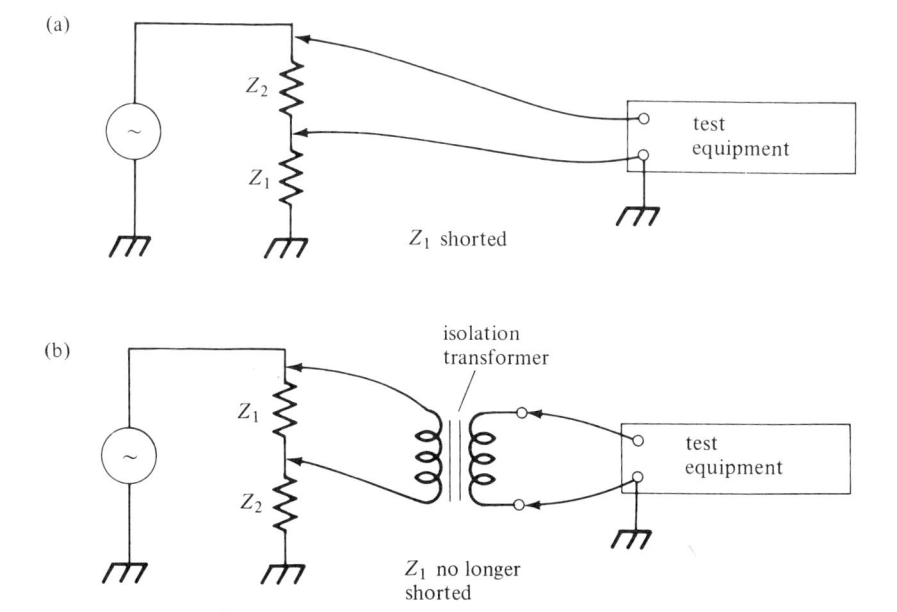

FIGURE 21.11 Use of Isolation Transformer to Raise a Potential Above Ground.

(a) center–tapped pri. and sec.

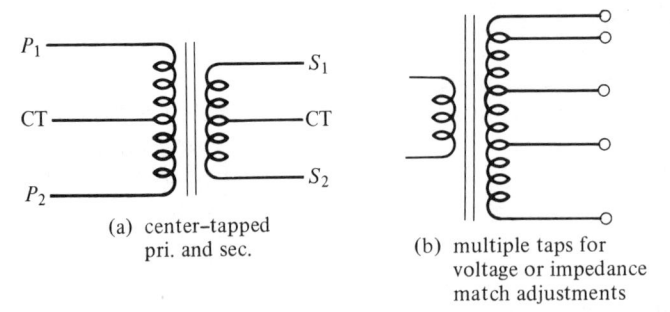

(b) multiple taps for voltage or impedance match adjustments

FIGURE 21.12 Tapped Transformers.

Autotransformers

The iron-core transformer in Figure 21.13, part (a), is designed to deliver 20 V(18 A) = 360 VA to a load without overheating. Suppose that we could disassemble the transformer and reassemble it as shown in part (b) of the figure, with the undotted secondary terminal connected to the dotted primary terminal. Let us consider the resulting voltage, current and volt-ampere capabilities.

Assuming that the load continues to draw 18 A from the secondary, the primary will have 3 A through it and induce 20 V across the secondary. However, because of series addition of E_{pri} and E_{sec}, load voltage is now 120 + 20 = 140 V. Since maximum secondary current remains at 18 A, the transformer now delivers an apparent power to the load of 140(18) = 2520 VA. This is an increase of 2520/360 = 7 times. The transformer that achieves this increase in apparent power by having its primary and secondary connected in series is called an *autotransformer*.

Although autotransformers have the advantage of increased VA capability, they have a distinct disadvantage in their direct connection between input and output. This

(a)

$VA_{pri} = 120(3) = 360$ VA
$VA_{sec} = 20(18) = 360$ VA

(b)

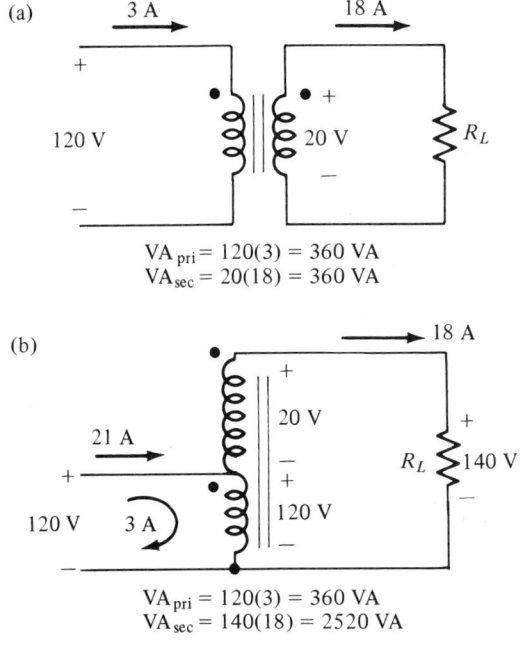

$VA_{pri} = 120(3) = 360$ VA
$VA_{sec} = 140(18) = 2520$ VA

FIGURE 21.13 Comparing Conventional Transformer and Autotransformer.

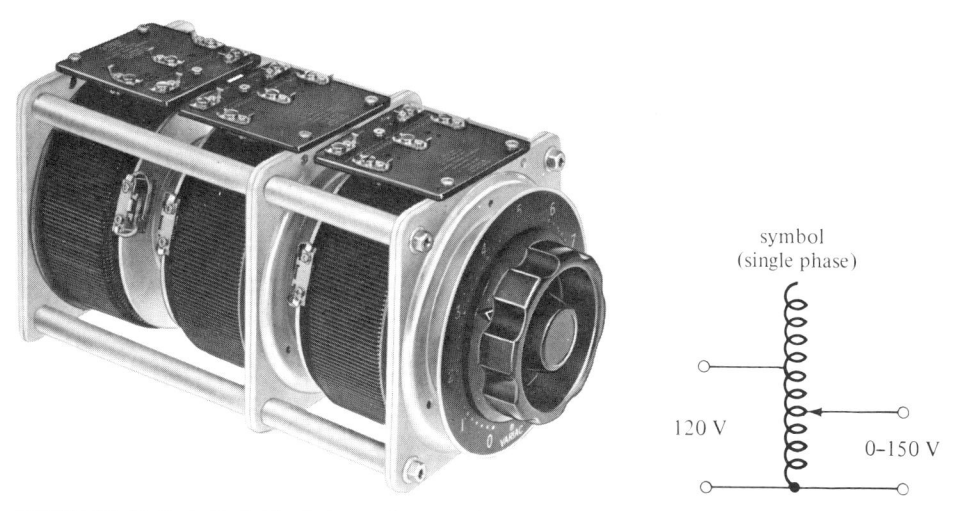

FIGURE 21.14 3-ϕ Variable Autotransformer. (Courtesy of Technipower)

eliminates the isolation provided by two-winding transformers and can be the cause of short circuits if care is not taken during wiring.

Autotransformers develop the high-voltage needed to operate TV picture tubes (flyback transformers) as well as generate sparks for gas engines (spark coils). By exposing the windings to contact by a movable carbon brush, a variable transformer is obtained. Sold under the trade name *Variac*, these units provide output from 0 to 150 V with a 120-V input. The diagram and a photo are shown in Figure 21.14. Compare the arrow indicating variability to the symbol for fixed taps in Figure 21.12.

An Assignment: Study the pole transformers in your neighborhood. They are various-sized gray cans (filled with cooling oil) approximately 30 to 40 in. high and 15 in. in diameter. Look for a thin wire going to a large insulator (the high-voltage, low-current primary lead; the other side may be grounded) and three thicker wires going from the smaller insulators to the houses (low-voltage, high-current secondary leads; three-wire circuit). Note possible cooling fins welded to the cans and perhaps a red overload light on during peak-load evening hours. Observe the ground wire from the can to a ground rod at the base of the pole. Look for high- and low-voltage circuits and cooling methods on the transformers in a high-tension switching yard.

EXERCISES

Section 21.1

1. Fill in the blanks.

	N_p	N_s	E_p/I_p	E_s/I_s	Step-Up or Step-Down
a.		150	320 mV	12 mV	
b.	490		10 mA	129 mA	
c.	300	45,000		1.8 kV	
d.	630	1837	2.1 A		

2. Explain why a transformer will not work on dc.

3. Economics says: transport electricity at high voltage; safety says: deliver it to consumers at low voltage. Therefore, 14 kV on a street's primary line is reduced to 120-V secondaries by regularly spaced transformers. If these units are rated at 25 kVA, they would be called a 25-kVA 14-kV/120 V transformer. Consider such a unit ideal. Find its: a. turns ratio; b. $I_{p_{max}}$; c. $I_{s_{max}}$.

4. Repeat Exercise 3 if the transformer is rated 15 kVA, 4 kV/120 V.

5. A transformer drops line voltage of 120 V to 15 V to run a battery charger. If the primary has 485 turns, how many are on the secondary?

6. Inverters are available that convert a car's 12 V dc to 110 V ac. The step-up action is done by a transformer. How is this possible since transformers do not transform dc? *Hint:* When does core flux change if dc is applied to a primary?

7. Fill in the blanks.

	Z_p (Ω)	a	Z_{load} (Ω)	Step-Up or Step-Down
a.		5	580 Ω	
b.	950 Ω		594 k	
c.	375 Ω	1.3		
d.	50 Ω		5 k	

8. A transformer is rated 60 kVA, 4 kV/220 V.
a. Determine Z_{load} to fully load the transformer.
b. Determine the value of this impedance when it has been reflected to the primary.

Section 21.2

9. Connect all four secondaries to obtain each of the voltages: a. 80 V; b. 0 V; c. 10 V; d. 70 V. Put one voltmeter terminal on pin C, the other on either I or J.

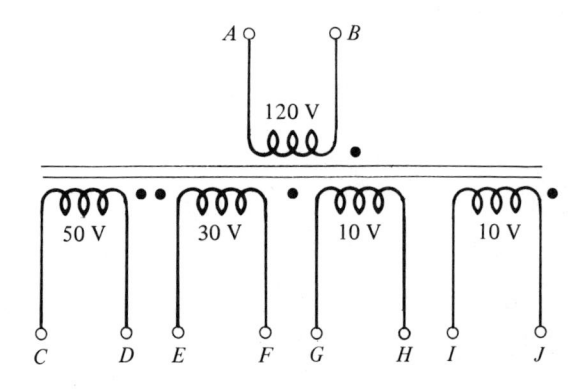

10. For a transformer rated 65 kVA, 14 kV/240 V:
a. Find the turns ratio.
b. Find $I_{p_{max}}$.
c. Find $I_{s_{max}}$.
d. Reverse the transformer, putting the 14 kV in on what was the secondary but is now the primary. Assuming that the insulation can stand the potential, find the new E_s and $I_{s_{max}}$.

EXERCISES

Section 21.3

11. The transformers in this circuit are ideal. Find I_1, I_2, and I_3.

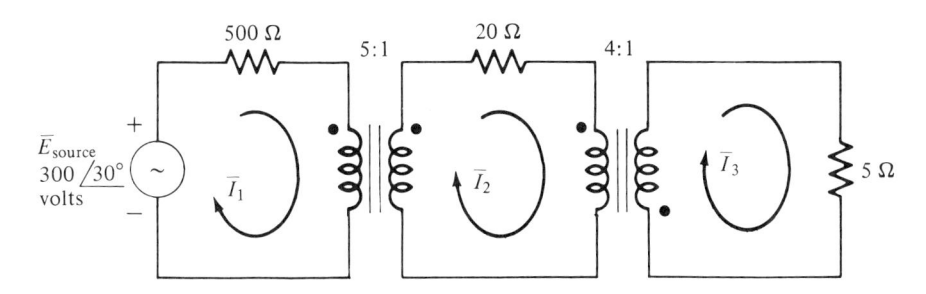

12. Repeat Exercise 11, changing the 5-Ω resistive load to a reactance of $j2$ Ω.

13. Find I_p and I_s by first reflecting the secondary circuit into the primary (the transformer is ideal).

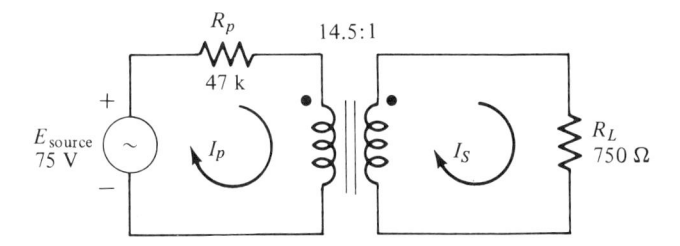

14. Repeat Exercise 13 if Z_{load} is changed to a reactance of $-j400$ Ω and the turns ratio is changed to 6.7:1.

15. Repeat Exercise 13 to find I_s by reflecting the primary into the secondary.

16. Find the load current by reflecting the primary circuit into the secondary.

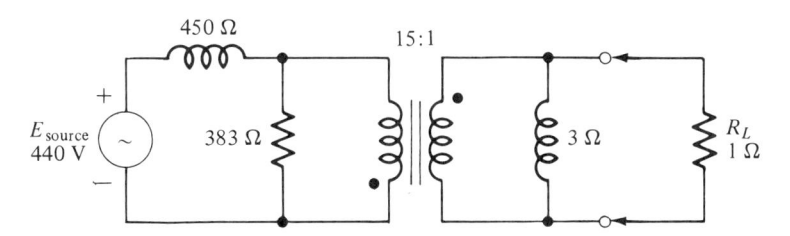

Section 21.4

17. For a current of 0.8 A, a transformer develops a flux of 15.3 W, 10.7 W of which passes through the secondary. Find:

a. The mutual flux.   b. The leakage flux.   c. The coefficient of coupling.

18. How many magnetomotive forces are there in a. an unloaded transformer; b. a loaded transformer? Repeat for the number of fluxes (consider total mutual flux as one).

19. A transformer having negligible winding resistance and primary and secondary turns of 75 and 22 turns, respectively, develops a primary flux of 12.3 Wb. The flux changes 2.6 Wb/100 ms and the leakage is 4.9 Wb.

a. Find the mutual flux.
b. Find k.
c. Find the voltage induced across the primary and open-circuited secondary.

20. Repeat Exercise 19 if the leakage is reduced to 1.2 Wb.

Section 21.5

21. A transformer with primary and secondary inductances of 5 H and 12 H, respectively, has a positive mutual inductance of 3 H. If the primary current changes 3 A/s and the secondary current, 1.5 A/s, what are E_p and E_s? The winding resistance is negligible.

22. Repeat Exercise 21 if M is negative.

23. Fill in the blanks.

	L_1 (H)	L_2 (H)	L_{total} (H)	M (H)	K
a.	5 H	3 H	4 H		
b.	9 H			−5 H	0.43
c.	3 mH	6 mH		+	0.94
d.	L_1 and L_2 equal valued			17.4 H	0.87
e.	35 H		63 H	5 H	

24. Explain why an ideal transformer draws power from the line only when it is delivering power to its load.

25. Two coils have a total inductance of 22 H when connected in series aiding, and 10 H when connected in series opposing. Find the mutual inductance.

26. Repeat Exercise 25 for 9 H and 8 H.

27. Determine the total inductance of this series arrangement:

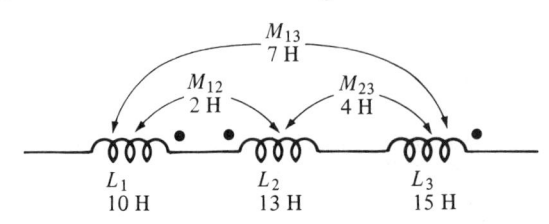

M_{13} 7 H
M_{12} 2 H
M_{23} 4 H
L_1 10 H
L_2 13 H
L_3 15 H

28. Explain why mutual inductance is negative when a transformer's secondary supplies current to a load.

Section 21.6

29. What are the three types of losses in a practical transformer? How is each one accounted for in the equivalent circuit?

30. Describe how the three capacitors in the transformer equivalent circuit are formed. What serves as the dielectric of each? What forms the plates of each?

31. What would be the effect (if any) on the frequency response curve in Figure 21.8 of each of the following.
   a. Decrease L_m.
   b. Decrease C_s.
   c. Increase N_s.
   d. Decrease R_c.
   e. Increase R_p and decrease R_c.

32. What component(s) in the equivalent circuit determine the low-frequency drop-off? The high-frequency drop-off?

Section 21.7

33. A transformer secondary has a resistance and inductance of 10 Ω and 6 mH, respectively. If its load consists of an 8-Ω resistance in series with 125 μF, and the mutual inductance is 19 mH, what is the impedance reflected into the primary at 500 rad/s?

EXERCISES

34. How does an open-circuited secondary reflect into the primary of a loosely coupled (nonideal) transformer? Repeat for an ideal transformer.

35. For this circuit, find $\bar{I}_p$, $\bar{E}_s$, and $\bar{I}_s$.

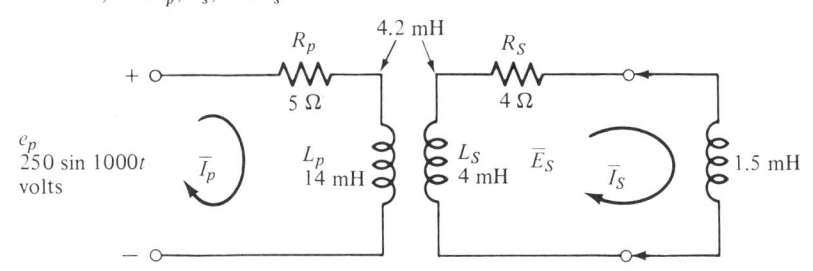

36. Redo Exercise 35 if Z_{load} is changed to series resonate the secondary circuit.

Section 21.8

37. Given: the multiple secondary transformer and loads shown.
   a. Find I_{source}.
   b. By Ohm's law, find the total resistance seen by the source.
   c. Find the total resistance seen by the source by reflecting loads into the primary.
   d. How do reflected loads combine to form R_{total}? Write this as a rule of multiload transformer operation.

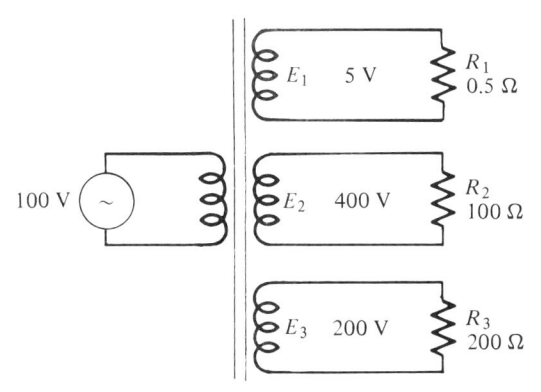

38. Show that two signals 180° out of phase are obtained by grounding the center tap of a transformer's secondary.

39. The multitapped secondary matching transformer shown has 1200 primary turns and is to match a 7500-Ω source to the loads indicated. At which turn should taps b and c be located?

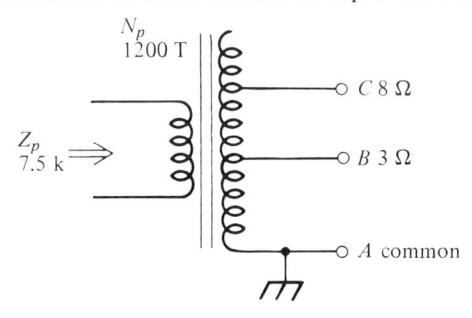

40. The terminals of the secondary in the autotransformer of Figure 21.13 are reversed. Calculate the new V_{load} and VA_{load}.

41. A 25-kVA 440/120-V ideal transformer is connected in the usual two-winding manner.
   a. Calculate maximum I_p and I_s.

b. Connect the secondary to the primary to form an autotransformer like that shown in Figure 21.13 and, assuming that I_s and I_p are at the maximum values of part a, calculate the new load voltage, load kVA, and source current.

c. The autotransformer load resistance increases, causing I_{load} to fall to 130 A. Find the new I_p and I_{source}.

42. Repeat Exercise 41, parts a and b, for a 65-kVA 14-kV/240 V transformer.

CHAPTER HIGHLIGHTS

Iron-core transformers have such low losses they can often be analyzed as though they were ideal.

If voltage is stepped up, current is stepped down, and vice versa.

Transformers are excellent impedance-matching devices. They also serve to isolate one circuit from another and block the passage of dc.

Primaries can be wired in series and parallel as well as secondaries. The dot convention indicates points having the same instantaneous polarity.

Transformers come in many different sizes designed for a wide variety of jobs.

Transformer circuits are regularly analyzed by reflecting the secondary circuit into the primary, and vice versa. However, this reflecting is done differently for tightly coupled (approximately ideal) transformers than it is for loosely coupled ones.

Mutual flux is flux generated by one coil that passes through another.

Three sources of transformer losses are copper, eddy currents, and hysterisis.

Transformer windings are tapped to provide a selection of turns ratios.

Autotransformers have windings in series aiding. They have considerably greater kVA handling ability than that of conventional transformers and are more efficient but have a common connection between input and output.

Formulas to Remember

$\dfrac{N_p}{N_s} = a$ turns ratio

$\dfrac{E_p}{E_s} = \dfrac{N_p}{N_s}$ voltage transformation ratio

$\dfrac{I_s}{I_p} = \dfrac{N_p}{N_s}$ current transformation ratio

$Z_p = a^2 Z_s$ impedance transformation ratio

$k = \dfrac{\phi_m}{\phi_{coil}} = \dfrac{M}{\sqrt{L_1 L_2}}$ coupling coefficient

$E_1 = I_1 R_1 + L_1 \dfrac{dI_1}{dt} + M \dfrac{dI_2}{dt}$ total voltage across coil 1

$L_{total} = L_1 + L_2 \pm 2M$ total inductance of two mutually coupled coils in series aiding (+) or series opposition (−)

$Z_{in} = Z_p + \dfrac{(\omega M)^2}{Z_s + Z_{load}}$ input impedance of loosely coupled transformer

CHAPTER 22

Ac Measurements

Ac measurements are more involved than dc measurements. Besides the usual problem of loading errors, electronics personnel must consider the bandwidth of the test equipment being used. Always remember:

Ac test equipment must have a bandwidth at least as wide as the frequency of the signals being measured.

In other words, do not try to measure a 3-MHz signal with a meter only capable of reading to 20 kHz.

Another problem in ac measurements is that of meters which do not respond to ac voltage or current. For them, the parameter must first be converted, or *rectified*, to dc, a step which causes a marked decrease in meter sensitivity. Finally, there is the need to measure frequency and phase, which is of no concern in dc (impedance measurements were covered in Section 17.2 and will not be discussed further). We study these and other subjects in this chapter and are introduced to the oscilloscope, which, together with the multimeter, are probably the two most frequently used pieces of test equipment in electronics.

At the completion of this chapter, you should be able to:

- Choose the right ac measuring tool for a particular job.
- Explain the need for rectification in ac meters and the difference between full- and half-wave rectification.
- Tell why iron-vane and electrodynamometer ac meters read effective values without the need of rectifiers.
- Describe the action of a thermocouple in measuring ac quantities.
- Measure single- and three-phase ac power.
- Interpret signals displayed on an oscilloscope to determine time, frequency, voltage, and current.
- Use Lissajous patterns to measure frequency and phase.
- Describe the theory behind the basic frequency counter.

22.1 USING THE D'ARSONVAL MOVEMENT FOR AC MEASUREMENTS

Our studies of Chapter 9 showed that torque was developed in the D'Arsonval movement by the interaction of two magnetic fields. These, you will recall, are the fixed field of a permanent magnet and the electromagnetic field created when the current being measured passes through the meter's movable coil. Since torque is proportional to current, a low-frequency ac, perhaps 5 or 10 Hz, causes the needle to swing back and forth. As the frequency increases, however, inertia prevents faster movement and the needle quivers at zero or does nothing. In other words, the meter responds only to current in one direction.

Ac can be made to go in only one direction by converting it to dc in a process called *rectification*. The device that does this is sometimes called a *rectifier*, but the more common name is *diode*.

Diodes are studied in detail in electronics courses. For our purposes, it is enough to say that ideal diodes have no resistance to current going one way, infinite resistance to current going the opposite way. They are somewhat like a valve that allows current in only one direction.

Figure 22.1 [part (a)] shows a meter with a diode in series with its coil resistance and its terminals connected to an unknown ac voltage. On positive alternations [part (b)], when the upper terminal is positive, current flows clockwise because this is the diode's zero resistance direction (current with the arrow). The negative alternations, however, try to push current against the arrow when the diode is an open circuit [part (c)]. The result, plotted in part (d) for three periods, is the positive half of every wave being passed with gaps in between for the missing negative halves. Note that the polarity for this *half-wave* rectified waveform is always positive; meter current now flows in only one direction. The average or dc value of the half-wave rectified waveform is also indicated in Figure 22.1, part (d). Determined by calculus, this

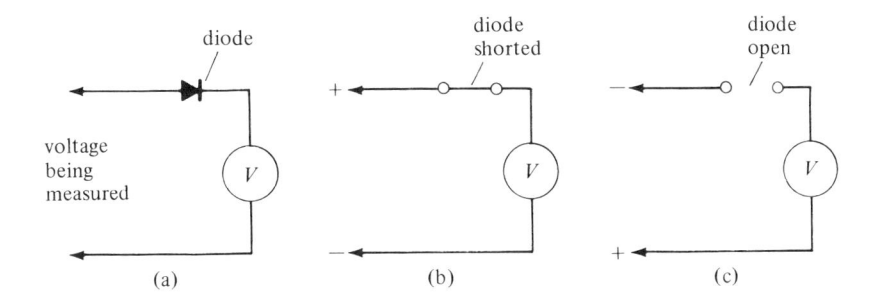

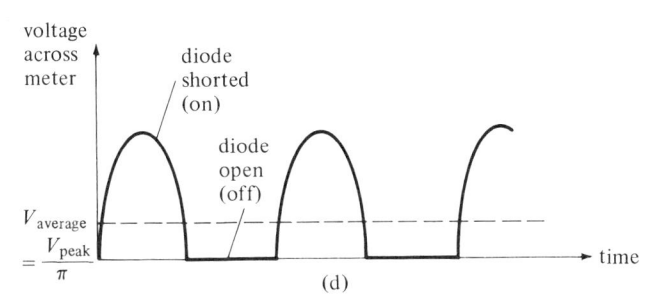

22.1 USING THE D'ARSONVAL MOVEMENT FOR AC MEASUREMENTS

FIGURE 22.1 Half-Wave Rectification.

value is:

For half-wave: $V_{\text{average}} = V_{dc}$
$$= \frac{V_{\text{peak}}}{\pi} \qquad (22.1)$$

The gaps between alternations of Figure 22.1 are filled and twice the average voltage obtained by use of the *full-wave bridge* rectifier seen in Figure 22.2. Follow the currents in this diagram carefully, noting that meter current is always in the same direction regardless of input terminal polarity. The average voltage of this waveform is:

For full-wave: $V_{\text{average}} = V_{dc}$
$$= \frac{2V_{\text{peak}}}{\pi} \qquad (22.2)$$

With twice the area, it is not surprising that this wave has twice the average value of half-wave units.

Reading ac with a D'Arsonval meter is not just a simple case of adding a rectifier between voltage and meter, however. Additional steps must be taken to express the new ac meter's readings in the usual rms form. This is easily done.

From Equation 22.1, we have
$$V_{\text{peak}} = \pi V_{dc}$$

We also know that
$$V_{\text{peak}} = \sqrt{2}\, V_{\text{rms}}$$

Therefore,
$$\sqrt{2}\, V_{\text{rms}} = \pi V_{dc}$$

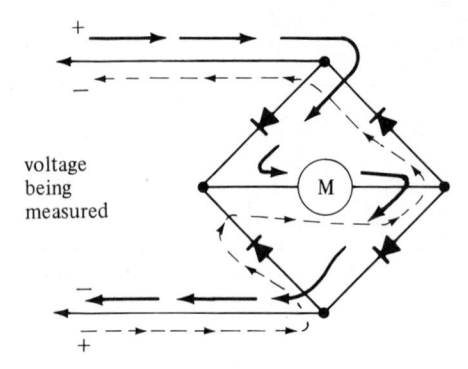

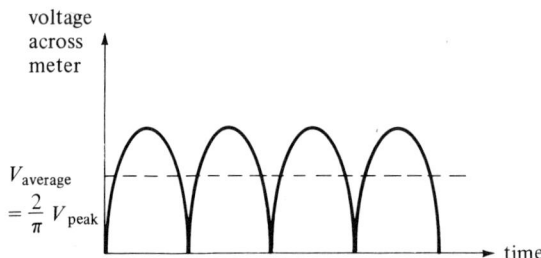

FIGURE 22.2 Full-Wave Bridge Rectification.

Solve for V_{rms}:

$$V_{rms} = \frac{\pi}{\sqrt{2}} V_{dc}$$
$$= 2.22 V_{dc} \qquad (22.3)$$

This equation tells us that with half-wave rectification, the rms value of an ac is 2.22 times the reading on the meter's dc scale. To avoid the inconvenience of multiplying every reading by 2.22, a new ac scale is added to the meter face or compensation made in the series/shunt resistances used to increase scale ranges.

Using the same steps for full-wave, we find that

$$V_{rms} = 1.11 V_{dc} \qquad (22.4)$$

EXAMPLE 22.1

An unknown ac voltage is half-wave rectified by an ideal diode and applied to the 0- to 50-V dc scale of a D'Arsonval meter. The meter reads 18.6 V.

a. Find the unknown's rms value.

$$\begin{aligned} V_{rms} &= 2.22 V_{dc} \\ &= 2.22(18.6) \\ &= 41.3 \text{ V} \end{aligned}$$

b. Find the peak value of the unknown voltage.

$$\begin{aligned} V_{peak} &= \pi V_{dc} \\ &= \pi(18.6) \\ &= 58.4 \text{ V} \end{aligned}$$

c. The rectifier is changed to full-wave. Repeat parts a and b.

$$V_{rms} = 1.11(18.6)$$
$$= 20.7 \text{ V}$$

$$V_{peak} = \frac{\pi}{2} 18.6$$
$$= 29.2 \text{ V}$$ ■

EXAMPLE 22.2

A 20-V battery is measured at 22.3 V by the ac scale of a D'Arsonval meter. What type of rectifier does the meter have?

Since $22.3/20 = 1.12 \approx 1.11$, the rectifier is full-wave. ■

If equations are rewritten in terms of I_{dc}, we find for the half-wave,

$$I_{dc} = \frac{\sqrt{2}}{\pi} I_{rms} = 0.45 I_{rms} \qquad (22.5)$$

which tells us the ac system has only 45% the sensitivity of the dc system. For the full-wave rectifier,

$$I_{dc} = \frac{2\sqrt{2}}{\pi} I_{rms} = 0.9 I_{rms} \qquad (22.6)$$

which is much better but still 10% below dc sensitivity. These figures are further reduced by nonideal diodes and instrument reactances. As a result, the 20,000 Ω/V dc rating of a popular brand of VOM drops to 5000 Ω/V on ac. The bandwidth of such meters is also limited. The unit mentioned above, a typical example, is only useful for frequencies below 20 kHz. The calculations to determine series and shunt range-extending resistances follows the procedures outlined in Chapter 9.

22.2 OTHER TYPES OF AC METERS

The Iron-Vane Movement

The iron-vane meter consists of a doughnut-shaped magnet (see Figure 22.3) and two soft-iron paddles or vanes in the doughnut hole. One of the vanes is fixed, the other rotates on an axle to which a return spring and pointer are attached.

The magnetic field created by coil current induces a field in each vane. Since both of these are in the same direction, the ends of the vanes have the same magnetic polarity. The vanes therefore repel one another with a force proportional to coil current, causing the movable vane to swing like a door on its hinges. Movement continues until balance is reached between spring and magnetic forces.

Because the vane's induced fields have like polarity regardless of coil current direction, the iron-vane movement always reads upscale. Therefore, it not only reads dc but the effective value of ac current and voltage, as well as other waveforms, with no need for a rectifier. Relatively high currents, up to about 5 A, can be measured by winding the coil with a few turns of heavy wire. For voltages to about 150 V, the coil is wound with many turns of fine wire. Instrument transformers (discussed below) are used to extend readings beyond these limits.

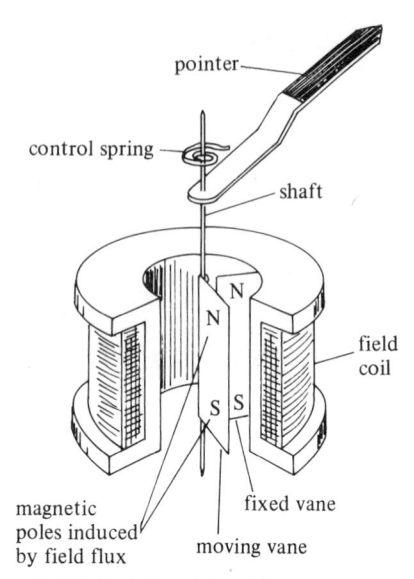

FIGURE 22.3 Iron-Vane Movement.

The disadvantages of the iron-vane meter are: low sensitivity; requires a relatively high power for its own operation; and is limited to use below a few hundred kilohertz because of unavoidable coil inductance.

The Electrodynamometer Movement

As noted in Chapter 9, rotation in the electrodynamometer meter is the result of opposition between magnetic fields of fixed and movable coils. This opposition is a function of the effective value of coil current and, like the iron-vane meter, does not change direction with polarity. Hence this meter also reads the effective value of ac as well as dc. In fact, with small modifications, it will read power factor or frequency.

The meter is also not as sensitive as the D'Arsonval movement, requires fairly high operating power, and is limited to use below about 500 Hz.

Instrument Transformers

Because iron-vane and electrodynamometer meters have coil resistances in the milliohms range, it is difficult to construct and connect low-resistance shunts with which to extend a meter's current range. Hence *current transformers* are used, such as the one illustrated in Figure 22.4. It reduces the various primary currents shown across the top to drive a 5-A movement. In a similar manner, *potential transformers* drop high voltages for reading on low-voltage meters. For safety reasons, this is often done to isolate meters from dangerous potentials.

Thermocouple Meters

Although seriously challenged by electronic techniques, the *thermocouple meter* is still used to measure the effective value of currents and voltages having frequencies above several hundred hertz. The heart of the meter is a *thermocouple* made of two different metals (for example, antimony and bismuth) which develops a voltage proportional to

22.2 OTHER TYPES OF AC METERS

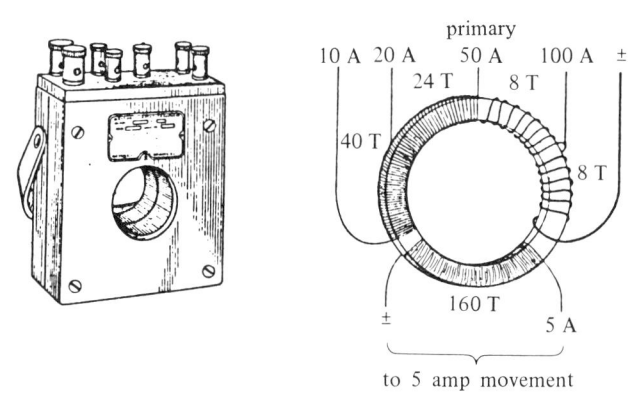

FIGURE 22.4 Instrument Transformer.

the temperature at the junction of the two metals (see Figure 22.5). The signal under test heats a resistance which, in turn, heats the thermocouple. A conventional D'Arsonval meter with a properly calibrated scale reads the resulting voltage.

Although capable of excellent results to 100 MHz or more, these meters have poor low-frequency response, change readings at a sluggish rate that can be annoying, and are easily burned out because heat is proportional to the square of applied voltage or current.

True RMS Meters

In Section 22.1 we learned that the rms value of rectified sinusoids is $2.22 V_{dc}$ and $1.11 V_{dc}$ for half- and full-wave, respectively. A meter's ac scales are calibrated to reflect this. Note, however, that this calibration is for sinusoids only. Should any other nonsinusoidal, periodic waveform be applied to the meter, the diodes will rectify it and the meter will give a reading. But that reading will not represent the actual rms (effective) value of that particular waveform (except by coincidence).

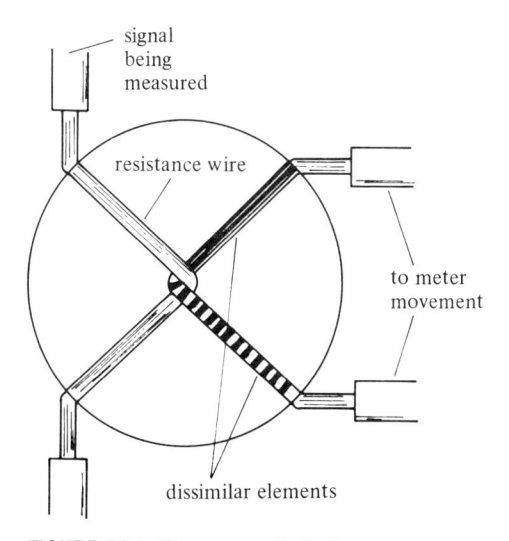

FIGURE 22.5 Thermocouple Meter.

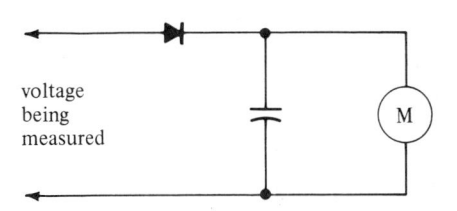

FIGURE 22.6 Peak-Reading Voltmeter.

To measure the rms value of any waveform requires a meter that reads *true rms*. Although iron-vane, electrodynamometer, and thermocouple units do this, electronic techniques using high-reliability, low-cost integrated circuits are incorporated in modern digital multimeters (DMMs) and other test equipment. These solid-state "chips" make true rms measurements by squaring an input signal instant by instant, taking the mean of these squares, then determining the root of that mean and displaying it.

Peak-Reading Voltmeters

The diode in Figure 22.6 passes the positive alternation of the input sinusoid, charging the capacitor to the wave's peak value. If the meter's input resistance is sufficiently high to prevent rapid capacitor discharge between cycles, the meter provides a reliable indication of input peak voltage.

22.3 AC POWER MEASUREMENTS

Single-Phase Wattmeters

As in the case with dc circuits, ac power can be determined by measuring load current and voltage and calculating their product. Measurements may be made with D'Arsonval or iron-vane meters; however, the load must be purely resistive or the circuit power factor known; otherwise, the calculations produce apparent power rather than true power.

The same results are obtained without calculations by using the two-coil electrodynamometer-type wattmeter. With one coil carrying load current, the other connected across the voltage, coil rotation is proportional to the product EI, or power. By placing one coil 90° in relation to the other, rotation becomes a function of the in-phase, resistive components only. The meter thus gives a true power reading regardless of circuit power factor.

For frequencies beyond the range of magnetic meters, thermocouple wattmeters are used.

Three-Phase Wattmeters

Three-phase power can be measured with a wattmeter in each phase. Two meters will do the job, however, if connected as shown in Figure 22.7. With this system, current is read in any two of the lines, with voltage being read between those lines and the third line. Total power = $P_1 + P_2$, and the load may be balanced or unbalanced.

There are numerous other techniques for three-phase measurements that are beyond the scope of this book.

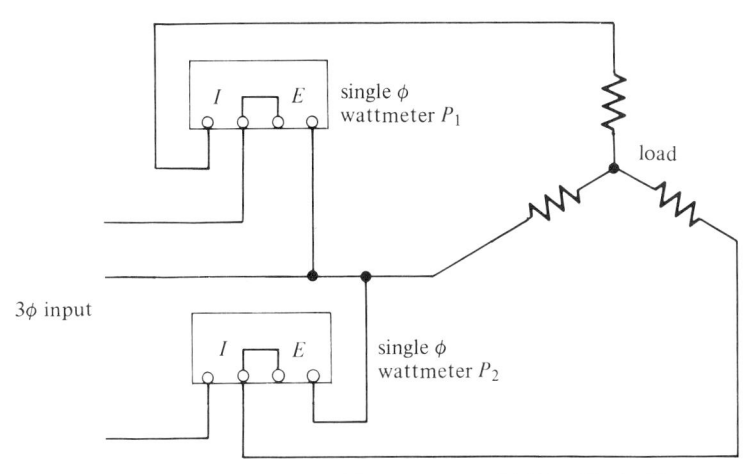

FIGURE 22.7 Three-Phase Power Measurement with Two Single-Phase Wattmeters.

22.4 BASIC AC MEASUREMENTS ON THE OSCILLOSCOPE

The *cathode-ray oscilloscope*, or simply the *oscilloscope*, is probably the most useful and versatile piece of test equipment in electronics. With it we can measure voltage, including peak and peak-to-peak readings, current, frequency, and phase of signals in the hundreds of megahertz. It allows us to observe waveforms and, if desired, store them on film or in digital memory. It also makes possible time measurements of events lasting only billionths of a second.

The heart of the oscilloscope is the *CRT* (*cathode ray tube*), mentioned briefly during the static electricity discussion of Chapter 2 and shown in Figure 2.2 [repeated for convenience in Figure 22.8, part (a)].

The electron gun in a CRT shoots a stream of electrons that causes special chemicals, called *phosphor materials*, on the screen to glow at the beam's point of impact. These materials determine glow color and *persistence* (the time needed for glow to fade after the electron beam moves away; high-persistence materials fade very slowly).

During the trip between gun and screen, the beam passes between two parallel metal plates called *vertical deflection plates*. If the upper plate is polarized positive relative to the bottom plate, the beam is bent upward by the forces of Coulomb's law. This causes the glowing spot to move to point *A*. Reverse the polarity, making the upper plate negative, and the beam bends downward to *B*. When a sine wave is applied between the plates, the beam moves up and down at the wave's frequency over a distance determined by peak-to-peak voltage. The result is a vertical line of little value. Something else is needed if we are to see the sine wave.

Let us try a simple experiment. Move a pencil up and down a constant distance on a sheet of paper and a line like that on the scope screen appears. Next, pull the paper smoothly and steadily beneath the moving pencil and, with a little practice, a surprisingly good sine wave appears. The point of our experiment is this: In order to see sine waves (or any other waves for that matter) on the CRT, the beam must be moved horizontally and vertically at the same time. Therefore, the CRT requires a second set of deflection plates, the *horizontal plates* [Figure 22.8, part (b)].

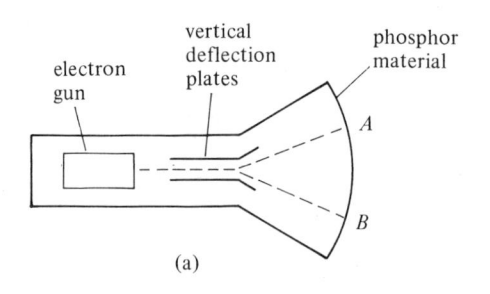

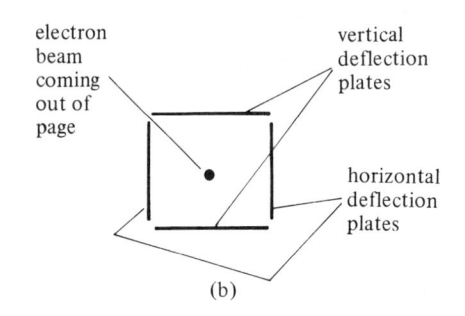

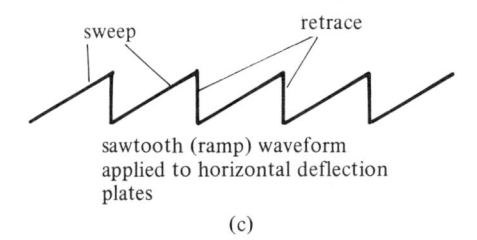

FIGURE 22.8 Cathode Ray Tube—Heart of the Oscilloscope.

Applied between the horizontal deflection plates is a voltage *sawtooth* or *ramp function* [Figure 22.8, part (c)] generated by circuits within the oscilloscope. As this sawtooth increases in amplitude during the relatively slow sweep portion of the wave, it causes one horizontal plate to become steadily more positively charged than the other. This forces the beam to move across or *sweep* the screen at a constant rate. The sharp drop to zero during *retrace* returns the beam to the left very rapidly and the next cycle begins. The movement is much like that of the eye scanning a printed page.

A more descriptive name for the horizontal sweep signal is *time base* because it makes time measurement possible with the oscilloscope. Let us see how.

The speed at which the sawtooth climbs determines how fast the beam sweeps the screen. In other words, sweep speed is proportional to sawtooth slope. That slope is set by a multiposition *time-per-division* switch on the oscilloscope panel. Its units, seconds/centimeter, indicate the time it takes the beam to move between lines 1 cm (1 division) apart on the CRT face. Etched on the inner screen surface of the CRT, these lines form a grid called the *graticule*.

To measure time, the signal being investigated is fed to the vertical plates by connecting it to the oscilloscope's *vertical* or *Y input* on the front panel. The signal's width is then measured on the graticule in centimeters. That width is multiplied by the

22.4 BASIC AC MEASUREMENTS ON THE OSCILLOSCOPE

setting of the time/div switch. The result is

$$\text{centimeters}\left(\frac{\text{seconds}}{\text{centimeters}}\right) = \text{seconds}$$

that is, units of time.

EXAMPLE 22.3

A wave measures 4.6 cm on a CRT graticule. The time-base switch setting is 20 μs/cm. What is the wave's period?

$$T = \frac{20\ \mu s}{cm}(4.6\ cm)$$
$$= 20(4.6)10^{-6}$$
$$= 92\ \mu s$$ ■

Voltage readings are made on a scope by the same two-step procedure used to determine time, namely:

- Measure a length in centimeters against the graticule.
- Multiply this length by a control setting.

The front-panel control in this case is the *vertical gain* or *vertical sensitivity*. Its units are volts/centimeter. As before, the signal being measured is applied to the oscilloscope's vertical input.

EXAMPLE 22.4

Perform the indicated operations for the sine wave seen on this oscilloscope screen.

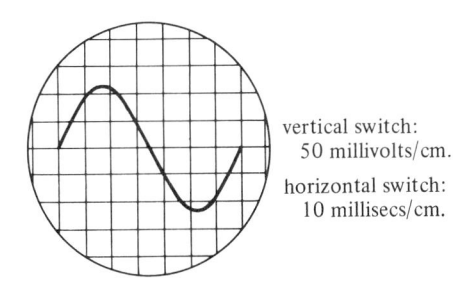

vertical switch: 50 millivolts/cm.
horizontal switch: 10 millisecs/cm.

a. Determine the peak amplitude.

$$e_{peak} = 2.2\ cm\ tall\left(\frac{50\ mV}{cm}\ \text{switch setting}\right)$$
$$= 2.2(50 \times 10^{-3})$$
$$= 110\ mV\ peak$$

b. Determine the peak-to-peak amplitude.

$$e_{\text{peak-to-peak}} = 4.4 \text{ cm} \frac{50 \text{ mV}}{\text{cm}}$$
$$= 220 \text{ mV}$$

c. Determine the period.

$$T = 7 \text{ cm wide} \frac{10 \text{ ms}}{\text{cm}}$$
$$= 70 \text{ ms}$$

d. Determine the frequency.

$$f = \frac{1}{T} = \frac{1}{70} \text{ ms}$$
$$= 14.3 \text{ Hz}$$

e. Write the wave's equation.

$$e_{(t)} = E_{\text{peak}} \sin 2\pi f t \text{ volts}$$
$$= 110 \sin 89.8t \quad \text{mV}$$

EXAMPLE 22.5

The switch settings in Example 22.4 are changed to 2 V/cm and 1 μs/cm. Redo parts a and d.

a. $e_{\text{peak}} = 2.2 \text{ cm} \left(\frac{2 \text{ V}}{\text{cm}}\right)$

$\quad = 4.4 \text{ V}$

b. $f = \frac{1}{T}$

$\quad = \frac{1}{7 \text{ cm}(1 \text{ } \mu\text{s/cm})}$

$\quad = \frac{1}{7 \mu s}$

$\quad = 12.8 \text{ MHz}$

22.5 MEASURING PHASE AND FREQUENCY ON THE OSCILLOSCOPE: THE LISSAJOUS PATTERN

Phase and frequency measurements can be made on a scope with Lissajous patterns (pronounced *liss-uh-jew*). Their use requires the input of two sinusoids, one to each set of deflection plates. Therefore, the internal sweep generator is not used, one of the few cases when it isn't.

To see how phase is measured by this method, consider what happens when two sinusoids of the same frequency and in phase are applied to a scope's X and Y plates (see Figure 22.9). Pausing at point O on each wave, we see that both voltages are zero.

22.5 MEASURING PHASE AND FREQUENCY ON THE OSCILLOSCOPE

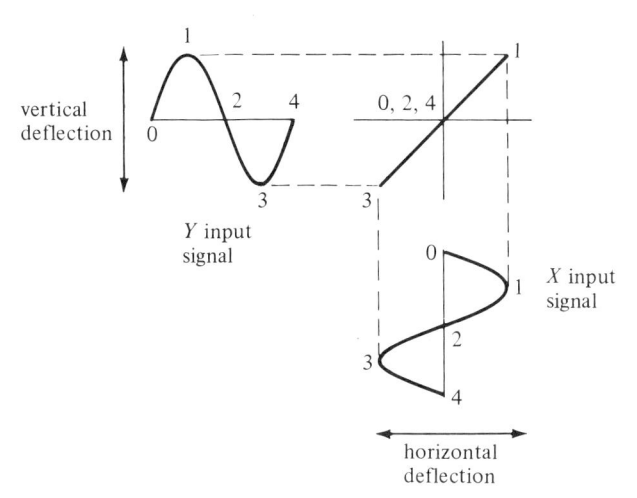

FIGURE 22.9 Lissajous Pattern of Two In-Phase Sinusoids.

The beam, therefore, is not deflected in either direction and lies at position O in the center of the screen. At point 1, the vertical signal has moved the beam up while the horizontal has moved it to the right. Note the position of point 1 on the resultant.

At 2, both inputs are back to zero and so is the resultant. Continuing to 3, we see that the Y and X signals have moved the point down and to the left, respectively. Point 4 finds all three at zero again.

Had other points been plotted, at every 30 or 45° perhaps, and all of these joined, we would see the resultant is a straight line. That line, extending from lower left to upper right, is the characteristic Lissajous pattern of two in-phase sine waves.

The procedure is repeated in Figure 22.10 for signals 90° out of phase. At point O, the Y signal causes no vertical deflection but the X input has the beam at the far right. Point 1 finds the spot at its maximum vertical position, but zero horizontal input causes no X displacement. Continue through the waves, connect the points, and a circle develops, the characteristic Lissajous pattern for signals in quadrature.

Figure 22.11 shows the patterns for a few significant angles between 0 and 180°.

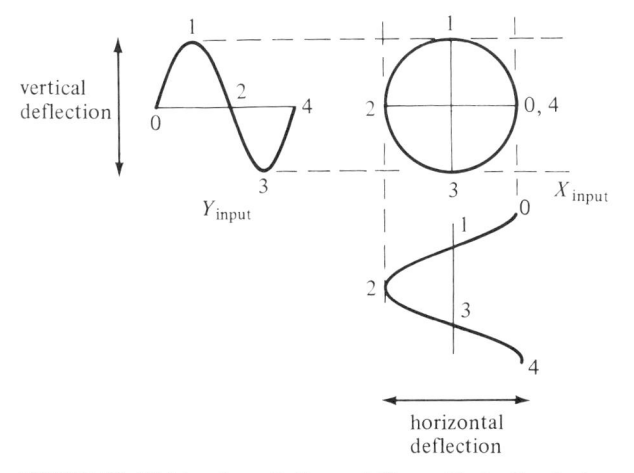

FIGURE 22.10 Lissajous Pattern of Sinusoids in Quadrature.

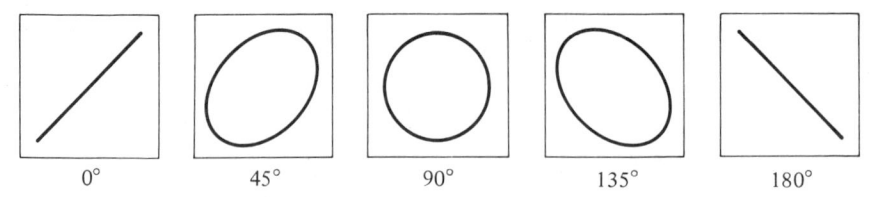

FIGURE 22.11 Lissajous Patterns for angles between 0° and 180°.

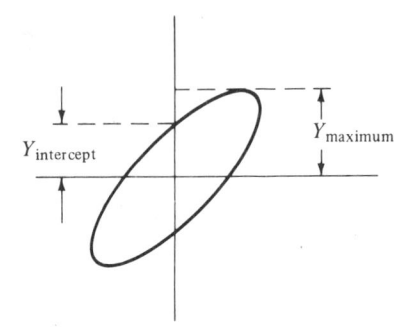

FIGURE 22.12 Determining Phase by Lissajous Pattern.

To determine phase angle between sinusoids by Lissajous patterns:

- Measure the length of the Y intercept [distance from the pattern's center to the point where it intercepts or crosses the Y axis (see Figure 22.12)].
- Measure the length of Y maximum.
- Use the formula:

$$\phi = \sin^{-1} \frac{Y_{\text{intercept}}}{Y_{\text{maximum}}} \qquad (22.7)$$

EXAMPLE 22.6

The Y intercept and Y maximum of a Lissajous ellipse are 2.2 and 5.5 cm, respectively. What is the phase angle between the sinusoids causing the pattern?

$$\phi = \arcsin \frac{2.2}{5.5}$$
$$= \arcsin 0.4$$
$$= 23.6°$$

■

Like phase measurements, frequency measurement by Lissajous patterns requires the input of two sinusoids to the oscilloscope. One of these is a reference signal of known frequency. The other is the unknown.

By the same process of point-to-point plotting with the two sinusoids of Figure 22.13, we see the familiar figure-8 pattern of signals having frequencies in a 2:1 ratio.

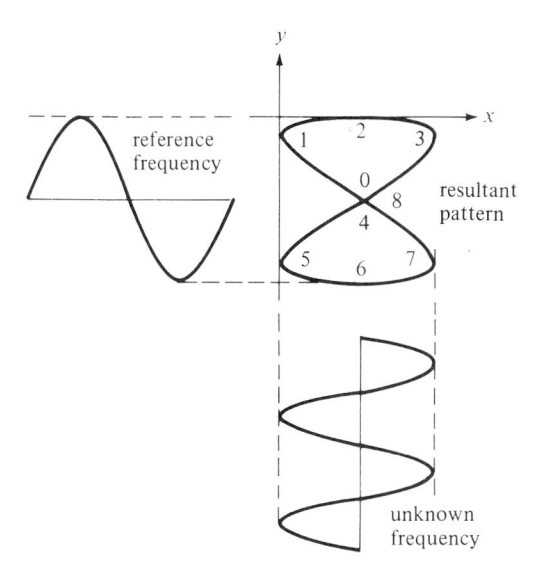

22.5 MEASURING PHASE AND FREQUENCY ON THE OSCILLOSCOPE

FIGURE 22.13 Lissajous Pattern of Signals with Frequencies in a 2:1 Ratio.

Note the X and Y lines drawn to the left of and above the resultant in Figure 22.13. Note also that two of the waveform's loops touch the vertical line, while only one touches the horizontal line. A mathematical relation exists between the frequencies and number of loops touching each line. It is

$$f_x L_x = f_y L_y \tag{22.8}$$

where f_x and f_y = frequency inputs to X and Y inputs, respectively
L_x and L_y = loops touching X and Y lines, respectively

Since the loops are easily counted and one of the frequencies is known, the unknown frequency can be determined. The technique works best with low frequencies.

Figure 22.14 shows the Lissajous patterns of a few common frequency ratios.

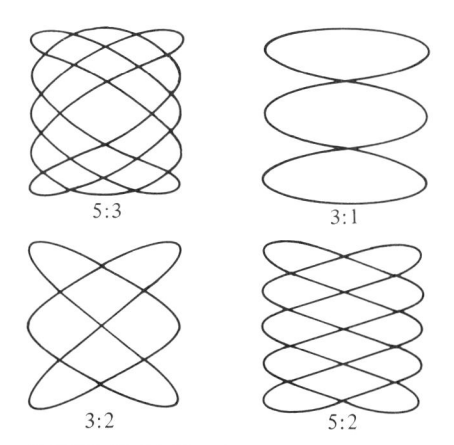

5:3   3:1
3:2   5:2

FIGURE 22.14 Lissajous Patterns of Common Frequency Ratios.

EXAMPLE 22.7

Two sinusoids, one at 150 Hz, the other unknown, are applied to an oscilloscope's vertical and horizontal inputs, respectively. If the resulting Lissajous pattern is the 3:2 diagram of Figure 22.14, what is the unknown frequency?

Count loops on the diagram:
$$L_x = 2$$
$$L_y = 3$$

By Equation 22.8,
$$f_x = \frac{L_y}{L_x} f_y$$
$$= \frac{3}{2} 150$$
$$= 225 \text{ Hz} \qquad \blacksquare$$

EXAMPLE 22.8

The 225-Hz signal of Example 22.7 remains on the horizontal input and the vertical input frequency is adjusted to get the 5:2 ratio pattern of Figure 22.14. What is the new vertical input frequency?

From the pattern:
$$L_x = 2$$
$$L_y = 5$$

and f_x is known to be 225 Hz. Therefore,
$$f_y = \frac{L_x}{L_y} f_x$$
$$= \frac{2}{5} 225$$
$$= 90 \text{ Hz} \qquad \blacksquare$$

22.6 FREQUENCY COUNTERS

Using digital electronics, modern *frequency counters* (see Figure 22.15), capable of measuring frequencies from a few hertz to the gigahertz range, are available as off-the-shelf items.

The basic block diagram of a counter, shown in Figure 22.16, begins with processing circuits that convert the input signal to the digital form (pulses) understood by the other sections of the system. This digital output is allowed through the gate for 1 s, counted, and displayed.

FIGURE 22.15 Frequency Counter Capable of Counting to Gigahertz Range. (Courtesy of Leader Instruments Corp.)

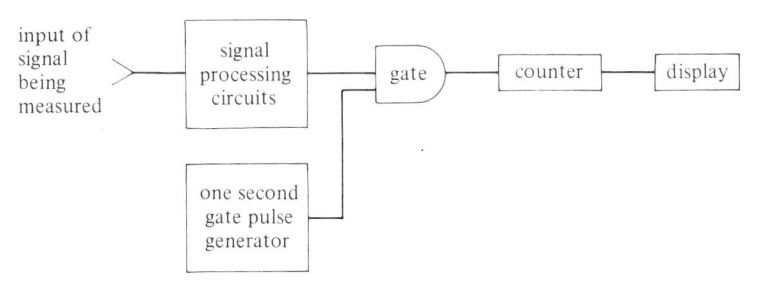

FIGURE 22.16 Simplified Block Diagram of Frequency Counter.

EXERCISES

Section 22.1

1. An unknown ac voltage is half-wave rectified with an ideal diode and read by a D'Arsonval meter. The meter reads 75.6 V.
  a. Find the unknown's rms value.   b. Find the peak value of the unknown.
  c. Repeat for full-wave rectification.

2. Repeat Exercise 1 for an unknown ac of 47.3 V.

3. A dc of 75.6 V measures approximately 168 V on the ac scale of a D'Arsonval meter. What type of rectifier does the meter have?

4. The half-wave rectification circuit of Figure 22.1 is simplified but not practical because it stops load current during half of each cycle. The following circuit permits that current throughout the entire cycle. Trace current around the loops in both directions. Compare on and off times for D_1 and D_2. Assuming ideal diodes and a zero-resistance meter, what is the load voltage waveform?

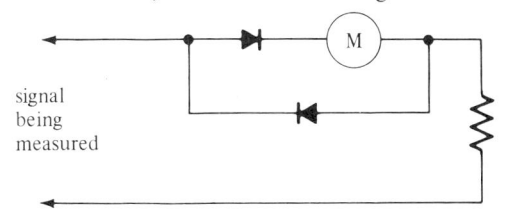

5. A current, 28 cos (750t) amperes, is sent to a D'Arsonval meter. What is the reading if the diodes are ideal and the rectification is: a. half-wave; b. full-wave?

6. Consider the bridge rectifier of Figure 22.2. Is there a time when: a. only one diode is on; b. all four diodes are on?

Section 22.2

7. Why does the iron-vane meter always read upscale regardless of current direction?

8. Repeat Exercise 7 for the electrodynamometer meter.

9. We wish to measure the effective value of a nonsinusoidal periodic signal. What three characteristics must the meter choosen for the job have?

10. Should a meter with low input resistance be used in a peak-reading voltmeter? Explain.

Section 22.4

11. A sinusoid is 7.3 cm wide on an oscilloscope's graticule. Its positive alternation is 4.8 cm high. For each set of switch settings, find e_{peak}, $e_{peak-to-peak}$, period, and frequency.

	Vertical gain	Time Base
a.	1 V/cm	2 s/cm
b.	20 mV/cm	100 s/cm
c.	500 mV/cm	500 ms/cm
d.	5 V/cm	1 s/cm

12. Explain why a CRT needs two sets of deflection plates.

13. What is the equation of this waveform monitored on an oscilloscope?

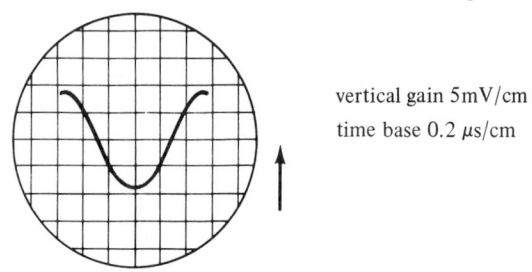

vertical gain 5mV/cm
time base 0.2 μs/cm

14. Redo Exercise 13 for the switch settings

vertical gain = 20 mV/cm
time base = 100 μs/cm

15. Describe the view on an oscilloscope screen if the deflection plate signals were reversed; that is, a sine wave was applied to the horizontal plates and the sawtooth to the vertical.

16. Oscilloscopes only respond to voltage. How, then, can one be used to measure current? *Hint:* Think Ohm's law.

Section 22.5

17. Find sin ϕ and ϕ for each of the following Lissajous patterns.

	$Y_{intercept}$ (cm)	Y_{max} (cm)
a.	1.3	8.7
b.	0	5.6
c.	5.1	6.4
d.	3.9	3.9

18. How do we tell if the angle determined by a Lissajous pattern is in the first or second quadrant?

19. With the help of the following Lissajous pattern, find each of the missing frequencies in the table.

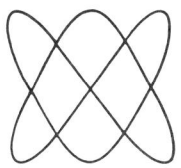

	f_x (Hz)	f_y (Hz)
a.	33	
b.		420
c.	510	
d.		197

20. Repeat Exercise 19, parts a and b, for the 5:3 pattern of Figure 22.14.

CHAPTER HIGHLIGHTS

Measuring equipment must always have a frequency capability equal to or greater than the frequency of the signal being measured.

A D'Arsonval movement, converted to read ac with rectification, suffers a sharp loss in sensitivity.

Know the various meter types. Some are much better suited than others for a particular job.

A meter that says rms volts reads rms for sine-wave inputs only. It will say "true rms" if it reads the rms value of any waveform.

Oscilloscope readings are a two-step procedure:
　Measure a length.
　Multiply that length by the setting of a panel switch.

Know the Lissajous patterns of 0°, 90°, and 180° phase differences.

Formulas to Remember

$$V_{dc} = \frac{V_{peak}}{\pi} \qquad \text{dc component of half-wave rectified output}$$

$$V_{dc} = \frac{2V_{peak}}{\pi} \qquad \text{dc component of full-wave rectified output}$$

$$\phi = \sin^{-1} \frac{Y_{intercept}}{Y_{maximum}} \qquad \text{phase angle by Lissajous pattern}$$

$$f_x L_x = f_y L_y \qquad \text{frequency measure by Lissajous patterns}$$

CHAPTER 23

Periodic Non-Sinusoidal Waveforms: The Fourier Series

Section 3.13 listed reasons for the importance of the sine wave. One of them was that all periodic waveforms, such as the square and sawtooth wave, are actually a combination of superimposed sine waves. Which sinusoids combine at what frequency, phase, and amplitude to form a specific waveform is described by a special math statement for that waveform. The statement is called a Fourier series (pronounced *four-ree-ay*), named after the French mathematician Jean Fourier (1768–1830).

We study the Fourier series in this chapter as well as a few of the problems arising from the sine-wave nature of periodic signals.

At the completion of this chapter, you should be able to:

- Define fundamental and harmonic frequencies.
- Explain why circuits carrying nonsinusoidal periodic signals must have a bandwidth much greater than the signal's fundamental frequency.
- Give a general description of the Fourier series.
- Discuss the effects of symmetry on the components in a waveform's series.
- Calculate effective voltage and current and power transfer ability of a non-sinusoidal periodic signal.
- Describe the Fourier series of common waveforms in electronics.
- Analyze basic circuits driven by non-sinusoidal periodic signals.

23.1 AN EXPERIMENT

Note the following about each of the series-connected generators in Figure 23.1, part (a):

■ The frequency of the bottom generator, ωt, is the lowest of the four. It is called the *fundamental frequency*, or simply the *fundamental*.
■ The next generator frequency, $3\omega t$, is three times the fundamental. It is called the *third harmonic* (a harmonic is a whole-number multiple of the fundamental). The next source operates at the fifth harmonic, $5\omega t$, and the top generator at the seventh harmonic of the fundamental.
■ Harmonic amplitude decreases as frequency increases.

If the fundamental and third harmonic are added instant by instant, the plot of their sum [dashed line in part (b)] is not a sine wave. Instead, it begins to look like a crude square wave.

Part (c) shows the results of adding generator outputs at the fifth and seventh harmonics. The summation looks even more like a square wave. It continues to improve with the addition of each *odd* harmonic of the correct amplitude.

It should come as no surprise that the superposition of an infinite number of these harmonics creates a perfect square wave. The math statement describing that superposition is the *Fourier series* of a square wave:

$$v = 1\sin\omega t + \tfrac{1}{3}\sin 3\omega t + \tfrac{1}{5}\sin 5\omega t + \tfrac{1}{7}\sin 7\omega t + \cdots \qquad (23.1)$$

fundamental   third harmonic   fifth harmonic   seventh harmonic

Although the series extends to infinity, we deal with only the first three or four harmonics because their amplitudes drop off so sharply that adding more makes little difference.

23.1 AN EXPERIMENT

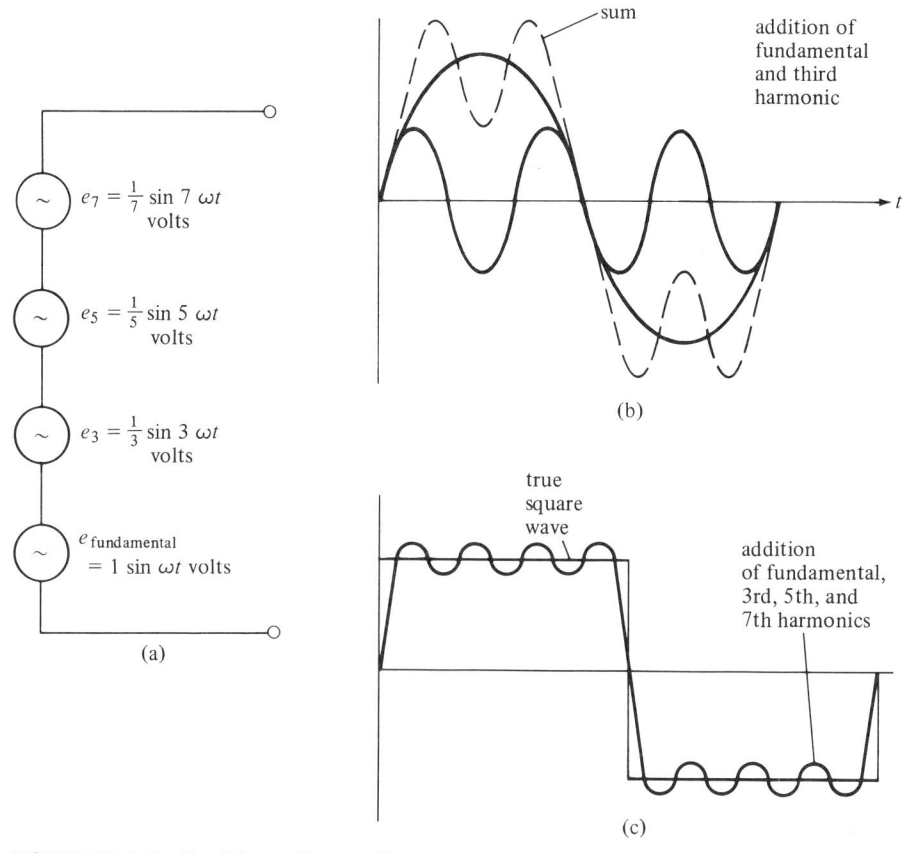

FIGURE 23.1 Synthesizing a Square Wave.

The Fourier series of any nonsinusoidal periodic waveform is similar to that of the square wave. It might have only even harmonics perhaps or both even and odd, but some combination of harmonics will always be there. That is the most important lesson of this chapter:

> All periodic nonsinusoidal signals contain high-frequency components extending far beyond the signal's fundamental frequency.

Any circuit carrying such signals must have the bandwidth necessary to pass these high-frequency components. If it does not, a periodic waveform at the input appears distorted at the output.

But there is another problem. High-frequency harmonics from the pulses in digital equipment are easily radiated as radio signals by the wires carrying them. When "received" by nearby equipment, these signals interfere with, even drown out, legitimate signals. The result is *RFI* (*radio-frequency interference*), a problem common to digital equipment or, for that matter, any device with a sharply rising waveform, such as household light dimmers. Preventing RFI requires careful attention to shielding, impedance matching, and proper grounding techniques.

23.2 THE FOURIER SERIES

Equation 23.1 showed the series of a square wave. Let us look next at the general Fourier series that applies to all periodic signals. It is written as a voltage but could just as well have been in current form:

$$v = A_0 + A_1 \cos \omega t + A_2 \cos 2\omega t + \cdots$$
$$+ A_n \cos n\omega t + B_1 \sin \omega t + B_2 \sin 2\omega t$$
$$+ \cdots + B_n \sin n\omega t \tag{23.2}$$

where n = any number.

Figure 23.2 is the circuit corresponding to Equation 23.2.

A_0, the only term in the series with no frequency, is the *dc component*. It does not have to be present. When it is, the wave described by that series is shifted above or below the time axis to an average value of A_0. From earlier discussions, we know that the wave will have equal areas above and below this value.

$A_1 \cos \omega t$ and $B_1 \sin \omega t$ are fundamental components. All other terms are sine or cosine harmonics of that fundamental. The coefficients A_1 through A_n and B_1 through B_n are determined by formulas obtained with calculus. Since this is beyond our present math level, we will be given the coefficients or the formulas with which they are determined.

An inspection of Equation 23.1 shows that our square wave has no dc component and only odd-numbered sine components. Vary the wave's amplitude or frequency or add a dc level, and the series changes accordingly. Replace the square wave with a triangular or sawtooth waveform and an entirely different series is needed to describe

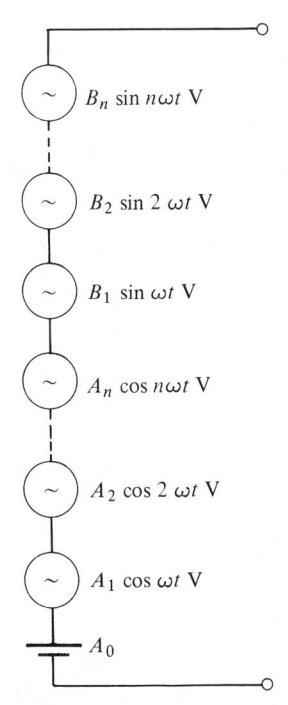

FIGURE 23.2 Fourier Series Connection.

23.3 SYMMETRY

it. The point is that every periodic wave has a unique combination of terms and coefficients that make up its Fourier series. Some waves have only sine terms; others, only cosine terms. Some have every harmonic; others skip some. There are infinite possibilities.

23.3 SYMMETRY

Although it is not our purpose to get deeply involved with the mathematics of Fourier series, let us briefly examine a characteristic of periodic waveforms that tells a great deal about a wave's series. That characteristic is *symmetry*. There are three types.

Even Symmetry

Imagine that the vertical axis in the diagram of Figure 23.3, part (a), was the back of a book. Everything to the left of that axis is then on the left page, everything to the right on the right page. When the book is closed, the wave's two sides fall on top of each other. For reasons that we need not go into, waveforms with this characteristic are said to have *even symmetry*. Note that all waves in Figure 23.3 are even functions.

The characteristic of even functions described above is expressed mathematically by saying that

$$f(t) = f(-t) \tag{23.3}$$

for an even function. That is, a wave with even symmetry has the same value for a given time whether that time is negative (before $t = 0$) or positive (after $t = 0$).

It can be shown that all B_1 through B_n coefficients in the Fourier series (Equation 23.2) equal zero for even functions. In other words, there are no sine terms in the series describing an even function, only cosine terms (and possibly a dc component). Note that the dc component of the waveforms in Figure 23.3 is zero for parts (a) and (b), positive for part (c), and negative for part (d).

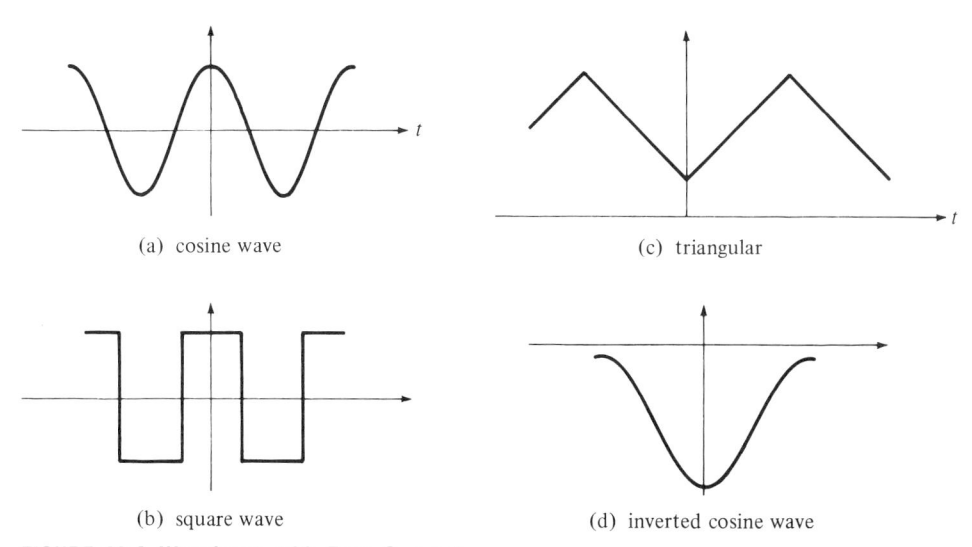

(a) cosine wave

(c) triangular

(b) square wave

(d) inverted cosine wave

FIGURE 23.3 Waveforms with Even Symmetry.

Odd Symmetry

If the sine function in Figure 23.4 is plotted to the left and right of our book hinge, the parts do not match when the book is closed. But turn the negative half upside down, that is, rotate everything left of the Y axis around the negative X axis to the dotted wave, and symmetry returns. All the waves in Figure 23.4 have this characteristic. It is called *odd symmetry*.

Odd symmetry is described by the math statement

$$f(t) = -f(-t) \tag{23.4}$$

In English this says: A waveform with odd symmetry has values of equal but opposite sign at points the same distance to the right and left of the Y axis.

It can be shown that all A_0 through A_n coefficients in the Fourier series equal zero for odd functions. In other words, there are no cosine terms in the series describing an odd function, only sine terms. A_0 is also zero. If it wasn't, $f(t)$ could not equal $-f(-t)$.

Half-Wave Symmetry

The value of the square wave in Figure 23.5 at t_1 is 5 V. A half-period later, at $t_1 + T/2$, the value is -5 V. For such a wave we can say that

$$f(t) = -f\left(t + \frac{T}{2}\right) \tag{23.5}$$

This characteristic, possessed by both waveforms in Figure 23.5, is called *half-wave symmetry*.

Waveforms with half-wave symmetry can be shown to have only odd harmonics in their Fourier series. This applies to both even and odd functions.

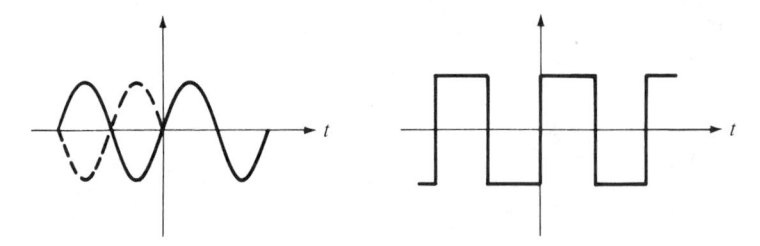

FIGURE 23.4 Waveforms with Odd Symmetry.

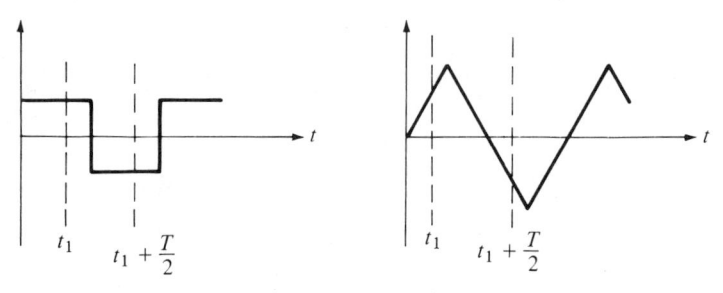

FIGURE 23.5 Waveforms with Half-Wave Symmetry.

23.4 FOURIER SERIES OF COMMON ELECTRONIC WAVEFORMS

Now that we have been introduced to Fourier characteristics, let's examine each of the most common waveforms found in electronics for dc component and symmetry (if any). With this information and the coefficients, we will be able to write the series describing that waveform.

The Square Wave

Our first example is the square wave introduced in Section 23.1 and redrawn in Figure 23.6, part (a). The wave has equal areas above and below the X axis. Therefore, the dc component or average value, A_0, in the series describing this wave must be zero. The wave also has odd symmetry since $f(t) = -f(-t)$; therefore, there are no cosine terms in the series; their A_n coefficients are all zero. Finally, the wave has half-wave symmetry, meaning that even-ordered coefficients, B_2, B_4, and so on, equal zero and are also absent from the series. The only coefficients remaining are those for odd B_n values. These are determined using the formula

$$B_n = \frac{4V_m}{n\pi} \quad \text{where } V_m = V_{\max} \tag{23.6}$$

obtained with calculus procedures not discussed. Solve for the first three of these:

$$\text{for } n = 1: \quad B_1 = \frac{4V_m}{1\pi} = \frac{4V_m}{\pi}$$

$$\text{for } n = 3: \quad B_3 = \frac{4V_m}{3\pi}$$

$$\text{for } n = 5: \quad B_5 = \frac{4V_m}{5\pi}$$

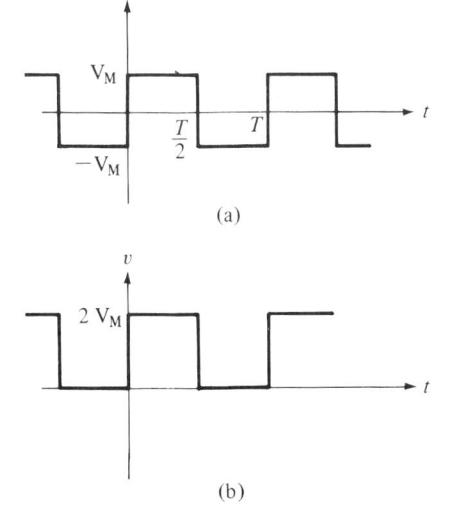

FIGURE 23.6 Square Waves With and Without a dc Component.

Substituting these coefficients into the general Fourier series of Equation 23.2, we get the specific series for this square wave:

$$v = \frac{4V_m}{\pi}\sin \omega t + \frac{4V_m}{3\pi}\sin 3\omega t$$
$$+ \frac{4V_m}{5\pi}\sin 5\omega t + \cdots$$
$$+ \frac{4V_m}{n\pi}\sin n\omega t \text{ V}$$

where the n are odd values to infinity. Factoring $4V_m/\pi$, we have

$$v = \frac{4V_m}{\pi}\left[\sin \omega t + \frac{1}{3}\sin 3\omega t + \frac{1}{5}\sin 5\omega t\right.$$
$$\left. + \cdots + \frac{1}{n}\sin n\omega t\right] \text{ V} \qquad (23.7)$$

The first three terms of this series were used earlier (Equation 23.1) with $\frac{4V_m}{\pi} = 1$ V.

When the square wave is combined with a DC voltage of $+V_m$ volts the entire waveform is shifted upwards (Figure 23.6, part b). As a result, the series now includes an A_0 term but is otherwise unchanged:

$$v = V_m + \frac{4V_m}{\pi}\left(\sin \omega t + \frac{1}{3}\sin 3\omega t + \frac{1}{5}\sin 5\omega t\right.$$
$$\left. + \cdots + \frac{1}{n}\sin n\omega t\right) \text{ V} \qquad (23.8)$$

Suppose that $V_m = 5$ V in Figure 23.6, part (a). Substitute that value into Equation 23.7:

$$v = \frac{4(5)}{\pi}\left(\sin \omega t + \frac{1}{3}\sin 3\omega t + \frac{1}{5}\sin 5\omega t\right.$$
$$\left. + \cdots + \frac{1}{n}\sin n\omega t\right) \text{ V}$$
$$= 6.4\left(\sin \omega t + \frac{1}{3}\sin 3\omega t + \frac{1}{5}\sin 5\omega t\right.$$
$$\left. + \cdots + \frac{1}{n}\sin n\omega t\right) \text{ V}$$

The Sawtooth

The sawtooth, introduced in Chapter 22 as the horizontal sweep signal for cathode ray tubes, is illustrated in Figure 23.7. With equal areas above and below the X axis, its dc component is zero. The waveform value a given distance to the right of the Y axis equals minus that value the same distance to the left of the axis. The waveform thus has odd symmetry; its series has no cosine terms. Although it might seem that the wave has half-wave symmetry, it does not. This is verified by noting that the wave's value at time t

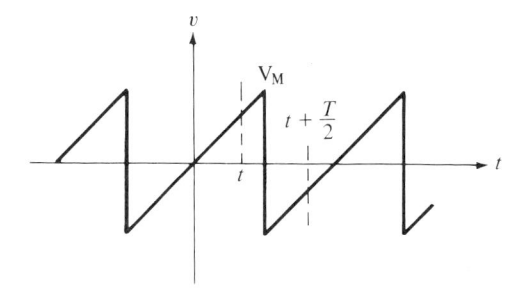

23.4 FOURIER SERIES OF COMMON ELECTRONIC WAVEFORMS

FIGURE 23.7 Sawtooth Waveform.

in Figure 23.7 is not equal to minus its value a half-period later, at $t + \frac{T}{2}$. The series therefore includes both odd- and even-ordered terms. From the calculus derivations for the coefficient equations, we get

$$B_n = \pm \frac{2}{n\pi} \quad (+ \text{ for odd values of } n) \tag{23.9}$$

Calculating the first four coefficients and substituting in the general Fourier series of Equation 23.2 yields

$$v = \frac{2V_m}{\pi}\left[\sin \omega t - \frac{1}{2}\sin 2\omega t + \frac{1}{3}\sin 3\omega t - \frac{1}{4}\sin 4\omega t + \cdots\right] \text{V} \tag{23.10}$$

The Triangular Wave

The triangular wave of Figure 23.8, part (a), has a dc or average component, A_0, which is obtained by dividing triangle area by period:

$$A_0 = \tfrac{1}{2}(\text{base})(\text{height})$$
$$= \frac{[(T/2) - (-T/2)]V_m}{2T}$$
$$= \frac{V_m}{2} \text{ V}$$

The wave has even symmetry. That appears to be its only symmetry. But remove the dc

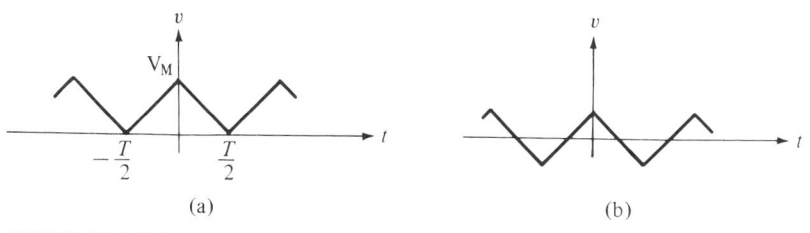

FIGURE 23.8 Triangular Wave.

component temporarily [part (b)] and it is easy to see that there is half-wave symmetry. The coefficient equation,

$$A_n = \frac{4}{\pi^2 n^2} \tag{23.11}$$

therefore, equals zero for even-numbered n values, leaving only odd-numbered components. When the first three of these are calculated and the series written, we have

$$v = \frac{V_m}{2} + \frac{4V_m}{\pi^2}\left[\cos \omega t + \frac{1}{9}\cos 3\omega t + \frac{1}{25}\cos 5\omega t + \cdots\right] \text{V} \tag{23.12}$$

Note that coefficient values decrease as $1/n^2$ ($1/3^2$, $1/5^2$, etc.), a much faster rate than that of other waves examined. This tells us the low-frequency components play the most important role in determining the wave's triangular shape. The accuracy of that statement is easily tested by evaluating the wave's instantaneous value at $t = 0$ using only the dc component, fundamental, and first two harmonics:

$$\begin{aligned} v &= \frac{V_m}{2} + \frac{4V_m}{\pi^2}\left[\cos 0 + \frac{1}{9}\cos 0 + \frac{1}{25}\cos 0\right] \\ &= V_m\left[\frac{1}{2} + \frac{4}{\pi^2}\left(1 + \frac{1}{9} + \frac{1}{25}\right)\right] \\ &= 0.97 V_m \end{aligned}$$

Since the value from Figure 23.8 is V_m at $t = 0$, we conclude that the contribution of all remaining harmonics would be only 3%.

Rectified Waveforms

The half-wave rectified waveform in Figure 23.9, part (a), consists of the same upper-half sinusoids introduced in Chapter 22. The wave has no symmetry; therefore, its

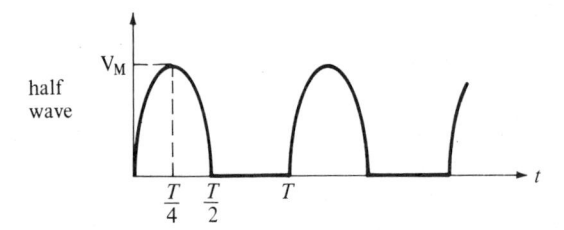

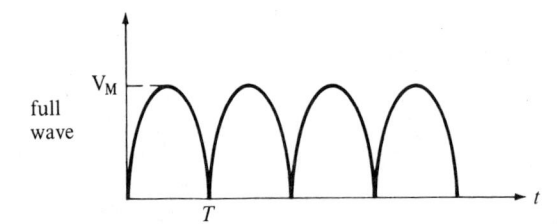

FIGURE 23.9 Half- and Full-wave Rectified Waveforms.

23.5 RMS VALUES AND AVERAGE POWER

series contains both sine and cosine terms. Its series is

$$v = \frac{V_m}{\pi}\left[1 + \frac{\pi}{2}\sin \omega t - \frac{2}{3}\cos 2\omega t - \frac{2}{15}\cos 4\omega t \right.$$
$$\left. - \frac{2}{35}\cos 6\omega t - \cdots \right] \text{V} \qquad (23.13)$$

Note that the dc component is V_m/π, as we learned in Chapter 22, Equation 22.1.

To show that the series really does describe the wave, let us use it to determine instantaneous voltage at time $t = T/4$. Replacing ω with $2\pi/T$ and substituting the time, $T/4$, into the equation, we find

$$v = \frac{V_m}{\pi}\left[1 + \frac{\pi}{2}\sin\frac{2\pi}{T}\frac{T}{4} - \frac{2}{3}\cos 2\frac{2\pi}{T}\frac{T}{4} - \frac{2}{15}\cos 4\frac{2\pi}{T}\frac{T}{4} - \frac{2}{35}\cos 6\frac{2\pi}{T}\frac{T}{4} - \cdots\right]$$
$$= \frac{V_m}{\pi}\left[1 + \frac{\pi}{2}\sin\frac{\pi}{2} - \frac{2}{3}\cos \pi - \frac{2}{15}\cos 2\pi - \frac{2}{35}\cos 3\pi - \cdots\right]$$
$$= \frac{V_m}{\pi}\left[1 + \frac{\pi}{2} + \frac{2}{3} - \frac{2}{15} + \frac{2}{35} - \cdots\right]$$
$$= 1.01 V_m$$

which is 1% off the value in Figure 23.9 at that instant.

Figure 23.9, part (b) of is the full-wave rectified waveform. It has even symmetry and thus no sine terms. Its series is

$$v = \frac{2V_m}{\pi}\left[1 - \frac{2}{3}\cos 2\omega t - \frac{2}{15}\cos 4\omega t - \frac{2}{35}\cos 6\omega t - \cdots\right] \text{V} \qquad (23.14)$$

As is to be expected with twice the number of loops, the dc component is now twice what it was for half-wave rectification. It is also interesting to note that the lowest harmonic in this full-wave series has twice the frequency of the lowest harmonic in the half-wave series.

23.5 RMS VALUES AND AVERAGE POWER

Power transferred by a nonsinusoidal periodic voltage equals the sum of the powers from the individual voltages in the waveform's Fourier series. This summation is

$$P_{\text{total}} = \frac{V_0^2}{R} + \frac{V_1^2}{R} + \cdots + \frac{V_n^2}{R} \qquad \text{watts} \qquad (23.15)$$

where all voltages are rms. Factor $1/R$ and we get:

$$P_{\text{total}} = \frac{1}{R}[V_0^2 + V_1^2 + \cdots + V_n^2]$$

But P_{total} can also be written

$$P_{\text{total}} = \frac{V_{\text{rms}}^2}{R}$$

in which V_{rms} is the waveform's effective value. From these it can be seen that the effective voltage of a nonsinusoidal waveform is

$$V_{rms} = \sqrt{V_0^2 + V_1^2 + \cdots + V_n^2} \qquad \text{volts} \qquad (23.16)$$

EXAMPLE 23.1

Find the rms value of this series of voltages.

$$v = 4.5 + 6.2 \cos 100t - 3.9 \cos 300t + 5.5 \sin 100t - 2.7 \sin 300t$$

All voltages except the dc must be converted to RMS. Therefore,

$$V_{rms} = \sqrt{(4.5)^2 + \left(\frac{6.2}{\sqrt{2}}\right)^2 + \left(\frac{-3.9}{\sqrt{2}}\right)^2 + \left(\frac{5.5}{\sqrt{2}}\right)^2 + \left(\frac{-2.7}{\sqrt{2}}\right)^2}$$
$$= \sqrt{65.9}$$
$$= 8.1 \text{ V}$$

∎

Power transferred by a nonsinusoidal periodic current is also the sum of the powers from the individual current components:

$$P_{total} = I_0^2 R + I_1^2 R + \cdots + I_n^2 R \qquad \text{watts} \qquad (23.17)$$

where all currents are rms. Factor R:

$$P_{total} = R[I_0^2 + I_1^2 + \cdots + I_n^2] \qquad \text{watts}$$

But P_{total} can also be written

$$P_{total} = I_{rms}^2 R$$

in which I_{rms} is the waveform's effective value. From these it can be seen that the effective current of a nonsinusoidal waveform is

$$I_{rms} = \sqrt{I_0^2 + I_1^2 + \cdots + I_n^2} \qquad \text{amperes} \qquad (23.18)$$

EXAMPLE 23.2

Find the power delivered to a 0.5-Ω load by the waveform described by the series.

$$i = 16.4 - 9.8 \cos 377t + 5.1 \cos 754t$$
$$- 11.3 \sin 377t + 2.6 \sin 754t \text{ amps}$$

Convert peak values to rms and finding the total effective current:

$$I_{rms} = \sqrt{(16.4)^2 + \left(\frac{-9.8}{\sqrt{2}}\right)^2 + \left(\frac{5.1}{\sqrt{2}}\right)^2 + \left(\frac{-11.3}{\sqrt{2}}\right)^2 + \left(\frac{2.6}{\sqrt{2}}\right)^2}$$
$$= \sqrt{397.2}$$
$$= 19.9 \text{ A}$$
$$P_{total} = I_{rms}^2 R = (19.9)^2 (0.5)$$
$$= 198 \text{ W}$$

∎

23.6 USING FOURIER SERIES IN CIRCUIT ANALYSIS

If both the current series and voltage series are known for a particular waveform, power delivered by that waveform can be determined by

$$P_{total} = V_0 I_0 + V_1 I_1 \cos \phi_1 + V_2 I_2 \cos \phi_2 \\ + \cdots + V_n I_n \cos \phi_n \tag{23.19}$$

in which ϕ_1 is the angle between fundamental voltage and current, ϕ_2 is the angle between the second harmonic voltage and current, and so on.

23.6 USING FOURIER SERIES IN CIRCUIT ANALYSIS

Fourier series make it possible to analyze circuits driven by nonsinusoidal periodic inputs. The signals are broken down into Fourier components and a separate analysis done for each (this sounds worse than it is; remember, the first few components are usually sufficient for acceptable accuracy). Final results are obtained by combining these components with superposition just as we did in Section 17.5. Power is calculated with effective values obtained by Equations 23.16 and 23.18.

In the exercises for this section and the following examples, we consider any RC or RL transients as over and solve for steady-state solutions only.

EXAMPLE 23.3

A voltage described by the statement.

$$e = 10 + 5 \sin 6t \text{ V}$$

is applied across a 5-Ω resistance. Perform the following steps.

a. Draw the circuit with separate sources for each input component.

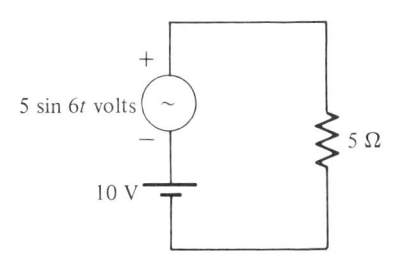

b. Find the time domain expression for the current.

For the 10-V, A_0, dc component:

$$I = \frac{E}{R} \\ = \frac{10}{5} \\ = 2 \text{ A}$$

For the sinusoid:

$$i_{peak} = \frac{e_{peak}}{R}$$
$$= \frac{5}{5}$$
$$= 1 \text{ A}$$

The current expression in the time domain is

$$i = 2 + 1 \sin 6t \text{ A} \qquad \blacksquare$$

EXAMPLE 23.4

Add a 2-H inductor in series with the resistance of Example 23.3. Find the time domain expressions for the current and the voltages, v_R and v_L.

The new circuit is

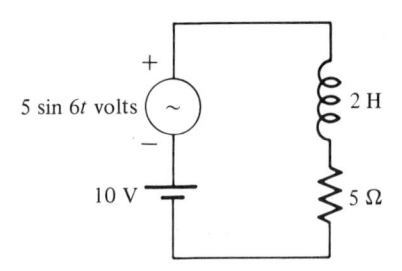

For the dc component:

$$I = \frac{E}{R}$$
$$= \frac{10}{5}$$
$$= 2 \text{ A}$$

For the sinusoid:

$$X_L = \omega L$$
$$= 6(2)$$
$$= j12 \text{ }\Omega$$
$$\bar{Z} = 5 + j12 \quad \text{or} \quad 13\,\underline{/67.4°}\text{ }\Omega$$
$$\bar{I} = \frac{\bar{E}}{\bar{Z}}$$
$$= \frac{5/\sqrt{2}}{13\,\underline{/67.4°}}$$
$$= \frac{0.385}{\sqrt{2}}\,\underline{/-67.4°}\text{ A}$$

23.6 USING FOURIER SERIES IN CIRCUIT ANALYSIS

Write the two currents as a Fourier statement in the time domain:

$$i = 2 + 0.385 \sin(6t - 67.4°) \text{ A}$$

Multiplying each part by 5, we get v_R:

$$\begin{aligned} v_R &= iR \\ &= 5[2 + 0.385 \sin(6t - 67.4°)] \\ &= 10 + 1.9 \sin(6t - 67.4°) \quad \text{V} \end{aligned}$$

There is no drop across the inductor at dc; therefore, the statement for v_L is

$$\begin{aligned} v_L &= iX_L \\ &= 12 \underline{/90°} \, [0.385 \sin(6t - 67.4°)] \\ &= 4.6 \sin(6t - 67.4° + 90°) \\ &= 4.6 \sin(6t + 22.6°) \quad \text{V} \end{aligned}$$ ■

EXAMPLE 23.5

The voltage

$$v = 100 + 80 \sin \omega t - 50 \sin 3\omega t \quad \text{volts}$$

is applied to a series-connected 10-Ω resistance and a 1-μF capacitor. If the fundamental radian frequency is 20×10^3 rad/s, find the Fourier expressions for the indicated quantity.

a. The current.

For the A_0 component: Current from the 100 V A_0 component is zero because the capacitor is an open.
For the fundamental:

$$\begin{aligned} X_{C_1} &= \frac{1}{\omega_1 C} \\ &= \frac{1}{(20 \times 10^3)(10^{-6})} \\ &= -j50 \, \Omega \\ \bar{Z}_1 &= 10 - j50 \quad \text{or} \quad 51 \underline{/-78.7°} \, \Omega \\ \bar{I}_1 &= \frac{\bar{E}_1}{\bar{Z}_1} \\ &= \frac{\frac{80}{\sqrt{2}} \underline{/0°}}{51 \underline{/-78.7°}} \\ &= \frac{1.6}{\sqrt{2}} \underline{/78.7°} \text{ A} \end{aligned}$$

Convert to time domain:

$$i_{C_1} = 1.6 \sin(20{,}000t + 78.7°) \text{ A}$$

For the third harmonic: X_{C_3} must be $-j50/3 = -j16.7 \ \Omega$

$$\bar{Z}_3 = 10 - j16.7 \quad \text{or} \quad 19.4 \underline{/-59°} \ \Omega$$

$$\bar{I}_3 = \frac{\bar{E}_3}{\bar{Z}_3}$$

$$= \frac{\frac{50}{\sqrt{2}} \underline{/180°}}{19.4 \underline{/-59°}}$$

$$= \frac{2.6}{\sqrt{2}} \underline{/239°} \ A$$

In the time domain:

$$i_3 = 2.6 \sin(60{,}000t + 239°) \ \text{A}$$

The final Fourier expression for current is

$$i = i_1 + i_3$$
$$= 1.6 \sin(20{,}000t + 78.7°) + 2.6 \sin(60{,}000t + 239°) \ \text{A}$$

b. The voltage across the resistance.

The expression for v_R has two components, one for each current; there is no dc component.

$$v_R = R(i_1 + i_3)$$
$$= 10[1.6 \sin(20{,}000t + 78.7°) + 2.6 \sin(60{,}000t + 239°)]$$
$$= 16 \sin(20{,}000t + 78.7°) + 26 \sin(60{,}000t + 239°) \ \text{V}$$

c. The voltage across the capacitor.

The expression for v_R has two components, one for each current; there is no dc component.
the fundamental:

$$\bar{V}_1 = \bar{I}_1 X_{C_1}$$

$$= \frac{1.6}{\sqrt{2}} \underline{/78.7°} \ (50 \underline{/-90°})$$

$$= \frac{80}{\sqrt{2}} \underline{/-11.3°}$$

In the time domain:

$$v_1 = 80 \sin(20{,}000t - 11.3°) \ \text{V}$$

For the third harmonic:

$$\bar{V}_3 = \bar{I}_3 X_{C_3}$$

$$= \frac{2.6}{\sqrt{2}} \underline{/239°} \ (16.7 \underline{/-90°})$$

$$= \frac{43.4}{\sqrt{2}} \underline{/149°}$$

In the time domain:

$$v_3 = 43.4 \sin(60{,}000t + 149°) \ \text{V}$$

The final Fourier expression for v_c:

$$v_C = 100 + 80 \sin(20{,}000t - 11.3°) + 43.4 \sin(60{,}000t + 149°) \ \text{V} \quad \blacksquare$$

EXERCISES

Section 23.1

1. Given: a fundamental at 625 Hz.
   a. What is its period?
   b. Find the frequency and period of the third harmonic.
   c. Repeat part b for the eleventh harmonic.

2. Explain why digital equipment can be the source of buzzing sounds in a nearby radio.

3. A rule of thumb says that at least the first five components of a square wave (fundamental and third, fifth, seventh, and ninth harmonics) should be passed to ensure a square shape. By this rule, what is the maximum pulse frequency that can be observed on an oscilloscope with a 500-kHz bandwidth?

4. Using the rule of thumb from Exercise 3, what oscilloscope bandwidth is needed to observe a 7.5 MHz pulse train?

Sections 23.2 and 23.3

5. Write a Fourier series consisting of a 12.3-V dc component, an 8.5-V rms sinusoid at 300 Hz, and its first two even harmonics whose rms amplitudes are 7.1 V and 2.9 V, respectively.

6. The Fourier series of a sine wave has how many components? Repeat for a cosine wave and for a dc.

7. Determine whether the Fourier series describing each of the following waveforms will have: a. a dc component; b. a cosine terms; c. sine terms; d. odd terms only.

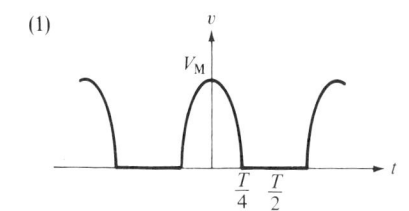

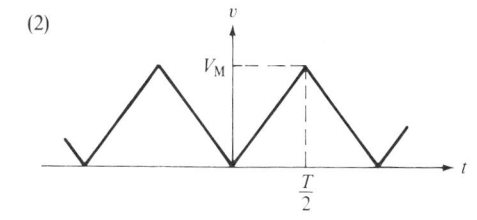

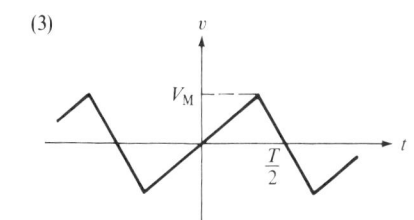

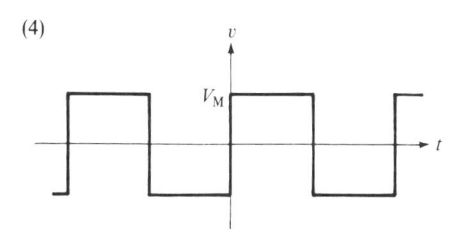

8. Repeat Exercise 7 for these waveforms:

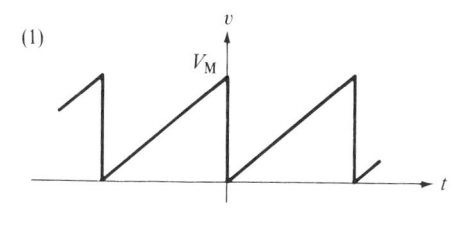

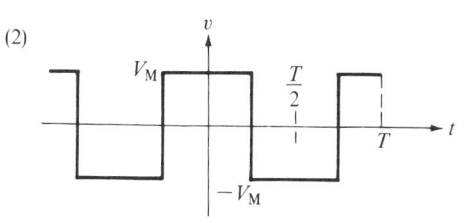

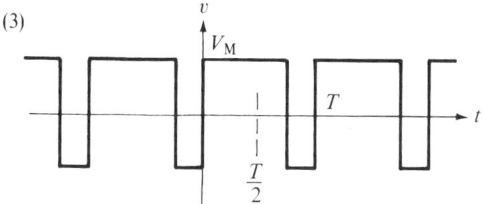

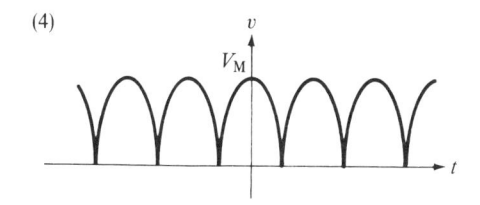

Section 23.4

9. Evaluate the square wave of Figure 23.6, part (a), at $t = T/4$, for the fundamental and first four harmonics. Compare the result to the plot's value at the same time. *Hint:* Follow the technique used to evaluate the half-wave rectifier output wave.

10. Given: a square wave with the Y axis drawn in position A, then in position B. Comment on the symmetry changes resulting from this move.

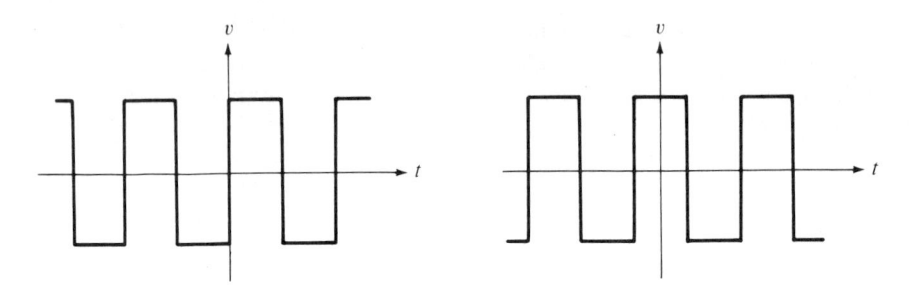

Section 23.5

11. Find the average and effective values of the signals described by the following equations. Determine the power dissipated by each signal in a 4.5-Ω resistance.
  a. $v = 3 + 5 \sin 50t + 1.8 \sin 150t$ V
  b. $i = -4.8 + 3.6 \cos 200t - 1.1 \cos 400t$ A

12. Repeat Exercise 11 for the following equations.
  a. $v = 50 \sin 377t - 37.5 \sin 1131t + 22 \sin 1885t - 9.2 \sin 2639t$ V
  b. $i = 33 + 19 \cos 5t + 11.3 \cos 10t + 8.1 \cos 15t + 4.7 \cos 20t$ A

13. A voltage waveform, described by the Fourier statement

$$e = 37 + 23 \sin \omega t + 18 \sin 3\omega t + 11.2 \sin 5\omega t + \cdots \text{ V}$$

is applied to a 12-Ω resistance.
  a. Find the time-domain current expression.
  b. Find the power dissipated by the resistance.

14. A current waveform described by the statement

$$i = 6.7 \sin 200t - 4.1 \sin 400t + 2.9 \sin 600t \text{ A}$$

is applied to a 33-Ω resistance.
  a. Find the time-domain voltage expression.
  b. Find the power dissipated by the resistance.

15. The input signal to the circuit is described by the Fourier series

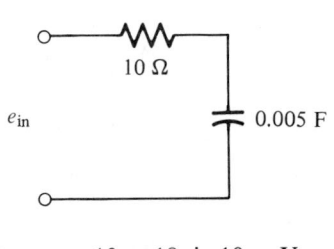

$$e_{in} = 10 + 18 \sin 10t \text{ V}$$

Find the time-domain expressions for circuit current and the voltage across each element.

16. Repeat Exercise 15 for this input signal:

$$e = 16 + 20.5 \sin 5t - 13.2 \sin 10t + 6.8 \sin 15t - 4.2 \sin 20t \text{ V}$$

17. The input signal to this circuit is described by the Fourier series

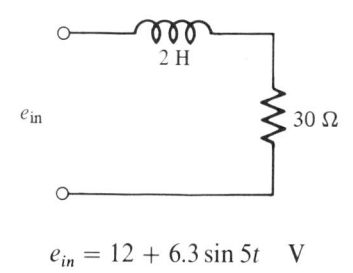

$$e_{in} = 12 + 6.3 \sin 5t \quad V$$

Find the time-domain expressions for circuit current and the voltage across each element. Calculate the power dissipated by the resistance.

18. Repeat Exercise 17 for this input signal:

$$e = 21 + 22 \sin 5t - 14.4 \sin 10t + 8.5 \sin 15t - 5 \sin 20t \quad V$$

19. *Extra Credit.* The output of a full-wave rectifier, described by the series

$$v = 100[1 - 2\cos 2\omega t - 2\cos 4\omega t - 2\cos 6\omega t - \cdots] \quad V$$

where $\omega = 377$ rad/s, is applied to the input of this filter circuit:

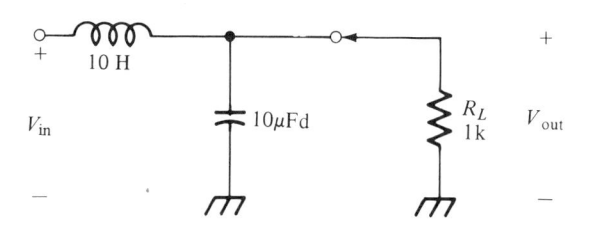

a. Determine the Fourier series of V_{out}.   b. Determine the effective value of V_{out}.
Carry results to five places beyond the decimal point. *Hint:* Using voltage division, derive an equation for output voltage.

CHAPTER HIGHLIGHTS

Every nonsinusoidal periodic waveform can be described by a unique Fourier series consisting of a sinusoid or cosinusoid at a fundamental frequency, combined with an infinite combination of harmonic components and possibly a dc component.

Equipment carrying nonsinusoidal periodic signals must have sufficient bandwidth to pass the signal's high-frequency components; otherwise, distortion occurs.

Harmonics are multiples of a fundamental frequency.

Checking a wave's symmetry is a quick way to determine the components in that wave's Fourier series.

Circuit analysis can be done on circuits with nonsinusoidal periodic inputs by using superposition on the input's Fourier components.

Formulas to Remember

$$v = A_0 + A_1 \cos \omega t + A_2 \cos 2\omega t + \cdots + A_n \cos n\omega t \\ + B_1 \sin \omega t + B_2 \sin 2\omega t + \cdots + B_n \sin n\omega t$$

the Fourier series:

$$I_{\text{rms}} = \sqrt{I_0^2 + I_1^2 + I_2^2 + \cdots + I_n^2} \quad \text{amperes} \quad \text{rms current}$$

$$V_{\text{rms}} = \sqrt{V_0^2 + V_1^2 + V_2^2 + \cdots + V_n^2} \quad \text{volts} \quad \text{rms voltage}$$

$$\begin{aligned} P_{\text{total}} &= V_0 I_0 + V_1 I_1 \cos\theta_1 \\ &\quad + V_2 I_2 \cos\theta_2 + \cdots \\ &\quad + V_n I_n \cos\theta_n \\ &= \frac{V_{\text{rms}}^2}{R} \\ &= I_{\text{rms}}^2 R \quad \text{watts} \quad \text{total power} \end{aligned}$$

APPENDIX

Derivation of Π-to-T and T-to-Π Conversion Formulas of Sections 6.5 and 17.3

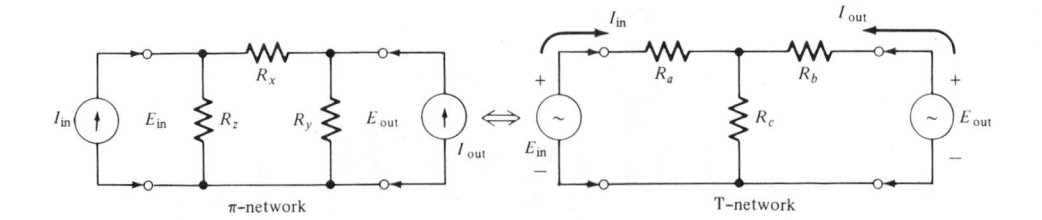

Writing nodal equations for the π network:
At input:
$$\frac{E_{in}}{R_z} + \frac{E_{in} - E_{out}}{R_x} = I_{in}$$

or

$$\left(\frac{R_x + R_z}{R_x R_z}\right) E_{in} - \frac{1}{R_x} E_{out} = I_{in} \qquad (A.1)$$

At output:
$$\frac{E_{out}}{R_y} + \frac{E_{out} - E_{in}}{R_x} = I_{out}$$

or

$$-\frac{1}{R_x} E_{in} + \left(\frac{R_x + R_y}{R_x R_y}\right) E_{out} = I_{out} \qquad (A.2)$$

Writing loop equations for the T network: At input:
$$E_{in} = (R_a + R_c) I_{in} + R_c I_{out} \qquad (A.3)$$

At output:
$$E_{out} = R_c I_{in} + (R_b + R_c) I_{out} \qquad (A.4)$$

Converting Π into T

Solve nodal equations A.1 and A.2 for E_{in} by Cramer's rule:

$$E_{in} = \frac{\begin{vmatrix} I_{in} & -\dfrac{1}{R_x} \\ I_{out} & \dfrac{R_x + R_y}{R_x R_y} \end{vmatrix}}{\begin{vmatrix} \dfrac{R_x + R_z}{R_x R_z} & -\dfrac{1}{R_x} \\ -\dfrac{1}{R_x} & \dfrac{R_x + R_y}{R_x R_y} \end{vmatrix}}$$

$$= \frac{\dfrac{R_x + R_y}{R_x R_y}}{D_\pi} I_{in} + \frac{1/R_x}{D_\pi} I_{out} \qquad (A.5)$$

APPENDIX

where

$$D_\pi = \frac{(R_x + R_z)(R_x + R_y)}{R_x^2 R_y R_z} - \frac{1}{R_x^2}$$

$$= \frac{R_x^2 + R_x R_y + R_y R_z + R_x R_z}{R_x^2 R_y R_z} - \frac{1}{R_x^2}\frac{R_y R_z}{R_y R_z}$$

$$= \frac{R_x^2 + R_x R_y + R_x R_z}{R_x^2 R_y R_z}$$

$$= \frac{R_x + R_y + R_z}{R_x R_y R_z}$$

Similarly, solving nodal equations A.1 and A.2 for E_{out} by Cramer's rule:

$$E_{out} = \frac{\begin{vmatrix} \frac{R_x + R_z}{R_x R_z} & I_{in} \\ -\frac{1}{R_x} & I_{out} \end{vmatrix}}{D_\pi}$$

$$= \frac{\frac{R_x + R_z}{R_x R_z}}{D_\pi} I_{out} + \frac{1/R_x}{D_\pi} I_{in} \quad \text{(A.6)}$$

If the networks are equivalent, the input/output currents and voltages for one network must equal the corresponding input/output currents and voltages for the other network. Therefore, the coefficient of I_{out} in Equation A.3 must equal the same coefficient in Equation A.5; that is,

$$R_c = \frac{1/R_x}{D_\pi} \quad \text{(A.7)}$$

Substitute for D_π:

$$R_c = \frac{1}{R_x \left(\frac{R_x + R_y + R_z}{R_x R_y R_z} \right)}$$

$$= \frac{R_y R_z}{R_x + R_y + R_z} \quad \text{(A.8)}$$

Similarly, the coefficients of I_{in} in Equations A.3 and A.5 must be equal:

$$R_a + R_c = \frac{\frac{R_x + R_y}{R_x R_y}}{D_\pi}$$

$$= \frac{R_x + R_y}{R_x R_y D_\pi}$$

$$= \frac{1}{R_y D_\pi} + \frac{1}{R_x D_\pi}$$

Subtracting Equation A.7 from both sides eliminates R_c and leaves

$$R_a = \frac{1}{R_y D_\pi}$$

Substitute for D_π:

$$R_a = \frac{R_x R_y R_z}{R_y(R_x + R_y + R_z)}$$
$$= \frac{R_x R_z}{R_x + R_y + R_z} \quad (A.9)$$

Finally, the coefficients of I_{out} in Equations A.4 and A.6 must be equal:

$$R_b + R_c = \frac{\frac{R_x + R_z}{R_x R_z}}{D_\pi}$$
$$= \frac{R_x + R_z}{R_x R_z D_\pi}$$
$$= \frac{1}{R_z D_\pi} + \frac{1}{R_x D_\pi}$$

Subtracting Equation A.7 from both sides eliminates R_c and leaves

$$R_b = \frac{1}{R_z D_\pi}$$

Substitute for D_π:

$$R_b = \frac{R_x R_y R_z}{R_z(R_x + R_y + R_z)}$$
$$= \frac{R_x R_y}{R_x + R_y + R_z} \quad (A.10)$$

Equation A.8 corresponds to Equations 6.4 and 17.9, Equation A.9 to Equation 6.2 and 17.7, and Equation A.10 to Equations 6.3 and 17.8.

Converting T into Π

Solving loop equations A.3 and A.4 for I_{in} by Cramer's rule:

$$I_{\text{in}} = \frac{\begin{vmatrix} E_{\text{in}} & R_c \\ E_{\text{out}} & R_b + R_c \end{vmatrix}}{\begin{vmatrix} R_a + R_c & R_c \\ R_c & R_b + R_c \end{vmatrix}}$$
$$= \frac{R_b + R_c}{D_T} E_{\text{in}} - \frac{R_c}{D_T} E_{\text{out}} \quad (A.11)$$

where

$$D_T = (R_a + R_c)(R_b + R_c) - R_c^2$$
$$= R_a R_b + R_b R_c + R_c R_a$$

APPENDIX 727

Similarly, solving loop equations A.3 and A.4 for I_{out} by Cramer's rule:

$$I_{out} = \frac{\begin{vmatrix} R_a + R_c & E_{in} \\ R_c & E_{out} \end{vmatrix}}{D_T}$$

$$= -\frac{R_c}{D_T}E_{in} + \frac{R_a + R_c}{D_T}E_{out} \qquad (A.12)$$

If the networks are equivalent, the input/output currents and voltages for one network must equal the corresponding input/output currents and voltages for the other network. Therefore, the coefficient of E_{out} in Equation A.1 must equal the same coefficient in Equation A.11; that is;

$$-\frac{1}{R_x} = -\frac{R_c}{D_T}$$

or

$$\frac{1}{R_x} = \frac{R_c}{D_T} \qquad (A.13)$$

Invert and substitute for D_T:

$$R_x = \frac{R_a R_b + R_b R_c + R_c R_a}{R_c} \qquad (A.14)$$

Similarly, the coefficients of E_{out} in Equations A.2 and A.12 must be equal:

$$\frac{R_x + R_y}{R_x R_y} = \frac{R_a + R_c}{D_T}$$

or

$$\frac{1}{R_y} + \frac{1}{R_x} = \frac{R_a}{D_T} + \frac{R_c}{D_T}$$

Subtracting Equation A.13 from both sides eliminates $1/R_x$ and leaves

$$\frac{1}{R_y} = \frac{R_a}{D_T}$$

Therefore,

$$R_y = \frac{R_a R_b + R_b R_c + R_c R_a}{R_a} \qquad (A.15)$$

Finally, the coefficients of E_{in} in Equation A.1 and A.11 must be equal:

$$\frac{R_x + R_z}{R_x R_z} = \frac{R_b + R_c}{D_T}$$

or

$$\frac{1}{R_x} + \frac{1}{R_z} = \frac{R_b}{D_T} + \frac{R_c}{D_T}$$

Subtracting Equation A.13 from both sides eliminates $1/R_x$ and leaves

$$\frac{1}{R_z} = \frac{R_b}{D_T}$$

Therefore,

$$R_z = \frac{R_a R_b + R_b R_c + R_c R_a}{R_b} \qquad (A.16)$$

Equation A.14 corresponds to Equations 6.5 and 17.10, Equation A.15 to Equations 6.6 and 17.11 and Equation A.16 to Equations 6.7 and 17.12.

Answers to Odd-Numbered Exercises

CHAPTER 1

1. **a.** 5.7×10^4, **b.** 9.23, **c.** 3.7×10^{-6}, **d.** 1.015×10^3, **e.** 6.85×10^{-2}, **f.** 1.86×10^5, **g.** 9.2×10^{-10}, **h.** 2.35×10^6, **i.** 7.7×10^{-5}, **j.** 1.56, **k.** 7.55×10^2, **l.** 7.55×10^{-2}, **m.** 2.2×10^1, **n.** 1.57×10^4, **o.** 5.3×10^{-4}. 3. **f.** 0.000186×10^9, **g.** 0.00092×10^{-6}, **h.** 0.000235×10^{10}, **i.** 0.00077×10^{-1}, **j.** 0.000156×10^4. 5. **a.** 2.2×10^7, **b.** 2.9×10^6, **c.** 4.7×10^{-10}, **d.** 1.4×10^6, **e.** 1.2×10^{-2}, **f.** 9.3×10^5, **g.** 6.3×10^{-2}, **h.** 6.8×10^2, **i.** 2×10^{16}, **j.** 6.5×10^{-9}. 7. **a.** 914.4 mm/yd, **b.** 904,977 g/ton, **c.** 1.712 cents/min, **d.** 63.245×10^6 pints/leap year, **e.** 0.015 in., **f.** 47 kilodollars, **g.** 6.8 l, **h.** 14×10^4 picobushels, **i.** 9.3×10^4 tons, **j.** 107 gigadollars. 9. 33 picoseconds. 13. 1 kiloday. 15. **a.** increases 25 times, **b.** increases 4 times, **c.** increases 121 times, **d.** decreases to $\frac{1}{64}$ of original value, **e.** decreases to $\frac{1}{3}$ original value. 17. **a.** X directly proportional to F, **b.** X inversely proportional to F. 19. 33%. 21. 15.62 times. 23. energy increases 4 times. Consider this concept with reference to braking a car moving at various speeds.

CHAPTER 2

3. **a.** 8.24×10^{22} electrons, **b.** 4.66×10^{15} electrons, **c.** 3.12×10^{14} electrons, **d.** 2.4×10^{-6} C, **e.** 9.9×10^{-9} C, **f.** 0.32 C. 5. Since the rod lost electrons it has become positively charged by 50 C with respect to the silk; or we can say that the silk, having gained electrons, is negatively charged by 50 C with respect to the rod. 7. -4972.5 N, attraction. 9. -5 C. 11. **b, c, e,** and **j** attract; **a, d, f, g, h,** and **i** repel; **k** has no effect. 13. **a** and **e** negative; **b** and **d** positive; **c** neither. 15. by forcing them apart. 17. Yes; push two like poles together or pull unlike poles apart. 19. **a** and **c** rise; **b** and **d** drop. 21. **a** and **d** store; **b** and **c** release. 23. **a.** Incorrect; between M and where? **b.** correct, **c.** incorrect; charge passed through the victim. 25. **a.** 9×10^{-6} C, **b.** 26.9 V, **c.** 0.005 J, **d.** 337 J, **e.** 568×10^{-3} C, **f.** 4×10^8 V. 27. 15 J. Double to 30 J; energy transfer directly proportional to voltage. 29. 110 V—wall outlet, 12 V—car battery, 9 V—radio battery, 1.5 V—flashlight cell. 31. **a.** 40 A, **b.** 7400 C, **c.** 314.8 s, **d.** 10 A, **e.** 619.8×10^{-9} C, **f.** 2 h. 33. no, I = 54.4 A. 35. 31.9 A. 37. **a.** 6 mA, **b.** 50 A, **c.** 7.5 μA.

CHAPTER 3

3. source when supplying energy to starter; load when being recharged by alternator. 5. **a** and **d** loads; **b** and **c** sources. 7. **a.** input negative, **b.** negative when at ground floor, **c.** negative at target, **d.** negative on ground. 9. insulator: **a, b, c, g**; conductor: **d, e, f, h**. 11. **a.** 10,000 Ω, **b.** 47,000 Ω, **c.** 330,000 Ω, **d.** 560,000 Ω, **e.** 1,200,000 Ω,

ANSWERS TO ODD-NUMBERED EXERCISES

f. 4,700,000 Ω, **g.** 2,200,000 Ω. **13. a.** 3.03×10^{-6} S, **b.** 6.67×10^{-7} S, **c.** 45.5 μS, **d.** 667 μS. **15. a** and **c**. **17. a.** 430 Ω 5%, 408.5 and 451.5 Ω, **b.** 5600 Ω 10%, 5040 and 6160 Ω, **c.** 22 k 20%, 17600 and 26400 Ω, **d.** 5.1 Ω 10% 1% reliability, 5.61 and 4.59 Ω, **e.** 3.9M 5%, 3.71 and 4.1 M, **f.** 82 Ω 10% 0.01% reliability, 73.8 and 90.2 Ω. **19.** b false; all others true. **21. b.** false; all others true. **23. a.** 652 Ω, **b.** 243 mV, **c.** 3.4 μA, **d.** 23.1 k, **e.** 7.9 mA, **f.** 7.5 V. **25. a.** 29.4 μA, **b.** 1.5 V, **c.** 108 μA, **d.** 2 M. **27. a.** 6×10^{-6} S, **b.** 149 V, **c.** 4 nA. **29.** 60 W: 0.55 A, 75 W: 0.68 A, 100 W: 0.91 A, 200 W: 1.8 A. I varies directly with P. Yes. 60 W: 201.7 Ω, 75 W: 161.3 Ω, 100 W, 121 Ω, 200 W: 60.5 Ω. R varies inversely with P. Yes. **31. a.** 40 A, 1 Ω, **b.** 3 V, 111 Ω, **c.** 7.5×10^{-13} W, 50 nA, **d.** 268.3 V, 3.7 A. **33.** 0.09 Ω 15.2 W, 1.2 V. **35. a** and **b** watts; **c** amperes. **37.** 61%. **39. a.** 32.2 W, **b.** 34.8 W, **c.** as heat. **41. a.** 0.43, **b.** 0.55, **c.** 0.89. **45.** heat generated increases with the square of the current. **47.** all connected together. **49. a.** 25 mA down, **b.** 3.3 mA down, **c.** 0.3 A down.
53.

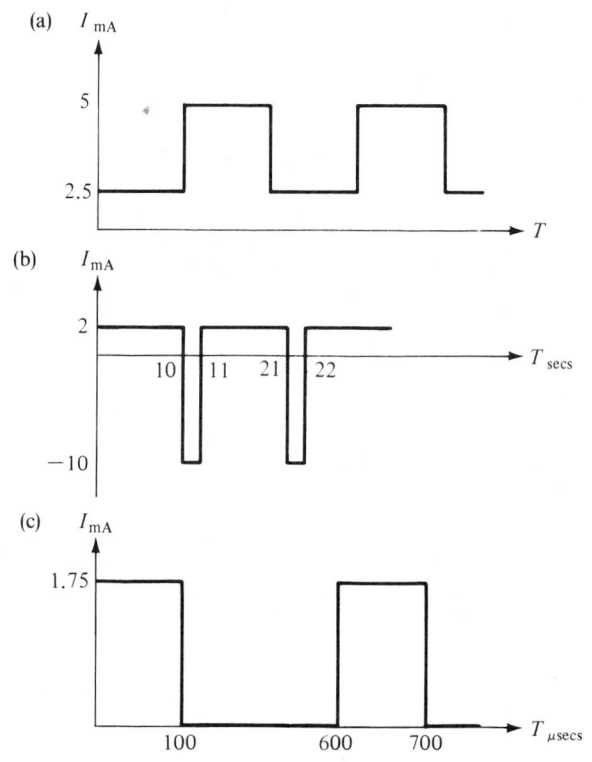

55. Current defined as positive in clockwise direction; since it is actually flowing counterclockwise, it is negative.

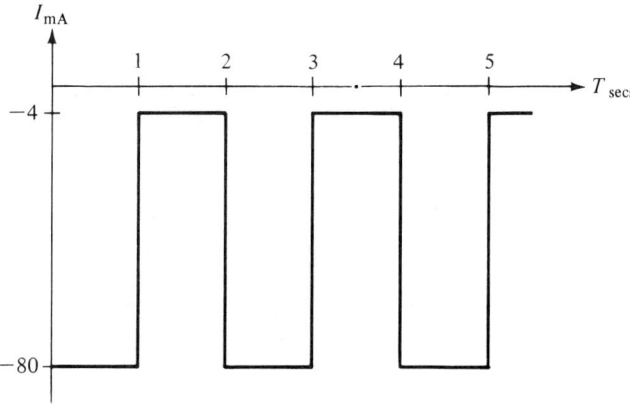

ANSWERS TO ODD-NUMBERED EXERCISES

CHAPTER 4

1. If the same 5 mA passed through all three. They would not be in series if each had a different 5 mA.   3. 2 and 4 have, **a**, 1 and 3 have **b** and **c**.   5. **a.** R_1, R_2, source,   **b.** R_1, source,   **c.** R_a, R_b, source,   **d.** none,   **e.** R_1, R_7, source. R_2 and R_6,   **f.** R_d, R_h, source,   **g.** none,   **h.** R_1, R_2, R_3. R_5 and source. R_6 and R_{11}. R_7, R_8 and R_{10}.   7. **a.** 1 or 3,   **b.** 2,   **c.** 4.
9. **a.** $(5\,V + 12\,V) - (8\,V + 6\,V)$,   **b.** $(12\,V + 8\,V) - (5 + 6\,V)$,   **c.** $(5\,V + 6\,V + 8\,V) - 12\,V$.
11. **a.** 12 V, A+,   **b.** 4 V, A+,   **c.** 6 V, B+,   **d.** 0 V (27 V each way),   **e.** 38 mV, A+.
13. **a.** 4 V, B+,   **b.** 3 V, B+,   **c.** 6 V, C+,   **d.** 5 V, A+,   **e.** 1 V, C+,   **f.** 9 V, B+.
15. **a.** 35 V,   **b.** 38 V,   **c.** 8 mV,   **d.** 0 V.   17. **a.** 27 V,   **b.** 9 mV.   19. **a.** 80 k,   **b.** 140 k,   **c.** 154.2 Ω,   **d.** 70 Ω.   21. **a.** 15 k,   **b.** 3 k.   23. **a.** 1.6 A,   **b.** 4.8 μA,   **c.** 59.4 μA.   25. **a.** 25 Ω,   **b.** 5 Ω,   **c.** five times,   **d.** twice the drop,   **e.** 2.5 times,   **f.** yes.
27. **a and b.** split source voltage equally because resistances equal,   **c.** equal resistances have equal drops,   **d.** both double, still equal,   **e.** relative resistance values.   29. **a.** 800 Ω,   **b.** 300 Ω,   **c.** 200 Ω,   **d.** 100 Ω,   **e.** 30 Ω.   31. **a.** I: 2 A, R_1: 24 V, 48 W, R_2: 40 V, 80 W,   **b.** I: 84.9 μA, R_t: 2.1 M, R_1: 18.7 V, 1.6 mW, R_2: 39.9 V, 3.4 mW, R_3: 30.6 V, 2.6 mW, R_4: 57.7 V, 4.9 mW, R_5: 33.2 V, 2.8 mW,   **c.** R_t: 52 k, I: 1.8 μA, R_1: 23 mV, 40.7 nW, R_2: 69 mV, 122.1 nW,   **d.** actually separate circuits because of ground at junction of 75 k and 240 k. I (160 k and 75 k): 93.6 μA, 160 k: 15 V, 1.4 mW, 75 k: 7 V, 0.7 mW, I (240 k): 70.8 μA, 240 k: 17 V, 1.2 mW.
33. **a.** 750 Ω,   **b.** 330 k.   35. **a.** 3 A, 27 V, 12 V,   **b.** 4 mA, 56 k.   37. current square wave: 0 to 213 mA every 5 mS, voltage square wave across R_1: 0 to 7 V, across R_2: 0 to 10 V, same times as current.   39. sawtooth, 0 to 4.53 V.   41. **a.** 2.17 A,   **b.** 2.9 mA,   **c.** 0.99 A.
43. **a.** 20 V, + on right,   **b.** 75 V, + on left,   **c.** 95 V, + on right.   45. **a.** 5 V, B+,   **b.** 40 V, A+,   **c.** 150 V, A+.   47. **a.** V_{R_2}: 150 V, E_s: 650 V, R_{total}: 43.3 k, V_{ab}: 450 V, A+,   **b.** 830 V.   49. 20 Ω, yes.   51. 200 Ω, 50 W.   53. **a.** A: 24.5 V, B: 39.1 V, C: 75 V,   **b.** A: −10 mV, B: 10 mV, C: 25 mV,   **c.** A: −264.6 V, B: −126.6 V, C: 46 V, D: 130.4 V.
55. **a.** 5 k: W and B, 15 k: W and A, 20 k: A and B,   **b.** halfway.   57. Batteries often run under oc conditions but sc conditions would force all of E_{oc} to be dropped across R_{int}: resulting heat might cause battery to explode.   59. **a.** E_{oc}: 200 V, I_{oc}: 0 A, E_{sc}: 0 V, I_{sc}: 400 A,   **b.** E_{oc}: 18.3 mV, I_{oc}: 0 A, E_{sc}: 0 V, I_{sc}: 44.6 μA.   61. 5.91 Ω.   63. **a.** 33.3 V, 333A,   **b.** 66.7 V, 267 A,   **c.** 100 V, 200 A,   **d.** 160 V, 80 A,   **e.** 190 V, 19 A,   **f.** As R_{load} increased, V_o increased.
65. **a.** 1.44 V,   **b.** V_{bulb} 1 V,   **c.** testing under oc conditions gives no indication of state of R_{int}.
67. 410 Ω, 0.5 Ω.   69. E_{oc}: 108 V, R_{int}: 52 Ω.   71. requires breaking the circuit.
73. dirty or corroded contacts.   75. **a.** 90 V,   **b.** 0 V,   **c.** 0 V,   **d.** 150 V.

CHAPTER 5

1. **a.** no,   **b.** yes.   3. **a.** 15 k,   **b.** drops equal,   **c.** currents equal,   **d.** 3 k.   5. nodes: **a** and **c**, none; **b**, **d**, and **g**, two; **e**, **f**, and **h** four. branches: **a** and **c**, one; **b**, **d**, and **g**, three; **e**, six; **f** eight; and **h** seven.   7. Circuit b: **a.** V_a,   **b.** V_b,   **c.** V_c,   **d.** E_{source}, Circuit e: **a.** V_a,   **b.** V_b,   **c.** V_c,   **d.** V_d,   **e.** zero,   **f.** V_{R_2},   **g.** V_{R_1},   **h.** $V_F - V_C$.   9. **a.** 20 A right,   **b.** I_1: 4 A down, I_2: 5 A down. I_3: 4 A down.   11. **a.** I, D,   **b.** D, I.   13. **a.** 1.3 k,   **b.** 118.5 Ω.
15. **a.** any $R < 10$ k,   **b.** any $R > 1$ M,   **c.** 100 k,   **d.** impossible.   17. R_1: 270 k, R_2: 330 k.
19. A: 4 A, B: 17 A, C: 13 A, D: 5 A.   21. **a.** 5.1 Ω: 32.7 mA, 5.5 mW. 16 Ω: 10.4 mA, 1.7 mW. 8.6 Ω: 19.4 mA, 3.2 mW,   **b.** 0.0005 S: 255 mA, 130 W. 0.003 S: 1.5 A, 780 W.
23. 36 k.   25. square wave, 5.1 k: −2.2 to 6.9 mA, 8.6 k: −1.3 to 4.1 mA.   27. R_1: 5 Ω, R_2: 12 Ω.   29. 1.3 A.   31. **a.** impossible situation in both cases,   **b.** yes, both cases,   **c.** R_{internal} now present.   33. **a.** 0 A, 16.5 kV, 0 W,   **b.** 5 A, 0 V, 0 W,   **c.** 1.9 A, 10.4 kV, 19.2 kW,   **d.** 3300 Ω,   **e.** when $R_{\text{load}} \ll R_{\text{int}}$. 10-to-1 rule says $R_{\text{load}} \leq 330$ Ω,   **f.** when $R_{\text{load}} \gg R_{\text{int}}$.
35. **a.** I_{sc}: 200 μA, R_{int}: 105 Ω,   **b.** E_{oc}: 45 V, R_{int}: 900 k,   **c.** I_{sc}: 3.7×10^5 A, R_{int}: 0.01 Ω,   **d.** E_{oc}: 86 V, R_{int}: 2 k.   37. **a.** $I_{\text{load}} = E_{oc}/(R_{\text{load}} + R_{\text{int}})$,   **c.** I_{load} constant,   **d.** $R_{\text{int}}, R_{\text{load}}$,

e. $I_{load} = I_{sc}(R_{int})/(R_{int} + R_{load})$, **f.** $R_{internal} \gg R_{load}$. **39. a.** 30.1 V, 4.3 k, **b.** 29.3 V, 95.8 k, **c.** 0 A, 5.1 Ω, **d.** 24.8 V, 5.5 k. **41. a.** 250 V, **b.** 20 V. **43.** M_1 and M_3: D, M_2: NC. **45. a, c** and **d**: NC, **b, e** and **f**: D. **47.** M_1, M_4, M_5, M_6 and M_7: NC, M_2 and M_3 decrease. **49.** R_2 branch open.

CHAPTER 6

1. a. 10 Ω, **b.** 40 Ω, **c.** 38.8 Ω. **3. a.** 13 Ω, **b.** 6.6 Ω, **c.** 9 Ω, **d.** 13 Ω, **e.** 10 Ω, **f.** 5.1 Ω. **5. a.** no, **b.** R_1 drop subtracts from source voltage to reduce V_{R_3}. Therefore, I_{R_3} decreases too. **7. a.** R_1: 20 V, 4 A, 80 W. R_2: 20 V, 2 A, 40 W, R_3: 2 A, 10 V, 20 W, $R_4 = R_5$: 2 A, 5 V, 10 W, **b.** $R_1 = R_2$: 133.3 mA, 66.7 V, 8.9 W. R_3 = 66.7 mA, 66.6 V, 4.4 W. R_4: 33.3 mA, 66.6 V, 2.2 W. $R_5 = R_7$: 33.3 mA, 16.7 V, 554 mW. R_6: 33.3 mA, 33.3 V, 1.1 W. **9.** 133 V. **11.** R_1: 100 V, 25 A, 4 Ω, R_2: 100 V, 5 A, 20 Ω, R_3: 50 V, 25 A, 2 Ω, R_4: 40 V, 20 A, 2 Ω, R_5: 60 V, 15 A, 4 Ω, R_6: 25 V, 5 A, 5 Ω, R_7: 20 V, 5 A, 4 Ω, R_8: 15 V, 5 A, 3 Ω. **13. a.** 18.5 mW, **b.** 0.29 mA, **c.** 48.6 V, B+. **15. a, b,** and **d**, I; **c**, D. **17. a.** 19 V, **b.** 13.9 V, **c.** increase. **19.** 5.1 k. **21. a.** balance won't change as batteries are depleted, **b.** lower battery drain, **c.** put in as one leg of the bridge an element whose resistance changes as a function of barometric pressure. **23. a.** not balanced, **b.** balanced. **25.** 60 Ω. **27.** 682 mW. **29.** when the board breaks from overload. **31.** 0.62 A up. **33.** 282.2 W. **35.** 2 A. **37. b.** 3 mA, **c.** V_{R_2}: 60 V, I_{R_3}: 6 mA, **d.** 54.4 V. **39. a.** 1212 V, **b.** 0.9 A, **c.** 774.3 V, **d.** 7.2 A, **e.** 4545 V. **41. j**, D; **a, c, d** and **i**, B; **b, e, f**, and **h**, G; **g**, U.

CHAPTER 7

1. a. −2, **b.** 9, **c.** 10, **d.** 23. **3. a.** 3/2, 2/3, **b.** 2, 3/5, **c.** −36/5, −88/5. **5. a.** 158, **b.** 26, **c.** 70. **7. a.** −1, 3, −2, **b.** 2, −3, 1. **9. a.** 3, **b.** 2, **c.** 2, **d.** $15I_1 - 10I_2 = 40$, $-10I_1 + 20I_2 = 0$. I_{R_1}: 4 A, I_{R_2}: 2 A, I_{R_3}: 2 A. **11.** $13I_1 + 5I_2 = -5$, $5I_1 + 8I_2 = 5$, V_{R_1}: 1.6 V, V_{R_2}: 3.4 V, V_{R_3}: 6.6 V. **13.** I_{in}: 1.6 A, I through 3 Ω: 0.6 A. **15.** 2 A. **17. a.** $6I_1 + 2I_2 + 2I_3 = 5$, $2I_1 + 7I_2 + 6I_3 = 8$, $2I_1 + 6I_2 + 8I_3 = 16$, **b.** I_{R_a}: 1.8 A, I_{R_b}: 0.33 A, I_{R_c}: 1.5 A, I_{R_d}: 1.7 A, I_{R_e}: 3.2 A. **19.** 6.9 A. **21. a.** 2 A, **b.** 5 A, **c.** 26 A. **23. a.** $(V_a - V_b)/5$, **b.** $(V_a + 10 - V_b)/5$, **c.** $(V_a - 10 - V_b)/5$. **25. a.** 2, **b.** 1, **c.** $(V - 40)/5 + V/10 + V/10 = 0$, **d.** I_{R_1}: 4 A, $I_{R_2} = I_{R_3} = 2$ A. **27.** 1.1 A. **29.** 2.4 A. **31. a.** 7.7 A, **b.** 0.7 V. **33. a.** 4 k, **b.** 0.9 mV, **c.** yes. **35.** (1) E_{oc}: 10 V, R_{int}: 6.7 Ω, 10 Ω: 0.6 A, 6 V, (2) E_{oc}: 10 V, R_{int}: 4 Ω, 10 Ω: 0.71 A, 7.1 V, (3) E_{oc}: 14 V, R_{int}: 6 Ω, 10 Ω: 0.88 A, 8.8 V. **37. a.** 6.7 Ω, **b.** 4 Ω, **c.** 6 Ω. **39. a.** I_{sc}: 1.5 A, R_{int}: 6.7 Ω, **b.** I_{sc}: 2.5 A, R_{int}: 4 Ω, **c.** I_{sc}: 2.3 A, R_{int}: 6 Ω. **41. a.** I_{sc}: 2 A, R_{int}: 3 Ω, I_{R_3}: 1 A, **b.** I_{sc}: 1.5 A, R_{int}: 1.7 Ω, I_{R_2}: 0.3 A, **c.** I_{sc}: 1.2 A, R_{int}: 5 Ω, I_{R_4}: 1 A.

CHAPTER 8

3. a. doubles, **b.** doubles, **c.** no change, **d.** quadruples, **e.** one-ninth original value, **f.** multiplied by 16. **5.** 0.13 Ω. **7.** 0.08 Ω. **11. a.** 2.16×10^{-3} Ω, **b.** 7×10^{-3} Ω. **13. a.** 1008 W, **b.** 121 Ω, **c.** 2039°C. **15. a.** 1,206,000 Ω, **b.** 150,825 Ω, **c.** 46,300 Ω, **d.** 792 Ω. **17. a.** 4.094 Ω, approximately 4 × 10 gage value, **b.** four jumps of 3 gage values from 18 to 30 gage. Therefore, 30 gage resistance $2^4 \times$ 18 gage value or 16(6.51) = 104 Ω, **c.** Table 8.3 lists 30 gage resistance as 105.2 Ω. **19. a.** 2.57 Ω, **b.** 16.9 Ω CM/ft., **c.** 17.3×10^5 ft, **d.** 2 mils. **21.** R_{copper}: 95.8×10^{-3} Ω, R_{steel}: 0.393×10^{-3} Ω. Chassis

ANSWERS TO ODD-NUMBERED EXERCISES 733

resistance lower because of larger cross-sectional area. **25.** reliability and low noise; use metal film; avoid carbon. **27. a.** reduce it, **b.** layer resistance in shunt. **29. a.** P and A, **b.** neither, **c.** A but no P, **d.** P but no A. **31.** It increases. Greater care and testing during production. **33.** more values needed to prevent gaps. **37.** 13, 17, 23 and 31 Ω. **39. a.** 2.5 V, 10 k, **b.** 1.25 V, 10 k, **c.** 0.625 V, 10 k, **d.** 10 k. 1/2, 1/4, 1/8. Constant $R_{internal}$. V_o always 5 divided by a power of 2, $V_o = 5/2^{3-N}$, where n = switch number. **41. a.** 10 k, 10%, **b.** 9.1 M, 2%, **c.** 5.7 Ω, 1%, **d.** 7.6 k, 5%. **45.** a and d, B; b and c, A. **47.** low current, high resistance for potentials below protection voltage. High current, low resistance for potentials above protection voltage. **49.** generate electrical signal changing with variations in some physical parameter. **a.** CRT or light bulb, **b.** microphone, **c.** loudspeaker, **d.** resistance. **55.** Control light from two locations.

CHAPTER 9

1. Not likely. Rule: Test equipment weight must be very much less than weight of tested device. **3. a.** 88.3 V, **b.** 113.7 V. **5.** ammeter resistance raises total circuit resistance. **7. a.** $R_{total} < 750$ Ω, **b.** $R_{total} > 750$ Ω. **11. a.** 6%, **b.** 0%, **c.** 20.8%. **13.** choose meter so that reading falls in upper third of scale. **15. a.** 0.2 Ω, **b.** 0.06 Ω, **c.** 0.03 Ω. **17.** R_{shunt} gets lower and lower. Practical limit: difficult to make R_{shunt} to very low values. **19. a.** 50 k, **b.** 68 mV, **c.** 21.6 k, **d.** 2.5 M. **21. a.** 2 k, **b.** 50 k, **c.** 200 k, **d.** 1 M. **25.** zero or infinite ohms. Defective switch has resistance when closed which varies as the switch is operated. **27.** needle position and range-switch setting. **29.** higher input resistance. **33. a.** 99999, 9999.9, 999.99, 99.999, 9.9999, 0.99999, **b.** 199999, 19999.9, 1999.99, 199.999, 19.9999, 1.99999, 0.199999 V.

CHAPTER 10

1. No, but to outside circuit effect is as though it did. **3.** unbalanced charge. Discharge capacitor. **5. a.** 667 V, **b.** 3 μC, **c.** 20 μF. **7.** (1)**a.** R_1: 50 V, 25 A. R_2: 0 V, 0 A, **b.** $v_{R_1} = v_{R_2} = 25$ V, $i_{R_1} = i_{R_2} = 12.5$ A, (2)**a.** $v_{R_1} = v_{R_2} = 50$ V, $i_{R_1} = i_{R_2} = 25$ A, **b.** R_1: 50 V, 25 A, R_2: 0 V, 0 A. **9.** equal currents. **11.** 1.2 mA, 100 to 125 ms, 0 mA, 125 to 200 ms, 0.2 mA, 200 to 300 ms, 0 mA, 300 to 350 ms, −5 mA, 350 to 360 ms. **13.** 890 pF. **17. a.** 0.75 μF, **b.** 150 kV/m, **c.** 225 μC. **19.** Increasing thickness reduces capacitance. **21. a.** 1.35 J, 77 × 10⁻³ C, **b.** 22.1 × 10⁻² J, 38 nF, **c.** 4.8 C, 32 V, **d.** 86 pF, 50 mV. **23. a.** 0.5 F, **b.** 175 μF, **c.** 2 pF, **d.** 7500 pF, **e.** 3027 pF. **25. a.** R_{total} approximately equal to larger valued resistance, **b.** C_{total} approximately equal to smaller valued capacitance. **27.** C_1: 75 V, C_2: 20 V, C_3, C_4, C_5 and C_6 all have 7.5 V, C_7: 40 V. **29. a.** 202.7 pF, **b.** 200 pF, **c.** 199.9 pF. **31.** Observe polarity markings. **33. a.** 100 ms, **b.** $v_C = 120(1 - e^{-t/100ms})$ volts, $v_R = 120e^{-t/100ms}$ volts, $i = 60e^{-t/100ms}$ mA, **c.** 34.8 mA, **d.** 118 V, **e.** 0 V. **35. a.** $5\tau = 5$ s, **b.** charge to 3001 V, **c.** $5\tau = 5$ s, **d.** capacitor discharges to 3 kV in 5 τ or 5 s, **e.** it takes 5τ for a capacitor to charge or discharge completely through any voltage change. **37. a.** 35 V, **b.** 0.21 mA.

CHAPTER 11

1. a. A: 3.8 × 10⁹ G, B: 3 × 10⁹ G, C: 164 × 10⁹ G, **b.** A: 3.8 × 10⁵ Wb/m², B: 3 × 10⁵ Wb/m², C: 164 × 10⁵ Wb/m². **3. a.** out, **b.** in. **5.** 180 turns. **7.** Present value of B depends on past value. **9.** when B–H curve flattens so that increase in H causes no increase in B.

11. a. 50 mA, **b.** 0.065 Wb/A-m.   **13.** 1 A entering terminal A.   **15.** 6.0 AT.
17. 21.4 × 10⁻⁶ Wb.   **19.** 3000 turns with gap versus 19.7 without.   **21. a.** decrease,
b. increase, **c.** no change.   **23.** move loop faster, increase B, wind loop into N turns.
25. 5 kV.   **27.** 200 turns.   **29.** current creates opposing field.

CHAPTER 12

1. increases because more linkages changed per unit time.   **3.** 0.75 H.   **3.** 880 mH.
7. 0 to 2 s: 36 V, 2 to 6 s: −36 V, 6 to 10 s: 36 V, etc.   **9.** both currents zero at $t = 0^+$ s.
At $t = \infty$, 2 H: 4 A, 5 H: 2 A.   **11.** both 0 J at $t = 0^+$ s. At $t = \infty$, 2 H: 16 J, 5 H: 10 J.
13. a and f, L; b and d, R; c and e, C.   **15.** they "choke" current changes.   **17.** 605 turns.
19. core currents developed by linkages between core and its own flux field. Increase core
resistance.   **21. a.** 200 µH, 10%, **b.** 7.5 µH, 5%, **c.** 3900 µH, 20%, **d.** 0.16 µH, 5%.
23. Reduce inductance below that with air core.   **25.** 12 H.   **27.** unavoidable resistance,
capacitance, and inductance of components. High frequencies. Low frequencies.   **31. a.** 10 s,
b. 4 A, **c.** 8.9 kiloamps. Still requires 10 s, **d.** I_L climbs from zero to max value in 5τ, no matter
what the value of i_{max}.   **33. a.** 2.5 ms, **b.** $v_L = 60\,e^{-\tau/(2.5 \times 10^{-3})}$ V, $v_R = 60(1 - e^{-\tau/(2.5 \times 10^{-3})})$ V,
$i = 60(1 - e^{-\tau/(2.5 \times 10^{-3})})$ mA, **c.** 51 mA, **d.** 0.6 V, **e.** 60 V.   **35.** 3.5 mJ.
37. a. $v_L = 0$ V, $i = 4.7$ mA, $v_{9k} = 42$ V, **b.** i remains at 4.7 mA, $v_L = -56.4$ V, v_{9k} remains
at 42 V, $v_{12k} = 56.4$ V, **c.** v_L to 42 V then exponential decay to zero, then to −56.4 V,
then decay to zero, i rises exponentially to 4.7 mA, then decays exponentially to 2 mA.
39. a. $v_L = 5\,e^{-\tau/(150 \times 10^{-3})}$ V, $i = 500(1 - e^{-\tau/(150 \times 10^{-3})})$ mA, **b.** $v_{8\Omega} = 4(1 - e^{-\tau/(150 \times 10^{-3})})$ V.

CHAPTER 13

1. a. $V_x = 0.8$ and $V_y = 2.9$ ft/s, **b.** $V_x = -400$ and $V_y = -692.8$ m/s.   **3. a.** no flux lines
being cut, **b.** maximum rate of flux cutting or, we can say, sin θ has its maximum value, 1, at
90°, **c.** $v_y = V \sin \theta$, **d.** flux lines cut in opposite direction.   **7. a.** all units rads.
(1) π, (2) π/9, (3) π/3, (4) π/4, (5) π/12, **b.** all units rad. (1) 1.13, (2) 0.31, (3) 0.49,
(4) 1.43, **c.** (1) 75°, (2) 11°, (3) 58°, (4) 38°, **d.** (1) 45°, (2) 72°, (3) 225°, (4) 84°.
9. (1)**a.** 2 rad, **b.** 0.91; (2)**a.** 6.3 rad; **b.** 0.02; (3)**a.** 8.5 rad; **b.** 0.8.   **11. a.** 5200 s,
b. 1.24 × 10⁶ rotations.   **13. a.** no maximum value, **b.** ±1, **c.** +75 to −75 V, **d.** 1:
radians, 2: no units, 3: volts.   **e.** infinite number of possibilities. One is $t = 7.48$ ms,
f. infinite possibilities. One is $t = 3.74$ ms,   **g.** $i(t) = 3 \sin 420t$ A.   **15. a.** 750 V,
b. 6720 rad/s, **c.** 1070 Hz, **d.** 706 V, **e.** −175 V, **f.** $i(t) = 3 \sin 6720t$ mA, **g.** −650 µA.
17. a. 12.6 × 10⁶ rad/s, 2 MHz, **b.** 756 Hz, 1.32 ms, **c.** 848 × 10⁶ rad/s, 7.4 ns.
19. a. $i(t) = 340 \sin 25000t$ µA, **b.** $e(t) = 66.2 \sin 377t$ V.   **21. a.** 16.7 ms, **b.** 7.7 A,
c. 15.4 A, **d.** 5.3 A.   **23. a.** 5 A, **b.** 4 A, **c.** 9 A, **d.** no.   **25.** $e(t) = -8.1 +$
$8.4 \sin(22 \times 10^6 t)$ V.   **27. a.** 1.25 µA, **b.** 1.9 V.   **29. a.** 30, 10.6, 0, **b.** 37, 26.2, 0,
c. 28.3, 56.6, 0, **d.** 75, 53, 0, **e.** 5.4, 10.8, 0.   **31.** 0.7 A.   **33.** 110.2 W.

CHAPTER 14

1. a, b, and c, e_2; d and e, e_1; f, e_1 and e_3.   **3. a.** $e(t) = 15 \sin 377t$ V, **b.** $e(t) = 13 \sin(157t +$
$135°)$ mV.   **5. a.** −30° lagging, **b.** 100° leading, **c.** 15° leading, **d.** no phase relation
because frequencies differ.   **7. a.** $\bar{V} = 55.2\,\underline{/0°}$ V, $\bar{I} = 21.9\,\underline{/-30°}$ A, **b.** $\bar{I} = 2.1\,\underline{/120°}$ A, $\bar{E} =$
$5\,\underline{/20°}$ V, **c.** $\bar{E} = 9.2\,\underline{/50°}$ V, $\bar{I} = 6.9\,\underline{/65°}$ A, **d.** $\bar{I} = 49.5\,\underline{/20°}$ A, $\bar{E} = 123.7\,\underline{/-83°}$ V.
9. a. $e = 14.1 \sin(377t + 63°)$ V, $i = 28.3 \sin(377t - 15°)$ mA, **b.** $e_1 = 46.7 \sin(2512t + 47°)$ V,
$e_2 = 70.7 \sin(2512t - 55°)$ V, $i = 5.7 \sin(2512t + 20°)$ A.   **11.** a and d, real; b and

ANSWERS TO ODD-NUMBERED EXERCISES 735

c, imaginary; **e.** complex. **13.** **a.** $10.9 + j5.1$, **b.** $2.3 - j7.2$, **c.** $-14.3 - j4.4$,
d. $-2.7 + j8.8$. **15.** **a.** $13\,\underline{/67.4°}$ **b.** $5\,\underline{/-126.9°}$ or $5\,\underline{/233.1°}$ **c.** $6.6\,\underline{/112.3°}$ **d.** $13.5\,\underline{/-35.7°}$.
17. **a.** $12 - j12.5$, **b.** $2.9 + j16.5$, **c.** $-19.8 - j10.8$, **d.** $19 - j13.5$, **e.** $15 - j3.6$,
f. $22.4 + j3.6$, **g.** $1.9 + j21.9$, **h.** $-33.3 - j4.9$, **i.** $-16.2 - j5.1$. **19.** **a.** $31.6\sin(100t - 2.7°)$,
b. $46.1\sin(377t - 37°)$, **c.** $9.6\sin(50t - 39.1°)$, **d.** $61.3\sin(1000t + 134.4°)$. **21.** **a.** $10\,\underline{/227°}$,
b. $28\,\underline{/107°}$, **c.** $888\,\underline{/146°}$, **d.** $240\,\underline{/147.1°}$, **e.** $612\,\underline{/16°}$, **f.** $13{,}375\,\underline{/153°}$, **g.** $48{,}675\,\underline{/-70.7°}$,
h. $119248\,\underline{/88°}$. **23.** **a.** $10(-3 + j1)$, **b.** 146, **c.** $24(2 - j1)$, **d.** $7 - j199$, **e.** $3(3 - j1)$,
f. $15(4 - j3)$, **g.** $25(5 - j2)$, **h.** $67 - j101$. **25.** **a.** $3\,\underline{/18°}$ **b.** $2.2\,\underline{/55°}$, **c.** $2\,\underline{/235°}$,
d. $17\,\underline{/113°}$, **e.** $2.5\,\underline{/36°}$, **f.** $13.2\,\underline{/-290°}$, **g.** $1.9\,\underline{/-8.1°}$, **h.** $0.9\,\underline{/299°}$. **27.** **a.** $-4.5 - j3$,
b. $-1.8 - j0.1$, **c.** $5.5 - j1.5$, **d.** $-15.4 + j6.1$, **e.** $-26.5 - j41.4$, **f.** $-0.7 + j0.1$,
g. $-1 + j0.6$, **h.** $-0.7 - j1.3$.

CHAPTER 15

1. **a.** 4, **b.** 2, **c.** 5, **d.** 1. **3.** For sine wave: **a.** 0 to $90°$ and 270 to $360°$, **b.** 90 to $270°$;
for cosine wave: **a.** $180°$ to $360°$, **b.** 0 to $180°$. **5.** **a.** $4500\sin(300t + 90°)$, **b.** $12.3 \times 10^6 \sin(40{,}000t + 131°)$, **c.** $0.8\sin(11.5t + 75°)$. **7.** **a.** $9.4\,\underline{/90°}$, $j9.4$, **b.** $310.9\,\underline{/90°}$, $j310.9$,
c. $47.1 \times 10^3\,\underline{/90°}$, $j47.1 \times 10^3$, **d.** $1{,}055{,}040\,\underline{/90°}$, $j1{,}055{,}040\,\Omega$. **9.** $470\,\Omega$. resistance not
affected by change in frequency. **11.** **a.** $i = 2\sin(1000t - 60°)$ A, $\bar{I} = 1.4\,\underline{/-60°}$ A, **b.** $i = 640\sin(400t + 90°)$ mA, $\bar{I} = 452\,\underline{/90°}$ mA, **c.** $i = 24.5\sin(377t - 45°)$ mA, $\bar{I} = 17.3\,\underline{/-45°}$ mA,
d. $i = 112\sin(500t + 7°)$ μA, $\bar{I} = 79.3\,\underline{/7°}$ μA **13.** a and c, L; b and f, R; d and e, C.
15. **a.** $1902 + j618$, **b.** $2.3 - j2.9$, **c.** $172.6 - j811.9$. **17.** **a.** $3.4\,\underline{/-72.7°}$ lead,
b. $26\,\underline{/57.5°}$ lag, **c.** $4264\,\underline{/-39.3°}$ lead, **d.** $844\text{ k}\,\underline{/36.3°}$ lag. **19.** a and e, NC; b, c, d, and
f, I. **21.** a and e, NC; b and c, I; d, M; f. D. **23.** **a.** $0\,\Omega$, **b.** $R < X$, **c.** $R \gg X$,
d. $R = X$, **e.** $R \ll X$, **f.** $R > X$. **25.** **a.** $f \ll 5310$ Hz, **b.** $f < 5310$ Hz, **c.** $f = 5310$ Hz,
d. $f > 5310$ Hz, **e.** $f \gg 5310$ Hz. **27.** **a.** $\bar{I} = 1.4\,\underline{/54.7°}$ A, $i = 1.9\sin(3400t + 54.7°)$ A,
b. $29.7°$, **c.** 2.1 A, **d.** 3.8 A. **29.** 724 Hz. **31.** **a.** 1.5 kvar, **b.** 184 megavar, **c.** 0 var.
33. **a.** 500 A; **b, c,** and **d.** $8\,\Omega$; **e.** 1.43 megavar. **35.** **a.** 0.5, 400 VA, 346.4 var, 200 W,
b. 5 A, $-12°$, 0.98, -36.4 var, **c.** 22 V, 0.33, 160.6 VA, -151.9 var, **d.** 25 A, $45°$, 0.71, 100 W,
e. 10 V, $82°$, 0.14, 16 VA, **f.** 5.8 A, $-34°$, 67 VA, 55.6 W. **37.** **a.** $2\,\underline{/-69.3°}$ A,
b. $226\,\underline{/-69.3°}$ VA, **c.** 79.9 W, **d.** -211 var, ind. **e.** 35% lagging.

CHAPTER 16

1. **a.** $10\,\Omega$, 670 μF, **b.** 1.1 k, 140 mH. **3.** **a.** $3 - j5$, **b.** $5 + j20$, **c.** $32 - j22$,
d. $1\text{ k} + j1.2\text{ k}$, **e.** $3\,\Omega$. **5.** **a.** $69 - j177\,\Omega$, **b.** $1\,\underline{/68.7°}$ A, **c.** $\bar{V}_R = 69\,\underline{/68.7°}$ V,
$\bar{V}_C = 177\,\underline{/-21.3°}$ V, **e.** $v_R = 97.6\sin(9420t + 68.7°)$ V, $v_C = 250.3\sin(9420t - 21.3°)$ V,
$i = 1.4\sin(9420t + 68.7°)$ **f.** $P_W = 69$ W, $P_Q = 177$ var, $P_A = 190$ VA, **g.** 36% lead.
7. **a.** $X_L = j1131\,\Omega$, $\bar{Z} = 1000 + j1131\,\Omega$, **b.** $110\,\underline{/-48.5°}$ mA, **c.** $\bar{V}_R = 110\,\underline{/-48.5°}$ V,
$\bar{V}_L = 124.4\,\underline{/41.5°}$ V, **e.** $v_R = 155.5\sin(377t - 48.5°)$ V, $V_L = 175.9\sin(377t + 41.5°)$ V,
$i = 155.5\sin(377t - 48.5°)$ mA, **f.** $P_W = 12.1$ W, $P_A = 13.7$ var, $P_A = 18.3$ VA, **g.** 66% lag.
9. $452.5\,\Omega$, 0.005 μF. **11.** **a.** $20 + j9.3\,\Omega$, **b.** $\bar{I} = 2.3\,\underline{/-24.9°}$ A, $\bar{V}_R = 46\,\underline{/-24.9°}$ V,
$\bar{V}_L = 52\,\underline{/65.1°}$ V, $\bar{V}_C = 30.6\,\underline{/-114.9°}$ V, **d.** $P_W = 104.3$, $P_{Q\text{ind}} = -j119.6$ var, $P_{Q\text{cap}} = j70.4$ var, $P_A = 115$ VA. **13.** **a.** 5 V, **b.** $-j25\,\Omega$, **c.** 5 V. **15.** b, f, h, and j, 1; d and i, 2;
a, c, e, and g, 3. **17.** d and f, 1; a, c, g, and h, 2; b and e, 3. **19.** **a.** 4 kHz, **b.** 250 μH,
c. $126\,\Omega$. **21.** **a.** $10\,\underline{/0°}$ V, **b.** $7\,\underline{/-45°}$ V, **c.** $1\,\underline{/-90°}$ V. **23.** **a.** 11 k, **b.** 0.001 μF,
c. 964 kHz. **25.** **a.** across the capacitor, **b.** across the resistor, **c.** $10\,\underline{/0°}$ V, $7\,\underline{/-45°}$ V,
$0.01\,\underline{/-90°}$ V, **d.** low-pass, **e.** $0.1\,\underline{/90°}$ V, $7\,\underline{/45°}$ V, $10\,\underline{/0°}$ V. **27.** **a.** $7.1\,\underline{/45°}\,\Omega$,
b. $860\,\underline{/-29.1°}\,\Omega$, **c.** $2.9\,\underline{/-12.4°}\,\Omega$. **29.** **a.** 277 μS, **b.** 900 Hz, **c.** 27 mH.
31. **a.** 0.1 S, **b.** $-j10^{-3}$ S, **c.** $j10^{-4}$ S, **d.** $-j3.3$ mS, **e.** 33.3 mS, **f.** $j32$ kS.
33. **a.** $12.6 \times 10^{-3}\,\underline{/37.6°}$ S, **b.** 1.3 A, **c.** $j1$ A, **d.** $(1.3 + j1)$ A, **e.** $37.6°$, **f.** 213 VA,

g. 169 W, **h.** 130 var, **i.** 79.4% lead. **35. a.** $G = 0.05$ S, $B_C = j0.1$ S, $\bar{Y}_{total} = 0.05 + j0.1$ S, **c.** $\bar{I}_{total} = 3.8\,\underline{/63.4°}$ A, $\bar{I}_R = 1.7\,\underline{/0°}$ A, $\bar{I}_C = 3.4\,\underline{/90°}$ A, **f.** 57.8 W, 115.6 var, 129.2 VA.
37. a. $2.42 \times 10^{-3}\,\underline{/-52.6°}$ S, **b.** $607\,\underline{/0°}$ mA, **c.** $793\,\underline{/-90°}$ mA, **d.** $1\,\underline{/-52.6°}$ A, **e.** $-52.6°$, **f.** 413 VA, **g.** 251 W, **h.** 328 var, **i.** 60.8% lag. **39.** $15\,\Omega$, $20\,\mu$H. **41. a.** 4.55×10^{-6} S, $-j5 \times 10^{-6}$ S, $6.75 \times 10^{-6}\,\underline{/-47.7°}$ S, **b.** $295\sin(14.4 \times 10^3 t + 47.7°)$ V, **c.** $i_R = 1.34\sin(14.4 \times 10^3 t)$ mA, $i_L = 1.48\sin(14.4 \times 10^3 t - 90°)$ mA. **43. a.** $0.18\,\underline{/56.3°}$ S, **b.** $19.8\,\underline{/56.3°}$ A, **c.** $\bar{I}_R = 11$ A, $\bar{I}_L = -j5.5$ A, $\bar{I}_C = j22$ A, **f.** $P_W = 1210$ W, $P_{QL} = -605$ var, $P_{QC} = 2420$ var, **g.** 2178 VA, 56% lead. **45.** $33.3\,\Omega$, $250\,\mu$H. **47.** $\bar{I}_R = 30.6\,\underline{/34.3°}$ mA, $\bar{I}_L = 20.9\,\underline{/-55.7°}$ mA. **49. a, d,** and **i**, 1; **e** and **g**, 2; **b, c, f** and **h**; 3. **51. d** and **h**, 1; **b, e, g,** and **i**, 2; **a, c, f,** and **j** 3. **53. a.** $85\,\Omega, j21.3\,\Omega$, **b.** $47.1\,\Omega, -j68.8\,\Omega$. **55. a.** $19.3\,\Omega, -j14.6\,\Omega$, **b.** $1.7\,\Omega, j0.7\,\Omega$. **57. a.** $325\,\Omega, j40.6\,\Omega$, **b.** both changed, **c.** no. **59.** $Q = 4$, PF = 24.3% lag.

CHAPTER 17

1. (1)**a.** $1.9 - j3\,\Omega$, **b.** $6.9\,\underline{/57.7°}$ A, **c.** $57.7°$, **d.** $5.6\,\underline{/94.6°}$ A, **e.** $4.1\,\underline{/4.6°}$ A. (2)**a.** $3.6 - j7.8\,\Omega$, **b.** $8.7\,\underline{/45.2°}$ A, **c.** $65.2°$, **d.** $7.4\,\underline{/77.2°}$ A, **e.** $4.6\,\underline{/-12.8°}$ A. (3)**a.** $8 + j2.5\,\Omega$, **b.** $14.3\,\underline{/-17.4°}$ A, **c.** $-17.4°$, **d.** $12.8\,\underline{/9.2°}$ A, **e.** $6.4\,\underline{/-80.8°}$ A. **3. a.** $21\,\underline{/-53.1°}$ A, **b.** $7.7\,\underline{/-89.1°}$ A. **c.** $46.3\,\underline{/-36°}$ V, **d.** $\bar{V}_C$ lags $\bar{E}_s$ by $126°$. **5. a.** 46.6 kW, **b.** $54.5\,\underline{/75.8°}$ A. **7.** $5 - j12\,\Omega$. **9. a.** with the a terminal as positive, $V_{ab} = 15.7\,\underline{/125°}$ V, **b.** $\bar{V}_{ab}$ now $15.7\,\underline{/-55.1°}$ V, **c.** $16.2\,\underline{/-127.7°}$ V. **11. a.** 45.4% lag, **b.** 82% lag, **c.** 397 V across a 220-V motor, **d.** $24.4\,\mu$F **e.** 220 V. Connect them in shunt. **13.** $24.9\,\mu$F. **15.** bridge unbalanced; R_x should be $850\,\Omega$. **17.** No, bridge is balanced. **19.** inductance: $R_4 = R_2 R_3/R_1$, $L_4 = L_2 R_3/R_1$. Hay: $R_4 = (\omega^2 R_1 C_1^2 R_2 R_3)/(1 + \omega^2 R_1^2 C_1^2)$, $L_4 = C_1 R_2 R_3/(1 + \omega^2 R_1^2 C_1^2)$. Hay bridge is frequency dependent. **23. a.** $30\,\underline{/40°}\,\Omega$, **b.** same as a, **c.** elements $45\,\underline{/40°}\,\Omega$, **d.** Yes. **25.** $\bar{Z}_a = 4.4\,\underline{/7.2°}\,\Omega$, $\bar{Z}_b = 1.5\,\underline{/14.1°}\,\Omega$, $\bar{Z}_c = 2.8\,\underline{/44.1°}\,\Omega$. **27.** $8.7\,\underline{/43.8°}$ A. **29. a.** 25 V, 0 V, **b.** $5\,\underline{/53.1°}$ A, 0 A, **c.** $26.9\,\underline{/29.6°}$ V, $2.7\,\underline{/-30.4°}$ A, **d.** $5\,\underline{/53.1°}$ A, **e.** 52.1 W. **31. a.** $\bar{E}_{oc} = 1.7\,\underline{/-56.3°}$ V, $\bar{Z}_{int} = 0.8 - j1.2\,\Omega$, **b.** $\bar{I}_{sc} = 10\,\underline{/-45°}$ A, $B_{int} = -j0.2$ S. **33.** $\bar{I}_{total} = 2.6\,\underline{/43.5°}$ A, $\bar{Y}_{total} = 0.05 - j0.08$ S. **35.** $\bar{E}_{oc} = 212.1\,\underline{/-25.6°}$ V, $\bar{Z}_{internal} = 30 + j40\,\Omega$. **37. a.** $11\,\underline{/17.9°}$ A, **b.** 2.3 A. **39.** $3.9\,\underline{/25.9°}$ A. **41. a.** $2\,\underline{/-30.3°}$ A, **b.** ≈ 50.2 V. **43. a.** $2\,\underline{/48.3°}$ A, **b.** 96.8 W, **c.** $29.4\,\underline{/6.2°}$ V, **d.** $5.6\,\underline{/-41.7°}$ A.

CHAPTER 18

1. $1.3\,\underline{/89.5°}$ A, right to left. **3.** 188.5 var, inductive. **5.** $3.3\,\underline{/-8.4°}$ A. **7.** $1.2\,\underline{/-75.6°}$ A. **9.** $1.1\,\underline{/12.1°}$ A **11. a.** $7.9\,\underline{/18.4°}$ V, **b.** $4.9\,\underline{/-9.5°}$ V. **13.** $11.1\,\underline{/18°}$ A. **15. a.** 0.64 var, **b.** 10.7 var, **c.** 3.7 var. **17.** $16.9\,\underline{/-55.3°}$ A. **19. a.** $8\,\underline{/35°}$ V, $j4\,\Omega$, **b.** $13.9\,\underline{/41.5°}$ V, $4.5\,\underline{/63.4°}\,\Omega$. **21. a.** 5 V, $10\,\Omega$, **b.** $15.9\,\underline{/7.5°}$ V, $5 - j4\,\Omega$. **23. a.** $10\,\Omega$, **b.** $5 + j4\,\Omega$. **25. a.** $3\,\underline{/45°}$ A, $18.5\,\underline{/22.6°}\,\Omega$, $\bar{I}_{cap} = 3.3\,\underline{/67.6°}$ A, **b.** $3.1\,\underline{/-21.7°}$ A, $4.5\,\underline{/63.4°}\,\Omega$, $\bar{I}_{cap} = 3.1\,\underline{/104.8°}$ A. **27. a.** 0.5 A, $10\,\Omega$, **b.** $2.5\,\underline{/46.2°}$ A, $5 - j4\,\Omega$. **29.** $6.1\,\underline{/35°}$ A, $1.8\,\underline{/5.2°}\,\Omega$, $I_{source} = 1.4\,\underline{/151.5°}$ A. **31.** $0.6\,\underline{/43.1°}$ A, $3.6\,\underline{/-23°}\,\Omega$.

CHAPTER 19

1. a. 1.6 MHz, **b.** 11 mH, **c.** 142 pF, **d.** 9.2 kHz, **e.** $38\,\mu$H, **f.** $14.8\,\mu$F **3. a.** 1 and 8, **b.** 3 and 5, **c.** 2, 4, 6, and 7. **5. a.** $82\,\Omega$, **b.** 14.7, **c.** $4.5\,\mu$H, **d.** 35 Hz, **e.** 2300, **f.** $0.003\,\Omega$. **7.** $Q = 13.3$, $V_{cap} = 400$ V. **9. a** and **b**, 100 V; **c** and **d**, 20 kV; **e.** 0 V. **11. a.** 8, **b.** $95\,\Omega$, **c.** 480 mH, **d.** $0.01\,\mu$F. **13. a, b, e, f, g, j,** and **k** are H;

ANSWERS TO ODD-NUMBERED EXERCISES

c, d, h, and i are L.   **15. a.** 1050 kHz, 950 kHz, 10,   **b.** 14.3 MHz, 14.25 MHz, 143,
c. 235.7 Hz, 2.6 Hz, 91.2,   **d.** 74.1 kHz, 69.5 kHz, 4.6 kHz.   **17. a.** 384 kHz,
b. 24,   **c.** 16 kHz,   **d.** 392 and 376 kHz,   **e.** 392.1 and 376 kHz.   **19. a.** no,   **b.** 453 kHz,
c. 12.5 mA,   **d.** 230 mA.   **21. a.** 20,   **b.** 100 k,   **c.** $R_p \approx 100$ k, $X_{L_p} \approx X_L = 5$ k   **d.** the same,
e. 5 mA,   **f.** 100 mA both ways,   **g.** 10.   **23.** circuit detuned to 637 kHz.   **25. a.** 690 kHz,
b. 31.3,   **c.** 22.1 kHz,   **d.** 701 kHz and 679 kHz,   **e.** 85.7 k.   **27.** Set inductor to 780 μH;
shunt with 319 k.   **31.** $C_s = 345$ pF, $L_p = 10.3$ μH.   **33.** $L_s = 5$ μH, $C_p = 100$ pF.
35. $C_s = 0.001$ μF, $L_{total} = 14.9$ μH.   **37. a.** 103.8 and 96.2 kHz,   **b.** 5 Ω,   **c.** 2.2 mH, 1160 pF,
d. 6.4 kHz.   **39.** $f_{center} = f_s = 784$ kHz, $f_2 = 819$ kHz, $f_1 = 749$ kHz, $V_{out_{max}} \approx 100$ V, $V_{out_{min}} =$
4.7 V. Reject region from 749 to 819 kHz. All frequencies above and below these are passed.
41. $L_s = 26.4$ μH, $L_p = 634$ μH.   **43.** $L_s = 100$ μH, $L_p = 1.5$ mH.
45. $f_{pass} = 1/2\pi\sqrt{L(C_s + C_p)}$.

CHAPTER 20

3. a. 3031 V,   **b.** 311.8 kV.   **5. a.** 549.1 V,   **b.** 317 V,   **c.** 31.7 A,   **d.** 31.7 A.
7. a. $\bar{I}_a = 92.8 \underline{/68.2°}$ A, $\bar{I}_b = 92.8 \underline{/-51.8°}$ A, $\bar{I}_c = 92.8 \underline{/-171.8°}$ A,   **c.** $P_A = 278$ kVA,
$P_W = 103.4$ kW, $P_Q = 258.6$ kvar.   **9. a.** 11.5 A,   **b.** 8952 W,   **c.** 15.7 A, 12,263 W.
11. 1334 Ω.   **13. a.** $\bar{I}_a = 27.7 \underline{/0°}$ A, $\bar{I}_b = 18.5 \underline{/-120°}$ A, $\bar{I}_c = 12.6 \underline{/-240°}$ A,   **b.** 16,276 W,
c. 0 var,   **d.** $P_A = P_W$, PF = 100%,   **e.** 13.2 $\underline{/-22.8°}$ A.   **15.** $\bar{I}_a = 12 \underline{/-80°}$ A,
$\bar{I}_b = 8 \underline{/-130°}$ A, $\bar{I}_c = 4.8 \underline{/85°}$ A, $\bar{I}_n = 13.4 \underline{/-101.3°}$ A.   **17. a.** 480 V,   **b.** 17.8 A,   **c.** 30.8 A,
d. 25,599 W.   **19. a.** $\bar{I}_{ab} = 201.7 \underline{/-60°}$ A, $\bar{I}_{bc} = 201.7 \underline{/-180°}$ A, $\bar{I}_{ca} = 201.7 \underline{/60°}$ A,   **b.** $\bar{I}_{line} =$
349.3 A.   **c.** $P_W = 732.2$ kW, $P_{Q_{total}} = -1270$ kvar, $P_{A_{total}} = 1466$ kVA, PF = 49.9% lag.
21. a. $\bar{I}_{ab} = 24.3 \underline{/120°}$ A, $I_{bc} = 58.4 \underline{/-180°}$ A, $\bar{I}_{ca} = 29.2 \underline{/150°}$ A,   **b.** $\bar{I}_a = 14.6 \underline{/26.1°}$ A,
$\bar{I}_b = 50.8 \underline{/-155.5°}$ A, $\bar{I}_c = 36.2 \underline{/23.8°}$ A,   **c.** $P_{W_{total}} = 8526$ W, $P_{Q_{total}} = -9967$ var, $P_{A_{total}} =$
13,116 VA, PF = 65% lagging.

CHAPTER 21

1. a. 4000, SD,   **b.** 38, SD,   **c.** 12 V, SU,   **d.** 720 mA, SU.   **3. a.** 117:1,   **b.** 1.8 A,
c. 208.3 A.   **5.** 60.6.   **7. a.** 14.5 k SD,   **b.** 0.04 SU,   **c.** 222 Ω SD,   **d.** 0.1 SU.
9. a. C, (D to F), (E to G), (H to I), J.   **b.** C, (D to E), (F to G), (H to J), I.   **c.** C, (D to E),
(F, G, and J together), (H and I together),   **d.** C, (D to F), (E, H, and I together), (G and J together).
11. $\bar{I}_1 = 100 \underline{/30°}$ mA, $\bar{I}_2 = 500 \underline{/30°}$ mA, $\bar{I}_3 = 2 \underline{/-150°}$ A.   **13.** $I_p = 366$ μA, $I_s = 5.3$ mA.
15. 5.3 mA.   **17. a.** 10.7 Wb,   **b.** 4.6 Wb,   **c.** 0.7.   **19. a.** 7.4 Wb,   **b.** 0.6,
c. $e_p = 1950$ V, $e_s = 343.2$ V.   **21.** $E_p = 19.5$ V, $E_s = 27$ V.   **23. a.** -2 H, 0.52.
b. 15 H, 14 H.   **c.** 17 mH, 4 mH.   **d.** 20 H, 20 H, 74.8 H.   **e.** 18 H, 0.2.   **25.** 3 H.
27. 40 H.   **29.** copper loss by R_p and R_s, eddy current and hysterisis losses by R_c.
31. a. frequency of low end of flat region increases,   **b.** frequency of resonant rise increases,
c. no effect,   **d.** frequency of low end of flat region decreases,   **e.** reduce height of flat region.
33. $33 + j2.4$ Ω.   **35.** $13 \underline{/-61.4°}$ A, $54.7 \underline{/-151.4°}$ V, $8.1 \underline{/-205.4°}$ A.   **37. a.** $I_s = 18.5$ A,
b. 5.4 Ω,   **c.** 5.4 Ω,   **d.** reflected loads combine in parallel.   **39.** b at 24 turns, c at 39.2 turns.
41. a. $I_p = 56.8$ A, $I_s = 208$ A.   **b.** 648 V, 135 kVA, 264.8 A.   **c.** $I_p = 35.5$ A, $I_{source} = 165.5$ A.

CHAPTER 22

1. a. 167.8 V,   **b.** 237.5 V,   **c.** 83.9 V, 118.8 V.   **3.** half-wave.   **5. a.** 8.9 A,   **b.** 17.8 A.
9. frequency range, voltage capability, and true rms.   **11. a.** 4.8 V, 9.6 V, 14.6 s, 68×10^{-3} Hz,
b. 96 mV, 192 mV, 730 μs, 1369.9 Hz,   **c.** 2.4 V, 4.8 V, 3.7 s, 27×10^{-2} Hz,   **d.** 24 V, 48 V, 7.3 μs,

137 kHz.   **13.** $e(t) = 9\cos(5.2 \times 10^6\, t)$ mV.   **15.** sine-wave axis vertical; positive maximum to the left, negative maximum to the right.   **17. a.** 0.15, 8.6°,   **b.** 0, 0°,   **c.** 0.8, 52.8°,   **d.** 1, 90°.   **19. a.** 49.5 Hz,   **b.** 280 Hz,   **c.** 765 Hz,   **d.** 131.3 Hz.

CHAPTER 23

1. a. 1.6 ms,   **b.** 1875 Hz, 533.3 μs,   **c.** 6875 Hz, 145.5 μs.   **3.** 55.5 kHz.   **5.** $v = 12.3 + 12\sin 1884t + 10\sin 3768t + 4.1\sin 7536t$ volts.   **7.** (1)**a** and **b**, yes; **c** and **d**, no. (2)**a** and **b**, yes; **c** and **d**, no. (3)**a** and **b**, no; **c** and **d**, yes. (4)**a** yes; **b, c,** and **d**, no.   **9.** 1.06 V_m.
11. a. $V_{\text{ave}} = 3$ V, $V_{\text{eff}} = 4.8$ V, $P = 5.1$ W.   **b.** $I_{\text{ave}} = -4.8$ A, $I_{\text{eff}} = 5.5$ A, $P = 136.1$ W.
13. a. $i = 3.1 + 1.9\sin \omega t + 1.5\sin 3\omega t + 0.9\sin 5\omega t$ A,   **b.** $P = 156$ W.   **15.** $i = 0.8\sin(10t + 63.4°)$ A, $v_R = 8\sin(10t + 63.4°)$ V, $v_C = 10 + 16\sin(10t - 26.6°)$ V.   **17.** $i = 0.4 + 0.2\sin(5t - 18.4°)$ A, $v_R = 12 + 6\sin(5t - 18.4°)$, V, $v_L = 2\sin(5t + 71.6°)$ V, $P = 5.4$ W.
19. a. $v_{\text{out}} = \frac{100}{\pi}[1 - 0.01183\cos(754t - 172°) - 0.00059\cos(1508t - 176°) - 0.00011\cos(2262t - 177°)]$ V,   **b.** $V \approx 100/\pi$ volts.

Index

ac generator, 376–381
Admittance, 482, 485
Air gaps, 331–332, 351
Air-core transformer, 671
American Wire Gage (AWG), 236
Ammeters, 85–86, 261–262
Ampere, 22–23
Ampère, André Marie, 22
Ampere turns, 322, 324, 651, 654
Ampere's circuital law, 327–328
Analog meters, 258, 271–274
 bargraph, 273–274
 comparison with digital meters, 271–272
 electrodynamometer, 272
 parallax error, 261
Angular velocity, 381
Anti-resonance. See Parallel resonance
Apparent power, 453–457
Armature, 376–379
Atom, 14–15
 Bohr model of, 14
 electrons, 14–15
 neutrons, 14
 nucleus, 14
 proton, 14
Audio transformer, 674
Autotransformer, 676–677
Average power, 632–633
Average value, 389–391

Balance equation
 ac, 527–528
 dc, 152–154
Balanced load
 in three-phase systems, 636–638
Ballpark estimates
 analyzing parallel circuits, 116–117
 analyzing series circuits, 70–71
Band-pass filter, 614, 615–617, 620

Band-reject filter, 614, 617–618, 620
Bandwidth
 parallel resonant circuits, 605–607
 series resonant circuits, 591–596
 signal, 596
Bargraph, 273–274
B-H curves, 324–327
Billiard ball effect, 24
Branch, 102, 188–190
Breakdown, insulation, 24
Breakdown voltage, 291
Bridges
 ac, 526–530
 ac balance equation, 527–528
 capacitance, 529
 dc, 152–156
 dc balance equation, 152–154
 Hay, 549
 inductance, 549
 Maxwell, 528–529
 Owens, 549
 Schering, 549
 Wheatstone, 155–156
 Wien, 529–530, 549, 550
Bridge rectification, 685

Calculators, required features of, 2
Capacitance, 280–282, 286–287
 bridge, 529
 parasitic, 357
 phase relations, 438–439
 stray, 299
 in transformers, 670
Capacitive reactance, 438–442
Capacitive susceptance, 482–483, 493, 495
Capacitors, 281–282
 current/voltage relations, 283–286

dielectric, 280
dielectric absorbtion (soaking), 291
dielectric breakdown, 291
dielectric constant, 289–290
dielectric permittivity, 290
dissipation factor, 499–500
energy storage, 281, 291–292
farad, 282
general characteristics, 296–297
in parallel, 292–293, 295–296
polarization of, 296
in series, 294–296
symbol, 281–282
time constant, 302
transients, 300–311
types, 297–299
Carbon-composition resistors, 33–36, 241
Cartesian form. See rectangular form
Cathode-ray tube (CRT), 18
 see also oscilloscope
Cascade connection, 149
CGS system, 6, 319
Charge, 14–17, 20, 22, 23, 280–282, 286–289, 291–292
Charge, induced, 289
Circuit
 basic, 30–32
 closed, 31
 driven, 149
 driving, 149
 energy transfers in, 31
 parts of, 30–31
 open, 43–44
 short, 43–44
Circular mils, 236–238
Coefficient of coupling, 663, 669
Color code (EIA), 34–36
Complex conjugate, 421, 537
Complex numbers, 413–415

739

Complex numbers (*cont'd*)
   complex conjugate, 421, 537
   j operator, 413–414
   math operations with, 417–421
   phasor representation, 415–416
   polar form, 415–416
   polar/rectangular conversions, 416
   rationalizing denominators, 420–421
   rectangular form, 416
   rectangular/polar conversions, 416–417
Components, linear, bilateral, 163
Composite signal, 388–389
Conductance, 33, 38, 482
   in parallel, 107
Conductors, 23, 235–238
   ideal, 31
   resistance of, 31, 33, 230–232
   types, 235–236
Conservation of energy, 20
Continuity, 83
Controlled sources, 123
Conventional current, 25
Conversions
   Norton/Thévenin
     ac, 539–540
     dc, 119–120
   T/Π networks
     ac, 531–535
     dc, 156–162
Copper losses, 326–327, 352, 670
Cosign wave, 386–388, 407
Coulomb, 16
Coulomb, Charles A., 16
Coulomb's law, 16–18, 20
Counter EMF, 344–345, 347
Cramer's rule, 185–187
Crystals, 620
Current, 21, 22–23, 74
   branch, 102
   conventional (Ben Franklin), 25
   eddy, 352, 671
   electron, 23–25
   full-scale deflection, 260, 261–262
Current division formula
   ac, 492
   dc, 114–115
Current source, ideal, 117, 535, 539, 568
Cycle, 379

D'Arsonval meter movement, 258–261
   ac measurements, 391, 684–687
   extending current range, 261–262
   ohmmeter, 265–267
   pivot and jewel suspension, 259
   taut-band suspension, 260
   voltmeter, 263–264
Deflection plates, 18, 691–692
Delta connection, 632
   generator, 636
   load, 642–645
Delta-wye networks. *See* T-Π networks
Dependent sources. *See* Controlled sources
Dependent variable, 9
Derivative, 430–432
Determinants, 184–187
   Cramer's rule, 185–187
Diamagnetic material, 318, 320
Dielectric, 280, 288–291
   absorbtion, 291
   constant, 289–290
   permittivity, 287, 290
   soaking, 291
Digital meters, 269–272
Diodes, 684
Dipoles, 288
Direct proportionality, 9
Dissipation factor, 499–500
Dot system, 657–659, 666–668, 672
Double-tuned filters, 618–620
Dynamic electricity, 15

Eddy currents, 352, 671
Effective value, sinusoid, 392–397
Efficiency, 41–43
Electric charge, 14–17, 20, 22, 23, 280–282, 286–289, 291–292
Electric field, 16, 24, 286–290
   in dielectric, 288–289
   intensity, 286–287
Electrodynamometers, 272, 688, 690
Electromagnets, 321–322
Electromagnetic devices, 336–338
Electromagnetic induction, 333–335

Electromagnetism, 321–322
   air gaps, 331–332, 351
   Ampere's circuital law, 327–328
   B-H curve, 324–327
   flux, 321–322, 323, 328–330, 333, 334, 336
   flux density, 319, 324–327, 332
   hysteresis, 326–327
   induction, 333–335
   leakage flux, 322, 650–651, 663
   Lenz's law, 335
   magnetizing force, 324, 326
   magnetomotive force, 322–323
   normal magnetizing curve, 327
   Oerstad's law, 321
   relative permeability, 320
   reluctance, 323
   right-hand rule, 321, 322
   solenoid, 321–322
Electromotive force (EMF). *See* Voltage
Electron, 14–15, 280–282
   drift, 24
   flow, 24
   free, 23–24
   speed in conductor, 24
Electrostatic force, 16–18
Energy, 19–20
   compared to power, 38
   electric, 20
   storage in capacitance, 291–292
   storage in magnetic field, 350
English units, 6
Equations, linear
   solution techniques, 184, 191–192, 193–194
Equivalent circuits, 495–501
   iron-core transformer, 670–671
Even functions, 707
Even symmetry, 707
Exponents, laws of, 3
   effects on math statements, 9

Farad, 282
Faraday, Michael, 282
Faraday's law, 334–335
Ferrite, 354–355
Ferromagnetic, 318, 320
FETVOM, 267–268

INDEX

Fields
  electric, 16, 286–290
  magnetic, 318–320
Filters
  band pass, 615–617
  band reject, 617–618
  double tuned, 618–620
  high pass, 480
  low pass, 476
  resonant, 614–620
Flux, 318–322, 323, 328–330, 333, 334, 336
  density, 319, 324–327, 332
  leakage, 332, 650–651, 663
  linkage, 333–335, 358
  mutual, 662–664
Force, electrostatic, 16–17
Fourier series, 704–707
  analyzing circuits with, 715–718
  bandwidth requirements, 705
  generation of RFI, 705
  rectified waveforms, 712–713
  sawtooth, 710–711
  square wave, 709
  symmetry, 707–709
  triangular wave, 711–712
Frequency, 384–386
  fundamental, 704
  half-power, 478–480, 591–593, 614–616
  harmonics, 704
  measuring on oscilloscope, 696–698
Frequency counter, 698
Frequency response, of audio transformer, 671
Frequency response curve, 478, 479
Full-wave rectification, 685–687
Fundamental frequency, 704
Fuse, 44

Generator, 333, 335
  ac, 376–379
Graticule, 692
Ground, 44–45
  chassis, 45
  earth, 45

Half digit, 270–271
Half-power frequency, 478–480, 591–593, 614–616
Half-wave rectification, 684–687

Half-wave symmetry, 708
Hall effect, 338
Harmonic frequency, 704
Hay bridge, 549
Henry, 346
Henry, Joseph, 346
Hertz, 384
Horsepower, 38
Hysteresis, 326–327
Hysteresis losses, 326–327, 352

Impedance, 443–449
  diagram (triangle), 444–446
  internal, 535–541, 542, 609
  reflected, 660–662, 672–673
  parallel, 480–481
  series, 466
Impedance matching
  with resonant circuits, 609–614
  with transformers, 655–656
Independent variable, 9
Indirect proportionality, 9
Induced voltage, 333–335
Inductance, 346
  mutual, 665–670
  parasitic, 357
  phase relations, 435
Inductive kick, 350
Inductive reactance, 435–438
Inductive susceptance, 482–483, 494–495
Inductors
  core construction, 352
  counter-EMF, 344–345, 347
  Henry, 346
  inductive kick, 350
  Lenz's law, 344–345
  losses in, 352
  parallel, 355–357
  practical considerations, 350–355
  self inductance, 345–346
  series, 355
  stored energy, 350
  time constant, 361
  transients, 359–365
Input resistance, 149–150
Instrument transformer, 688
Insulation, 235
Insulator, 23
Insulator, resistance of, 32
Insulator breakdown, 24
Internal impedance, 535–541, 542, 609

Internal resistance, 77, 80, 81, 117–122, 163–164, 207, 208, 602, 603
Ion, 15, 25
Iron-core transformer. *See* Transformer
Iron-vane meter, 687–688
Isolation transformer, 675

j operator, 413–414
Joule, 19

Kirchhoff's current law (KCL)
  ac, 480, 483, 486
  dc, 103–104
Kirchhoff's voltage law (KVL)
  ac, 466–467
  dc, 60–62
  use in finding unknown voltage, 71–74
Kirchhoff, Gustav, 61

Ladder method
  ac, 545–546
  dc, 169–171
Lagging power factor, 458, 476, 494
Laminations, 352
L/C ratio, 590–591
Leading power factor, 458, 475, 493
Leakage flux, 322, 663
Lenz's law, 335
Line current, 633, 638, 640, 643, 644–645
Line loss, 33
Line voltage, 633–636, 640, 642
Linear relationship, 243
Lissajous patterns, 694–698
L-networks, 609–614
Load, 30, 32
Loading, 79–81, 149–152, 256–258
Loop analysis
  ac, 556–559
  dc, 188–197
Loop equation, 72–74
  finding unknown voltage with, 71–74
Loudspeaker, 336–337

Magnetic circuits, 322–323
  analysis, 328–333
  Ohm's law for, 323–324
Magnetic field, 318–320
Magnetic recording, 336
Magnetism
  diamagnetic, 318, 320

Magnetism (cont'd)
 electromagnetism, 321–322
 ferromagnetic, 318, 320
 flux, 318–320
 flux density, 319
 flux lines, 318–320
 in meter movements, 258–259
 paramagnetic, 318, 320
 permeability, 320, 351
 remanent, 326
 see also Electromagnetism
Magnetizing force, 324, 326
Magnetomotive force, 322–323
Matched conditions, 537–538, 609–610, 655
Maximum power transfer
 ac, 537–539, 609–610
 dc, 80–81
 see also Matched conditions
Maxwell bridge, 528–529
Mesh analysis. *See* Loop analysis
Meters
 ammeters, 85–86, 261–262
 analog, 258, 272–274
 bargraph, 273–274
 D'Arsonval movement, 258–267, 391, 684–687
 digital, 269–272
 electrodynamometer, 688, 690
 FETVOM (TVOM), 268
 full-scale deflection current, 260, 261–262
 iron vane, 687–688
 loading by, 256–258
 multi (VOM), 267–268
 ohmmeter, 83, 265–267
 ohms/volt rating, 264–265
 peak reading, 690
 true rms, 689–690
 voltmeters, 125–126, 263–265
 wattmeter, 690
Mho, 33
Mil, 236
Millman's theorem
 ac, 541
 dc, 121–122
Multimeter. *See* VOM
Multimeter, digital, 271
Mutual inductance, 665–670

Network transformations, T-Π
 ac, 531–535
 dc, 156–162
Networks, T-Π, 156, 531–535

Networks, T-Π, derivation of equations, Appendix A
Neutral, 630, 632, 633, 636, 638
Neutron, 14
Nodal analysis
 ac, 559–564
 dc, 197–206
Nodal equation, 104
Node, 102
Normal magnetization curve, 327
Norton analysis
 ac, 568–571
 dc, 214–220
Norton equivalent source
 ac, 536–541
 dc, 117–119,
 combining, 121–122
 converting to Thévenin
  ac, 539–540
  dc, 119–120
 in superposition, 163–164
Norton's theorem, 214

Odd function, 708
Odd symmetry, 708
Oerstad's law, 321
Ohm, 32–33
Ohm, Georg Simon, 36
Ohmmeter, 265–267
 continuity checker, 83
 how to use, 267
Ohm's law, 36–38
 for ac circuits, 447
 in conductance terms, 38
 for magnetic circuits, 323–324
Ohms/volt rating, 264–265
Open, ideal, 43–44
Open circuit, 43, 83, 123–124
Oscilloscope, 691–695
 deflection plates, 691–692
 graticule, 692
 Lissajous patterns, 694–698
 measuring frequency, 696–698
 measuring phase, 694–696
 persistance, 691
 sweep signal, 692
 time base, 692
Output characteristics, Thévenin source, 78
Owens bridge, 549

Parallax error, 261
Parallel circuit
 ac, 480–492

current division formula, 114–115, 492
 dc, 102, 113–117
 define "in parallel," 102
 equivalent ac series circuit, 495–501
 response to changing frequency, 493–495
 troubleshooting, 123–124
 using approximations in analysis, 116–117
Parallel resonance, 494–495, 597–609
 bandwidth, 605–607
 in filters, 617, 618
 frequency, 598–599
 half-power frequencies, 605
 Q, 601–605
Paramagnetic, 318, 320
Parasitic components, 357–358
Peak-reading voltmeter, 690
Period, 47, 385–386
Permanent magnet, 259, 318, 326, 336–337
Permeability, 320
Phase, 404–411
 cause of, 433–442
 measuring on oscilloscope, 694–696
 in parallel circuits, 493–495
 in phasor domain, 408–412
 in series circuits, 474–477
 in time domain, 405–408
Phase current, 633, 637, 643
Phase voltage, 633–636, 637, 642
Phasor domain, 409
 converting from time domain, 410–412
 representing by complex numbers, 415–416
 vertical projection, 380
Phasors, 380, 408–409
Π-T conversions
 ac, 531–535
 dc, 156–162
Piezoelectric effect, 620, 622
Pivot and jewel suspension, 259
Polar form, 416
Polarity, 32
 transformer dot system, 657–659, 666–668, 672
Polarized capacitors, 296
Polyphase circuits, 630–633
Potential, 21
 difference, 21
 drop, 20

INDEX

energy, 20
rise, 20
see also Voltage
Potentiometer, 76–77
Power
ac, 449–457
apparent, 453–457
average, 632–633
dc, 38–40
of nonsinusoidal waveforms, 714–715
reactive, 452–453
in three-phase circuits, 632, 637, 640, 641–642, 643
triangle, 454–456
true, 450–451
Power factor, 457–459, 475, 479, 493, 494
correction of, 523–526
Power triangle, 454–455, 456
Primary, transformer, 650
Printed circuit boards, 238–240
Product-over-sum (POS) rule, 111–112, 480–481
Proportionality, direct and indirect, 9
Proton, 14
Pulse transformer, 675
Pure components, 357

Q-factor (figure), 499, 585–591
parallel resonant circuit, 601–604
series resonant circuit, 585–591
Q multiplication, 588–589, 604–605

Radian, 381–382, 384, 387
Radio frequency interference (RFI), 705
Rationalizing denominators, 420–421
Reactance
capacitive, 438–442
inductive, 435–438
Reactive factor, 458
Reactive power (VAR), 452–453
Rectangular form, 416
Rectification, 684–687
Rectifier. *See* Diode
Reflected impedance, 660–662, 672–673
Relative permeability, 320
Reluctance, 323

Remanent magnetism, 326
Resistance, 32–33, 36, 230–232, 238
dynamic, 244–246
internal, 77–78, 80, 81, 117–122, 163–164, 207, 208, 602, 603
in parallel, 105–113
phase relations, 433–435
POS rule, 111–112
in series, 62–64
in series-parallel, 138–140
skin effect, 358
temperature effects, 232–235
Resistive networks, 242
Resistivity, 231, 237–238
Resistors, 33, 34, 40, 240–243
carbon composition, 33–36, 241
determining value, 242–243
EIA color code, 34–36
general characteristics, 240–241
military coding, 243
power ratings, 40
reliability, 34
tolerance, 33, 242–243
types, 33, 241–242
Resonant circuits
parallel, 494–495, 597–609
series, 475, 477, 581–583
tank circuit as ac generator, 608–609
Resonator, ceramic, 620, 622
Rheostat, 74
Rho, 231, 237–238
Right-hand rule, 321, 322, 377, 658
rms value, 392–397, 713–715
Root mean square value, 392–397, 713–715
Rotor, 376–377

Saturation, 325–327
Sawtooth waveform, 48, 50, 710–711
Scalars, 374
Schematic diagram, 44–45, 66–67
ground symbols, 44–45
Schering bridge, 549
Scientific notation, 3–5
Secondary, transformer, 650
Self inductance, 345–346
Semiconductors, 23
Series circuits
ac, 466–474

approximations in analysis, 70–71
dc, 64–71
define "in series," 58
equivalent ac parallel circuit, 495–501
finding unknown voltage in, 71–74
Kirchhoff's voltage law (KVL), 60–62, 71
response to changing frequency, 474–475
ten-to-one rule, 71
troubleshooting, 82–83
voltage division formula, 65–68, 473–474
Series-parallel circuits
ac, 512–526
dc, 140–146
Series-parallel reduction (SPR) technique
ac, 512–515
dc, 140–146
Series resonance, 581–585
bandwidth, 591–597
half-power frequencies, 591–593
L/C ratio, 590
Q, 585–591
Q multiplication, 588
Short circuit, 43–44
ideal, 43–44
troubleshooting parallel circuits, 123–124
troubleshooting series circuits, 82–83
Shunt, meter, 261–262
Shunt connection. *See* Parallel
SI system of units, 6
Siemen, 33, 482
Signal, 47
Single-phase circuits, 630
Sinusoidal waveform, 47
alternations, 380
angular velocity, 381–382
average value, 389–392
on dc level, 388–389
derivative, 430–432
effective value, 392–397
frequency, 384–386
generation of, 376–380, 381
peak value, 379
peak-to-peak value, 386
period, 385–386
phase, 404–411
phasor representation of, 380
polarity definitions, 380
rms value, 392–397

INDEX

Sinusoidal waveform (cont'd)
  sketching of, 387–388
Skin effect, 358
Slope, 244–246
Solenoid, 321–322
Sources, 30
  combining Thévenin and Nortons, 541
  controlled, 123
  conversions, 119–120, 539–540
  ideal current, 117
  ideal voltage, 77
  Norton model, 117–119
  in parallel, 121–122
  polarity of voltage across, 32
  in series, 59–60
  Thévenin model, 77–81
SPR. See Series-parallel reduction
Square wave, 704–705, 709–710
Star connection. See Wye connection
Static electricity, 15
  uses of, 18
Stator, 376–377
Step-down transformer, 652
Step-up transformer, 652
Stray capacitance, 299
Superposition
  ac, 542–545
  dc, 163–169
  in Fourier series, 704
Susceptance, 482–483
  capacitive, 482
  inductive, 482
Switches, 246–247
Symmetry, 707–709
  even, 707
  half-wave, 708
  odd, 708
Systems of units, 6–8

T-Π conversions
  ac, 531–535
  dc, 156–162
Tank circuit, 608–609
Tapped, resistor, 76
  transformer, 675
Temperature coefficient, 232–235
Ten-to-one rule
  parallel resistances, 112–113
  series resistances, 71, 80
Thermistor, 155, 233, 246
Thermocouple, 688–689, 690
Thévenin analysis
  ac, 564–568
  dc, 206–214
Thévenin equivalent source
  ac, 536–541
  combining, 81, 121–122
  converting to Norton
    ac, 539–540
    dc, 119–120
  dc, 77–81
  output characteristics, 78
  in superposition, 163–164
  volt/ampere characteristics, 78
Thévenin's theorem, 207
  in RC-transient analysis, 309–311
  in RL-transient analysis, 364–365
Three-phase circuits
  balanced load, 636
  current in neutral line, 638
  delta connected sources, 636
  delta connection, 632, 642–645
  line current, 633
  line voltage, 633–636
  measuring power, 690
  neutral line, 632
  phase current, 633, 643
  phase voltage, 633–636, 642–643
  power in, 631–633, 637
  unbalanced load, 636, 640–642
  wye connected load, 636–642
  wye connected source, 633–639
  wye connection, 632
Time base, 692
Time constant
  RC circuit, 302
  RL circuit, 361
Time domain, 383
  converting from phasor domain, 410–411
Tolerance, 33
Toroid, 322
Transducers
  capacitive, 30
  inductive, 359
  resistive, 246
Transformations, T-Π
  ac, 531–535
  dc, 156–162
Transformers, 334, 650–656
  air core, 671
  audio, 674
  auto, 676–677
  coefficient of coupling, 663, 669
  copper loss, 670
  dot convention, 657–659, 666–668, 672
  eddy current losses, 671
  equivalent circuit, iron core, 670–671
  frequency response of audio, 671
  hysteresis losses, 671
  ideal, 650–651, 654–655, 657, 659–662
  impedance matching with, 655–656
  instrument, 688
  iron core, 650–656
  isolation, 675
  leakage flux, 650–651, 663
  loosely coupled, 671–674
  mutual flux, 662–664
  mutual inductance, 665–670
  primary, 650
  pulse, 675
  reflected impedance, 660–662, 672–673
  secondaries in series/parallel, 657–659
  secondary, 650
  step down, 652
  step up, 652
  tapped, 675
  turns ratio, 652, 654
Transients
  capacitive, 300–311
  inductive, 359–365
Triangular waveform, 48, 707, 708, 711–712
Troubleshooting
  parallel circuits, 123–124
  series circuits, 82–83
True rms meter, 689–690
Tuned circuits. See Resonant circuits
Turns ratio, 652, 654
Two-phase circuits, 630
Two-wattmeter method, 690

Unbalanced load, 636–638
Units
  CGS, 6
  conversions, 6–8
  English, 6
  standard international (SI), 6
  systems of, 6

INDEX

Variables, dependent and independent, 9
Variac, 677
Vectors, 374–376
 in reference position, 374
 resolving components, 375–376, 379
Volt, 21
Volt-ampere (VA) characteristics, 243–246
Volt-ampere-reactive (VAR), 452–453
Volt-amperes (VA), 454–457
Volta, Alessandro, 22
Voltage
 induced, 333–335
 using KVL to find unknown, 71–74
Voltage divider
 loaded, 149–152
 unloaded, 75–77
Voltage division formula
 ac, 473–474
 dc, 65–68
Voltage drop, 21–22, 31
Voltage regulation, 80–81, 118
Voltage rise, 21–22, 31
Voltage source, ideal, 77, 535, 539, 564
Voltmeter, 125–126, 263–264
 input resistance, 264–265
VOM, 267–268

Watt, 38
Wattmeters, 690
Waveforms, 46–50
 amplitude, 47
 aperiodic, 47
 exponential, 304, 359
 Fourier series of, 709–713
 period, 47
 periodic, 47
 sawtooth (ramp), 692
 sinusoid, 379–381
 square, 704–705, 709–710
 symmetry in, 707–709
 triangular, 48, 707, 708, 711–712
Webers, 319
Wheatstone bridge, 155–156
Wien bridge, 529–530, 549, 550
Wire table, 236
Work. *See* Energy
Wye connected load, 636–642
Wye connected source, 633–642
Wye connection, 632
Wye delta. *See* T-II